THE INVERTEBRATE TREE OF LIFE

THE
INVERTEBRATE
TREE
OF LIFE

Gonzalo Giribet

Gregory D. Edgecombe

PRINCETON UNIVERSITY PRESS

PRINCETON AND OXFORD

Published by Princeton University Press
41 William Street, Princeton, New Jersey 08540
6 Oxford Street, Woodstock, Oxfordshire OX20 1TR

press.princeton.edu

Library of Congress Control Number: 2019937612
ISBN 978-0-691-17025-1

British Library Cataloging-in-Publication Data is available

Editorial: Alison Kalett, Kristin Zodrow, and Abigail Johnson
Production Editorial: Mark Bellis
Text and Jacket Design: C. Alvarez-Gaffin
Production: Jacqueline Poirier
Publicity: Matthew Taylor and Katie Lewis
Copyeditor: Margery Tippie

Jacket Credit: images courtesy of the authors

This book has been composed in Palatino with Gotham display
by Westchester Publishing Services

Printed on acid-free paper. ∞

Printed in China

1 3 5 7 9 10 8 6 4 2

CONTENTS

PREFACE AND NOTES

The idea of writing a textbook on invertebrate zoology had crossed our minds for some years, but committing to the task required the encouragement of our editor, Alison Kalett. Invertebrate textbooks are a difficult breed, as for many students the one chosen for their undergraduate class may be the only invertebrate text they will ever see. Many of these students may have barely studied invertebrates in high school, making it all the more challenging, as such textbooks should introduce many new and complex concepts, names, and classifications and cover an endless number of body plans, fossils, and so on. In addition, the vibrant field of invertebrate zoology has disappeared from the curriculum in many universities. Nevertheless, invertebrate textbooks are constantly used by academics to continue learning about the different animal groups. A geneticist probably does not care to have the latest genetics textbooks on their desk unless that person teaches the genetics course in their university. Invertebrate zoologists, on the other hand, treasure their invertebrate textbooks as the first port of call for facts about a particular group of animals. Hence the challenge of writing a book that meets the needs of a readership that spans beginning undergraduate students to the already learned.

We also thought to write a book that incorporates novel aspects not treated in other texts, and that was able to synthesize a specific item that tends to frustrate nonspecialists the most: the classification of animals. This is one of the questions we most often face, sometimes with accompanying accusations of having changed the phylogeny of this or that group. And indeed, we have been working on animal phylogenetics for over two decades, having witnessed several paradigm changes, among them the cladistic revolution, the genetic revolution, and now the genomic revolution. We have published hundreds of scientific papers using morphological data matrices, a handful of genes, and later transcriptomes and genomes to investigate the phylogenies of all Metazoa and at finer levels, down to a genus of, for example, small centipedes. We have experienced these revolutions first-hand and have participated in many of the debates that have appeared in the literature about subjects like Ecdysozoa versus Articulata, the position of Ctenophora and Porifera, the systematic position of controversial groups like xenoturbellids and acoelomorphs, and the resolution of the main clades of Spiralia and Lophotrochozoa. We thus aim to bring our expertise in this field to provide what we hope is a coherent account of the evolution of metazoans, without shying away from debate where it exists.

While our book is fundamentally about understanding invertebrates in the light of their evolutionary history—captured by the metaphor of a tree—this book could not just be about phylogenetics. Instead we wanted to place phylogenetic discussions in the broader context of the characters of the animals, the features that make this enterprise interesting and worthy of study. The relevance of whether sponges or ctenophores are the sister group to all other animals involves not just the details of phylogenetic inference methods (which are themselves a lively field of research, especially as data sets have become genomic in scale), but also with a fundamental idea of the evolution of complexity. Was the first animal already complex morphologically, with nerves and muscles, or only genetically (we know that genetic complexity comparable to that of animals with tissues is already present in sponges)? Resolving these relationships is essential to answer this and other questions in animal evolution. We also thought it fundamental to bring a detailed discussion of fossils to this book, something rarely covered in an invertebrate textbook, unless it is a book on invertebrate paleontology, which then tends to downplay the living forms. Valentine's *On the Origin of Phyla* is quite unique in covering this niche.

But, of course, we are not alone in this universe of textbooks. Two of our models are Nielsen's *Animal Evolution* (currently in its third edition, published in 2012) and Brusca et al.'s *Invertebrates* (also in its third edition, published in 2016). Other books that have been constant sources of inspiration are Westheide and Rieger's *Spezielle Zoologie* (in its third edition), Schmidt-Rhaesa's *The Evolution of Organ Systems*, and Ax's *Multicellular Animals* series. Of these last, *The Evolution of Organ Systems* is an outstanding source to learn about a particular organ system and then find information about each individual phylum, in an unprecedented way for a zoology textbook. But it is difficult to amass all knowledge about invertebrates in just a few brains. The third edition of *Invertebrates* has, in addition to its three lead authors, many chapters that were completely rewritten by specialists. This model has also been followed by two of our favorite and most cited modern treatments in this book, *Structure and Evolution of Invertebrate Nervous Systems* (edited by A. Schmidt-Rhaesa, S. Harzsch, and G. Purschke) and the six-volume series *Evolutionary Developmental Biology of Invertebrates* (edited by A. Wanninger). These are monumental volumes that are impossible to replicate by a single author. The new volumes of the *Handbook of Zoology* and *Treatise on Zoology* also cover some of the groups here discussed in much greater detail. And of course, we must acknowledge other series that have been at our fingertips constantly, including the *Microscopic Anatomy of Invertebrates* series, *Reproductive Biology of Invertebrates*, and *Meiofauna Marina*, among others.

But like the Brusca series, all these books are too massive to follow in an introductory course on invertebrates. In Brusca's third edition, each chapter could almost be turned into multiple lectures or some even into an entire course or degree program (e.g., Entomology), a privilege we teachers often don't have. We obviously cannot compete with the level of detail of such masterpieces. Our book is therefore more along the lines of Nielsen's first edition of *Animal Evolution*, hopefully a groundbreaking treatment of animal relationships and relevant aspects of those phyla that concern the evolution of their body plans. Claus Nielsen has been an inspiration to so many of us because of this book. His revised editions have introduced changes to the phylogenetic tree of animals based on new theories and discoveries, thus changing the structure of the book. But we felt that it had not yet fully integrated the genomic revolution or the fossil record at the level we have strived to accomplish here. We also wanted a tree based more on explicit analyses of data than on evolutionary scenarios, which have informed many of Nielsen's hypotheses. We do not claim our approach is better, but it explains why we have placed less emphasis on some of the aspects that drove his books, such as the trochaea theory, among others. Yet we follow him in designing our book as the pages of the animal tree, navigating through nodes, and making arbitrary decisions as to whether to take one branch or another at each bifurcation.

That said, in some cases, the decision is not so arbitrary. The first split in Nephrozoa is between protostomes/deuterostomes or deuterostomes/protostomes, depending on how one draws the tree. Most books leave the deuterostomes to the end, with the chordates (and vertebrates) coming in the very last chapter or volume. Like in Piper's elegant *Animal Earth*, we chose not to do so, trying to make a statement that avoids all connections to the old concept of the ladder of life, which is often followed even when using a phylogenetic scheme—chordates at the end. We also understand that the text sometimes reads like a scientific paper, with what may seem a large number of references. The reader does not need to check them unless the topic is of special interest, but we needed to check most of the statements that are given as for granted in so many places. Some of these, even involving facts as basic as whether or not tardigrades are eutelic, required considerable investigation from our side, and these are the references we ended up using to reach our conclusions.

Writing a textbook covering the vast zoological literature of invertebrates can thus be a formidable task, yet we have attempted to check nearly all the approximately three thousand references that appear in this one, even if citing them because prior scholarly work had done so. Because our intention was to check every paper we cite, our citations are biased toward the present (indeed we have drawn most heavily on very recent literature to try and capture the state-of-play in the field up to mid-2019), yet in many cases we wanted to credit original and unparalleled work by many extraordinary zoologists of the late nineteenth and early twentieth centuries. A large part of that literature was written in German, which we could not attempt to fully read. Many of our references to early German workers are secondary citations, and thus we may have mischaracterized some of this seminal work. We preferred to do this rather than to ignore such fundamental contributions. If our literature is biased toward English-language works, we can only apologize.

READING PHYLOGENIES

Reading phylogenetic trees quickly becomes second nature to biologists. This book, as has been done, inspiringly, earlier (e.g., Ax, 1984; Hennig, 1994; Nielsen, 1995), navigates through the dense Animal Tree of Life, with discussion framed around its main nodes. We here thus introduce some notes to basic nomenclatural issues associated with reading phylogenies. It is well known that phylogenetic trees are a metaphor for the nested hierarchies of the living entities represented in their branches, and that when groups are monophyletic, based on evidence in the form of shared derived characters, phylogenetic trees do not reflect ancestor–descendant relationships. We therefore often refer to taxa as sets of sister-group relationships. For example, we can say that Echinodermata is the sister group (and closest relative) of Hemichordata, the two together constituting the taxon Ambulacraria, all of which are monophyletic. Neither Echinodermata nor Hemichordata is derived from the other, and at most we can infer a common ancestor using a diversity of methods for optimizing characters on trees.

Because trees reflect sister-group relationships and not ancestor–descendant relationships, it is also important to be precise when using the term "basal." The improper usage of the term has been elegantly reviewed elsewhere (Krell and Cranston, 2004), a position to which we also subscribe. Their title, "Which Side of the Tree Is More Basal?" is clear, and authors often (and incorrectly) refer to a taxon being basal when it is the less diverse of two sister groups and sometimes depicts plesiomorphic features. But sponges or ctenophores are not basal to Planulozoa (see fig. 1); rather, they are simply their potential sister group. The term "basal" thus needs to be carefully applied, and then qualified. For example, sponges are topologically more basal than arthropods in the animal tree, as the latter are deeply nested within Planulozoa, but sponges are not more basal than their sister group, Planulozoa (likewise, we could say—although no one would—that Planulozoa is more basal than a particular derived sponge taxon, such as Hexactinellida). "Basal" may be useful when referring to the first offshoots in a grade of taxa. (As a note for clarification, when we later refer to "the 'base' of the Animal Tree of Life," we do not refer to any taxon we think is more basal than others but to the deepest nodes in the tree, that is, the first splits of the animal phylogeny.)

We thus employ a widely used tree-based method for describing the relationships of fossils that avoids the pitfalls of "basal" taxa. This makes a distinction between stem groups and crown groups, a convention first developed by German zoologist Willi Hennig. A crown group is composed of the most recent common ancestor of the extant/living members of a clade and all of its descendants. The crown group of arthropods includes the most recent common ancestor of the reciprocally monophyletic sister groups Chelicerata (e.g., a horseshoe crab) and Mandibulata (e.g., a fruit fly) and all species descended from that ancestor. The latter would of course include all fossil chelicerates and mandibulates; they are crown-group Arthropoda. Fossil species that branched off between the divergence of the most recent common ancestor of Arthropoda and its living sister group—Onychophora—are members of the arthropod stem group. Stem groups are composed entirely of extinct species. The arthropod stem group and crown group together are called "total-group Arthropoda."

We can used "stemward" and "crownward" as ways of describing the shapes of phylogenetic trees, orienting toward regions of a tree occupied by living diversity and that represented by extinct diversity. Some see the distinction between stem and crown groups as an arbitrary convention based on whether or not a taxon happens to have survived to the Recent rather than having gone extinct, and it is true that extinct and extant taxa are not ontologically different from each other; in either case they are monophyletic groups based on shared derived characters. The distinction is epistemological: in most cases, extant species can have their phylogenetic position established based on vastly larger character samples than extinct ones (e.g., genome scale molecular data), so taxonomic nomenclature based on extant diversity recognizes this difference in amount of evidence.

PHYLA AND LINNEAN RANKS

When Ernst Haeckel introduced the new rank "phylum," based on the Greek word *phylon* ("tribe" or "stock"), between Linnaeus' "kingdom" and "class," he intended it to represent major lineages that display a common body plan not shared with any other species. In the modern usage, "phylum" is traditionally employed to designate a major clade of organisms; typically between 35 and 40 phyla are recognized for metazoans. Whether phylum or any other Linnaean ranks are useful has been debated elsewhere, and we adhere to a position that phyla must represent monophyletic groups of animals that share a body plan that is radically different from that of others (Giribet et al., 2016). We do not support paraphyletic taxa (groups that include some but not all descendants of their most recent common ancestor), and hence Acanthocephala must be recognized as a subtaxon of Rotifera to avoid paraphyly of the latter (chapter 39), Myxozoa as a subtaxon of Cnidaria (chapter 8), and former phyla, such as Diurodrilida, Echiura, Sipuncula, Pogonophora, and Vestimentifera are now subclades of Annelida (chapter 47)—to provide just some notable examples.

We do not necessarily stand on whether Xenacoelomorpha (chapter 10) is a phylum or a superphylum containing two or three phyla—even though we choose to treat Xenoturbellida and Acoelomorpha as two phyla in our chapters—as this changes nothing of its topology (we take an admittedly arbitrary stand here), and treat Chordata as a single phylum, although we could equally use three phyla instead. Likewise, we make no assumption of equating phyla with their time of origin, as one phylum can be sister group to one, two, or even all other phyla. Our only assumption is thus that two sister taxa, whether they are both phyla or one is a phylum and the other includes the remaining phyla, have the same age.

A more substantial debate is whether the concept of "phylum" has a real biological meaning, since this goes beyond classification systems, as proposed by some authors who have suggested that the phylotypic period (a stage in the middle of embryonic development) is conserved among species within some phyla. This has been termed the "developmental hourglass model" (Kalinka et al., 2010) and has led to the proposal that a phylum may be defined as a collection of species whose gene expression at the mid-developmental transition is both highly conserved among them and divergent relative to other species (Levin et al., 2016). While appealing, this concept is not supported experimentally (Hejnol and Dunn, 2016).

We thus understand that a phylum is just a rank, a useful label used to compare and refer to large clades of animal diversity without implying that the same rank has an evolutionary equivalence. Indeed, many of our chapters correspond to these phyla. Operationally it allows us to draw a line for the lowest taxonomic rank for which we write chapters (as is traditionally done in most invertebrate textbooks) and to make certain comparisons such as how much species diversity there is in a phylum, that is, Arthropoda is more diverse than Mollusca. While some advocate that this does not require that we call them "phyla," we argue that by being phyla they cannot belong to any other phylum, and this, in itself, contains information (e.g., Platnick, 2009).

NOMENCLATURAL NOTES

We shall not try here to provide an exhaustive discussion about taxonomic nomenclature and rules. Some are well known to zoologists and the general public, as for example how binomials require the generic epithet to be capitalized but not the specific epithet, or how both should be in italics (underlined in some older fonts), as, for example, *Trichoplax adhaerens* or <u>Trichoplax adhaerens</u>. There are nonetheless many other common rules that are often overlooked or misinterpreted.

While common names are not available for all taxa—and often these come from translations of other languages and are not really used or useful—each Latin scientific name can easily be translated into English. We refer to members of Nematoda as "nematodes," and a common name may be "roundworm" (there are many others for specific species of nematodes). While the rules of nomenclature dictate that scientific names for genera and above must be capitalized, the English version should not. This is a common mistake found in many scientific papers. In order to reflect the taxonomic rank of the English name, one must adhere to certain nomenclatural rules. Drosophilidae therefore becomes "drosophilid" rather than "Drosophilid," and Synthetonychiidae becomes "synthetonychiid." "Drosophilidae" (and "drosophilid") uses a single "i" because the nominal genus, *Drosophila*, does not have an "I" in its last syllable, while *Synthetonychia* does—hence "Synthetonychiidae." The English for "subfamilies" should add the suffix–*ne*, for example, the beetle subfamily Rutelinae would become "ruteline."

It is well known among systematists, but perhaps not among other scientists, that a taxon name is often accompanied by its authority. Biologists often cite the author and publication year of the taxon, although in some cases shortcuts are taken. Higher taxa are referred to the authority without parentheses, which thus cannot be confused with a citation. But parentheses as regards species authorities have a specific meaning, that the species has been transferred to a different genus from that originally assigned by its author. Thus, *Nautilus pompilius* Linnaeus, 1758, is a species described by Linnaeus in his 1758 edition of *Systema Naturae* in the genus *Nautilus*, while *Nautilus scrobiculatus* (Lightfoot, 1786) is a species described by Lightfoot in 1786, under a different genus, and later transferred to Linnaeus' genus, *Nautilus*. Because zoological textbooks often do not provide authorship to all named species (few know the author of *Drosophila melanogaster*, perhaps the most cited invertebrate), and because a large fraction of animals are now in a genus different from when they were originally described, it is often difficult to distinguish a species author from a citation following a species name. Thus, to avoid confusion we generally add "see" before a citation following a species name, for example, *Nautilus pompilius* (see Ward et al., 2016) instead of *Nautilus pompilius* (Ward et al., 2016), to avoid possible ambiguities between authorship and a scientific citation.

▆ TRENDS IN MORPHOLOGICAL SYSTEMATICS

Because this book is in part about invertebrate systematics, we mention here some areas of morphological phylogenetics that have had considerable influence in our interpretation of animal relationships. The first of these is what has been known as "spermiocladistics," or the study of sperm ultrastructure for phylogenetic purposes. This field has had influence in many metazoan clades, and a true landmark was the publication of the volume *Advances in Spermatozoal Phylogeny and Taxonomy* (Jamieson et al., 1995), which summarized much of the comparative sperm ultrastructural work up to that time. Special areas in which spermiocladistics has been applied are Platyhelminthes (e.g., Hendelberg, 1986), Gastrotricha (e.g., Marotta et al., 2005), Annelida (e.g., Jamieson and Rouse, 1989), Mollusca (e.g., Franzén, 1955; Bieler et al., 2014), and Arthropoda (e.g., Alberti, 1995; Jamieson et al., 1999). In molluscs it has been shown that the information content of sperm ultrastructure (and shell morphology) is closest to that of molecular data, at least for bivalves (Bieler et al., 2014). It is also evident, though, that some major convergences exist, such as the biflagellate sperm of Acoela and many Platyhelminthes, or a positioning of the mitochondrial midpiece between the nucleus and axoneme that was proposed as a synapomorphy of Onychophora and Oligochaeta (Jamieson, 1986). An excellent review of sperm ultrastructure across animal phyla can be found in Schmidt-Rhaesa (2007)

Neurophylogeny (see Richter et al., 2010) is a more recent but flourishing area in animal systematics, especially in groups such as arthropods, where the nervous system has been the basis of detailed comparative work (Strausfeld, 2012; Wolff and Strausfeld, 2016). In addition to molecular data, the first field to explain the close relationship between hexapods and their crustacean counterparts was neuroanatomy (Strausfeld, 1998; Strausfeld and Andrew, 2011). Arthropod visual systems have also been used to link and explain the Cambrian diversity and the recent taxa (Strausfeld et al., 2016a; Strausfeld et al., 2016b), and thus, unlike spermiocladistics, neurophylogeny has had an impact on paleontologists trying to reconstruct appendage homologies (Ortega-Hernández et al., 2017). These two areas are integrated into the relevant chapters.

ACKNOWLEDGMENTS

Writing a book is never an easy task, and if such books attempt to summarize the body of knowledge of virtually all animals, it quickly becomes daunting. It is for this reason that our editor Alison A. Kalett has been so instrumental in ensuring that this book would see the light of day. From her initial contact at Gonzalo's Harvard office to the many subsequent visits, emails and phone conversations, she checked in at the right times and always encouraged us about the next steps. The book would definitely not have happened without her. Gonzalo would also like to acknowledge support from the Guggenheim Foundation, which provided the fellowship that allowed him to stay in London for six months to work with Greg on what would become the first seeds of this final product.

For reading chapters and providing corrections and suggestions, we are most grateful to our close colleagues and friends Jaume Baguñà (Universitat de Barcelona), Federico Brown (University of São Paulo), Tauana Cunha (Harvard University), Reinhardt Møbjerg Kristensen (Natural History Museum of Denmark), Christopher Laumer (European Bioinformatics Institute), David Lindberg (University of California, Berkeley), Mark Martindale (Whitney Marine Laboratory), Juan Moles (Harvard University), Ricardo Neves (University of Copenhagen), Vicki Pearse (Monterey, CA), Winston Ponder (Australian Museum), Ana Riesgo (The Natural History Museum, London), Greg Rouse (SCRIPPS Institution of Oceanography), Martin Vinther Sørensen (Natural History Museum of Denmark), Malin Strand (Swedish University of Agricultural Sciences), Per Sundberg (University of Gothenburg), Robert Woollacott (Harvard University), and most especially to Margery Tippie, for her superb copyediting of such a complicated text, and to Mark Bellis, for guidance and help during production and proofing.

Many doubts and clarifications benefited from our exchanges with many other colleagues, especially Thomas Bartolomaeus (Universität Bonn), Richard Brusca (University of Arizona), Marcelo Carrera (Universidad Nacional de Córdoba), Cassandra Extavour (Harvard University), Pat Hutchings (Australian Museum), Sebastian Kvist (Royal Ontario Museum), Carsten Lüter (Museum für Naturkunde, Berlin), Kristen Marhaver (CARMABI, Curaçao), Claus Nielsen (Natural History Museum of Denmark), John Pearse (Monterey, CA), Andreas Schmidt-Rhaesa (Universität Hamburg), Mansi Srivastava (Harvard University), Paul Taylor (The Natural History Museum, London), Owen Wangensteen (The Arctic University of Norway), and Katrine Worsaae (University of Copenhagen).

For images we relied on an army of friends and colleagues, who are the true specialists on many of these groups. Thanks are extended to Loren Babcock, Thomas Bartolomaeus, Christoph Bleidorn, Glenn A. Brock, Jean-Bernard Caron,

Phil Donoghue, Blanca Figuerola, Peter Funch, Paul Gonzalez, Noemí Guil, Steven Haddock, Ken Halanych, Conrad Helm, Diying Huang, Sören Jensen, Collin Johnson, Ulf Jondelius, Andrew Knoll, Reinhardt Møbjerg Kristensen, Christopher Laumer, Juliana Leme, Alex Liu, Xiaoya Ma, Gabriela Mángano, Ricardo Neves, Akiko Okusu, Cruz Palacín, Tae-Yoon Park, John Paterson, Fredrik Pleijel, Susannah Porter, Lars Ramsköld, Greg Rouse, Miguel Salinas Saavedra, Derek Siveter, Christian Skovsted, Martin Smith, Andy Sombke, Martin Vinther Sørensen, Mansi Srivastava, Mark Sutton, Paul Taylor, Timothy Topper, Kira Treibergs, Jakob Vinther, Xing Wang, Katrine Worsaae, Samuel Zamora, and Alexander Ziegler. (We note here that any images not otherwise credited in the captions are the authors' own.)

And Mary Sears, from the Ernst Mayr Library in the Museum of Comparative Zoology was incredibly helpful with bibliography.

This book has not always been easy to write, and we would like to acknowledge our partners, Christine and Zerina, for always being there, listening to book-related issues that emerged continuously. As always, they provided encouragement and support when needed. Our students and postdocs also had to be extra patient, reading chapters or constantly hearing about "the book this" or "the book that," and it certainly was a fierce competitor for their deserved time. We hope that they now understand a bit better.

THE
INVERTEBRATE
TREE
OF LIFE

BEFORE ANIMALS

1

Despite anecdotal results from the early days of molecular phylogenetics (e.g., Field et al., 1988), all extant animals (Metazoa) unite as a monophyletic group, sharing a common ancestor that evolved from unicellular organisms in the Precambrian (Sebé-Pedrós et al., 2017). The nature and age of this ancestor are a matter of intense debate, one that may not be resolved anytime soon for many reasons. Nonetheless, progress has been made in terms of the genomic complement of such an ancestor by comparing the genomes of metazoans and their closely related unicellular holozoans (choanoflagellates, ichthyosporeans and filastereans) (fig. 1.1) with those of other outgroups (e.g., King et al., 2008; Sebé-Pedrós et al., 2017; Paps and Holland, 2018; Richter et al., 2018) and reconstructing the common repertoire of genes found across metazoans (see Lewis and Dunn, 2018).

This has shown that the addition of novel groups of genes at the node that leads to Metazoa is considerably larger than the novel genes at any nodes surrounding it. Indeed, 25 groups of metazoan-specific genes have been established as essential for this clade (Paps and Holland, 2018), facilitated by the complete genome sequences of four unicellular holozoans (Sebé-Pedrós et al., 2017): two choanoflagellates (*Monosiga brevicollis* and *Salpingoeca rosetta*), a filasterean (*Capsaspora owczarzaki*), and an ichthyosporean (*Creolimax fragrantissima*) (King et al., 2008; Fairclough et al., 2013; Suga et al., 2013; de Mendoza et al., 2015). This data set enables reconstructing the gene content of the unicellular ancestor of animals at an unprecedented level of detail—including the so-called multicellularity genes that have roles in cell–cell recognition, signaling, and adhesion. The study of these genomes resulted in a quite surprising result; although there has been gene innovation at the origin of Metazoa (see Paps and Holland, 2018), the unicellular ancestor of animals already had a rich repertoire of genes that are required for cell adhesion, cell signaling, and transcriptional regulation in modern animals (Sebé-Pedrós et al., 2017).

Another recent study, sampling transcriptomes of nineteen additional choanoflagellates, also suggested that a large number of gene families were gained at the stem of Metazoa (Richter et al., 2018). However, whereas Paps and Holland (2018) estimated that the number of gains was much larger than the number of losses, Richter et al. (2018) found that these numbers are very similar, which has been portrayed as evidence for an "accelerated expansion of gene families" versus an "accelerated churn of gene families" along the metazoan stem (Lewis and Dunn, 2018). Perhaps most important, the new study thoroughly sampling choanoflagellate transcriptomes has provided evidence that hundreds of gene families previously thought to be animal-specific, including Notch, Delta, and homologs of

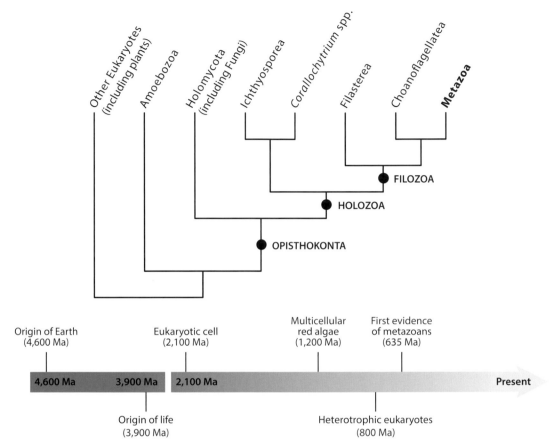

FIGURE 1.1. Top: phylogenetic position of Metazoa among Holozoa and other eukaryotes. Bottom: a timeline of major events leading to the origins of metazoans. Based on Sebé-Pedrós et al. (2017).

the animal Toll-like receptor genes, are also found in choanoflagellates (but not in the two highly derived, previously sequenced genomes) and thus predate the choanoflagellate–metazoan divergence. It is anticipated that the early history of the animal gene repertoire will continue to be refined as the genomes of more closely related holozoans are brought into the picture.

The history of the origins of metazoans goes back to Haeckel and Metschnikoff (see a recent historical account in Nielsen, 2012a). Among historical hypotheses, Remane (1963) argued explicitly for a colonial spherical choanoflagellate as an ancestor to Metazoa, instead of the hypothesis of a multinucleated plasmodial cell (e.g., Hadži, 1953), a hypothesis that at least is supported from a sister group perspective between Choanoflagellatea and Metazoa. However, from a traditional morphological perspective, reconstructing the nature of the oldest metazoan requires optimization of characters on phylogenetic trees. Optimizing characters on a well-resolved phylogeny is especially difficult when few characters are shared between the deepest nodes. Supposing that groups like Ctenophora, Porifera or even

Placozoa were the first offshoots of animal evolution, meaningful character optimization would be reduced to a handful of molecular markers and subcellular structures, something that would not help us in reconstructing the external morphology of an ancestor.

Likewise, such characters are unlikely to be recognized in the fossil record and thus if the last common ancestor of all animals looked like a comb jelly, a sponge, or a placozoan they would be recognized as stem groups of each of those three lineages, but probably not as the so-called Urmetazoan. Only one scenario, that of sponge paraphyly at the base of the animal tree, would provide the necessary power to say something about such an ancestor, as proposed by Nielsen (2008) in his "choanoblastaea" hypothesis. Sponge paraphyly is, however, disfavored in most recent phylogenetic analyses of sponges and metazoans.

Two facts are important for this book. First is the position of Metazoa in the broader tree of life within a clade of Opisthokonta named Holozoa. Holozoa includes, in addition to animals, choanoflagellates, filastereans and ichthyosporideans. Metazoa is well supported in all molecular phylogenetic analyses as sister group of Choanoflagellatea (e.g., Torruella et al., 2015) [fig. 1.1]. The resemblance of choanoflagellates to sponge choanocytes is striking and has been used a synapomorphy for the clade containing choanoflagellates and metazoans (= Choanozoa), reinforced in those topologies that suggest sponge paraphyly at the base of animals (Nielsen, 2012a). However, few real comparisons have been made between choanoflagellates and choanocytes until recently (Mah et al., 2014), and these authors indicated that although these cells are similar in some aspects, they differ in others, concluding that homology cannot be taken for granted. Similarities in collar-flagellum systems separated by 600 million years of evolution, whether homologous or convergent, suggest that these form important adaptations for optimizing fluid flow at microscale levels (Mah et al., 2014).

Irrespective of whether or not these two cell types are homologous, animal biologists have much to learn from animals' closest relatives. The first choanoflagellate genome, for the unicellular species *Monosiga brevicollis*, was thus sequenced to better understand the transition to multicellularity and tissue integration in metazoans. This genome, consisting of approximately 9,200 intron-rich genes, includes genes that encode for cell adhesion and signaling protein domains that were thought to be restricted to metazoans (King et al., 2008), but abundant domain shuffling followed the separation of the choanoflagellate and metazoan lineages. Nonetheless, a series of molecular synapomorphies of metazoans is still supported in the presence of special signaling, adhesion, and transcriptional regulation factors, including Wnt, Frizzled, Hedgehog, EGFR, classical cadherin, HOX, ETS, and POU, or the exclusive metazoan extracellular matrix components such as collagen type IV, nidogen, and perlecan. A list of core animal-specific gene families is given in Richter et al. (2018).

Second is the age of the oldest metazoan fossils, a much more controversial matter (see Sperling and Stockey, 2018, for a recent review). We begin this section by discussing some key paleontological facts and hypotheses in relation to the origin of metazoans.

WHAT IS A METAZOAN?

Defining Metazoa as a term is not trivial, and we now mostly recognize monophyletic groups as they are defined in phylogenies. Metazoa therefore includes any organism that shares a common ancestor with Ctenophora, Porifera, and Bilateria but excludes Choanoflagellatea. We do not consider here therefore the plethora of "protozoan" groups that used to be included in some textbooks as "unicellular animals," for these are not necessarily the closest sister groups of animals. Metazoans are organisms of multicellular organization, as opposed to unicellular or colonial ones, which means that there are special cell–cell junction molecules (Leys and Riesgo, 2012). That said, multicellularity is not exclusive to metazoans, as it occurs in multiple lineages of eukaryotes, even within Opisthokonta (Ruiz-Trillo et al., 2007). This has allowed division of labor, and even the simplest extant metazoans have multiple cell types.

All metazoans are also ingestive heterotrophic, but that is not equivalent to having a mouth, as pinocytosis and phagocytosis are the sole feeding mechanism of sponges and extracellular digestion with endocytosis by the lower epithelium occurs in placozoans. Virtually all other free-living metazoans ingest food through a mouth, with some exceptions of parasitic or symbiotic species. Nevertheless, the ability of metazoans to phagocytize food is unique among the multicellular eukaryotes (Mills and Canfield, 2016). A prevalent hypothesis is that the first metazoans—the common ancestor of all living metazoans—likely subsisted on picoplankton (planktonic microbes 0.2–2 μm in diameter) and dissolved organic matter, as sponges do nowadays and that therefore, through their feeding, helped bridge the strictly microbial food webs of the Proterozoic Eon (2.5–0.541 billion years ago) to the more macroscopic, metazoan-sustaining food webs of the Phanerozoic Eon (the past 541 mllion years) (Mills and Canfield, 2016). This hypothesis, however, relies upon a similarity between the last common ancestor of modern metazoans and modern sponges. Alternative phylogenetic hypotheses placing ctenophores more basally than sponges have been informally criticized for requiring carnivory to have evolved at the base of the animal tree, but this is not necessarily the case, as extant ctenophores seem to have diversified relatively recently, and there could have been other ecologies earlier in the stem ctenophore lineages. Furthermore, it is difficult to predict the feeding mode of possible extinct stem metazoans, but it is not unlikely that they would have fed on phytoplankton, as many animal larvae do nowadays.

Because multicellular animals must begin as unicellular, metazoan development shares some basic principles in sexual animals. While most nonsexual species tend to have sexual sister species, a few lineages of long-term asexual (mostly parthenogenetic) animals are supposed to exist, for example, bdelloid rotifers (Mark Welch and Meselson, 2000). This phenomenon has, however, recently been disputed, suggesting that bdelloids may have some sort of infrequent or atypical sex, in which segregation occurs without requiring homologous chromosome pairs (Signorovitch et al., 2015). Therefore, the presence of eggs and sperm cells could be considered the typical metazoan condition. After fertilization, metazoan zygotes develop from one of the four cells resulting from meiosis, whereas the other three cells become polar bodies and often degenerate. Embryogenesis in animals is, however, extremely diverse, and polar bodies can carry information or have specific functions. For example, they are key in fertilization of eggs in parthenogenetic animals, or have a role as extra-embryonic tissue in some parasitic wasps (Schmerler and Wessel, 2011).

Some authors have also attempted to identify metazoan-specific markers, including special glycoproteins such as collagens (a large family of proteins found in the extracellular matrix of metazoans), protein kinase C for cell signaling, or even specific neurotransmitters, but many of these molecules are now known from the genome of *Monosiga brevicollis*, suggesting a premetazoan history of protein domains required for multicellularity (King et al., 2008), and even neurotransmission may have a common origin with the primordial secretion machinery of choanoflagellates (Burkhardt et al., 2011; Hoffmeyer and Burkhardt, 2016). Some recent research may indicate that while fibrillary collagen motifs evolved in the common ancestor of choanoflagellates and metazoans, fibrillary collagen with covalent cross-links between individual fibrils are metazoan synapomorphies (Rodriguez-Pascual and Slatter, 2016). From these, type IV collagen or a type IV–like form (spongin short chain collagen) is present in the basement membrane of all metazoans (Leys and Riesgo, 2012). TGF-ß is also found in all animals but nowhere outside animals, although there are some differences in the complement of genes between sponges and ctenophores with the rest of animals (Pang et al., 2011).

In addition to protein-coding genes, animal *cis*-regulatory complexity (i.e., distinct enhancers and transcription factor binding sites for different genes that regulate their spatial and temporal expression), once thought to be the trademark of complex animals, is now known to be present in sponges (Gaiti et al., 2017; Hinman and Cary, 2017), but this has not been studied in ctenophores or placozoans.

Germ cells play a unique role in gamete production and thus in heredity and evolution. They can be specified either by maternally inherited determinants (preformation) or by inductive signals (epigenesis) (Extavour and Akam, 2003). At the molecular level, metazoans seem to share a germline multipotency program (GMP) with 18 GMP genes present in representatives of sponges, ctenophores, cnidarians, and bilaterians (Fierro-Constaín et al., 2017), showing that some of them evolved in Metazoa. Likewise, while homologizing germ layers across animals may be difficult, the expression of the transcription factor GATA in the sponge inner-cell layer suggests a shared ancestry with the endomesoderm of other metazoans and that the ancestral role of GATA in specifying internalized cells may precede the origin of germ layers (Nakanishi et al., 2014). This may imply that germ layers and gastrulation evolved early in eumetazoan evolution from developmental programs used for the simple patterning of cells in the first multicellular animals (Nakanishi et al., 2014).

▰▰▰ THE EARLIEST METAZOAN FOSSIL RECORD— THE PRECAMBRIAN

Most of the characters discussed above as likely apomorphies of Metazoa have negligible fossilization potential and, as is often the case, fossils may be assigned to a

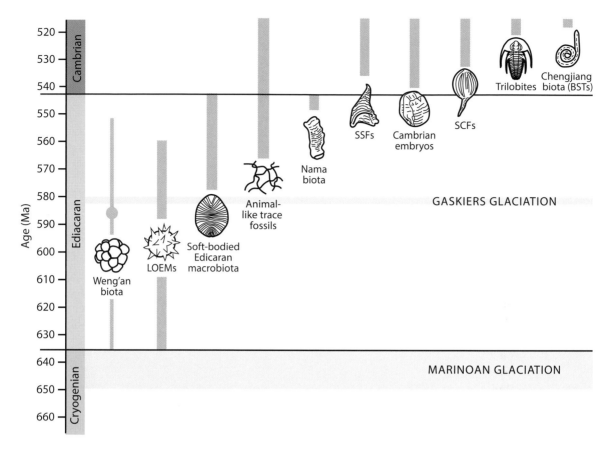

FIGURE 1.2. Temporal occurrence of key Ediacaran and early Cambrian fossil assemblages and preservational styles (based on Cunningham et al., 2017). Abbreviations: BSTs, Burgess Shale-type biotas; LOEMs, Large ornamented Ediacaran microfossils; SCFs, Small carbonaceous fossils; SSFs, Small shelly fossils.

group because they possess diagnostic characters of one or more of its subgroups. A review of metazoan characters proposed by Ax (1996), for example, noted that radial cleavage was the only one with reasonable fossilization potential (Cunningham et al., 2017). This particularly applies to the body fossil record of the earliest metazoans and impedes chances of identifying a stem-group metazoan, very few compelling candidates of which are known from the Proterozoic Eon (the late Precambrian), and these are confined to its latest period, the Ediacaran (635–541 Mya) (fig. 1.2).

Evidence from behavior in the form of trace fossils (=ichnofossils) has generally been regarded as providing the strongest evidence of Metazoa before the Ediacaran–Cambrian boundary (fig. 1.3). Trace fossils from the late Ediacaran (ca. 555 Ma) have the most widespread acceptance as having been made by animals, and some of these traces are generally ascribed to Bilateria (Gaidos et al., 2007; Gehling et al., 2014; Oji et al., 2018), some likely even made by bilaterians with paired appendages (Chen et al., 2018). That said, the situation is less straightforward than it might seem, as locomotory traces that have classically been attributed to Bilateria, such as bilobed trails with a median groove, have been observed to have been made by large protists (Matz et al., 2008), and other claimed bilaterian traces (Chen et al., 2013) have been attributed to nonanimal behavior, such as by slime molds (Retallack, 2013). Various sources have argued for bilaterian locomotory and feeding traces from earlier sediments, for example >585 Ma (Pecoits et al., 2012). Well-dated locomotory traces from the deep-water Ediacaran deposits in Newfoundland at least indicate that relatively large organisms (trace widths up to 13 mm) were motile by 565 Ma, and the form of the traces is consistent with muscular locomotion as in metazoans (Liu et al., 2010).

The trace fossil record for Ediacaran metazoans is potentially supplemented by biomarkers. At the center of the debate is the discovery of 24-isopropylcholestanes (24-ipc), the hydrocarbon remains of sterols interpreted as having been produced by marine demosponges, derived from rocks as early as 635 Myr old and extending into the Cambrian (Love et al., 2009). This hypothesis was later questioned, as several modern marine algae are also able to produce compositional isomers that are identical to the claimed sponge biomarker (Antcliffe, 2013) and a recent reexamination of the sponge fossil record was unable to unambiguously assign any Precambrian fossil to Porifera (Antcliffe et al., 2014). However, the hypothesis of the sponge biomarker again gained support based on molecular analyses of the origin of the sponge and plant 24-isopropylcholestanes (Gold et al., 2016a), suggesting indeed a gap on the order of 100 Myr for siliceous sponge spicules in the fossil record (Sperling et al., 2010). An additional demosponge-specific biomarker, 26-methylstigmatane (26-mes), has also been identified in Ediacaran sediments (see a summary of demosponge biomarkers in Sperling and Stockey, 2018). However, both 24-ipc and 26-mes have been found to be abundantly biosynthesized

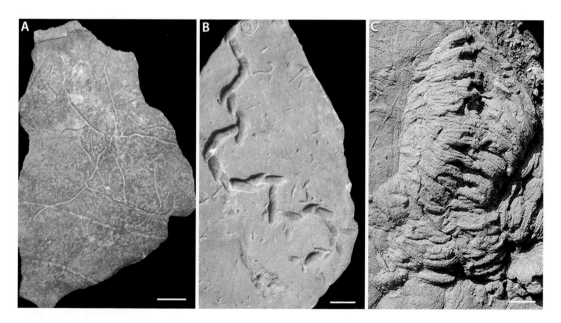

FIGURE 1.3. Metazoan trace fossils from the Ediacaran–Cambrian transition. A, *Helminthoidichnites* from the Ediacaran, scale 2 cm; B, *Treptichnus pedum*, index fossil for the base of the Cambrian, scale 1 cm; C, *Rusophycus burjensis*, a middle Cambrian arthropod resting trace, scale 5 mm. Image credits: B, Sören Jensen; C, Gabriela Mángano.

by unicellular protists in the clade Rhizaria, and the appearance of this group in the fossil record matches the geological appearance of the biomarkers better than does that that of sponges (Nettersheim et al., 2019). Irrespective of this debate over demosponges, lipid biomarkers from organic films on Ediacaran dickinsoniid megafossils have been identified as cholesteroids (Bobrovskiy et al., 2018), discussed below as evidence for total-group Metazoa.

A series of phosphatized and silicified microfossils (Muscente et al., 2015a) from the Doushantuo Formation, South China, known as the Weng'an biota, has yielded an abundance of three-dimensional fossils, including what have been interpreted as metazoan embryos (Chen et al., 2000; Xiao and Knoll, 2000; Chen et al., 2006; Chen et al., 2009b) and small postembryonic stages of metazoans (Xiao et al., 2000; Wang et al., 2008). However, subsequent reexamination or reinterpretation of the Doushantuo specimens has in several cases disputed the proposed evidence for metazoan identities, with many of the putatively biological structures being reinterpreted as geological in origin (reviewed by Crosby and Bailey, 2018). With regards to the putative embryos (fig. 1.4 A), some workers have accepted an identity as embryos but questioned their identification as metazoans, suggesting alternatives such as non-metazoan Holozoa, whereas others have proposed that these fossils are algal or bacterial (Huldtgren et al., 2011; Bengtson et al., 2012; Cunningham et al., 2012; Cunningham et al., 2015).

Another proposed indicator of metazoans in the Ediacaran are acritarchs, known as large ornamented Ediacaran microfossils (fig. 1.4 B, C), some of which have been interpreted as the encysted resting stages of Metazoa (Yin et al., 2007; Cohen et al., 2009). These acritarchs have a temporal range from approximately 635 to 560 Ma, but their precise affinities remain unclear and appear to be phylogenetically varied (Liu et al., 2014b), although some encase embryo-like Doushantuo fossils (Yin et al., 2007).

Perhaps the most fascinating Precambrian biota is the famous Ediacaran megafossils, a series of large (sometimes more than a meter long), soft-bodied, mostly sessile organisms originally described from the Flinders Ranges in Australia (fig. 1.4 G–J) but later extended to be a globally distributed marine biota, now known from all continents but Antarctica (Fedonkin et al., 2007a). Three different assemblages— the Avalon, White Sea Ediacaran, and Nama—are identified, the differences between them reflecting a mix of temporal, biogeographic, and especially environmental/biofacies differences (the Avalon, named for a classic site in Newfoundland, being the oldest and deepest water, and the Nama, first described from Namibia, being the youngest and shallowest). Together they span the interval from ca. 571–542 Mya, and span biofacies from lower-energy inner shelf settings (Avalon-type), wave- and current-agitated shoreface (White Sea Ediacaran) to high-energy distributary systems (Nama) (Grazhdankin, 2014).

Ediacaran organisms like *Dickinsonia, Mawsonites, Rangea,* and *Charniodiscus,* among others, have been discussed in the context of modern metazoan taxa, especially Cnidaria, but a plethora of other interpretations have been proposed, including that of a protozoan affinity (Seilacher et al., 2003) or being a wholly extinct radiation of nonanimal life, collectively grouped as Vendobionta. *Dickinsonia* (fig. 1.4 J), historically allied to annelids or cnidarians, exemplifies the diversity of current

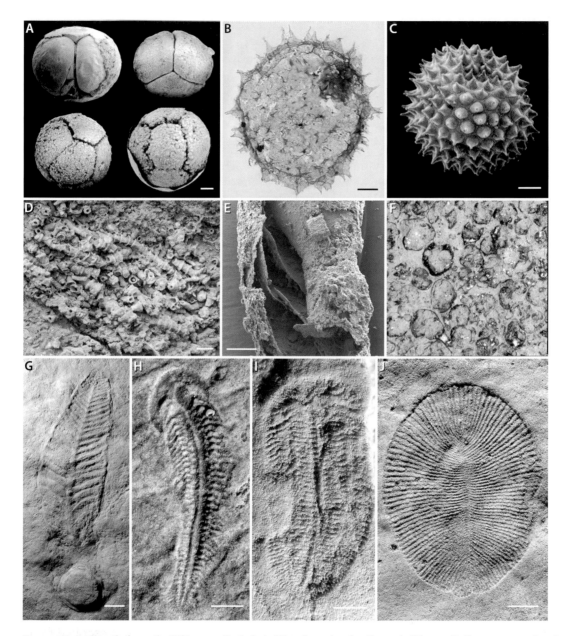

FIGURE 1.4. Fossils from the Ediacaran Period. A, Weng'an microfossils, scale 0.1 mm; B, C, ornamented acritarchs, B, *Aliceosphaeridium lappaceum*, scale 50 μm; C, *Meghystrichosphaeridium reticulatum*, scale 50 μm; D, *Cloudina carinata*, field context, scale 5 mm; E, *Cloudina hartmanae*, SEM, scale 200 μm; F, *Namacalathus hermanastes*, field context; G–J, South Australian Ediacaran macrofossils; G, *Charniodiscus arboreus*, scale 3 cm; H, *Spriggina floundersi*, scale 5 mm; I, *Marywadea ovata*, scale 5 mm; J, *Dickinsonia costata*, scale 1 cm. Photo credits: A–C, F, Andrew Knoll; D, E, Iván Cortijo; G–J, John Paterson.

opinion on the question of affinities for Ediacaran macrofossils. Its feeding traces (grazing on a microbial mat by absorption across its ventral surface) characteristic of external digestion have been argued to support a placozoan affinity (Sperling and Vinther, 2010). Its inferred mode of growth has alternatively been interpreted as involving repeated units being added at one pole and compared to terminal addition in metazoans, prompting an assignment to Bilateria (Gold et al., 2015) or instead involving preterminal addition of new units as in Metazoa more generally (Hoekzema et al., 2017; Dunn et al., 2018). *Dickinsonia* is confidently identified as being a mobile organism (Evans et al., 2015), but like the spectrum of Ediacaran macrofossils it lacks any structures that can be convincingly interpreted as a mouth or gut. Cholesteroids as the dominant lipids in *Dickinsonia* and the allied *Andiva* (Bobrovskiy et al., 2018) support an affinity of these organisms with Filozoa, the clade that unites metazoans with Filasterea and Choanoflagellatea (see fig. 1.1). This evidence, combined with large size, motility and mode of growth, amplifies the case for such dickinsoniids (and, by association, other Ediacaran megafossils with similar preterminal addition of new units and their subsequent inflation; Dunn et al., 2018) being total-group Metazoa. There is evidence for some frondose Ediacaran-type organisms surviving until the Cambrian (Jensen et al., 1998; Shu et al., 2006; Hoyal Cuthill et al., 2018).

The most widely endorsed body fossil evidence for Ediacaran metazoans comes from its terminal part, in the Nama Group (550–541 Ma) of Namibia and coeval, terminal Ediacaran biotas in Brazil, Russia, China, and other parts of the world. The key fossils are biomineralized and assume quite variable forms. Collectively they represent the earliest experiment in likely animal skeletonization. Such early skeletal fossils include *Cloudina*, a tiny tubular fossil organized as nested calcareous cones (fig. 1.4 D, E), and *Namacalathus*, which has a goblet-shaped calyx with several apertures (fig. 1.4 F), attached to a stalk (see chapters 3 and 44). Both of these as well as the tubular *Corumbella* have been compared to Cnidaria (see chapter 8) as well as other metazoan phyla.

To summarize, early skeletal fossils and a substantial body of indirect evidence suggests a Precambrian origin of Metazoa, in addition to support from nearly all analyses of molecular dating. Indeed, the latter converge on a minimal age for crown-group Metazoa in the Tonian, more than 720 Ma (Sperling and Stockey, 2018). While unambiguous consensus on the animal identity of any of these fossils is yet lacking, biomarkers and developmental modes are consistent with some Ediacaran megafossils being total-group metazoans. In the chapter on Cnidaria (chapter 8), the most likely cases of Ediacaran-age metazoan body fossils are discussed.

▬ THE PRECAMBRIAN–CAMBRIAN BOUNDARY

The dearth of definite metazoans in the Precambrian may of course be the result of multiple factors. Soft-bodied pelagic animals do not fossilize well, and neither do the smallest meiofaunal animals (although Cambrian loriciferans preserved as small carbonaceous fossils provide a fascinating exception; Harvey and Butterfield, 2017). Being small, thin, rare, and of low population density would severely impede the possibility of preservation (Sperling and Stockey, 2018). It is not out of the realm of possibility that some of the earliest animals may have looked like sponge larvae or like many of the microscopic meiofaunal taxa inhabiting water bodies all over the world nowadays.

Most of the animals that today branch out early in the Animal Tree of Life are either pelagic (most ctenophores and jellies, plus several other cnidarians) or, if benthic, are sessile (sponges and many cnidarians). Placozoans, another early animal lineage probably related to cnidarians (Laumer et al., 2018), have been observed swimming but are often collected on glass slides, crawling on surfaces. None of these early animals make burrows or penetrate marine sediments. That said, in contradiction to predictions from molecular dating, there is a sound basis for doubting that animals existed deep into the Proterozoic but failed to be preserved because of deficiencies or preservational biases in the fossil record. Precambrian sediments do in fact preserve fossils in cellular detail (Brasier, 2009), and it is increasingly understood that exceptionally preserved biotas in the late Ediacaran and the Cambrian have similar taphonomy (Daley et al., 2018).

The Ediacaran provides a fossil record in phosphorites, in cherts, as organic fossils, and as compression fossils, yet in striking contrast to the early Cambrian, the abundant and often exquisitely preserved fossil remains from the late Ediacaran are not crown-group animals (Daley et al., 2018). The sequential increase in complexity and diversity of trace fossils across the Ediacaran to middle Cambrian interval (Mángano and Buatois, 2014) is an especially powerful indication that the early Cambrian records key aspects of animal behavior and ecology evolving in real time.

Originally defined by the first appearance of fossil metazoans with mineralized skeletons, such as trilobites, the dating of the Precambrian–Cambrian boundary has changed considerably in the past decades. In the 1970s, with the discovery of small shelly fossils (SSF) below the oldest Cambrian trilobites—a fauna composed of spicules, sclerites, and ossicles of animals plausibly interpreted as sponges, molluscs (including extinct scleritome-bearing groups such as halkieriids), stem-group brachiopods, and so on—the base of the Cambrian was revised. Nowadays it is defined by the first occurrence of the trace fossil *Treptichnus pedum* (fig. 1.3 B), one of the first penetrative burrows, suggesting a metazoan tracemaker able to move between layers of sediments and probably also causing a great deal of bioturbation that oxygenated and mixed the sediment at a depth not previously attained—the so-called agronomic revolution that provides one of the possible ecological explanations for the Cambrian "explosion" of animal life (Brasier, 2009).

The nature of the animal that left the trace fossil *T. pedum* has been a matter of speculation, but it must have been a relatively large (macroscopic) animal based on the size of the burrows. Comparisons with the feeding traces of extant priapulans are consistent with *T. pedum* having been produced by a priapulan-like animal (Vannier et al., 2010), and exceptionally preserved treptichnid traces from the early Cambrian of Sweden closely replicate the morphology of priapulans as well as actualistic observations on burrowing behavior by *Priapulus caudatus* (see Kesidis et al., 2019). *Treptichnus pedum* has also often been associated with the presence of some sort of hydrostatic skeleton, such as a coelomic or a large pseudocoelomic cavity, that allowed the animals to burrow but probably lacked a well-developed cuticle.

Although the presence of a cuticle is quite widespread across animals, a cuticle is in fact difficult to define (Rieger, 1984; Ruppert, 1991b) but may be characterized as an apical extracellular matrix secreted by and covering the epidermis. Cuticles can be a simple glycocalyx or complex ones consisting of an organic matrix with protein fibers, such as collagen or keratin, or polysaccharides, such as chitin and cellulose. The cuticle can also be mineralized to form spicules or shells. The fossilization potential of different macroscopic coelomate worms can thus be radically different in animals with complex cuticles (e.g., annelids) and those with a simpler ciliated epidermis (e.g., nemerteans), as the latter decay rapidly after death. A bias in sediments that affects fossilization need also be acknowledged. Trace fossils in the earliest Cambrian are often in coarser-grained siliciclastic rocks that typically lack body fossils, particularly nonshelly ones, whereas nonbiomineralized body fossil preservation almost requires a lack of burrowing and disruption/oxygenation of the sediment. This means the two styles of preservation—the traces and the bodies—are almost exclusive of each other.

THE CAMBRIAN EXPLOSION

The term "Cambrian explosion" refers to the relatively sudden appearance of a large number of mostly large-bodied animals during the early Cambrian, including the first records of disparate animal phyla and, ultimately, the rise of metazoan-dominated marine ecosystems. It seems clear now that animal diversification during the Cambrian, giving origin to many animal phyla, was not an explosive radiation per se but that such diversity was made apparent to us through a series of narrow fossilization windows, giving the appearance of a sudden origin of so many animal lineages. Mineralization of course plays a key role in recognizing such diversity (Kouchinsky et al., 2012; Briggs, 2015), given that in several lineages (such as chaetognaths, represented by their organophosphatic grasping spines known as "protoconodonts"), mineralized microfossils precede the first appearance of nonbiomineralized parts of the body. For some of the major groups of animals, the latter coincides with the opening of a taphonomic window known as "Burgess Shale-type preservation" (Gaines, 2014), after the eponymous Burgess Shale in British Columbia, Canada, in which recalcitrant nonbiomineralized tissues are preserved as largely two-dimensional carbonaceous compressions.

This style of preservation is known from the late Proterozoic (Dornbos et al., 2016), exemplified by the Miaohe biota in China, which preserves abundant and diverse macroscopic algae (An et al., 2015), but it is especially characteristic of offshore siliciclastic rocks from a temporal window spanning Cambrian Stage 3 (ca. 518 Ma) to the Early Ordovician (fig. 1.5). For groups like arthropods and priapulans, the skeletal fossil record commences with Burgess Shale-type preservation in the Chengjiang and Qingjiang biotas in China (Hou et al., 2017; Fu et al., 2019) and the nearly synchronous Sirius Passet biota in Greenland (Cambrian Series 2), but the trace fossil record indicates that the lineages extend to or close to the base of the Cambrian.

Other styles of fossil preservation provide vital information about the diversity and ecology of Cambrian animal communities, such as secondarily phosphatized fossils known as Orsten. First and most comprehensively documented from the Guzhangian–Furongian (late Cambrian) of Sweden, Orsten-style preservation is

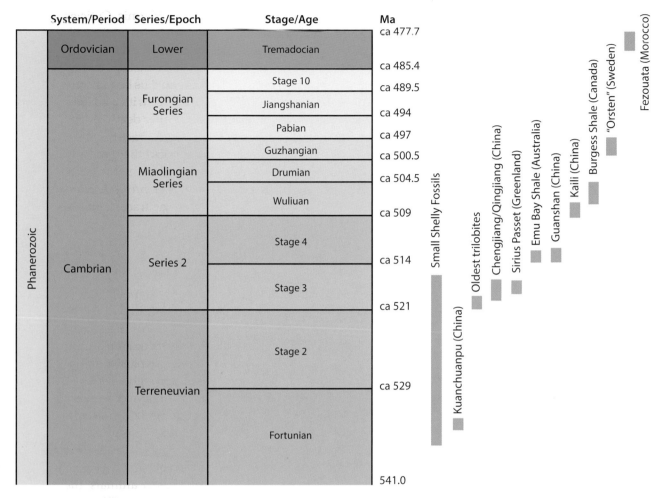

Figure 1.5. Temporal occurrence of key Cambrian and earliest Ordovician fossil assemblages relative to the Cambrian time scale (modified from Erwin and Valentine, 2013).

also known from the early Cambrian, permitting embryos, early postembryonic and larval stages to be documented for various groups of animals, including ec-dysozoans and cnidarians (Donoghue et al., 2015). Taken together, exceptionally preserved fossil biotas from the Cambrian include a large diversity of animals in extant and extinct lineages. Although it is often said that all major animal phyla except bryozoans appear already in the Cambrian fossil record, this is not true; about half of the currently recognized phyla, including some with numerous species and large sized-animals, like Platyhelminthes and Nemertea, do not have a confirmed Cambrian fossil record.

Many hypotheses have been discussed in the context of the Cambrian explosion of animal life, including a series of intrinsic (genetics, arms race hypothesis) and extrinsic abiotic (changes in ocean circulation patterns, and sea level rise that induced continental flooding and regolith erosion on a vast scale, the Great Unconformity; Peters and Gaines, 2012) and biotic (availability of oxygen and/or food) factors. For some, the timing of the onset may be constrained by the environment, whereas the duration may be conditioned by developmental innovation (Marshall, 2006). Understanding why this event was unique is often explained by factors ranging from developmental constraints to ecological saturation (Erwin and Valentine, 2013). Conversely, a fundamental distinction between the Cambrian "explosion" and earlier (Ediacaran) events of evolutionary radiation at the base of Metazoa has been questioned, these being seen as part of a longer phase in which animals evolved in response to changes in the Earth's biogeochemical cycles (Wood et al., 2019). It is inescapable that the Cambrian explosion is in fact a pattern of profound but sequential ecological change that resulted from an interconnected set of biotic and abiotic factors (Smith and Harper, 2013), including carnivory and food availability (Sperling et al., 2013).

PHYLOGENETICS AND THE BASE OF THE ANIMAL TREE OF LIFE

This book is structured around the Animal Tree of Life. We thus begin with a brief discussion of the recent history of metazoan phylogenetics, as well as addressing some methodological and technological developments that have influenced our views of the tree. Additionally, we outline some of the most heated discussion about how the base of the animal tree is shaped and review the most prevalent hypotheses about early animal evolution.

INFERRING THE ANIMAL TREE OF LIFE— A HISTORICAL INTRODUCTION

The first formulations of the Animal Tree of Life were mostly based on authoritative—although often not unfounded—knowledge of animal anatomy and morphology. Perhaps the most famous of these trees were the ones published by Prussian/German zoologist Ernst Heinrich Philipp August Haeckel in his *Generelle Morphologie der Organismen* (Haeckel, 1866). Haeckel was a prolific biologist, philosopher, and artist who coined, among many others, the terms "phylogeny," "ecology," and "phylum." Discussions among the relationships of animal groups are rooted in the principle of evolution, although the implicit notion of relationship, translated into classification systems, is as old as zoology itself. Here we focus on three major developments in systematics. First is the introduction of explicit methodologies to identify clades: the concept of synapomorphy and the principles of cladistics, which were put forward by Willi Hennig (1950, 1966). Numerical cladistics then played a key role in the publication of the first explicit morphological data matrices to explore metazoan relationships (e.g., Meglitsch and Schram, 1991; Schram, 1991; Eernisse et al., 1992; Nielsen et al., 1996; Zrzavý et al., 1998; Peterson and Eernisse, 2001). Molecular methods applied to phylogenetics are the third key development and one that has had a profound impact on the field of zoology.

Even before the first explicit parsimony-based cladistic hypotheses about metazoan phylogeny based on morphological data were published, molecular sequence data erupted in the zoological community. Since the discovery of the structure of DNA by Watson and Crick (1953), scientists attempted to sequence the genetic code, but the first efficient method was that proposed by Sanger et al. (1977). A decade later, DNA sequencing was beginning to be applied to animal phylogenetics. The first analyses of metazoan relationships using information from nucleic acids started with a series of seminal papers using partial or the nearly complete sequences of the *18S rRNA* gene (Field et al., 1988; Raff et al., 1989; Turbeville et al., 1991; Turbeville et al, 1992; Halanych et al., 1995). These early studies sequenced either RNA (initially) or DNA amplified by polymerase chain reaction or PCR (later). The introduction of the PCR method for amplifying DNA (Mullis and Faloona, 1987; Saiki et al., 1988) popularized molecular work to an unprecedented level. DNA sequencing until then used radioactivity and polyacrylamide gels and was mostly restricted to molecular laboratories that were often not accessible to zoologists. Substitution of radioactivity for the much cleaner labeled fluorochromes detected by a laser ("automated sequencers"), and subsequently the introduction of capillary electrophoresis to substitute the tedious polyacrylamide gels, facilitated DNA sequencing, making it much more accessible to a broader range of scientists.

This series of technological breakthroughs allowed for the first sequencing of complete eukaryotic genomes (Goffeau et al., 1996), including the first animal, the nematode *Caenorhabditis elegans* (*C. elegans* Sequencing Consortium, 1998), using the so-called Sanger technology. Soon after, new animal genomes were sequenced using brute-force capillary sequencing, including the model organism *Drosophila melanogaster* by 2000 (Adams et al., 2000), and the two first human genomes in 2001 (Lander et al., 2001; Venter et al., 2001). The technology and sequencing facilities assembled to sequence the human genomes soon found vast numbers of DNA facilities offering cheap sequencing projects to researchers without the capacity to sequence at that scale in their home institutions. Molecular phylogenies of the animal tree and its major branches were soon produced by the thousands.

During this time, the first combined analyses of morphological and molecular data sets were also published (e.g., Zrzavý et al., 1998; Giribet et al., 2000; Peterson and Eernisse, 2001; Glenner et al., 2004), but few new morphological data sets were developed to study deep metazoan phylogeny, the last morphological data sets for animals dating to the mid-2000s (Almeida et al., 2003, 2008; Jenner and Scholtz, 2005). Three reasons seem most apparent for this halt in producing metazoan morphological data sets. First is a pragmatic one: molecular data sets grew rapidly with the general acceptance of molecular phylogenetics by the broader community.

Second, a seminal paper exposed problems with the common treatment of terminals in morphological data sets as ground patterns rather than as exemplars (Prendini, 2001), partly inspired by how taxa are "coded" in molecular analyses. Until then, metazoan morphological data matrices used higher taxa (often phyla) as terminals and coded a combination of character states that were often based more on theory than on observation. The taxon Annelida, for example, was coded as segmented and possessing parapodia and chaetae despite the fact that many annelids lack some or all of these characters (Andrade et al., 2015). Mollusca were likewise coded with a shell, radula, and a molluscan cross, and Kinorhyncha were often coded as nonsegmented, despite having external cuticular plates that correspond with internal musculature. But all these characters, which were inferred to be present in the ground plan of these phyla were not real observations on real animals, not allowing a test of the monophyly of the group (since then many groups have been classified within Annelida), and not including in the data matrix the variation of characters that may have an impact on the tree inference as well as on character optimization.

Third was a series of papers by Ronald Jenner that criticized common practices in coding metazoan data matrices, especially with respect to recycling characters and the common recognition of nonhomology as a character state (e.g., Jenner, 1999, 2000, 2001, 2002; Jenner and Schram, 1999). As an example, this refers to coding characters such as spiral cleavage as "absent" or "present." While the *presence* may indicate homology, the amount of different patterns (radial, duet spiral, etc.) cannot be lumped into a single homologous state. When extending this principle to many other characters applicable to only parts of a data matrix, it becomes impossible to establish homologies across many characters in the animal kingdom when comparing a sponge to an arthropod, for example, as most characters that can be coded across taxa will refer to molecular markers and cellular and subcellular processes. Following the new recommendations of Lorenzo Prendini and Ronald Jenner made it intractable to produce a data matrix across the metazoans based on observations for every cell. In addition, the generalized use of large molecular data sets made the enterprise of coding such matrices less attractive (Giribet, 2010).

In parallel, a series of technological developments changed the landscape of animal phylogenetics. Traditional Sanger-based phylogenies (those that required specific primers to amplify a preselected gene) of metazoans had been based on just one or a few genes, this number only growing considerably for the best-studied groups of animals, such as arthropods (Giribet et al., 2001; Regier et al., 2010). Soon the presence of complete genomes for a handful of metazoans allowed broad phylogenetic comparisons using hundreds to thousands of genes (Blair et al., 2002; Dopazo et al., 2004; Wolf et al., 2004; Philip et al., 2005)—these were the first papers to use the term "phylogenomics" in a phylogenetic context. Series of orthologous genes were selected using different algorithms and these genes were used for subsequent phylogenetic comparisons, realizing that genes did not necessarily needed to be preselected (as in PCR-based approaches) for phylogenetic comparison.

Together with these available genomes, researchers changed their sequencing strategy and shifted efforts from PCR-based amplification to produce *expressed sequence tags* (ESTs), the sequencing of the mRNA after total RNA extraction, with the aim of obtaining a representation of the transcriptome of a tissue or an organism. ESTs were being produced for many purposes, especially for evo-devo studies, and soon the first EST-based Sanger phylogenies appeared (Philippe et al., 2005; Delsuc et al., 2006). This launched the field of phylogenomics (Delsuc et al., 2005). The first complete metazoan phylogenies using a combination of genomes and ESTs, and using automated methods for orthology selection, followed, and to date these remain the most cited phylogenomic analyses of metazoans (Dunn et al., 2008; Hejnol et al., 2009).

Subsequent changes followed, with the appearance of new technologies for parallel massive sequencing that allowed producing hundreds of millions of sequences at a very low cost (Dunn et al., 2014; Wheeler and Giribet, 2016). Current phylogenies are now based on entire genomes (e.g., Clark et al., 2007) or large numbers of fairly complete transcriptomes (Misof et al., 2014), opening new doors into broader questions in animal phylogenetics. A new generation of phylogenomic analyses of large fractions of the Animal Tree of Life are presently emerging (e.g., Laumer et al., 2015a; Laumer et al., 2019; Marlétaz et al., 2019). The specific results and interpretation of these phylogenies are discussed in the appropriate sections of this book.

THE BASE OF THE ANIMAL TREE OF LIFE

Resolution at the base of the animal phylogenetic tree remains contentious despite huge recent efforts using phylogenomic and genomic data. Many often-contradictory hypotheses have been supported by different sets of data and methodologies. Traditional views followed a scheme in which complexity increased in a ladder-like fashion, with Porifera appearing as the sister group to all other animals, followed by Placozoa, Cnidaria, and Ctenophora, among the nonbilaterians, and some authors even placed Ctenophora higher up in the tree, nested within Bilateria (e.g., Nielsen et al., 1996). The position of Placozoa and Porifera was often interchanged, or the two formed a clade (e.g., Schram and Ellis, 1994), and the non-poriferans/non-placozoans were often termed Eumetazoa (Bütschli, 1910)—or the metazoans with tissues and organs (also called Histozoa).

The name Neuralia has also been applied more recently to those animals with specialized communication cells, neurons with axons, electrical synapses known as "gap junctions," chemical synapses with various neurotransmitters, and signal propagation via action potentials using sodium/potassium ion channels (Nielsen, 2012a). This scheme, prevalent for most of the twentieth century and mostly based on morphology, was challenged by the introduction of molecular data, which often placed Placozoa higher up in the animal tree than previously anticipated. Thus placozoans are now interpreted as being secondarily reduced. This is also supported by the presence in Placozoa of a ParaHox cluster, also found in cnidarians and bilaterians (hence the name "Parahoxozoa"), but absent in Porifera (Fortunato et al., 2014; Pastrana et al., 2019) and Ctenophora (Ryan et al., 2010).

Two hypotheses remain with respect to the candidate sister group to all other animals, the "Porifera-sister" hypothesis, in which sponges occupy this position (e.g., Pisani et al., 2015; Feuda et al., 2017; Simion et al., 2017), and the "Ctenophora-sister" one, in which ctenophores occupy it (e.g., Dunn et al., 2008; Hejnol et al., 2009; Ryan et al., 2013; Whelan et al., 2015; Whelan et al., 2017) (fig. 2.1). The "Porifera-sister" scenario is almost compatible with the proposed clade Epitheliozoa, which includes Ctenophora, Cnidaria and Bilateria (but not Placozoa), and is grounded on the idea that sponges lack an epithelium and a basement membrane. An epithelium is especially important to maintain a controlled internal milieu and distinguish a multicellular animal from a colony.

However, recent morphological, functional, and molecular characterization of epithelia in sponges show that they have claudin-like genes that align most closely with sequences from *Drosophila melanogaster* that have a barrier function in septate

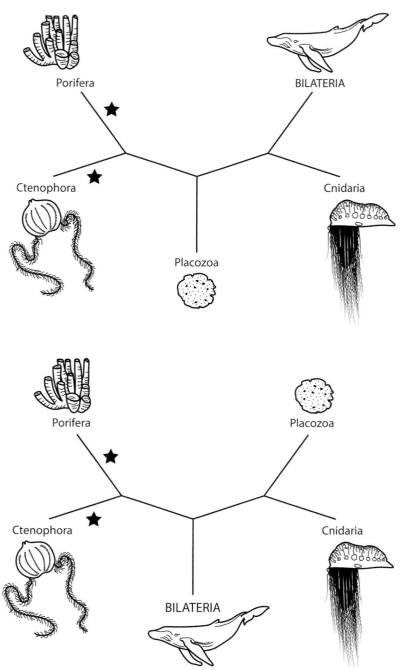

FIGURE 2.1. Schematic representation of two unrooted animal trees. Stars indicate possible rooting of the trees.

junctions. They also present type IV collagen, the main component of the basement membrane (BM) in calcareous sponges, or type IV-like collagen (spongin short chain collagen) in other sponges (Leys and Riesgo, 2012). Thus, the main character of a functional epithelium—the ability to seal and control the ionic composition of the internal milieu—is a property of even the simplest sponge epithelium (Leys and Riesgo, 2012). The putative clade Epitheliozoa, although partially supported in some phylogenetic analyses (barring Placozoa), is defined by an incorrect synapomorphy and will therefore not be used here.

"Ctenophora-sister" is difficult to support by any clear morphological character, but several molecular synapomorphies have been proposed, including paired domains linked to homeodomains, NR2A genes including a DNA binding domain, and Drosha microRNA processing (Dunn et al., 2014). It has also been suggested that ctenophores lack many genes that are essential for nervous system development and function in cnidarians and bilaterians, suggesting that ctenophores have nervous systems that are quite different from those of other animals (Ryan et al., 2013; Moroz et al., 2014). While some authors do not find this argument satisfactory because differences cannot in themselves obviate homology, virtually all supported topologies of the Animal Tree of Life require either multiple gains or multiple losses of nerves, as no topology based on phylogenomic analysis supports Neuralia—a sister-group relationship or a basal grade of the nerveless Porifera and Placozoa.

Resolution of this debate, which has been ongoing for a decade, requires methodological consensus, and it is unlikely to be resolved anytime soon, despite claims to the contrary (e.g., Dunn et al., 2008; Hejnol et al., 2009; Philippe et al., 2009; Pick et al., 2010; Feuda et al., 2012; Nosenko et al., 2013; Ryan et al., 2013; Feuda et al., 2014; Moroz et al., 2014; Pisani et al., 2015; Whelan et al., 2015; Arcila et al., 2017; Feuda et al., 2017; Shen et al., 2017; Simion et al., 2017; Whelan et al., 2017; Laumer et al., 2018). The effect of taxon selection (and possible amino acid compositional biases in ctenophores and fungal outgroups), model choice, gene representation, and other factors are advocated to be the artefactual basis of one topology or the other, but subsequent analyses correcting for each of these effects continue without producing fully convincing evidence (see Giribet, 2016a; King and Rokas, 2017). Genome-scale analyses using presence/absence of either ortholog groups or homologous protein families continue to differ in the placement of Ctenophora (the former placing them as sister group of all metazoans apart from Porifera and the latter uniting them as sister group of Cnidaria) (Pett et al., 2019). Ultimately, massive extinction or extremely rapid evolution at the base of the animal tree may be insurmountable issues leading to the current conflicting resolution of these nodes.

Forcing Placozoa to a basal position, another possibility proposed by some authors, however, has resulted in odd topologies, where Porifera, Cnidaria, and Ctenophora form the sister clade to *Trichoplax*, this larger group being the sister clade to Bilateria (Schierwater et al., 2009). This hypothesis forces a clade of animals with all levels of complexity and has not been obtained in any other phylogenomic analyses of deep animal relationships and is therefore not further considered here. However, a recent analysis including genomes of four placozoans found strong support for a sister-group relationship of Placozoa and Cnidaria (Laumer et al., 2018), contradicting most previous phylogenomic analyses that supported Planulozoa when only the genome of *T. adhaerens* was sampled (e.g., Srivastava et al., 2008; Hejnol et al., 2009; Pisani et al., 2015; Simion et al., 2017; Whelan et al., 2017). Because this topology was thoroughly tested by Laumer et al. (2018), it is largely followed in this book.

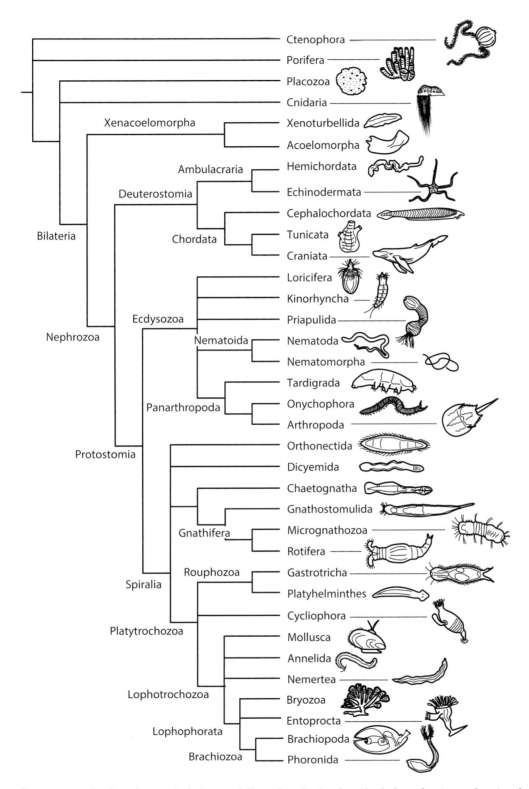

Ctenophora
Porifera
Placozoa
Cnidaria
Xenacoelomorpha
  Xenoturbellida
  Acoelomorpha
Bilateria
Deuterostomia
  Ambulacraria
    Hemichordata
    Echinodermata
  Chordata
    Cephalochordata
    Tunicata
    Craniata
Nephrozoa
Ecdysozoa
  Loricifera
  Kinorhyncha
  Priapulida
  Nematoida
    Nematoda
    Nematomorpha
  Panarthropoda
    Tardigrada
    Onychophora
    Arthropoda
Protostomia
  Orthonectida
  Dicyemida
  Chaetognatha
  Gnathifera
    Gnathostomulida
    Micrognathozoa
    Rotifera
  Rouphozoa
    Gastrotricha
    Platyhelminthes
Spiralia
  Cycliophora
Platytrochozoa
  Mollusca
  Annelida
  Nemertea
Lophotrochozoa
  Bryozoa
  Entoprocta
  Lophophorata
    Brachiopoda
  Brachiozoa
    Phoronida

FIGURE 2.2. Preferred animal phylogeny followed in this book with phyla at the tips and major clades indicated at nodes. This tree is derived from a synthesis of works, most of which are discussed in these pages.

Studies of the anatomy and development of sponges, ctenophores and placozoans have flourished as a consequence of this debate, providing new insights into the biology of these interesting animals (Dunn et al., 2015). One of these studies has postulated a primitive sensory organ in sponges (Ludeman et al., 2014). Another has revisited the question of homology between choanocytes and choanoflagellates (Mah et al., 2014). Others have shown complex behaviors in placozoans requiring some sort of cell integration and possible neurotransmitters (Smith et al., 2015; Senatore et al., 2017) or have analyzed opsins to infer how ciliary, rhabdomeric, and Go-coupled phototransduction evolved at the origins of animal vision (Feuda et al., 2012; Schnitzler et al., 2012; Feuda et al., 2014). As advised by Dunn et al. (2015), "It is now clear that the phylogenetic placement of Porifera and Ctenophora are not independent questions, and must be addressed together."

We are just beginning to assemble the genomic data sets to attempt to resolve this conundrum, and the response is not a simple one that can find its only explanation in a model of sequence evolution or the inclusion/exclusion of certain outgroups. Development of phylogenetic methods and careful analysis of data—including a thoughtful choice of key taxa, not only those available in current databases—will be required to continue shedding light into the early evolution of animals. Porifera and Ctenophora remain as candidates for the sister group to all other animals (= Planulozoa; see chapter 5), but it is also true that a greater genomic diversity of placozoans may also help to settle this debate (e.g., Laumer et al., 2018), as we know that the group includes enormous genetic diversity (Voigt et al., 2004; Pearse and Voigt, 2007; Eitel and Schierwater, 2010; Eitel et al., 2013).

CTENOPHORA

Ctenophora (comb jellies or sea gooseberries) are mostly macropelagic (from a few cm to almost 2 m in length), exclusively marine animals of a gelatinous consistency, nearly all of which spend their lives in the water column. A handful of benthic species exist, including the sedentary *Tjalfiella tristoma* and one of the two species of *Lyrocteis*. There are between 150 and 250 described valid species, but due to their gelatinous nature they are difficult to collect and study, and thus many remain to be formally described. Ctenophores have been recorded from the water surface, where they can be locally abundant, to depths exceeding 7,000 m, as shown by a report of a benthic species discovered at a depth of 7,217 m at the Ryukyu Trench off the coast of Japan (Lindsay and Miyake, 2007).

Ctenophores are carnivorous, feeding upon other small pelagic and planktonic animals, often captured by a pair of tentacles loaded with a special type of sticky cells called "colloblasts,"[1] or by engulfing the prey, including smaller ctenophores, with their large mouth, as is the case of *Beroe*. Their exclusively predatory lifestyle is difficult to reconcile with the hypothesis placing them as sister group to all other metazoans—unless the stem ctenophore had a different feeding ecology.

Ctenophores are the largest animals to swim by ciliary action. For this they use their characteristic rows of iridescent ciliary plates (hence their name, "combs" or "ctenes") that are present at least during some part of their life cycle. In addition to ciliary action, ctenophores can also move by muscular contraction of their lobes, or they can use their musculature to withdraw the retractable tentacles into tentacular sheaths.

Ctenophores have nerves, muscle cells, symmetry (with an infinite number of planes of rotational symmetry), and a through-gut, properties that have often been associated with animals placed higher up in the phylogenetic scale. They also share similarities with cnidarians, especially in their unipolar mode of early cleavage (Scholtz, 2004)—a mode of total cleavage in which the cleavage furrow starts at one side of the cell and progresses in a zipper-like mode and one of the characters used to support the putative clade Coelenterata, together with the supposed structure of the digestive tract (but see recent interpretations of the ctenophore anus below). Overall, this unique combination of characters makes it difficult to place ctenophores morphologically.

As seen in chapter 2, ctenophores have received special attention due to the recent debate about whether they or sponges are sister group to all other animals (e.g., Dunn et al., 2008; Hejnol et al., 2009; Ryan et al., 2013; Moroz et al., 2014; Pisani

---

[1] The term "collocyte" is also applied to these cells, but as a more general term it is applied to glue cells in other invertebrates, including tunicates, and thus we prefer the ctenophore-exclusive "colloblast."

et al., 2015; Whelan et al., 2015; Feuda et al., 2017; Simion et al., 2017; Laumer et al., 2018; Laumer et al., 2019). This debate has invigorated ctenophore research and has led to key genomic resources and to careful comparisons of the body plans of ctenophores and sponges with respect to other animals and outgroups (Dunn et al., 2015; Halanych, 2015; Jékely et al., 2015).

SYSTEMATICS

Ctenophores are currently classified into 9 orders, 27 families, and roughly 150 to 250 species (Mills, 1998–2016), but the traditional system conflicts with molecular phylogenetic work using traditional Sanger sequencing of a handful of markers (Podar et al., 2001; Simion et al., 2014). Results from a large phylogenomic analysis, including more than 35 species of ctenophores (Whelan et al., 2017), are comparable to those of the more densely sampled published Sanger-based phylogenetic analyses (Simion et al., 2014), which show that the larger orders Cydippida and Lobata are paraphyletic, while many of the smaller orders are probably monophyletic, although this remains largely untested. Until a comprehensive phylogeny of Ctenophora including a large diversity within its main groups becomes available, it seems futile to repeat here the classification system currently in use but widely recognized to be artificial.

CTENOPHORA: A SYNOPSIS

- Pelagic or benthic gelatinous, macroscopic marine animals
- Two embryonic layers, endoderm and ectoderm; the endoderm and ectoderm often separated by a cellular mesenchyme with muscle and macrophage-like mesenchymal cells
- Oral–aboral main body axis and a so-called rotational symmetry along three main planes of symmetry
- Special type of adhesive exocytotic cells called "colloblasts" (absent in some species)
- Branching digestive cavity, the mouth on the oral side and two anal pores on the aboral side, constituting a through-gut
- Pair of long tentacles (missing entirely in beroids), with branching tentilla loaded with colloblasts used for prey capture; the tentacles are extended by the water current but can be retracted into their sheath by muscular action
- Eight rows of ciliary comb plates (reduced in adults of benthic species) controlled by a unique apical sense organ
- Simple aboral apical sense organ, consisting of a mineral statolith supported by four balancers (one for each bifurcating row of comb plates) inside a ciliated cellular statocyst capsule; this organ is often associated with a pair of polar fields
- Nerve cells and muscle cells
- Monomorphic life cycle, with direct development (the "cydippid larva" is just a juvenile stage that transforms into an adult progressively)

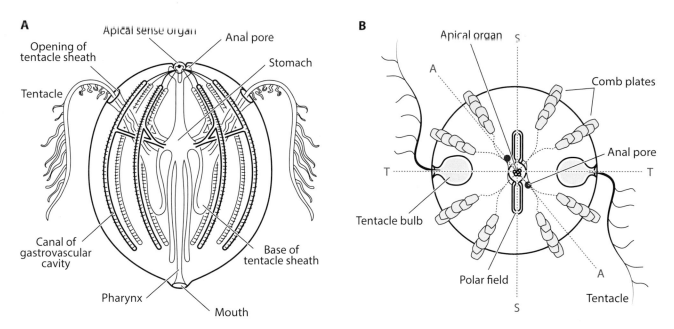

A Schematic image of an idealized ctenophore with main structures illustrated. Labels: Opening of tentacle sheath, Apical sense organ, Anal pore, Stomach, Tentacle, Canal of gastrovascular cavity, Base of tentacle sheath, Pharynx, Mouth.

B Aboral schematic view labels: Apical organ, S, A, Comb plates, Anal pore, T, T, Tentacle bulb, Polar field, A, Tentacle, S.

FIGURE 3.1. A. Schematic image of an idealized ctenophore with main structures illustrated. B. Aboral schematic view of an idealized ctenophore with the tentacular (T), sagittal (S) and anal (A) planes of symmetry. B modified from Martindale et al. (2002).

- Most species hermaphroditic
- Distinctive stereotyped early cleavage; unilateral cleavage during early embryogenesis; all major cell types determined by the 60-cell stage

Symmetry is often used to describe metazoan body plans. While some animals, like Porifera and Placozoa, lack a defined plane of symmetry, most animals are bilaterally symmetrical—their body can be divided in two subequal halves, left and right mirror images, along a longitudinal axis. This symmetry can be highly modified in the adult forms of echinoderms (said to present pentaradial symmetry) and many gastropods, which undergo a developmental process called "torsion." Symmetry in ctenophores is difficult to define. Three main planes of symmetry can be identified, none of them generating two exact mirror images: a tentacular plane is defined by the tentacles, a sagittal (or stomodeal) plane is defined by the stomodeum, and an anal plane is defined by the position of the anal pores, which connect the anal canals to the gut. It is for this reason that a rotational plane of symmetry has been postulated (fig. 3.1) (Martindale et al., 2002).

The digestive cavity of ctenophores is complex, with a mouth opening at the end of a muscular ciliated pharynx, where extracellular digestion happens, and the digested food passes to a complex system of gastrovascular canals. Ctenophores are considered to have a pair of anal pores opening on their aboral side, defining the anal plane of symmetry, 45° with respect to the tentacular and sagittal planes. The nature of the anal pores in ctenophores has been a matter of debate since their discovery by German zoologist Carl Chun (1880), but it has recently been suggested that they have muscular sphincters and that they are able to evacuate the wastes of the digestive cavity, while regurgitation from the mouth does not occur in normal conditions (Presnell et al., 2016). It has further been shown that in the North Atlantic species *Mnemiopsis leidyi* defecation occurs following an ultra-dian rhythm—a recurrent period or cycle repeated throughout a 24-hour day—and only through one of the two anal pores, always the same in a given animal, and that this pore forms transiently (Tamm, 2019).

The tentacular plane is defined by the eponymous pair of long tentacles, which may have branching tentilla loaded with colloblasts used for prey capture. Such tentacles are found in adults of many species (most sources refer to tentacles being present in Cyddipida, but the group is paraphyletic) and the juveniles of most other forms. The retractable tentacles have a muscular core; they are extended by the water current but can be retracted into their sheaths by muscular action. In some groups with short tentacles, the prey is trapped in mucus of the body surface and transported to the mouth by ciliary action. Beroids entirely lack tentacles, and in-stead they engulf their prey, which consists of other ctenophores.

Colloblasts are adhesive exocytotic cells exclusive to the members of the phy-lum Ctenophora. Each colloblast consists of a single cell that contains a secretory adhesive portion and a spiral filament, surrounding a straight filament; the coiled filament then uncoils upon discharge, trapping prey. Colloblasts do not penetrate the skin of prey, but, unlike previously thought, the tentacles can secrete toxins (Babonis et al., 2018). A common embryological origin for colloblasts and neurons has recently been demonstrated (Babonis et al., 2018), as both cell lineages appear to be the descendants of a single micromere in *Mnemiopsis leidyi*. A common em-bryological precursor of cnidocytes and neurons has also been shown for cnidar-ians (see chapter 8), but this does not necessarily mean that cnidocytes are ho-mologous to colloblasts.

A few species of ctenophores lack tentacles and/or colloblasts. *Haeckelia rubra*, from the Pacific and the Sea of Japan, has long been known to have nematocysts—derived from the characteristic stinging cells of Cnidaria—instead of colloblasts in its tentacles. The discovery of nematocysts in ctenophores in the mid- to late 1800s reinforced the idea of the clade Coelenterata (uniting Ctenophora with Cni-daria based on their supposed digestive cavity, called the "coelenteron"). It is, however, now known that these nematocysts are acquired secondarily, as the ctenophore preferentially eats the tentacles of a medusa, *Aegina citrea* (see Mills and Miller, 1984), a phenomenon known as "kleptocnidism" (Carré et al., 1989), most commonly known from nudibranchs. In this case, the nematocysts are trans-ferred from parent to offspring, as the outer layer of the egg contains nematocysts which are ingested by the larva after gastrulation; these nematocysts make up the initial stock of exogenous nematocysts that will allow the ctenophore to catch its first cnidarian (Carré and Carré, 1989b).

An additional kleptocnidate species, *Haeckelia bimaculata*, is known from the Mediterranean Sea (Carré and Carré, 1989a). In this case, the hatching larva has no nematocysts, but its tentacles are lined by a special type of glandular cells named "pseudocolloblasts," composed of a secretory head bearing regularly interspaced, striated structures and an anchoring stalk (Carré and Carré, 1989b). Pseudocollo-blasts eject their contents progressively during successive phases of exocytosis, as colloblasts do, but differ morphologically from real colloblasts. They, however, allow the larvae of *H. bimaculata* to catch cnidarian larvae and thus acquire the exogenous nematocysts present in postlarval and adult stages.

Another apomorphy of ctenophores is their characteristic rows of ciliary comb plates, although these structures are lost in the benthic Platynectida. These are laid down in eight meridional rows of the giant locomotory comb plates that converge apically in four pairs and are controlled by a unique apical sense organ. Each comb plate is composed of thousands of tightly packed 9+2 cilia, each with its own membrane, forming stiff plates that push the animal through the water by a series of metachronal waves. The power stroke is oriented aborally, making the animals swim mouth first, although ciliary reversal also occurs in all eight rows, providing for a fine motor control of swimming behavior (Tamm, 2014). Excitation or inhibition of ciliary beating provides adaptive locomotory responses, and global reversal of beat direction causes escape swimming.

The beating cilia produce the characteristic iridescence that is perhaps the most conspicuous feature of ctenophores. This color is structural, rather than a pigment. The iridescence is thus caused by many thousands of tightly packed cilia, whose dimensions, combined with their highly regular arrangement, form a structure known as a "photonic crystal." In the case of ctenophores, this photonic crystal can reflect ambient light to generate bright coloration across the visible spectrum, but it also transmits light of wavelengths around that of the organism's bioluminescent organ (Welch et al., 2005).

In contrast to other early-branching metazoans (i.e., sponges and placozoans), ctenophores have both a complex nervous system and muscles, but these are best understood from developmental and genomics perspectives (see the discussions below). The nervous system of ctenophores is characterized by two distinct subepidermal diffuse nerve nets, a mesogleal nerve net that is loosely organized throughout the body mesoglea, and a much more compact "nerve net" with polygonal meshes in the ectodermal epithelium, organized as a plexus of short nerve cords (Jager et al., 2011). There is a concentration of neurons below the ciliary plates, around the mouth, at the base of the tentacles, and especially in the aboral neurosensory complex comprising the apical organ and polar fields. This neurosensory complex includes nerve nets underlying the apical organ and polar fields, a tangential bundle of actin-rich fibers (interpreted as a muscle) within the polar fields, and distinct groups of neurons labeled by anti-FMRFamide and anti-vasopressin antibodies, within the apical organ floor (Jager et al., 2011), illustrating the complexity of such sensory structures.

The ctenophore nervous system is thus quite complex, with specialized sensory cells, peripheral nerve nets, a centralized apical organ, and neurons in the epidermis, gastrodermis, and mesoglea, and with extensive regional specialization (Simmons and Martindale, 2016). However, the genomic content of neural associated genes—neural transcription factors, pre- and postsynaptic scaffold, neurotransmitter enzymes, and axon guidance genes—is the most reduced of any animal (Simmons and Martindale, 2016). In spite of this, other animals with a more elaborate set of neural genes, such as sponges and placozoans, lack a nervous system as elaborate as that of ctenophores, and it has been said that the ctenophore neurons and synapses have originated independently from those of other animals (Moroz and Kohn, 2016).

▮ REPRODUCTION AND DEVELOPMENT

Sexual reproduction is the norm in ctenophores. Most adult ctenophores (*Beroe* being the exception) have a high regenerative capacity, with the ability for whole-body regeneration (Martindale and Henry, 1997a). The benthic Platynectida can reproduce asexually by fragmentation, as fragments of the animal that break off while crawling about are able to regenerate the entire adult organism.

Hermaphroditism with the possibility of self-fertilization is common in ctenophores (Sasson and Ryan, 2016), but cross-fertilization also happens in broadcast spawners. Polyspermy is common in ctenophores, unlike in most other animals. Fertilization is internal in some groups, which rear their juvenile stages, the so-called cydippid larva. Some species have been described as undergoing a reproductive pattern called "dissogeny" ("larval" reproduction), where functional gametes are generated within days after development is complete but before reaching adulthood, as observed for *Mnemiopsis leidyi* (see Martindale, 1987). This has also been documented in *Mertensia ovum*, a species widely distributed in the Arctic region, but in the Baltic Sea its population consists solely of small larvae (less than 1.6 mm). Despite the absence of adults, the larvae are reproductively active, being the first documented account of a ctenophore population entirely recruiting through larval reproduction (paedogenesis) (Jaspers et al., 2012).

Ctenophores have been excellent models for experimental embryology and evo-devo, *Mnemiopsis leidyi* being one of the best-studied species. Much of this work has been led by American developmental biologists Mark Q. Martindale and Jonathan Q. Henry (see, e.g., Martindale and Henry, 1997a, b, 2015; Henry and Martindale, 2004). Their elegant work has shown the origin of every cell type in certain species of ctenophores. For example, the cilia from the comb plates are derived from two embryonic lineages, which include both daughters of the four e_1 micromeres (e_{11} and e_{12}) and a single daughter of the four m_1 micromeres (the m_{12} micromeres). Although the e_1 lineage is established autonomously, the m_1 lineage requires an inductive interaction from the e_1 lineage to contribute to comb plate formation (Henry and Martindale, 2004)—"induction" being the process by which the identity of certain cells influences the developmental fate of surrounding cells. These structures thus originate by a combination of noninduction and induction.

Ctenophore embryos have a unique and highly stereotyped cleavage program (recently summarized in Martindale and Henry, 2015), including unipolar cleavage—a mode of cell cleavage, also found in some cnidarians, where the cleavage furrow ingresses from only one side of the cell. The early stages of embryogenesis are well understood in *Mnemiopsis leidyi*, with all major cell types determined by the 60-cell stage (Martindale and Henry, 1999).

All ctenophores that have been studied to date undergo direct development to rapidly give rise to a miniature functional juvenile (called a "cydippid larva") in less than one day. Although a planula larva (the typical larva of cnidarians) has been mentioned in the literature (Komai, 1922), this has not been confirmed by other authors and is not currently accepted. Ctenophore embryogenesis generates a body plan that is essentially the same as in their adult stages, so they are basically direct developers (Martindale and Henry, 2015).

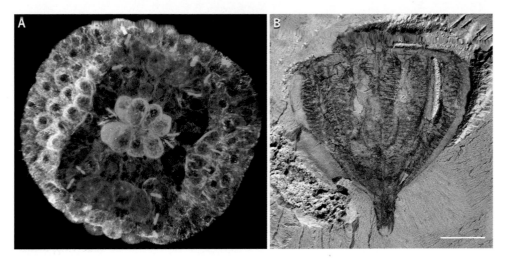

FIGURE 3.2. A, ctenophore "mesoderm" or ctenoderm in a maximum projection of Z-stack diagonal cut of a *M. leidyi* embryo. Nuclei stained with DAPI. Image courtesy of Miguel Salinas Saavedra. B, the Cambrian stem-group ctenophore *Galeactena hemispherica*, scale 2 mm. Image courtesy of Derek Siveter.

A host of contractile and mesenchymal cells reside in the mesoglea between the outer ectodermal epidermis and the inner gastrodermis in ctenophores, the latter derived from endoderm. Although often called "mesoderm" (e.g., in Martindale and Henry, 1997a), this embryonic layer does not seem to be homologous to bilaterian mesoderm because ctenophore genomes do not possess most of the developmental regulatory genes required to form mesoderm in bilaterians (Ryan et al., 2013; Moroz et al., 2014). We thus prefer to use the term "ctenoderm" to refer to cells residing in this layer (fig. 3.2 A).

ECOLOGY

Most ctenophores are holopelagic and thus can contribute, along with other gelatinous particulate organic matter, to the fast export of carbon to the deep ocean (e.g., Lebrato et al., 2013). Ctenophore blooms can be caused by anthropogenic effects and have deep impact on ecosystems, including eutrophication (Purcell, 2012). The ctenophore *Mnemiopsis leidyi* is one of the best-documented examples of a successful marine invasion (Kideys et al., 2005; Bayha et al., 2014). Native to the Atlantic coast of the Americas, *M. leidyi* invaded the Black, Caspian, and Mediterranean Seas beginning in the late 1980s, followed by the North and Baltic Seas starting in 2006, with major concomitant alterations in pelagic ecology, including fishery collapses (Bayha et al., 2014). The Black Sea invasion has been best characterized in a series of studies, *Mnemiopsis* biomass levels peaking by the end of the 1980s. By the end of the 1990s, the accidental introduction of another ctenophore to the Black Sea, *Beroe ovata*, known to feed exclusively on other ctenophores, reduced the population of introduced *M. leidyi*, leading to beneficial effects to the local ecosystem and the fisheries (Kideys et al., 2005).

A clade of ctenophores, the "order" Platynectida, so far monophyletic (Simion et al., 2014), is composed of benthic forms, often associated with corals or other invertebrates (Glynn et al., 2014). Platynectida species are dorsoventrally flattened and have the general appearance of a flatworm, rather than a pelagic ctenophore. Platynectida develop ciliated comb plates during embryogenesis but lose them after hatching, and all have a pair of tentacles bearing tentilla equipped with colloblasts for feeding, which are extended to capture the prey (Eeckhaut et al., 1997). The most commonly encountered shallow-water benthic ctenophores are placed in the genus *Coeloplana*, members of which move slowly by everting their pharynxes and pulling themselves forward, similar to leeches. The feeding ecology as well as the predators of some of these species have recently been elucidated (Glynn et al., 2018).

Tjalfiella tristoma, from Umanak Fjord, in West Greenland, was described as a "sessile" ctenophore species (Mortensen, 1910, 1912), living permanently attached to substrate (on the sea pen *Umbellula encrinus*) at around 500 m depth and without the ability to swim. This species was only found once after the original collection of 1908, in 1931, and only on sea pens at the exact type locality (see Gravili et al., 2018). Another species resembling *Tjalfiella*, also said to be sessile, *Lyrocteis imperatoris*, was described from the Sagami Sea, in Japan (Komai, 1941). However, the second species of the genus, *L. flavopallidus*, from McMurdo Sound, Antarctica, is described as sedentary instead of as sessile, as it is able to move 1–2 m per day (Robilliard and Dayton, 1972). Another species, *Lobatolampea tetragona*, seems to have a preference for a benthic lifestyle as well, but it can swim until it finds appropriate substrate, landing on seaweed and hanging on there (Horita, 2000; Uyeno et al., 2015). *Tjalfiella*, *Lobatolampea*, and *Coeloplana* have internal fertilization and brood their embryos. *Lampea pancerina* (referred to in the old literature as *Gastrodes parasiticum*) lives parasitically embedded in the body of pelagic tunicates of the genus *Salpa* (Komai, 1922).

Feeding in ctenophores is thus quite diverse for such a small group of invertebrates in numbers of species, and these diverse types of prey and feeding mechanisms have been suggested to be related to their radiation in body form and morphology (Tamm, 2014). Feeding ecologies involve thus the use of tentacles or lobes or by engulfing their prey, and these characters, being the basis of the current taxonomic groups, have an adaptive component that does not reflect the phylogenetic history of the group. While most species are generalists, others are specialists, feeding exclusively on salps, larvaceans, or other ctenophores. For a review of the feeding ecology of ctenophores, see Haddock (2007).

GENOMICS

Due to the debate about their phylogenetic position in the Animal Tree of Life, ctenophore genomics was a natural area of interest and the genomes of *Mnemiopsis leidyi* and *Pleurobrachia brachei* have been sequenced (Ryan et al., 2010; Ryan et al., 2013; Moroz et al., 2014). These studies have shown that ctenophore genomes are unique in their paucity of known bilaterian neurogenic, immune, and developmental genes, including the apparent absence of Hox genes, canonical microRNA machinery, and a reduced immune complement (Ryan et al., 2010; Moroz et al., 2014). Surprisingly, the sets of neural components present in the genomes of ctenophores and sponges are quite similar, which has been interpreted as suggesting that sponges have the necessary genetic machinery for a functioning nervous system but may have lost these cell types (Ryan et al., 2013). Although two distinct nervous systems are well recognized in ctenophores, many bilaterian neuron-specific genes and genes of "classical" neurotransmitter pathways are either absent or unrecognizeable or, if present, not expressed in neurons. Thus, it has been suggested that ctenophore neural systems, and possibly muscle specification, evolved independently from those in other animals (Ryan et al., 2013; Moroz et al., 2014), as supported by new theories about the multiple gains and losses of complex traits during metazoan evolution (Dunn et al., 2015). These views are, however, not universally accepted.

With respect to the debate about the mesodermal origin of the ctenophore muscle cells, most genes coding for structural components of mesodermal cells (e.g., tropomyosin) have been found in these genomes, but ctenophores lack many other genes involved in bilaterian mesodermal specification (Ryan et al., 2013).

Ctenophore mitochondrial genomes are unique, much shorter than those of almost all other animals, at 10–11 Kb (Pett et al., 2011; Kohn et al., 2012; Arafat et al., 2018), as shown in recent reports of the mitochondrial genomes of *Mnemiopsis leidyi*, *Pleurobrachia brachei* and three benthic species. These mitogenomes have a reduced number of intron-less protein-encoding genes, with some typical mitochondrial genes like *atp6* and *atp9* being located in the nuclear genome. For noncoding genes, there is also a major reduction and/or fragmentation in its ribosomal RNAs and a lack of tRNAs (Arafat et al., 2018). Ctenophore mitochondrial genomes are thus a clear example of the dynamic evolution of mitochondrial translation in metazoans (Pett and Lavrov, 2015).

▬ FOSSIL RECORD

The "Ctenophora-sister" hypothesis would predict an early appearance of Ctenophora in the fossil record. A fossil of Ediacaran age, *Eoandromeda octobranchiata*, preserved as two-dimensional compressions in shale in southern China, and from less compacted material from Australia, has been interpreted as a stem-group ctenophore (Tang et al., 2011). *Eoandromeda* resembles ctenophores in having octoradial symmetry (equated with the four pairs of comb plates), and its "arms" bear transverse bands homologized with ctenes (Tang et al., 2011). The significance of octoradial symmetry is, however, problematic because likely relatives of *Eoandromeda* in the Ediacaran—for example, *Tribrachidium*—have differing numbers of "arms" (triradial symmetry in the case of *Tribrachidium*). Another candidate for an Ediacaran total-group ctenophore, *Namacalathus*, is discussed below.

The confidence in assigning fossils to Ctenophora increases in the Cambrian. The oldest Cambrian ctenophore is material identified as embryos from the Terreneuvian of China (Chen et al., 2007b), in which the comb rows exhibit diagnostic morphology. In the early Cambrian Chengjiang biota, six genera, among them *Maotianascus* and *Gemmactaena*, share eight sclerotized struts that support an equal number of soft lobes or flaps. Based on the distinctive rigid struts or "spokes," these taxa have been united as a clade named Scleroctenophora (Ou et al., 2015) (fig. 3.2 B). Their ctenophore identity is upheld by having the distinctive oral–aboral body axis and the octameral lobes that each bears a pair of comb plates. In the Chengjiang fossils the comb plates are represented by what is likely the muscular pad.

Some other taxa from the Chengjiang biota that have been proposed to belong to Ctenophora are contentious or have been refuted (Hu et al., 2007); *Yunnanoascus*, for example, has been reinterpreted as a cnidarian (Ou et al., 2015, 2017; Han et al., 2016a). One of the controversies regarding Cambrian ctenophores is the number of comb rows, which can vastly exceed the conserved number of eight in living forms, various taxa having 16, 24, or as many as ca. 80 in the Burgess Shale species *Fasciculus vesanus* (Conway Morris and Collins, 1996). In convincingly attributed (and counted) species the number appears nonetheless to be in multiples of eight. The strategy of feeding with two elongated tentacles had been known from species from the Early Devonian (Stanley and Stürmer, 1983, 1987), and has recently been identified in a Cambrian ctenophore (Fu et al., 2019).

A series of Cambrian taxa, including *Daihua* from the Chengjiang biota, *Dinomischus* from the Burgess Shale and Chengjiang, and the "tulip animal" *Siphusauctum* from the Burgess and Spence Shales, have been interpreted as stem-group ctenophores (Zhao et al., 2019). They have a calyx with 18 tentacles that surround the mouth, the tentacles lined with pinnules bearing structures interpreted as macrocilia of comparable dimensions to the compound cilia of ctenophores. *Dinomischus* and *Siphusauctum* are stalked, and the "dinomischid" grade as a whole are interpreted as sessile suspension feeders. The most stemward taxa, *Daihua* and *Xianguangia*, suggest that the ctenophore total-group comprised polypoid forms at its origin. A possible terminal Ediacaran record of ctenophores is provided by the biomineralized *Namacalathus* (fig. 1.4 F), which resembles *Siphusauctum* in having a stalk/calyx organization, the calyx with an apical opening and ring of six lateral openings (Zhao et al., 2019).

A frondose Chengjiang fossil, *Stromatoveris psygmoglena*, has been interpreted as an Ediacaran survivor into the Cambrian and simultaneously as a stem-group ctenophore (Shu et al., 2006). The latter hypothesis is undermined by the speculative nature of the key apomorphy; branches on the body are suggested to be ciliated, but there is no direct evidence for this organization. Restudy of *Stromatovermis* in the context of a phylogenetic analysis is consistent with it being part of a mostly Ediacaran clade named Petalonamae (Hoyal Cuthill and Han, 2018).

Some authors have argued from molecular divergence dates that extant ctenophores are a relatively recent radiation, when compared to Porifera or Cnidaria (Podar et al., 2001; Dohrmann and Wörheide, 2017; Whelan et al., 2017), which would be consistent with the interpretation of early ctenophore fossils being members of the stem-group, instead of crown-group ctenophores, as evidenced by the long branch separating the extant diversity from the divergence of the clade from its sister group (e.g., Pick et al., 2010; Ryan et al., 2013; Whelan et al., 2015). This is consistent with the ecology of the putative stem ctenophores being very different from that of the modern group, as predicted by the hypothesis that the earliest ctenophores were suspension feeders (Zhao et al., 2019). This would eliminate the incongruence in the carnivorous habits of the purported first animals under the "Ctenophora-sister" hypothesis.

PORIFERA

Porifera (sponges) constitutes one of the dominant groups in many aquatic environments with ca. 8300 extant marine and ca. 50 freshwater species living from the subtidal to the deepest depths. Adult sponges are generally sessile, but the larvae of most species swim actively, serving as a dispersal phase—there are also unciliated larvae that are transported passively by currents. Most sponges are filter feeders, as they actively pump water through their bodies, helped by the beating of the cilia of a unique type of cells called "choanocytes," at rates of about 1 liter per cm^3 an hour (Reiswig, 1971). Sponges are unlike any other animals in that they can disassociate all their cells and reaggregate to constitute a new adult individual (Lavrov and Kosevich, 2014).

Ecologically, sponges play many roles, fighting for substrate by means of becoming active chemical competitors, and massive sponge grounds in the deep sea have a role analogous to that of scleractinians in shallow coral reefs (Maldonado et al., 2015; Maldonado et al., 2016). The giant barrel sponge *Xestospongia muta* dominates shallow Caribbean coral reef communities, where it increases habitat complexity while filtering large volumes of seawater. The largest specimens have been estimated to live for up to 2,300 years (McMurray et al., 2008). Contrary to common supposition that most of the silicon (Si) in marine coastal systems is recirculated under the biological control of planktonic diatoms, it has been shown that in tropical habitats sponge biogenic Si stocks surpass those of diatoms (Maldonado et al., 2010), and it is suggested that the role of sponges in the benthopelagic coupling of the Si cycle is much more important than previously thought (Maldonado et al., 2005; Maldonado et al., 2010).

Sponges are also an important component of many reef environments, where much marine biodiversity is sustained in oligotrophic waters—which has been termed "the reef's paradox." This large accumulation of biodiversity is in part facilitated by what has been termed the "sponge loop," where coral reef sponges are part of a highly efficient recycling pathway for dissolved organic matter, converting it, via rapid sponge-cell turnover, into cellular detritus that becomes food for reef consumers (de Goeij et al., 2013). This dissolved organic matter transfer through the sponge loop approaches the gross primary production rates required for the entire coral reef ecosystem.

In addition to their ecological importance, sponges play a key role in our understanding of the evolution of the most-basal lineages of animals, as it is widely accepted that choanocytes are homologous to choanoflagellates. This hypothesis is, however, not universally accepted, and similar collar cells have been reported in a variety of animal groups, including cnidarians (Goreau and Philpott, 1956;

Lyons, 1973), hemichordates (Nørrevang, 1964), echinoderms (Nørrevang and Wingstrand, 1970; Crawford and Campbell, 1993), and ascidians (Milanesi, 1971), and it has been claimed that these cells could be more widespread among animals than initially thought (Nørrevang and Wingstrand, 1970).

Comparisons of sponge anatomy to that of other animals is difficult. Some authors have proposed a link through the sponge larvae, arguing that they may be more comparable to other simple metazoans (Maldonado, 2004). Whereas our former understanding of sponge function and evolution was largely based on a morphological perspective, the availability of the first sponge genome, that of *Amphimedon queenslandica* (see Srivastava et al., 2010), and ample transcriptomic resources for a diversity of sponge species (e.g., Riesgo et al., 2014a; Simion et al., 2017), provide new ways to understand sponges by their molecular components.

In this chapter we highlight some of the newest discoveries in sponge anatomy, ecology, and evolution, directing the reader to the most recent treatments for further details (Leys and Hill, 2012; Brusca et al., 2016b).

SYSTEMATICS

Although currently considered monophyletic by most authors (e.g., Pick et al., 2010; Whelan et al., 2015; Simion et al., 2017), a series of Sanger-based molecular phylogenetic analyses in the 1990s and early 2000s suggested paraphyly of sponges at the base of the animal tree (e.g., Lafay et al., 1992; Cavalier-Smith et al., 1996; Collins, 1998; Borchiellini et al., 2001; Medina et al., 2001; Sperling et al., 2007; Sperling et al., 2009; Nosenko et al., 2013; Riesgo et al., 2014b). The implications of sponge paraphyly are especially relevant for understanding the last common ancestor of Metazoa, which has been proposed to be a "choanoblastaea," a pelagic sphere consisting of choanocytes (Nielsen, 2008). This hypothesis is the basis of the classification system proposed by Nielsen (2012a), where Silicea (Hexactinellida + Demospongiae) is the sister group to the clade Euradiculata (sensu Nielsen, 2012a) (= Calcarea, Homoscleromorpha + all other metazoans; this clade was previously named Laminazoa by Leys and Riesgo, 2012), with Calcarea as sister group to Proepitheliozoa (sensu Nielsen, 2012a) (= Homoscleromorpha + all other metazoans), and Homoscleromorpha as sister group to the remaining Metazoa. The segregation of Homoscleromorpha from Demospongiae, where it had been traditionally placed (Hooper and Van Soest, 2002a, b), is now broadly accepted (Gazave et al., 2012). This system is, however, rooted in phylogenetic results obtained prior to the latest generation of phylogenomic analyses and is also based on a misinterpretation of the lack of epithelia in sponges, an area of active recent research (see Adams et al., 2010; Leys and Riesgo, 2012).

Here we follow a system in which Porifera is monophyletic and composed of four classes, Calcarea, Hexactinellida, Demospongiae, and Homoscleromorpha (e.g., Erpenbeck and Wörheide, 2007; Voigt et al., 2008; Wörheide et al., 2012; Thacker et al., 2013; Riesgo et al., 2014b; Simion et al., 2017). Most authors further support a clade named Silicea, grouping Hexactinellida with Demospongiae (e.g., Pick et al., 2010; Belinky et al., 2012; Nosenko et al., 2013; Simion et al., 2017), the two groups that share siliceous spicules. For a detailed modern framework of sponge systematics, see Van Soest et al. (2012).

Phylum Porifera
    Class Calcarea
        Subclass Calcinea
        Subclass Calcaronea
    Class Homoscleromorpha
  Silicea
    Class Hexactinellida
        Subclass Amphidiscophora
        Subclass Hexasterophora
    Class Demospongiae
        Subclass Keratosa
        Subclass Myxospongiae
        Subclass Haploscleromorpha
        Subclass Heteroscleromorpha

PORIFERA: A SYNOPSIS

- Most sponges lack any form of symmetry, but others show a characteristic cylindrical body plan and a fractal pattern of choanocyte chambers
- Unique type of cells called "choanocytes," featuring an apical collar of tightly spaced, rodlike microvilli that surround a long flagellum; they constitute the choanoderm
- External pinacoderm and a middle layer, or mesohyl, containing the skeletal elements and free cells
- Skeletal elements can be organic (collagen fibers) or inorganic spicules, made of silica or calcium carbonate
- Sessile animals as adults, aquatic, without differentiated tissues but with an epithelium and type IV collagen
- Filter feeders, the particles transported by water currents being ingested by phagocytosis or pinocytosis; carnivorous nonfilter feeders exist, with intracellular digestion
- Many totipotent and pluripotent cells
- Adult tissues with the capacity to disassociate and reassociate into small individuals
- Without organs, including nerves

Sponges are unique in so many respects that their anatomy and physiology differ profoundly from those of any other animal, as they lack typical tissues, but their 16 cell types present some tissue-level integration and communication. Major aspects of their anatomy are discussed here, but for more detailed accounts on the specific cell types, diversity of body plans, as well as the multiple types of larvae one can find in sponges we direct the reader to recent anatomical summaries (Leys and Hill, 2012; Brusca et al., 2016b).

The sponge body (fig. 4.1) is characterized by an external monolayer of flat cells, or pinacocytes, that form the pinacoderm, and an internal monolayer of choanocytes, forming the choanoderm, which line the internal sponge chambers. The choanocytes of hexactinellid sponges are syncytial, and the structure is thus referred to as a choanosyncitium. The pinacoderm and choanoderm layers are traversed by cylindrical hollow pore cells, or porocytes, which connect the external milieu with the spongocoel—the cavity inside the sponge. Between the choanoderm and the pinacoderm is a mesohyl containing the supporting skeleton of spicules and, often, an elastic network of fibers and an extracellular matrix with collagen (and other molecules). The mesohyl may be thin in encrusting sponges or massive and thick in larger species. The number of cells that secrete the skeleton is quite varied, with sclerocytes, which secrete the spicules—a process that can take hours and that involves biomineralization (Voigt et al., 2017)—and several types of ameboid cells (collencytes, lophocytes, spongocytes) that secrete fibrillar collagen including spongin, a modified type of collagen that forms the fibrous skeleton of sponges. In addition, myocytes are a type of cell of possible contractile function, although most contractions seem to be mediated by the pinacoderm (Nickel et al., 2011).

Archaeocytes are the undifferentiated stage or stem cell population and are able to differentiate into many cell types in the sponge, including pinacocytes and choanocytes. In the early days of sponge science, distinct populations within what we now call archaeocytes were recognized (e.g., Weltner, 1907): amoebocytes when they transport the nutrients; thesocytes when they are the undifferentiated state emerging during gemmule hatching containing large storage products; tokocytes, which were thought to be the precursors of gametes; and finally archaeocytes, the stem cells' cell lineage. From all these cells, we can morphologically identify archaeocytes in their transport duties (therefore amoebocytes, with their anucleolate nucleus), in their stem state (nucleolate nucleus and large endoplasmic reticulum), and thesocytes when they emerge from gemmules with large yolk bodies. In the late 1980s, several authors proposed a terminological homogenization and unified archaeocytes, amoebocytes, and tokocytes under the name "archaeocyte" (e.g., Simpson, 1984). A subpopulation of tokocytes with germline markers and a defensive stage within archaeocytes that express lectins can also be detected. Finally, spherulous cells are often found storing secondary metabolites. Whether most cells in the mesohyl are undifferentiated and therefore stem or not remains a matter of debate.

Choanocytes are the most characteristic cells in sponges (fig. 4.1 B). They are often described as specialized "sieve" cells, as they have a collar of tightly spaced, rodlike microvilli that surround a long flagellum. The beat of the flagellum draws water through this collar, but how particles caught on the collar are brought to the cell surface is unknown, and multiple mechanisms and additional cells are ultimately responsible for the assimilation of the particles transported by the water currents, the ingestion mechanism related to particle size and seasonality (Reiswig, 1971;

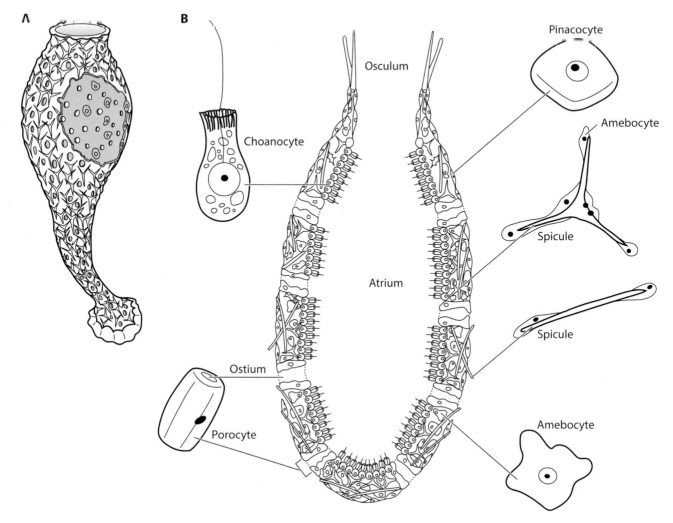

Figure 4.1. An olynthus of an asconoid sponge (A) and (B) a cross-section showing the details of a sponge's anatomy with the different cell types, including choanocytes forming the choanoderm, pinacocytes constituting the pinacoderm, and different types of cells and skeletal elements. After Brusca and Brusca (1990).

Ribes et al., 1999; Leys and Eerkes-Medrano, 2006). Choanocytes are also important because of their similarity to choanoflagellates, which are often discussed as homologous cells and are thus the most important character used to support a basal phylogenetic position of sponges in the Animal Tree of Life (see chapter 2). This homology is, however, difficult to test (see Mah et al., 2014) and has been rejected by some workers (Dunn et al., 2015).

A few demosponges inhabiting nutrient-poor environments are carnivorous, trapping prey using hook-shaped spicules (Vacelet and Boury-Esnault, 1995). These species have partially or entirely lost the characteristic apomorphy of sponges, their aquiferous system with choanocytes (Riesgo et al., 2007). Research on carnivorous sponges has flourished in recent years with the improved ease for sampling in remote environments, especially in the deepest parts of the ocean, where filter-feeding organisms do not find optimal conditions (see Lee et al., 2012). It has been shown recently that carnivory in sponges had a single origin (Hestetun et al., 2016).

Most species, with the exception of hexactinellids, are now thought to have symbiotic relationships with bacteria, a fact that was reported relatively recently (Vacelet et al., 1995). New research has demonstrated that microbial symbionts can comprise over 40 percent of the total sponge–microbe holobiont volume, providing supplementary nutrition, secondary metabolites, enhancing the rigidity of their skeleton, and conferring protection from radiation (Thacker and Freeman, 2012). The presence of zooxanthellae has also been documented in sponges (Sarà and Liaci, 1964; Carlos et al., 1999).

The same secondary metabolites that provide sponges with chemical warfare to defend their territory can be used by other organisms for their own defense. Such is the case of the nudibranch *Chromodoris quadricolor*, which feeds on the sponge *Negombata magnifica* (formerly in the genus *Latrunculia*) from the Red Sea, incorporating the icthyotoxic latrunculin B in its mucus secretion (Cimino et al., 1999).

The sponge body plan therefore relates to the choanocytes and its filter-feeding habits, with the exceptions outlined above. Sponges are arranged in a way that they constitute an internal aquiferous system that brings oxygenated water and food particles through the sponge and expels excretory and digestive wastes. The aquiferous system also has a role in reproduction. Researchers have recorded sponge pumping rates that range from 0.002 to 0.84 mL of water per second per cm^3 of sponge body, with large sponges filtering their own volume of water every 10 to 20 seconds (Brusca et al., 2016b).

Sponge anatomy can be extremely complex, but textbooks typically use a model of increasing complexity to explain their body plan with respect to the aquiferous system (fig. 4.2). In what it is known as the "asconoid condition," a sponge is depicted as a vase, the pinacoderm lying outside, and the choanoderm delimiting the internal surface. The two layers are connected by ostia ("ostium" in singular) that draw water into an internal atrium or spongocoel. The water then exits the sponge via the larger opening or osculum, taking the products of cellular respiration and excretion, the particles that have not been digested, as well as the gametes.

This body plan becomes more complex in the syconoid condition (fig. 4.2 B), where the choanocytes, instead of lining the atrium, line the interior of the so-called choanocyte chambers. Each choanocyte chamber could be understood as an individual asconoid sponge, each opening by an apopyle to the atrium of the vase-shaped sponge, then leading to the osculum. Dermal pores lead to incurrent canals traversing the cortex of the sponge, and the prosopyles lead to each of the choanocyte chambers. In this condition, water particles travel through the dermal (incurrent) pore → incurrent canal → prosopyle → choanocyte chamber → apopyle → atrium → osculum.

The asconoid condition is typically found in the olynthus stage of some calcareous sponges, while the syconoid stage, produced by the simple folding of the pinacoderm and the choanoderm, is found in many adult calcareous sponges, including the eponymous genus *Sycon*. In the sylleibid condition of some calcareous sponges (fig. 4.2 C), the choanocyte chambers form clusters, each cluster opening directly to the atrium.

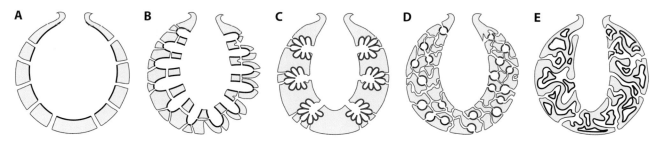

FIGURE 4.2. Schematic cross sections of the five types of aquiferous systems in sponges, the thick line representing the choanoderm: (A) asconoid, (B) syconoid, (C) sylleibid, (D) leuconoid, and (E) solenoid. After Cavalcanti and Klautau (2011).

This architecture can get much more complicated, with the next level of organization, the leuconoid condition (fig. 4.2 D), found in all sponge classes. Here many choanocyte chambers embedded in the thickened mesohyl open to a common chamber leading to a series of anastomosing excurrent canals, ultimately reaching the osculum or multiple oscula. Other types of conditions are also discussed by some authors, including the solenoid condition (fig. 4.2 E), a different type of aquiferous system of calcareous sponges composed of anastomosed (interconnected) choanocyte tubes with an atrium entirely lacking a choanoderm (Cavalcanti and Klautau, 2011).

Sponge taxonomy relies primarily on morphological external features such as overall body structure, shape, size, and color, and on internal traits, especially skeletal traits such as the shape and size of spicules or spongin fibers, these especially important in species that do not possess spicules, as is the case of Halisarcidae (Alvizu et al., 2013). Spicules are a key character in sponge taxonomy and can be composed of calcium carbonate in the form of calcite or aragonite in the members of the class Calcarea or of hydrated silicon dioxide in Silicea (Hexactinellida and Demospongiae); sponges are the only animals that use hydrated silica as a skeletal material. Silicatein is a protein often associated with the formation of silica, most likely derived from a gene expansion in the cathepsin family (Riesgo et al., 2015)—the expansion of a gene cluster is the duplication of genes that leads to larger gene families, giving some of these copies the possibility of neofunctionalization. However, not all sponges with a siliceous skeleton express silicatein, while this enzyme is expressed in some sponges that lack a skeleton (Riesgo et al., 2015). Another protein involved in the process of biosilification is glassin, found in some Hexactinellida (Shimizu et al., 2015). The exact biosilification mechanism of homoscleromorphs, however, remains unknown.

The range of spicule morphology includes simple cylinders with pointed ends (monaxons; these can be monactinal or diactinal, depending on whether they have one or two pointed ends) that can be covered in spines (acanthoxea and acanthostrongyles); other names apply to different morphologies. Triaxons have three axes, tetraxons four, and polyaxons, multiple axes. Sigma C-spicules are shaped like the letter C. Depending on their size, spicules can be megascleres (often from 60 to 2,000 μm) and often function as the main support elements in the skeleton. Microscleres are small spicules (10 to 60 μm), often scattered throughout the mesohyl and pinacoderm. A selection of these spicules can be found in fig. 4.3 (Van Soest et al., 2012).

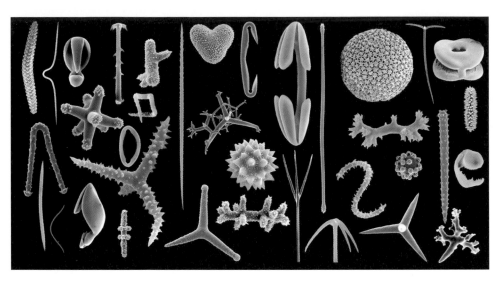

FIGURE 4.3. SEM images of a selection of microscleres and megascleres, not to scale, sizes vary between 0.01 and 1 mm. From Van Soest et al. (2012) PLoS One 7(4): e35105. doi:10.1371/journal.pone.0035105, CC BY 2.5, https://commons .wikimedia.org/w/index.php?curid=34595288

Sponges lack muscles and nerves or any other type of discrete sensory structures, yet they are able to sense and respond to environmental changes and possess synaptic elements homologous to those of other animals with nerves (Ryan and Grant, 2009). They also express striated muscle *myhc* orthologues, suggesting a functional diversification of these genes before the origin of muscles (Steinmetz et al., 2012), if they were sister group to all other animals. It is now clear that at the molecular level they possess the typical neurogenic circuit found in bilaterians (Richards et al., 2008; Ryan and Grant, 2009).

From an anatomical point of view, it has recently been demonstrated that sponges possess arrays of nonmotile cilia in the oscula that act as flow sensors (Ludeman et al., 2014). These authors have gone further to suggest that the osculum may be considered a sensory organ that is used to coordinate whole animal responses in sponges, and that arrays of primary cilia like these could represent a first step in the evolution of sensory and coordination systems in metazoans (Ludeman et al., 2014). The presence of sensory systems in the osculum and in other parts of the aquiferous system is consistent with the idea that sponges may be able to actively control the flow to compensate for changes in the environment and for repairing and cleaning the aquiferous system. Sponges thus have behavior, including contractions of myoid cells, but responses are too slow to be controlled by rapid neurotransmitter-based signaling mechanisms. They use glutamatergic signaling to coordinate slow contractions, and one group, glass sponges, uses calcium/ potassium action potential (electrical signaling) for rapid arrest of feeding (Dunn et al., 2015). In addition, sponges may present other elements of more sophisticated nervous systems, including phototactic behavior in larvae.

Coordination systems show that sponges are largely epithelial animals, with sensory cells that are epithelial, effectors that are contractile epithelial cells, as well as flagellated collar bodies lining the feeding chambers of glass sponges. Signaling pathways also seem to use the epithelia (Leys, 2015).

REPRODUCTION AND DEVELOPMENT

Sponges reproduce both sexually and asexually (by budding or fragmentation) and can produce gemmules and reduction bodies. Gemmules are small spherical structures with a thick collagenous coating and siliceous spicules typical of overwintering forms in limnic species. Much evo-devo work on sponges has been based on these resistance bodies, as they are easier to manipulate than reproductive adult individuals. Similar to these gemmules are the reduction bodies of some marine species. Reproduction by fragmentation is probably possible in most or all sponges.

Sponges display considerable diversity with respect to reproductive and developmental biology (Riesgo et al., 2014b; Koutsouveli et al., 2018). Some species are gonochoristic while others are hermaphroditic (mostly sequential, in the form of protogyny or protandry, and with single or multiple sex reversals). Cross-fertilization is thus probably the norm, and mature sperm cells and sometimes oocytes are generally released into the environment through the aquiferous system, fertilization taking place either internally or in the water column, depending on the species. Gametogenesis is poorly understood in most species, but eggs and sperm appear to develop in clusters in the carnivorous sponge *Lycopodina occidentalis*, which may explain how fertilization happens in this species and why embryos appear to develop in synchronous clusters (Riesgo et al., 2007).

In other cases, it has been observed that oogenesis results in modifications of the external morphology and of the aquiferous system, where a large part of the sponge body becomes filled with oocytes with the consequent disappearance of choanocyte chambers in the reproductive portion of the sponge (Sidri et al., 2005). Developmental work on the homoscleromorph *Oscarella lobularis* has shown that 11 germline multipotency program (GMP) genes are expressed throughout the life cycle of the sponge in pluri/multipotent progenitors, during gametogenesis, embryogenesis and during wound healing, reinforcing the hypothesis of an ancestral multipotency program (Fierro-Constaín et al., 2017).

The reproductive condition can be either oviparity or viviparity, but in general it involves a planktonic larva that allows the dispersal of these sessile animals as adults. A few species are direct developers, without an intermediate larval form. Developmental work on sponges has recently been summarized (Ereskovsky, 2010; Degnan et al., 2015). However, early embryogenesis and cell fates continue to be poorly understood for most species, as it often occurs embedded within the sponge tissues.

Sponge larvae can be of many different types (amphiblastula and calciblastula in calcareous sponges; trichimella in hexactinellids; disphaerula, parenchymella, and coeloblastula in demosponges; and cinctoblastula in homoscleromorphs) (Degnan et al., 2015).

Reproduction in carnivorous sponges has received special attention due to the lack of the aquiferous system (Riesgo et al., 2007; Riesgo, 2010). Development and gene expression patterns are best characterized in the demosponge *Amphimedon queenslandica* and the calcareous *Sycon ciliatum*, two species with well-annotated genomes.

GENOMICS

Due to their simple morphology, sponges were one of the first animal phyla with a sequenced genome, that of the Australian Great Barrier Reef demosponge species *Amphimedon queenslandica* (Srivastava et al., 2010). Despite their radical morphological differences with other animals, the genome of this sponge is remarkably similar to those of other animals with respect to gene content, structure, and organization (Srivastava et al., 2010). Genomic work on *A. queenslandica* has also revealed that the typical neurogenic circuit found in bilaterians, comprising proneural atonal-related bHLH genes coupled with Notch–Delta signaling, was already present in sponges to generate an ancient sensory cell type (Richards et al., 2008). The genomes of the calcareous sponge *Sycon ciliatum* (see Fortunato et al., 2012) and the homoscleromorph *Oscarella carmela* (see Nichols et al., 2012) have also been available for a few years but not formally published. Finally, the genomes of the first hexactinellid, *Oopsacas minuta*, and a second species of Homoscleromorpha, *Oscarella lobularis*, have recently been published (Belahbib et al., 2018).

A relatively large number of sponge transcriptomes has recently become available (e.g., Nichols et al., 2006; Riesgo et al., 2012; Pérez-Porro et al., 2013; Riesgo et al., 2014a; Guzman and Conaco, 2016). These transcriptomes have corroborated early observations from the genome of *A. queenslandica* in terms of high gene complexity across all sponge classes (Riesgo et al., 2014a).

Long noncoding RNAs (lncRNAs or lincRNAs) play regulatory roles during animal development, and it has been hypothesized that RNA-based gene regulation was important for the evolution of developmental complexity in animals (Bråte et al., 2015). A recent analysis of RNA-Seq data derived from a comprehensive set of embryonic stages in the calcareous sponge *Sycon ciliatum* identified hundreds of developmentally expressed intergenic lncRNAs with a dynamic spatial and temporal expression during embryonic development, providing insights into the noncoding gene repertoire in sponges and suggesting RNA-based gene regulation early in animal evolution (Bråte et al., 2015).

Complete mitochondrial genomes are available for a large number of sponge species (e.g., Lavrov et al., 2005; Lavrov and Lang, 2005; Erpenbeck et al., 2007; Wang and Lavrov, 2007; Rosengarten et al., 2008; Plese et al., 2012; Sperling et al., 2012; Lavrov et al., 2013), and they differ considerably from those of most other metazoans, especially in the case of calcareous sponges. The mt genomes of most sponges are circular and exceed the typical length of metazoan mt genomes—which is approximately 16 kb—and they possess long noncoding stretches of DNA, have no identifiable control region, and bear additional open reading frames (ORFs)—typically *atp9* in demosponges (Erpenbeck et al., 2007).

Sponge mitochondrial genes often show a slow rate of evolution, which makes the use of the so-called universal barcode, a fragment of cytochrome *c* oxidase subunit I, difficult (e.g., Wörheide, 2006; Sperling et al., 2012). However, in the case of the calcareous sponge *Clathrina clathrus* mitochondrial DNA (mtDNA) consists of

6 linear chromosomes 7.6–9.4 kb in size and encodes at least 37 genes: 13 protein-coding, 2 ribosomal RNAs (rRNAs), and 24 transfer RNAs (tRNAs) (Lavrov et al., 2013), with its rRNAs encoded in 3 (*srRNA*) and >6 (*lrRNA*) fragments distributed out of order and on several chromosomes, in addition to several other extreme re-arrangements. Contrary to other sponge groups, calcareous sponges' mitochondrial genes show high rates of evolution. Extensive mRNA editing (a variety of unrelated biochemical processes that alter RNA sequence during or after transcription) has been documented in most members of the calcarean subclass Calcaronea (Lavrov et al., 2016).

FOSSIL RECORD

Sterol biomarkers attributed to demosponges in Neoproterozoic sediments are discussed in chapter 1. Although there are many putative Precambrian sponges, a comprehensive review of the early fossil record of sponges disputes all of them as either lacking diagnostic poriferan characters or being incorrectly dated (Antcliffe et al., 2014; see also Muscente et al., 2015b). Candidates from the Ediacaran nonetheless continue to be suggested as possibly of sponge grade, such as the ca. 600 Ma *Eocyathispongia qiania* (Yin et al., 2015), though this identity has been disputed on functional and morphological grounds (Botting and Muir, 2018).

The oldest reliable sponge remains are siliceous spicules from the early Cambrian (*Protohertzina anabarica* Zone) of Iran (Antcliffe et al., 2014). Archaeocyatha, which have a calcareous, aspiculate skeleton, are the most diverse group of Cambrian sponges and constituted the framework of the earliest metazoan reefs. Archaeocyaths are not known from the basal Cambrian; rather, they appeared and radiated as late as the middle of the Fortunian Stage of the Cambrian. The gradual assembly of their skeletal structure through this time is coincident with the evolution of other metazoan groups, a pattern suggestive of sponges diversifying as part of the Cambrian explosion rather than being a vastly older (Precambrian) radiation (Antcliffe et al., 2014). Their affinities have long been debated, but their growth and reproductive patterns make demosponge affinities most likely (Botting and Muir, 2018). Archaeocyaths went extinct in Cambrian Stage 4.

Cambrian sponges are often known only from disarticulated spicules retrieved from residues after dissolution of sedimentary rocks, but some systematically novel character combinations are provided by fossils known from articulated remains. For example, the Burgess Shale sponge *Eiffelia globosa* depicts characters of both Calcarea and Silicea that are unknown in extant sponges (Botting and Butterfield, 2005). Notably, the spicules are bilayered, a silica layer surrounding a calcium carbonate core, and the spicule geometries are a mix of hexactine (as in Silicea, but potentially a general character of Porifera as a whole: Botting and Muir, 2018) and hexaradiate (as in Calcarea). Likewise novel is the discovery of early Cambrian (Stage 4) spicules with triaxine geometry as found in hexactinellids but with an organic sheath as in calcareans (Harvey, 2010), or the report of a transitional form between Demospongiae and Hexactinellida from the Late Ordovician, *Conciliospongia anjiensis*, with a reticulate, tufted skeleton of minute monaxon spicules, with hexactine spicules and a globose body form (Botting et al., 2017). Cambrian Burgess

Shale–type biotas worldwide have been recognized as sharing a distinctive assemblage of demosponges and hexactinellids that are largely exclusive to that temporal and taphonomic window, though some genera persist into the Ordovician. They are mostly thin walled, with loose spicules, and they inhabited offshore environments. The geographic uniformity of Cambrian sponges contrasts sharply with greater geographic provincialism in the Ordovician (Muir et al., 2013). The fossil record of the stem and crown groups of the four extant classes of sponges are reviewed by Botting and Muir (2018). Of these, demosponges are most confidently identified as having crown-group members in the Cambrian, as early as the Sirius Passet Konservat-Lagerstätte in Cambrian Stage 3 (Botting et al., 2015).

The sponge fossil record includes some geographically and temporally wide-ranging groups that have long been difficult to place, among them stromatoporoids, chaetitids, and sphinctozoans. These share the property of having calcareous skeletons, and the term "hypercalcified sponges" is applied to them.

Stromatoporoids are sponges that played a major role in constructing the frameworks of Silurian and Devonian reefs. They secreted a laminated, aspiculate calcareous skeleton with rodlike pillars perpendicular to the laminae, and they have homologues of the stellate exhalant canals of sponges, called "astrorhizae." "True" stromatoporoids ranged through the Ordovician to the Devonian, whereas younger fossils informally attributed to the group are probably an unrelated polyphyletic assemblage (Stock, 2001). Since the rediscovery of sclerosponges in the 1970s, this has widely been regarded as their systematic affinity.

Chaetitids are a group of fossil sponges with a skeleton of fused tubules. They were long classified as cnidarians, bryozoans, or algae, but some members have spicule pseudomorphs, and they are now usually allied with Demospongiae (West, 2011). Some chaetitids closely resemble stromatoporoids, particularly those with astrorhizae, and both of these groups are regarded as grades of construction of hypercalcified sponges (Wood, 1991).

Sphinctozoans are sponges, most of them aspiculate, organized as a series of stacked chambers. They are widely regarded as being polyphyletic, the earliest members, from the early Cambrian, being calcareans, whereas others are demosponges (Senowbari-Daryan and Rigby, 2011). They were especially diverse in the Permian and Triassic, but the extant demosponge *Vaceletia progenitor* has been identified as a sphinctozoan (Vacelet, 2002).

Sponge neontology and paleontology intersect in the discovery of lithistid demosponges forming reeflike buildups in the Mediterranean (Maldonado et al., 2015). Such monospecific lithistid sponge grounds have an extensive fossil history, being most common in Jurassic and Cretaceous buildups in the Tethyan region, the extant deep-sea habitat being seen as relictual.

Finally, Chancelloriida represents another fossil group of debated systematic position, but among the alternatives an affinity to sponges, or indeed an identity as sponges, is a recurring theme. Chancelloriids are known from whole-body compression fossils in many Cambrian Burgess Shale-type biotas (Bengtson and Collins, 2015). Chancelloriids have radial symmetry and, like sponges, they present an apical opening. The body is shaped like a sac, with an "integument" embedded with distinctive, usually multipronged sclerites. These sclerites are commonly found isolated as microfossils, their hollow construction closely resembling those of halkieriid molluscs, a similarity that has had them united as "coeloscleritophorans" (discussed in chapter 44) (Porter, 2008). The mode of addition of the sclerites in an apical zone has been compared to extant calcarean sponges (Cong et al., 2018), although even in the best preserved material, no evidence of ostia is present. Chancelloriid sclerites depict some similarities in their inferred mode of secretion with the spicules of Cambrian protomonaxidid sponges, leading some to assign Chancelloriida to that group and thus total-group Porifera (Botting and Muir, 2018).

5

PLANULOZOA

Planulozoa is a clade composed of Cnidaria, Placozoa, and Bilateria (Wallberg et al., 2004), characterized by the presence of serotonin (Dunn et al., 2014) and perhaps by the presence of the blastopore-associated axial organizer based on Wnt/ß-catenin signaling, an organizer being a cell or set of cells in an embryo that induce development of a certain structure. This signaling is known (and plausibly maintained from common ancestry) from cnidarians, deuterostomes and, possibly, several protostome lineages (Kraus et al., 2016), but this needs further testing in placozoans and non-planulozoans. The name Planulozoa refers to the planula larva of cnidarians, which was said to resemble an acoelomorph in its vermiform shape and in the polar development of the nervous system (Gröger and Schmid, 2001; Wallberg et al., 2004). While the favored resolution within Planulozoa based on phylogenomic analyses found Cnidaria closer to Bilateria than either is to Placozoa (e.g., Srivastava et al., 2008; Hejnol et al., 2009; Pisani et al., 2015; Simion et al., 2017; Whelan et al., 2017), the addition of three placozoan genomes has found strong support for a clade uniting Placozoa and Cnidaria (Laumer et al., 2018).

The same clade of Placozoa, Cnidaria, and Bilateria was named Parahoxozoa (originally ParaHoxozoa) by Ryan et al. (2010), based on the apparent absence of Hox/ParaHox, S50 and K50 PRD and HNF class homeodomains in Porifera and Ctenophora (see also Dunn and Ryan, 2015). Subsequent genomic work identified one ParaHox gene, *Cdx*, in a calcareous sponge (Fortunato et al., 2014), which would render the name Parahoxozoa uninformative. However, the identity of the sponge *Cdx* with ParaHox genes is questionable (Pastrana et al., 2019).

This ParaHox cluster seems to have originated from a duplication event of an ancestral ProtoHox gene cluster, which may have given origin to the paralogous Hox and ParaHox clusters, facilitating an increase in body complexity during the Cambrian Explosion (Brooke et al., 1998; Holland, 2013). It remains likely that the ParaHox cluster is found in all animals but sponges and ctenophores (Ryan et al., 2010), but that it consisted of two genes (as in cnidarians) instead of three typical of bilaterians, *Gsx*, *Xlox*, and *Cdx* (Chourrout et al., 2006).

Some authors, including ourselves, have used Planulozoa to refer to Cnidaria and Bilateria to the exclusion of Placozoa (e.g., Dunn et al., 2014; Giribet, 2016b), but this contradicts the original concept of the group introduced by Wallberg et al. (2004), and a cnidarian–bilaterian clade is contradicted by the alliance of placozoans with cnidarians (Laumer et al., 2018).

Planulozoa as used herein reflects most modern phylogenetic resolutions placing either Porifera and Ctenophora (in either order) as the basalmost grade in the animal tree (see a discussion in Dunn et al., 2014). Few modern analyses contradict

Planulozoa, and those that do propose hypotheses at odds with other analyses or with zoological paradigms: these involve either a split between Bilateria and all other animals, which form a clade including Cnidaria, Ctenophora, Placozoa, and Porifera (Schierwater et al., 2009; Nosenko et al., 2013), or monophyly of Coelenterata (Philippe et al., 2009). Despite these early anomalous results, Planulozoa is one of the most stable nodes at the base of the animal tree (e.g., Ryan et al., 2013; Pisani et al., 2015; Whelan et al., 2015).

The clade Planulozoa is incompatible with the putative clade Acrosomata, proposed by Ax (1995), which united Ctenophora and Bilateria based on the presence of an acrosome in the spermatozoon. In fact, acrosomes are also found in multiple clades of Porifera, including Homoscleromorpha (Harrison and De Vos, 1991) and Demospongiae (Riesgo and Maldonado, 2009). Therefore, whether the spermatozoon has a discrete acrosome or an acrosomal vesicle seems to have little relevance to resolve the base of the animal tree.

Planulozoa and Acrosomata are also in conflict with Coelenterata (also called Radiata), grouping Cnidaria and Ctenophora, a clade often appearing in the old literature that has received little support (Martindale et al., 2002). Some embryological features have been put forward as potential apomorphies for Coelenterata, including the presence of unipolar cleavage (see chapter 3), and the site of gastrulation (Scholtz, 2004). Morphological parsimony analysis including fossil ctenophores and the embryological characters noted above recovered Coelenterata (Ou et al., 2017; Zhao et al., 2019), but with respect to molecular phylogenies, this group was supported only in some of the earliest phylogenomic analyses (Philippe et al., 2009) and, more recently, when presence/absence of homologous gene families are used as characters (Pett et al., 2019). The clade Neuralia (Cnidaria, Ctenophora, and Bilateria, to the exclusion of Porifera and Placozoa) (Nielsen, 2008) (= Eumetazoa of many authors), supported by the presence of nerve cells, is likewise incompatible with the current view of animal phylogeny. We do not follow Acrosomata, Coelenterata, or Neuralia/Eumetazoa in this book.

DIPLOBLASTY AND TRIPLOBLASTY?

One concept, the embryonic germ layers resulting from gastrulation, is pervasive in the metazoan literature and has often been used to explain animal evolution and to define major animal clades. A germ layer is a group of cells, usually layered one on top of the other in an embryo, that gives rise to different tissues in the adult animal. All authors agree that bilaterians have three germ layers, from the inside to the outside, namely, endoderm, mesoderm, and ectoderm. For example, in insects the endoderm gives rise to the gut; the mesoderm gives origin to the muscle, heart and blood; and the ectoderm gives origin to the cuticle, the nervous system, and all their derivatives. It is for this reason that bilaterians are often called "triploblasts," while traditionally cnidarians and ctenophores have been referred to as "diploblasts", [2] as their adult tissues have traditionally been thought to derive from

---

[2] While some texts refer to Porifera and Placozoa as diploblasts, these two animal groups lack identifiable germ layers, and thus, since Coelenterata is not monopyletic, the term "diploblast" is used

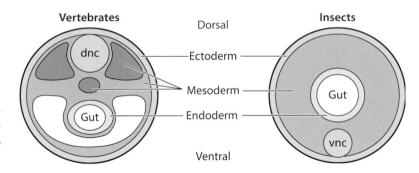

FIGURE 5.1. Idealized schematic of the germ layers of chordates and arthropods. Abbreviations: dnc, dorsal nerve cord; vnc, ventral nerve cord.

just two germ layers, endoderm and ectoderm (but see the discussion of a putative third layer in ctenophores in chapter 3). In cnidarians the endoderm gives origin to the gastrodermis (the lining of the gastrovascular cavity) and the ectoderm gives rise to the outer epithelium; cnidarian ectoderm and endoderm are epithelial monolayers during development and throughout the life of the animal (Technau et al., 2015) and are separated by a common extracellular matrix, called the "mesoglea." The mesoglea is secreted by the epithelial cells and is composed of laminin, fibronectin, and collagen IV, similar to the basement membrane of Bilateria.

In triploblastic animals the musculature is the adult tissue that defines mesoderm, but not all animals with muscles are strictly triploblastic. For example, ctenophores have muscle cells (see, e.g., Martindale et al., 2002), but not mesoderm, although they do have a third embryonic layer (see chapter 3). In addition, none of the genes that are involved in bilaterian muscle development are present in the ctenophore genome. In the cnidarian *Nematostella vectensis* the musculature is largely formed by myoepithelial cells of endodermal origin, but the origin of the myoepithelial cells from the ectoderm (e.g., in the tentacles) is less clear. The endoderm, however, expresses all the mesodermal markers found in bilaterians. More recently, a study by Salinas-Saavedra et al. (2018) has shown that by disrupting the apicobasal cell-polarity complex in *Nematostella vectensis*, cells from the endomesodermal epithelium could be converted into mesenchymal-like cells, translocate Nvß-catenin, and emulate EMT-like processes (apical constriction and individual cell-detachments) in the ectodermal epithelium. This implies that the mechanisms necessary to segregate individual germ layers, such as mesoderm, are also present in cnidarians despite not forming such a layer.

Given the current resolution of the tree of life, especially with respect to the position of ctenophores, cnidarians or even placozoans, it seems clear that triploblasty (as in Bilateria) may be a unique derived character, but "diploblasty" may not. One idea is that all metazoans were primitively triploblastic, with subsequent reduction or fusion of embryonic layers, but this does not seem to be supported by molecular markers. Another solution may be that the diploblastic condition of cnidarians and ctenophores may have originated independently, with the ctenophores deriving a third body layer that is not homologous to mesoderm.

exclusively to describe a level of organization, and not as the vernacular of a putative taxon "Diploblastica."

PLACOZOA + CNIDARIA CLADE

A clade grouping Placozoa + Cnidaria has recently been suggested based on phylogenomic analyses using multiple placozoan genomes (Laumer et al., 2018). In fact, some earlier phylogenetic analyses had suggested a close relationship of placozoans to cnidarians (e.g., Borchiellini et al., 2001; Glenner et al., 2004). The study of Laumer et al. (2018) shows that using site-heterogeneous substitution models for their phylogenetic analyses can generate full support for either a sister-group relationship of Cnidaria + Placozoa, or for the more traditional Cnidaria + Bilateria (fig. 2.1), depending on the orthologues selected to construct the matrix. Sensitivity tests designed to detect compositional heterogeneity in amino-acid usage among taxa—a known cause of artifacts in phylogenetic reconstruction—show that support for Cnidaria + Bilateria can be pinpointed to this source of systematic error. When amino acids are recoded using a Dayhoff-6 matrix, which is supposed to reduce compositional effects, Placozoa always appears as the sister group of Cnidaria. We thus favor this position in the current book (but it is left unresolved in fig. 2.2), recognizing that many features recognized in cnidarians and bilaterians, including bilateral symmetry, nervous systems, basement membrane-lined epithelia, musculature, embryonic germ-layer organization, and internal digestion, may have been lost in placozoans.

7

PLACOZOA

Placozoa constitutes a group of anatomically simple, disc-shaped or branching marine metazoans that adhere to substrates and move by ciliary gliding. They were known from a single valid described species for more than a century, *Trichoplax adhaerens*, discovered in a seawater aquarium at the Graz Zoological Institute in Austria in the late nineteenth century (Schulze, 1883), but the phylum now includes two additional species, *Hoilungia hongkongensis* (Eitel et al., 2018) and *Polyplacotoma mediterranea* (Osigus et al., 2019). Placozoans have played a key role in discussions about early metazoan evolution. Most early work on *Trichoplax* is owed to Karl Gottlieb Grell, who, together with colleagues, studied its ultrastructure and reproduction and named the phylum Placozoa, based on Bütschli's "Placula"—a hypothetical two-layered benthic "Urmetazoon" (Eitel et al., 2013). Most of what we know about placozoans has been based on aquarium specimens (see Grell and Ruthmann, 1991). More recently, a series of marine surveys have shown a wide distribution and astonishing molecular diversity of Placozoa (Voigt et al., 2004; Signorovitch et al., 2006; Pearse and Voigt, 2007; Eitel and Schierwater, 2010; Guidi et al., 2011; Eitel et al., 2013).

Several salient aspects of placozoans, in light of their apparent simple morphology, refer to their body patterning, cell-to-cell communication, behavior (integration of ciliary movement, feeding, sensing the environment, etc.), and reproduction. Their phylogenetic position has been in flux, with some authors suggesting a sister-group relationship to all other metazoans—a position rarely supported in any formal phylogenetic analysis—sister group to Neuralia (sensu Nielsen, 2008), or in a series of alternative positions in relation to sponges, ctenophores, and cnidarians. Recent phylogenomic analyses including a diversity of early metazoan lineages suggest that placozoans are not primitively simple, but rather that they have acquired their simple body plan secondarily, as they nest higher up in the metazoan tree than Porifera and Ctenophora, constituting the possible sister group of Cnidaria + Bilateria (e.g., Chang et al., 2015; Pisani et al., 2015; Whelan et al., 2015) or Cnidaria (Laumer et al., 2018) (fig. 2.1).

These positions contrast with the lack of neural structures but are congruent with the bioinformatic prediction of many genes essential for neuronal function (Nikitin, 2015), with the expression by gland cells of several proteins typical of neurosecretory cells, including an FMRFamide-like neuropeptide (Smith et al., 2014), and with coordinated feeding behavioral responses (Senatore et al., 2017). Also congruent with this hypothesis is the presence of paraHox genes in Placozoa (Mendivil Ramos et al., 2012) and other Planulozoa, but not in Porifera and Ctenophora (Ryan et al., 2010; Pastrana et al., 2019). The issue of their anatomical simplicity is thus a conundrum that requires further attention (Miller and Ball, 2005).

While placozoans have become an emerging developmental model system (Schierwater and Eitel, 2015), with predicted distributions in temperate to warm seawaters (Paknia and Schierwater, 2015), their ecology and biogeography remain largely unknown, and only a handful of studies have documented interactions with other species (Pearse and Voigt, 2007). Riedl (1959) showed *Rhodope veranii* feeding on placozoans, and more recently it has been proposed that another rhodopemorph mollusc, *Rhodope placozophagus*, may also exhibit a specific predator-prey interaction (Cuervo-González, 2017).

SYSTEMATICS

The systematics of Placozoa is simple, as the group has only three formalized species in three genera: *Trichoplax adhaerens*, *Hoilungia hongkongensis*, and *Polyplacotoma mediterranea*. An earlier species described by Monticelli (1893), *Treptoplax reptans*, is currently unaccepted. It is also unclear whether Grell's specimens and those studied subsequently as *T. adhaerens* are actually conspecific with Schulze's species (Eitel et al., 2013). However, the species-level diversity of placozoans is much higher than what has been formalized, as understood now from morphological and molecular perspectives (Guidi et al., 2011; Eitel et al., 2013). The most recent phylogenetic analyses of the mitochondrial ribosomal *16S rRNA* has identified 19 haplotypes constituting seven distinct clades divided into two main groups, and many of these show ultrastructural differences (Eitel et al., 2013). Placozoans are found in all oceans sampled to date, including the Atlantic, Pacific, and Indian Oceans, inhabiting a variety of substrates. We suspect that ongoing metagenomic sampling may unravel an unsuspected diversity of these previously rare metazoans.

PLACOZOA: A SYNOPSIS

- Small (usually up to 3 mm, one species >10 mm), dislike, or with polytomous branching, asymmetrical metazoans consisting of two cell layers and six cell types
- Three described species and seven or more clades with a distribution in all temperate to tropical oceans
- Without a nervous system, but able to secrete neuropeptides and coordinate behavior
- Extracellular digestion
- Regular clonal proliferation by fission, fragmentation, and budding from the surface
- Genetic evidence for sexual reproduction having occurred in the evolutionary history of the group; fragmentary observations have been interpreted as indicating (poorly characterized) sexual reproduction

The asymmetrical body plan with a lack of anteroposterior or dorsoventral axes makes it difficult to describe the animals, but their usual disc-like shape allows us to refer to the upper and lower sides, the latter being that side facing the substrate (irrespective of the orientation of the substrate). Upper and lower main cells are

Figure 7.1. Schematic illustration of a cross section of *Trichoplax*. Below (facing the substrate) is a layer of lower epithelial cells, each with a cilium and microvilli; lipophil cells that contain large lipophilic inclusions, including a large spherical inclusion near the distal end; and gland cells, with secretory granules near the margin. Opposing are the upper epithelial cells. In between are fiber cells with branching processes that contact the other cell types. A shiny sphere containing a birefringent crystal lies under the epithelium. After Smith et al. (2014).

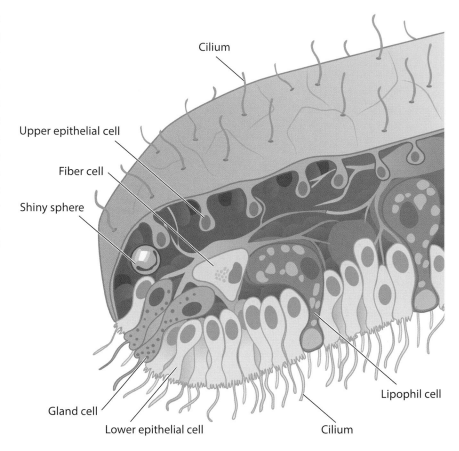

monociliated but different, the lower ones being mostly columnar and the upper being flattened (fig. 7.1). Fiber cells are located between these two layers and may play a role in the behavior of the animals, as they have transverse branches that contact other fiber cells and the cell bodies of a special type of secretory cells, called "lipophils," and resemble the electrically conductive syncytial junctions of glass sponges (Smith et al., 2015). Special structures unique to placozoans, called "shiny spheres" and containing lipid inclusions, are found at the surface of the interstices of the upper epithelial cells. These cells, which can be dislodged, have shown to have an antipredator function (Jackson and Buss, 2009).

Despite consisting of only six cell types (Smith et al., 2014) and lacking synapses, *Trichoplax adhaerens* can coordinate a complex sequence of behaviors, its feeding behavior elegantly characterized by Smith et al. (2015). These authors showed how the animal is able to integrate the beating of cilia and secretion to be able to conduct external digestion of algae. When *Trichoplax* glides over a patch of algae, its cilia stop beating so it ceases movement. A subset of one of the cell types, the lipophils—and only those located near the algae—simultaneously secretes granules whose content rapidly lysates algae. The animal pauses while the algal content is ingested, and then resumes gliding. This clearly shows that global control of gliding is coordinated with precise local control of lipophil secretion, suggesting the presence of mechanisms for cellular communication and integration. How this can be done in the absence of a discrete nervous system seems to be explained by the existence of a certain level of neurosecretion (Smith et al., 2014). Nonetheless, the lack of gap junctions or genes associated with gap junctions makes them beat their cilia asynchronously, unlike any other known metazoan (Smith et al., 2015).

Placozoans lack a basement membrane but express genes that in other animals encode for extracellular matrix (ECM) proteins, including collagen IV, laminin-α, -β and -γ, nidogen, heparan sulphate proteoglycans (including two glypicans) and a matrilin-2-like gene. The presence of these genes and their transcripts has led investigators to postulate that an ECM may be present but that it is not detected using traditional histological stains (Srivastava et al., 2008).

A few studies have tried to characterize gene expression patterns in placozoans (Martinelli and Spring, 2003, 2004; Monteiro et al., 2006), but these are difficult to interpret in light of their lack of anatomical features, as several genes are expressed along the periphery of the animal, which has been interpreted in the context of the Placula theory as being a precursor to the rim of the main body opening in cnidarians (Schierwater et al., 2009). More recently, a study focusing on expression patterns in a diversity of genes has postulated that the bottom surface of *Trichoplax* may be equivalent to cnidarian oral and endomesodermal tissue, and thus that placozoans have an oral–aboral axis, similar to cnidarians (DuBuc et al., 2019).

The placozoan life cycle is not well understood. While it has been claimed that there is an alternation of reproductive modes between bisexual and vegetative reproduction (Eitel et al., 2011), the evidence for this is tenuous. The regular observation of oocytes and expressed sperm markers is not sufficient to support that placozoans reproduce sexually, at least in the field. Embryonic development has been observed up to a 128-cell stage, during which new ultrastructural features not observed in adults develop (Eitel et al., 2011), but again, cleavage could occur without fertilization, as it does in some other animals. It has been claimed that predation upon placozoans can contribute to asexual reproduction (Cuervo-González, 2017), but regular clonal proliferation by fission, fragmentation, and budding certainly occurs spontaneously without the stimulus of predation (V. Pearse, pers. comm.).

As described above, placozoans are able to control feeding behavior and do this by responding to endomorphin-like peptides. A recent study has shown that signal amplification by peptide-induced peptide secretion may explain how a small number of sensory secretory cells lacking processes and synapses can evoke a wave of peptide secretion across the entire animal and between individuals that are not in contact to globally arrest ciliary beating and allow pausing during feeding (Senatore et al., 2017).

GENOMICS

Due to their phylogenetic significance, the nuclear genome of putative *T. adhaerens* was sequenced relatively early, compared to many other, better-known metazoans (Srivastava et al., 2008). Five additional genomes have been released a decade later (Eitel et al., 2018; Kamm et al., 2018; Laumer et al., 2018), placozoans being the only animal phylum with more genomes sequenced than species described.

The compact nuclear genome of *T. adhaerens* (ca. 98 million base pairs) shows conserved gene content, gene structure, and synteny in relation to the human and other complex eumetazoan genomes, something originally unexpected due to its apparent cellular and organismal simplicity. However, its genome encodes a rich array of transcription factors and signaling pathway genes that are typically associated with diverse cell types and developmental processes in other metazoans (Srivastava et al., 2008). This has later been corroborated by proteomic analysis of 2,800 *T. adhaerens* individuals showing a large, complex, protein-mediated, regulatory-process activity and high activity of tyrosine phosphorylation, in line with the hypothesized burst of tyrosine-regulated signaling at the instance of animal multicellularity (Ringrose et al., 2013).

The mitochondrial genome of *T. adhaerens* also offers interesting insights about these mysterious metazoans. Typical metazoan mitochondrial genomes are 15- to 24-kb circular molecules that encode a nearly identical set of 12 to 14 proteins for oxidative phosphorylation and 24 to 25 structural RNAs (*16S rRNA, 12S rRNA*, and *tRNAs*; lack significant intragenic spacers; and are generally without introns. In contrast, placozoan mitochondrial genomes are much larger (32- to 43-kb), highly variable circular molecules with numerous intragenic spacers, several introns and ORFs of unknown function, and protein-coding regions that are generally larger than those found in other animals (Dellaporta et al., 2006; Signorovitch et al., 2007; Miyazawa et al., 2012). Because these mitochondrial genomes are said to be more similar to those of known non-metazoan outgroups, such as chytrid fungi and choanoflagellates, some authors have suggested that mitochondrial genomes provide strong support for the placement of placozoans as the sister group to all other animals (Dellaporta et al., 2006), a position contradicted by virtually all molecular phylogenetic analyses published to date.

FOSSIL RECORD

Although an affinity to the Ediacaran *Dickinsonia* was proposed based on the feeding traces left by *Dickinsonia* and *Trichoplax adhaerens* (Sperling and Vinther, 2010), no accepted fossil can be assigned to Placozoa, as expected from their simple morphology.

CNIDARIA

8

If the reader is lucky enough to snorkel or dive in a tropical reef, cnidarians will be, no doubt, the dominant animals (probably followed by sponges and echinoderms) in terms of biomass and abundance. They will compose the basis of the reef, with its majestic hermatypic (= reef-building) corals, while other cnidarians may form a carpet of soft-bodied zooanthideans and corallimorpharians, covering most of the substrate. Frondose gorgonians and soft corals may then be exposed to the stronger water currents. But cnidarians, with their over 12,300 described species[3] (Daly et al., 2007; Zhang, 2011; Brusca et al., 2016b), can also be found in all other ocean regions and depths, dominating the water column and deep-sea reefs, as well as in most limnic environments.

In fact, cnidarians may be sessile, as in the so-called polyp-stage, which can be solitary, colonial, or free living as a swimming medusa. They can range from microscopic-size polyps and medusae to gigantic proportions, some medusae having tentacles many meters long. Likewise, some coral colonies are massive. The members of some clades show an alternation of an asexual polyp and a dispersal phase or medusa, which typically reproduces sexually. This life cycle is often described as a "biphasic life cycle,"[4] due to the presence of two adult phases. In this life cycle, one form reproduces asexually (the polyp) while the other reproduces sexually (the medusa). But not all cnidarians follow this pattern, and some exist only in one of these two stages (e.g., anthozoans only as polyps). Yet others, the myxozoans, can have vermiform stages in an obscure parasitic lifestyle.

The typical cnidarian body plan (we discuss the myxozoan body plan in a separate section below) resembles that of ctenophores in having a main oral–aboral axis and multiple planes of symmetry. In this case, cnidarians are said to have radial symmetry, but some authors interpret them as being bilaterally symmetrical, as shown in their gastrovascular cavity (or coelenteron). Some body parts have tetraradial (some medusae), hexaradial or octoradial symmetry, the latter two constituting the basis for the taxonomy of some groups (i.e., Hexacorallia and Octocorallia; see below).

---

[3] To the 10,100+ species reported by Zhang (2011), we add here the 2,200 described myxozoan species (see Brusca et al., 2016b).

[4] Some use the term "biphasic" to refer to life cycles with an adult and a larva; in the case of cnidarians, they can have the two phases discussed here plus a larva, which would make them "triphasic." We thus distinguish between a biphasic life cycle, with two adult forms, and refer to the animals with larvae as having indirect development. This is opposed to direct developers, those animals giving birth to juveniles that are similar to the adults except for size, gonads, and other sex-related characters.

Cnidarians are predators (some are parasites), being able to paralyze their prey with an arsenal of neurotoxins delivered via a unique type of stinging cells called "cnidocytes," the best-known being the nematocysts. In fact, in some oceans of the world, cnidarians are among the most dangerous animals one can encounter, the stings of some box jellies (also called "sea wasps") (class Cubozoa) being extremely painful and even fatal to humans. Entire beaches in Queensland, Australia, are often closed due to these almost invisible animals. The discharge mechanism of cnidocytes has remained a mystery for centuries, but recent research has shown that light may trigger the discharge, thus proposing a nonvisual function for opsin-mediated phototransduction (Plachetzki et al., 2012).

In addition, many lineages inhabiting shallow tropical waters have acquired zooxanthellae (dinoflagellates) to form one of the most successful and spectacular symbioses of the animal–vegetal kingdoms, giving origin to coral reefs—the most productive and diverse of marine ecosystems. While sessile cnidarians in the polyp form, especially colonial ones, can be the dominant group of benthic organisms in tropical waters, medusae dominate the water column at nearly all latitudes and depths. Some medusae have also developed intimate symbioses with zooxanthellae, including the members of the genus *Cassiopea*. In addition to the zooxanthellae, corals can also host symbiotic nitrogen-fixing cyanobacteria within the coral cells (Lesser et al., 2004). The presence of this prokaryotic symbiont in a nitrogen-limited zooxanthellate coral suggests that nitrogen fixation may be an important factor for this symbiotic association.

Understanding of the clade of parasitic cnidarians called "myxozoans," their development and life cycles, is also of broad interest to parasitologists and fisheries. Recent work on the group is revitalizing the field of evolution of these parasites (Okamura et al., 2015) and now have to be dealt with in a chapter on cnidarians, despite their major differences in anatomy and lifestyle. When we talk about "cnidarians" in a morphological or ecological sense, we tend to refer to the classical group, while using "myxozoans" to refer exclusively to this parasitic group of cnidarians.

Typical cnidarians are also notorious for their spectacular reproductive strategies. Who has not heard of the synchronized coral spawning or the bizarre pedal laceration, a mode of asexual reproduction by which walking anemones leave small fragments of their pedal disc behind, those giving origin to new individuals, in a mechanism similar to that described for platyctenid benthic ctenophores (Martindale and Henry, 1997a)? Clonality and asexual perpetuation are thus fascinating topics of research, scientists even having proposed that some cnidarians are immortal. Such is the case of the small hydrozoans of the genus *Turritopsis*, a jellyfish capable of reverse development via cell transdifferentiation (Piraino et al., 1996) and thus called "the immortal jellyfish." While most animals with indirect development cannot reverse their larval development, cnidarians have the ability of reversible metamorphosis of their planula larva (Richmond, 1985).

Another odd reproductive behavior is the so-called polyp bailout of corals, where under conditions of stress the soft parts of the polyp may detach and disperse to a new site (Sanmarco, 1982; Serrano et al., 2017). But perhaps the most bizarre of all cnidarians are boloceroidid sea anemones, which are able both to crawl and swim actively and reproduce both sexually and asexually by fission, pedal laceration, oral laceration, or even by sheding and swallowing their own tentacles, which develop into new polyps in the coelenteron (Pearse, 2002).

SYSTEMATICS

The systematics of Cnidaria has been investigated using morphological and molecular markers, including phylogenomic data (Zapata et al., 2015; Lin et al., 2016; Quattrini et al., 2017; Kayal et al., 2018). Most studies agree on a system with two main clades—Anthozoa, for the polyp-only stages, and Medusozoa—but resolution based on mitochondrial genes conflicts with respect to the monophyly of Anthozoa (Kayal et al., 2013). Our modern understanding of the systematics of Cnidaria was synthesized by Daly et al. (2007). We here follow this system with some additions from recent phylogenetic analyses of *Polypodium* and Myxozoa, which form a clade, Endocnidozoa, that is sister group to Medusozoa (Jiménez-Guri et al., 2007b; Chang et al., 2015; Kayal et al., 2018). Structure within Medusozoa has long been debated, but the phylogenomic analysis of Kayal et al. (2018) rescues the old clade Acraspeda, uniting Cubozoa, Scyphozoa, and Staurozoa as the sister group of Hydrozoa. Our system resembles that recently proposed by Brusca et al. (2016b), and given the extensive history of the study of cnidarian systematics, we refer to Collins (2009).

Phylum Cnidaria
    Subphylum Anthozoa
        Class Anthozoa
            Subclass Hexacorallia
                Order Actiniaria (sea anemones)
                Order Antipatharia (black corals)
                Order Corallimorpharia
                Order Scleractinia (stony corals)
                Order Zoanthidea
            Subclass Ceriantharia
                Order Penicillaria
                Order Spirularia
            Subclass Octocorallia (soft corals, gorgonians, red coral, sea pens, etc.)
                Order Alcyonacea (red coral, soft corals, gorgonians)
                Order Helioporacea (blue corals)
                Order Holaxonia
                Order Pennatulacea (sea pens)
    Subphylum Endocnidozoa
        Class Polypodiozoa
        Class Myxozoa

Subclass Myxosporea

Subclass Malacosporea

Subphylum Medusozoa

Class Hydrozoa

Subclass Hydroidolina

Order Anthoathecata

Order Leptothecata

Order Siphonophora (siphonophores)

Subclass Trachylina

Order Actinulida

Order Limnomedusae

Order Narcomedusae

Order Trachymedusae

Superclass Acraspeda

Class Cubozoa (box jellies or sea wasps)

Order Carybdeida

Order Chirodropida

Class Scyphozoa (jellyfish)

Order Coronatae

Order Semaestomeae

Order Rhizostomeae

Class Staurozoa (stalked jellyfish)

Order Stauromedusae

The taxonomy of Hexacorallia, especially with respect to the internal relationships of the stony corals, Scleractinia, may require further attention (Fukami et al., 2008) and novel genomic developments may provide further resolution (Quattrini et al., 2017). Within Octocorallia, the order Alcyonacea now includes a diversity of groups that once were considered separate orders, but their relationships are poorly understood (Collins, 2009; Kayal et al., 2018), and a rapid radiation of the members of this order has been proposed as an explanation for such poor resolution (McFadden et al., 2006). In addition, the current lower-level taxonomy of corals is in strong conflict with genetic data in many cases (e.g., Addamo et al., 2016), but this is extensible to the highly calcified members of other cnidarian classes (e.g., Ruiz-Ramos et al., 2014).

Medusozoan relationships have received special attention due to the importance of understanding the evolution of the cnidarian biphasic life cycle (e.g., Collins, 2002; Collins et al., 2006b; Kayal et al., 2013; Zapata et al., 2015; Kayal et al., 2018), all agreeing on its monophyly. The newest data sets divide Medusozoa into two clades, one comprising Staurozoa, Cubozoa and Scyphozoa, collectively named Acraspeda (Kayal et al., 2018), and another one with Hydrozoa.

Hydrozoans display the most morphological disparity within Cnidaria but have perhaps received the least broad systematic attention, and sister-group relationships among its main orders remain largely unresolved (Collins, 2009). A recent study using mitochondrial genomes has addressed these deficiencies within the larger clade Hydroidolina, showing monophyly of Siphonophora, which constitutes the sister group of all other hydroidolines, as well as monophyly of Leptothecata, but Anthoathecata is paraphyletic with respect to Leptothecata (Kayal et al., 2015). This study also shows that Filifera, one of the subclades of Anthoathecata, is polyphyletic, and thus a new system for Hydroidolina should be established. An additional development within Hydrozoa is the recognition of the recently described *Dendrogramma* (Just et al., 2014) as the detached bract of a benthic siphonophore (O'Hara et al., 2016).

The most drastic change in recent classifications of cnidarians is the inclusion of *Polypodium hydriforme* and Myxozoa within Cnidaria (e.g., Collins, 2009; Brusca et al., 2016b), a position long anticipated by morphological and molecular data (Siddall et al., 1995) and more recently shown also in multiple phylogenomic analyses (Nesnidal et al., 2013a; Feng et al., 2014; Chang et al., 2015). Given the molecular and morphological support for this clade (see Okamura and Gruhl, 2016), we use the name Endocnidozoa (sensu Zrzavý and Hypša, 2003) and assign to it a rank equivalent to that of the other two cnidarian classes. We further support a sister-group relationship of Endocnidozoa to Medusozoa (one could consider the free-living stage of *Polypodium* a medusa), as suggested in recent phylogenomic analyses (Chang et al., 2015), and by the tetraradial arrangement of the musculature in Endocnidozoa and Medusozoa (Gruhl and Okamura, 2012). Earlier analyses placing Endocnidozoa within Bilateria or as basal bilaterians are thus certainly artefactual (see Evans et al., 2008).

Myxozoans are intracellular parasites of vertebrates, annelids, and bryozoans, with myxospores (cnida-like structures containing nematocyst-like filaments) that had been treated as protozoans by some authors due to their obligate parasitism until phylogenetic analyses suggested a metazoan affinity (Smothers et al., 1994), a relationship also proposed by others based on their ontogeny and morphology. Some have a life cycle with multiple hosts, such as myxosporeans, with a myxosporean phase in vertebrates and an actinosporean phase within polychaete annelids and other coelomate worms. Another interesting player in this debate is *Buddenbrockia plumatellae*, a nematode-like parasite of freshwater bryozoans (Okamura et al., 2002), with polar capsules like those of the malacosporean myxozoans, parasites of bryozoans. The homology of the polar capsules to the nematocysts of the other cnidarians has been defended on both morphological and molecular grounds (Holland et al., 2011). It is now clear that *Buddenbrockia* is a member of the clade Myxozoa (Monteiro et al., 2002; Zrzavý and Hypša, 2003; Jiménez-Guri et al., 2007a; Jiménez-Guri et al., 2007b), and thus a cnidarian.

Polypodium hydriforme is the sole member of the class Polypodiozoa, an animal adapted to intracellular parasitism in oocytes of acipenserid and polyodontid fishes, with a complicated life cycle (Raikova, 1994).

Molecular systematics has also helped resolve the position of a few odd cnidarians. *Tetraplatia* is a genus composed of two species of pelagic cnidarians of curious morphology, with a vermiform shape and four swimming flaps. Since their discovery in the mid-1800s, a number of prominent cnidarian workers have argued whether *Tetraplatia* is an aberrant hydrozoan or a scyphozoan, a discussion resolved on the side of Hydrozoa (more specifically Narcomedusae) based on analyses of ribosomal RNA gene data (Collins et al., 2006a). Nonetheless, this and *Polypodium* have significance with regards to the ability to modify a supposedly radial body plan into a wormlike animal.

CNIDARIA: A SYNOPSIS

- Endoderm and ectoderm (diploblastic) separated by a cellular mesoglea or partly cellular mesenchyme
- Oral–aboral axis and radial symmetry (often hexaradial or octoradial)
- Blind gastrovascular cavity (coelenteron); the mouth facing generally downward in free-living medusae or upward in attached solitary polyps; the mouth orientation in large colonial organisms varies depending on the position of the zooid
- Solitary, clonal, or colonial; when colonial, with or without zooid specialization; specialization peaks in the members of Siphonophora
- Unique type of cell called "cnidocyte," which produces a cnida; cnidae are adhesive or stinging structures, the most common ones being nematocysts
- Neurons forming a simple nerve net
- Musculature formed largely of myoepithelial cells
- Simple sensory structures that can become complex in some Medusozoa, which have highly complex rhopalia
- Biphasic life cycle consisting of an asexual polypoid phase and a sexual medusoid stage (fig. 8.1); the life cycle can miss either of these stages, and if missing the medusoid stage, the polyp develops gonads to complete the sexual life cycle
- Typically with a ciliated planula larva
- Parasitic myxozoans an exception to virtually all these characteristics

THE CNIDARIAN BODY-PLAN: POLYPS AND MEDUSAE

Anthozoans are restricted to the polyp phase and can be solitary, clonal, or colonial (e.g., most scleractinian corals, gorgonians, soft corals, pennatulaceans, and zoanthideans are colonial). Gorgonians are colonial anthozoans (Alcyonacea) that can grow in bushy shapes, whereas others are planar, size and shape of the colony often defined by the hydrodynamics of the local surges and currents. In the clonal sea anemona *Anthopleura elegantissima*, central anemones have larger gonads, peripheral anemones investing more in the acrorhagi used in intraspecific aggression (Francis, 1976). While the polyps are not truly polymorphic, as in some other cnidarians, they are not identical.

Cnidarian polyps are formed by a column arranged along the oral–aboral axis and a crown of tentacles around the mouth. They are generally sessile, attached

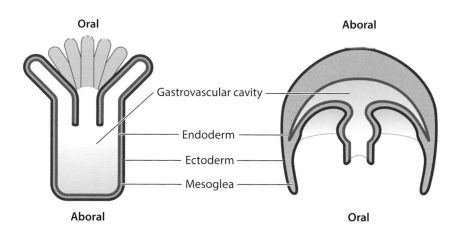

Oral

Aboral

Gastrovascular cavity

Endoderm

Ectoderm

Mesoglea

Aboral

Oral

FIGURE 8.1. Schematic drawing of cross sections of the polyp (left) and medusa (right) of cnidarians.

by a pedal disc at the aboral end, although some have the ability to walk. The aboral end may also be rounded and adapted for digging in the sediment, as in the burrowing anemones, or the polyp may arise from a common stalk or stolon in colonial species. In anthozoans the mouth is a slitlike structure on a flat oral disc, leading to a muscular pharynx that extends into the gastrovascular cavity (in anthozoans). In hydrozoans the mouth may be set on an elevated hypostome. The pharynx usually bears one to multiple ciliated grooves called "siphonoglyphs," which drive water into the gut cavity. The gastrovascular cavity is lined by the gastrodermis, while the external surfaces are covered by the epidermis.

In hydrozoan polyps, the gastrovascular cavity is a single, uncompartmentalized tube. In scyphozoan polyps, it is partially subdivided by four longitudinal, ridgelike mesenteries. In anthozoans the coelenteron is extensively compartmentalized by mesenteries, with the free inner edge below the pharynx presenting a thickened, cordlike margin armed with cnidae, cilia, and gland cells and is called the "mesenterial filament." These can form special structures called acontia in some sea anemones (see below). In soft corals the gastrovascular cavities of its polyps are interconnected by canals called "solenia."

Marine hydroids are colonial and most are surrounded, at least in part, by a nonliving protein–chitin exoskeleton secreted by the epidermis called "perisarc"—as opposed to the living tissue, or "coenosarc." The outer covering is absent in freshwater hydroids.

A complex terminology has been developed to describe hydrozoan polyps or zooids (called "hydroids"). The reason for this special nomenclature is that hydroid colonies are usually polymorphic, containing more than one kind of polyp. The term "hydranth," or "gastrozooid," refers to feeding polyps, which typically bear tentacles and the mouth. Other commonly occurring polyp types include defense polyps (dactylozooids) and reproductive polyps (gonozooids or gonangia). Each polyp typically arises from a stalk or hydrocaulus. In most colonial hydrozoans, the individual polyps arise from the rootlike stolon, or hydrorhiza (Brusca et al., 2016b).

Gastrozooids capture and ingest prey and provide energy and nutrients to the rest of the colony, including all the nonfeeding polyps. Dactylozooids are heavily armed with cnidae and often surround each gastrozooid, serving for both defense and prey capture. Gonozooids produce medusa buds, or gonophores, that are either released or retained. Whether released as free medusae or retained as gonophores, they produce gametes for the sexual phase of the hydrozoan life cycle. The coenosarc of the gonozooid is called the "blastostyle," from which the gonophores arise. When a gonotheca surrounds the blastostyle, the zooid is called a "gonangium."

The perisarc may extend around each hydranth and gonozooid as a hydrotheca and gonotheca, respectively. When this occurs, the hydroids are said to be thecate, while hydroids whose perisarcs do not extend around the zooids are athecate; these conditions serve as the bases for hydroid taxonomy.

The most dramatic examples of polymorphism among cnidarian colonies are seen in Siphonophora (Hydrozoa) and Pennatulacea (Anthozoa). Pennatulaceans may have two types of polyps, autozooids, bearing tentacles for feeding, and the smaller siphonozooids, with reduced tentacles that create water currents through the colony. Siphonophores are hydrozoan colonies composed of highly polymorphic individuals, with as many as a thousand zooids in a single colony. The gastrozooids of siphonophores are highly modified polyps with a large mouth and one long, hollow tentacle that bears many cnidae. This long feeding tentacle reaches lengths of 13 m in the Portuguese man-o-war (*Physalia physalis*). The nonfeeding dactylozooids also bear one long (unbranched) tentacle. The gonozooids are usually branched; they produce sessile gonophores that are never released as free medusae.

Siphonophores use propulsive zooids, sometimes in the shape of a swimming bell (nectophore), a gas-filled float (pneumatophore), or both, to help with navigation. The nectophore shows many of the structures common to free-swimming medusae, although it lacks a mouth, tentacles, and sense organs, and this is why some authors thought it derived from a medusoid individual. The pneumatophore, once also thought to be a modified medusa, derives directly from the larval stage and probably represents a highly modified polyp. Pneumatophores are double-walled chambers lined with chitin. Each float houses a gland that secretes gas to control flotation and keep the colony at a particular depth.

Siphonophores can have colonies with swimming bells but no float; can have a small float and a long train of swimming bells; or can have a large float and no bell. Calycophorans have a long tubular stem extending from the swimming bell, from which various types of zooids bud in groups called "cormidia." Each cormidium acts as an individual colony, usually composed of a shieldlike bract, a gastrozooid, and one or more gonophores that may function as swimming bells. As discussed above, the mysterious *Dendrogramma* has been recognized as the detached bract of a benthic siphonophore (O'Hara et al., 2016). The cormidia can break loose from the parental colony; the detached cormidia are termed "eudoxids." Physonectans have an apical float with a long stem bearing a series of nectophores followed by a long train of cormidia. Cystonectans, including *Physalia*, usually have a large pneumatophore with a prominent budding zone at its base, which produces the various polyps and medusoids.

Because these colonies usually contain individuals at different developmental stages, they have recently attracted the interest of developmental biologists (Dunn, 2005; Siebert et al., 2011). The oldest cormidium defines the posterior part of the colony, but the body axes are not intuitive and a standard nomenclature has been developed for consistency (Haddock et al., 2005).

Velella velella (the "by-the-wind-sailor"; Hydroidolina, Anthoathecata) is a colorful oceanic organism that drifts about on the sea surface sometimes in large groups drawn by the wind and prevailing current, occasionally washing ashore to coat the beach with their bluish-purple bodies, sometimes for hundreds of kilometers. These are short lived and do not possess a functional mouth or gut, probably relying on their combined ingestion of small planktonic prey supplemented with autotrophic nutrition from symbiotic zooxanthellae (Purcell et al., 2015).

The medusa phase, a free-living, noncolonial, often planktonic phase, is found in members of the clade Medusozoa. It is generally oriented with the mouth toward the bottom, and due to the difficulties in preserving and collecting mid-water animals, they are often less studied than polyps. Many species were originally described as different taxa from the polyp and the medusa before closing their life cycle. In a few hydrozoan taxa, medusae remain attached to the colonies as sessile gonophores, producing gametes for the sexual phase of the hydrozoan life cycle. The "stalked medusae" (Staurozoa, Stauromedusae) are benthic and solitary, but most others are free living, even when some lie on the bottom, as do the upside-down jellies of the genus *Cassiopea*, which rest with the oral side facing up while photosynthesizing.

Medusae are generally umbrella- or bell-shaped (from deep to flat), and usually have a thick, mesogleal layer of a jellylike consistency (hence the name "jellyfish"), reminiscent of the body plan of ctenophores, with the oral end facing generally downwards. The mouth is located in the center of the umbrella, at the end of a tubular extension called the "manubrium." The manubrium is often present in hydromedusae (the medusae of hydrozoans) but can be reduced or absent in scyphomedusae.

The gastrovascular cavity occupies the central region of the umbrella and extends radially via radial canals. In most hydromedusae, a marginal ring canal within the rim of the bell connects the ends of the radial canals. The presence of four radial canals and of tentacles in multiples of four (in hydromedusae) and the division of the stomach by mesenteries into four gastric pouches (in scyphomedusae) give most jellyfish a tetraradial symmetry. Most hydromedusae have a circular flap of tissue, the velum, within the margin of the bell. Such medusae are termed "craspedote." Those lacking a velum, such as scyphomedusae, are said to be acraspedote. As in polyps, the external surfaces of medusae are covered with epidermis, and the internal surfaces (coelenteron and canals) are lined with gastrodermis. The gelatinous middle layer is either a largely acellular mesoglea or a partly cellular mesenchyme.

In cnidarians, the endoderm gives origin to the gastrodermis (the lining of the gastrovascular cavity) and the ectoderm gives rise to the outer epithelium; cnidarian ectoderm and endoderm are epithelial monolayers during development and throughout the life of the animal (Technau et al., 2015) and are separated by a common extracellular matrix, called the "mesoglea." The mesoglea is secreted by the epithelial cells and is composed of laminin, fibronectin, and collagen IV, similar to the basement membrane of Bilateria.

In the actiniarian *Nematostella vectensis* the musculature is largely formed by myoepithelial cells of endodermal origin, but there are also myoepithelial cells derived from the ectoderm in the tentacles. The endoderm expresses all the mesodermal markers found in bilaterians and thus is homologized by some authors to both, the endoderm and the mesoderm of bilaterians. A typical cnidarian myoepithelial cell (called epitheliomuscular if in the epidermis and nutritive-muscular if found in the gastrodermis) is columnar, bearing flattened contractile basal extensions called "myonemes" that interconnect with those of neighboring cells to form longitudinal and circular sheets of muscle (Brusca et al., 2016b). Similar epitheliomuscular cells are also found in the mesenteries of an inarticulate brachiopod and in the coelomic channels of cephalochordates (Storch and Welsch, 1974).

With very few exceptions, polyps have smooth muscle, but medusae also have striated, mononuclear muscle cells in the subumbrella (Technau et al., 2015), which express striated muscle *myhc* orthologues. However, the striated muscle of cnidarians (and ctenophores) lacks crucial components of bilaterian striated muscles, such as genes that code for titin and the troponin complex. For these reasons, some authors interpret the striated muscles in cnidarians and bilaterians to have evolved convergently (Steinmetz et al., 2012).

The worm-shaped parasitic *Polypodium* has only smooth muscle cells situated within the mesoglea, not epithelial muscle cells (Raikova et al., 2007), and *Buddenbrockia* has four well-defined blocks of longitudinal muscle running the length of the body (Jiménez-Guri et al., 2007b), consisting of independent myocytes embedded in the extracellular matrix between inner and outer epithelial tissue layers. This has led some authors to postulate a possible support for the presence of mesoderm in cnidarians (Gruhl and Okamura, 2012). The embryological origin of these muscles is, however, unknown, and it has been suggested that they may be analogous to anthozoan mesenteries or folds/concentrations of cells (M. Q. Martindale, pers. comm.).

The digestive cavity of cnidarians is a blind gastrovascular cavity with a single opening, although exceptions, with a flow-through system, have been noted among deep-sea corals (Schlichter, 1991) and a hydrozoan medusa (Aria and Chan, 1989). Further observations of additional openings have been made on the scyphozoan *Phyllorhiza punctata* (Jean Jaubert, pers. comm.).

The gastrodermis contains numerous enzyme-producing cells that facilitate extracellular digestion, and after distribution of the nutritive soup, digestion is completed intracellularly. The gastrovascular cavity can be highly partitioned and branched, to facilitate digestion—the larger the animal, the more partitioned it is. In some sea anemones (the so-called Acontiaria), there are acontia, free coiled threads at the end of the thickened edge of mesenteries, near the pedal disc, and loaded with cnidae that float in the gastrovascular cavity to subdue prey (Rodríguez et al., 2012). The acontia may be shot out through the mouth or through pores in

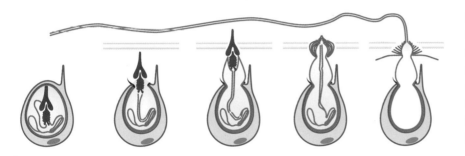

the body wall when the animal contracts. In *Exaiptasia pallida*, the animal possibly controls a network of body muscles and manipulates water pressure in the gastrovascular cavity to eject and retract acontia (Lam et al., 2017), instead of resynthesizing acontia after each ejection. This mechanism is also probably found in other acontiarians.

Cnidae (including nematocysts; fig. 8.2) are complex organelles produced and contained within special cells called "cnidocytes" derived from the epidermis or gastrodermis, and the synapomorphy that gives the name to Cnidaria. Fully developed cnidocytes may have a variety of functions, including prey capture, defense, and environmental sensing. They also play a role in extracoelenteric digestion in anthozoans (Schlesinger et al., 2009).

Cnidocyte development is well known in hydrozoans, but not fully understood in other groups of cnidarians. Hydrozoans, best characterized in the model species of the genera *Hydractinia* and *Hydra*, possess a population of stem/progenitor cells—interstitial cells, or i-cells—defined morphologically rather than functionally (Gahan et al., 2016). I-cells are multipotent (in *Hydra*) or totipotent (in *Hydractinia*). In fact, these are the first stem cells described in the biological literature, and their detection by Weismann in 1893 gave rise to his germline and "germ plasm" theory (see a review in Frank et al., 2009). I-cells are small and reside in the interstitial spaces of epithelial cells, primarily in the epidermis, are often found to be proliferative and migratory and express genes whose bilaterian homologues are known to be involved in stem cell and germ cell biology (Gahan et al., 2016). Little is known about the stem cell system in other cnidarians.

Recent work has focused on understanding cnidocyte development in the model anthozoan species *Nematostella vectensis* (Wolenski et al., 2013; Babonis and Martindale, 2017). Cnidocyte development requires the conserved transcription factor *PaxA* (but not *PaxC*), a gene often involved in eye development in other animals. Other transcription factors also have a role in cnidocyte development, including *Mef2* (*myocyte enhancer factor-2*), which coexpresses in a subset of the *PaxA*-expressing cells. A third transcription factor, *SoxB2*, regulates expression of *PaxA*, *Mef2*, and several cnidocyte-specific genes. Because *PaxA* is not coexpressed with *SoxB2* at any time during cnidocyte development, a simple model for cnidogenesis has been proposed whereby a *SoxB2*-expressing progenitor cell population (equivalent to hydrozoan i-cells) undergoes division to give rise to *PaxA*-expressing cnidocytes, some of which also express *Mef2* (Babonis and Martindale, 2017).

A diversity of animal groups is able to incorporate unfired nematocysts as a defense mechanism, in the phenomenon known as "nematocyst sequestration" or "kleptocnidism" (e.g., Salvini-Plawen, 1972; Greenwood, 2009). This has evolved multiple times, having been documented between 10 and 18 times among the members of Ctenophora, Porifera, Acoelomorpha, Platyhelminthes, and Mollusca (Goodheart and Bely, 2017; Schellenberg et al., 2019). Structures homologous to nematocysts were once believed to occur among dinoflagellate protists, although these have recently been shown to be convergent (Gavelis et al., 2017). However, not all cnidarians use nematocysts for feeding, as in the case of the tropical corallimorpharian *Amplexidiscus fenestrafer*, which captures prey directly with the oral disc without the use of nematocysts.

NERVOUS SYSTEM

Due to their phylogenetic position as an outgroup to Bilateria, the cnidarian nervous system has attracted broad interest, its structure having been thoroughly reviewed by Leitz (2016). Knowledge gleaned from the nervous system of freshwater hydrozoans in the genus *Hydra* has been especially influential, this traditionally having been considered to be composed of a diffuse noncentralized nerve net made of nerve cells (Lentz and Barrnett, 1965). There are also the cnidocytes, which are highly specialized mechanoreceptor cells. However, using novel calcium imaging of genetically engineered animals, Dupre and Yuste (2017) measured the activity of essentially all of the neurons in *Hydra vulgaris*, to conclude that instead of a simple nerve net, there is a series of functional networks that are anatomically nonoverlapping and are associated with specific behaviors. The nervous system of *Hydra* is therefore composed of three major functional networks that extend through the entire animal and are activated selectively during longitudinal contractions, elongations in response to light, and radial contractions, whereas an additional network is located near the hypostome and is active during nodding (Dupre and Yuste, 2017).

Anthozoan nervous systems have received little attention until the emergence of the starlet anemone *Nematostella vectensis* as a developmental model. *Nematostella*'s nervous system has been described as a diffuse nerve net with both ectodermal sensory and effector cells and endodermal multipolar ganglion cells. This nervous system, however, consists of several distinct neural territories along the oral–aboral axis, including pharyngeal and oral nerve rings and the larval apical tuft. Interestingly, these neuralized regions correspond to expression of conserved bilaterian neural developmental regulatory genes, including homeodomain transcription factors and neural cell adhesion molecules (Marlow et al., 2009). All these neural cells are derived from endoderm (which derives from the animal pole of the embryo), and only nematocyst neurons derive from ectoderm. The idea of a simple diffuse nerve net in cnidarians does not, therefore, seem to be accurate.

Nervous systems in Medusozoa have received attention, especially in relation to their sophisticated visual structures. Cnidarian eyes span from simple eyespots, pigment-cup ocelli of varying complexity, to camera-type eyes (Martin, 2002).

A recent large-scale analysis has suggested that eyes originated at least eight times within Cnidaria, with complex, lensed eyes having an independent origin from other eye types (Picciani et al., 2018). The multiple origin of eyes within Medusozoa also correlates with the evolutionary history of opsins. Eyes are well developed in some medusae, especially in Scyphozoa and Cubozoa, with complex rhopalia—each of the four eye-bearing structures in these groups—representing the most complex sensorial structures in cnidarians, and a structure well integrated with the central nervous system (Garm et al., 2006). Each rhopalium often has a statocyst, sensory epithelia, and ocelli.

In Cubozoa each rhopalium contains two lensed eyes and four bilaterally paired pigment-cup eyes (Parkefelt and Ekström, 2009), a structure that allows cubozoans to actively capture prey by luring them with the twitching of their extended tentacles (Courtney et al., 2015). Cubozoan ocelli can be morphologically complex, with cornea, lens, and a multilayered retina with pigmented ciliary photoreceptor cells (Nakanishi et al., 2010), and may contain a finely tuned refractive index gradient producing nearly aberration-free imaging (Nilsson et al., 2005). Rhopalia thus show us how relatively simple nervous systems can become highly complex and support stereotyped and well-coordinated behaviors.

SKELETAL STRUCTURES AND SUPPORT

Soft-bodied cnidarians have developed a whole series of mechanisms for support, from a water-filled coelenteron, to the presence of stiffened mesenchymes containing fibers. In addition, some incorporate hard inorganic materials into their bodies, greatly increasing support. Many cnidarian groups have developed specific structures, from the chitinous perisarc of colonial hydrozoans, to the many hard skeletal structures found mostly in some anthozoan groups. These are mainly organic axial skeletal structures, calcareous sclerites, and massive calcareous skeletons. Organic axial structures formed of protein–mucopolysaccharide complexes of gorgonin are found in holaxonian gorgonians (Alcyonacea) and black corals (Antipatharia), the latter used for jewelry due to its density. Scleraxonian gorgonians have an axis composed predominantly of sclerites that may be either unfused or fused with calcite, and soft corals often have calcareous sclerites embedded in their tissues. In some cases, the sclerites may fuse to form more rigid skeletons, such as in the precious corals of the genera *Corallium* and *Paracorallium* or in the organ-pipe coral (*Tubipora musica*), all in the non-monophyletic Alcyonacea. The sclerites are generally secreted by mesenchymal cells called "scleroblasts."

Perhaps the most conspicuous or charismatic cnidarians are scleractinian corals, magnificent organisms able to form massive calcareous skeletons. In some Scleractinia, epidermal cells secrete a skeleton of calcium carbonate that, when the coral is active, appears covered by a thin layer of tissue that in many cases can be withdrawn. Corals come in many shapes and forms, and their difficult taxonomy is often based on the structure of this internal skeleton, each fundamental unit called a "corallite." In solitary corals, such as in the genus *Caryophyllia*, the corallite shows a series of septa that increase the surface of the individual. Colonial corals are formed by many such fused polyps with different degrees of fusion, including a branching pattern of individual polyps (e.g., *Dendrophyllia*), to tightly packed colonies, such as in *Porites*, with intermediates such as *Acropora*. These are the typical hermatypic corals, in the order Scleractinia, that build reefs.

Corals that do not contribute to coral reef development are referred to as "ahermatypic" species. Scleractinians are more or less evenly split between zooxanthellate and azooxanthellate species. Most azooxanthellate species are solitary and nearly absent from reefs but have wider geographic and bathymetric distributions than hermatypic corals, which tend to be colonial. While the traditional classification of corals divided them between solitary and colonial, recent analyses have shown that symbiosis has been lost at least three times in the evolutionary history of scleractinian corals; coloniality has been lost at least six times; and there have been at least two instances in which both characters were lost (Barbeitos et al., 2010).

Some Hydrozoa also have calcareous skeletons and contribute to reef formation, especially those in the families Milleporidae (fire corals) and Stylasteridae. In this case, the skeleton is not surrounded by tissue, as in the scleractinian corals, but is penetrated by tissue through a series of canals (Brusca et al., 2016b).

The ecological importance of cnidarians, especially those with massive skeletons, and the chemistry relating the formation of calcium carbonate in relation to atmospheric carbon, temperature and pH, have resulted in corals becoming perhaps the most emblematic organism to study global climate change (e.g., Mayfield et al., 2011; Toth et al., 2012; Serrano et al., 2018) and heat stress tolerance (Barshis et al., 2013). The production of reef coral skeletons has important physiological demands. It is now known that some of the largest invertebrates require symbionts to help them grow. Among these, dinoflagellate algal symbionts are prominent in shallow clean waters of the tropics, allowing true invertebrate giants, like the giant clams of the genus *Tridacna* (Mollusca, Bivalvia, Cardiidae), the upside-down jellyfish of the genus *Cassiopea* (Scyphozoa), and, perhaps most conspicuously, hermatypic corals—a relationship that seems to have originated in the Mesozoic during the emergence and ecological success of reef-building corals (LaJeunesse et al., 2018). *Symbiodinium* is thus fundamental for the growth of corals (Shoguchi et al., 2013), and their loss causes the increasingly common disease known as "coral bleaching" (DeSalvo et al., 2010), which is also correlated with above-the-norm seawater temperatures (McWilliams et al., 2005). Therefore the effects of temperature on the symbiosis (Baird et al., 2009) as well as on the physiological demands of calcification (Vidal-Dupiol et al., 2013) have a dual effect on the viability of coral reefs in a changing ocean.

In addition to tentacular feeding on small planktonic animals, corals can filter-feed with the aid of mucus threads they secrete. Mucus feeding may be the prevalent direct feeding mode in some hard corals (e.g., Agariciidae). The mucus is also a nutritive food source for some reef fishes, and the reefs also act as a refuge for reproduction or settlement of many marine organisms, to the point that larvae of a multitude of animals, mostly fishes and crustaceans, are attracted to the sounds of reefs when searching for a place to settle (Montgomery et al., 2006; Simpson et al., 2008), as do coral larvae (Vermeij et al., 2010).

REPRODUCTION AND LIFE CYCLES

Cnidarian reproduction and the multiple cnidarian life cycles, including those of the parasitic Endocnidozoa, are diverse and have been discussed in several key revisionary works (Campbell, 1974; Martin, 1997; Martin and Koss, 2002), and more recently in two seminal reviews summarizing the state of the art of free-living cnidarian reproduction and development (Fautin, 2002; Technau et al., 2015) and one on myxozoans (Eszterbauer et al., 2015). Traditionally, cnidarian life cycles are divided into asexual and sexual phases whereby the animals can multiply by numerous asexual mechanisms sometimes throughout most of their life (e.g., fission, pedal laceration, budding, or fragmentation), but under certain conditions they reproduce sexually. Some species have well-defined biphasic life cycles, with the sessile polyp stage and the pelagic medusa that returns to the benthic life stage after a planula larva settles. The transition from polyp to medusa is less understood in most species, but in *Aurelia aurita* the mechanism known as "strobilation" (see below; fig. 8.2) is a well documented example of this biphasic life cycle. Many other life cycles exist, including skipping either the polyp phase or the medusa phase (Martin and Koss, 2002).

Asexual reproduction, colony growth, and regeneration play an important role in maintaining the life cycle of cnidarians and are mediated by their stem cells. *Hydra* is famous for its regenerative abilities, being able to regenerate fully after having been dissociated into single cells and recombined into a heterogeneous aggregate (Gierer et al., 1972). Transverse fission has been documented in members of Anthozoa, Hydrozoa, and Scyphozoa, and its absence in Cubozoa has been ascribed to poor knowledge (Fautin, 2002). The morphogenetic mechanisms across the classes of cnidarians require further study.

Budding is a main mode of growth that can result in increase of colony size or in the generation of new unattached, clonal individuals, but both processes require transformation of existing tissue and in some cases, the origin of secondary axes of growth—requiring what has been termed an organizer. A defining characteristic of an organizer region is its ability to induce the formation of a secondary axis when transplanted to another region of an embryo (Broun and Bode, 2002), and these organizers are well studied in vertebrates, where they are found only in the embryo and localized exclusively in the head region. In *Hydra* there is a head activation gradient that works throughout the life of the polyp and is not localized in any specific region. This gradient consists of two components, the hypostome, which has the inductive capacity of an organizer, and the self-organizing ability of the column tissue to form the second axis, processes that are mediated by the Wnt signaling pathway (Lengfeld et al., 2009)—as in the case of the vertebrate organizer.

Asexual reproduction in scyphozoans occurs at different levels. The asexual polyp of scyphozoans is called the scyphistoma, and it may produce new scyphistomae by budding from the column wall or from stolons. At certain times of the year, sexual medusae are produced by transverse fission of a modified scyphistoma (called "strobila"), the process of strobilation (Martin and Koss, 2002). Medusae may be produced one at a time in what is called "monodisc strobilation," or numerous

FIGURE 8.3. The biphasic life cycle of the scyphozoan *Aurelia aurita*, illustrating the process of strobilation and all the life cycle phases (planula, scyphistoma, ephyra and adult).

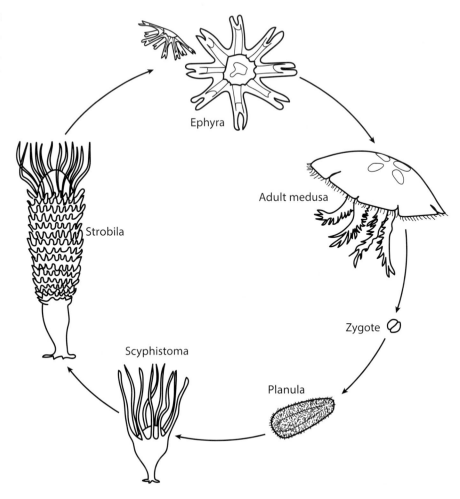

immature medusae may stack up one on top of another and then be released singly as they mature (polydisc strobilation), as is the case of *Aurelia aurita* (fig. 8.3). Immature and newly released medusae are called ephyrae.

During sexual reproduction, most cnidarians release their gametes into the water column. However, a variety of reproductive methods and strategies exist, including numerous cases of internal fertilization by uptake of sperm with different mechanisms of sperm storage (Marques et al., 2015), or the odd case of mating and even "copulation," as in the sea anemone *Sagartia troglodytes* (see Brusca et al., 2016b). But the most complex sexual behaviors are seen in the tripedaliid cubozoans. In *Copula sivickisi* and *Tripedalia cystophora*, two species of tiny box jellyfish, sexual dimorphism is evident and courtship behavior in which spermatophores are formed by coalescing sperm packets ejaculated from "seminal receptacles" of the male testes. During pair formation the male uses his tentacles to control the female, bringing her in close to transfer the spermatophore that she invariably ingests (Lewis and Long, 2005). The sperm bundles, sometimes of multiple males, are accommodated into her gastrovascular cavity (within several hours) and transferred to specialized structures that function in sperm storage.

Coral reproduction has attracted the attention of scientists and the public, especially through the episodes of mass spawning, and there has been a renaissance of coral reproductive biology, with information on sexual reproduction now available for more than 440 scleractinian species (Harrison, 2011). It is now clear that hermaphroditic broadcast spawning is the dominant pattern among coral species studied to date, while relatively few brooding species (whether hermaphroditic or gonochoristic) have been recorded—although most work on larval settlement has been done in these species. Brooding species in general transmit their dinoflagellate symbionts vertically while spawners get them from the environment, in which case the development of the mouth and coelenteron play important roles in symbiont acquisition (Harii et al., 2009).

In broadcasting spawning species, individuals must synchronize gamete release to ensure gamete encounter and successful fertilization (Harrison et al., 1984). Monthly larval release by some brooding species is documented to occur in coincidence with the lunar cycle, and in many reefs, mass coral spawning involving multiple species has been recorded. The simultaneous release of gametes from multiple species probably conditions coral biology, a group where hybridization between species is well documented (Vollmer and Palumbi, 2002; Combosch et al., 2008; Combosch and Vollmer, 2015). This is probably one of the reasons for the strong disagreement between morphological and molecular species delineation in corals, in addition to other factors like morphological convergence (see van Oppen et al., 2001) and phenotypic plasticity.

A recent phenomenon of larval cloning in corals has been documented, as early embryonic stages probably break by the physical action of waves and wind. These broken embryos are, however, able to complete development and settle (Heyward and Negri, 2012).

After completing development, the planula larva of corals (whether brooded or not; feeding or nonfeeding) becomes competent and starts probing the substrate while undergoing a series of morphological transformations. Quite a lot of recent research has focused on aspects of coral larval settlement and the effects of numerous physical and biological factors (Grasso et al., 2011; Strader et al., 2018).

Probing behaviors include body elongation and shifting of the phototactic behavior. During time of competency, larvae react to a wide range of environmental cues, including light intensity (e.g., Maida et al., 1994; Mundy and Babcock, 1998), light color (Strader et al., 2015), and reef sounds mostly generated by fish and crustaceans (Vermeij et al., 2010), as well as chemical signals from crustose coralline algae (Heyward and Negri, 1999) and microbial biofilms (Sneed et al., 2014). While the morpho/anatomical changes larvae undergo to settle and metamorphose are relatively well understood (Harrison and Wallace, 1990; Hirose et al., 2008), there is little known about the underlying molecular responses that drive these major changes. For example, in *Acropora* distinct aboral and oral expression patterns were observed during the transition from larva to polyp (Grasso et al., 2011), and patterns of upregulation in sensory and signal transduction genes were found while following development from embryos to settled larvae (Strader et al., 2018).

The probing period of the larva is crucial for the coral holobiont, and until recently it was believed that only the coral larval behavioral clues played a role in selecting the place for colony settlement. However, via differential gene expression analyses it has been shown that for the Caribbean widespread brooding species *Porites astreoides*, where symbionts are transmitted vertically, many more genes are differentially expressed in the symbiont than in the host, thus assigning a key role to the dinoflagellate symbiont in choosing the optimal place to establish the new colony (Walker et al., submitted). Once established, the planula metamorphoses into a corallite, calcification starts, and the animal becomes sessile.

Larval settlement, however, does not always follow the canonical pattern, and a few unconventional phenomena have been observed in some species of corals. For example, floating corallites have been observed in aquarium settings (Vermeij, 2009), in which a subset of planulae excreted a buoyant skeleton, causing the individual to float upside down on the water surface as tentacles developed pointing downward. Coral "rambling" after settlement, where a planula metamorphoses into a polyp but does not start calcifying, remaining moveable for a few days, has been observed in situ (Vermeij and Bakm, 2002), in a type of movement previously known only from anemones. The reversible metamorphosis of planulae (Richmond, 1985) and the bizarre polyp bailout (Sanmarco, 1982; Serrano et al., 2017) are discussed above. All these phenomena clearly indicate that coral reproduction can be more complex than previously thought.

The first myxozoan life cycle—that of *Myxobolus cerebralis*, which causes salmonid whirling disease—was closed by Wolf and Markiw (1984), showing that it involves two spore forms, which develop in alternating vertebrate and invertebrate hosts (fig. 8.4). This finding constituted a milestone in myxozoan research and significantly accelerated studies on myxozoan transmission and development (Eszterbauer et al., 2015). Since 1984 dozens of life cycles have been studied; however, most of the studies either replicate only partial life cycles (typically fish-to-worm transmission) or identify transmission of myxozoans based on DNA sequence data, and only have been completed for five species. The best-understood life cycle involves a waterborne myxospore, which is produced through asexual reproduction in the fish host and infects an intermediate invertebrate coelomate host, where the myxospores reproduce sexually to produce actinospores that infect the final host. For a recent review and details of the myxozoan life cycles, see Eszterbauer et al. (2015).

The life cycle of *Polypodium hydriforme* begins as a binucleate cell in the oocyte of the host and undergoes a series of changes until emergence from the host. The free-living stage is a stolon carrying buds with outer tentacles that, upon entering the water, fragments into pieces, which then take on the semblance of small jellyfish and acquire 24 tentacles. These individuals then form a mouth and feed on oligochaetes and other small invertebrates. The mature individuals develop gonads and complete the life cycle with a binucleate cell in the free-living organism, although no one has been able to close it (Raikova, 1994). The presence of cnidocytes during its life cycle has, however, always allied *P. hydriforme* with cnidarians (Ussow, 1887).

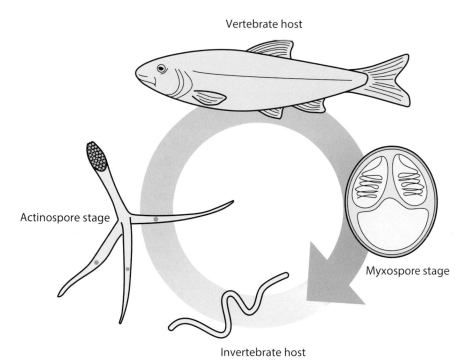

Vertebrate host

Actinospore stage

Myxospore stage

Invertebrate host

FIGURE 8.4. Life cycle of the myxozoan *Myxobolus cerebralis*. Based on an illustration from S. Atkinson for fishpathogens.net.

DEVELOPMENT

Cnidarian development is well studied, especially in the group's model organisms. The sperm is of the primitive type (ectaquasperm), but with acrosomal vesicles instead of an acrosome (Ehlers, 1993; Franzén, 1996).

The animal pole of the egg corresponds to the sites of first cleavage and gastrulation, and becomes the oral pole of the adult, both in hydrozoans (Freeman, 1980) and anthozoans (Lee et al., 2007). Cnidarian cleavage is generally radial and holoblastic but with quite some variation in the type of cleavage (which can be total or plasmodial), and the types of gastrulation can be diverse (Nielsen, 2012a). In most cnidarians, cell division is unipolar during cleavage—the cleavage furrow ingresses from only one side of the cell, which will become the oral pole.

Development is best known in *Nematostella vectensis*, for which early embryogenesis and gene expression can be mapped simultaneously (e.g., DuBuc et al., 2014). Early cleavage stages present substantial variability from embryo to embryo yet invariably lead to the formation of a coeloblastula, which then undergoes a series of unusual broad invaginations–evaginations (Fritzenwanker et al., 2007). Gastrulation has been previously described as a combination of invagination and unipolar ingression, but ingression could not be observed in a subsequent study (Magie et al., 2007). In the end, there is a clear correspondence of the animal–vegetal axis of the egg with the oral–aboral axis of the embryo. In *N. vectensis* embryos the animal hemisphere of the embryo contains organizing activity to form a normal polyp (Fritzenwanker et al., 2007). The end result of early development is the planula larva, which may settle and turn into a polyp or give rise to a medusa in those groups that lack the polyp stage. Planulae are well studied in corals (e.g., Hirose et al., 2008) and can be feeding or nonfeeding; it is at this stage that the larva can acquire their zooxanthellate endosymbionts in some species. The planula is elongated and fully ciliated, often with an apical tuft of longer cilia. Some planulae can undergo reverse development (Richmond, 1985).

While gastrulation defines the germ layers relatively early in development, Yuan et al. (2008) proposed that in *Aurelia* sp., during metamorphosis from planula into a polyp, cells in the planula (primary) endoderm undergo apoptosis, so the polyp (secondary) endoderm derives from the planula ectoderm. This implies a second reorganization of germ layers, what they called "secondary gastrulation." This hypothesis was not supported by the authors of a transmission electron microscopy study suggesting a germ-layer continuity from larva to polyp in *Aurelia aurita* (Mayorova et al., 2012), but transmission electron microscopy does not allow for following cell movements. A third study, designed to settle this dispute, performed a pulse-chase experiment to trace the fate of larval ectodermal cells, observing that prior to metamorphosis, ectodermal cells that proliferated early in larval development concentrate at the future oral end of the polyp. During metamorphosis, these cells migrate into the endoderm, extending all the way to the aboral portion of the gut, rejecting the hypothesis that larval endoderm remains intact during metamorphosis (Gold et al., 2016b).

The "secondary gastrulation" hypothesis seems to be a unique way of embryonic layer morphogenesis among animals—the only described case where an animal derives its endoderm twice during normal (noncatastrophic) development. This adds evidence that germ layers can be dramatically reorganized in cnidarian life cycles, favoring *de novo* generation of structures over the reorganization of structures from the larva, including nervous systems, musculature, and tentacles (see a discussion in Gold et al., 2016b).

Brooding is well known in some coral species (see above), but it can also be frequent in sea anemones. In the latter, brooding occurs in two primary modes, internal and external, in which offspring may be produced either sexually or asexually (Larson, 2017), and it is frequent at the poles.

GENOMICS

Due to their pivotal position among Planulozoa, and especially after the establishment of the actiniarian *Nematostella vectensis* as a model organism (e.g., see Layden et al., 2016), cnidarian genomics took off early, the first completed genome being that of *N. vectensis* (Putnam et al., 2007). This was followed by the genomes of the hydrozoan *Hydra magnipapillata* (Chapman et al., 2010), a model to study the nervous system, and the scleractinian coral *Acropora digitifera*, due to interest in coral conservation (Shinzato et al., 2011). A second actiniarian, *Aiptasia pallida*, soon followed (Baumgarten et al., 2015), and more recently the pennatulacean *Renilla reniformis* was also sequenced (Kayal et al., 2018). Additional draft genomes are available in NCBI, including the coral *Porites rus*.

These genomes are as complex as those of the best-known bilaterians and have unique features, such as a different arrangement of the Hox cluster, even among different anthozoans. Anthozoan genomes are known to have genomic rearrangements and taxonomically restricted genes that may be functionally related to the symbiosis, aspects of host dependence on alga-derived nutrients, a novel cnidarian-specific family of putative pattern-recognition receptors that might be involved in the animal–algal interactions, and extensive lineage-specific horizontal gene transfer (Baumgarten et al., 2015).

Genomic resources are also availalbe for the myxozoans *Kudoa iwatai* and *Myxobolus cerebralis*, as well as the reduced cnidarian parasite, *Polypodium hydriforme*. The myxozoan genomes are among the smallest metazoan genomes, both in size and in gene content, with ca. 5,500 genes, lacking key elements of signaling pathways and transcriptional factors important for multicellular development (Chang et al., 2015). In addition, they show a depletion of expressed genes related to development, cell differentiation, and cell–cell communication. These characteristics are common in the reduced genomes of other extreme parasites (e.g., Mikhailov et al., 2016).

Cnidarian mitochondrial genomes have been studied for a long time, and while anthozoans have a typical circular genome, all known medusozoans have a linear genome (Bridge et al., 1992), as in Calcarea (Porifera). As in sponges and other deep-branching metazoans, mitochondrial evolutionary rates are considerably slower than in bilaterians (Shearer et al., 2002), although the mitochondrial genomes of myxozoans and *Polypodium hydriforme* are highly reduced and fast evolving (Takeuchi et al., 2015; Lavrov and Pett, 2016).

Foundation species such as corals are often long-lived and clonal, but the age of colonies is difficult to estimate. Genets may consist of many (hundreds) of members (ramets) that originated hundreds to thousands of years ago (Devlin-Durante et al., 2016). During these long periods, as climate change and other factors exert selection pressure, the demography of populations changes. In order to assess the age of colonies, Devlin-Durante et al. (2016) correlated somatic mutations with genet age to provide the first estimates of genet age in the emblematic Caribbean species *Acropora palmata*, which has been largely decimated throughout its range. The age of such colonies could be on the order of hundreds to thousands of (up to 6,500) years old (depending on the mutation rate utilized).

FOSSIL RECORD

The cnidarian fossil record is extensive, especially for the skeleton-bearing species. Cnidaria provide the most likely candidates for body fossils of Metazoa from the Ediacaran Period, and they are well represented from the early Cambrian onward.

Frondose Ediacaran megafossils have repeatedly been compared to Cnidaria, particularly with members of Pennatulacea, and superficial similarities between some hermatypic corals in the family Fungiidae and Ediacaran fossils such as *Dickinsonia* have also been noted. Little hard evidence exists to support such affinities, and with respect to rangeomorph fronds such as *Charnia* and *Charniodiscus* (fig. 1.3 G), the growth polarity is in fact the opposite of that for Pennatulacea (Antcliffe and Brasier, 2008), undermining homology. As well, the attachment of lateral branches in rangeomorphs would not have permitted the water flow required for octocoral-style filter feeding (Xiao and Laflamme, 2009). Radial Ediacaran fossils also have a long history of comparison with Cnidaria, often with porpitid chondrophorine Hydrozoa. These fossils are morphologically and taphonomically inconsistent with cnidarian medusa (Young and Hagadorn, 2010), and the oldest reliable porpitids are Paleozoic (Fryer and Stanley, 2004).

A more recently described Ediacaran fossil from Mistaken Point, Newfoundland, *Haootia quadriformis* (fig. 8.5 A), was proposed to be a total-group cnidarian based on its twisted impressions of bundles of fibers interpreted as muscle extending into four corners of a polyp-shaped organism (Liu et al., 2014a). The musculature of *Haootia* was particularly compared with that of Staurozoa, but a survey of muscles in extant taxa of that group noted some conflicts with what is seen in the fossils, and an alternative reconstruction of gross myoanatomy was proposed (Miranda et al., 2015).

Among putative Proterozoic cnidarians is a suite of taxa including *Sinocyclocylicus*, *Ramitubus*, *Crassitubus* and *Quadratitubus* from the Weng'an biota in China, known from secondarily phosphatized tubular microfossils. These have closely spaced cross walls that have been compared with the tabulae of Paleozoic tabulate corals, and at least some taxa have tetrameral symmetry. However, the case for cnidarian affinities has been challenged based on evidence that cellular tissue was likely present in the space between the cross walls, and algal or cyanobacterial affinities are more likely than with Cnidaria (Cunningham et al., 2015; Landing et al., 2018).

Corumbella werneri (fig. 8.5 B), known from the latest Neoproterozoic of Brazil, Paraguay, and the United States, represents a strong candidate for an Ediacaran cnidarian. The fossils consist of elongate, annulated tubes with a square cross-section conferring tetraradial symmetry, with evidence for budding (Babcock et al., 2005). The tube was organic, organized as polygonal plates with pores and papillae (Warren et al., 2012). These plates and a lamellar microfabric (Pacheco et al., 2015) resemble those of conulariids, a fossil group widely endorsed as scyphozoans (see below). The morphology and likely reproductive mode have prompted comparisons between *Corumbella* and coronate Scyphozoa, thus making it plausible that crown-group Cnidaria crossed the Ediacaran–Cambrian boundary.

Some evidence has also been marshaled for Scyphozoa in the Ediacaran in the form of conulariids, a fairly common group of fossils throughout the Paleozoic, ranging at least from the late Cambrian to the Late Triassic. Affinities of conulariids to Scyphozoa have been recovered in phylogenetic analyses of Medusozoa (Van Iten et al., 2006). An origin of the group in the Neoproterozoic is indicated in two taxa from the terminal Ediacaran, *Vendoconularia triradiata*, from the terminal Ediacaran in Russia (Van Iten et al., 2005) and fossils from Brazil compared to the Paleozoic genus *Paraconularia* (Van Iten et al., 2014). An alternative position within Medusozoa allies conulariids and Cambrian fossil groups such as the tubular carinachitids, anabaritids (triradiate fossils that are known from the basal Cambrian and are among the earliest cnidarian "small shelly fossils"), and olivooidids (see below), these taxa being grouped with Staurozoa rather than Scyphozoa (Han et al., 2016b; Han et al., 2018).

Thus, some reasonable fossil evidence supports a latest Ediacaran fossil record for a few cnidarian lineages, but the phylum is known with greater confidence from

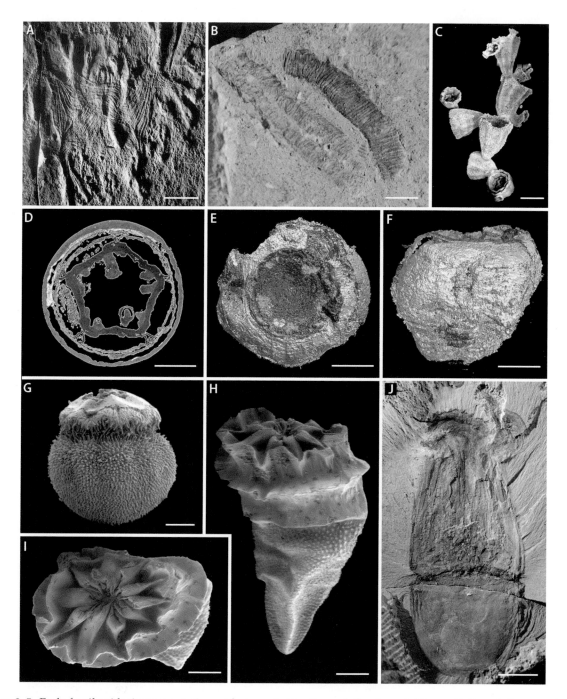

Figure 8.5. Early fossil cnidarians or putative cnidarians. A, *Haootia quadriformis*, scale 20 mm; B, *Corumbella werneri*, scale 3 mm; C, *Cambroctoconus orientalis*, scale 5 mm; D–F, *Sinaster petalon*. D, horizontal microtomographic section, scale 200 μm; E–F, reconstructed microtomographic models, scale 150 μm; G–I, *Olivooides* sp., G, embryo, scale 150 μm; H–I, post-embryonic stage, scale 250 μm; J, *Xianguangia sinica*, scale 5 mm. Image credits: A, Alex Liu; B, Loren Babcock; C, Tae-Yoon Park; D–F, Xing Wang; G–I, Phil Donoghue; J, Derek Siveter.

the Fortunian Stage, the earliest Cambrian. The Kuanchuanpu Formation of South China is the source of three-dimensional, secondarily phosphatized microfossils ca. 535 million years old, including developmental series. The recently described *Sinaster petalon* (see Wang et al., 2017b) (fig. 8.5 D–F) in particular shares numerous characters that indicate membership in Medusozoa. These include coronal muscles, interradial and accessory septa, gonad lamellae, and tentacle buds. The suggested phylogenetic position of this species is within a putative cubozoan-scyphozoan clade (Wang et al., 2017b). The same formation yields pre-hatching–stage embryos identified as pentamerous Cubozoa (Han et al., 2013).

One of the best associations of Cambrian embryos and postembryonic stages is a spherical, externally stellate organism known as *Olivooides* (Han et al., 2016b) or *Punctatus* (Yasui et al., 2013), also known from Kuanchuanpu microfossils (fig. 8.5 G–I). The arrangement of the mesenteries, tentacles, and gonad lamellae indicates pentaradial symmetry in some of the best-known cases, though tetraradial symmetry has also been documented (Han et al., 2016b). Affinities to "cycloneuralian" ecdysozoans have been proposed for *Olivooides/Punctatus*, but this can be dismissed based on the lack of a through-gut or introvert in all known ontogenetic stages (Dong et al., 2016). Although some recent studies have questioned its membership in the Cnidaria (Yasui et al., 2013), analyses of internal anatomy strengthen an identification as a medusozoan (Han et al., 2016b), and more specific affinities to Scyphozoa are recovered in other phylogenetic analyses (Dong et al., 2016). This draws especially on similarities in the ornament of the theca to coronate Scyphozoa and structures that resemble strobilating ephyrae.

Additional members of Cnidaria are first represented as compression fossils in Burgess Shale-type biotas. A polypoid animal from the Chengjiang biota, *Xianguangia sinica* (fig. 8.5 J), interpreted in earlier studies as an actinarian-like crown-group cnidarian, has been resolved in phylogenetic analyses as a stem-group cnidarian (Ou et al., 2017) or as the most deeply branching fossil in the ctenophore stem group (Zhao et al., 2019). Irrespective of the contested position of *Xianguangia*, medusae with rhopalia are present in the Chengjiang biota (Han et al., 2016a).

Summing up data from Fortunian microfossils and younger Burgess Shale-type macrofossils from Cambrian Stage 3, the major lineages of Cnidaria had diversified by the early Cambrian (Van Iten et al., 2014; Han et al., 2016b). Slightly younger medusozoan jellyfish from the middle Cambrian preserve the tentacles and muscles and provide characters shared with particular extant families of Hydrozoa and Scyphozoa that suggest they are likely crown-group members of those classes (Cartwright et al., 2007). One of these taxa is suggestive of Cubozoa (Cartwright et al., 2007), though such medusae are not identified with confidence until the Carboniferous (Young and Hagadorn, 2010). The latter work summarized anatomical and taphonomic criteria for recognizing authentic cnidarian medusa among "medusoid" fossils and reviewed the stratigraphic and systematic distribution of medusae.

Originally interpreted as a possible stem-group cnidarian, the middle Cambrian *Cambroctoconus*, which has a cuplike calcareous skeleton with an octoradial arrangement of its septa (Park et al., 2011) (fig. 8.5 C), has now been found in several parts of the world in the lower and middle Cambrian and is reinterpreted as an anthozoan, likely allied to octocorals (Peel, 2017).

A variety of either solitary or colonial coral-like fossils grouped under the concept of coralomorphs (Jell, 1984) appear from the middle part of the early Cambrian (Landing et al., 2018). They are interpreted as multiple events of calcification by anemones and have uncertain affinities to post-Cambrian corals. The earliest members of this grade are isolated tubes, some of which have tabulae reminiscent of those of tabulate corals (see below), but by the late early Cambrian modular colonial taxa have more diverse forms, and some of them show diagnostic zoantharian budding (Landing et al., 2018).

The fossil record of Cnidaria in the post-Cambrian Paleozoic is dominated by rugose and tabulate corals, both of which are key components of the Great Ordovician Biodiversification Event (GOBE). Tabulata are exclusively colonial and mainly calcitic; they have a low-diversity, patchy record in the Early Ordovician but radiated prolifically in the Middle Ordovician (Webby et al., 2004). They remained diverse, abundant and widely distributed throughout the Paleozoic. Rugosa are variably solitary or colonial, united by a distinctive radiobilateral symmetry depicted by their septa. They originated somewhat later in the Ordovician than tabulates, and the first solitary members (commonly known as "horn corals") appeared by the Late Ordovician. Like tabulates, rugose corals were eliminated in the end-Permian mass extinction.

Extant—indeed all post-Paleozoic—corals are members of Scleractinia. The first fossil scleractinians are of Middle Triassic age, and by the end of that period taxa that are known (in the Recent) to have associations with zooxanthellae ("z-like corals") are present. Species-level diversity patterns for scleractinians since the Cretaceous depict "boom and bust" cycles of diversification and extinction that correspond to the Early Cretaceous, the Late Cretaceous, the Paleogene, and the Neogene (Rosen, 2000). Molecular dating suggests that the deepest divergence between extant scleractinians is deep in the Paleozoic (Stolarski et al., 2011). If this is accurate, it bears on the question of the cryptic origins of Scleractinia, which has variably been thought to either involve an origin in rugose corals or in Paleozoic corals that are called "scleractiniamorphs." The latter range as far back as the Ordovician, and in light of molecular dates for Ordovician–Silurian origins of crown-group Scleractinia, may be members of the stem group (Stolarski et al., 2011).

9

BILATERIA

Bilateria or Triploblastica is a clade of animals whose adult tissues are derived from three embryonic germ layers—endoderm, mesoderm, and ectoderm—and whose main body axis is anteroposterior (A–P). This type of axial patterning makes individuals divisible into two symmetrical halves, the left and right sides of the animal. Some bilaterians have secondarily lost this bilateral symmetry in the adult form, as is the case for echinoderms (see chapter 17), which present pentaradial symmetry, or gastropods (see chapter 46), which undergo the developmental process of torsion. The term Bilateria was introduced by Hatschek (1888) and recently redefined, among others, by Baguñà et al. (2008), who included Cnidaria as part of Bilateria. We favor the more traditional definition of the group, including all triploblasts with primitively bilateral symmetry, but not cnidarians. Mesozoans (dicyemids and orthonectids), which lack bilateral symmetry, are also part of Bilateria from a phylogenetic perspective and are not treated separately, as is done by some authors (Cavalier-Smith, 1998). The clade Bilateria is supported in virtually every phylogenetic analysis of morphological or molecular data, whether using PCR-based target-gene approaches, transcriptomes, or genomes.

The acquisition of bilateral symmetry is traditionally considered the main characteristic of bilaterians, but it has been argued that the anemone *Nematostella vectensis* uses the same genes as bilaterians to achieve its bilateral symmetry, with multiple Hox genes being expressed in a staggered fashion along the primary body axis of the animal (Finnerty et al., 2004). This suggests that the molecular mechanism for bilateral symmetry may have been in place prior to the divergence between cnidarians and bilaterians, and hence conforms to Baguñà et al.'s (2008) interpretation.

With respect to the acquisition of specific axial properties, it has been suggested that the oral–aboral axis of cnidarians probably corresponds to the A–P axis of bilaterians, although this has not been demonstrated (Martindale, 2005). Bilateral symmetry and the axial properties of bilaterians have at least one notable consequence in their morphology, the concentration of nerves and sensory structures in the anterior body region in a process often referred to as as "cephalization." The nervous system of the earliest bilaterians, such as xenacoelomorphs (see chapter 10), is relatively simple but increases in complexity phylogenetically, especially in acoels (Perea-Atienza et al., 2015) and in parallel becomes highly complex in multiple lineages of nephrozoans. Indeed, most nonsessile nephrozoans have more conspicuous concentrations of ganglia, forming anterior brains often with fusion of multiple ganglia, such as in arthropods or in the craniates. The anterior brain is thus interpreted as a consequence of directionality in bilaterian animals, and this directionality is probably responsible for also positioning the mouth anteriorly, as is found in most nephrozoans (except in the majority of Platyhelminthes), but not in the xenacoelomorpha. Developmentally, cephalization occurs by elaboration of tissues born at the animal pole of the embryo.

While it has traditionally been hypothesized that a condensed nervous system with a medial ventral nerve cord is an ancestral character of Bilateria, others suggest that the similarities in dorsoventral patterning and trunk neuroanatomies among many bilaterian groups evolved independently (Martín-Durán et al., 2017).

The A–P axis is a conserved feature of bilateral animals and is defined in the anterior by a head and in the posterior by a trunk. The secreted family of Wnt proteins and their antagonists play a conserved role in setting up this axis during early development in many nephrozoans (Darras et al., 2018) and control A–P axis during whole-body regeneration in acoels (Srivastava et al., 2014). However, the role of these proteins and antagonists has not been studied during development in the acoels. The bilaterian central nervous system is usually asymmetrically localized along the dorsoventral axis under the control of Bmp (Bone morphogenetic protein) signaling. A high cell–type diversity and centralization of the nervous system with spatial restriction along the dorsoventral axis are characteristics of the bilaterian CNS.

Within animals 11 classes of homeodomain proteins have been defined; these include ANTP (with HOXL and NKL subclasses), PRD, LIM, POU, HNF, SINE, TALE, CUT, PROS, ZF, and CERS domain proteins. From these, all are found in bilaterians, including Xenacoelomorpha and Nephrozoa (Brauchle et al., 2018). In addition, bilaterians are characterized by having three classes of Hox genes—anterior, central, and posterior—and this set of Hox genes increases complexity in the node leading to Nephrozoa, with the addition of new classes (Hejnol and Martindale, 2009a) (fig. 9.1). Hox genes may also control the development of germ layers; it has been suggested that the different *Antennapedia* (*Antp*) Hox genes control each germ layer, the patterning and differentiation of the mesoderm being under the control of the NKL cluster. This is, however, weakly supported by empirical evidence (see a discussion in Derelle and Manuel, 2007).

Mesoderm is the main synapomorphy of Bilateria, permitting the evolution of often larger and more complex body plans than in non-bilaterian animals (Martindale et al., 2004). The embryological origin of mesoderm is well established in a diversity of bilaterians, and some genes have been studied in detail with respect to their role in mesoderm formation, especially *twist* and *snail*, members of the helix-loop-helix and zinc-finger protein families, respectively, that determine the development of the mesoderm in *Drosophila melahogaster* (Leptin, 1991). However, these (and other) "mesoderm" genes are also found in non-bilaterians, expressed in the gastrodermis (endodermal) of the anthozoan *Nematostella vectensis*, where they play a role in germ layer specification (Martindale et al., 2004).

During the process of gastrulation, the embryo has to go from being a single layer of cells to become a multilayered structure, which can be done in many ways. Gastrulation thus results in the formation of germ layers, the mesoderm originating through ingression or invagination from ectoderm (thus called "ectomesoderm" or "mesectoderm") or most commonly from endoderm ("endomesoderm" or

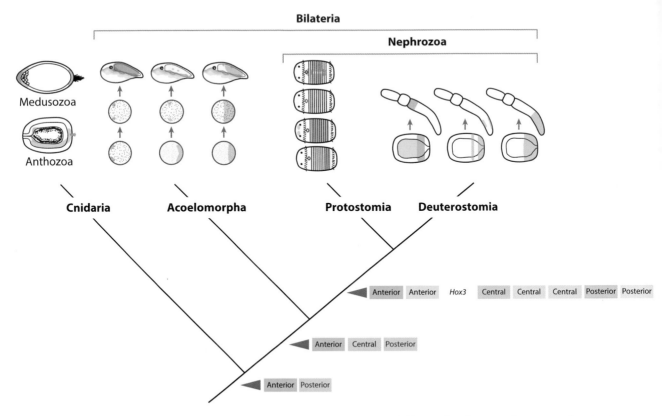

FIGURE 9.1. Evolutionary scenario of the expression of Hox genes across different developmental stages and animal groups. Two Hox genes—one anterior and one posterior—are present in the cnidarian–bilaterian ancestor. Based on Hox gene expression data of the acoelomorph *Convolutriloba longifissura*, three Hox genes (a central gene is added) are expressed in a spatial staggered pattern along the anterior–posterior axis. The spatial staggered expression however, is ancestral for Bilateria, but *Hox3* evolved in the ancestor of Nephrozoa. The protostome Hox expression is represented by the annelid *Capitella teleta*, the deuterostome expression by the hemichordate *Saccoglossus kowalevskii*. After Hejnol and Martindale (2009).

"mesendoderm"). Two main types of mesoderm formation are found among bilaterians (mesoderm forms from derivatives of a mesentoblast, or by archenteric pouching), and these are discussed in the relevant chapters.

Early mesoderm specification seems to be evolutionarily conserved among bilaterians, at least among the studied groups of nephrozoans (Brunet et al., 2013). Cell lineage studies of a diversity of bilaterians clearly show that the mesoderm gives rise to a series of derivative tissues, including muscle cells. There is, however, some level of misunderstanding about the significance of having muscle cells and having mesoderm (see chapter 5 heading "Diploblasty and triploblasty?"). In the previous chapters we have seen that cnidarians (except for the highly derived parasitic endocnidozoans) lack muscle cells and instead have myoepithelial cells, but ctenophores have isolated muscle cells, although they lack many of the genes involved in bilaterian mesodermal specification (Ryan et al., 2013). But most important, they do not have the third embryonic germ layer patterned as in bilaterians, and thus we cannot talk about mesoderm (but see the discussion of a ctenophore-specific third body layer in chapter 3). Some authors, however, consider that ctenophores have well-defined mesoderm because they have muscle cells (e.g., Derelle and Manuel, 2007; Nielsen, 2012a), but we follow a more strict definition of mesoderm.

▆▆▆ BODY CAVITIES AND COELOMS

Mesoderm also gives origin to the heart and blood, and in a functional group of nephrozoan animals called "coelomates," mesoderm is associated with the formation and definition of the main coelomic body cavity (also called "secondary body cavity," as opposed to the pseudocoelomic or "primary body cavity"; see Jenner, 2004 for a discussion on these terms), mainly with a hydrostatic function. A coelomic cavity can be recognized on the ultrastructural level because its lining is a true epithelium with polarized cells interconnected by apical adherens junctions (Bartolomaeus, 2001). All coeloms, by definition, are thus lined by mesodermally derived tissues although they can be formed in different ways, including schizocoely or enterocoely. In schizocoely the mesoderm originates as a solid mass of cells that ends up forming a coelomic cavity followed by a proliferation of new coelomic spaces, often associated with segmental units. Schizocoely is common in protostomes. In enterocoely, common in deuterostomes, the coelomic cavities form from a process called "archenteric pouching."

Typical coelomate animals are annelids, nemerteans, phoronids, brachiopods, bryozoans, chaetognaths, and all the deuterostome phyla, while the coelomate nature of molluscs is debatable. While some have considered arthropods to lack a coelom, this has been refuted in a recent study showing the formation of a coelom in chelicerates and insects (Koch et al., 2014), but this may turn into a mixocoel (i.e., a fusion of primary and secondary body cavities), as it does in onychophorans (Mayer et al., 2004). At least one species of priapulan has also been shown to be coelomate (Storch et al., 1989a).

The circulatory systems of animals (a system of blood vessels and a muscular pump or heart) normally define a coelomic cavity. In general, animals with coelomic cavities can grow larger than those with compact bodies, as the latter lack circulatory and respiratory structures that can distribute nutrients and gases through the body. All animals with a closed circulatory system are thus considered coelomates.

In other animals the body cavity is not surrounded by mesoderm and we then talk about pseudocoelomate (or blastocoelomate) animals (see Ruppert, 1991b). According to some authors, different types of these primary body cavities exist, ranging from narrow interstitial spaces in gnathostomulids (which some consider acoelomates) to a more spacious cavity lined by ECM and bordered by the basal surfaces of the epidermis and gut, as in some priapulans (Jenner, 2004). The function of these primary body cavities, especially if they are large, is similar to the coelomic cavities of coelomates. Typical pseudocoelomate animals are nematodes and priapulans, but in the case of nematodes some small species obliterate their pseudoceolomic cavities, becoming acoelomates, and, as mentioned earlier, one priapulan species is known to have a coelomic cavity.

In other animals the body is compact, with no body cavities (other than the digestive tube), and these are called "acoelomates." Many small interstitial animals are acoelomate, but some large animals, such as polyclad flatworms, are also acoelomate, thus having a very flat body in which most cells are in close proximity to the gut and the exterior, since they lack circulatory and respiratory systems.

This threefold distinction (acoelomates, pseudocoelomates, and coelomates) was based primarily on light microscopical data, but electron microscopic studies required a revision of these concepts (Ruppert, 1991b), primarily by showing that acoelomate and pseudocoelomate organizations can be arbitrary points along an ultrastructural continuum (Jenner, 2004).

Bilaterians thus include an incredible disparity of animal body plans, probably appearing in the terminal Ediacaran, which have been able to colonize an ever-increasing diversity of habitats and environments, including multiple instances of invasion of land and air.

FOSSIL RECORD

The early fossil record of bilaterians is vast, with about half of the bilaterian phyla already being represented in the Cambrian (Valentine, 2004). However, nearly all of these are recognized as Bilateria because they possess diagnostic characters of particular bilaterian subclades (such as Mollusca, Brachiopoda, Arthropoda, etc.) and to date, no clear stem-group bilaterian has been identified. It is well accepted, however, that bilaterally symmetrical animals with brains and ganglionated nerve cords had evolved by the early Cambrian (e.g., Edgecombe et al., 2015). Relevant fossils are thus discussed in subsequent chapters. Perhaps the Ediacaran fossil that has engendered the most discussion as a candidate for belonging to total-group Bilateria is *Kimberella quadrata*. This is based on its bilateral symmetry, differentiation of the body along an A–P axis, apparent differentiation of tissues, and associated feeding traces attesting to motility (Cunningham et al., 2017). As many advocates of *Kimberella* as a bilaterian have specifically allied it to Mollusca, it is discussed in more detail in that chapter.

The trace fossil record has been used to make inferences about the likely presence of Bilateria in terminal Ediacaran or earliest Cambrian (Fortunian) sediments that lack body fossils identifiable as bilaterian. As discussed in chapter 1, late Ediacaran sediments provide stronger evidence for bilaterians via trace fossils than via body fossils. An annulated vermiform organism, *Vittatusivermis annularis*, from the Fortunian of South China (Zhang et al., 2017b), provides a rare example of a macroscopic organism (body width ca. 1 cm and length to 26 cm) associated with burrows plausibly made by it, allowing similar ichnofossils to be attributed, plausibly, to Bilateria. At the opposite end of the size scale, ichnofossils have been reported from the terminal Ediacaran–Cambrian boundary interval in Brazil, attributed to meiofaunal, possibly nematoid-like worms, this based on the inferred undulating style of locomotion (Parry et al., 2017). With the caveat that relatively simple burrows are difficult to ascribe to a tracemaker systematically, this behavior provides a strong indication for meiofaunal organisms crossing the Ediacaran–Cambrian boundary.

XENACOELOMORPHA

Xenacoelomorpha is a taxon composed of three clades of simple but poorly understood predominantly marine worms, Xenoturbellida, Acoela, and Nemertodermatida (fig. 10.1). The latter two were previously considered part of Platyhelminthes (e.g., Ax, 1984; Ehlers, 1985a) until they were removed from this protostome phylum, first Acoela (Ruiz-Trillo et al., 1999) and later Nemertodermatida (Jondelius et al., 2002; Ruiz-Trillo et al., 2002).

Because in some of the earlier Sanger-based molecular analyses Acoela and Nemertodermatida did not form a clade (e.g., Jondelius et al., 2002; Ruiz-Trillo et al., 2002; Telford et al., 2003), they were subsequently treated as separate phyla (Wallberg et al., 2007; Nielsen, 2012a). More recent phylogenomic analyses, however, suggested that they do form a well-supported group (Hejnol et al., 2009; Cannon et al., 2016), and due to their many morphological affinities, we prefer to use the name Acoelomorpha for the two groups (Ruiz-Trillo and Paps, 2015) but treat them separate from Xenoturbellida, for historical reasons (the name Acoelomorpha has been used to refer to these taxa by many authors, but Nielsen, 2012 uses it to refer to our Xenacoelomorpha; see, e.g., Nielsen, 2010). Others treat the three taxa as the single phylum Xenacoelomorpha (e.g., Philippe et al., 2011; Brusca et al., 2016b), but since each of Xenacoelomorpha, Xenoturbellida, Acoelomorpha, Acoela, and Nemertodermatida is monophyletic, the question of whether they represent one, two, or three phyla may be a matter of semantics (see Giribet et al., 2016).

Xenacoelomorphs range in size from less than 1 mm to several cms. The smallest species are interstitial, while the largest xentoturbellids are suprabenthic. Acoels can often be found swarming on algae and corals, and a good number of species hosts symbiotic algae. One species of nemertodermatid is parasitic on holothuroids. They are all direct developers and live primarily in marine environments (two species of Acoela inhabit limnic environments).

Xenacoelomorphs share several characteristics that many perceive as "primitive," including a primarily simple brain and a blind gut with a single opening. However, the structure of the anterior nervous system is highly variable, as is the digestive cavity. Their epidermis is completely ciliated, the cilia forming a complex system of ciliary rootlets (Lundin, 1997, 1998), although epidermal ultrastructure may not be well studied in many phyla to compare with that of xenacoelomorphs. Xenacoelomorphs also share a mode of withdrawing and resorbing worn ciliated epidermal cells (Lundin, 2001), also known as "restitution cells" in nemertodermatids and "pulsatile bodies" in acoels (Lundin and Hendelberg, 1996). As in the case of the ciliary rootlets, data on non-xenacoelomorph species are scarce,

Figure 10.1. A, The acoel *Philactinoposthia* sp.; B, the nemertodermatid *Flagellophora apelti*; C, the xenoturbellid *Xenoturbella churro*. Photo credits: A–B, Ulf Jondelius; C, Greg Rouse.

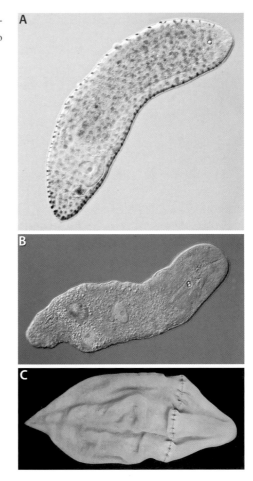

but studies in molluscs have not identified this kind of withdrawal mechanism (e.g., Lundin et al., 2009).

Acoelomorphs lack a typical basement membrane (sometimes called "basal membrane" or "basal lamina"), which is reduced to small islands of some sort of extracellular material that can be found between the epidermal cells and underlying muscle (Tyler and Hooge, 2004) or around the statocyst (Martínez et al., 2017). A thick multilayered basement membrane termed "subepidermal membrane complex" is present in xenoturbellids, separating the epidermis from the gastrodermis (Martínez et al., 2017).

Xenoturbellids and acoelomorphs possess a simple nervous system that generally is a basiepidermal or intraepithelial nerve net; however, in some cases this net is condensed into basiepidermal neurite bundles at different parts of the body or is submerged under the epidermis, where a condensed brain and submuscular cords form (Achatz and Martinez, 2012; Hejnol, 2016). Some nemertodermatids and acoels have a ring-shaped brain, which to some authors has been used to question the presence of a brain in xenacoelomorphs (the so-called skin brain of Holland, 2003). However, other species do have an elaborate brain, with paired ganglia complete with neuropil and rind (Achatz and Martinez, 2012; Achatz et al., 2013; Perea-Atienza et al., 2015), as is typically found in other bilaterians.

Acoelomorphs have a frontal gland complex that opens at the animal's anterior tip, the so-called frontal organ (Smith and Tyler, 1985b, 1986; Ehlers, 1992b), which is now also reported in xenoturbellids (Nakano et al., 2017) and may consitute a synapomorphy of Xenacoelomorpha.

The direct development of xenacoelomorphs with a hatchling being a completely ciliated worm without a digestive tract or mouth opening (although acoels hatch as juvenile worms with all major organ systems but the reproductive tract present) (Hejnol, 2015a) has revived discussion about the "urbilaterian"[5] nature and the planula-acoeloid theory (e.g., Hejnol and Martindale, 2008b; promoted by Hyman 1951a), which gains support now that aceolomorphs are not considered Platyhelminthes.

Research on xenacoelomorph anatomy, including extensive work on neuroanatomy and development (including modern evo-devo approaches) has bloomed in the past decade or so due to the key position of xenacoelomorphs as sister group to Nephrozoa (e.g., Hejnol et al., 2009; Cannon et al., 2016). Although this relationship has been challenged by a series of studies placing either Xenoturbellida or Xenacoelomorpha within or as sister group to Deuterostomia (Bourlat et al., 2006; Perseke et al., 2007; Bourlat et al., 2009; Mwinyi et al., 2010; Philippe et al., 2011; Robertson et al., 2017), a result that was mainly driven by mitochondrial data, it has been refuted by most subsequent studies using phylogenomic analyses. However, a study of the presence of the gene UDP-GlcNAc 2-epimerase/N-acetylmannosamine kinase (GNE) in metazoans has shown that GNE is encoded in the genomes of deuterostomes, acoelomorphs, and *Xenoturbella*, whereas it is absent in protostomes and non-bilaterians, suggesting this rare genome change as a putative synapomorphy for xenacoelomorphs and deuterostomes (de Mendoza and Ruiz-Trillo, 2011). A similar result, with Xenoturbellida as sister group to Ambulacraria (but resulting in paraphyletic Deuterostomia) was found in a recent phylogenomic analysis (Marlétaz et al., 2019), but this is only found when acoelomorphs were excluded from the analyses.

---

[5] Urbilateria is a theoretical organism representing the first triploblastic animal with A–P and D–V axes and predecessor of the protostome–deuterostome ancestor.

11

XENOTURBELLIDA

Xenoturbellida has remained one of the most mysterious animals, due to their simple morphology and contentious phylogenetic affinities (Reisinger, 1960; Nakano, 2015; Gee, 2016). The first known species, *Xenoturbella bocki*, was described by Westblad (1949) as a "basal platyhelminth" and subsequently considered a "basal" metazoan (Jägersten, 1959), the sister group of Bilateria (Ehlers and Sopott-Ehlers, 1997), or a deuterostome possibly related to Ambulacraria (Reisinger, 1960; Pedersen and Pedersen, 1986). Other studies have been more specific, placing them as ingroup flatworms related to Acoelomorpha (Franzén and Afzelius, 1987), as a derived triploblast probably related to bryozoans (Zrzavý et al., 1998), or even as a highly derived spiralian, with affinities to bivalve molluscs (Israelsson, 1997, 1999; Norén and Jondelius, 1997). After decades of debate and uncertainty about their phylogenetic position (see Ax, 1996), molecular data have, after initial issues with DNA contamination, universally agreed upon certain aspects. First, xenoturbellids are not Platyhelminthes or related to any other protostomes. Second, they are closely allied to Acoelomorpha, as justified by their epidermal ciliary rootlets and mode of withdrawal of worn epidermal cells. Disagreement, however, exists on whether they have a deuterostome affinity, with or without Acoelomorpha, a result based mostly on mitochondrial-based phylogenies (Bourlat et al., 2003; Bourlat et al., 2006; Perseke et al., 2007; Bourlat et al., 2009; Philippe et al., 2011; Robertson et al., 2017; but see also Marlétaz et al., 2019), or whether they constitute, again, with Acoelomorpha, the sister group to Nephrozoa (Hejnol et al., 2009; Cannon et al., 2016; Rouse et al., 2016). The latter position, supported by large phylogenomic data sets (Laumer et al., 2019), is favored here, especially as most recent phylogenomic analyses do not place Xenacoelomorpha within Deuterostomia but as its sister group with low support (Robertson et al., 2017), or result in deuterostome paraphyly (Marlétaz et al., 2019).

A phylum of a single species for decades, known only from the coasts of Scandinavia and Scotland, it now includes six distinct species from the Atlantic and the Pacific Oceans, ranging in size from a few mm to more than 20 cm and in depth from 20 to 3,700 m (Ehlers and Sopott-Ehlers, 1997; Rouse et al., 2016; Nakano et al., 2017). A species sympatric with the nominal species, *X. westbladi*, was described by Israelsson (1999), but it is considered a junior synonym of *X. bocki* (Rouse et al., 2016).

SYSTEMATICS

Xenoturbellida is composed of six species in the genus *Xenoturbella* and divides into two clades, a shallow-water clade, including the nominal species from the west coast of Sweden and two species from ca. 400–600 m in the Monterey Submarine Canyon, California and in Japan, and a deep-water clade, including three large species, from waters deeper than 1,700 m, from the East Pacific coast (Rouse et al., 2016).

Phylum Xenoturbellida

Family Xenoturbellidae

Shallow-water clade

Xenoturbella bocki

X. hollandorum

X. japonica

Deep-water clade

X. monstrosa

X. profunda

X. churro

XENOTURBELLIDA: A SYNOPSIS

- Dorsoventrally flattened acoelomate worms with an anterior ring furrow
- Body completely ciliated (multiciliated epidermal cells) with special cilia forming a unique network of interconnecting rootlets and a unique mode of withdrawing and resorbing worn ciliated epidermal cells
- Without excretory system or organized gonads
- Midventral mouth pore leading to a digestive cavity; without an anus
- Commissural brain and an intraepidermal nerve plexus
- Anterior statocyst and a frontal pore organ
- Direct development

Externally, *Xenoturbella* is a flattened pinkish worm, with a conspicuous anterior ring furrow and a flaccid consistency that is reflected in longitudinal folds, at least in the larger species (fig. 10.1 C). A side furrow and a conspicuous epidermal ventral glandular network have been described in the Pacific species. A simple diamond-shaped mouth pore is located ventrally in the anterior half of the animal. Specimens of *X. bocki* have been observed opening the simple mouth pore and protruding the aciliated gastrodermis, with the help of the body wall musculature, which also assists in the withdrawal of the gastrodermis (Ehlers and Sopott-Ehlers, 1997). The body wall musculature is well developed, with outer circular muscles, a prominent layer of longitudinal muscles, and radial muscles that extend from the outer circular myocytes to the musculature surrounding the gastrodermis (Ehlers and Sopott-Ehlers, 1997; Raikova et al., 2000b).

Xenoturbella's nervous system consists of a diffuse intraepithelial (basiepithelial) nerve net, resolving previous confusion stemming from the original description (Westblad, 1949) over whether it is subepithelial or basiepithelial (see a historical account in Stach, 2016b). The latter was subsequently corroborated (Pedersen and Pedersen, 1988; Raikova et al., 2000b; Stach et al., 2005; Stach, 2016b), as was the lack of ganglia (Raikova et al., 2000b). The metabolism of serotonin was investigated by Squires et al. (2010).

The statocyst of *Xenoturbella* is situated between the epidermis and the muscular layers, in the anterior end of the body, and contains motile flagellated cells (Ehlers, 1991). Its ultrastructure has recently been described, and although it contains a statolith, it also contains secretory vesicles, thus questioning its georeceptor nature (Israelsson, 2007). A modern detailed account of *Xenoturbella*'s nervous system is provided by Stach (2016b).

A frontal pore, continuous with a ventral glandular network, may be homologous with the frontal organ of acoelomorphs (Nakano et al., 2017) (see chapter 12).

A PCR-based Hox gene survey in *X. bocki* identified one anterior, one posterior and three central class Hox genes (Fritzsch et al., 2008), lacking *Hox3*, characteristic of the members of Nephrozoa but with more central class genes than acoelomorphs (Hejnol and Martindale, 2009a). However, no functional analysis of these putative Hox genes has been conducted to date.

Earlier results placing *Xenoturbellida* with molluscs (Norén and Jondelius, 1997) were shown to be due to food contamination (Bourlat et al., 2003 Bourlat et al., 2008), despite claims to the contrary (Israelsson, 1997, 1999; Israelsson and Budd, 2005). Likewise, three out of four recently described species also yielded sequences of co-occurring bivalves (Rouse et al., 2016), indicating a preference for this food source.

Little is known about the embryonic development of Xenoturbellida (Hejnol, 2015a), other than a few observations of eggs and that they have direct development, where the hatchling lacks a gut (Nakano et al., 2013). This nonfeeding juvenile stage is thus similar to that of nemertodermatids.

Spermiogenesis has been studied in detail for *X. bocki* (Obst et al., 2011)—and a few observations have been made in the Pacific species *X. profunda* (see Rouse et al., 2016)—and it has been seen that xenoturbellids have the so-called primitive mature sperm cells, the plesiomorphic type for metazoans, which probably fertilizes the eggs externally. Sperm cells resembling those of *Nucula* have been observed in *X. bocki* by Nielsen (2012a), thus reinforcing the hypothesis that they feed preferentially on these bivalves, which are common in the same sediments.

GENOMICS

Given its key phylogenetic position to understand the origins and subsequent diversification of Bilateria, researchers have embarked on a genome project, but this is yet unpublished. Transcriptomic resources are, however, available for multiple individuals (Bourlat et al., 2006; Dunn et al., 2008; Cannon et al., 2016).

Complete mitochondrial genomes are now available for five of the six described species of *Xenoturbella* (Bourlat et al., 2006; Perseke et al., 2007; Rouse et al., 2016), the gene order being consistent across species with only minor variation in amino acids and length of the control region (Rouse et al., 2016). It has been suggested that the mitochondrial gene order of xenoturbellids is typical of deuterostomes, but a detailed analysis of gene order data concluded that it is unlikely that gene order alone can be used to distinguish between a "basal bilaterian position" (being part of the sister group to Nephrozoa) or a deuterostome affiliation (Perseke et al., 2007).

FOSSIL RECORD

There is no fossil record attributed to Xenoturbellida, but the recent discovery of large (> 20 cm in length) deep-sea species off the coast of California and Mexico could indicate the possibility of finding relevant fossils.

ACOELOMORPHA

Acoelomorpha is composed of two singular groups of worms, Acoela and Nemertodermatida, traditionally considered as related to "turbellarians"[6] (e.g., Tyler and Rieger, 1977; Ehlers, 1985a), although Haszprunar (1996a, c) already recognized non-monophyly of the then Platyhelminthes, placing Acoelomorpha as the sister group to other Bilateria (admittedly with Rhabditophora and Catenulida forming the next branches and thus maintaining "Platyhelminthes" as a grade at the base of Bilateria). While acoels were known since the turn of the nineteenth century and were considered to be a group of turbellarians, the discovery of nemertodermatids was based on a small worm dredged by Otto Steinböck and Eric Reisinger in Disko Bay, Greenland, that led Steinböck (1930–1931) to claim having found the most primitive bilaterian. Later these two groups were recognized to share key synapomorphies and treated as the clade Acoelomorpha, within Platyhelminthes.

With the advent of molecular phylogenetics, however, a plethora of analyses suggested that Acoela and Nemertodermatida were instead earlier branching bilaterians unrelated to true flatworms (e.g., Carranza et al., 1997; Ruiz-Trillo et al., 1999; Jondelius et al., 2002; Ruiz-Trillo et al., 2002; Telford et al., 2003; Ruiz-Trillo et al., 2004). Although early phylogenetic analyses of Sanger-based data sets suggested paraphyly of Acoelomorpha (e.g., Wallberg et al., 2007; Paps et al., 2009a), subsequent analyses based on phylogenomic data clearly showed a clade composed of Acoela and Nemertodermatida, both being reciprocally monophyletic (e.g., Hejnol et al., 2009; Cannon et al., 2016). Thus, whether one wants to call Acoelomorpha a phylum, or Acoela and Nemertodermatida two phyla, is not that relevant. We follow here the decision of using Acoelomorpha as a phylum (as in Baguñà and Riutort, 2004a, b), to reflect many of the similarities in the body organization of these worms, and because this clade has been broadly recognized at that rank in the modern literature.

Acoelomorphs are acoelomate flatworm-like animals with a body completely covered by locomotory cilia. As in other acoelomates, the space between the gut and body of acoels is filled with parenchymal cells that occasionally contain the so-called chordoid vacuoles and the insunk bodies of epidermal and gland cells (Achatz et al., 2013). These chordoid vacuoles, sometimes called "chordoid cells," are absent in nemertodermatids (Rieger et al., 1991). Acoelomorphs have either a transient or a permanent blind gut cavity with a ventral mouth, and they lack excretory organs. The nervous system is ganglionated but relatively simple, with cephalization and a statocyst as its main sense organ.

---

[6] "Turbellaria" is an old term referring to a paraphyletic assemblage of Platyhelminthes, the so-called free-living platyhelminths, from which the parasitic Neodermata originate.

Acoelomorphs are mostly marine and free living. Of the approximately four hundred described species of acoelomorphs, only two are known from limnic environments: *Limonoposthia polonica* occurs in lakes in Poland, while *Oligochoerus limnophilus* has been found in a variety of such environments as lakes, irrigation ditches, and rivers in continental Europe and the United Kingdom (Vila-Farré et al., 2013). Interestingly, the six other species of the genus *Oligochoerus* occur exclusively in the Caspian Sea, suggesting that the brackish waters of the Pontic–Caspian have acted as a stepping-stone for *O. limnophilus* to colonize freshwater (Ax and Dörjes, 1966).

Recently a metabarcoding study has discovered new lineages of acoels in the deep sea as well as previously undocumented planktonic diversity (Arroyo et al., 2016). One species of nemertodermatid, *Meara stichopi*, commonly found off Bergen, is endocommensal with the holothurian *Parastichopus tremulus*, where it is generally found in the foregut, immediately behind the mouth, and only rarely in the body cavity (Sterrer, 1998b). A second yet undescribed *Meara* has been collected from the intestine of a different holothurian host in the Bahamian *Holothuria lentigenosa* (Lundin, 1999). The two *Meara* species and at least one specimen of the free-living *Nemertoderma westbladi* have been found with extraepidermal bacteria among the epidermal cilia (Lundin, 1999), having a possible symbiotic role.

Some species of acoels harbor photosymbiotic algae, important for the carbon budget of the animals, which have thus developed mechanisms to protect the symbionts as well as previously optimize algal growth. For example, *Convolutriloba longifissura* is a striking red acoel with white dots that harbors unicellular green algae within its body (Hirose and Hirose, 2007). Its red pigment is located in round cells and is soluble in ethanol or acetone, whereas the white pigment, contained in the crystalline (retractile) platelets of amoeboid-shaped cells, is soluble in 1% ammonium hydroxide (NH_4OH). These two types of pigment cells form the body coloration and are probably involved in light protection of the algal symbionts, as many algal cells are distributed beneath the body wall and some are in the highly vacuolated parenchyma (Hirose and Hirose, 2007). It has also been shown that other acoels can migrate to obtain the maximum benefit for their algal symbionts (Serodio et al., 2011). An undescribed acoel from Sardinia has been shown to sequester nematocysts and store them on the dorsal surface (C.E. Laumer, pers. comm.).

Acoelomorphs are direct developers, and acoels in particular show a special type of cleavage pattern called "duet spiral cleavage." The two main groups, acoels and nemertodermatids, differ in some fundamental characters, including the structure of the digestive cavity, the statocyst, and the sperm flagella.

▰▰▰ SYSTEMATICS

Acoelomorpha is composed of two clades, Acoela and Nemertodermatida, variably ranked as orders (e.g., Valentine, 2004) to phyla (e.g., Nielsen, 2012a). Historically, Westblad (1937) placed nemertodermatids in Acoela, at that time considered an order of Platyhelminthes. Several morphological differences led Karling (1940) to recognize Nemertodermatida as a separate clade from Acoela. Acoelomorpha was later erected as an order of Platyhelminthes by Ehlers (1985a) in recognition of special similarities in their epidermal cilia. Molecular data analyses in the late 1990s and early 2000s finally removed Acoelomorpha from Platyhelminthes (see above), a proposal that was formalized in a series of review articles (e.g., Baguñà and Riutort, 2004a, b).

The phylogenetic relationships of the twenty or so species of Nemertodermatida (Meyer-Wachsmuth et al., 2014) have been studied in detail using both morphological (Sterrer, 1998b; Lundin, 2000) and molecular (Wallberg et al., 2007; Meyer-Wachsmuth and Jondelius, 2016) characters. The most comprehensive molecular study of Meyer-Wachsmuth and Jondelius (2016) supports Sterrer (1998b) in dividing the group into Ascopariidae and Nemertodermatidae.

Acoels include ca. 400 known species, and the relationships of about a third of these have been investigated by a combination of Sanger-based nuclear and mitochondrial markers and morphological characters (Jondelius et al., 2011). This study proposed an unranked phylogenetic classification and was used to infer the anatomy of the ancestral acoel worm, which had frontal glands, mixed hermaphroditic gonads, a tubular antrum, a copulatory stylet, a posteroterminal gonopore, and a posteroterminal muscular pharynx. We represent the acoel system here mostly without assigning Linnean ranks.

Phylum Acoelomorpha
  Class Acoela
    Diopisthoporidae
    Bitesticulata
      Paratomellidae
      Bursalia
        Prosopharyngida
        Crucimusculata
          Families Dakuidae, Isodiametridae, Otocelididae, Proporidae
          Aberrantospermata

  Class Nemertodermatida
    Family Ascopariidae
    Family Nemertodermatidae

ACOELOMORPHA: A SYNOPSIS

- Dorsoventrally flattened acoelomate worms
- Body completely ciliated (multiciliated epidermal cells), with special cilia forming a unique network of interconnecting rootlets and a unique mode of withdrawing and resorbing worn ciliated epidermal cells
- Without a typical basement membrane
- Without excretory system or organized gonads
- Midventral mouth and permanent or transient digestive system; without an anus

- More or less ganglionated brain and a subepidermal or intraepidermal nerve plexus
- Anterior statocyst; some with simple ocelli
- Frontal organ opening at the anterior tip
- Posterior gonopore
- Direct development
- Uniflagellate (Nemertodermatida) or biflagellate (Acoela) sperm cells
- Set of anterior, central, and posterior Hox genes, and only anterior and posterior ParaHox genes
- Population of neoblasts able to produce all other cell types

Several characters of Acoelomorpha are shared with Xenoturbellida and are discussed in chapter 10 (Xenacoelomorpha). We focus here on a series of characters that are unique to acoelomorphs or that are important to understand the differences between Acoela and Nemertodermatida, as they differ in a number of morphological attributes, including the digestive cavity, the statocyst, the reproductive organs, the type of sperm cells and the type of early cleavage.

Acoelomorphs have a unique digestive system (fig. 12.1), with a ventral mouth, generally in the median region (but some have anterior or posterior mouths), and without an anus. While the digestive cavity of nemertodermatids has a distinct gut lumen, in acoels the mouth leads to a central syncytium surrounded by peripheral intestinal cells or has a gut lumen lined by a double layer of cells (in *Paratomella*), and lacks glandular digestive cells (Smith and Tyler, 1985a; Ax, 1996). Some acoels have a pharynx simplex.[7]

Consistent with their simple nervous system, few sense organs are found in acoelomorphs, the statocyst being the most conspicuous one. Statocyst-like organs are common in cnidarians and ctenophores, as well as in other animals that live in three-dimensional environments. The typical acoel statocyst contains a single statolith surrounded by two parietal cells, while nemertodermatids often have a statocyst with two statoliths and several parietal cells, but some have one to four statoliths. Acoelomorphs also have a frontal gland complex, with an accumulation both of different body wall glands and ciliated sensory receptors, that opens at the anterior tip of the animals, the so-called frontal organ (Smith and Tyler, 1985b, 1986; Ehlers, 1992b), possibly homologous to a similar structure found in xenoturbellids (Nakano et al., 2017). In addition, acoels, but not nemertodermatids, present photoreceptive ocelli that lack rhabdomeres or cilia (Yamasu, 1991), probably used to regulate the productivity of algal endosymbionts, but others have reported rhabdomeric eyes (Bailly et al., 2014). A recent revision of the nervous system of acoelomorphs can be found in Hejnol (2016).

Another characteristic of some acoels is a unique extrusome called a "sagittocyst," a needle-shaped secretory product that can be extruded upon physical or chemical stimulation (Gschwentner et al., 1999), like nematocysts of cnidarians and rhabdites of turbellarian flatworms. Sagittocysts form an extrusion apparatus with a muscle cell (Gschwentner et al., 2002) and seem to be restricted to a particular clade of acoels, the "convolutimorphs" (Mamkaev and Kostenko, 1991).

---

[7] This is a term applied typically to the pharynxes of Platyhelminthes.

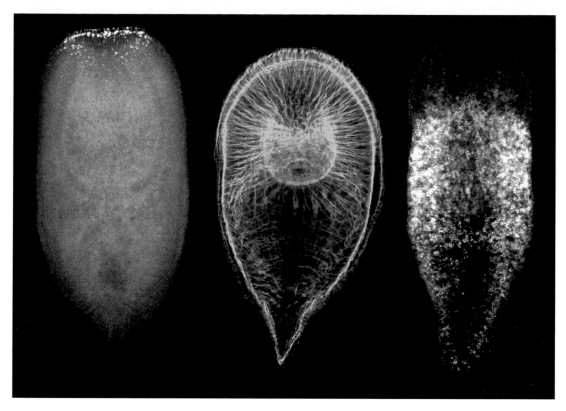

FIGURE 12.1. Inmunostaining images of *Hofstenia miamia*. Left, patterning information (expression of sfrp-1 [secreted frizzled related protein 1]); middle, phalloidin staining showing musculature; Right, expression of piwi-1 revealing distribution of stem cells (neoblasts). Photo credits: Mansi Srivastava.

Sperm ultrastructure has played an important role in understanding animal relationships. Few animals have biflagellate sperm cells, and acoels are one of them (e.g., Raikova et al., 1998a; Tekle et al., 2006). Coincidentally, Platyhelminthes, the group where acoels were thought to belong until the advent of molecular data, also have mostly biflagellate sperm cells. However, nemertodermatids show uniflagellate sperm (Tyler and Rieger, 1975), considered closer to the basic metazoan sperm cell (Lundin and Hendelberg, 1998).

Early cleavage and development in acoelomorphs are best characterized in acoels (Bresslau, 1909; Apelt, 1969; Boyer, 1971; Ladurner and Rieger, 2000), while much less is known from nemertodermatids (Jondelius et al., 2004; Børve and Hejnol, 2014). The cleavage of *Neochildia fusca* is perhaps the best known among acoels (Henry et al., 2000), and all acoels investigated show a consistent pattern in their cleavage program and gastrulation that has been termed "duet cleavage" (Hejnol, 2015a) and are able to regulate blastomere deletions (Boyer, 1971). The investigated nemertodermatid species deviate from this duet pattern by having four micromeres instead of two and being less stereotypic after the 16-cell stage (Børve and Hejnol, 2014).

Acoels and nemertodermatids have caught the attention of evolutionary developmental biologists (e.g., Hejnol and Martindale, 2008a, b, 2009a; Moreno et al., 2009;

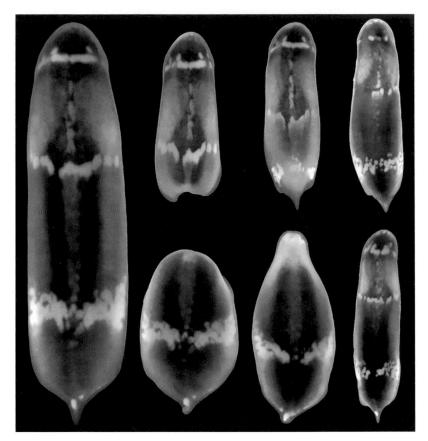

FIGURE **12.2.** The model acoel *Hofstenia miamia*. Left, adult uncut individual; top, cut individuals regenerating the tail after 1 day, 8 days and several weeks; bottom, cut individuals regenerating the head after 1 day, 8 days and several weeks. Photo credits: Mansi Srivastava.

Moreno et al., 2010; Chiodin et al., 2013; Børve and Hejnol, 2014) due to their simple anatomy and pivotal phylogenetic position (Arboleda et al., 2018). Immunohistochemical studies have explored the anatomy of the nervous system, musculature, and copulatory organs of a diverse set of acoelomorph species (Raikova et al., 1998b; Reuter et al., 1998; Hooge and Tyler, 1999; Ladurner and Rieger, 2000; Raikova et al., 2000a; Raikova et al., 2004a, b; Raikova et al., 2006; Semmler et al., 2010; Chiodin et al., 2011; Achatz and Martinez, 2012). Acoels have also recently been used to study regeneration (fig. 12.2) at the molecular level, which interestingly is not very different from that of planarians (Srivastava et al., 2014), with which they also share a similar type of stem cells or neoblasts (Gschwentner et al., 2001). As in planarians, Wnt signaling controls regeneration of the anterior–posterior body axis, and Bmp-Admp signaling controls regeneration of the dorso–ventral axis (Srivastava et al., 2014). Unlike in planarians, however, acoels have the ability to invert the A–P axis during bud formation and modify the midline of a bilateral organism to become the right or left body sides during asexual reproduction (Sikes and Bely, 2008, 2010).

Hox genes are critical for patterning the bilaterian anterior–posterior axis, where they are typically expressed in a colinear fashion. However, a study on acoels shows that expression of anterior, central, and posterior class Hox genes begins contemporaneously after gastrulation, only later resolving into staggered domains along the anterior–posterior axis (Hejnol and Martindale, 2009a). This lack of temporal colinearity in Hox gene expression in acoels has been interpreted as if the ancestral set of Hox genes were involved in the anterior–posterior patterning of the nervous system of the last common bilaterian ancestor and was later co-opted for patterning in diverse tissues in the bilaterian radiation, or, alternatively, temporal colinearity may have arisen in conjunction with the expansion of the Hox cluster in the members of Nephrozoa. While there are anterior and posterior ParaHox genes (*Xlox* and *Cdx*), the central one is missing in acoels and nemertodermatids (Jiménez-Guri et al., 2006; Moreno et al., 2009).

GENOMICS

Given their key phylogenetic position for understanding the origins and subsequent diversification of Bilateria, several genome projects of acoelomorphs have been undertaken, but just one, that of the species *Hofstenia miamia*, developed as a model to study regeneration (see Srivastava et al., 2014), has been released (Gehrke et al., 2019), containing ca. 22,600 gene predictions. Transcriptomic resources are, however, available for a relatively large number of species (Hejnol et al., 2009; Srivastava et al., 2014; Cannon et al., 2016; Laumer et al., 2019).

Complete and partial mitochondrial genomes are now available for a handful of species including one nemertodermatid and five acoels (Ruiz-Trillo et al., 2004; Mwinyi et al., 2010), differing considerably among them and with respect to those of other bilaterians; for example, there is little noncoding sequence in the compact mitochondrial genome of *Paratomella rubra*, while in *Aarchaphanostoma ylvae* and *Isodiametra pulchra* there are many long noncoding sequences between genes, likely driving mitogenome-size expansion (Robertson et al., 2017).

FOSSIL RECORD

There is no fossil record attributed to Acoelomorpha, although some have argued that acoelomorph-like animals would have been the first to leave horizontal borrows in the early trace-fossil record (Valentine, 2004), yet others think that they could be related to the animals that made the three-dimensional burrows of *Treptichnus pedum* at the base of the Cambrian. As discussed in chapter 1, it is more likely that such trace fossils could only have been produced by animals with some sort of hydrostatic skeleton, that is, a pseudocoelomate or a coelomate, and a vermiform scalidophoran conforms to details of the burrows (Kesidis et al., 2019).

13

NEPHROZOA

Excretion of ammonia and osmoregulation are fundamental cellular processes that must be done by cellular diffusion in all the animals examined in previous chapters; osmoregulatory contractile vacuoles have evolved in parallel in some protists. However, as animals became larger, more voluminous, and/or more complex, discrete excretory organs are needed to evolve to control these processes. Thus, a clade of bilaterians with discrete excretory organs was proposed by Jondelius et al. (2002) and named Nephrozoa, although that analysis did not include Xenoturbellida and found paraphyly of Acoelomorpha. The names Eubilateria and Eutriploblastica have also been applied to a similar clade but sometimes excluding particular taxa (e.g., Zrzavý et al., 1998, excluded dicyemids and orthonectids from its Eutriploblastica). Modern textbooks like Nielsen (2012a) use Eubilateria as a strict synonym of Nephrozoa. Traditionally, Nephrozoa was equated to Bilateria, but with the removal of Acoelomorpha from Platyhelminthes and the proposal of a sister-group relationship of the former to Xenoturbellida, together as sister group to the remaining Bilateria (e.g., Hejnol et al., 2009), a new clade name was necessary.

Nephrozoa is found in virtually all phylogenetic analyses of molecular data, with the exception of a series of studies placing Xenoturbellida or Xenoturbellida + Acoelomorpha within deuterosomes (Bourlat et al., 2003; Bourlat et al., 2006; Bourlat et al., 2009; Philippe et al., 2011; Marlétaz et al., 2019). This result was, however, poorly supported and has been rejected in most analyses using well-sampled phylogenomic data sets (e.g., Hejnol et al., 2009; Cannon et al., 2016; Rouse et al., 2016; Laumer et al., 2019).

The name Nephrozoa does not necessarily imply homology among all the varied excretory structures in metazoans, but rather it is intended to identify a physiological necessity due to body plan complexity. Excretory organs are unknown in sponges, ctenophores, placozoans, cnidarians, and xenacoelomorphs, but Ehlers (1992a) described a specialized type of cells in the acoel *Paratomella rubra* with a possible excretory function, although this function has not really been tested.

Among nephrozoans, many discrete excretory structures have been described, but several groups, including Chaetognatha, Bryozoa, and Tunicata, lack discrete excretory structures; excretion is often assumed to take place through the body wall or the gut. In adults of Echinodermata there are podocytes in the coelomic epithelium of the axial organ, which were postulated to have an excretory function in echinoids and asteroids (Welsch and Rehkämper, 1987), but the whole body surface and the respiratory tree have been suggested to play an excretory role in holothurians (Ruppert et al., 2004; Schmidt-Rhaesa, 2007).

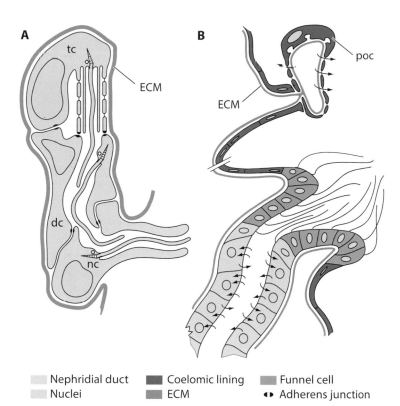

FIGURE 13.1. Schematics of filtration structures: A, protonephridium showing the terminal cell (tc), duct cell (dc) and nephropore cell (nc). The structural integrity of the protonephridium is maintained by adherens junctions connecting adjacent cells; B, metanephridium showing the direction of the ultrafiltrate and the podocytes (poc) resting on the coelomic face of a blood vessel and acting as a supporting structure. ECM = extracellular matrix. Based on Koch et al. (2014).

It is clear that no homology exists between the gut-associated Malpighian tubules (excretory organs of most terrestrial arthropods and tardigrades) and the renette cells and excretory canal system of nematodes, or between these and the two main excretory structures in nephrozoans, protonephridia, and metanephridia (fig. 13.1). All these structures filter and subsequently modify intercellular fluids and have a role in the water budget and therefore in osmoregulation. A summary of all the excretory organs in nephrozoans can be found in Schmidt-Rhaesa (2007) and a detailed anatomical discussion of some of these organs in annelids and arthropods in Koch et al. (2014).

Protonephridia (fig. 13.1 A) are simple, ectodermically derived excretory organs closed up by a terminal cell with filtering function. For example, the protonephridium of *Limnognathia maerski* (Micrognathozoa) epitomizes a simplicity of this excretory organ, being composed of four terminal cells with a flagellum, two canal cells, and a single nephridiopore cell, all monociliated. The terminal cell forms a cylindrical weir apparatus or ciliary pit, which consists of nine to ten stiff microvilli (inner rods of the weir) and the single flagellum in the middle (Kristensen and Funch, 2000). The cilia in the lumen of the cell generate an inward flow, allowing filtration across the extracellular matrix (ECM), but some authors have questioned that ciliary movement is the mechanism driving filtration (Nielsen, 2012a).

Protonephridia are independent of a circulatory system and thus are typically present in animals without such a system or in the larvae or certain structures of coelomates. They occur in some acoelomates (e.g., Platyhelminthes, Gastrotricha), pseudocoelomates (e.g., Priapulida, Kinorhyncha) and in larvae (e.g., Annelida, Mollusca) and adults (e.g., Annelida, Mollusca, Nemertea) of some coelomates. Protonephridia are thought to be homologous across taxa (Bartolomaeus and Ax, 1992). In annelids and molluscs larval protonephridia may persist in the adult or be replaced by metanephridia (Bartolomaeus and Quast, 2005; Baeumler et al., 2012). Given our current understanding of animal phylogeny, protonephridia, once thought to be part of the ground plan of bilaterians, are now restricted to protostomes.

Metanephridia (fig. 13.1 B) are mesodermally derived organs opening into an associated coelomic cavity, and thus are known only from coelomate protostomes, including annelids, phoronids, arthropods and onychophorans. Metanephridia are complex organs with special cells (podocytes, rhogocytes, etc.) and the filtration occurring through the ECM. The transition between the coelomic cavity and the canal of the metanephridium can be a more or less conspicuous ciliated funnel (e.g., in annelids and onychophorans), but in arthropods there is no ciliation in the funnel. The discharge through the nephridiopore is generally far from the site of ultrafiltration, in the case of some annelids both ends opening in different segments. The metanephridial canal is often used for releasing gametes in those animals that store them in the coelom.

The nephridial system of adult molluscs is highly modified, as the main body cavity is the pericard and filtration occurs through rhogocytes (probably a modified podocyte; Haszprunar, 1996b), the filtration product traveling through pericardioducts to the body wall. This whole structure is often referred to as a "kidney." There seems to be an ontogenetic transformation of protonephridial terminal cells into rhogocytes (Stewart et al., 2014). In craniates, a kidney composed of smaller subunits (nephrons) is found, each nephron being considered a metanephridial system (Schmidt-Rhaesa, 2007).

Two main questions about the origin and homology of nephridial structures are whether protonephridia and metanephridia are connected through a transformation series (e.g., Haszprunar, 1996b) and whether metanephridia are homologous in different phyla (e.g., Bartolomaeus and Ax, 1992; Koch et al., 2014). Hasse et al. (2010) observed three generations of nephridia in the annelid *Platynereis dumerilii*, with a transitory unciliated head kidney; a transitory larval nephridium, which undergoes a morphological transition from a protonephridium to a funneled metanephridium concomitant with the development of the coelomic cavity; and, finally, the serially arranged metanephridia. Development of these three consecutive sets of nephridia with different morphologies, including a direct link of a protonephridium with a metanephridium, reveals an interesting multistep process in the development of excretory structures in annelids. A developmental link between protonephridia and metanephridia is also the case in the chiton *Lepidochitona caprearum* (Baeumler et al., 2012; as *L. corrugata*).

Although highly convergent in some instances, several other characters evolved in Nephrozoa. We have briefly discussed the nature of coeloms, which are now considered to have had multiple origins. A through-gut has evolved twice, once in Ctenophora and again in Nephrozoa, probably independently (ctenophores are the only animals with two anal openings). The guts we have seen until now (excluding Ctenophora) consisted of a single opening that acted both as mouth and as anus, but nephrozoans, save some exceptions, evolved a through-gut with the mouth and the anus at opposite ends of the body —although this changes drastically in many sessile animals. Platyhelminthes, with a blind gut, are an exception to the rule. Some brachiopods also lack an anus in the adult, and other interesting observations have been noted in the literature, such as the transient anus of Gnathostomulida (but see a discussion in Hejnol and Martín-Durán, 2015).

The origin of a through-gut has often been explained by the process of amphistomy, in which the blastopore elongates and both openings originate by the lateral closure of a slitlike blastopore (Nielsen, 2001). Recent research suggests that the through-gut may have independent origins in different nephrozoan lineages (Hejnol and Martindale, 2008a, 2009b; Hejnol and Martín-Durán, 2015), although additional interpretations have also been provided with respect to the fate of the blastopore and its relation to body axes, as well as the multiple origins of deuterostomous development in metazoans (Nielsen et al., 2018).

The evolution of circulatory systems and haemal systems is also necessary linked to the evolution of nephrozoans, since these are the only animals with coeloms. In spite of this, no commonality is found here for all nephrozoans, and the different systems will be discussed in the relevant chapters. Furthermore, we do not find justification for a common egg-cleavage pattern in nephrozoans, and certain stereotypical patterns, such as spiral cleavage, are discussed in the relevant clades. Likewise, we do not think that a ciliated larva is characteristic of nephrozoans because ciliated larvae occur in Porifera and Cnidaria, and because the many ciliated larvae found in Nephrozoa are certainly nonhomologous. In addition, direct developers are the norm in ecdysozoans as well as in many basal spiralians (e.g., Gnathostomulida, Micrognathozoa, Gastrotricha, nonpolyclad Platyhelminthes).

Nephrozoans are also characterized by having increased complexity in the three main classes of Hox genes found in bilaterians (i.e., anterior, central and posterior), as well as the acquisition of *Hox3*, not present in xenacoelomorphs (Hejnol and Martindale, 2009a). The three main classes of Hox genes show an expression pattern of temporal colinearity, which helps pattern the different body regions.

FOSSIL RECORD

Many nephrozoan phyla have an extensive fossil record. Excretory organs including nephridia and podocytes, on which the concept of the group are founded, have low fossilization potential, but guts are commonly fossilized. The early fossil record has identified many fossils with a through-gut, though they are confined to the Phanerozoic. Although guts are labile organs and are observed to be lost relatively early in decay experiments, preservation potential of the gut is enhanced by its microbes commonly serving as templates for authigenic mineralization (Butler et al., 2015), often in the form of calcium phosphate. Guts are thus often preserved with greater three-dimensionality than more decay-resistant tissues such as cuticle.

14

DEUTEROSTOMIA

Deuterostomia constitutes a well-supported metazoan clade phylogenetically, although striking body plan differences among its constituents made it difficult to recognize until recent progress in molecular genetics and embryology identified common patterns across the deuterostome groups before their body plans begin to diverge (Lowe et al., 2015). This is why its composition has changed quite drastically through the years. For example, many textbooks have traditionally treated taxa such as Phoronida, Brachiopoda, Bryozoa or Chaetognatha as deuterostomes (but note that Grobben, 1908, placed the lophophorates with Protostomia in his Molluscoidea), although they are now universally placed within Protostomia (e.g., Nesnidal et al., 2013b; Laumer et al., 2015a). Likewise, some recent molecular phylogenetic analyses have placed Xenoturbellida and/or Acoelomorpha within deuterostomes (e.g., Telford, 2008), but this is most probably an analytical artifact (see Cannon et al., 2016). The unstable history of these phyla is discussed in detail in their relevant sections.

Nowadays our understanding of Deuterostomia includes two extant clades, Ambulacraria, with the phyla Echinodermata and Hemichordata, and Chordata, with Cephalochordata, Tunicata, and Vertebrata (e.g., Nielsen, 2012a; Stach, 2014; Brusca et al., 2016b), as supported by most phylogenetic analyses of molecular data (see a review in Schlegel et al., 2014).

The name Deuterostomia refers to the fate of the blastopore during early embryogenesis, literally meaning that the blastopore becomes the anus in the adult. The mouth is thus formed de novo from the bottom of the archenteron. However, it is now known that this pattern is not entirely conserved and there are members of Protostomia with deuterostomous development, including nematomorphs (Montgomery, 1904) and priapulans (Martín-Durán et al., 2012a), among others (see Nielsen, 2015). In addition, in some deuterostomes the anus does not originate from the blastopore (see a review in Hejnol and Martín-Durán, 2015). The presence of radial cleavage is also often used to define deuterostomes, but radial cleavage exists in non-bilaterians, as well as in some non-deuterostome bilaterians.

Choanocyte-like cells, such as those of sponges, are found in multiple deuterostome groups—including echinoderms, hemichordates, and ascidians (see the discussion in Nørrevang and Wingstrand, 1970; Crawford and Campbell, 1993)—in addition to cnidarians.

The "diplozoon" hypothesis proposes that the mouth and the anus originate from different individuals in a putative colonial ancestor of deuterostomes (Lacalli, 1997), but no evidence can be provided to support this hypothesis.

Early development with coelom formation via enterocoely—or the formation of coelomic pouches from the archenteron—giving origin to a tripartite coelom is typical of deuterostomes. The coelomic cavities can form via archenteric pouching, in which each coelom forms from the gut, pinching off and becoming a coelomic cavity, whether forming a ring surrounding the gut or a sphere. But schizocoely has also been reported in deuterostomes, including scattered mesodermal cells in the tornariae of some enteropneusts (Nielsen, 2012a), the mesodermal pericardium in *Saccoglossus kowalevskii*, which is ectodermal in origin and develops via schizocoely (Kaul-Strehlow and Stach, 2011), and probably in pterobranchs (Dilly, 2014). These three coelomic compartments are called "protocoel," "mesocoel," and "metacoel" (also "prosome," "mesosome," and "metasome"), from anterior to posterior, respectively, and are found in the adults of several deuterostome groups. It has thus often been argued that fossils with three distinct body regions show the deuterostome trimeric body plan (also called "archimery"), despite not being able to observe the coelomic cavities.

The condition of coelom formation from separate evaginations of the archenteron is found in enteropneusts, the larva of amphioxus, and in crinoids. Therefore, coelom formation from separated pouches, rather than from a single apical pouch with eventual subdivision, has been suggested as the ancestral type of coelom formation in deuterostomes (Kaul-Strehlow and Stach, 2013). Other authors have tried to homologize a single echinoderm arm with the chordate axis, based on a positional similarity between the origin of the hydrocoel in echinoderm development and the origin of the notochord in chordate development, and a positional similarity between the respective origins of the coelomic mesoderm and chordate mesoderm in echinoderm and chordate development (Morris et al., 2009; Morris, 2012).

Except for extant echinoderms—which have undergone major body plan reorganization (Smith, 2008)—deuterostomes typically have pharyngeal gill slits, and, in fact, pharyngeal gill slits are also found in some extinct stem-group echinoderms (e.g., Dominguez et al., 2002). The homology between the gill slits of hemichordates and chordates has been shown from both morphological (Schaeffer, 1987) as well as developmental points of view (see Cannon et al., 2009). A widely conserved deuterostome-specific cluster of six ordered genes, including four transcription factors expressed during the development of pharyngeal gill slits and the branchial apparatus, are thus considered the most prominent morphological synapomorphy of deuterostomes (Simakov et al., 2015).

The evolution of the complex gill slit and endostyle system has often been cited as an example of stepwise evolution of a transformational series (Nielsen, 2012a), where the ciliary filter-feeding slits of enteropneusts with iodine-binding cells in the epithelia evolve into the mucus ciliary filter-feeding systems of cephalochordates and tunicates. In these animals, the endostyle has cells that secrete the mucus and cells that secrete iodinated proteins. In vertebrates the gills become respiratory and nonfeeding, and the endostyle transforms into the endocrine thyroid gland.

Some authors have considered echinoderms to be highly apomorphic anatomically and genetically, and thus of little value to contribute to the understanding of deuterostome evolution (Holland, 2003), and instead have focused on homologies between hemichordates and chordates, as exemplified by the gill slit and endostyle system described above. However, the nervous systems of Hemichordata and

Chordata are radically different, making it difficult to infer an ancestral deutero-stome condition for neuroanatomical organization.

In adult enteropneusts, the entire nervous system is basiepidermal and divisible into localized main nerve tracts and a pervasive nerve net comprising the perikarya and neurites of interneurons and motor neurons, as well as the neurites of epidermal sensory cells (Holland, 2003; Lowe et al., 2003). This contrasts with the localized dorsal nerve cord of chordates, in which centralization was added by neural–epidermal segregation (Lowe et al., 2006), and a dorsoventral inversion of a main body axis has been postulated (e.g., Arendt and Nübler-Jung, 1994; Nübler-Jung and Arendt, 1999; Lowe et al., 2006). This has led some authors to propose a homology between the dorsal notochord of chordates and the enteropneusts' posteroventral pygochord (Nübler-Jung and Arendt, 1999). Developmental genetic data have thus provided unique insights into early deuterostome evolution, revealing a complexity of genetic regulation previously attributed only to vertebrates (Lowe, 2008). It is thus logical that evo-devo research has flourished within this clade, trying to understand better the transition to the vertebrate body plan.

A conserved neuroectodermal signaling center has been suggested for deuterostome brains (Pani et al., 2012), as it is conserved in Hemichordata (until now only tested in Enteropneusta) and in Vertebrata, but these were not conserved in Cephalochordata and Tunicata.

There is experimental evidence for a conserved role of Wnt signaling in the early specification of the A–P axis during deuterostome body-plan diversification (Darras et al., 2018).

GENOMICS

No other clade of animals has genomic data comparable to that of deuterostomes, both in terms of species numbers and phylogenetic diversity, and thus information about genomic novelties is better supported than in other animals. Well-annotated genomes are now available for all major deuterostome lineages, including Echinodermata (Sodergren et al., 2006; Baughman et al., 2014), Hemichordata (Simakov et al., 2015), Cephalochordata (Putnam et al., 2008), Tunicata (Seo et al., 2001; Dehal et al., 2002; Small et al., 2007), and Vertebrata (more than a hundred genomes already available by 2015; Dunn and Ryan, 2015). At the genomic scale there is extensive conserved macrosynteny among deuterostomes (Simakov et al., 2015). It is also known that numerous gene novelties, including large gene families, are shared among deuterostomes. Some of these gene families consist of proteins with putative functions that imply physiological, metabolic, and developmental specializations for filter feeding (Simakov et al., 2015)—the putative ancestral feeding mode in deuterostomes.

FOSSIL RECORD

Multiple deuterostome lineages were already present in the early Cambrian, including echinoderms, hemichordates, tunicates, and vertebrates. A number of extinct groups known from Cambrian Burgess Shale-type biotas—notably vetulicolians, yunnanozoans, and cambroernids—are widely identified as deuterostomes, but each of them has alternatively been excluded from the clade by some workers, and even when they are assigned to Deuterostomia, their phylogenetic positions have been debated.

Vetulicolians (fig. 14.1 B) are known from 14 species from various early and middle Cambrian Konservat-Lagerstätten, although their anatomy is best understood from species from the Chengjiang biota (Aldridge et al., 2007; Shu et al., 2010). Their bipartite body was originally interpreted in the context of them being arthropods with a bivalved carapace and a segmented trunk, but the lack of appendages and the discovery of pharyngeal gill slits in the anterior body region shifted most discussion about their affinities to Deuterostomia (Shu et al., 2001b; Ou et al., 2012a; Smith, 2012a). The bipartite body, with a barrellike anterior part and segmented posterior, has invited particular comparison with tunicates (Lacalli, 2002; García-Bellido et al., 2014), but they have alternatively been assigned to the chordate (Mallatt and Holland, 2013) or deuterostome (Ou et al., 2012a) stem groups. Some investigators have left deuterostome affinities uncertain and made comparisons with protostomes (Aldridge et al., 2007), whereas others explicitly ally them with ecdysozoan protostomes (Chen, 2008; Bergström, 2010).

Yunnanozoans are known from *Yunnanozoon lividum* (fig. 14.1 C) and the apparently synonymous (Cong et al., 2015) *Haikouella lanceolata* from the Chengjiang biota. Various soft anatomical characters have been interpreted in different ways, and even fundamental features are disputed, such as whether the trunk depicts chordate-like segmental muscle or ecdysozoan-like segmental sclerites. Interpretations favoring an identity of serially repeated branchial bars with filamentous gills have allied yunnanozoans with Chordata, where they have either been placed in the chordate stem group or variably grouped with cephalochordates, emphasizing the alternation of serialized organs interpreted as gonads (Chen et al., 1995; Shu et al., 2003a), or with vertebrates (Holland and Chen, 2001; Chen, 2008). The latter draws on the interpretation of specific sensory organs (including eyes and nostrils) and a complex brain. Some deuterostome assignments have allied these animals with ambulacrarians rather than with chordates (Shu et al., 2003a). As is the case for vetulicolians, the affinities of yunnanozoans to Deuterostomia have also met opposition, some students of the group leaving a protostome identity open (Cong et al., 2015). It has been noted that vetulicolians and yunnanozoans are likely related to each other, this argument being used to ally both to deuterostomes (Ou et al., 2012a) or to ecdyozoans, including arguments for ecdysis (Bergström, 2010).

Cambroernids are an extinct group composed of the vermiform *Herpetogaster collinsi* from the Burgess Shale (Caron et al., 2010) and a congener from the early Cambrian of Nevada (Kimmig et al., 2018), *Phlogites* from the Chengjiang biota (Hou et al., 2006; Caron et al., 2010), and a suite of discoidal fossils known from several Cambrian Konservat-Lagerstätten known as "eldoniids" (see, e.g., Zhu et al., 2002). Characters cited to unite these fossils are prominent feeding tentacles and a conspicuous gut housed in a coiled coelomic sac suspended by mesenterial elements (Caron et al., 2010). *Phlogites* and eldoniids had been compared to lophophorates in various previous studies, but the former had also been identified as a tunicate (the synonym *Cheungkongella*; Shu et al., 2001a), and eldoniids had been compared to echinoderms. Deuterostome affinities for the cambroernids are suggested by

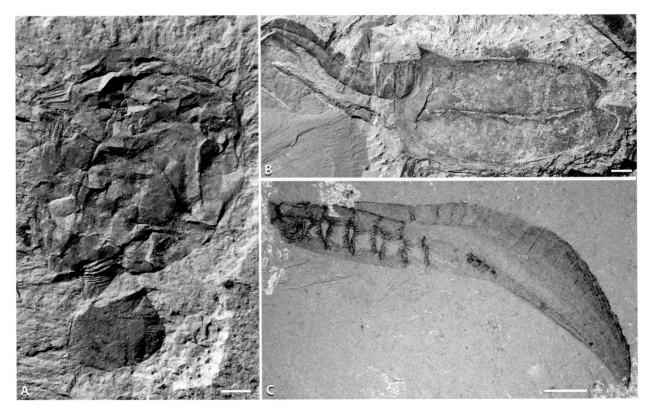

FIGURE 14.1. Early Cambrian fossils from the Chengjiang biota widely identified as deuterostomes. A, *Vetulocystis catenata*, a vetulocystid, scale 1 mm; B, *Vetulicolia cuneata*, a vetulicolian, scale 5 mm; C, *Yunnanozoon lividium*, a yunnanozoan, scale 3 mm. Photo credits: Derek Siveter.

comparisons with pterobranchs in particular, and while a position in the ambulacrarian stem group was cautiously advocated (Caron et al., 2010), the exact affinities of cambroernids are uncertain. An ambulacrarian model for interpreting *Herpetogaster* has been criticized based on inconsistencies with hemichordate morphology, such as whether the arms arise from the collar or the anterior part of the head and whether the oral tentacles are dendritic or bilaterally symmetrical (Maletz, 2014b).

Vetulocystida (Shu et al., 2004) was established for two species from the Chengjiang biota (fig. 14.1 A) originally assigned to total-group Ambulacraria but implicated in the origin of echinoderms. A representative from the middle Cambrian of Utah was documented subsequently (Conway Morris et al., 2015). Having a bipartite body described as an inflated theca and a tail that may be segmented, these animals depict similarities with vetulicolians, but, more contentiously, homologies have been proposed with asymmetrical stem-group echinoderms. The non-biomineralized theca bears two radial cones, one near the inferred anterior end and the other near the juncture with the theca, presumably housing the mouth and anus, respectively. This interpretation underpinned a claimed homology with echinoderms, restriction of the digestive tract to the anterior part of the body. The lack of definite echinoderm or ambulacrarian autapomorphies in vetulocystids

(Swalla and Smith, 2008; Clausen et al., 2010) makes it difficult to assign them more precisely than as likely relatives of vetulicolians.

Predating all of the putative deuterostomes discussed above is *Saccorhytus coronarius*, known from millimetric fossils from Fortunian sediments in South China that have been attributed to Deuterostomia based on the presence of a mouth with lateral openings, interpreted as precursors of the pharyngeal gill slits (Han et al., 2017), but *Saccorhytus* was originally described as lacking an anus.

15

AMBULACRARIA

A clade composed of Hemichordata and Echinodermata has been almost unanimously found in molecular analyses and has been recognized morphologically since the late nineteenth century (Metschnikoff, 1881; Hatschek, 1888). Sometimes referred to as Ambulacralia (Grobben, 1908) or Coelomopora (Marcus, 1958), the clade is now termed Ambulacraria (Peterson and Eernisse, 2001) and includes Hemichordata and Echinodermata, two well-corroborated sister clades from a molecular perspective, which also have a unique mitochondrial genetic code, with AUA coding for isoleucine instead of methionine. Although several morphological characters have been proposed as apomorphies of Ambulacraria (e.g., Nielsen, 2012a), the radically distinct body plans of Enteropneusta, Pterobranchia, and the five extant echinoderm classes, make it difficult to define the group anatomically. Some characters proposed include their trimeric architecture—but this is also supposedly found in some chordates, while being difficult to define in most echinoderms (see chapter 14: Deuterostomia). The axial organ and dipleurula larva have been proposed as putative apomorphies (e.g., Nielsen, 2012a), but the first is difficult to observe in adult echinoderms, and other authors have basically used the same characters as synapomorphies for Deuterostomia as a whole (Ax, 2003).

The dipleurula larva sensu Nielsen (other authors have used this term for other concepts) is an ancestral larval type with a perioral, upstream-collecting ciliary ring named "neotroch," which is formed of monociliated cells (Nielsen, 2012a). Whether the enteropneust tornaria or some of the echinoderm larvae (i.e., auricularia and bipinnaria) are variations of the hypothetical dipleurula (Nielsen, 1995; Lacalli and Gilmour, 2001; Nakano et al., 2003) remains controversial, as they have fundamental differences in the presence or absence of ganglia (Nezlin, 2000).

An incompatible clade uniting Enteropneusta and Chordata, named Cyrtotreta, based on the branchial gut with U-shaped slits, was proposed by Nielsen (1995) and followed by subsequent authors (e.g., Ax, 2003). Because hemichordates are more closely related to echinoderms than to chordates, any homologous features shared between hemichordates and chordates must have been present in the last common ancestor of deuterostomes (Cannon et al., 2009).

FOSSIL RECORD

A group of mostly Cambrian tentaculate fossils named "cambroernids" are likely deuterostomes and have been discussed in the context of being potential stem-group Ambulacraria (Caron et al., 2010). They are treated in Deuterostomia (chapter 14).

HEMICHORDATA

16

Hemichordata constitutes a relatively small clade of marine deuterostomes containing about 140 described species with two extreme morphologies: the wormlike, free-living Enteropneusta (fig. 16.1 A–B; often referred to as "acorn worms") and the sessile, colonial Pterobranchia (fig. 16.1 C), now considered members of the largely fossil Graptholithoidea. Graptolites (fig. 16.1 D) are represented by hundreds of fossil species spanning the middle Cambrian to Late Carboniferous, and because of their widespread geographic distributions (many species being planktonic) and rapid rate of evolution they provide one of the most valuable index fossil groups for biozonation of Ordovician–Lower Devonian shales.

Hemichordates have played a key role in understanding the evolution of the chordate body plan, as many authors identify enteropneusts as the closest morphology to that of the last common ancestor of chordates. It is for this reason that *Saccoglossus kowalevskii* has emerged as a prominent developmental model (e.g., Pani et al., 2012; Röttinger and Lowe, 2012; Kaul-Strehlow and Stach, 2013). All known hemichordates are restricted to the marine realm, where they occur from the in-

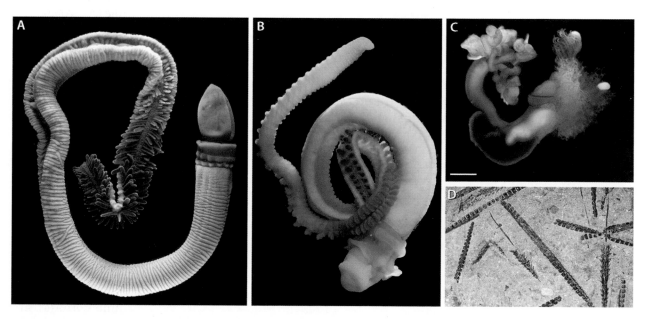

Figure 16.1. A, adult female *Schizocardium californicum* (Enteropneusta); B, adult gravid female of *Ptychodera bahamensis* (Enteropneusta); C, male zooid of *Cephalodiscus fumosus* (Pterobranchia), scale 1 mm; D, a Silurian graptoloid graptolite, *Monograptus tumescens*. Photo credits: A, Paul Gonzalez; C, Kenneth Halanych.

tertidal region to the deep-sea, with records down to 4,000 m. The knowledge of the group has exploded in the past decade with the discovery of new deep-sea taxa displaying new morphologies and life habits (Holland et al., 2009; Osborn et al., 2012), and by the description of the smallest deuterostome species, a meiofaunal enteropneust (Worsaae et al., 2012).

SYSTEMATICS

Traditionally treated as a phylum, Hemichordata was found by Nielsen et al. (1996) to be paraphyletic in a cladistic analysis of morphological data of metazoans, and thus Pterobranchia and Enteropneusta were treated as separate phyla in Nielsen (1995) and in subsequent editions of his book (Nielsen, 2001, 2012a). Early phylogenetic analyses including a diverse assemblage of hemichordates based on morphological (Cameron, 2005) and Sanger-based markers, however, found monophyly of Hemichordata (e.g., Halanych, 1996a; Osborn et al., 2012; Cannon et al., 2013; Osborn et al., 2013), but often with Pterobranchia nesting within Enteropneusta (Cannon et al., 2009; Worsaae et al., 2012), and thus the use of two phyla seemed unjustified. The most recent phylogenomic analysis supports reciprocal monophyly of Enteropneusta and Pterobranchia (Cannon et al., 2014), and thus it should not matter whether we treat Hemichordata as one (e.g., Brusca et al., 2016b) or two (e.g., Nielsen, 2012a) deuterostome phyla.

Recent taxonomic developments in hemichordates include the placement of the so-called lophenteropneusts within Enteropneusta, in the new family Torquaratoridae (Holland et al., 2005). In addition, molecular analyses have also shown that the highly modified deep-sea species *Saxipendium coronatum*, the only representative of the former Saxipendiidae, is nested within Harrimaniidae (Cannon et al., 2009; Osborn et al., 2012; Worsaae et al., 2012).

Planctosphaeroidea has been referred to as another class of Hemichordata to accommodate *Planctosphaera pelagica*, a large-size "larva" described by Spengel (1932) and not observed alive until 1992 (Hart et al., 1994). It is now interpreted as the larva of an unknown enteropneust (see Hadfield and Young, 1983), perhaps the larva of the deep-sea Torquaratoridae with large eggs (Osborn et al., 2012), although this remains to be tested.

It is now widely accepted that the fossil graptolites are related to Pterobranchia (Mitchell et al., 2013; Maletz, 2014a; Beli et al., 2018), and recently Pterobranchia has been synonymized with Graptolithoidea, composed of seven orders, including the two extant Rhabdopleuroidea (with 5 accepted species) and Cephalodiscoidea (with 19 accepted species).

> Phylum Hemichordata
>> Class Enteropneusta
>>> Family Harrimaniidae
>>> Family Spengelidae
>>> Family Ptychoderidae
>>> Family Torquaratoridae

Class Graptolithoidea

   Order Rhabdopleuroidea

      Family Rhabdopleuridae

   Order Cephalodiscoidea

      Family Cephalodiscidae

   Order Crustoidea †

   Order Dendroidea †

   Order Graptoloidea †

   Order Tuboidea †

   Order Camaroidea †

Incertae sedis: Planctosphaera

HEMICHORDATA: A SYNOPSIS

- Triploblastic, bilaterally symmetric deuterostomes with a trimeric body architecture, with proboscis (prosome), collar (mesosome) and trunk (metasome), each with its corresponding coelomic compartment (protocoel, mesocoel, and metacoel); mesocoel extending into the tentacular arms in pterobranchs
- Pharyngeal gill pores
- Well-developed open circulatory system
- Unique excretory system, the heart–glomerulus complex
- Complete through-gut, with a preoral gut diverticulum
- Short, dorsal, mesosomal nerve cord
- Gonochoristic, with external fertilization and indirect development; asexual reproduction not uncommon
- Tornaria larva
- About 140 extant species, strictly marine, adults benthic, free living (enteropneusts), or sessile colonial (pterobranchs)

Hemichordates are deuterostome coelomate animals with a tripartite body plan composed of a proboscis, collar, and trunk and displaying characteristic deuterostome features, including an anus originating from the blastopore, pharyngeal gill pores, and a dorsal nerve cord, at least in the free-living enteropneusts. This trimery corresponds to each of the three coelomic compartments that characterize deuterostomes, or at least ambulacrarians. The prosome houses an unpaired protocoel, the mesosome a pair of mesocoels, and the metasome a pair of metacoels (Stach, 2016a). The prosome of enteropneusts forms the "acorn-shaped" proboscis that gives the common name to the group, a highly muscularized structure used for burrowing in sediment. The prosome houses the unique excretory system called the "heart–glomerulus complex" (Balser and Ruppert, 1990; Benito and Pardos, 1997; Kaul-Strehlow and Stach, 2011; Merker et al., 2014). The glomerulus is composed of a vessel overlain by mesodermal podocytes, specialized cells of the protocoelic lining (Balser and Ruppert, 1990). Ultrafiltration podocytes are also found in the protocoel duct (Ruppert and Balser, 1986) and are associated with the gill structures (Pardos and Benito, 1988).

While the proboscis, collar, and trunk of enteropneusts correspond to the prosome, mesosome, and metasome, in pterobranchs the prosome constitutes the cephalic shield, which also holds a heart–glomerulus complex (Merker et al., 2014). It is from a gland in the cephalic shield that pterobranchs secrete the mucus to build their collagenous tubes, or coenecia. While the presence of tubes was thought to be a unique feature of pterobranchs, some deep-sea torquaratorids also build tubes (Halanych et al., 2013), and tubes are also known from Cambrian fossils (Caron et al., 2013). The pterobranch collar has the tentacle-bearing arms that characterize pterobranchs. The metasome then constitutes the saclike trunk that contains a pair of gill clefts, a U-shaped gut, and the gonads and tapers into a long stalk connecting individuals from the same colony and containing budding tissue (Stach, 2016a).

A recent study of the coelomogenesis of *Saccoglossus kowalevskii* showed that all main coelomic cavities (single protocoel, paired mesocoels and metacoels) derive from the endoderm via enterocoely (Kaul-Strehlow and Stach, 2013), but the pericardium develops from the ectoderm via schizocoely (Kaul-Strehlow and Stach, 2011). Schizocoely seems to be the main mode of coelom formation in pterobranchs (Dilly, 2014), but more detailed studies are required.

The nervous system of enteropneusts (Nielsen and Hay-Schmidt, 2007) and pterobranchs (Rehkämper et al., 1987; Stach et al., 2012) has received recent attention in the context of understanding the unique nervous systems of deuterostomes (Holland, 2003). The nervous system of enteropneusts has been characterized since the mid- to late nineteenth century, when it was already described to be basiepithelial and to have a subepidermal dorsal collar cord (Stach, 2016a). More recently it has been shown that juvenile and adult enteropneusts have dense agglomerations of neurons associated with a neuropil, forming two cords, ventral and dorsal, the latter being internalized in the collar region as a chordate-like neural tube (Nomaksteinsky et al., 2009), but nonetheless it is highly diffuse. In pterobranchs the brain is part of the epidermis (Stach, 2016a). The interpretation of this nervous system and the nature of the nerve center have implications for understanding chordate evolution. The branchial nerves in Pterobranchia have been suggested to be homologous to branchial nerves in craniates (Stach et al., 2012).

A study focusing on neuroectodermal signaling centers in enteropneusts has shown molecular conservation with chordates, thus suggesting deuterostome-wide brain signaling centers (Pani et al., 2012), although these were, however, not conserved in cephalochordates and tunicates.

■ REPRODUCTION AND DEVELOPMENT

Recent work has shown that the enteropneust species *Ptychodera flava* possesses the remarkable capacity to regenerate its entire nervous system, including the dorsal neural tube and an anterior headlike structure, or proboscis, this being a unique case among deuterostomes (Luttrell et al., 2016), although asexual reproduction by paratomy had already been reported in a meiofaunal species (Worsaae et al., 2012), probably involving similar regeneration mechanisms.

Enteropneust reproduction can thus be asexual by paratomy, but sexual reproduction via spawning of mucoid egg masses and shedding of sperm seems to be the norm. Brooding has also been described in an Arctic torquaratorid species (Osborn et al., 2013). Development can be direct or indirect, through a planktonic tornaria larva, which is now considered a "swimming head," which lacks the conserved, Hox-patterned posterior territory of adults (Gonzalez et al., 2017). The mode of neurogenesis is, however, highly conserved irrespective of the developmental mode (Kaul-Strehlow et al., 2015). Adult hemichordates have few sensory structures and lack eyes, but tornaria larvae have rhabdomeric eyes (Braun et al., 2015). These larvae display a negatively phototactic behavior.

Juvenile and small enteropneust species inhabit the mesopsammon, but large adult individuals inhabit U-shaped burrows with accessory openings in the anterior end, digesting organic matter that is directed via ciliary action to the mouth, located between the proboscis and the collar. The collar acts as a sorting area for the rejected material, principally inorganic particles. The animals typically accumulate fecal coils at the other extreme.

Acorn worms have become important developmental models (e.g., Röttinger and Lowe, 2012; Kaul-Strehlow and Röttinger, 2015; Gonzalez et al., 2017; Darras et al., 2018), as they are thought to resemble the chordate ancestor more closely than any other group of animals (Tassia et al., 2016), but much less is known about the development of pterobranchs, particularly due to their small size and predominance in cold or deep-sea waters. With the discovery of more accessible shallower species in temperate and tropical seas (Dilly, 1975, 1985), developmental work on pterobranchs is now catching up (Sato et al., 2008; Dilly, 2014).

GENOMICS

The nuclear genome is currently available for the enteropneust species *Saccoglossus kowalevskii* and *Ptychodera flava* (Simakov et al., 2015), but no pterobranch genome is yet available. These genomes have been shown to exhibit extensive conserved synteny with amphioxus and other bilaterians, and to possess a deuterostome-specific genomic cluster of four ordered transcription factor genes associated with the development of pharyngeal gill slits (Simakov et al., 2015). Comparative analyses have revealed numerous deuterostome-specific gene novelties whose putative functions can be linked to physiological, metabolic, and developmental specializations for filter feeding.

Four mitochondrial genomes are currently available, including three enteropneusts—*Balanoglossus carnosus*, *Balanoglossus clavigerus*, and *Saccoglossus kowalevskii*—and one pterobranch, *Rhabdopleura compacta* (Castresana et al., 1998; Perseke et al., 2010; Perseke et al., 2011). The enteropneust mitochondrial genomes are typical of deuterostomes, but the pterobranch genome is unique in showing a very GT-rich main-coding strand, representing the strongest known strand-specific mutational bias in the nucleotide composition among deuterostomes (Perseke et al., 2011). The protein-coding genes have been affected by a strong strand-specific mutational pressure, showing unusual codon frequency and amino acid composition. Perseke et al. (2011) have argued that the strong reversed asymmetrical mutational constraint in the mitochondrial genome of *Rhabdopleura compacta* may have arisen by an inversion of the replication direction and adaptation to this bias in the protein sequences. This may have given rise to arguably the most unusual deuterostome mitochondrial geonome and their unique mitochondrial code.

FOSSIL RECORD

The hemichordate fossil record is almost entirely restricted to peridermal skeletons of pterobranchs, notably graptolites (Harvey et al., 2012; Caron et al., 2013; Mitchell et al., 2013), and a few enteropneusts (Arduini et al., 1981; Frickhinger, 1999; Alessandrello et al., 2004; Caron et al., 2013; Maletz, 2014b), the latter including *Spartobranchus tenuis*, from the middle Cambrian (Series 3, Stage 5) Burgess Shale (Caron et al., 2013). When discovered, it was thought to differ from all known enteropneusts in that it is often associated with a fibrous tube, but Antarctic torquaratorids are known to form tubes (Halanych et al., 2013). Another Burgess Shale fossil, *Oesia disjuncta*, has upon study of hundreds of newly collected specimens been reidentified as an epifaunal, tube-dwelling hemichordate (Nanglu et al., 2016). Based on its vermiform morphology, *Oesia* is identified as an enteropneust, but the sessile life habits and fibrous filaments of its perforated tube are comparable to those of pterobranchs. Enteropneusts also have a potential trace fossil record (reviewed by Maletz, 2014b).

While the oldest definite pterobranchs are the fragments of rhabdosomes from the Cambrian Stage 5 Kaili Formation in China (Harvey et al., 2012), a possibly earlier Cambrian form is *Malongitubus kuangshanensis* from Chengjiang (Hu et al., 2018). This would be the earliest known pterobranch, as another putative Chengjiang hemichordate zooid, *Galaeoplumosus abilus* (see Hou et al., 2011), has been reinterpreted as part of the stem-group cnidarian *Xianguangia sinica* (see Ou et al., 2017).

Among the two extant pterobranch groups, the noncolonial cephalodiscids have sparse fossil representation in the middle Cambrian and Ordovician (reviewed by Maletz, 2014b), succeeded by a long gap to their Recent members.

Most of the diversity of graptolites has traditionally been classified as Dendroidea (containing benthic, sessile forms) and the much more speciose Graptoloidea (planktonic forms; fig. 16.1 D). Dendroidea includes nearly all Cambrian graptolites, but they are joined in the Ordovician by a few other scarce, sessile groups that are sometimes ranked as orders (Camaroidea, Crustoidea, Stolonoidea, and Tuboidea). The pelagic graptoloids emerged in the Great Ordovician Biodiversification Event, when they diversified into varied morphologies, including coiled, tuning fork, and petaloid colonies. Graptoloids went extinct in the Early Devonian, whereas Dendroidea survived until the Late Carboniferous.

The fossil record of graptolites is dominated by their organic housing structures, the rhabdosome. Zooids are very rare as fossils, though examples are known as far back as the Cambrian (see Maletz, 2014a). This is consistent with experimental decay of *Rhabdopleura* (Briggs et al., 1995) in that zooids decay in a matter of days, whereas the peridermal rhabdosome and stolon system are much more resistant to decay.

ECHINODERMATA

Echinoderms—literally "spiny-skinned animals"—include sea urchins, sea stars or starfish, sea cucumbers, and their relatives and constitute a clade of exclusively marine deuterostomes that is characterized by the presence of a dermal skeleton formed mostly of calcium carbonate, as well as by an ambulacral or water vascular system. They are among the most conspicuous marine invertebrates, often with vivid colors, achieving large sizes and huge densities in many marine environments. Their extant diversity is divided into five major groups or classes (Crinoidea, Asteroidea, Ophiuroidea, Echinoidea, and Holothuroidea), but they include numerous extinct high-ranking lineages (see below). Ecologically, echinoderms encompass an array of feeding modes, including carnivory (e.g., sea stars), herbivory (e.g., sea urchins), sediment ingestion (e.g., sea cucumbers), and particulate suspension feeding (e.g., sea lilies, feather stars, sea cucumbers). Likewise, echinoderms can be functionally sessile (although with the possibility of swimming, as in crinoids) or can move about. While most species are benthic as adults, some pelagic sea cucumbers are known to swim using swimming appendages, as can the feather stars.

Echinoderms are one of the most diverse animal phyla, with ca. 7,300 extant described species, as well as 15,000 fossil species (Appeltans et al., 2012; Mooi, 2016), and a total estimated diversity of 9,617 to 13,251 living species (Appeltans et al., 2012). Among the extant classes (fig. 17.1), Ophiuroidea is the largest group, with 2,064 described species of brittle stars and basket stars found in all oceans, from the intertidal to the greatest depths (Stöhr et al., 2012); Asteroidea includes 1,900 described species of sea stars and concentricycloids, some reaching a meter between the tips of two arms (Mah and Blake, 2012); Holothuroidea includes 1,693 species of sea cucumbers that range in length from 2 mm to more than 3 m (Miller et al., 2017); Echinoidea includes slightly more than 1,000 species of sea urchins, pencil urchins, sand dollars, and heart urchins (Kroh and Mooi, 2018); Crinoidea includes about 620 species, of which more than half occur at depths of 200 m or less (Rouse et al., 2013).

This number of living crinoids pales in comparison to the more than 6,000 known fossil species, some of which were true giants that probably attached to floating structures and hung with the oral end looking down. Sea cucumbers and sea urchins are not only important fisheries in many places around the world, especially in Asia and Europe, but are becoming a global fishery. Sea stars and urchins are also harvested in large numbers in some tropical regions to be sold as souvenirs, and sea urchins are, furthermore, model organisms for understanding fertilization, embryology, cell biology, genomic regulatory systems, and genomics (Kober and Bernardi, 2013). Sea cucumbers themselves have become a model for studying nervous system regeneration. Because of this extraordinary capacity for regeneration, some holothuroids are able to quickly undergo complete evisceration when disturbed (García-Arrarás and Greenberg, 2001).

Echinoderms are bizarre animals in many senses, as they are the result of what has been termed "maximal indirect development," a type of development in which the adult arises from set-aside cells in the larva that are held out from the early embryonic specification processes but retain extensive proliferative capacity (Peterson et al., 1997). Because of their radial (pentaradial or pentamerous) symmetry, they lack the typical body axes of other bilaterian animals, and instead the terms "oral" and "aboral" apply, as in some non-bilaterians. Their anus can open either on the oral or on the aboral side of the animal, and the oral end may face upward (as in crinoids), toward the substrate (in asteroids, echinoids and ophiuroids), or their oral–aboral axis can be parallel to the substrate, as in the secondarily, bilaterally symmetrical holothuroids. Echinoderms exhibit such surprising properties as anal suspension feeding (Jaeckle and Strathmann, 2013); or, as in some holothurian species, may have anal teeth; or their anal sphincter can be stimulated to open by pearlfish (Carapidae), which reside in the digestive or coelomic cavities of sea urchins and sea stars, apparently as a refuge (Meyer-Rochow, 1979).

Echinoderms have developed commensal associations with a suite of invertebrates, including crustaceans, annelids, platyhelminths, and nemertodermatids, and are often parasitized by annelids (e.g., myzostomes) and molluscs, especially gastropods of the family Eulimidae, some of which even lose their shells and become endoparasites, as is the case of *Entoconcha mirabilis*, a parasite in sea cucumbers. The thick spines of cidaroids are home to sessile invertebrates, including bryozoans, bivalves, sponges, and hydroids. A recent account of many of these interactions with other invertebrates can be found in Byrne and O'Hara (2017).

All the world's oceans are home to echinoderms, and a handful of species can tolerate brackish waters (Turner and Meyer, 1980), although they have no discrete excretion/osmoregulation organs and therefore have not been able to adapt to freshwater environments. They are found from the upper intertidal—sea stars and sand dollars often being exposed during low tide among algae or humid sand—to some of the deepest parts of the oceans. While all echinoderm classes have deepsea representatives, the deepest record is for the holothuroid *Prototrochus bruuni*, which has been collected to depths of 10,687 m in the Philippine Trench (C. Mah, Echinoblog). Their distribution ranges from the tropics to the poles, abounding in both cold and warm waters. They are also among the most abundant animals in many ecosystems, with ophiuroid densities of hundreds of *Ophiura sarsii* individuals registered per m^2 (Fujita and Ohta, 1990), although their distribution can be extremely patchy (Berkenbusch et al., 2011). Ophiuroids are also commonly found on sponges in tropical waters, where they capture particles driven by their host's active filtering, as well as on gorgonians, which are often exposed to steady currents, again driving food particles to these echinoderms.

Although living echinoderms are often described as pentaradial, fivefold symmetry of the adult ambulacra describes only four of the five living classes (grouped together as Eleutherozoa), as crinoids and many related extinct groups have a bi-

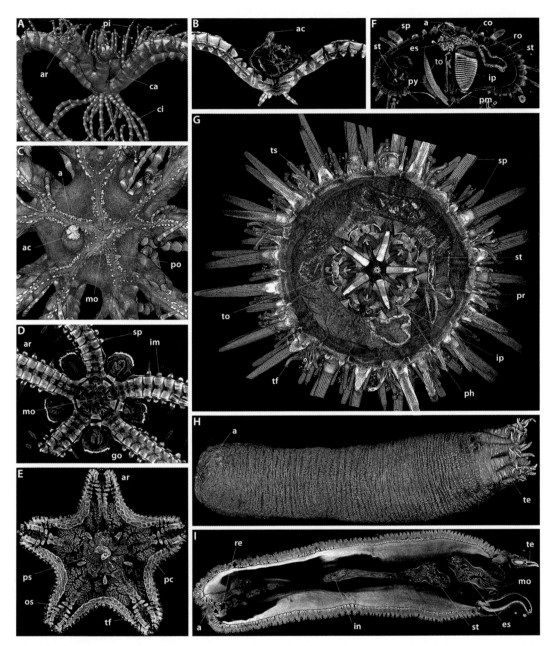

Figure 17.1. Contrast-enhanced micro-CT scanning of selected echinoderms. A–C, *Antedon mediterranea* (Crinoidea): A, volume rendering in lateral view; B, Virtual sagittal section at the level of mouth and anal cone; C, volume rendering, oral view. D, virtual transverse section of *Ophiothrix fragilis* (Ophiuroidea) at the level of the gonads. E, virtual transverse section of *Asterina gibbosa* (Asteroidea) at the level of the pyloric caeca. F–G, *Paracentrotus lividus* (Echinoidea): F, virtual sagittal section at the mid-level of the pharynx; G, virtual transverse section at the level of the stomach. H–I, *Oestergrenia digitata* (Holothuroidea): H, volume rendering, lateral view; I, virtual sagittal section at the mid-level of the pharynx. Abbreviations: a, anus; ac, anal cone; ar, arm; ca, calyx; ci, cirrus; co, compass; es, esophagus; go, gonad; im, intervertebral muscle; in, intestine; ip, interpyramidal muscle; mo, mouth; os, ossicle; pc, pyloric caecum; ph, pharynx; pi, pinnule; pm, peristomial membrane; po, podium; pr, protractor muscle; ps, pyloric stomach; py, pyramid; re, rectum; ro, rotula; sp, spine; st, stomach; te, tentacle; tf, tube foot; to, tooth; ts, test. Photo credits: Alexander Ziegler and publicly available in Morphobank, Project #3268, doi:10.7934/P3268.

laterally symmetrical 2-1-2 arrangement, with three ambulacra radiating from the mouth, the two opposed ones later bifurcating (Sumrall and Wray, 2007). Indeed, the pentaradial symmetry is achieved in the course of development from a bilaterally symmetrical larva, and different mechanisms can give rise to the adult morphology. Fossil stem-group echinoderms additionally depict triradial and bilateral symmetry as well as asymmetry (Zamora and Rahman, 2015). Due to their rich fossil record, echinoderms have featured prominently in the study of macroevolutionary patterns (e.g., Hopkins and Smith, 2015; Whittle et al., 2018).

SYSTEMATICS

Extant echinoderms are among the most familiar animals to most invertebrate biologists. They are classified in five classes: Crinoidea, Asteroidea, Ophiuroidea, Holothuroidea, and Echinoidea. A new discovery in 1986 claimed to have found (and described) the first new class of living echinoderms since 1821. The discovery of the deep-sea *Xyloplax medusiformis* from sunken wood off the coast of New Zealand resulted in the description of Concentricycloidea as the sixth class of echinoderms (Baker et al., 1986). A second species, *X. turnerae*, soon followed, also from sunken wood, this time from 2000 m depths in the Bahamas (Rowe et al., 1988). In the latter paper, the status as a new class was maintained, although recognizing that *Xyloplax*—despite its autapomorphic disc-shaped morphology with concentrically arranged skeletal ossicles and what at the time was believed to be a double-ring form of the water vascular system (Baker et al., 1986; Janies and McEdward, 1994)— is derived from an asteroid ancestor.

A third species was collected almost three decades later, this time from the North Pacific (Voight, 2005) and later described as *Xyloplax janetae* by Mah (2006). Additional *Xyloplax* species are known but remain undescribed (G.W. Rouse, pers. comm.). The status of *Xyloplax* as a distinct echinoderm class was eventually rejected on cladistic grounds (Janies and Mooi, 1999; Janies, 2001), its water vascular system reinterpreted as having the ambulacra splayed out to the periphery of the body (e.g., Janies and Mooi, 1999; Janies et al., 2011). Its modern classification in Asteroidea reflects its position as a derived asteroid (Janies et al., 2011; Linchangco et al., 2017). Phylogenetic discussions of the five extant echinoderm classes have largely agreed on the sister-group relationship of Crinoidea and Eleutherozoa, the clade that includes the other four classes of echinoderms with the oral surface directed toward the substrate (secondarily modified in holothuroids) (for a review of hypotheses and classification systems, see Smith, 1984). The relationships among the eleutherozoan classes have thus been the focus of a series of papers, including morphological (Smith, 1984), molecular (Smith et al., 1993), and a seminal study by Littlewood et al. (1997) combining adult and larval morphology and two nuclear rRNA markers.

Littlewood et al. (1997) concluded that Echinozoa (Echinoidea and Holothuroidea) was monophyletic, as in most subsequent analyses (but see Pisani et al., 2012), although they could not distinguish between the monophyly or paraphyly of Asterozoa (Asteroidea and Ophiuroidea). This therefore resolved an old discussion of whether the classes with pluteus larvae (Echinoidea and Ophiuroidea) formed a clade, leaving the door open for Ophiuroidea being the sister group of Echinozoa—the Cryptosyringida hypothesis of Smith (1984), on the basis of closed ambulacral grooves. However, these relationships were contradicted by mitogenomic data (Perseke et al., 2010), which instead supported Asteroidea as the sister group of Echinozoa. This debate seems to have settled after a series of independent phylogenomic analyses using transcriptomic data recognized monophyly of both Echinozoa and Asterozoa (Cannon et al., 2014; Telford et al., 2014; Reich et al., 2015; Simakov et al., 2015); this resolution is reflected here.

Internal relationships within each of the five echinoderm classes have received considerable attention from both morphological as well as molecular points of view, including several phylogenomic studies. Recent phylogenetic hypotheses are thus available for Crinoidea (e.g., Lanterbecq et al., 2006; Rouse et al., 2013); Ophiuroidea (e.g., Smith et al., 1995b; O'Hara et al., 2014; Thuy and Stöhr, 2016); Asteroidea (e.g., Linchangco et al., 2017); Holothuroidea (e.g., Miller et al., 2017); and Echinoidea (e.g., Littlewood and Smith, 1995; Smith et al., 1995a; Smith et al., 2004; Smith et al., 2006; Hopkins and Smith, 2015). Due to length limits we do not discuss these internal phylogenies here.

Subphylum Crinozoa
    Class Crinoidea
        Order Comatulida
        Order Cyrtocrinida
        Order Hyocrinida
        Order Isocrinida
Eleutherozoa
    Subphylum Asterozoa
        Class Asteroidea
            Order Forcipulatida
            Order Bridingida
            Order Spinulosida
            Order Notomyotida
            Order Paxillosida
            Order Valvatida
            Order Velatida
        Class Ophiuroidea
            Order Ophiurida
            Order Euryalida
    Subphylum Echinozoa
        Class Echinoidea
            Subclass Cidaroidea
            Subclass Euechinoidea

Class Holothuroidea
  Order Apodida
  Order Aspidochirotida
  Order Dendrochirotida
  Order Elasipodida
  Order Molpadida

ECHINODERMATA: A SYNOPSIS

- Adults with pentaradial symmetry, at least in some body structures
- Main body axis oral–aboral
- Mesodermally derived dermal endoskeleton of free or fused ossicles and plates as an open meshwork (stereom) with the interstices filled with living tissue (stroma)
- Water vascular system derived from the coelomic hydrocoel (if development is indirect); composed of a series of internal fluid-filled canals and external tube feet lined in ambulacra
- Body wall containing mutable collagenous tissue
- Through-gut (sometimes without an anus)
- Without discrete excretory organs
- Sometimes with a hemal system derived from coelomic cavities
- Decentralized nervous system, consisting of subepithelial nerve net, nerve ring, and radial nerves
- Mostly gonochoristic; development indirect with a diversity of larvae, or secondarily direct
- Cleavage radial, enterocoelous, with mouth not derived from blastopore; mesoderm derived from endoderm

The echinoderm body plan is difficult to generalize, as it differs enormously between at least some of the classes. This is why each of the five body plans almost requires an independent section. We refer the reader to more extensive textbooks, especially the echinoderm chapter by Mooi in Brusca et al. (2016b), on which the anatomical account below draws, and in some instances we address only some key features using examples from one or two of the classes.

Crinoids have multiple arms that radiate from a central calyx, with the oral surface directed upward. The aboral end of the calyx may be attached by a stem (in sea lilies) or by cirri (in feather stars) that can detach and swim away to a new location. The arms bear numerous lateral branches called "pinnules." Asteroids and ophiuroids, the stellate forms, normally have five arms, although more arms exist in both classes; some sea stars have up to forty, and some ophiuroid species have arms that bifurcate, in some cases indefinitely. In these the mouth is generally directed toward the substrate, with the exception of the basket stars and other ophiuroids when they are filter-feeding. The oral disc is clearly demarcated in ophiuroids, which have thin, "snakelike" arms, whereas asteroid arms touch at their bases and the disc is not clearly demarcated.

Echinoids and holothuroids are "globose" forms, lacking arms. Echinoids may be flattened and discoidal, the mouth directed toward the substrate. The stereom, sutured by connective tissue and interdigitating calcitic plates, forms a rigid test. The jaw apparatus, or "Aristotle's lantern" (fig. 17.3) and movable spines mounted on tubercles are characteristics of this group. Finally, holothuroids are fleshy, often elongated animals (along the oral–aboral axis) with the skeleton reduced to ossicles embedded in the body wall. They usually move forward along this oral–aboral axis, which is parallel to the substrate. The oral tube feet become feeding tentacles and preserve the pentaradial symmetry. Tube feet may also be present along the body.

One of the synapomorphies of echinoderms is their water vascular system, which derives from the five lobes of the hydrocoel and helps us understand the body plan of each echinoderm class. During gastrulation, echinoderms form the three pairs of coelomic compartments that characterize deuterostomes, the protocoels, mesocoels, and metacoels (or somatocoels). The mesocoel is dominated by the left compartment, the hydrocoel. During larval development, the hydrocoel and left somatocoel become a unit called the "rudiment." This rudiment becomes what is known as the "axial region," which includes the adult mouth, water vascular system, tube feet, and ambulacral skeleton. The non-rudiment part of the larva is called the "extraxial region" and makes up most of the central disc, although it is almost entirely lost in echinoids. The anus, genital pores, and openings of the water vascular system are part of the perforate extraxial region.

In asteroids, the water vascular system is designed around a ring canal that surrounds the esophagus and five lateral canals (fig. 17.2). It opens to the exterior through the madreporite, a skeletal sieve plate on an interambulacrum on the aboral surface. The madreporite is perforated by hydropores, covered by epithelium, and allows water into the water vascular system, as demonstrated in asteroids and echinoids (e.g., Ferguson, 1990, 1996), as well as in the perivisceral coelom of asteroids, but not echinoids. The madreporite is connected to the ring canal through the stone canal but also to the perivisceral coelom (hemal system). Interradial blind pouches known as "Tiedemann's bodies" and "Polian vesicles" attach to the ring canal. Tiedemann's bodies receive high-pressure water from the stone canal to generate perivisceral coelomic fluid (Baccetti and Rosati, 1968; Ferguson, 1990; Mashanov et al., 2010) and are thought to regulate internal pressure of the water vascular system.

Each arm has lateral canals that originate at each side of the radial canal, and each lateral canal terminates in a tube foot–ampulla system. The asteroid tube feet are contractile, muscular tubes that protrude from the oral side, usually ending in a sucker that may bear ossicles and mucus glands. At the other end, internally, the tube feet end in a bulb-like ampulla. Tube feet are also well innervated, and they serve as the primary tactile, locomotory, attachment, and manipulating organs of echinoderms. In many sea stars, the terminal unpaired tube foot of each arm is probably a photoreceptor organ. Photosensitivity is also localized in the tube feet of sea urchins (see below).

The tube feet and ampullae (ambulacral system) with the two sets of muscles and the fluid inside the water vascular system constitute a hydrostatic system. Each

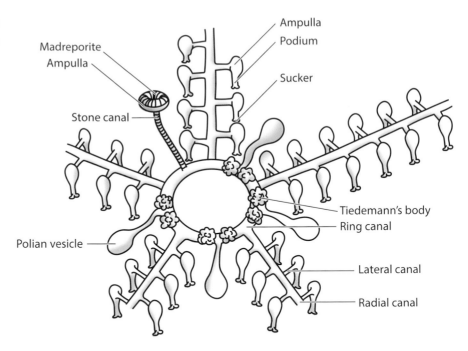

Figure 17.2. Water vascular system of an asteroidean. Based on Brusca and Brusca (1990).

unit composed of a tube foot and ampulla can be isolated from the rest of the system by a valve in the lateral canal. When the valve closes, the ampullar muscles contract so that fluid enters the lumen of the tube foot, which consequently extends and relaxes the longitudinal musculature. The sucker then comes into action, using a mix of physical muscular suction and adhesive secretions to adhere to the substrate or prey. When the longitudinal muscles of the tube foot contract, it shortens, forcing fluid back into the relaxed ampulla.

Variations of this water vascular system are found across the echinoderm classes. In ophiuroids, the madreporite is on the oral side of the central disc, rather than the aboral, and in some cases has only two hydropores. The tube feet lack suckers but instead use sticky mucus. In crinoids the water vascular system operates entirely on coelomic fluid, without a madreporite, and, instead, up to hundreds of stone canals emerging from the ring canal open to the coelomic channels, with water entering the perivisceral coeloms. The radial canals bifurcate into all branches of each crinoid arm, and extensions of the radial canals go into the pinnules or side branches of the crinoid arms. Tube feet are often in groups of three and, as in ophiuroids, lack suckers.

In echinoids the madreporite is a special periproctal plate (one of the skeletal plates around the anus) from the so-called apical system. The madreporite leads to an ampulla–axial gland complex. The stone canal extends to the ring canal atop the jaw apparatus, which in turn gives rise to the five radial canals in the ambulacra. The tube feet protrude through pores in the skeletal plates; they may or may not bear suckers and can be extended beyond the length of most spines, at least around the oral side. In addition to attachment, feeding, and locomotion, tube feet function in gas exchange.

In holothuroids the stone canal usually extends from beneath the pharynx in an interradial position, and the madreporite opens into the coelom. The esophagus is encircled by the ring canal, which bears between one and fifty Polian vesicles. From the ring canal, five radial canals project into the oral tentacles. These are modified terminal tentacles and tube feet, homologous with the radial canal of other echinoderms. They are the only part of the water vascular system in Apodida—the sister group to all other holothurians. In the other groups, extensions from the ring canal develop, during postlarval growth, to form interradial longitudinal canals that lie along the inside of the larval body wall and form the tube feet (called "tube footlike structures" by Mooi, 2016).

This does not seem to follow the Ocular Plate Rule of Mooi et al. (1994), which refers to the ontogenetic pattern by which new tube feet are laid down in the axial region of a sea star (Mooi et al., 1994; Mooi et al., 2005; Mooi, 2016). As the radial canals of the developing rudiment expand outward from the ring canal along the rays, new tube feet appear just proximal to the ocular tentacle (the terminal tentacle). A new tube foot appears first on one side of the radial canal, and then on the other, in a zigzag pattern, new ambulacral plates forming as the new tube feet appear. Therefore, the oldest tube feet are toward the mouth and the youngest at the tip. Mooi et al. (1994) therefore distinguish between tube feet around the mouth and tube footlike appendages in the body of holothurians. The latter can be laid down longitudinally in five ambulacral lines (or ambulacral-like *fide* Mooi), can be missing, or can be condensed toward the part of the animal closest to the substrate or lower part of the animal (holothuroids have the mouth facing forward, although some species that bury themselves inside the sediment have the mouth directed upward while collecting food particles from the water column).

The body wall of echinoderms is also quite special, as it consists of an epidermis that overlies a mesodermally derived dermis containing the skeletal elements of the stereom (ossicles and plates). These can be fused, as in echinoids, or scattered within the dermis, as in holothuroids. The stereom is mostly calcitic or may also contain small amounts of magnesite. Each ossicle or plate begins to form as a separate spicule-like element to which additional material is added in a specific pattern, the final ossicles being porous, the pore spaces containing dermal cells and fibers, the so-called stroma.

The genes responsible for the biomineralization process are highly conserved among the echinoderm classes (Dylus et al., 2018), but class-specific characteristics are also found, including an independent duplication of the *msp130* class of genes in different classes and a unique occurrence of spicule matrix genes in echinoids. External plates may bear tubercles of different sizes, some supporting spines and other structures, such as the pedicellariae of asteroids and echinoids, which have their own neuromuscular responses independent of the central nervous system. Each pedicellaria is basically an effector organ with its own set of muscles, neuropils, and sensory receptors. Curiously, pedicellariae were first described as a new genus of parasitic polyps, *Pedicellaria*, with three species, *P. globifera*, *P. triphylla*, and *P. tridens*, reflecting the different types of pedicellariae. There are different morphologies, which have defensive and protective functions; some are like pincers that hold on to debris for camouflage, while others have evolved into venomous structures, as in the globigerous pedicellariae of the sea urchin family Toxopneustidae. Some species of this group, such as the Pacific species *Toxopneustes pileolus*, are extremely venomous.

Because much of the external tissue of echinoderms is isolated from the coelom and digestive system (i.e., epidermis, spines, and pedicellariae), it is thought that nutrient uptake directly from the seawater occurs at this level, although probably not in a significant manner (Ferguson, 1980). Nutrient uptake by the epidermis has also been reported in many other invertebrates (Gomme, 1982).

Below the dermis lie the muscle fibers and the peritoneum of the coelom. The muscles are well developed in active species, but are weakly developed in echinoids, since they have a rigid skeleton, or test. In the arms of the stellate forms, adjacent skeletal plates articulate through bands of muscles connecting such plates.

Another bizarre characteristic of echinoderms is their unique so-called mutable collagenous (or connective) tissue, which has the ability to undergo rapid and reversible changes in passive mechanical properties that are initiated and modulated by the nervous system (Ribeiro et al., 2011). The normal state of this tissue is rigid, but it can be temporarily softened to the familiar flaccid state of sea cucumbers, which can change its tensile properties in a matter of seconds. Likewise, the bases of the spines of sea urchins have a muscular ring and a similar ring of mutable collagenous tissue that "freezes" the spine into place. While it has been argued that muscle may have a function in maintaining turgidity, this has been refuted (see a review in Wilkie, 2002). The mutable collagenous tissue can disintegrate, resulting in autotomy (Byrne, 1985, 2001).

As in most other deuterostomes, the coelom develops as a tripartite series of paired pouches, the protocoels, mesocoels and metacoels (or somatocoels). The main adult body coeloms derive from the left and right somatocoels, and the water vascular system, from the hydrocoel. The main body cavity or perivisceral coeloms are lined by ciliated peritoneum and their coelomic fluid, containing coelomocytes, plays a major circulatory role. In some holothuroids and ophiuroids, the coelomocytes contain hemoglobins. The coelomic fluid is isosmotic with seawater. This and the lack of active osmoregulatory organs seem to be the reason for the lack of echinoderms in freshwater, as discussed earlier, although some osmoregulation can obviously occur at the cellular level (Diehl, 1986). The buffer capacity of the coelomic fluid seems to be inversely related to the gas exchange capacity of the different groups (Collard et al., 2013).

NERVOUS SYSTEM

As in every other structure in echinoderms, the bilateral larval nervous system has to transform into a radically different pentaradial system that differs from that of the larva, in both its complexity and function. The changes undergone by the nervous system as the larva metamorphoses remain largely understudied, yet it has been shown that most larval structures are eliminated during metamorphosis (Chia and Burke, 1978). The adult nervous system of echinoderms was thoroughly reviewed by Cobb (1987), and more recently by Mashanov et al. (2016) and Díaz-Balzac and García-Arrarás (2018). We mostly follow the latter sources for the description provided here. The neuromuscular system in echinoderms described by Cobb and Laverack (1967) reported the presence of "muscle tails" approaching axons, similar to the muscle arms of Nematoda and Cephalochordata.

The adult nervous system follows the typical pentaradial symmetry of the other body parts and is fundamentally decentralized, lacking a brain. The best studied representatives to date are echinoid embryos and larvae (see below) and adult holothurians, which have become a model for studying nervous system regeneration (e.g., Mashanov et al., 2008). Although often referred to as a neural net, the echinoderm nervous system consists of well-defined neural structures. The adult nervous system is composed of a central nervous system made up of a nerve ring connected to a series of radial nerve cords. Peripheral nerves extending from the radial nerve cords or nerve ring connect with the peripheral nervous system, located in other organs or effectors, including the viscera, podia, body wall muscles, and mutable collagenous tissue (Díaz-Balzac and García-Arrarás, 2018).

In general, each radius of the nervous system is supplied with its own radial nerve cord, which runs along the entire length of the oral–aboral axis and terminates either at the tip of the arm (in the stellate classes), or near the aboral end (in the globose forms). At the oral end, all five individual radial nerve cords coalesce into a circumoral nerve ring. This combination of the radial nerve cords with the nerve ring constitutes the central nervous system of echinoderms, an anatomically and histologically distinct agglomeration of neurons and glial cells associated with an extensive neuropil, which occurs nowhere else in the body and is responsible for initiation and coordination of various body-wide responses (Mashanov et al., 2016).

Adult echinoderm neuroanatomy is unique in being composed of distinct superimposed domains or layers of nervous tissue at different levels relative to the oral–aboral axis (Mashanov et al., 2016). These domains are referred to as the "ectoneural," "hyponeural," and "entoneural" (or "aboral") systems and are unrelated to the embryonic origin of these tissues. The ectoneural system is predominantly sensory in function and occupies the oral-most position, as it lies either within or directly beneath the oral epidermis. It is always present both in the nerve ring and radial nerve cords of all echinoderms, showing the most consistent organization across the phylum, and constitutes the predominant part of the nervous system in all classes except crinoids. The hyponeural system may or may not be present as a component of the nerve ring or radial nerve cords. Its organization varies in different classes and generally shows a correlation with the degree of development of large muscles. When the hyponeural system is present, it directly overlays the aboral surface of the respective ectoneural cords as a second (usually thinner) layer of nervous tissue, largely motor in function. Crinoids are different from the other four classes in that they possess a well-developed third component, the entoneural system, which is the main part of their central nervous system.

There are two different designs of the ectoneural system in echinoderms. In crinoids and asteroids, it is integrated into the oral epidermis in a superficial position, separated from the external environment only by an apical cuticle. In ophiuroids, echinoids, and holothuroids, the ectoneural cords are located inside the body, being completely immersed within the connective tissue of the body wall and overlain by a narrow cavity called the "epineural canal" (Mashanov et al., 2016). The hypo-neural cords are located deeper in the body and always have a tubular organization with an inner cavity called the "hyponeural sinus." The ectoneural and hyponeural cords parallel each other in such a way that the basal surfaces of the two neuroepithelia are closely opposed to one another, extensively connected by numerous short neural bridges (Mashanov et al., 2006; Mashanov et al., 2007; Hoekstra et al., 2012; Mashanov et al., 2013). The entire complex of the ectoneural and hyponeural cords is surrounded by a continuous basal lamina, unifying them while being separate from the surrounding tissues. The two subsystems, therefore, are anatomically and functionally interconnected components of a single continuous anatomical entity, rather than two distinct and independent nervous systems, as previously thought (Mashanov et al., 2016).

The central nervous system is extensively connected to various parts of the body by peripheral motor nerves and receives a variety of sensory inputs as well. The nerve ring, besides connecting the radial nerve cords together, sends nerves to peri-stomial appendages (e.g., tentacles in holothuroids, buccal podia in ophiuroids, spines in asteroids) and extensively innervates the anterior regions of the digestive tube. The radial nerve cords give off a series of nerves that innervate various types of effectors, such as the podia, muscles, body wall peritoneum, epidermis, and elements of the mutable collagenous tissue. These peripheral nerves can be purely ectoneural, hyponeural, or mixed (Mashanov et al., 2016). For example, the podia of holothuroids are innervated ectoneurally (Díaz-Balzac et al., 2010).

Both the central and peripheral nervous systems are composed of complex and diverse subdivisions mainly characterized by the expression of neurotransmitters and neuropeptides (Díaz-Balzac and García-Arrarás, 2018).

Echinoderms are highly responsive animals with behaviors that include being able to track prey, quickly escape by the action of tube feet or swimming (in feather stars), direct spines toward possible attackers (in sea urchins), or quickly folding arms to drop more rapidly (in ophiuroids). However, discrete sense organs are rare among echinoderms. Therefore, the sensory perception in the phylum is thought to be due mostly to receptors for various sensory modalities scattered throughout the body and either initiate local reflexes or eventually communicate to the central nervous system (Cobb, 1987; Heinzeller and Welsch, 2001); the sensory biology of asteroids has recently been reviewed by Garm (2017).

Echinoids, asteroids, and ophiuroids, however, have photosensory organs. The iridophores of diadematid sea urchins—the typical bluish lines found on the test of these species—have been proposed to be eyes, but they are not photosensitive (Ullrich-Lüter et al., 2011). Instead, light-sensitive cells of echinoids localized at the base and distal disc of each tube foot form what might be a compound eye, capable of directional vision (Ullrich-Lüter et al., 2011). In asteroids the distal tip of the arm shows a photoreceptor together with surrounding epithelial cells, forming what is known as an optic cushion. These photoreceptors contain the visual pigment r-opsin (Johnsen, 1997; Ullrich-Lüter et al., 2011), and in echinoids originate from regions of the embryo, which express a homologue of *pax6*, a transcription factor with a conserved role in eye formation in other animals. Asteroid compound eyes consist of up to two hundred ommatidia, are capable of low-resolution image formation (Garm and Nilsson, 2014), and have a role in photonavigation and orientation (Garm and Nilsson, 2014; Sigl et al., 2016).

In light-sensitive ophiuroids, presumed photoreceptors are buried in the dermis at the aboral surface of the body and are positioned at the focal points of calcitic mounds that have been called "microlenses," which form the outer surface of the plates of the dorsal arm endoskeleton. The ability of these calcareous ossicles to polarize and focus light has been proposed experimentally, it having been suggested that the entire light-sensitive surface of the body functions as a compound eye (Hendler and Byrne, 1987; Byrne, 1994; Aizenberg et al., 2001). However, more recent investigation of species of the ophiuroid *Ophiocoma* suggest that photoreception is extraocular, effected by opsin-reactive cells scattered all over the body surface, whereas the supposed "microlenses" on the arm plates that were interpreted as the basis for a compound eyelike visual system are not associated with the photoreceptive cells (Sumner-Rooney et al., 2018).

Statocysts are found in some holothuroids (Ehlers, 1997). In apodids, five pairs of statocysts are found, each pair associated with the junction between each of the five radial nerve cords and the circumoral nerve ring. The statocysts are hollow, with the central cavity lined with monociliated sensory cells that project their axons into the ectoneural part of the radial nerve cord and provide information about the position of the animal.

Minute skeletal appendages of bright appearance called "sphaeridia" are found on tubercles of the test in all Recent echinoids except cidaroids. Myocytes stretch from the tubercle to the sphaeridium. The tubercles are encircled by a basiepithelial nerve ring containing neurons that are provided with a cilium that lies close to the contact region with the myocytes and thus probably have a proprioceptor function (Märkel et al., 1992).

FEEDING AND DIGESTION

Feeding and digestion are highly variable in echinoderms. Crinoids have the oral surface facing upward and extend the arms and its pinnules to trap particles in suspension in areas of strong current. The food particles are trapped by the tube feet, which draw them into a ciliated groove; the food is then driven toward the mouth, which opens into a short esophagus. A long intestine loops into the calyx and opens on the oral surface, on top of an anal cone. Suspension feeding is also common in ophiuroids (including basket stars), some holothuroids, and even in some bathyal asteroids, which use their arms and pedicellaria to trap small crustaceans.

Suspension feeding in brittle stars is achieved in many ways, a common one involving clinging on to sponges, which draw particles through the action of their choanocytes. Other brittle stars and basket stars often cling to other invertebrates or a hard substrate in areas of high current. Basket stars display their continuously bifurcating, fractal-like, branching arms to trap small animals, which are then drawn to the mouth by flexing the arms. Many brittle stars use the tube feet, spines, and mucus secretion to trap small animals and particles and form a bolus that is directed to the mouth by the tube feet. Some species are also predators and even

deposit feeders. The mouth of ophiuroids is surrounded by modified jawlike ambulacral ossicles. From there food passes through a short esophagous to a large
blind stomach, the site of digestion.

Sea stars include a great variety of feeding modes, many being predators, scavengers, or even deposit feeders, whereas *Xyloplax* depends on symbiotic bacteria
to derive nutrients from wood. Asteroids can play an important role as predators
in many environments. While most species are generalists, a few are well-known
specialists, as for example the crown-of-thorns sea stars of the *Acanthaster planci*
species complex, which feed on coral polyps (although not of a single species) and
is a known cause of coral mass mortality in some Indo-Pacific reefs due to both
population booms and removal of natural predators. In this case, vision and chemoreception play a role in prey detection, but vision seems to be more important
at long distances (Sigl et al., 2016).

Feeding in sea stars is quite bizarre, everting part of the stomach (called the "cardiac" stomach) to obtain food, digesting it outside, as for coral tissue, or prying
bivalve shells open and deploying the stomach into the shell to secrete the digestive enzymes and digest the soft parts of the bivalve in situ. Smaller prey items are
often swallowed and slowly digested inside the stomach. As discussed above, suspension feeding is also used, sometimes as a supplementary form of feeding. In
some species, pedicellaria help trap small prey, even fishes. The cardiac stomach
leads to the pyloric stomach, which extends into the arms to form pyloric caeca;
the pyloric stomach, through a short intestine, opens to the anus on the aboral side.
External digestion is facilitated by the proteases and other digestive enzymes secreted by the pyloric caeca and the cardiac stomach.

Many holothurians live buried in sediment, extending their oral tentacles to trap
food particles and small animals and constantly drawing the tentacles inside the
mouth one at a time. Deposit feeding is also common in sea cucumbers, which ingest sediment as they move, thus acting as an important bioturbator (e.g., Mactavish
et al., 2012). The anterior mouth of holothurians is surrounded by buccal tentacles
and leads into the esophagus, which passes through a ring of calcareous plates
supporting the foregut and the ring canal of the water vascular system. The intestine
has an enlarged anterior area and a posterior area that loops forward and backward
to extend its length, continuing into an expanded rectum, and ending in a posterior
anus (the elongate body along the oral–aboral axis can be referred to as anterior and
posterior, unlike in other echinoderms). The respiratory trees (if present) attach in
the rectal area.

The Cuvierian tubules of some holothuroids are clusters of sticky, blind tubules
that arise from the respiratory trees, discharged by a contraction of the body wall
as an effective defense mechanism. This discharge prompts rupture of the hindgut, everting the tubules and entangling a predator (or an annoying human hand)
in a mass of sticky secretion. As for mostly everything else in the sea cucumbers,
the Cuvierian tubules are easily regenerated. In addition, holothurians can evis-
cerate part or even the entire digestive tract, respiratory trees, muscles, and gonads,
sometimes as an almost immediate response to human contact. Whether this is seasonal (Swan, 1961; Byrne, 1985) or just a defense mechanism is not well understood, but interest in gut regeneration has led to recent research on this topic, which

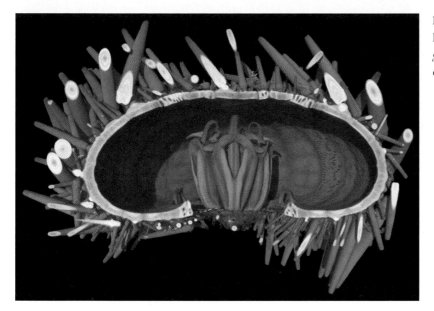

Figure 17.3. MicroCT detail of Aristotle's lantern in situ of the purple sea urchin *Strongylocentrotus purpuratus*. Photo credit: Alexander Ziegler.

is now well understood from developmental and molecular points of view (Mashanov and García-Arrarás, 2011).

Echinoids also fill a diversity of feeding niches, from deposit feeding or detritivory to being a main group of algal consumers, and a few are active predators. Their voracious appetite is facilitated by one of the most unique organs in the animal world, their jaw apparatus, or Aristotle's lantern of the globose species (fig. 17.3). The lantern has been studied in detail by many authors (e.g., Candia Carnevali et al., 1993; Ziegler et al., 2012). It consists of 40 skeletal elements—5 teeth, 5 rotulae, 10 hemipyramids, 10 epiphyses, and 10 compass elements—as well as numerous soft tissue structures, among them a large number of unpaired and paired muscle groups, including protractor and retractor muscles (Ziegler et al., 2012). The lantern is oriented along the oral–aboral axis and "suspended" on a soft peristomial membrane, which allows the teeth to be directed in different directions.

The five pairs of calcareous, trapezoidal elements (the hemipyramids) connect to form five triangular pyramids, each of which supports a single tooth in a groove called the "tooth slide." The pyramids are positioned in the interambulacral region and attach to each other by muscles that pull the pyramids and the teeth together to close the jaws. Sharpening of the teeth is accomplished by the constant addition of platelets where the tooth emerges from the top of the pyramid. The lantern is used for scraping algae from the substrate, for biting chunks of macroalgae, or for excavating burrows. While the lantern is absent in some irregular urchins (at least in the adult), in sand dollars (Euechinoidea: Clypeasteroida) the lantern serves as a crushing device that grinds ingested sediment into finer material.

In addition, echinoids can also use the spines and tube feet to trap organic matter. The irregular spatangoids burrow and ingest sediment so as to digest its organic matter, and many other feeding modes exist in the secondarily bilateral irregular urchins (see Mooi, 2016). Cidaroid sea urchins have been shown to feed on live stalked crinoids (Baumiller et al., 2010). Perhaps more surprising is that the stalked crinoids were able to detach and crawl away with their arms, dragging along the long stem.

The digestive system of echinoids is formed by a simple tube uniting the mouth, on the oral side, and anus, at the aboral end. In the irregular forms, which adopt a secondarily bilateral symmetry, the mouth is shifted "anteriorly" and the anus has migrated toward the "posterior" side of the animal. The pharynx is located inside the lantern, and the intestine loops around until it narrows down and opens at the anus.

RESPIRATION, EXCRETION, CIRCULATION

Gas exchange in echinoderms relies on thin surfaces, especially in crinoids, in which respiration occurs through all exposed thin parts of the body wall, especially the tube feet. In asteroids, gas exchange occurs via the tube feet and in special outgrowths of the body wall called "papulae," which are also used for excretion. In ophiuroids, ten invaginations of the body, called "bursae," open to the outside via ciliated slits. Water circulates through the bursae by ciliary action, or in some cases through muscular action. Echinoids typically use the tube feet for respiration, and these can be modified and leaflike in irregular urchins. The so-called gills of the peristomal area of echinoids may not have a respiratory function or may supply only the muscles of Aristotle's lantern. The respiratory trees of certain holothurians, located in the hindgut area and extending toward the oral side, are used for gas exchange through the active pumping in and out of the hindgut via the anus. Gas exchange may also involve the tube feet in holothurians.

Despite the lack of specific excretory organs, nitrogenous waste in the form of ammonia needs to be eliminated somehow, and this seems to happen also through the tube feet and papulae (see below) in asteroids and probably through the respiratory trees in holothuroids. Excretion thus occurs mostly via precipitation of nitrogenous material in the coelomocytes and then discharged by various methods in the different structures involved.

Internal transport in echinoderms relies on the perivisceral coeloms, hemal system and water vascular system. The hemal system is a complex array of canals, mostly within the perihemal sinuses, best developed in crinoids and holothuroids. In asteroids there are two oral and aboral hemal rings, paralleling the water vascular system, connected by an axial sinus. The axial gland of the axial sinus is thought to produce coelomocytes. Radial hemal channels of the oral side extend with the radial canal of the water vascular system, while in the aboral region the hemal channels reach the gonads. A third hemal gastric ring associated with the digestive system is found in many echinoderms. Echinoids have a pulsatile dorsal sac in the aboral ring, next to the axial sinus, which, in addition to ciliary action, aids in hemal fluid movement. The hemal system of holothurians is intimately associated with the digestive tract and the respiratory trees (when present) and is pumped by a multitude of "hearts."

The integrative nature of the echinoderm body plan, with its multiple coelomic compartments and their role in nutrient uptake, was elegantly studied by Ferguson

(1984), who showed that radioactively labeled food entered the sea star hemal system by the aboral ring and axial organ and from there extended to the gonads and the radial hemal channels and the tube feet. These observations suggest that these materials came from the colelomocytes rather than from the digestive system.

REGENERATION, ASEXUAL REPRODUCTION AND DEVELOPMENT

Most echinoderms are capable of regenerating lost body parts (podia, spines, arms, digestive system, Cuverian tubules, etc.) to a degree much higher than that of most other invertebrates, with the exception of acoels and flatworms (see reviews in Candia Carnevali, 2006; Arnone et al., 2015). This may have a role in wound regeneration, as well as in defense or escape mechanisms, and can serve for asexual reproduction (Conand, 1996). Arm regeneration in crinoids and ophiuroids is an epimorphic blastemal process (i.e., it involves dedifferentiation of adult structures to form an undifferentiated mass of cells from which the new structures eventually develop). In contrast, morphallaxis (i.e., cellular reorganization with limited production of new cells) is the main process involved in regeneration in asteroids and echinoids (Arnone et al., 2015), although both modes of regeneration can occur in the same individual. Arm regeneration and visceral regeneration have received considerable attention, whereas other regenerative processes are less well understood. Among these, the regeneration of the holothurian nervous system has received some attention due to possible medical applications (Mashanov et al., 2008; Mashanov et al., 2013). In addition, an extreme mode of "regeneration" is the so-called spontaneous larval cloning, a form of asexual reproduction in larvae, which has been observed in members of all echinoderm classes (Eaves and Palmer, 2003).

Sea urchins have been a favorite animal for studying development because of the ease of obtaining billions of gametes for synchronous embryo culture, the transparency of the embryo, and the ability to introduce foreign DNA or RNA into the embryos (Arnone et al., 2015). One species in particular, the North American Pacific urchin *Strongylocentrotus purpuratus*, has played a fundamental role in the field of evo-devo. From the original characterization of the dynamic changes of transcription in embryogenesis to recent analyses of gene regulatory networks, due to the availability of whole genome sequencing, this species has become a key model in our modern understanding of developmental processes. The gene regulatory network for endomesoderm specification in *S. purpuratus* (Davidson et al., 2002) is often used as an example of these gene interactions. More than a hundred years of intense research on echinoderm development have inspired a vast literature. We address the reader to some classic reviews (Giese et al., 1991; Wray, 1997), as well as to the recent summary by Arnone et al. (2015).

Development in echinoderms involves a transition from bilateral to radial symmetry. This is accomplished by a first phase of development, in which the zygote gives rise to a bilaterally symmetrical larva (direct development has evolved multiple times in echinoderms), and by a second phase, in which specific larval cells give rise to the adult with pentaradial symmetry. This convoluted process is known as "maximal indirect development" (see above). It requires, among other things, a change of polarity between the two developmental phases.

Most echinoderms are gonochoristic, although hermaphroditic species are known among asteroids, holothuroids, and ophiuroids. The gonads are usually in fives, but a single dorsal gonad is found in holothuroids, while crinoids lack distinct gonads. Most echinoderms are free spawners, fertilization occurring in the water column, often in synchrony and related to the lunar cycle. Mechanisms for preventing polyspermy are well understood, and fertilization biology is especially well studied in sea urchins.

Fertilization involves gamete fusion—the fusion of the sperm and egg plasma membranes—and pronuclear fusion—the fusion of the male and female haploid pronuclei. As the surfaces of the gametes approach each other, a specific interaction takes place between the sperm protein bindin and a receptor located on the egg surface. For this reason bindin has played a key role in understanding gamete recognition and in driving speciation in sea urchins (e.g., Geyer and Palumbi, 2003; Landry et al., 2003).

In early embryogenesis of the model sea urchin, cleavage is radial, holoblastic, and initially equal or subequal, leading to a spacious coeloblastula (fig. 17.4). First and second cleavages are both longitudinal and at right angles to each other, intersecting the animal and vegetal poles and resulting in four cells of the same size. The third cleavage, resulting in the eight-cell stage, is equatorial, perpendicular to the first two. During fourth cleavage, the upper four cells divide meridionally, forming eight equal mesomeres, while the lower four cells divide unequally and horizontally, producing four large macromeres and four small micromeres in the vegetal pole of the embryo.

At the fifth cleavage, the eight mesomeres in the animal hemisphere divide equally and horizontally to form two tiers of cells (an1 and an2), one staggered above the other. The four macromeres divide meridionally, forming a tier of eight cells, while the micromeres divide unequally once more, generating four large micromeres and four small micromeres. At sixth cleavage, all the cells divide horizontally, producing a sixty-cell–stage embryo. The four small micromere founder cells arise at the unequal fifth cleavage; they divide only once more during embryogenesis, being set aside and contributing to the coelomic pouch and adult rudiment (Juliano et al., 2010). The sister cells of the small micromeres give rise to the skeleton in the larva.

The blastula stage begins after the next cell division, the cells forming a hollow sphere surrounding the blastocoel and becoming organized as an epithelium. The cells in the vegetal pole begin to thicken, forming a vegetal plate. At the other end of the embryo, in the animal pole, a small tuft of cilia forms, which allows the embryo to rotate inside the fertilization membrane until the blastula hatches and starts to swim freely. The advanced blastula consists of a single outer layer of about five hundred cells, flattened and thickened at the vegetal side. Morphogenesis occurs at this stage, being understood in detail in sea urchin embryos.

Coelomogenesis is initiated early in echinoderms, shortly after gastrulation. The left coelom usually gives rise to the adult hydrocoel and somatocoel. The hydrocoel has five lobes that constitute the primary podia and give rise to the radial canals of

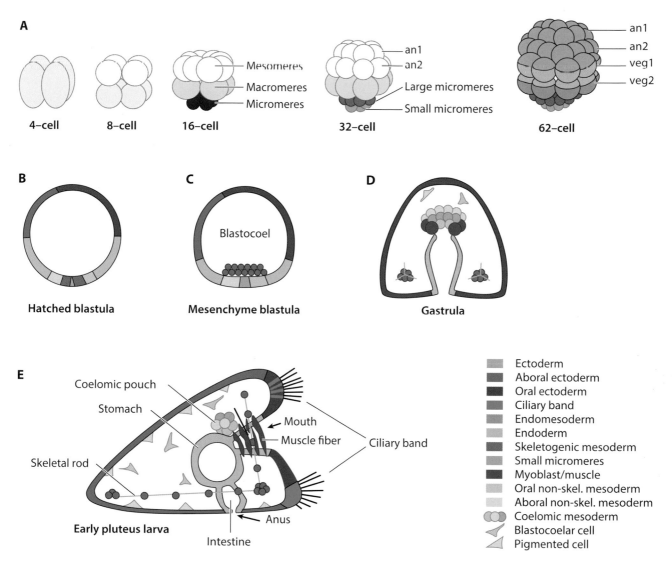

FIGURE 17.4. Early development of a sea urchin. A, early cleavage stages seen along the animal (an)–vegetal (veg) axis from 4- to 62-cell stages. At the 16-cell stage there are four micromeres (red) at the vegetal pole, four central macromeres (light yellow), and eight mesomeres (white) at the animal pole. At the 32-cell stage the micromeres have divided into large micromeres (purple) and small micromeres (green). The colors here and in the 62-cell stage indicate when the cells begin to be specified toward ectoderm, endoderm, and mesoderm in the following images. B–D, midsagittal sections of: B, hatched blastula stage where the ectoderm is already subdivided into oral ectoderm (dark blue) and aboral ectoderm (lighter blue) and the nonskeletogenic mesoderm (oral and aboral) has separated from the endoderm; C, mesenchyme blastula stage where primary mesenchyme cells (purple) have ingressed into the blastocoel while small micromeres (green) stay behind; D, mid-gastrula stage, showing the gut invaginating (dark yellow), the skeletogenic cells (purple) beginning to synthesize the skeleton, and nonskeletogenic mesoderm at the tip of the archenteron subdividing into domains (different cell types are indicated following the color key). (E) Pluteus larva, midsagittal view, showing the definite structures and cell types generated during embryogenesis. Based on Arnone et al. (2015).

the adult water vascular system. Each radial canal extends from a growth zone at the base of the primary podium (Morris, 2012). Coelomic development on the larval left side is the basis for constructing the adult, the left somatocoel becoming the body coelom of the adult, while the right coelom of the sea star larva also contributes to the adult body coelom (Morris et al., 2011). This model of development of the coelomic compartments may vary across species (Janies and McEdward, 1993).

Echinoderm development is highly regulative (Davidson, 1989), but cell lineages can be highly variable, even among sea urchin species (Wray, 1997).

Echinoderms generally have planktonic larvae, and these can constitute a large fraction of the zooplankton. Crinoids have nonfeeding (lecithotrophic) larvae, and their early development is not well understood. The stalked crinoids have two successive larvae, a nonfeeding auricularia stage with partially longitudinal ciliary bands, followed by a doliolaria larva with circumferential ciliary bands (Nakano et al., 2003). Other crinoids have only a doliolaria larva. The other echinoderm larvae are generally planktotrophic, but lecithotrophy has evolved in numerous lineages. Brooding is also important, especially among deep-sea and polar species. The eleutherozoan larvae may include feeding larvae with skeletal elements in the arms, such as the pluteus larva of Echinoidea (echinopluteus) and Ophiuroidea (ophiopluteus), or other forms of feeding auricularia and doliolaria in Holothuroidea; vitellaria in Ophiuroidea; or bipinnaria, brachiolaria, and vitellaria in Asteroidea.

The transformation of the larva into the adult is often explained in relation to the fate of the coeloms. Toward the end of larval life in the plankton, a portion of the coelom and the surrounding ectoderm form a region of rapidly proliferating cells, or imaginal adult rudiment, which produces the radially symmetrical adult, with an already functional water vascular system. It takes several days, however, until the mouth and anus form.

GENOMICS

The field of echinoderm genomics was launched relatively early, with the sequence of the nuclear genome of the purple sea urchin *Strongylocentrotus purpuratus* by Sodergren et al. (2006), and continues to be a growing field (Cameron et al., 2015), in part maintained by the online resource Echinobase (http://www.echinobase.org /Echinobase) (Kudtarkar and Cameron, 2017). Currently it includes draft genomes of a few other species, including two other echinoids (*Eucidaris tribuloides* and *Lytechinus variegatus*), one asteroid (*Patiria miniata*), one holothuroid (*Apostichopus parvimensis*) and one ophiuroid (*Ophiothrix spiculata*). Three other species, including a second ophiuroid (*Ophionereis fasciata*), a second asteroid (*Patiriella regularis*), and a second holothuroid (*Australostichopus mollis*), have become recently available, although with low coverage (Long et al., 2016). The genome of the ecologically important coral predator crown-of-thorns starfish, *Acanthaster solaris* (see Haszprunar et al., 2017), population outbreaks of which can cause substantial loss of coral cover, has been sequenced with the objective of understanding this coral predator (Hall et al., 2017). Also, the commercially important Japanese sea cucumber *Apostichopus japonicus* genome (Jo et al., 2017) and the sea urchin *Hemicentrotus pulcherrimus* genome (Kinjo et al., 2018) have become available.

Despite this considerable amount of information, most studies have focused on the statistics of the assemblies, and there is not much broadly available comparative data about echinoderm genomics as a whole, even in those cases where echinoderm genomics has been reviewed (Arnone et al., 2015). Nevertheless, the original sequencing of the *S. purpuratus* genome already highlighted a number of genes comparable to many other animals—23,300—and the presence in the genome of many genes previously thought to be vertebrate innovations or known only outside the deuterostomes (Sodergren et al., 2006). The genome of *A. solaris* has been analyzed in more detail than most other echinoderm genomes and shares many gene families and domain expansions with hemichordates and *S. purpuratus*. There is also extensive microsynteny with other deuterostomes, including conserved Hox, Para-Hox, and Nkx gene clusters (Hall et al., 2017), rejecting a prevailing hypothesis, based on the sea urchin genome, that the pentamerous symmetry had a relationship to chromosomal Hox gene arrangements in the chromosomes and the origin of pentamery (see Byrne et al., 2016).

Transcriptomic resources are abundant for echinoderms (Janies et al., 2016), in part due to recent phylogenomic work (Kober and Bernardi, 2013; Cannon et al., 2014; O'Hara et al., 2014; Telford et al., 2014; Reich et al., 2015; Simakov et al., 2015; Linchangco et al., 2017), as well as developmental (e.g., Samanta et al., 2006; Dylus et al., 2018) and regenerative (e.g., Mashanov et al., 2014) work.

Mitogenomics also had an early start in echinoderms (Jacobs et al., 1988; Smith et al., 1993), and complete mitogenomes are now available for a large number of species in all echinoderm classes (e.g., Scouras and Smith, 2001; Scouras et al., 2004; Scouras and Smith, 2006; Perseke et al., 2010; Cea et al., 2015; Bronstein and Kroh, 2019). Echinoderm mitogenomes show considerable variation across the different classes (Perseke et al., 2010; Cea et al., 2015), and are characterized by an alternative mitochondrial code (Osawa et al., 1992), in which AUA codes for isoleucine (as in hemichordates) in contrast to methionine, and AAA codes for asparagine instead of lysine (Perseke et al., 2011).

FOSSIL RECORD

Echinoderms are very diverse in the fossil record, with about 13,000 to 15,000 named species. Their record extends to at least Cambrian Series 2, Stage 3, and possibly as far back as the Fortunian. Recognizing echinoderms, even fragments, in the fossil record is facilitated by the unique property of their skeleton—the meshlike microstructure of the stereom (Bottjer et al., 2006). Stereom can be identified even in calcite skeletons that have been recrystallized or secondarily replaced by other minerals, as is common in geologically ancient fossils (Zamora and Rahman, 2015).

Larval development indicates that echinoderms pass through a stage of bilateral symmetry before assuming pentaradial symmetry (Smith, 2008). Cambrian stem-group echinoderms reveal that bilateral symmetry was likewise present in adults at the base of total-group Echinodermata (Zamora et al., 2012). Fossils such as *Ctenoimbricata spinosa* and early ctenocystoids such as *Courtessolea* from the middle Cambrian exemplify such bilaterally symmetrical echinoderms (fig. 17.5 A). In these taxa, the theca is encircled by a ring of marginal plates and the dorsal and ventral surfaces have different arrangements of ossicles and plates embedded in the variably calcified membrane.

The stratigraphically earliest taxon that may be assigned to Echinodermata is the recently redescribed *Yanjiahella biscarpa*, a Fortunian species from China resolved in phylogenetic analysis as the earliest branching member of the phylum (Topper et al., 2019). Like *Ctenoimbricata* and ctenocystoids, it exhibits bilateral summetry and a plated theca. *Yanjiahella biscarpa* has a muscular stalk reminiscent of that of hemichordates and has paired feeding appendages.

Phylogenetic analysis including the diversity of Cambrian stem-group echinoderms allows the transition from preradial to radial symmetry to be understood (Zamora and Rahman, 2015), although alternative phylogenetic schemes differ in whether or not asymmetrical forms are stemward or crownward of radial forms (David et al., 2000). As noted just above, the phylogenetically earliest echinoderms are bilaterally symmetrical (as is the sister group of echinoderms: hemichordates), whereas crownward of these are a grade of asymmetrical taxa, including cinctans (fig. 17.5 B), solutes (fig. 17.5 C), and stylophorans (cornutes and mitrates). These have sometimes been grouped as carpoids, or formally as a taxon Homalozoa, but the case for monophyly is dubious.

For many years these asymmetrical taxa were the focus of a debate over whether they were in fact echinoderms or were instead an assemblage of stem groups of Chordata and its constituent major taxa. This hypothesis, which grouped these animals as "calcichordates," was championed by British paleontologist R. P. S. Jefferies, most comprehensively documented in his book *The Ancestry of the Vertebrates* (Jefferies, 1986). Perhaps the strongest counterargument to the calcichordate theory is the genetic mechanisms of producing stereom that are unique to this calcite skeleton in echinoderms (and inferred to be the same in fossils that possess a calcitic stereom) and are not shared by the hydroxyapatite skeleton of chordates or aragonite endoskeletal elements of hemichordates (Bottjer et al., 2006; Cameron and Bishop, 2012). Paraphyly of the asymmetrical stem-group echinoderms is indicated by solutes but not cinctans, sharing a hydropore in their water vascular system, as in crown-group echinoderms (Zamora et al., 2012). Computational fluid dynamic reconstruction of feeding in a Cambrian cinctan suggests that these stem-group echinoderms employed pharyngeal filter-feeding, as in tunicates, cephalochordates and enteropneusts, rendering this the likely feeding mode of deuterostomes as a whole (Rahman et al., 2015). The discovery of soft parts in exceptionally preserved cornute stylophorans from the Ordovician revealed the presence of a hydrocoel and water vascular system in their unpaired appendage (Lefebvre et al., 2019). This confirms their identity as echinoderms (rather than hemichordates or chordates). Furthermore, a debate over whether the appendage was a locomotory stalk/tail or a feeding arm has swung in favor of the latter model, based on the similarity in the organization of the water vascular system (i.e., having an ambulacral canal and tube feet) to that of feeding arms in such taxa as crinoids.

Helicoplacoids are the phylogenetically first (and geologically earliest) radial echinoderms, and the first to exhibit ambulacra. They have a typically spindle-shaped theca, with spirally arranged plates and three ambulacra, and thus triradial symmetry (fig. 17.5 F). Their remains include arguably the most stratigraphi-

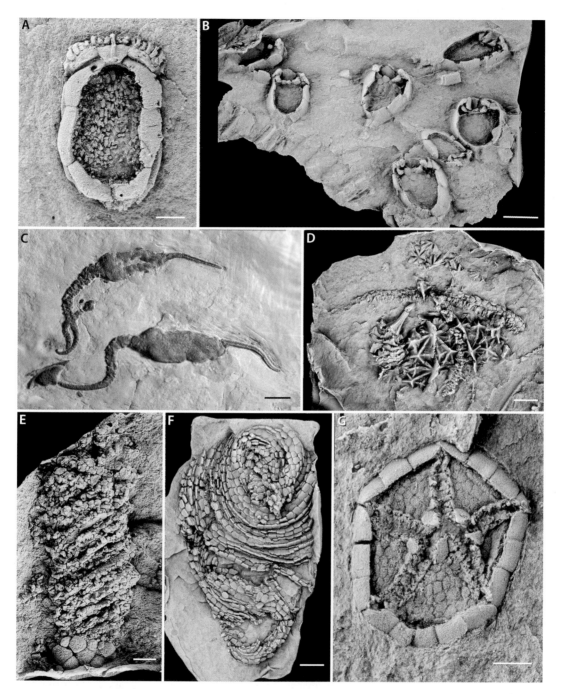

Figure 17.5. Cambrian echinoderm fossils. A, the ctenocystoid *Ctenocystis utahensis*, scale 1 mm; B, the cinctan *Gyrocystis platessa*, scale 5 mm; C, the solute *Coleicarpus sprinklei*, scale 5 mm; D, the rhombiferan *Dibrachicystis purujoensi*, scale 5 mm; E, the helicocystid *Helicocystis moroccoensis*, scale 1 mm; F, the helicioplacoid *Helicoplacus gilberti*, scale 5 mm; G, the edrioasteroid *Cambraster cannati*, scale 2 mm. Photo credits: Samuel Zamora except C, original.

cally early echinoderm fossils, known from Cambrian Series 2, Stage 3, their record restricted to North America (Sprinkle and Wilbur, 2005).

Helicocystids, established with the description of *Helicocystis moroccoensis* (fig. 17.5 E) from the early Cambrian of Morocco (Smith and Zamora, 2013), shed light on the origin of pentaradial symmetry. They provide a unique combination of characters shared on the one hand with helicoplacoids (such as a spiral body plating) and on the other with pentaradial forms. Most important, they exhibit torsion, such that the mouth is situated apically and the periproct laterally, versus the opposite in helicoplacoids.

Edrioasteroids, which usually have a disc-shaped theca, are among the oldest Cambrian echinoderms but persisted until the Permian. They are obviously pentaradial, with five ambulacra (fig. 17.5 G). Phylogenetic analysis of early pentaradial forms suggests that edrioasteroids may be para- or polyphyletic, with different members allied to different lineages of a group discussed in the next paragraph, Blastozoa (Kammer et al., 2013).

The Cambrian forms that depict non-pentaradial symmetry are just a subset of extinct diversity of echinoderms, because the Paleozoic includes some thirty major extinct groups (Sumrall and Wray, 2007). Most of these are stalked forms that are more closely related to crinoids than to Eleutherozoa, a relationship recognized by their grouping as Pelmatozoa (stemmed echinoderms) in traditional classifications. A grouping named Blastozoa unites many of the stalked Paleozoic taxa based on unique supposed respiratory structures and biserially plated feeding organs known as "brachioles" that lined the ambulacra (Sprinkle, 1973). At least some members, such as the middle Cambrian rhombiferan *Dibrachicystis purujoensis* (fig. 17.5 D), have feeding appendages that are thecal extensions of the body and are thus considered homologous with the arms of crinoids (Zamora and Smith, 2012), and some phylogenetic analyses recover crinoids nested within a paraphyletic blastozoan grade (Ausich et al., 2015).

Other common Paleozoic groups of blastozoans such as the Cambrian–Silurian eocrinoids have brachioles that arise from the ambulacra in the typical 2-1-2 pattern of crinoids. Unlike those of crinoids, however, these appendages are exothecal, lacking a connection to the body cavity. Blastozoa includes some diverse, widespread groups, such as cystoids and blastoids, both of which appeared in the Great Ordovician Biodiversification Event, a time of major radiation of echinoderms, and persisted until the Devonian and Permian, respectively. True crinoids are known from near the base of the Ordovician onward, and went on to dominate echinoderm diversity from the Late Ordovician to the Permian. Crinoid remains can be so common in Carboniferous limestones that they are significant rock-forming elements. Most of the major groups of crinoids had already appeared by the Ordovician (Ausich et al., 2015), and indeed, all five extant echinoderm classes have their first record in that period. Despite this ancient divergence, the four living eleutherozoan classes underwent only moderate diversification during the Paleozoic and predominantly radiated in the Mesozoic and Cenozoic. Echinoids diversified rapidly in the Cretaceous and Paleogene, whereas ophiuroids and asteroids were much less diverse in the Mesozoic and are largely Neogene radiations.

CHORDATA

A clade composed of the phyla Cephalochordata (also called Acrania), Tunicata (or Urochordata) and Vertebrata (or Craniota/Craniata) has been recognized since the nineteenth century by the presence of a notochord–neural tube complex in all its members (Stach, 2008; Nielsen, 2012a). This clade, named Chordata, has been well supported in most morphological and molecular phylogenetic analyses of animal relationships. We humans belong to it, hence the broad interest in this clade, as reflected in the vast literature related to development and morphogenesis of many chordate species.

While vertebrates are not specifically discussed in this book, we refer to them within the context of Chordata and its two invertebrate branches, Cephalochordata (commonly known as "lancelets" or "amphioxus") and Tunicata (sea squirts and their relatives). Cephalochordata, Tunicata, and Vertebrata are treated either as three individual phyla (e.g., Nielsen, 2012a; Satoh et al., 2014) or as subtaxa of the phylum Chordata (e.g., Brusca et al., 2016b)—a subjective matter, as they form a well-defined clade (see Giribet et al., 2016 for a recent discussion of Linnean ranks in this context). Since there is no scientific or biological justification for the use of one or three distinct "phyla" (Hejnol and Dunn, 2016)—other than the principles of monophyly and nestedness—we choose to treat Chordata as one entity. For a historical revision of the usage of different names, see Satoh et al. (2014).

Chordates are primarily marine, free-living, solitary (as opposed to colonial) animals, although multiple lineages of vertebrates have colonized freshwater and land, with striking cases of return to life at sea; and tunicates evolved both sessile lifestyles and coloniality. All the invertebrate chordates are restricted to the marine realm, with only some ascidians tolerating low-salinity environments but not freshwater. Cephalochordates include 33 or so described free-living benthic species, inhabiting shallow clean sand in temperate and tropical regions. Tunicates include ca. 3,000 recognized species (Zhang, 2011; Shenkar et al., 2018), with different ecologies. Sea squirts, or ascidians, are sessile (although some colonial ascidians can walk, as they can propagate and grow in different directions), and most are filter feeders (but a few deep-sea species are "carnivorous" or macrophagous). Thaliacea (pyrosomes, salps, and doliolids) are pelagic and mainly colonial, some members showing complex life cycles. Appendicularia (also called Larvacea) are free swimming and they produce a large and canalized shelter of animal cellulose for filter feeding. Ascidians can be solitary (some forming clumps from gregarious settlement) or colonial, while the pelagic groups can be solitary (appendicularians), colonial (pyrosomes), or temporarily colonial (salps and doliolids). They are found at all latitudes and depths.

While almost all chordates have sexual cycles (only some fishes and a few tetrapods show some form of parthenogenesis), colonial ascidians can have a mix of sexual and asexual phases, and thaliaceans use asexual reproduction for colony growth (Lemaire and Piette, 2015). Finally, vertebrates are the most diverse clade of Chordata, with nearly 65,000 species (Zhang, 2011), but they still represent a tiny fraction of the ca. 1.55 million animals currently accounted for (about 4% of animal diversity as a whole); it is this other 96% of animals with which we deal in this book.

The presence of pharyngeal gill slits during some part of the life cycle was once considered one of the apomorphies of Chordata, but it is now thought to be a plesiomorphy of Deuterostomia (see chapter 14), as pharyngeal gill slits are found in hemichordates and are also interpreted in several calcite-plated deuterostome fossils that are identified as stem-group echinoderms (e.g., Dominguez et al., 2002). The notochord–neural tube complex and the branchial basket with an endostyle, tight junctions, and myomery (Nielsen, 2012a) are possible autapomorphies of Chordata. Others include the post-anal tail—a tail that extends beyond the anus—found at least in the early developmental stages (and some adults) of all chordates (Brusca et al., 2016b).

The invertebrate chordates are perhaps not the most conspicuous of invertebrates, although ascidians can be common in many shallow marine environments and cephalochordates, although not commonly encountered, can be locally abundant, giving the name to a type of substrate formed by clean coarse sand (amphioxus sand). Tunicates have a direct impact on humans in that several species of solitary ascidians are harvested for food in some cultures, while amphioxus fisheries have existed in Asia, at least historically (Light, 1923). As do many other sessile animals, ascidians produce a large array of chemical compounds used for defense from predators and to compete for space. They are thus the focus of substantial research on marine natural products, with anticancer and antimalarial drugs having been developed from compounds derived from ascidians (see Palanisamy et al., 2017). Finally, the light-emitting pyrosomes—hence their name—have captured the interest of zoologists and sailors for centuries, and both pyrosomes and appendicularians constitute an important component of the marine bioluminescence (Martini and Haddock, 2017).

▬ SYSTEMATICS

While cephalochordates have been considered the closest relatives of vertebrates based on morphological characters (e.g., Nielsen et al., 1996; Ax, 2001; Stach, 2014)—a clade named Notochordata (Ax, 2001 uses the name Vertebrata for this whole clade)—molecular phylogenies have almost unanimously suggested a sister-group relationship of tunicates and vertebrates, in an unranked clade named Olfactores (Delsuc et al., 2006), a result first received with skepticism (Stach, 2008). Olfactory organs associated with the telencephalon have been interpreted as an apomorphic character of craniates (Stach, 2008), and putatively homologous organs were described in fossils assigned to the stem-Olfactores by Jefferies (1991). The clade Olfactores has been recovered in both Sanger-based phylogenies (e.g., Zrzavý et al., 1998; Giribet et al., 2000) and phylogenomic analyses (e.g., Delsuc et al., 2006;

Delsuc et al., 2008; Dunn et al., 2008; Putnam et al., 2008; Hejnol et al., 2009). This clade is, however, contradicted by some mitochondrial data (Bernt et al., 2013) and some Sanger-based analyses of protein-encoding genes (Paps et al., 2009a).

The phylogeny of cephalochordates has been investigated using whole mitochondrial genomes (Nohara et al., 2005; Kon et al., 2007; Igawa et al., 2017). These studies show monophyly of the genera *Asymmetron* and *Branchiostoma*,[8] the latter being the sister group to *Epigonichthys*. Calibration with transcriptomic data and using densely sampled mitochondrial genomes suggests a Cenozoic diversification of the extant species (Igawa et al., 2017).

Tunicate phylogeny has been notoriously difficult to resolve with molecular methods, as many of its members have exceptionally accelerated rates of evolution (Tsagkogeorga et al., 2010) and display remarkable genome reorganization (Berná and Alvarez-Valin, 2014). Until recently, tunicate molecular phylogenies were inferred using a single nuclear ribosomal RNA marker (e.g., Wada, 1998; Swalla et al., 2000; Govindarajan et al., 2011; Tatián et al., 2011) or one to a few mitochondrial genes (Turon and López-Legentil, 2004; Singh et al., 2009; Stach et al., 2010; Rubinstein et al., 2013). One study used a combination of a nuclear and a mitochondrial gene together with morphology (Stach and Turbeville, 2002).

Classically, tunicates have been divided into three classes: Ascidiacea (the benthic ascidians), Thaliacea (the pelagic salps, pyrosomes and allies), and Appendicularia or Larvacea (the solitary pelagic tunicates that retain the notochord as adults). An additional group of carnivorous abyssal ascidians was sometimes classified in the class Sorberacea (Monniot and Monniot, 1990), now known to be related to the molgulid ascidians (Tatián et al., 2011). The ascidians have been divided into Stolidobranchiata (also spelled Stolidobranchia), Aplousobranchiata (= Aplousobranchia), and Phlebobranchiata (= Phlebobranchia), the latter recognized as probably paraphyletic and giving rise to the other two groups (see a historical review in Stach and Turbeville, 2002). Aplousobranchiata and Phlebobranchiata are often grouped in the taxon Enterogona. Most molecular phylogenetic analyses published to date have found ascidians to be paraphyletic, with thaliaceans closest to phlebobranchiates. We thus use the term "ascidians" as a common name for a non-monophyletic group.

Two recent phylogenomic studies using a partially overlapping set of eighteen genomes and transcriptomes both found maximum support for the main splits within Tunicata (fig. 18.1), including a sister-group relationship of the pelagic Appendicularia to all other tunicates and a main division between Stolidobranchiata (one of the main clades of ascidians) and the other groups, Thaliacea + Enterogona (Delsuc et al., 2018; Kocot et al., 2018). Thaliacea appears as the sister group to Phlebobranchiata–Aplousobranchiata, but the monophyly of Phlebobranchiata, the clade that includes the genetic and developmental model species *Ciona robusta*, is still unresolved. While some of these groups had been resolved before, using *18S*

---

[8] The first cephalochordate was described by Pallas (1774), who classified it within molluscs and named the first species *Limax lanceolatus*. The name was later revised to *Branchiostoma lubricum* and to *Amphioxus lanceolatus*. Due to priority of the name *Branchiostoma*, the European species became *Branchiostoma lanceolatum*, while the term "amphioxus" has become a common name for cephalochordates.

rRNA data sets (Tsagkogeorga et al., 2009; Govindarajan et al., 2011), the phylogenomic framework provides stronger support for nearly all nodes and now allows identifying the presence of the adult notochord in appendicularians as a plesiomorphy of the group and not as a derived neotenic feature (Giribet, 2018b). Whereas in appendicularians the notochord cells continue to divide until they reach 120 to 160 cells in the adult, ascidians reach the maximum number of 40 notochord cells in the larval stage before being resorbed during metamorphosis (Soviknes and Glover, 2008).

Vertebrate relationships have been explored in great detail but are out of the scope of this volume. Several questions worth addressing phylogenetically are the relationships among the basal fish lineages (Braasch et al., 2016) or the position of unstable taxa such as Testudines (Fong et al., 2012). The evolutionary history of the group has been summarized in a number of places (e.g., Bertrand and Escrivá, 2014).

    Phylum Chordata
        Subphylum Cephalochordata
          Class Leptocardii
            Family Branchiostomatidae
              Asymmetron
              Epigonichthys
              Branchiostoma

       Olfactores
       Subphylum Tunicata (= Urochordata)
           Class Appendicularia (= Larvacea)
         Unranked taxon 1
           Pleurogona
             Class Stolidobranchiata

         Unranked taxon 2
           Class Thaliacea
             Order Doliolida
             Order Pyrosomatida
             Order Salpida
          Enterogona
            Class Aplousobranchiata
            Class Phlebobranchiata

      Subphylum Vertebrata (= Craniata)
       Superclass Agnatha (= Cephalaspidomorphi)
       Superclass Gnathostomata
         Class Chondrichthyes
         Class Osteichthyes
          Actinopterygii
          Sarcopterygii
           Actinistia
           Dipnoi
           Tetrapoda

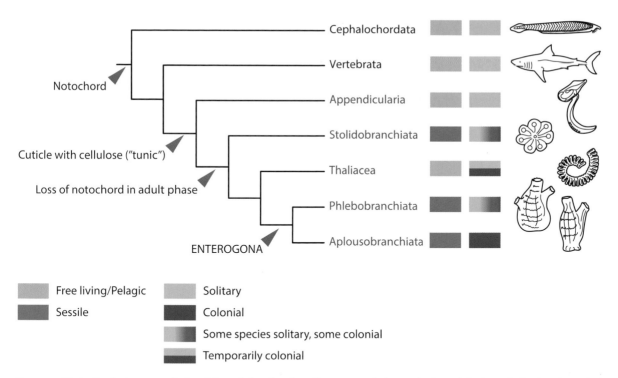

FIGURE 18.1. Evolutionary relationships of the chordate lineages and the evolution of their main body plans with respect to habitat and coloniality. Based on Giribet (2018b).

CHORDATA: A SYNOPSIS

- Bilaterally symmetrical, coelomate deuterostomes
- Pharyngeal gill slits present at some stage of development
- Mesodermally derived dorsal notochord present at some stage in development
- Dorsal, hollow nerve cord giving rise to an anterior brain
- Characteristic myomery, absent in tunicates
- Muscular post-anal tail at some stage in development
- Endodermal, pharyngeal endostyle (Cephalochordata and Tunicata) or the homologous thyroid gland (Vertebrata)
- Complete through-gut, straight or U-shaped, often regionalized
- Closed circulatory system (open in adult ascidians) with a contractile heart

Although most of the chordate diversity lies within vertebrates, we focus here on the two invertebrate groups, only making reference to some vertebrate characters when needed. Furthermore, the body plans of cephalochordates and tunicates differ substantially in many fundamental aspects, especially due to the adaptations of tunicates to their pelagic or sessile lifestyles, and we often refer to one or the other when describing their anatomy or their biology.

Cephalochordates live in coarse sediments where they can burrow but easily escape from possible predators, being able to swim equally well head-first or tail-first. This is accomplished by the distinctive, chevron-shaped blocks of muscle and by the notochord.

Tunicates can be solitary, gregarious (sometimes with vascular connections between individuals), or colonial, with benthic (mostly sessile) and pelagic species (fig. 18.1). The ascidian classes (Aplousobranchiata, Phlebobranchiata, and Stolidobranchiata) are composed of all benthic sessile animals, including solitary and colonial forms, some of the colonies measuring several meters in length. Colonial and compound ascidians are composed of individual zooids embedded in a common gelatinous matrix. The colonies grow by asexual reproduction after the larva settles. In some cases the colonial ascidians can form rosettes of individuals that share an atrial siphon but have individual oral siphons. The highly modified adult has lost some of the larval characters, including the notochord and the hollow dorsal nerve cord, and has a U-shaped gut with the two siphons pointing upward or sideways.

Thaliaceans are pelagic and have the two siphons at opposite ends of the body, the exiting water providing a means of jet propulsion. Thaliaceans can be colonial or temporarily colonial, and they often alternate sexual and asexual phases in their life cycle. Pyrosomes are large, gelatinous colonies of small zooids arranged along a common chamber, or cloaca that receives all the exhalant siphons. This water exits the colony through a single large opening, projecting the whole colony forward. Movement in doliolids and salps is partially or totally mediated by muscular action.

Appendicularians are peculiar chordates in some ways, although they retain the notochord and part of the dorsal nerve cord as adults, a feature that seems to be plesiomorphic for Tunicata. They are solitary animals encased in a gelatinous house they secrete and that they use for filtering food from the water column by a combination of tail wagging and a mesh of filters within the house. When clogged with food, they can generate a new house every two hours (B. Gemmell, pers. comm).

The most conspicuous characteristic of chordates is the notochord. It is a dorsal elastic rod of a likely primitive type of cartilage, present in all embryonic and larval stages as well as in the adults of cephalochordates, appendicularians, and some basal vertebrates.

The notochord derives from a middorsal strip of embryonic mesoderm, serving as a source of signals that pattern surrounding tissues and as a major skeletal element of the developing embryo (Stemple, 2005). For the cephalochordates, the notochord is essential for locomotion and persists throughout life. In this case, it is composed of longitudinally stacked discoidal lamellae surrounded by connective tissue. These lamellae are composed of muscle cells with their fibers oriented transversally and are responsible for stiffening the notochord during some kinds of movements, like burrowing. The notochord is associated with another chordate characteristic, the hollow dorsal neural tube. In cephalochordates, the notochord extends anteriorly to the brain, which characteristic gives the name to this clade, having a "chord" in the head.

The notochord is positioned centrally in the embryo with respect to both the dorsal–ventral (DV) and left–right (LR) axes, where it signals to all surrounding tissues. The notochord is therefore important for specifying ventral fates in the central nervous system, controlling aspects of LR asymmetry, inducing pancreatic fates, controlling the arterial versus venous identity of the major axial blood vessels, and specifying a variety of cell types in forming somites (see a review in Stemple, 2005). The notochord also plays a fundamental structural role, serving as the axial skeleton of the embryo until other elements form. In ascidians, the notochord exists during embryonic and larval free-swimming stages, providing the axial structural support necessary for locomotion (Satoh, 2003), and is resorbed during metamorphosis. In some vertebrate clades, such as cyclostomes (lampreys) as well as in particular gnathostomes, the notochord persists throughout life. In derived vertebrates the notochord becomes ossified and contributes to the center of the intervertebral discs.

The body of cephalochordates is completely covered by a simple columnar epithelium (the epidermis) on top of a thin dermis. The body wall muscles are a conspicuous characteristic of these fishlike animals, forming chevron-shaped blocks (myotomes) arranged along most of the length of the body, leaving little space for the coelom. The notochord extends beyond the myotomes, both anteriorly and posteriorly. The animals typically swim by contracting the myotomes, which bends the body laterally, as the notochord does not allow shortening of the body, and is aided by a series of finlike storage organs. Muscle arms, similar to those known in nematodes (and a few other animals) are also found in amphioxus (Flood, 1966). Epitheliomuscular cells, largely corresponding to those of cnidarians, have been found in various coelomic channels of *Branchiostoma lanceolatum* (see Storch and Welsch, 1974).

In tunicates, the secreted tunic, which is the main synapomorphy of the group, overlies the epidermis. The tunic has different degrees of development in the different groups, especially in ascidians and thaliaceans, and it can be gelatinous or tough and leathery, sometimes including calcareous spicules. The tunic contains fibers and is largely made of a cellulose polymer (tunicin) produced by a cellulose synthase gene thought to have been acquired by horizontal gene transfer from an actinobacterial genome (Matthysse et al., 2004), but this still required acquisition of binding sites for specific transcription factors (Sasakura et al., 2016). In addition, the tunic contains amebocytes, and in some cases blood cells and blood vessels, and thus is sort of a living cuticle. Some ascidians keep a diversity of symbiotic bacteria within the tunic, perhaps explaining the mechanism for acquisition of cellulose synthase. These bacteria are responsible for the production of numerous secondary metabolites, including the rare vanadium that is famously accumulated by the ascidians.

Tunicate musculature includes subepidermal muscle bands, longitudinal muscles that serve to pull the siphons, and the sphincters that control the closing of the siphons. Circular musculature used to pump water in and expel water is well developed in doliolids and salps. Appendicularians also have muscles that allow for movement of the tail.

Chordates have a through-gut that is U-shaped in the sessile tunicates. Feeding in cephalochordates and tunicates relies on a pharyngeal mucus net and the pharyngeal gill slits. Cephalochordates drive water into the mouth and pharynx via pharyngeal cilia in a brachial chamber and expel it through the pharyngeal gill slits, which can number up to two hundred. These slits are separated by gill bars supported by cartilaginous rods. The food from the inflowing water is then trapped by the mucus net secreted by the secretory Hatschek's pit located on the roof of the vestibule, and enters the mouth, located in the anterior vestibule in the so-called oral hood. The oral hood bears the oral cirri, or buccal cirri, which act as a filter that sorts out the particles that are not ingested. The mouth and its associated velar tentacles constitute a second filter. A ciliary wheel organ is found along the internal vestibular walls. When feeding, the animals are partially buried in sand, the vestibule facing upward.

The endostyle (or hypobranchial groove) is found in the ventral surface of the pharynx in cephalochordates, tunicates, and basal filter-feeding vertebrates such as lampreys. It secretes mucus and binds iodine, and it is often interpreted as a homolog of the thyroid gland of vertebrates (e.g., Cañestro et al., 2008). The digestive caecum, located near the junction of the pharynx with the esophagus, serves for glycogen storage and protein synthesis, and it is seen as a possible homolog of the vertebrate liver and pancreas.

In ascidians, water flows into the pharynx through an incurrent oral (or branchial) siphon, passes through the pharyngeal gill slits (also called "stigmata"), into a spacious atrium (the cloacal chamber) and exits through an excurrent atrial siphon—in a way that resembles that of bivalves, with which they "compete" in their filtration rates (Petersen, 2007). The anus opens into the excurrent flow of water. The oral siphon is generally anterior, and the atrial siphon is anterodorsal in ascidians or posterior in thaliaceans. Likewise, dorsoventral orientation is generally recognized by the position of the dorsal ganglion and the ventral endostyle.

Cephalochordates have a closed circulatory system formed by a series of vessels but lack a heart, the blood moving by peristaltic movements (Rähr, 1979). The blood does not carry oxygen-binding pigments or cells, and it is thought to only distribute nutrients. Gas exchange thus takes place mainly through diffusion, especially across the walls of the so-called metapleural folds.

The circulatory system is weakly developed in tunicates and is best understood in ascidians, which have a tubular heart without valves that is controlled by two myogenic pacemakers at each end of the tube. They are able to periodically reverse the flow of blood. The heart is surrounded by a pericardium, which is interpreted by some to be the remnants of a coelom (e.g., Schmidt-Rhaesa, 2007). The tubular heart delivers the blood into lacunae of the body toward the internal organs and the tunic. The blood is rich in amebocytes and other cell types and has high concentrations of vanadium and iron, among other heavy metals. Gas exchange occurs mostly across the pharyngeal epithelium and the cloaca.

The coelom is highly reduced to many small spaces in cephalochordates (Schmidt-Rhaesa, 2007) and perhaps to the epicardial sacs of tunicates. These epicardial sacs are sometimes involved in bud formation during asexual reproduction and may also serve for accumulation of nitrogenous wastes. Other than these and some nephrocytes, elimination of metabolic wastes is probably by diffusion. In contrast, cephalochordates have protonephridia of the solenocyte type as their excretory organs (Moller and Ellis, 1974). The protonephridia are ectodermally derived, segmentally arranged structures that extend throughout the pharyngeal region, forming clusters that collect the nitrogenous wastes into a common nephridioduct that opens to the atrium. In addition to the paired nephridia, a single large nephridium, called the "nephridium of Hatschek," is located above the roof of the pharynx (Ruppert, 1996). It resembles the paired nephridia in structure, opening at its posterior end into the pharynx, just behind the velum, and then passes anteriorly to end blindly just in front of Hatschek's pit. An excretory function is also attributed to the renal papillae found in the floor of the atrium, but this requires experimental testing.

NERVOUS SYSTEM

The central nervous system of cephalochordates has recently been reviewed (Wicht and Lacalli, 2005; Lacalli and Stach, 2016), and the neuronal types and neurotransmitters of the anterior central nervous system have been thoroughly investigated (Castro et al., 2015). It is relatively simple, with a hollow dorsal nerve cord that extends for most of the length of the body (fig. 18.2), with an anterior cerebral vesicle (see a discussion in Lacalli and Stach, 2016). Sensory organs are restricted to four types of photoreceptors, including an unpaired frontal eye, a lamellar body, dorsal ocelli (or Hesse organs), and Joseph cells, plus a series of sensory cell types found mostly on the epidermis.

The central nervous system of tunicates has also recently been reviewed, by Manni and Pennati (2016). In adults it is reduced, reflecting their sessile or not-so-active filter-feeding lifestyle. A small cerebral ganglion lies dorsal to the anterior end of the pharynx, innervating the muscles and a few other parts of the body, but it is not regionalized in the same way as the vertebrate brain. In tunicates there are homologs of the forebrain, hindbrain, and spinal cord, but not the midbrain (Cañestro et al., 2005). The dorsal nerve cord is present in the tails of tunicate larvae and is lost during metamorphosis, except in appendicularians. A neural or subneural gland located anterior to the brain is thought to be homologous to the pituitary gland of vertebrates. Sensory organs are mostly of the ciliary type and are largely inconspicuous in the sessile forms, though some are well characterized, especially in the pelagic forms. Although adult tunicates lack eyes or ocelli, they have clear responses to light, including siphon contraction and phototropism, which are probably mediated by the pigmented spots around the siphon apertures in some species, in the sperm duct epithelial cells, and in the brain (Manni and Pennati, 2016).

The larval central nervous system has received more attention, especially from an evo-devo perspective, since it can be more easily connected to the vertebrate structures on the basis of morphology and gene expression patterns, being considered a simplified model of the vertebrate nervous system, with 177 neurons in the CNS of the model species *Ciona robusta* (Manni and Pennati, 2016; Ryan et al., 2016). The small number of cells in the larval central nervous system of *C. robusta* and *Halocynthia roretzi* seems to be fixed, showing some level of conservation but probably not true eutely (Meinertzhagen, 2005). Eutely has been claimed for appendicularians (Ax, 2001).

FIGURE 18.2. Schematic drawing of a midsagittal section of a cephalochordate. Based on Brusca and Brusca (1990).

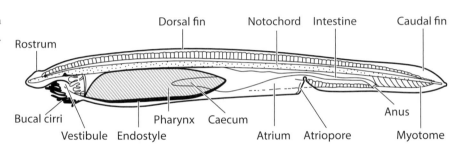

REPRODUCTION AND DEVELOPMENT

Cephalochordates are gonochoristic and reproduce seasonally. In mature specimens, rows of twenty-eight to thirty-five pairs of gonads are arranged along each side of the atrium. During the spawning season, the reproductive tissue occupies a large volume of the body to the point of impeding feeding. Spawning occurs externally, after the gametes exit the body by the rupture of the atrial wall.

Cephalochordate development has been thoroughly documented (Whittaker, 1997) and recently reviewed by Holland (2015). The field of cephalochordate developmental biology initiated, as for many other animal phyla, with the pioneering work of Kowalevsky (1867) on the Mediterranean *Branchiostoma lanceolatum*. This species has become a model organism (Holland et al., 2004), and the third invertebrate (after *Drosophila melanogaster* and *Caenorhabditis elegans*) for which the entire Hox cluster was deciphered, showing that although it was complete, it had not undergone the gene expansion found in vertebrates (Garcia-Fernàndez and Holland, 1994). Amphioxus was also instrumental for proposing the theory of homeobox evolution, with a supposedly ancestral ProtoHox cluster giving origin to the two clusters found in *B. lanceolatum*, the Hox cluster and its sister ParaHox gene cluster (Brooke et al., 1998). Developmental fate maps have been established by different authors, most recently by Holland and Holland (2007).

Cleavage in cephalochordates is radial, holoblastic and subequal, with first cleavage starting at the animal pole, second cleavage at right angles, and third cleavage dividing the embryo into an animal and a vegetal half. Blastomeres are not tightly adherent at the early developmental stages, often resulting in twins or quadruplets. The blastula gastrulates by simple invagination. At the late gastrula stage, each ectodermal cell becomes ciliated and the gastrula begins to rotate within the fertilization layer. Embryogenesis is especially well studied, especially in relation to the formation of the neural plate from the dorsal ectoderm, and the origin of the nerve cord. The blastopore at one point connects the archenteron to the lumen of the hollow nerve cord, forming a neuropore. Later the structures separate and the blastopore gives rise to the anus. The mouth appears secondarily. For a detailed description of the embryogenesis and state of the art of amphioxus evo-devo, see Holland (2015).

After embryonic development is completed, the asymmetrical free-swimming, feeding larva, with the mouth on the left side of head and the first gill openings on the right side, swims up to the plankton. The larvae can remain in the plankton for 75 to 200 days.

Tunicates combine sexual reproduction with asexual propagation/growth, with the exception of appendicularians, which probably lack asexual growth. In social and colonial (compound) ascidians, budding allows for colony growth. Some of these colonial ascidians can also regenerate an entire zooid from a small piece of vascular tissue. Budding is highly variable, but in most cases the first buds are generated by a sexually produced individual after settlement (termed an "oozoid"), which produces blastozooids, which in turn produce additional buds. In social species, like those in the genus *Clavelina*, the blastozooids arise from the stolons, but in others the germinal tissue can form in different parts, including the epicardial sacs described above, the gut, the gonads, and the epidermis.

Recent research on the colonial species *Botryllus schlosseri* has shown incredible properties of regeneration from stem cells circulating in the vascular system (e.g., Braden et al., 2014). Its budding and regenerative properties have made this species a model for studying aging and regeneration (Voskoboynik and Weissman, 2015). Regeneration has also been found in the tropical ascidian *Polycarpa mytiligera*, which uses evisceration as a probable defense mechanism (Shenkar and Gordon, 2015), as is commonly found in holothurians. Doliolids produce chains of buds that are released intact, but eventually each blastozooid becomes a separate individual.

Tunicates are hermaphroditic, with relatively simple reproductive systems containing a single ovary and a single testis (or an ovotestis, in some cases) or multiple gonads that open to the cloacal chamber. Fertilization is generally external, and embryonic development results in the so-called tadpole larva, which eventually metamorphoses into the adult. Other species, mainly colonial, brood the embryos until the larvae are released, and a few have a connection to the parental individual (placental development). A few species have direct development, or nonswimming larvae without a tail. Larvae may also be absent in other groups. In pyrosomes, each zygote develops into an oozoid. Doliolids produce larvae, but these are encased in a cuticular capsule and do not swim. In salps, which have internal fertilization, development is also placental-like. Appendicularians have external fertilization, the zygote developing into the tadpole-larva that does not lose the notochord after maturation.

Early cleavage is stereotypically bilateral, with some variation later in development. The developmental biology of the group has been thoroughly reviewed (Jeffery and Swalla, 1997; Stolfi and Brown, 2015). Detailed embryological studies include many classical ones (e.g., Conklin, 1905b), and a series of modern studies including detailed fate maps of ascidians (e.g., Kumano and Nishida, 2007; Lemaire, 2009) and the appendicularian *Oikopleura dioica* (Stach et al., 2008). Generally, tunicate cleavage is bilateral, holoblastic, and subequal (as in cephalochordates), leading to the formation of a coeloblastula, which gastrulates by invagination. The blastopore closes through development. Ascidians are a classical example of mosaic development (Conklin, 1905a)—meaning that each cell has its fate established as early as the two-cell stage and is not regulated by any other cell. Myogenesis and neurogenesis have been favorite areas in tunicate evo-devo and are known in great detail.

The tadpole larvae are varied in different groups, but in general they are short lived, with a free-swimming lifespan sometimes shorter than two days (fig. 18.3).

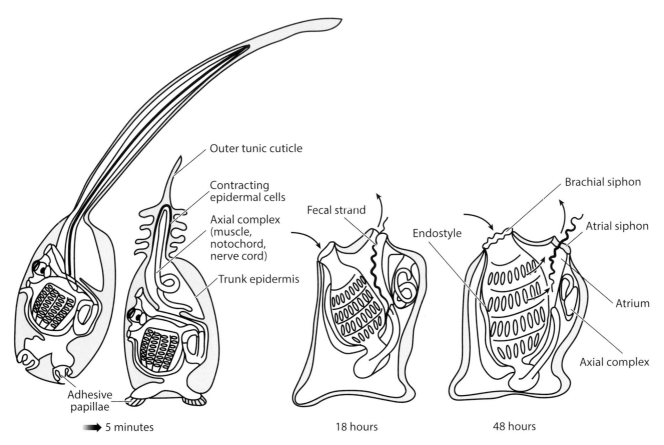

FIGURE 18.3. Metamorphosis of a tadpole larva into an adult ascidian. After the larva, using the anterior adhesive papillae, settles, resorption of the tail begins within the first five minutes, followed by the rotation of the viscerae and siphons until they attain the adult position. Metamorphosis continues with the enlargement of the pharynx and by shedding the outer layer of the cuticle and is completed in about 48 hours. Arrows indicate incurrent and excurrent currents. Based on multiple sources.

The larva, however, plays an important role in substrate selection. After settlement, metamorphosis occurs, which often involves probing with the anterior end and the secretion of a glue from the anterior adhesive papillae. This secretion triggers a series of events, including the resorption of the tail and a rotation of the viscerae and siphons until they attain the adult position. Metamorphosis continues with the enlargement of the pharynx and by shedding the outer layer of the cuticle.

The field of chordate evo-devo is especially rich, and since we could not attempt to summarize it here, we thus have limited ourselves to provide a few general references and examples. Many theories have attempted to explain the transition from an animal with a ventral nerve cord to the chordate body plan, with a dorsal nerve cord, in what has been interpreted as an inversion of a main body axis (see chapter 14). Many larval theories have been proposed for the origin of the chordate body plan (e.g., see Lacalli, 2005), a central one being Garstang's theory of the evolution of the chordate central nervous system (Garstang, 1894, 1928), which postulates that the CNS evolved through modification of the circumoral ciliary band of a dipleurula larva. Nielsen (1999) proposed an alternative theory in which the chordate neural tube evolved through lateral fusion of a ventral, postoral loop of the ciliary band in a dipleurula larva. A detailed discussion of these and other theories and some of their pros and cons are provided in Nielsen (2012a).

GENOMICS

Chordate genomes have flourished since the early sequencing of the human (*Homo sapiens*), mouse (*Mus musculus*), and fugu (*Takifugu rubripes*) genomes (released between 2001 and 2002), followed by a second batch including chicken (*Gallus gallus*), puffer fish (*Tetraodon nigroviridis*), and rat (*Rattus norvegicus*) (published in 2004). Hundreds of species and thousands of vertebrate genomes are now available, and worldwide consortia are producing large-scale genomic resources for a diversity of clades (G10KCOS, 2009). For the invertebrate chordates, currently available are the genomes of one cephalochordate, *Branchiostoma floridae* (Putnam et al., 2008); one appendicularian, *Oikopleura dioica* (Seo et al., 2001; Denoeud et al., 2010); and several ascidians: *Ciona robusta* (formerly *C. intestinalis* type A) (Dehal et al., 2002), *Ciona savignyi* (Small et al., 2007), three species of *Molgula* (Stolfi et al., 2014), *Botryllus schlosseri* (Voskoboynik et al., 2013), *Botrylloides leachii* (Blanchoud et al., 2018), and *Didemnum vexillum* (Velandia-Huerto et al., 2016).

The early sequencing of the *Ciona robusta* genome revealed 16,000 protein-coding genes, about half of those found in vertebrates (Dehal et al., 2002). Vertebrate gene families are typically found in simplified form in *Ciona*, suggesting that ascidians contain the basic ancestral complement of genes involved in cell signaling and development, but their genome is simplified when compared to those of cephalochordates and vertebrates (Putnam et al., 2008), thus supporting the idea of secondary simplification. This ascidian genome has also acquired a number of lineage-specific innovations, including a group of genes engaged in cellulose metabolism that are related to those in bacteria and fungi (Dehal et al., 2002).

The genome of the appendicularian *Oikopleura dioica* is also rapidly evolving, showing tremendous genome plasticity (Denoeud et al., 2010) and a total lack of synteny with other animal genomes at small scales (less than 30 genes) and modest conservation at larger distances. This miniaturized genome (Seo et al., 2001) has been interpreted as constituting "an unprecedented genome revolution," in that this animal preserves ancestral morphological features lost in other groups with more conservative genomes—as appendicularians maintain the chordate body plan in the adult stage. The genome of amphioxus seems to be closer to that of the supposed ancestral chordate in terms of gene content and structure, and even chromosomal organization (Garcia-Fernàndez and Benito-Gutiérrez, 2009), when compared to those of vertebrates (with numerous genome duplications) and tunicates (with many genomic simplifications and plasticity).

The mitochondrial genome of chordates is a typical small, circular molecule, about 15,000 (in amphioxus) to 18,000 base pairs in length (Naylor and Brown, 1998; Yokobori et al., 2003). The genome typically comprises 12 to 13 protein-coding regions, 2 *rRNA* genes, a replication control region, and 22 *tRNA* genes, although a control region may be missing in cephalochordates (Spruyt et al., 1998), which also show some gene rearrangement (Nohara et al., 2005). Gene order is broadly conserved across chordates, with the exception of tunicates, which show extensive gene rearrangements and particularly high evolutionary rates when compared to other chordates (Singh et al., 2009). The vertebrate and tunicate mtDNA genetic code differs from the "Universal" code—which is also the code found in cephalochordates—in several respects. In the case of tunicates, AGG and AGA code for glycine instead of the typical serine (Perseke et al., 2011), while these are stop codons in vertebrates; AUA codes for methionine, and UGA for tryptophan.

▬ FOSSIL RECORD

The first chordates appear in the fossil record with the opening of the Burgess Shale-type taphonomic window, near the base of Cambrian Series 2, Stage 3, ca. 519 million years ago. As discussed in the fossil section on Deuterostomia (chapter 14), several controversial groups of Cambrian fossils have been regarded by some workers as chordates, but this is disputed by others, the disagreement extending as far as offering protostome groups such as Ecdysozoa as alternatives. Even in the context of chordate affinities, groups such as yunnanozoans have been variably allied to cephalochordates, tunicates, or vertebrates. The discussion below thus concentrates on fossils that have not been part of that debate.

Pikaia gracilens from the Burgess Shale has been meticulously documented from study of more than a hundred exceptionally preserved specimens (Conway Morris and Caron, 2012). Sigmoidal myomeres, about a hundred along the length of the fusiform body, and a possible notochord were identified as characters that underpin an assignment to Chordata. *Pikaia* has been placed in the chordate stem group (Conway Morris and Caron, 2012), although phylogenetic analyses by Mallatt and Holland (2013) recover it in several alternative positions within Chordata, including within the crown group. *Cathaymyrus diadexis* from the Chengjiang biota was originally interpreted as *Pikaia*-like (Shu et al., 1996), and allied to Cephalochordata at a time when affinities to that group were considered likely for *Pikaia*. The later revision of *Pikaia*, as just noted, has left its position in the chordate crown group ambiguous. A second species, *C. haikouensis* (Luo et al., 2001), adds little more to this taxon. *Cathaymyrus* has been noted to preserve the most decay-resistant features of non-biomineralized chordates as a whole, and its few specimens could be decayed remains of any of the three subphyla of total-group Chordata (Sansom et al., 2010).

Comparisons with Cephalochordata have also been made for the Cambrian yunnanozoans (Chen et al., 1995), though as discussed in chapter 14, the systematic postion of this group is exceptionally labile. The only fossil identified as a cephalochordate is the Early Permian *Palaeobranchiostoma hamatogergum* (Oelofsen and Loock, 1981), from South Africa. The single known specimen of this species has been reconsidered in light of experimental decay of extant *Branchiostoma*, calling into question the identification of such features as a nerve cord and pharyngeal gill slits (Briggs and Kear, 1994), leaving its position within Chordata equivocal.

As discussed in Deuterostomia (chapter 14), a relationship between the early Cambrian vetulicolians and tunicates has been argued (García-Bellido et al., 2014).

One species from the early Cambrian Chengjiang biota, *Shankouclava anningense* (see Chen et al., 2003), has been identified as a tunicate and accepted as such by subsequent workers, including in reviews of early deuterostomes that have noted its similarity to aplousobranch ascidians (Swalla and Smith, 2008). Phosphatic microfossils from the late Cambrian in the United States named *Palaeobotryllus taylori* are one of a set of Cambro–Ordovician microfossils that have been compared to colonial ascidians (Müller, 1977; Lehnert et al., 1999), although alternative identities have been proposed, including Cnidaria (Baliński and Sun, 2017). The remainder of the Phanerozoic offers few fossil tunicates (reviewed by Valentine, 2004). An Ediacaran fossil from the White Sea Vendian, *Burykhia hunti* (see Fedonkin et al., 2012), classified together with previously described taxa from the Nama Group in Namibia, has been tentatively attributed to Tunicata. Only two specimens are known, and a crown-group chordate identity for any Ediacaran fossil demands further scrutiny.

The earliest well-corroborated verebrates are agnathans from the Chengjiang biota (Shu et al., 1999; Hou et al., 2002; Shu et al., 2003b; Shu et al., 2010). These have been assigned as many as four generic and specific names, but it has been suggested that most or all of these may be synonymous, representing taphonomic rather than biological variation (Hou et al., 2017). Most anatomical work on them uses the names *Haikouichthys ercaicunensis* and *Myllokunmingia fengjiaoa* (Shu et al., 1999). These jawless vertebrates reveal paired eyes and a notochord with separate vertebral elements, and evidence for possible nasal sacs (Shu et al., 2003b), as well as zigzag muscle blocks and a dorsal fin (Hou et al., 2002). A Burgess Shale species, *Metaspriggina walcotti*, has been shown to possess chordate and more specifically vertebrate characters as well, incuding a notocord, paired eyes, nasal sacs, W-shaped myomeres, and a post-anal tail (Conway Morris and Caron, 2014). In phylogenetic trees constrained with monophyly of cyclostomes (hagfish and lampreys), the Cambrian taxa are collapsed in a polytomy at the base of total-group Vertebrata, although a subset of the shortest trees unite them as a clade of stem-group vertebrates (Conway Morris and Caron, 2014).

19

PROTOSTOMIA

The traditional major division among bilateral animals has been between Protostomia and Deuterostomia (Grobben, 1908), a dichotomy referring to a fundamental mode of early development related to the fate of the blastopore, giving origin to the mouth (in protostomes), or whether the mouth originates from a new embryonic opening, the blastopore becoming the anus (in deuterostomes). Although the name Protostomia is widely used, it is now well known that the fate of the blastopore is highly variable, sometimes even within a single phylum (Martín-Durán et al., 2012a; Martín-Durán et al., 2016a), therefore concluding that the blastopore, the mouth, and the anus are developmentally independent. The name Protostomia is therefore broadly used for the non-deuterostome, non-xenacoelomorph bilaterians, that is, the members of the clades Ecdysozoa and Spiralia. This clade has been obtained in nearly all molecular phylogenetic analyses of traditional Sanger-based data as well as phylogenomics.

A group of similar composition, Zygoneura, in reference to the presence of a ventral nerve cord, was recognized by Hatschek (1888), and this is now considered a fundamental character of protostomes (Giribet, 2003; Nielsen, 2012b), although with important exceptions even in large groups like nemerteans and platyhelminths (Nielsen, 2012a). A historical account of Protostomia and names for groups within it was provided by Nielsen (2012b).

Protostomia is difficult to diagnose morphologically. Currently, phylogenetic criteria are used to define protosome animals, most of which possess a special type of central nervous system with an apical nerve center (brain) connected circumesophageally with a pair of longitudinal ventral nerve cords, but deviations from this model are extensive, and in some cases definition of ventral can be subjective (e.g., Priapulida).

According to the Nielsen's trochaea theory (Nielsen, 1985; Nielsen and Nørrevang, 1985), protostomes are derived from a holoplanktonic gastraea with a circumblastoporal ring of downstream-collecting compound cilia (archaeotroch) and a nervous system comprising an apical ganglion and a circumblastoporal nerve ring (Nielsen, 2012b), but this theory has difficulty explaining the evolution of many direct developing protostomes, as well as the entire clade Ecdysozoa. Ecdysozoa includes many direct developers as well as indirect developers without ciliated larvae. This theory was also developed before major reassignment of animal phyla from deuterostomes to protostomes, as is the case of the lophophorate phyla (e.g., Halanych et al., 1995; de Rosa, 2001) and chaetognaths (Shimotori and Goto, 2001; Helfenbein et al., 2004; Papillon et al., 2004; Marlétaz et al., 2006; Matus et al., 2006)—although Grobben

(1908) had already treated the lophophorates as protostomes (his "Molluscoidea") but placed chaetognaths with the deuterostomes.

Protostome animals include the two animal phyla with the highest number of described living species: arthropods (approximately 1.165 million species) and molluscs (over 117,000), along with other highly diverse phyla—nematodes (26,600), platyhelminths (30,000), and annelids (23,000). In all, protostomes constitute 95% of extant animal diversity; of these, 87% are arthropods (Giribet, 2014b).

Phylogenetically, protostomes include more than 20 phyla, 8 in the clade Ecdysozoa, and 14 in the clade Spiralia. Ecdysozoans include Arthropoda, Kinorhyncha, Loricifera, Nematoda, Nematomorpha, Onychophora, Priapulida, and Tardigrada, a clade of pseudocoelomate and coelomate animals, with or without segmentation. Spiralia includes a set of animals that mainly undergo spiral cleavage—a stereotyped cleavage pattern with highly conserved cell fate (Henry and Martindale, 1998)—plus a series of animals without this pattern. The currently accepted spiralian phyla are Annelida, Brachiopoda, Bryozoa, Chaetognatha, Cycliophora, Dicyemida, Entoprocta, Gastrotricha, Gnathostomulida, Micrognathozoa, Mollusca, Nemertea, Orthonectida, Phoronida, Platyhelminthes, and Rotifera, although some have considered Orthonectida as derived Annelida (Schiffer et al., 2018).

FOSSIL RECORD

The fossil record of protostomes is abundant and dates back to the early Cambrian, as explained in the relevant chapters (especially Mollusca, Brachiopoda, and Arthropoda, as major biomineralized phyla). Because the clade Protostomia is difficult to define morphologically, fossils are recognized as protostomes when they have diagnostic characters of particular protostome phyla or groups of phyla.

20 ECDYSOZOA

Among the most revolutionary changes in animal phylogeny in the twentieth century was the introduction of Ecdysozoa, a clade originally proposed based on an analysis of *18S rRNA* sequence data to comprise the members of Panarthropoda (Arthropoda,[9] Onychophora, and Tardigrada), Nematoida (Nematoda and Nematomorpha), and Scalidophora (Kinorhyncha and Priapulida) (Aguinaldo et al., 1997)—and later to also include Loricifera. Ecdysozoa is composed of animals that molt their cuticles during their life cycle (Aguinaldo et al., 1997; Giribet and Ribera, 1998; Schmidt-Rhaesa et al., 1998).

Since the proposal of Ecdysozoa, this grouping has been supported by numerous lines of both morphological and molecular evidence, although some criticisms of the hypothesis emerged soon after it was introduced. The arguments included scepticism about the data used to support Ecdysozoa (e.g., Wägele et al., 1999; Wägele and Misof, 2001; Pilato et al., 2005), as well as defenses of the homology of segmentation that panarthropods share with annelids (Scholtz, 2002, 2003). Segmentation was a key line of evidence underpinning the longstanding hypothesis that panarthropods and annelids unite as a monophyletic group, Articulata (Haeckel, 1866). In light of this conflict between molting and segmentation, attempts were made to reconcile the Articulata and Ecdysozoa hypotheses by providing intermediate evolutionary scenarios between these two groups (Nielsen, 2003). Some molecular studies placed additional (nonmolting) taxa within Ecdysozoa, such as Chaetognatha (e.g., Zrzavý et al., 1998; Paps et al., 2009a) and *Buddenbrockia* (Zrzavý et al., 1998). Both these have since been more convincingly positioned elsewhere in the animal tree, chaetognaths allying with Gnathifera (Fröbius and Funch, 2017; Laumer et al., 2019; Marlétaz et al., 2019) and *Buddenbrockia* reassigned with confidence to Myxozoa (Jiménez-Guri et al., 2007b).

Another theme that ran through early phylogenomic analyses restricted to a few available genomes was a rejection of Ecdysozoa by instead recognizing a putative clade named Coelomata (Wolf et al., 2004; Philip et al., 2005; Rogozin et al., 2007). This grouping was based on the premise that arthropods are more closely allied to chordates than to nematodes. Coelomata was discarded in favor of Ecdysozoa in subsequent analyses (e.g., Philippe et al., 2005; Irimia et al., 2007; Dunn et al., 2008; Hejnol et al., 2009). Support for Ecdysozoa has included evidence from mitogenomics (Podsiadlowski et al., 2008; Rota-Stabelli et al., 2010; Popova et al., 2016), though it need be acknowledged that this requires further testing, no mitochondrial genomes being as yet available for Nematomorpha or Loricifera.

[9] Euarthropoda sensu Ortega-Hernández (2016); see that paper for a historical account of the use of names such as Arthropoda, Euarthropoda, Tactopoda, and others.

Although Ecdysozoa was originally portrayed by some as an artifact of flaws in molecular systematics (Wägele and Misof, 2001), morphologists had already implicitly or explicitly questioned Articulata while instead placing arthropods within a molting clade. For example, Eernisse et al. (1992) published a morphological phylogenetic analysis resolving panarthropods with Nematoda and Kinorhyncha (Priapulida was left unresolved in a basal protostome trichotomy) but recognizing the annelid lineages as part of Spiralia. Even earlier, Crowe et al. (1970) noted similarity in the organization of the cuticles of tardigrades and nematodes, concluding, "On this basis a phylogenetic affinity of tardigrades for nematodes was supported." Similarities between tardigrades and other cycloneuralian phyla had also been noted, such as with Loricifera: "Annulation of the flexible buccal tube, telescopic mouth cone, and the three rows of placoids are found only in Tardigrada and Loricifera (Kristensen, 1987). Because tardigrades exhibit several arthropod characters (see Kristensen, 1976, 1978, 1981), this last finding supports a theory about a relationship between some aschelminth groups and arthropods (Higgins, 1961). That theory has recently gained support derived primarily from new ultrastructural data, e.g., the fine structure of the chitinous cuticular layer, molting cycle, sense organs, and muscle attachments" (Kristensen, 1991: p. 352).

Several authors have advanced a set of characters that could be apomorphic for Ecdysozoa, many of which are related to their cuticle (fig. 20.1). Some suggested that cuticular apomorphies for Ecdysozoa are a trilayered cuticle, formation of epicuticle from the tips of epidermal microvilli, annulation, molting (where known, using ecdysteroid-mediated hormones), and lack of locomotory cilia (Schmidt-Rhaesa et al., 1998). While the presence of a cuticle is not in conflict with the presence of epidermal ciliation, as is the case of Gastrotricha, it has been suggested that molting is correlated with the lack of epidermal ciliation (Valentine and Collins, 2000). A jaw replacement in a mechanism similar to molting has been described in one family of annelids (see chapter 47). Being so taxonomically restricted in Annelida, this observation thus does not constitute a major challenge for cuticular molting as an ecdysozoan apomorphy.

Another suggested ecdysozoan character is the terminal position of the mouth (Giribet, 2003). Recent developmental data, however, suggest that the terminal mouth of priapulans has a ventral origin, interpreted as the ancestral state in ecdysozoans (Martín-Durán and Hejnol, 2015). Irrespective of the developmental pathway followed to a terminal mouth, a terminal mouth in adults can still be interpreted as a putative synapomorphy of Ecdysozoa (Ortega-Hernández et al., 2019). An annulated cuticle and a terminal mouth are prevalent in Cambrian fossils assigned to the stem groups of lineages whose extant members lack one or both of these traits. These include stem-group arthropods such as *Kerygmachela* (Budd, 1998), stem-group onychophorans such as *Onychodictyon* (Ou et al., 2012b), and lobopodians near the base of Panarthropoda such as *Aysheaia*. The annulated cuticle does not, however, occur in many modern ecdysozoans; it is only present in Priapulida, Onychophora, and some Nematoda. Likewise, the mouth has shifted toward a ventral position in some Tardigrada (Nielsen, 2019), in Onychophora, and in most Arthropoda. Segmentation is found in four of the seven ecdysozoan phyla, but it is unclear how many times it evolved. The unstable position of Tardigrada

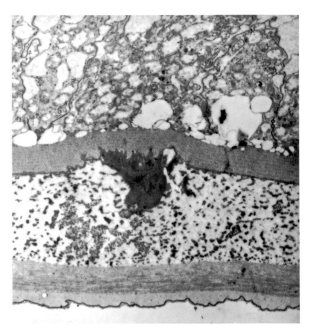

FIGURE 20.1. TEM image of the two cuticular layers of a nematomorph before molting (the two-electron-dense parallel layers). The section includes part of a natatory bristle, which attaches to the adult (second) cuticle. Photo credit: Andreas Schmidt-Rhaesa.

(discussed below) complicates this inference, but in any case it would at least have originated independently in Kinorhyncha and Panarthropoda.

A characteristic often cited for Ecdysozoa is the presence of α-chitin in their cuticle, but this has until now only been confirmed for Priapulida and Panarthropoda (Greven et al., 2016). The cuticle of Pentastomida, which are arthropods confidently nested in the crustacean–insect clade Pancrustacea (Abele et al., 1989; Giribet et al., 2005), contains ß-chitin (Karuppaswamy, 1977). Data are lacking on the type of chitin present in the other members of Ecdysozoa.

Some features of early cleavage and gastrulation are shared by cycloneuralians (particularly nematodes and priapulans) and arthropods (in pycnogonids and crustaceans), and stand as potentially apomorphic for Ecdysozoa (Ungerer and Scholtz, 2009). Such shared features are holoblastic, irregular radial, equal to subequal cleavage, and gastrulation involving one or two large division-retarded immigrating cells followed by smaller immigrating blastomeres that surround the large cells.

Molecular apomorphies have also been proposed for Ecdysozoa in the form of ecdysozoan tissue-specific markers, notably neural expression of horseradish peroxidase (HRP) immunoreactivity (Haase et al., 2001). A multimeric form of a *ß-thymosin* gene in arthropods and nematodes that was thought to be exclusive to ecdysozoans (Manuel et al., 2000) was later refuted (Telford, 2004).

The morphological arguments in defense of Ecdysozoa have been complemented by comparative morphological studies that have called putative homologies of Articulata into question (e.g., Kristensen, 2003; Giribet, 2004; Mayer, 2006a; Koch et al., 2014). Given that segmental nephridia or nephridial derivatives and coelomic cavities were parts of the concept of a segment in Articulata, these critiques undermine the homology between annelid and panarthropod segmentation (see the discussions in Scholtz, 2002, 2003; Giribet, 2003; Minelli, 2017).

Relationships within Ecdysozoa remain in flux (Giribet and Edgecombe, 2017), and unlike for many other branches of the animal tree, most authors rely on a mainly morphological hypothesis, dividing Ecdysozoa into three main clades: Scalidophora (= Cephalorhyncha sensu Nielsen; Kinorhyncha, Loricifera, and Priapulida), Nematoida (Nematoda and Nematomorpha), and Panarthropoda (e.g., Edgecombe, 2009; Dunn et al., 2014; Martín-Durán et al., 2016b). Despite its widespread acceptance, the topology implied by this classification is virtually unsupported by molecular trees. Whether Scalidophora and Nematoida form the clade Cycloneuralia (= Introverta sensu Nielsen) or a grade, with Nematoida as the sister group of Panarthropoda, is still unresolved. Molecular analyses excluding Loricifera have typically supported a sister-group relationship of Scalidophora to the other two clades (e.g., Petrov and Vladychenskaya, 2005; Mallatt and Giribet, 2006; Campbell et al., 2011; Pisani et al., 2013; Rota-Stabelli et al., 2013; Borner et al., 2014) or have favored a sister-group relationship of Priapulida to the remaining ecdysozoans (Hejnol et al., 2009; Laumer et al., 2019), a position consistent with some interpretations of priapulans as closest to the ancestral ecdysozoan (Webster et al., 2006).

Other analyses have contradicted Scalidophora by placing Loricifera closer to Nematomorpha than to Kinorhyncha and Priapulida (Sørensen et al., 2008) or closer to Nematoida and Panarthropoda than to Kinorhyncha and Priapulida (Yamasaki et al., 2015), or Loricifera appeared largely unresolved (Park et al., 2006), though each of these studies were based on only one or two markers. The first phylogenomic analysis to include data on Loricifera places *Armorloricus elegans* with Priapulida, albeit without significant support (Laumer et al., 2015a), but this study lacked data on Kinorhyncha.

In addition to the issues with Loricifera, discussions on ecdysozoan phylogeny have debated the position of tardigrades, which have been placed in Panarthropoda using a suite of morphological characters but are often drawn to Nematoida in molecular analyses. This labile positioning of tardigrades contrasts with that of Onychophora, which repeatedly resolve as the closest sister group of Arthropoda (Hejnol et al., 2009; Borner et al., 2014; but see Rota-Stabelli et al., 2013).

Discussion on the evolution of the ecdysozoan nervous system has centered on understanding the nature of the brain, which is circumoral and shares a characteristic ring neuropil with anterior and posterior neuronal somata in the non-panarthropods apart from Nematomorpha (Schmidt-Rhaesa and Rothe, 2014), and indeed this morphology constitutes the basis for the name Cycloneuralia for this putative clade. In contrast, the brain is ganglionar in the three panarthropod groups (Martin and Mayer, 2014; Martín-Durán et al., 2016b). Debate also surrounds the nature of the paired versus unpaired nerve cords (Martín-Durán et al., 2016b). According to these authors, the ancestral nervous system of the Ecdysozoa might have comprised an unpaired ventral nerve cord (seen in Priapulida, Kinorhyncha, Nematoda, and Nematomorpha), but the architecture of the brain in the ancestral ecdysozoan remains unclear. This is influenced by the debate over whether cycloneuralians are paraphyletic (in which case a collar-shaped brain is more likely the state in the last common ancestor of Ecdysozoa) or are monophyletic (in which case the transformation series between collar-shaped and dorsal ganglionar brains is ambiguous).

GENOMICS

Because they include model species, Ecdysozoa were at the forefront of animal genomics, *Caenorhabditis elegans* being the first published animal genome (*C. elegans* Sequencing Consortium, 1998), soon to be followed by that of *Drosophila melanogaster* (Adams et al., 2000). More than a hundred genomes from different ecdysozoan species have since been published (Dunn and Ryan, 2015), and thousands more sequenced, but genomes are entirely missing for Loricifera, Kinorhyncha, and Nematomorpha. High-coverage transcriptomes are at hand for most major ecdysozoan lineages (orders or equivalent).

FOSSIL RECORD

Given the shared presence of a cuticle and the fact that its molting allows for molts to outnumber carcasses as fossils, ecdysozoans have a rich fossil record from the early Cambrian onward (for a synopsis of Paleozoic vermiform ecdysozoans, see Maas, 2013). Cambrian diversity and the challenge of confidently establishing the phylogenetic position of early cycloneuralians is exemplified by Palaeoscolecida (Harvey et al., 2010), a group of often large-bodied worms that have a high preservation potential because of their robust cuticle. Although most speciose in the Cambrian, palaeoscolecids sensu stricto persisted into the Ordovician with moderate diversity (Botting et al., 2012), and their youngest members are Silurian. The cuticle is annulated, and it bears sclerites with taxon-specific geometries, arrangements, and ornament. Palaeoscolecid sclerites have an extensive microfossil record when preserved disarticulated from their scleritome. Burgess Shale-type compression fossils, as well as three-dimensionally preserved, secondarily phosphatized Orsten-type fossils, allow the sclerites to be associated with both the overall cuticular structure and other body parts, including paired terminal hooks and an introvert that bears radially arranged spines (Maas, 2013). A few features of palaeoscolecids have been identified as evidence for an affinity to Nematoida, such as the posterior hooks that resemble the posterior lobes of some male Nematomorpha (Hou and Bergström, 1994) and helically wound cross fibres in the innermost cuticular layer shared with nematoids more generally (Harvey et al., 2010). Several cladistic analyses have explored the relationships of palaeoscolecids and other vermiform ecdysozoans (for recent versions, see Harvey et al., 2010; Wills et al., 2012; Liu et al., 2014c; Zhang et al., 2015; Shao et al., 2016). Some of these, such as the Burgess Shale species *Ottoia prolifica*, are known from vast numbers of exceptionally preserved specimens that allow, in that example, for details of the eversion of the introvert to be documented (Conway Morris, 1977a) and gut contents studied to reveal a diversity of prey (Vannier, 2012).

The issues noted above with regard to uncertainty in the higher-level systematics of Ecdysozoa, especially whether Cycloneuralia is mono- or paraphyletic with respect to Panarthropoda and whether or not Scalidophora is a clade, impact on the classification of these fossils. For example, introvert morphology may support "cycloneuralian" affinities, but this may be a plesiomorphic character for Ecdysozoa rather than an apomorphy for Cycloneuralia. In published phylogenetic analyses, palaeoscolecids are allied either with nematoids or with priapulans, and *Ottoia* shifts from a position in the priapulan stem group to elsewhere in Cycloneuralia, depending on taxon sampling and character weighting.

Unique combinations of arthropod and cycloneuralian characters in fossil taxa are instructive for reconstructing ancestral states for deep nodes in Ecdysozoa. A radial mouth cone with overlapping plates and radially aranged, scalid-like pharyngeal teeth occurs, for example, in the stem-group arthropod *Pambdelurion*. These are interpreted as plesiomorphies shared by Panarthropoda and "Cycloneuralia," and thus characters of Ecdysozoa as a whole (Edgecombe, 2009; Vinther et al., 2016). Similarly, the Cambrian lobopodian *Hallucigenia*, which has been interpreted either as a stem-group onychophoran (Smith and Ortega-Hernández, 2014) or a stem-group panarthropod (Caron and Aria, 2017), has radially arranged circumoral lamellae and pharyngeal teeth that compare with putative homologues in tardigrades and cycloneuralians. These characters have thus been inferred to be possible autapomorphies of Ecdysozoa (Smith and Caron, 2015).

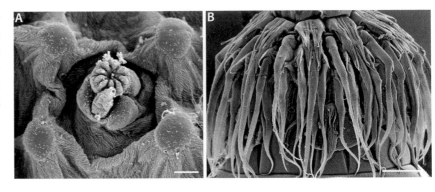

FIGURE 21.2. Scanning electron micrographs of scalidophorans: A, flosculus of a priapulan in the genus *Tubiluchus*, scale 2 μm; B, introvert of the kinorhynch *Echinoderes bookhouti* covered by scalids and spines, scale 10 μm. Photo credits: Martin Vinther Sørensen.

have been united as Vinctiplicata (Lemburg, 1999) based on details of their indirect development, including a larva enveloped by a longitudinally folded lorica, to the exclusion of the direct developing kinorhynchs.

Therefore, for the time being we continue to use Scalidophora, although recognizing that this clade may need to be dismantled, as reflected in fig. 2.2. Also, given our current understanding of ecdysozoan phylogeny, we choose not to provide any internal structure for this clade.

▬ FOSSIL RECORD

Treptichnid burrows dating to the base of the Cambrian are exceedingly likely to have been produced by a priapulan-like scalidophoran (Kesidis et al., 2019), providing a constraint on the group's age (see chapter 1). As discussed above, defining Scalidophora morphologically is difficult, as some characters, like the presence of a lorica or an annulated cuticle may be plesiomorphic for Ecdysozoa, hence making it difficult to assign fossils to a clade that is poorly supported, especially with molecular data. As discussed under the concept of Cycloneuralia in chapter 20, the phylogenetic position of many Cambrian vermiform ecdysozoans has been labile across analyses, often being sensitive to different character-weighting schemes.

Recent descriptions of secondarily phosphatized fossils from the Fortunian in South China have revealed that the fossil record of scalidophoran-like ecdysozoans extends to ca. 535 Ma, substantially earlier than the Burgess Shale-type compression fossils that had provided most of what was previously known about these taxa (Zhang et al., 2017a). *Eopriapulites sphinx*, for example, has been resolved as a stem-group scalidophoran of this age (Liu et al., 2014c; Shao et al., 2016). Its scalids and pharyngeal teeth are arranged hexaradially (18 longitudinal rows of scalids and 18 basal rows of pharyngeal teeth), and comparisons with other cycloneuralians open up the possibility that aspects of hexaradial symmetry may be a fairly general character for ecdysozoans (Liu et al., 2014c). Another putative scalidophoran from the same stratigraphic unit, *Eokinorhynchus rarus*, is discussed in chapter 23.

Markuelia is a grouping of five species of vermiform, annulated ecdysozoans known from several embryonic stages that are enrolled within their fertilization envelope. They range from the early Cambrian to the Early Ordovician, with records in Siberia, China, Australia, and the United States (Dong et al., 2010). The earliest species are among the earliest known ecdysozoan body fossils. Recent phylogenetic analyses of fossil and living ecdysozoans resolve *Markuelia* near the base of the scalidophoran stem group (Wills et al., 2012; Liu et al., 2014c), though it need be noted that coding of *Markuelia* is based on embryos.

Among the diversity of Cambrian cycloneuralians, a small but abundant group of dumb-bell shaped loricate taxa, represented by *Sicyophorus* and *Palaeopriapulites* in the Chengjiang biota, has attracted comparison with extant loricate scalidophorans, that is, Priapulida and Loricifera (= Vinctiplicata) (Maas et al., 2007a). These have an ovoid introvert about as large as the ovoid trunk. The latter is filled by an extensively coiled gut and, externally, is covered by around twenty longitudinal plates, as in larval priapulans and larval and adult loriciferans.

Priapulan-like scalids, teeth, and hooks are common as small carbonaceous fossils from Cambrian shales from various parts of the world, ranging back to the late Fortunian Stage of the early Cambrian (Slater et al., 2017). Some of these isolated microfossils can reliably be assigned to taxa known from whole-body preservation, such as ottoiids (Smith et al., 2015) (see chapter 20).

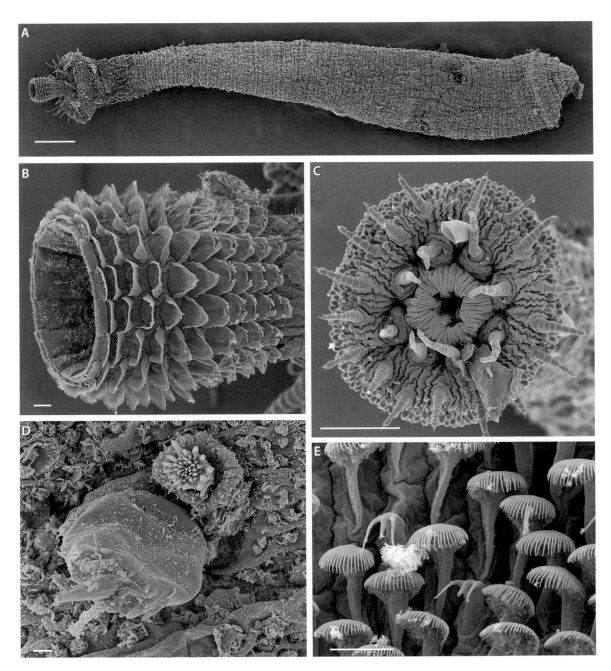

FIGURE 22.2. Scanning electron micrographs of anatomical details of the coelomate priapulan *Meiopriapulus fijiensis*. A, whole animal, scale 200 μm; B, detail of the everted pharynx, scale 10 μm; C, frontal view of the introvert and mouth, scale 100 μm; D, flosculus with tubule, scale 1 μm; E, locomotory scalids, scale 10 μm. Photo credits: Martin Vinther Sørensen.

innervated by intraepithelial longitudinal neurite bundles, running below each longitudinal row of scalids (Joffe and Kotikova, 1988; Rothe and Schmidt-Rhaesa, 2010). The brain may possess somata anterior and posterior of the neuropil, only anterior of the neuropil in the macroscopic species (see a summary in Schmidt-

Rhaesa and Rothe, 2014), or in the case of *Maccabeus tentaculatus*, somata are found medial of the neuropil (Por and Bromley, 1974).

Priapulans have a few types of sensory structures on their cuticle, most of them being mechano- or chemosensory. Scalids are perhaps the best-known sensory structures and generally include receptor cells, with a few exceptions in scalids that may have just a locomotory (fig. 22.2 E) or mechanical function (Storch, 1991). Receptor cells are also found in the buccal papillae of *Halicryptus spinulosus* (Storch et al., 1990). Flosculi (fig. 22.2 D) are another major sensory organ—a flower-shaped structure with a central cilium and seven surrounding microvilli—present in the neck region of larvae and on the trunk of adults and larvae (Schmidt-Rhaesa, 2013b).

Anatomically, the digestive system of priapulans is straight, divided into three regions, an ectodermal foregut, an endodermal midgut, and an ectodermal hindgut, ending in an anus subterminally, while the mouth is terminal. Recent developmental data, however, suggest that the anlage of the mouth is formed ventrally, by invagination, and shifts to a terminal anterior position as the ventral anterior ectoderm differentially proliferates (Martín-Durán and Hejnol, 2015). These authors thus propose that a midgut formed by a single endodermal population of vegetal cells, a ventral mouth, and the blastoporal origin of the anus (Martín-Durán et al., 2012a) are ancestral features in Ecdysozoa. This blastoporal origin of the anus had until recently been interpreted as a strictly deuterostome character.

The body cavity of priapulans can be large in the larger species, sometimes extending into the caudal appendage. While early studies suggested that the large body cavity of *Priapulus caudatus* was a coelom lined by a peritoneum (Shapeero, 1961), it is in fact a primary body cavity or pseudocoelom, with the exception of a small coelomic cavity around the foregut of *Meiopriapulus fijiensis*, a species unique in several other aspects (Storch et al., 1989a). While coeloms (or secondary body cavities) have been discussed earlier, a primary body cavity is an internal cavity lined by extracellular matrix (ECM) but not by an epithelium. The body cavity contains two types of free cells or coelomocytes: amoebocytes and erythrocytes. The latter contain hemerythrin, a respiratory pigment, although it has low affinity for oxygen in *P. caudatus* and thus may be related to their abundance in colder waters (Weber et al., 1979) or may not play an important role in respiration (Weber and Fänge, 1980). Whether the caudal appendage has a function in respiration remains a matter of debate (Schmidt-Rhaesa, 2013b), given the contradictory observations of Lang (1953) and Fänge and Mattisson (1961).

The excretory (protonephridia) and reproductive systems are combined into a urogenital system in which the excretory structure develops prior to the reproductive ones. Protonephidia are probably mesodemally derived (A. Hejnol, pers. comm.).

Priapulans are gonochoristic, but parthenogenesis is postulated in *Maccabeus tentaculatus*. Fertilization is external in the larger species that have typical ectaquasperm with round heads (Afzelius and Ferraguti, 1978b). It is probably internal in *Tubiluchus* species, with elongate sperm heads (Alberti and Storch, 1983; Storch and Higgins, 1989), and in the direct developer *Meiopriapulus fijiensis*, which also has round-headed sperm cells (Storch et al., 1989a) and has been shown to brood its embryos (Higgins and Storch, 1991).

The burrowing behavior has been studied in *Priapulus caudatus*, showing that both contraction of the longitudinal and circular muscles of the body wall and hydrostatic pressure are responsible for locomotion, often associated with feeding (Hammond, 1970, 1980), whereas movement in the larvae is limited by the presence of a lorica.

GENOMICS

Only a single unpublished low-quality genome is currently available, that of *Priapulus caudatus*, supplemented by just a few published transcriptomes (Borner et al., 2014; Laumer et al., 2015a), and EST libraries (Dunn et al., 2008). Mitochondrial genomes are available for two species (Webster et al., 2007; Rota-Stabelli et al., 2010).

FOSSIL RECORD

As discussed under Ecdysozoa (chapter 20), various Cambrian vermiform ecdysozoans known from Burgess Shale-type compression fossils have been placed in the priapulan stem group in cladistic analyses, but they have typically been shifted to alternative positions in "Cycloneuralia" under different analytical conditions (e.g., Wills et al., 2012). Despite the instability of such possible stem-group forms, and the likelihood that an extinct grouping of "archipriapulids" is polyphyletic (Wills et al., 2012), a few Cambrian species have been the focus of comparison with the priapulan crown group. In particular, the Chengjiang taxa *Xiaoheiqingella peculiaris* and *Yunnanpriapulus halteroformis* (Han et al., 2004; Huang et al., 2004a), *Paratubiluchus bicaudatus* (Han et al., 2004), and *Eximipriapulus globocaudatus* (Ma et al., 2014b) have nested within the priapulan crown group in some cladistic analyses (Ma et al., 2014b), although some of them are relegated to the priapulan stem group or assigned elsewhere in Scalidophora in other analyses (Wills et al., 2012). In the case of *Xiaoheiqingella/Yunnanpriapulus*, such characters as pentagonally arranged pharyngeal teeth and slender caudal appendage(s) have been argued to support affinities to Priapulidae and Tubiluchidae (Huang et al., 2004a). Whether *Paratubiluchus* is a member of Tubiluchidae (Han et al., 2004) or Priapulidae (Huang et al., 2006) has been debated, though in either case it would be a crown-group priapulan.

The fossil record of priapulans is very limited after the Burgess Shale-type preservational window closes, their post-Cambrian history consisting of *Priapulites koneconium* from the Upper Carboniferous of Mazon Creek, Illinois (Schram, 1973).

KINORHYNCHA

23

Discovered by French naturalist Félix Dujardin, as an intermediate between crustaceans and "vermes" (Dujardin, 1851), and with slightly over 270 described species to date, Kinorhyncha continues to be an understudied group. Kinorhynchs, or "mud dragons," are small (from 0.12 to 1.2 mm in body length), exclusively marine, meiofaunal ecdysozoans found around the world. They typically inhabit marine sediments from the intertidal zone to abyssal depths, the deepest records being from 7,800 m in the Atacama Trench off Chile (Danovaro et al., 2002; Adrianov and Maiorova, 2015). Live kinorhynchs resemble miniaturized segmented priapulans, with their protrusible introvert being continuously withdrawn and everted, while displaying their scalids and spines. Kinorhynchs, like the members of Panarthropoda, are segmented, although this segmentation is not seen in

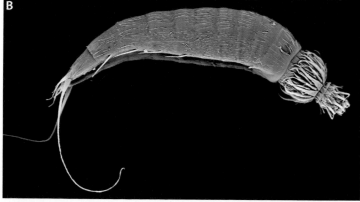

FIGURE 23.1. Illustrations of two cyclorhagid species: A, light micrograph of *Sphenoderes poseidon*; B, scanning electron micrograph of *Echinoderes charlotteae*. Photo credits: Martin Vinther Sørensen.

some internal structures, like the anterior longitudinal muscles. Nonetheless, they constitute an ideal system for studying independent origins of segmentation within ecdysozoans.

SYSTEMATICS

Since the earliest monograph by Zelinka (1928), kinorhynchs have received taxonomic treatment in a series of revisions (e.g., Sørensen and Pardos, 2008), and most authors divide them into two main clades, Cyclorhagida and Homalorhagida (e.g., Adrianov and Malakhov, 1995a). Phylogenetic work based on molecular data analysis of the nuclear genes *18S rRNA* and *28S rRNA* suggested polyphyly of Cyclorhagida and paraphyly of Homalorhagida (Yamasaki et al., 2013). A recent comprehensive phylogeny of the group, based on a combined analysis of morphology and nuclear rRNAs (Sørensen et al., 2015), provided the classification system followed here. Even more detailed analyses are available for certain kinorhynch subclades, such as Pycnophyidae (Sánchez et al., 2016).

Class Allomalorhagida
    Family Dracoderidae
    Family Franciscideridae
    Family Pycnophyidae
    Family Neocentrophyidae
Class Cyclorhagida
    Order Echinorhagata
        Family Echinoderidae
    Order Kentrorhagata
        Family Antygomonidae
        Family Cateriidae
        Family Centroderidae
        Family Semnoderidae
        Family Zelinkaderidae
    Order Xenostomata
        Family Campyloderidae

KINORHYNCHA: A SYNOPSIS

- Bilaterally symmetrical worms with introvert, neck, and eleven trunk segments as adults
- Chitinous cuticle and discrete cuticular plates; the cuticle is molted
- Protrusible introvert with a ring of scalids, with inner and outer retractors attached through a collar-shaped brain
- Sensorial spots

- Triradiate pharynx with radiating musculature
- Complete through-gut
- Pseudocoelomate or acoelomate; without circulatory or respiratory structures
- One pair of protonephridia
- Muscular system showing a mix of segmental and nonsegmental muscles
- Nervous system with a neuropil ring and a paired, ganglionated ventral nerve cord

The most salient feature of the kinorhynch body plan is its external segmentation (fig. 23.1) with repeated cuticular rings (Kozloff, 1972), sometimes called "zonites," but this term is not used in modern studies. While the cuticle contains chitin, it is not known what type of chitin is present. Adult kinorhynchs have a body divided into three regions, as in other scalidophorans: a head divided into a mouth cone and an introvert (fig. 23.2 A) that can be completely withdrawn into the trunk, a neck region that acts as a closure apparatus when the head is withdrawn, and 11 trunk segments, delimited by an external cuticular ring. Each cuticular ring (composed of up to four plates) is similar to those of arthropods, with thicker dorsal and ventral regions in each segment, separated by intersegmental regions with a thin cuticle. The anterior margin of segments 2–11 is thickened interiorly, as it also is paraventrally in many species, forming apodemes called "pachycycli" ("pachycyclus" in singular) (Neuhaus, 2013), for muscle attachment. In species with a thick cuticle, a ball-and-socket articulation may be expressed in the pachycyclus of segments 2–10 consisting of a sternal cuticular ring and a tergal cuticular process (Neuhaus, 2013).

According to some authors, segmentation in Kinorhyncha likely evolved from an unsegmented ancestor, as evidenced by the continuous longitudinal musculature in anterior segments 1–6 of the trunk, the continuous pharyngeal bulb protractors and retractors throughout the anterior segments, and the lack of segmentation in the digestive system, coupled with the absence of ganglia in most segments (Altenburger, 2016), but the anterior segments also have segmental dorsal, ventral, and diagonal muscles (Herranz et al., 2014).

Externally, the introvert has seven rows of scalids, but the nomenclature of the different types of scalids (spinoscalids, trichoscalids, etc.) has been debated by specialists (see Higgins, 1990; Neuhaus, 2013; Sørensen et al., 2015). It is now agreed that the first ring contains primary spinoscalids, that spinoscalids are found in rings 2–7, and that trichoscalids are found in the most posterior ring. The type and arrangement of the scalids is of taxonomic importance. The introvert bears the pigmented eyespots, which vary in number across species.

The protrusible mouth cone bears 9 outer oral styles and 20 inner oral styles in 3 rings, which become visible in the protruded introvert, although these numbers may vary. The neck may present a series of small, trapezoidal to triangular cuticular plates with rounded edges, termed "placids," also of taxonomic importance (Neuhaus, 2013).

The trunk may be triangular, circular, or oval in cross-section, with the internal musculature pulling the ventral cuticular plates toward the interior, conferring upon the animal the aspect of having a midventral depression. The cuticular rings

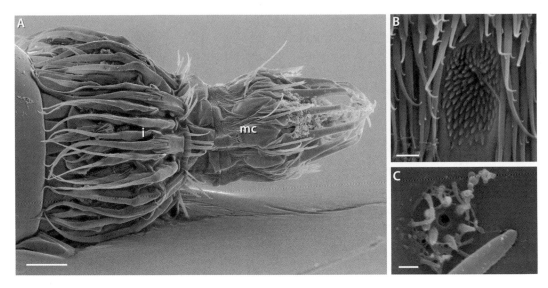

FIGURE 23.2. Scanning electron micrographs of *Echinoderes*: A, head with clearly delimited mouth cone (mc) and introvert (i) regions of *E. bookhouti*, scale 10 μm; B, detail of a regular sensory spot in *E. bookhouti*, scale 1 μm; C, flosculus-like sensory spot of *E. intermedius*, scale 100 nm. Photo credits: Martin Vinther Sørensen.

may be divided into four plates, with the ventral (= sternal) and dorsal (= tergal) plates being thickest, but some variation in the number of plates exists. There is a cuticular flap that overlaps in segments 1–10. The cuticle of the terminal segment may end in sternal and tergal spinose, bladelike, or truncate extensions (Neuhaus, 2013). In addition to pores and other cuticular structures, kinorhynchs present sensory spots (fig. 23.2 B–C), a type of ciliary receptor generally arranged symmetrically that is supposed to be homologous to the flosculi of other scalidophorans, but most species lack the typical flower shape that characterizes the flosculus. Each sensory spot appears as a round to oval or droplet-like area with one to three pores and numerous cuticular micropapillae; a cilium may jut out of a pore. An intracuticular conical to flask-shaped atrium or a subcuticular cuticular tube with or without a basal, funnel-like extension is usually found (Neuhaus, 2013). Multiple types of sensory spots have been characterized, and are thought to be chemoreceptors. There are also gland cells, sometimes showing sexual dimorphism. Appendages, in the form of several types of spines, are of great taxonomic significance.

Internally, kinorhynchs are characterized by longitudinal and dorsoventral segmental muscles and by circular musculature around the mouth cone (Müller and Schmidt-Rhaesa, 2003; Schmidt-Rhaesa and Rothe, 2006; Herranz et al., 2014; Altenburger, 2016). The trunk musculature consists of paired ventral and dorsal longitudinal muscles in segments 1–10, as well as dorsoventral and diagonal muscles in most trunk segments. Dorsal and ventral longitudinal muscles insert on cuticular apodemes within each segment. Strands of longitudinal musculature extend over segment borders in segments 1–6, while in segments 7–10 the trunk musculature is confined to the segments (Altenburger, 2016). Cell processes of the dorsoventral, longitudinal, and some circular and dilator muscles extending toward nerve cells, similar to the muscle arms of nematodes (see chapter 26), have been described (Neuhaus, 1994; Neuhaus and Higgins, 2002).

The musculature of the digestive system comprises a strong pharyngeal bulb with attached mouth-cone muscles, as well as pharyngeal bulb protractors and retractors, and shows no sign of segmentation (Altenburger, 2016).

Kinorhynchs possess one pair of complex protonephridia in the 8th trunk segment, each formed by up to 25 cells in the adult (Neuhaus, 1988), with the nephropore opening on the 9th trunk segment. These protonephridia have a peripheral cytoplasmic wall in each terminal cell, considered to be plesiomorphic by Neuhaus (1988), while the compound filter composed of several terminal cells and with a common filtration area, and the biciliarity of the protonephridial cells, are considered apomorphic for kinorhynchs.

The nervous system of kinorhynchs has been studied using transmission electron microscopy (Nebelsick, 1993) as well as immunohistochemistry (Herranz et al., 2013; Altenburger, 2016; Herranz et al., 2019). The nervous system comprises a neuropil ring anterior to the pharyngeal bulb and associated flask-shaped serotonergic somata extending anteriorly and posteriorly. A paired, ganglionated ventral nerve cord within segments 1–9 is connected to the neuropil ring and runs toward the anterior until an attachment point in segment 1, and from there toward the posterior.

Sensory receptors include scalids, sensory sensillae and sensory spots. The number of sensory structures increases through development (Neuhaus, 1993). Three types of these paired sensory spots are recognized in adults and juvenile stages. Microvillar pigmented eyespots with lenses are present on the pharyngeal nerve in some species (see above), but pigmentation disappears during fixation (Neuhaus and Higgins, 2002). A pair of cephalic sensory organs located at the base of the first ring of scalids has also been reported for some species (Neuhaus, 1997). Interestingly, a similar organ is also found in Gastrotricha.

The introvert works as a complex sensorial and locomotor apparatus with functional regionalization. The first row of scalids is highly innervated, playing a locomotory and sensorial role, sometimes carrying a variable number of eyes, whereas the remaining scalids are less innervated, combining secretory and sensorial functions. The mouth cone, also highly innervated, is used for feeding and food processing. The organization of the nervous system suggests that separate muscular systems are involved with feeding and locomotory behaviors.

Kinorhynchs have internal fertilization with direct transfer of spermatophores from the male to the female, although the process of fertilization remains unknown. The spermatophore is presumably extruded through the male gonopore and directed toward the female by the ductless penile spines, which are pressed directly against the cuticular plates of the female and usually cover the female gonopores, as observed by Brown (1983). The spermatophore contains a mass of intertwined spermatids and spermatozoa surrounded by clear material covered with a layer of debris. In the female, spermatozoa are found lodged in the seminal receptacle tissue applied to the dorsal aspect of posterior oocytes, where the spermatozoa complete their development. During this process, nuclei change from filiform to geniculate, and oval corpuscles surrounding the nuclei disappear, so that the spermatozoa are seen as densely packed, polyhedral cells (Brown, 1983), conferring on them the aberrant spermatozoan aspect previously reported in the literature (Nyholm and Nyholm, 1982).

Despite previous claims of indirect larval development, kinorhynchs have direct development, with the addition of two segments after hatching (Kozloff, 1972) through a series of five to six molts. Egg cleavage is, however, poorly known, with a single report having studied the development of *Echinoderes kozloffi* using light microscopy of unstained embryos (Kozloff, 2007).

GENOMICS

Little is known about kinorhynch nuclear genomes, with no size estimate or sequences publicly available. And only recently have two mitochondrial genomes been published (Popova et al., 2016): *Echinoderes svetlanae* (Cyclorhagida) and *Pycnophyes kielensis* (Allomalorhagida). Their mitochondrial genomes are circular molecules approximately 15 Kbp in size, with the typical bilaterian complement of 37 genes, which are all positioned on the major strand, but the gene order is distinct and unique among Ecdysozoa (Popova et al., 2016), including duplicated methionine tRNA genes.

FOSSIL RECORD

A putative fossil kinorhynch, *Eokinorhynchus rarus*, has been described from three-dimensional fossils that are among the oldest known bilaterian body fossils, from Fortunian strata in China dated to ca. 535 Ma (Zhang et al., 2015): two additional forms are also known from the same unit, and possible ontogenetic stages have been associated (Zhang et al., 2017a). Phylogenetic analysis resolved *E. rarus* as sister taxon to extant Kinorhyncha, but support for this grouping is weak, and the specimens lack clear kinorhynch synapomorphies. The relationship to kinorhynchs is based on the small number of elongated trunk annuli, but these are not like the cuticular rings of extant kinorhynchs, more closely resembling the cuticular annulation of priapulans or onychophorans. Identity as a total-group scalidophoran is at least secure (but see the discussion on scalidophoran monophyly). Three species of kinorhynch-like scalidophorans from the Qingjiang biota (Cambrian Stage 3) await formal description (Fu et al., 2019).

LORICIFERA

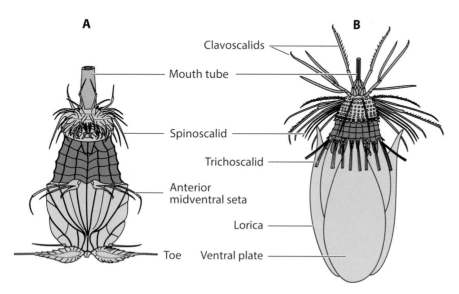

24

Loricifera, referred to by its discoverer as "jewel animals," is an exclusively marine, free-living meiofaunal group of miniaturized animals with complex life cycles (fig. 24.1). Described by Danish zoologist Reinhardt Møbjerg Kristensen in 1983 (Kristensen, 1983), this group now includes 38 named species (Higgins and Kristensen, 1986; Kristensen, 2016; Neves and Kristensen, 2016; Neves et al., 2018) and a handful of undescribed ones. The first specimens were found nearly simultaneously by Higgins and Kristensen in North Carolina (U.S.) and Helsingør (Denmark), respectively, in 1974 (Higgins and Kristensen, 1986), and the first Hamlet larva of a loriciferan was reported in 1975 (see a historical account of the discovery in Kristensen, 1984). The first Danish specimen was lost during preparation for transmission electron microscopy, but additional specimens, of other species, appeared in subsequent years from Greenland. At the time, these were juveniles and thus interpreted as some sort of larva, perhaps related to priapulans. The first adults were finally discovered in Roscoff (France), leading to the description of the phylum in 1983.

Most loriciferan specimens have been collected at depths between 20 and 450 m in coarse marine sediments, but one species, *Pliciloricus hadalis*, was collected in the Izu-Ogasawara Trench, western Pacific, at a depth of over 8,000 m (Kristensen and

Clavoscalids

Mouth tube

Spinoscalid

Trichoscalid

Anterior
midventral seta

Lorica

Toe   Ventral plate

FIGURE 24.1. Schematic drawings of two life stages of loriciferans in ventral view: A, the Higgins larva of *Nanaloricus mysticus* in ventral view; B, adult male *Armorloricus*. Green represents the mouth cone, blue the introvert, dark brown the thorax, and light brown the abdomen. Originals courtesy of Ricardo Neves.

Shirayama, 1988). Loriciferans have been shown to be the first metazoans without mitochondria and able to complete their life cycle in complete anoxia in the deep hypersaline anoxic basins of the Mediterranean Sea (Danovaro et al., 2010; Neves et al., 2014b)—a claim that has been disputed (Bernhard et al., 2015) but subsequently confirmed (Danovaro et al., 2016). This condition has been proposed to be possible through the presence of hydrogenosome-like organelles similar to those present in some obligate anaerobic unicellular eukaryotes (Mentel and Martin, 2010), expanding the universe of possible metazoan life (Levin, 2010).

A recent update of the 30 years of knowledge of the phylum and its terminology was provided by Neves et al. (2016), who stressed that many aspects of loriciferan ecology, such as their behavior and feeding habits, remain unknown, but they may feed upon bacteria or unicellular algae. Few scientists (including the authors of this book) have observed these minuscule animals alive, and this was first possible after applying a brief freshwater shock technique followed by reintroduction into seawater (Neves et al., 2016).

SYSTEMATICS

Few studies have focused on the internal relationships of Loricifera, and none have analyzed explicit data sets. Currently 38 species have been described in eleven genera (Neves et al., 2016; Neves et al., 2018), but many remain undescribed (Heiner Bang-Berthelsen et al., 2013; Pardos and Kristensen, 2013; Neves et al., 2014b; Neves et al., 2016). The described species are currently assigned to three families, with the genus *Tenuiloricus* remaining of uncertain affinity. The group is in need of phylogenetic work.

Order Nanaloricida
    Family Nanaloricidae
        Nanaloricus
        Armorloricus
        Phoeniciloricus
        Spinoloricus
        Culexiregiloricus
        Australoricus
    Family Pliciloricidae
        Pliciloricus
        Rugiloricus
        Titaniloricus
    Family Urnaloricidae
        Urnaloricus

    Incertae sedis
        Tenuiloricus

LORICIFERA: A SYNOPSIS

- Marine meiobenthic triploblasts, probably pseudocoelomate
- Body divided into a mouth cone, head, neck, thorax, and abdomen; abdomen covered by a lorica
- Head and neck with 8 to10 rows of scalids: 1 ring of clavoscalids and 8 rings of spinoscalids on the head and 1 ring of trichoscalids on the neck
- Complete through-gut
- Protonephridia combined into a urogenital system
- Dioecious, sometimes with sexual dimorphism; sometimes hermaphrodites
- Complex life cycle, generally with Higgins larva, sometimes with alternating sexual and asexual phases

Loriciferans are minuscule animals (85–800 µm in size) showing a highly regionalized body, with five main body regions. The head (introvert), which is everted by the body cavity pressure and retracted by musculature, bears a mouth cone, the anterior of which is telescopic, and multiple rows of scalids and spines. Loriciferans can bear 200 to 400 specialized sensory and locomotory appendages in the head region, each operated by tiny muscles and complex cuticular structures. These include 9 rows of scalids, the first bearing clavoscalids—club-shaped—of probable chemosensory function. These are followed by 2 to 8 rows of spinoscalids, of locomotory and mechanoreceptor functions. The characteristic Higgins larva bears 6 to 12 oral stylets.

The trunk region is formed by a thorax and abdomen; the abdomen is covered by a lorica formed by longitudinal folds (plicae) or multiple sclerotized plates. The Higgins larva bears setae and locomotory spines. The trunk is separated from the head region by the neck. The neck bears trichoscalids, a series of flattened single or double scalids, attaching to a basal plate, often characterized by a longitudinal median ridge and serrated lateral edges. They have either 15 single trichoscalids or 8 single plus 7 double. Trichoscalids can have a locomotory function, but no intrinsic muscles have been described, and thus it is suggested that their movement is produced by the circular musculature of the neck (Neves et al., 2013). The head, neck and thorax can withdraw into the lorica.

The lorica of loriciferans has attracted attention because it is thought to be homologous to the lorica of larval priapulans, and some have argued that loriciferans are just progenetic priapulans (Warwick, 2000), although the ultrastructure of these two loricas differs in some important aspects (Neuhaus et al., 1997b), and *Meiopriapulus* larvae lack a lorica.

The digestive tract is complete, with a long, cuticular buccal canal, a muscular triradiate pharynx, an esophagus, a long midgut, and a cuticular rectum, with the anus located on an anal cone. One pair of salivary glands opens dorsally into the buccal canal.

Ultrastructural data on the musculature of loriciferans are poor (Kristensen, 1991), but more recently immunohistochemical techniques have been applied to study the myoanatomy of adult *Nanaloricus* species and the Higgins larva of

Armorloricus elegans (Neves et al., 2013). Their somatic musculature is very complex and includes several muscles arranged in circular, transverse, and longitudinal orientations. In adult *Nanaloricus*, the introvert is characterized by a netlike muscular arrangement composed of thin circular fibers crossed by several thin longitudinal fibers with bifurcated anterior ends. Two sets of muscles surround the prepharyngeal armature: six buccal tube retractors and eight mouth cone retractors (Neves et al., 2013). Additionally, a thick, circular muscle marks the neck region, and a putative anal sphincter is found posteriorly. In the Higgins larva, two circular muscles are distinguished anteriorly in the introvert: a dorsal semicircular fiber and a thin ring muscle; the posterior-most region of the body is characterized by an anal sphincter and a triangular muscle (Neves et al., 2013). Loriciferans are unique among scalidophorans in having internal muscles in the spinoscalids (Kristensen, 1991). Overall, the myoanatomy of Loricifera is more similar to Kinorhyncha and Nematomorpha than to Priapulida (Neves et al., 2013).

The central nervous system of loriciferans has been studied using transmission electron microscopy (Kristensen, 1991), but no immunohistochemical data are publicly available. It includes a large circumpharyngeal brain located in the introvert, around the buccal tube or the esophagus (Schmidt-Rhaesa and Rothe, 2014; Schmidt-Rhaesa and Henne, 2016). Results of 5HT-immunoreactivity in the whole brain do not show any other ganglion or nerve cords (R. Neves, pers. comm.). Additional smaller ganglia are associated with the other various body regions, both anterior and posterior of the neuropil (Schmidt-Rhaesa and Henne, 2016). A large ventral nerve cord also bears ganglia. At least some scalids are probably sensory in function; however, it has been shown that at least the spinoscalids are involved in a jumping behavior (Neves et al., 2013; Kristensen, 2016). Flosculi and some setae (in the larvae) are also present as sensory structures (Heiner Bang-Berthelsen et al., 2013). A photoreceptor organ—two large, pigmented spots—has been found in the Higgins larva of *Australoricus oculatus* (Heiner et al., 2009). The fact that the eyespots disappeared after fixation may indicate that photoreceptors are present in other species of loriciferans.

LIFE CYCLES

Loriciferans show a diversity of life cycles that may include dioecy (with sexual dimorphism in their scalids) or hermaphroditism in the adult phase and a series of larval and postlarval stages. While a Higgins larva (with two to five instars) is found in most species, others are particular to certain life cycles, including the ghost larva of *Rugiloricus bacatus* and *Pliciloricus diva*, or the mega-larva of *Urnaloricus gadi*; the Shira larva of *Tenuiloricus shirayamai* putatively replaces the Higgins larva in this species (Neves et al., 2016). A postlarva is known from about half of the loriciferan species.

The first life cycle to be elucidated was that of the dioecious species *Nanaloricus mysticus* (fig. 24.2). All known Nanoloricidae have a simple life cycle with only sexual reproduction and dioecious adults. After fecundation the egg develops into a Higgins larva, with two caudal locomotory appendages referred to as "toes." This larva molts multiple times until becoming a postlarva, which gives

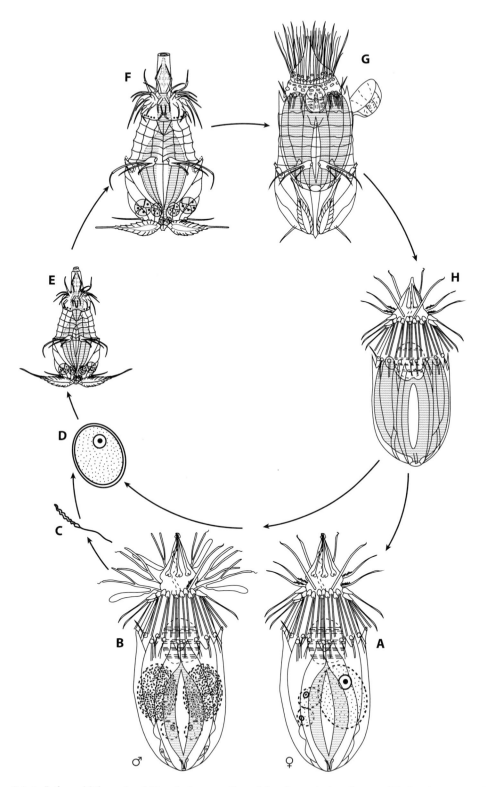

FIGURE 24.2. Inferred life cycle of *Nanaloricus mysticus*: After fecundation the egg (D) develops into a Higgins larva (E, F), which molts multiple times and metamorphoses (G) to become a postlarva (H), which gives rise to the adult female (A) or male (B), which produce eggs (D) and sperm (C), respectively. After Kristensen (1991).

rise to the adult male or female (Kristensen et al., 2007). Fertilization and embryogenesis have not been observed, but sperm cells have been found in the two seminal receptacles of a female (Kristensen, 1992; Kristensen and Brooke, 2002).

The members of Pliciloricidae have a variety of life cycles, including sexual and asexual phases, with internal or external maturation of a paedogenetic ghost larva (Heiner, 2008; Gad, 2009), and a hermaphroditic stage has been reported in a species of the family (Kristensen et al., 2013).

The life cycle of *Urnaloricus gadi* is difficult to understand and has been called viviparous paedogenesis, in which the larvae absorb the tissues from the maternal stage (i.e., a ghost larva). After escaping the maternal stage, the free-living Higgins larva undergoes a sequence of molts until a last Higgins larval stage is generated. This larval stage apparently gives rise to the pre mega-larva, probably by a process of metamorphosis, and the pre mega-larva, which contains a large ovary with few oocytes, presumably molts into a cyst-forming mega-larva. The maturation process of the ovary continues, while the cyst-forming mega-larva molts into a postlarva. This is a very brief stage because it molts immediately into a ghost larva, which rests inside the empty cuticles of both the postlarva and the cyst-forming mega-larva (the outermost cuticle layer). The eggs inside the ovary of the ghost larva mature into embryos, while feeding upon the cell tissue of the ghost larva. Subsequently, the Higgins larva hatches and escapes from the cystlike structure (Heiner and Kristensen, 2008; Neves et al., 2016).

GENOMICS

Only a relatively low-quality transcriptome of *Armorloricus elegans* is available (Laumer et al., 2015a). Information is lacking on any mitochondrial gene, making Loricifera probably the only animal phylum without even a single sequence of cytochrome *c* oxidase subunit I—the so-called universal barcode marker. In addition, a lack of mitochondria has been reported in some species inhabiting anoxic environments (Danovaro et al., 2010; Danovaro et al., 2016).

FOSSIL RECORD

The Shira larva of *Tenuiloricus shirayamai*, collected from depths over 3,000 m (Neves and Kristensen, 2014), shows some morphological similarities with the Cambrian fossil *Orstenoloricus shergoldii* (Maas et al., 2009). The latter was recognized as part of a putative loriciferan–priapulan clade in its original description, citing pairs of spines on the anterior and posterior parts of the lorica as a character most consistent with closest affinities to Loricifera. It has subsequently been accepted as a loriciferan by experts (Kristensen, 2017), although affinities to Priapulida have also been advocated (Harvey and Butterfield, 2017).

The least ambiguous Cambrian loriciferan is *Eolorica deadwoodensis*, a meiobenthic species extracted as small carbonaceous fossils from late Furongian (late Cambrian) sediments in western Canada (Harvey and Butterfield, 2017). This species exhibits the characteristic high number of scalids of Loricifera, typical spinoscalid morphology, and a loricate body. The latter particularly resembles the members of the extant family Pliciloricidae in its large number of plicae and opens up the possibility that *Eolorica* may be a crown-group loriciferan. The diversification of Loricifera in the meiofauna can thus be dated to at least 485 Ma.

Additional Cambrian fossils that may be allied to Loricifera are two species from the early Cambrian Sirius Passet Konservat-Lagerstätte of Greenland assigned to the genus *Sirilorica* (Peel, 2010a, b; Peel et al., 2013). They are represented by loricate fossils up to 80 mm in length, orders of magnitude larger than extant Loricifera. In the more common and better-known *S. carlsbergi*, the lorica consists of two interdigitating circlets each of seven elongate plates. In some specimens, longitudinal muscle bands are preserved at the margins of the plates, indicating that the lorica could be expanded and contracted. The introvert is conical and bears a single circlet of multicusped teeth in the usual position of a row of scalids. A maximum of six such teeth is observed, prompting a comparison to the hexaradial symmetry of the buccal tube retractors in Loricifera. The terminal parts of the body are represented by a slender, tapering mouth tube and a conical anal field. A few small plicate lorica in the Sirius Passet biota have been identified as larvae or postlarvae of *Sirilorica* (Peel et al., 2013). Identification of *Sirilorica* as a total-group loriciferan can be supported by a well-developed lorica in the adult and the extended mouth tube.

25

NEMATOIDA

Nematoda (roundworms) and Nematomorpha (horsehair worms) constitute a clade named Nematoida (Schmidt-Rhaesa, 1996c) (= Nematoidea of other authors). Nematoida was proposed based on clear synapomorphies of the cuticle and the body wall musculature and is almost universally supported by molecular data, both from Sanger-based methods and phylogenomics. Morphologically, adult nematodes and nematomorphs share a special cuticle, mostly without chitin—chitin is restricted to the cuticular pharynx of nematodes and to the juvenile cuticle of nematomorphs—with layers of crossing collagenous fibrils (Nielsen, 2012a). They also lack circular musculature in the body wall, and thus their erratic movement is dictated by the longitudinal muscles, sharing a similar type of musculature (Meyer-Rochow and Royuela, 2018). Protonephridia are also absent in all nematoids. The lack of cilia in their sperm cells has also been listed as a putative synapomorphy of the clade, but other similarities between the sperm cells are difficult to derive. Many features, however, distinguish the members of these two phyla. Nematomorphs include just a handful of species all with very similar parasitic life cycles that rely on different members of Arthropoda (marine decapods in the case of *Nectonema* and terrestrial arthropods for the gordioids) (Schmidt-Rhaesa, 2013a). Nematodes display incredible adaptability to nearly all habitats on Earth (marine and terrestrial), including an infinity of free-living forms and the many adaptations to parasitic lifestyles, both in plants and animals (Meldal et al., 2007). They are thus a classical example of a hyperdiverse taxon. Nematoda, with tens of thousands of described species and some estimates that they may contain millions, is sister group to a depauperate clade, Nematomorpha, with ca. 360 named species.

Despite its well-corroborated monophyly, the position of Nematoida with respect to the other ecdysozoan phyla remains contentious. Nematodes share a basic organization of the anterior nervous system with the so-called scalidophoran taxa, characters that have lent support for Cycloncuralia, but the ringlike neuropil that defines a "cycloneuralian" brain is not shared by nematomorphs (Schmidt-Rhaesa and Rothe, 2014). Furthermore, the central nervous system organization of cycloneuralians may actually be plesiomorphic for Ecdysozoa, as molecular evidence for the monophyly of Cycloneuralia is mostly lacking. For example, some molecular analyses support a relationship of Tardigrada and/or Loricifera to Nematoida (Borner et al., 2014), or a relationship of Nematoida to Panarthropoda (Campbell et al., 2011; Rota-Stabelli et al., 2013), both contradicting the clade Cycloneuralia.

GENOMICS

Nematoid genomics is well advanced, thanks to the pioneering work done with the model species *Caenorhabditis elegans*, which was the first animal genome to be sequenced (*C. elegans* Sequencing Consortium, 1998). Due to the medical and agricultural interest of many species, more than a hundred other nematode genomes are currently publicly available in NCBI, and an initiative to sequence large numbers of nematode genomes was already put in place in 2012 (Kumar et al., 2012; International Helminth Genomes Consortium, 2019). No nematomorph nuclear genome is yet publicly available, but a draft genome of *Gordionus alpestris* is underway by Efeykin and collaborators.

FOSSIL RECORD

The Cambrian–Silurian Palaeoscolecida are discussed under the fossil record of Ecdysozoa (chapter 20), their posterior hooks having been used as a possible apomorphy with Nematomorpha (Hou and Bergström, 1994), and a systematic position within Nematoida has been recovered in some phylogenetic analyses (Wills et al., 2012; Shao et al., 2016). Another character proposed for nematoid affinities of palaeoscolecids is a system of large, helically wound crosswise fibers in the innermost layer of the cuticle (Harvey et al., 2010). In nematomorphs and at least some nematodes, this is collagenous.

Some authors place the Cambrian "Orsten" *Shergoldana australiensis* in the stem group of Nematoida, based on similarities of its anterior region to that of nematomorphs (Maas et al., 2007b). The oldest unambiguous nematode, *Palaeonema phyticum*, from the Lower Devonian Rhynie Chert (Poinar Jr. et al., 2008; Dunlop and Garwood, 2018), is the earliest widely accepted crown-group nematoid.

NEMATODA

Nematodes, or roundworms, are among the most neglected animals, despite their ubiquity and abundance, as well as their agricultural and medical importance. They are agents of a large number of major tropical diseases, and the cause of more crop losses worldwide than all arthropods. Nematodes range in size from 82 μm in a free-living marine species, to over 8 m in a parasite of the placenta of the sperm whale (Decraemer et al., 2014), and can be found in most habitats, including the meiobenthos and soil, or within and on host animals and plants (Blaxter and Koutsovoulos, 2015). From a medical perspective, most of the human population experiences nematode parasitism during their lives, with estimates of one-quarter to one-third of the global population being infected at any time (Blaxter and Koutsovoulos, 2015).

There are many thousands of individual nematodes in a handful of garden soil, with some estimates of abundance suggesting that three-quarters or more of all animals on earth are nematodes (Moens et al., 2014). They are the dominant group of soil animals in grasslands (Wu et al., 2011). Metagenomic analyses of marine meiofaunal and tropical rainforest confirm that nematodes are indeed the most abundant "Operational Clustering of Taxonomic Units" in both environments (Creer et al., 2010). Nematodes are also the most abundant and the richest animal phylum in many marine sediments (Fonseca et al., 2010), including in the abyss (Schmidt and Martínez Arbizu, 2015). Some species of marine nematodes are covered in a coat of ectosymbiotic sulfur-oxidizing bacteria (Ott et al., 2004) (fig. 26.1 C–D).

We know that the nearly 26,646 named species (Hugot et al., 2001) constitute just a small fraction of their true diversity. For example, the 11,400 described marine species seem to be a mere 19% of the estimated marine diversity, with one of the lowest percentages of synonymy in marine animals (ca. 9%) (Appeltans et al., 2012). Estimates of marine species are ca. 61,400 (Appeltans et al., 2012), but these estimates probably do not take into account the large number of cryptic species recently discovered using genetic data (Moens et al., 2014). Reliable estimates of continental species are more difficult to obtain, and unsupported quotes often talk about half a million to more than a million species (Lambshead, 1993; Hugot et al., 2001), but these numbers have been questioned by other authors (Lambshead and Boucher, 2003). Yet some accounts even refer to 100 million species (Stock, 2016).

It seems that most terrestrial plants and larger animals are indeed associated with at least one species of parasitic nematode (Blaxter and Koutsovoulos, 2015). Estimates of the number of species of parasitic nematode per host suggest that there may be of the order of 24,000 nematode parasites of vertebrates alone (Dobson et al., 2008), most of which remain undescribed. Despite the supposed large numbers of insect parasites (obligate or facultative), limited taxonomic work on this group makes it nearly impossible to provide accurate estimates on the magnitude of nematode diversity.

Aside from the unaccounted megadiversity, or the medical and agricultural importance that have made Nematoda a highlight in invertebrate textbooks, one nematode, the free-living rhabditid *Caenorhabditis elegans*, is one of the best-known animals from genetic, developmental, neuroanatomical, and morphological points of view—and hence has earned the popular but unfortunate name "the worm" and is nearly always abbreviated as "*C. elegans*" in many scientific circles. Its simplicity, with 959 somatic cells in adult hermaphrodite individuals (one of the best-documented cases of eutely), promoted its status as a model organism in the 1970s by South African biologist Sydney Brenner, who, due to the low number of cells and the ability of doing forward mutations, established it as a genetic model for understanding the brain (Brenner, 1973, 1974). The complete cell lineage, from zygote to hatchling, was completed by Sulston et al. (1983). Brenner, Sulston and Horvitz shared the Nobel Prize in Physiology or Medicine in 2002 for their genetic and developmental work on *C. elegans*.

By the end of the twentieth century, *C. elegans* had become the first multicellular organism to have its entire genome sequenced (*C. elegans* Sequencing Consortium, 1998). This model organism also played a crucial role in the discovery of caspases (Yuan et al., 1993), RNA interference (Fire et al., 1998), and microRNAs (Lee et al., 1993). Nematodes are also prominent examples of two relatively uncommon biological phenomena, cryptobiosis, and eutely. The online resources available to understand this small organism are unparalleled apart from *Drosophila melanogaster*; Wormbase (wormbase.org), WormBook (wormbook.org), WormAtlas (wormatlas .org), and the Caenorhabditis Genetics Center (https://cbs.umn.edu/cgc/home) amass the global knowledge of *C. elegans*.

It is therefore ironic that one of the most poorly known animal phyla, at least from ecological and taxonomic points of view, contains what it is arguably the best-known animal at many levels, an animal amenable to transgenesis and a preferred study subject of neurobiologists, for which neuroscience has been able to understand neural circuits and behavior to an unparalleled degree (Sengupta and Samuel, 2009).

SYSTEMATICS

The difficulty to correctly identify and describe the rapidly increasing numbers of nematode taxa has discouraged many young scientists and diverted them toward more appealing disciplines (Coomans, 2000). This perception of the status of nematode taxonomy is not far from the truth, although the exact numbers of species described per year are difficult to compile, as descriptions are scattered throughout the zoological, biomedical, and agricultural literature.

Declining morphological expertise, in part due to the limited anatomical characteristics of nematodes, has left nematode systematics almost entirely in the hands of molecular biologists. Since the first comprehensive molecular phylogenetic studies of nematodes using *18S rRNA* (Aleshin et al., 1998; Blaxter et al., 1998), the

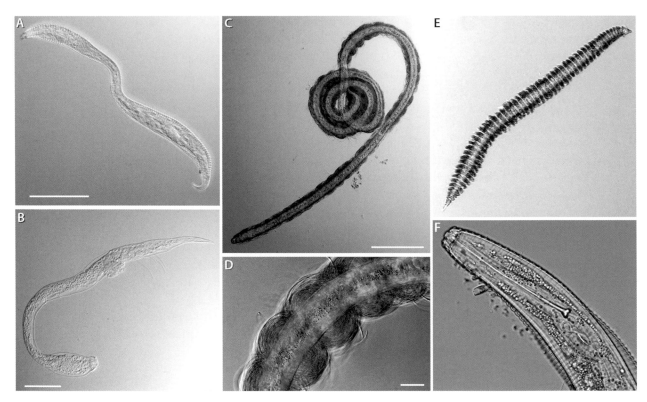

Figure 26.1. A, Chromadorida, *Epsilonema* sp., scale 20 μm; B, Chromadorida, Draconematidae sp., scale 40 μm; C, Desmodoridae, Stilbonematinae, notable for being completely covered in a coat of ectosymbiotic sulfur-oxidizing bacteria, scale 200 μm; D, detail of the coat of ectosymbiotic sulfur-oxidizing bacteria, scale 20 μm; E, Desmoscolecida, *Desmoscolex* sp.; F, detail of the anterior end of an unidentified nematode. Photo credits: A–B, E–F, Peter Funch; C–D, Katrine Worsaae.

number of available sequences has increased rapidly (Holterman et al., 2006; Meldal et al., 2007; Nadler et al., 2007; Holterman et al., 2009), and recent phylogenies have included ca. 1,200 terminals (van Megen et al., 2009).

Traditional classifications have used different characteristics of the group, including the system of Chitwood and Chitwood (1933), with a subdivision of the phylum into Aphasmidia and Phasmidia, which were later changed into Adenophorea ("gland bearers") and Secernentea ("secretors"), respectively, and for decades nearly all nematologists followed this main subdivision (see a recent historical account in van Megen et al., 2009). Lorenzen (1981) proposed the first phylogenetic scheme for nematodes based on cladistic principles, indicating the scarcity of informative characters for establishing a well-supported cladogram, and some other authors proposed alternative phylogenetic schemes (e.g., Malakhov, 1994). Since then, virtually all new schemes have been based on molecular data, or on a consilience of morphology with molecular information (De Ley and Blaxter, 2004), and have placed Secernentea as a derived clade within Chromadorea.

Many recent systems continue to use a division between the classes Enoplea and Chromadorea; the former is divided into the clades Enoplia and Dorylaimia and may not be monophyletic. Most studies support a division into three clades—

Chromadoria (= Chromadorea), Enoplia, and Dorylaimia—which subdivide into a number of clades, depending on the author. For example, Holterman et al. (2006) proposed a system of twelve clades on the basis of *18S rRNA* sequence data, each encompassing part, one or more, of the traditional nematode orders. A similar system based on clades and subclades has been endorsed by Mark Blaxter's work (e.g., Blaxter, 2011; Blaxter and Koutsovoulos, 2015). Additional subdivisions have been proposed by van Megen et al. (2009). Several traditional orders, including Desmodorida, Monhysterida, and Araeolaimida, have been shown to be polyphyletic (van Megen et al., 2009). While these molecular-based classification systems tend to be good for placing families in specific clades, a more useful higher-order classification may be needed.

An ordinal-level classification system with 21 orders that differs from the molecular classification has recently been proposed in the *Handbook of Zoology* (Schmidt-Rhaesa, 2014), the most up-to-date treatment on nematodes, but it does not provide any higher-level structure for these orders. More recently, a phylogenomic analysis based on 56 nematode genomes has provided a phylogenetic framework for a large part of the phylum (International Helminth Genomes Consortium, 2019), but fails to incorporate the diversity of the marine lineages. What we present here is a hybrid system, following the general structure of the phylogenetic analyses of molecular data—recognizing that they largely remain limited to a single marker—and some of the ordinal groups that have been used by nematode taxonomists. This system needs a great deal of refinement.

Class Enoplea[11]
    Subclass Enoplia (Clade 1 of van Megen et al., 2009)
        Order Enoplida
        Order Triplonchida
    Subclass Dorylaimia (Clade 2)
        Order Dorylaimida
        Order Mononchida
        Order Isolaimida
        Order Dioctophymatida
        Order Muspiceida
        Order Marimermithida
        Order Mermithida
        Order Trichinellida
  Class Chromadorea
    Subclass Chromadoria
        Order Chromadorida
        Order Desmodorida

[11] Enoplea may be diphyletic, but this is not well resolved.

Order Monhysterida

Order Araeolaimida

Order Plectida

Order Desmoscolecida

Spirurina

Myolaimina

Tylenchina

Rhabditina (Secernentea) (Clades 8 and 9)

▨ NEMATODA: A SYNOPSIS

- Pseudocoelomate or acoelomate unsegmented worms
- Body wall with longitudinal muscles; no circular musculature
- Eutely present at least in some organs
- Gut complete; mouth structures arranged radially
- Body typically round in cross-section and covered by a layered cuticle with collagen; chitin present in the pharyngeal cuticle of some species
- Unique cephalic sense organs (amphids) and some with caudal organs (phasmids)
- Unique secretory–excretory system; some with renette cells
- Without circulatory or gas exchange structures
- Ubiquitous; marine, limnic, terrestrial; parasitic on plants and animals

The body wall of nematodes is covered, as in other ecdysozoans, by a well-developed layered cuticle, which also covers the buccal cavity, the pharynx, and the rectum. The cuticle is an extracellular matrix consisting predominantly of small collagen-like proteins that are extensively crosslinked (Johnstone, 1994), in addition to other protein and nonprotein compounds, including lipids and mucopolysaccharides. The *C. elegans* genome contains between 50 and 150 collagen genes, most of which are believed to encode cuticular collagens (Johnstone, 1994). While collagen, but not chitin, is the main compound of the cuticle, chitin is, however, found in the egg-shell of probably all nematodes and in the buccal apparatus of at least a few species (Neuhaus et al., 1997a), in which case two different chitin synthases may be present (Veronico et al., 2001).

The cuticle changes configuration during development. In *Oesophagostomum dentatum*, the cuticle of the first three juvenile stages consists of a trilaminate epicuticle, an amorphous layer, and a radially striated layer, but this changes in the last juvenile stage and the adult worm, the radially striated layer being replaced by a fibrous layer with three sublayers of giant fibers and a basal amorphous layer (Neuhaus et al., 1996a).

This cuticle provides rigidity and homeostasis to the worm, and as in the case of the arthropods, it confers a protective layer that has allowed nematodes to thrive in all environments, including terrestrial environments and inside other organisms. Also, as in other ecdysozoans, nematodes must shed their cuticle in order to grow, a process they do typically four times during their lifetime, although in some cases there are three molts (Félix et al., 1999). The cuticle also displays a great deal of variation across species, being annulated in many free-living worms or displaying highly varied cuticular ornamentation, scales, striae, alae, and so on, as well as cuticular setae, very similar to those of arthropods.

The process of molting is best understood in *C. elegans*, for which many of the genes and pathways identified as important for molting are highly conserved with respect to those of vertebrates—in which ecdysteroid hormones have other roles (it is also known that ecdysone also has multiple roles in arthropods). These genes include regulators and components of vesicular trafficking, steroid-hormone signaling, developmental timers, and hedgehog-like signaling, among others (Lažetić and Fay, 2017). In insects and other pancrustaceans, the steroid hormone ecdysone triggers molting at each developmental phase and, likewise, steroid hormone signaling is thought to regulate molting cycles in *C. elegans* (Lažetić and Fay, 2017).

However, with the exception of one steroid hormone that regulates entry into the dauer stage (Matyash et al., 2004; Entchev and Kurzchalia, 2005; Li et al., 2013)—an alternative developmental stage whereby the juvenile goes into a type of stasis triggered by environmental conditions—the specific hormones that regulate normal molting cycles have not been identified (Lažetić and Fay, 2017). Evidence in support of a role for steroid-hormone signaling in *C. elegans* molting includes the finding that several nuclear hormone receptors are required for molting. In addition, worms that are deprived of cholesterol, which cannot be synthesized by *C. elegans*, undergo larval arrest, or display defects in molting (Yochem et al., 1999). A better understanding of the molting process should also aid with nematode infections and parasitism.

The epidermis of nematodes (called "hypodermis") is also unique in often being thickened in four equidistant longitudinal cords: ventral, dorsal, and two laterals. It can be cellular or syncytial, the latter in many parasitic species. The ventral and dorsal epidermal cords house the longitudinal unpaired nerve cords. The lateral epidermal cords may contain excretory canals and neurons.

The musculature of nematodes has been studied in detail in a few species, perhaps most prominently in the large human parasite *Ascaris lumbricoides* (which belongs to "clade 8," formerly in the polyphyletic Ascaridida), that can grow up to 35 cm in length. Nematodes' body wall musculature is limited to longitudinal muscles, which are divided into four quadrants: the two above the lateral cords are innervated by the dorsal nerve cord, and the two below the epidermal cords are innervated by the ventral nerve cord (fig. 26.2 A).

The muscle cells of nematodes are unusual in that they send multiple branches (called "muscle arms") to the neurons in contrast to the more usual situation in other animals where neurons send processes to the muscles (Stretton, 1976). The neuromuscular synapses are therefore made at the ends of the muscle arms (fig. 26.3). In *A. lumbricoides* the second-stage juvenile[12] has 83 muscle cells, while the adult has approximately 50,000 muscle cells (Stretton, 1976). This striking increase in cell numbers of the musculature is not matched by a corresponding increase in the

[12] The nematode literature often refers to "larvae," although other authors prefer to use the term "juvenile" for the intermolt stages before the last molt. Because the term "larva" is often associated with indirect development, we favor the term "juvenile," as in Schmidt-Rhaesa and Henne (2016).

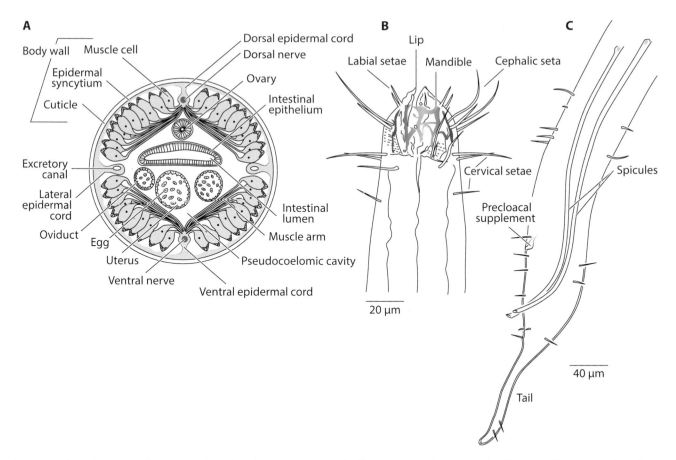

A

Body wall Muscle cell
Epidermal syncytium
Cuticle
Excretory canal
Lateral epidermal cord
Oviduct
Egg
Uterus
Ventral nerve
Ventral epidermal cord

Dorsal epidermal cord
Dorsal nerve
Ovary
Intestinal epithelium
Intestinal lumen
Muscle arm
Pseudocoelomic cavity

B

Lip
Labial setae   Mandible   Cephalic seta
Cervical setae

20 µm

C

Spicules
Precloacal supplement
40 µm
Tail

FIGURE 26.2. A, schematic cross section of an idealized nematode illustrating major anatomical features. B–C, an undescribed *Mesacanthion* species (Enoplida) illustrating, B, the anterior end with its sensory structures, scale 20 µm; C, posterior end illustrating the male tail and reproductive structures, scale 40 µm. A, based on Brusca and Brusca (1990); B–C, courtesy of Cruz Palacín.

number of cells in the nervous system, which is 250 in the adult worm (Stretton et al., 1978). Each neuron can therefore receive hundreds of muscle arms. While this system is highly unique among animals, it has also been observed in some of the wall muscles of kinohynchs (Neuhaus, 1994; Neuhaus and Higgins, 2002) (see chapter 23), as well as in platyhelminths, echinoderms, and cephalochordates (Chien and Koopowitz, 1972).

In addition to the musculature of the body wall, the other prominent set of muscles are those of the pharynx. The cylindrical pharynx of *Anguillicoloides crassus* consists of a short anterior muscular corpus and an enlarged posterior glandular and muscular postcorpus (Brunanská et al., 2007). The main cellular components of the pharynx include the muscle cells, the marginal cells, the nerve cells, and one dorsal and two subventral glands.

Nematodes have a through-gut that forms a straight inner tube, delimiting the inner portion of the body cavity (when present). It is divided into foregut (or stomodeum), midgut, and hindgut (or proctodeum). The midgut has an endodermal origin, whereas the other two regions have a mixed ecto- and mesodermal origin

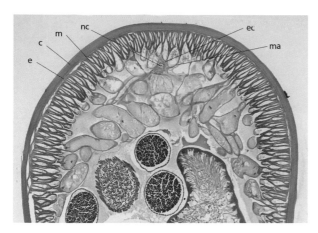

Figure 26.3. Cross section of *Ascaris lumbricoides* showing the extension of the muscles connecting to the longitudinal nerve cord. Abbreviations: c, cuticle; e, epidermis; ec, epidermal cord; m, muscle fiber; ma, muscle arm; nc, longitudinal nerve cord.

(Decraemer et al., 2014). The gut of *C. elegans* derives from all the progeny of the E blastomere, a cell of the eight-cell stage (Goldstein, 1993). Some taxa lack a mouth or have reduced digestive cavities, in some cases of symbiotic sulphur-oxidizing bacteria (Polz et al., 1992; Bayer et al., 2009). The pharynx is discussed in several other sections of this chapter; its lumen is triradiate, Y-shaped.

The nervous system of nematodes has recently been reviewed by Schmidt-Rhaesa and Henne (2016). The level of understanding of the nervous system in *C. elegans* is unparalleled by any other animal (Altun and Hall, 2011). The adult hermaphrodite has 302 neurons, 282 of which belong to the larger somatic nervous system and 20 to the largely independent stomatogastric pharyngeal nervous system. Many of the descriptions of nematode nervous systems are thus based on this model. The somatic nervous system comprises a broad network of neuronal elements, including the dense circumpharyngeal brain, typical of cycloneuralians, and several longitudinally and circularly oriented neurites or neurite bundles and their ascending cell clusters or ganglia.

The circumpharyngeal neuropil, or nerve ring, is usually slanted a few degrees with respect to the meeting point with the ventral nerve cord. Additional somata form clusters anterior and posterior to the neuropil. Anterior to the brain, six neurite bundles project from the brain neuropil to supply peripheral sense organs at the tip of the head. Posterior to the brain, a prominent ventral unpaired nerve cord and other neurite bundles extend longitudinally (the number of neurite bundles varies across species, but a dorsal neurite bundle is found). The ventral nerve cord separates into two branches in the region of the vulva.

Nematode sensory organs are of two main types, papillary or setiform, and can be mechanoreceptors, or chemoreceptors, if they bear a cuticular pore (Schmidt-Rhaesa and Henne, 2016). The diversity of sensory structures is described in Decraemer et al. (2014) and includes a series of external and internal sensory structures. Most prominent among these are the structures around the mouth (including radially arranged sensillae and a pair of amphids) and the reproductive organs (fig. 26.2 B–C). Amphids (or amphid sensillae) are the main sensory organs of nematodes, formed by a highly innervated invagination of the cuticle, each innervated by 12 neurons in *C. elegans*. Other sensory organs include phasmid sensillae (in the tail). Nematodes also may present internal sensory structures, including paired pigment spots, or ocelli, associated with the pharynx of some aquatic nematodes (Decraemer et al., 2014).

The body cavity of nematodes is lined by the epithelial cords (which are ectodermal), the body wall musculature, and the endodermal intestine, as nematodes lack intestinal musculature, and thus the body cavity is considered a pseudocoelom. The pseudocoelomic cavity can be ample in the large parasitic species but reduced or absent in smaller species (Ehlers, 1994), and thus some authors consider nematodes to be acoelomates and not pseudocoelomates (e.g., Ax, 1996; Schmidt-Rhaesa, 2007). Given the diminished emphasis on body cavities for understanding animal evolution and the fact that body cavities do vary within phyla (e.g., Priapulida can be pseudocoelomate or coelomate; see chapter 22), we consider nematodes both acoelomates or pseudocoelomates, depending on the species.

The excretory organ of nematodes is simple, consisting of one to five cells and differs considerably across taxa. The plesiomorphic excretory organ found in many marine species consist of one large glandular cell called the "ventral gland" or "renette cell," but the system may also have one or a pair of longitudinal extensions (Nielsen, 2012a). The renette cell of *Sphaerolaimus gracilis* occurs posterior to the esophageal–intestinal junction and opens through an ampulla to a ventral renette pore behind the nerve ring (Turpeenniemi and Hyvärinen, 1996). Interestingly, the renette cell and the two caudal gland cells are similar in structure, and both secrete a substance that is used by the worm to attach to the substrate.

Many members of the subclass Enoplia lack renette cells and instead have numerous unicellular units distributed along the entire length of the body, each opening via a duct and a pore. However, the excretory nature of these cells is not well understood, and some authors have proposed that they may represent nonciliated protonephridia (Stock, 2016).

The excretory system of *Caenorhabditis elegans* is one of the most complicated nematode systems, consisting of three unicellular tubes (a canal, duct, and pore), a secretory paired gland (formed of two cells), and two associated neurons and has been used as a model of tubular organogenesis involving a minimum of cells (Sundaram and Buechner, 2016). The excretory canal cell is the largest and one of the most distinctively shaped mononucleate cells in the animal. The cell body and large nucleus are located just ventral to the posterior bulb of the pharynx. The cell sends out four hollow tubules or "canals," two posterior and two anterior, forming an "H-shape." The canals are closed at their tips and extend the entire length of the adult animal. The canal cell is a seamless tube, with adherens junctions only at its connection to the excretory duct cell and gland (Sundaram and Buechner, 2016). The osmoregulatory function of the excretory system has long been established (Nelson and Riddle, 1984).

Many soil nematodes are able to undergo quiescence (suspension of activities, such as feeding and reproduction) to cryptobiosis (nearly complete suspension of their metabolism) including its specific anhydrobiotic state, when conditions become unfavorable—a phenomenon also well known in tardigrades and rotifers. In this resistant state the roundworms can survive the most extreme environmental conditions, such as the extreme cold and dryness of the Antarctic Dry Valleys

(Treonis and Wall, 2005) and return to life when favorable conditions return. The molecular mechanism of anhydrobiosis is, however, poorly understood but has been studied in the dauer juvenile of *C. elegans* (the only stage with this capacity) (Erkut et al., 2013) and in a handful of other anhydrobiotic nematode species, showing that a large number of genes and processes have a role in desiccation tolerance, including effective antioxidant systems, and genes involved in proteostasis, DNA repair, and signal transduction pathways (Reardon et al., 2010; Evangelista et al., 2017).

The phenomenon of eutely or cell constancy occurs when embryogenesis has a limit of cell divisions (generally referring to somatic cells only), further increase in size of the organism being possible only by cell growth and not by an increase in number of cells. Eutely is often suggested in other animals that coincidentally are also cryptobiotic (tardigrades and rotifers), but this is difficult to assess, at least in tardigrades (see chapter 29). Eutely can occur at the level of the organism or for some specific organs or tissues. For example, Goldschmidt (1908) established that the number of neurons in the central nervous system of *Ascaris lumbricoides* was very small (162 in the head ganglion), and reproducible from animal to animal.

The case of *C. elegans* is especially well studied, and consists of only about 1,000 somatic cells and 1,000 to 2,000 germ cells. More precisely, 959 somatic cell nuclei (of which 302 are neurons) and about 2,000 germ cells are counted in the hermaphrodite individuals and 1,031 somatic cell nuclei plus about 1,000 germ cells in the male. These constant numbers are largely responsible for the general idea that nematodes are eutelic, although it is now clear that, especially for the epidermis of large, free-living species (Cunha et al., 1999; Azevedo et al., 2000), not all organs present eutely, and, in fact, eutely may be the result of adaptation to small size (Rusin and Malakhov, 1998). And, of course, the germ-line stem cells are established early during embryogenesis and maintained throughout adulthood, but they essentially continue to divide throughout adult life.

DEVELOPMENT AND LIFE CYCLES

Few animal phyla are understood at the developmental level as are nematodes, as evidenced by a recent review by Schierenberg and Sommer (2014). Early embryogenesis has been studied in a number of species and is obviously best-understood in *C. elegans* (e.g., Sulston et al., 1983; Schnabel et al., 2006) (fig. 26.4), but the first model nematodes used were the gonochoristic (male/female) parasitic species *Parascaris equorum* (when horse manure was abundant in the streets, prior to the use of cars) and *Ascaris lumbricoides*, although their parasitic nature made them more intractable. During early development, the point of sperm entry determines A–P polarity (Goldstein and Hird, 1996), but there is no sperm contribution in root-knot nematode (Calderón-Urrea et al., 2016). First cleavage is already asymmetric and generates an anterior first somatic founder cell (AB) and smaller posterior germline cell (P1), with an early segregation of the germline. Ten minutes later, AB divides with a transverse spindle orientation, resulting in an anterior ABa and a dorsal ABp cells. P1 divides with longitudinal spindle into a larger somatic EMS cell and a posterior new germline P2 cell. The left–right axis is formally established at this time.

FIGURE 26.4. Early cleavage and fate map of *Caenorhabditis elegans* from the 4- to the 28-cell stage.

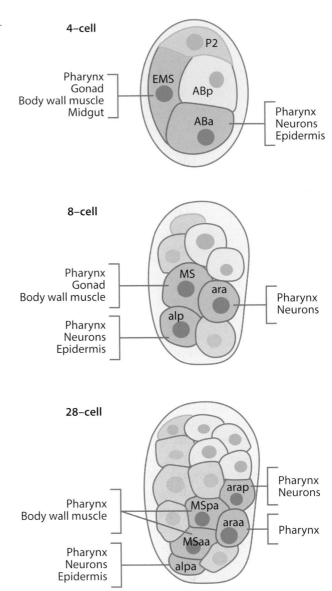

Gastrulation occurs when small groups of cells ingress at various times into the small blastocoel space. Complete development, from fertilization to hatching, takes 840 minutes.

The pattern of cleavage, spatial arrangement, and differentiation of cells diverged dramatically during the evolutionary history of nematodes without much change in the phenotype (Schulze and Schierenberg, 2011), and a good number of species have now been investigated (e.g., Schulze and Schierenberg, 2008; Calderón-Urrea et al., 2016). Developmental mechanisms can be very variable among species of nematodes. For example, early cleavage is asynchronous, unequal, and invariant in *C. elegans* while it is synchronous, equal, and noninvariant in *Enoplus brevis* (Félix, 1999). This variation in the fate and timing of cell fate in early blastomeres is beginning to be understood as additional cell lineages of nematodes are being established.

The diverse life cycles of nematodes have been summarized by Lee (2002), and, of course, that of *C. elegans* is known in great detail. As discussed above, *C. elegans* has two sexes, a hermaphrodite and a male, the former being viewed as a female that produces a limited number of sperm and that can reproduce either by self-fertilization or by cross-fertilization after mating with a male. It is now known that hermaphrodites of the wild-type laboratory reference strain N2 favor self-reproduction, whereas a wild isolate favors outcrossing (Bahrami and Zhang, 2013).

The ability to self-fertilize to produce homozygous progeny is one of the features that has contributed to make *C. elegans* the genetic model it is. Still, the predominant reproductive mode in nematodes is gonochory with males and females, but nematodes are known for the multiple types of reproduction that have evolved independently in several clades, including hermaphroditism, parthenogenesis, alternation of hermaphroditism or parthenogenesis with gonochory, and coexistence of males, females, and hermaphrodites, a mode called "trioecy" (Kanzaki et al., 2017). In one of these trioecic systems in the genus *Auanema*, there is convergence in some features of the complex life cycles of some parasitic nematodes, including the production of few males after outcrossing and the obligatory development of dauers into self-propagating adults (Kanzaki et al., 2017).

The complex life cycles of the parasitic forms have received special attention, especially in human enteric parasites, which include *Ascaris lumbricoides*, two major hookworms (*Ancylostoma duodenale* in the Old World and *Necator americanus* in the New World), *Strongyloides stercoralis*, *Trichuris trichiura*, and *Enterobius vermicularis*. Most enteric nematodes have established a well-balanced host–parasite relationship with the human host; humans tolerate these parasites well, and little disease manifests with light infection (Cross, 1996a). As worm load increases, a corresponding increase in disease usually occurs, producing different levels of irritation. The life cycles of these animals can differ in their complexity.

The whipworm *Trichuris trichiura* causes trichuriasis, an important neglected tropical disease, when it infects the human large intestine. It has a simple life cycle, the whipworms reproducing in the large intestine and passing the fertilized eggs (females produce between 2,000 and 10,000 eggs per day) with the feces until zygotes are ingested by humans (Cross, 1996a). Other cycles require intermediate hosts, as is the case of *Trichinella spiralis*, which produces trichinosis from ingestion of raw or inadequately cooked meat that contains encysted juveniles of the nematode. The parasite completes all stages of development in one host (human or other carnivorous mammals). Infective juveniles encyst in striated muscles and excyst in the gut of carnivores that eat the infected meat. The worms mature and reproduce in the small intestine until newborn juveniles migrate to the muscles, where they encyst (Kelsey, 1996).

Perhaps the most dreaded infections are those of the filarial nematodes, summarized by Cross (1996b). Filariae are threadlike parasites transmitted by arthropod vectors. The adult worms inhabit specific tissues, where they mate and produce microfilariae, the characteristic tiny, threadlike juveniles. The microfilariae infect vector arthropods through blood meals or through skin, in which they mature to infective juveniles. In some species, the female worms show a circadian periodicity in microfilaria production, the time of peak production usually corresponding to the peak feeding period of the vector. Filarial diseases are a major health problem in many tropical and subtropical areas. The exact type of disease produced by a filarial worm depends on the tissue locations preferred by adults and microfilariae.

The adults of the lymphatic filariae (*Wuchereria bancrofti* and *Brugia malayi*) inhabit lymph vessels, where blockage and host reaction can result in lymphatic inflammation and dysfunction, and eventually in lymphedema and fibrosis. Repeated, prolonged infection with these worms can lead to elephantiasis, a buildup of excess tissue in the affected area. Other filariae mature in the skin and subcutaneous tissues, where they induce nodule formation and dermatitis; migrating filariae of these species can cause ocular damage. Major filarial nematodes that infect humans are *Wuchereria bancrofti*, *Brugia malayi*, and *Onchocerca volvulus*; minor filariae are *Loa loa*, *Mansonella* species, and *Dirofilaria* species. The filariae of *Onchocerca volvulus* are transmitted by biting black flies and cause blindness through scarification in the eye. Onchocerciasis (river blindness) affects about 37 million people in Africa and causes blindness in hundreds of thousands of cases.

Many filarial nematodes, including those causing onchocerciasis, are known to harbor intracellular *Wolbachia* (see a review on the role of bacterial symbionts in onchocerciasis in Tamarozzi et al., 2011), which are often found in the female germline, epidermis, and gut (Ferri et al., 2011). However, some filarial species, including *Loa loa*, the African eyeworm, lack *Wolbachia* (Desjardins et al., 2013). The presence of intracellular bacteria in *Onchocerca volvulus* was first reported by Kozek and Marroquin (1977), and all individual worms and all life cycle stages contain the endosymbionts. They inhabit the lateral cords of adult worms and the reproductive system of females, where they are transmitted transovarially (Taylor et al., 2005). In nematodes, *Wolbachia* are obligate mutualistic endosymbionts of some members of the family Onchocercidae, but mostly in those that infect mammals (Ferri et al., 2011), unlike in most arthropod–*Wolbachia* associations, in which they are generally considered to be reproductive parasites (Taylor et al., 2005). *Wolbachia* are thus an important target for antifilarial therapy (Slatko et al., 2010), since clearance of the endosymbionts by antibiotic treatment causes inhibition of worm development, blocks embryogenesis and fertility, and reduces viability (Tamarozzi et al., 2011). Although our understanding of the biological basis of the symbiotic relationship is limited, it has been suggested that various *Wolbachia* biochemical pathways that are absent or incomplete in the nematode are candidates for *Wolbachia*'s contribution to nematode biology (Slatko et al., 2010).

The life cycles of plant parasites are also well understood in a few species, especially in crop-infecting species. The root-knot nematodes in the genus *Meloidogyne* infect plants, which as a result of nematode feeding, develop large galls or "knots" throughout the root system. Severe infections result in reduced yields of numerous crops. Root-knot nematodes begin their lives as eggs that rapidly develop into nonmotile first-stage juveniles (J1), which reside entirely inside the translucent egg case, where they molt into a J2 stage. The motile J2 is the only stage that can initiate infections by attacking growing root tips and moving to the area of cell elongation, where they initiate a feeding site by injecting esophageal gland secretions into root cells. These nematode secretions cause dramatic physiological changes in the parasitized cells, transforming them into giant syncytial cells that die if the nematode dies. In addition to the roots, nematodes feed on all parts of the plant, including stems, leaves, flowers and seeds and can cause many different plant diseases. Nematodes feed from plants in a variety of ways, but all use a specialized stylet to pierce plant tissue (Lambert and Bekal, 2002). Plant nematode control is an active area of research.

Entomopathogenic nematology is another broad area of research. Some parasitic nematodes have evolved an association with insect-pathogenic bacteria that constitute a lethal duo (Dillman and Sternberg, 2012). Essentially, the nematodes serve as mobile vectors for the bacteria. The nematodes invade potential hosts and release their pathogenic bacteria into the nutrient-rich hemolymph. Infected insects die quickly, the bacteria proliferate, the nematodes feed on bacteria and insect tissues, and then reproduce. The cooperation with bacteria and the speed with which they kill set entomopathogenic nematodes apart from other nematode parasites.

GENOMICS

Nematode genomics is well advanced, thanks to the pioneering work done with *Caenorhabditis elegans*, which was the first animal genome to be sequenced (*C. elegans* Sequencing Consortium, 1998). Due to the medical and agricultural interest of many species, the sequencing of nematode genomes flourished after the initial *C. elegans* draft genome. These genomes represent taxa closely related to the model organism (Stein et al., 2003; Haag et al., 2007), human parasites (e.g., Ghedin et al., 2007; Desjardins et al., 2013; Foth et al., 2014; Tang et al., 2014; Schwarz et al., 2015; Cotton et al., 2016), nonhuman mammal parasites (Jex et al., 2011; Godel et al., 2012; Laing et al., 2013; Foth et al., 2014; Jex et al., 2014), entomopathogenic species (Dieterich et al., 2008; Bai et al., 2013; Schiffer et al., 2013), and a plethora of species of agricultural importance (Abad et al., 2008; Opperman et al., 2008; Kikuchi et al., 2011; Cotton et al., 2014; Eves-van den Akker et al., 2016). Currently, genomes of about a hundred nematode species are publicly available between NCBI and Nematode.net, trying to follow an early initiative to sequence the nematode mythical number of *959* genomes, put in place by Kumar et al. (2012). Many of these novel parasitic nematode genomes have recently been released (International Helminth Genomes Consortium, 2019).

The *C. elegans* genome contains about 19,700 protein-encoding genes (Hillier et al., 2005), and nematode genomes as a whole are small, compared to mammalian genomes, even when containing almost the same number of genes (Sommer and Streit, 2011), making them highly tractable. Some parasitic nematodes have genomes that are not only compact but are also reduced in terms of numbers of genes. The lesion nematode *Pratylenchus coffeae*, a significant pest of banana and other staple crops in tropical and subtropical regions worldwide, encodes for only 6,712 protein-encoding genes (Burke et al., 2015). This is the smallest number of genes known for a metazoan genome with the exception of the myxozoan *Kudoa iwatai*, a highly derived cnidarian that parasitizes fishes (Chang et al., 2015). Interestingly, in the case of *P. coffeae*, no developmental or physiological pathways are obviously missing when compared to *C. elegans*. In the case of plant-parasitic nematodes, it has been suggested that horizontal gene transfer, gene family expansions, evolution of new genes that mediate interactions with the host, and parasitism-specific gene regulation are important adaptations that allow parasitism (Kikuchi et al., 2017). Understanding these genomes is a first step toward controlling parasitism (Eves-van den Akker et al., 2016).

FOSSIL RECORD

The fossil record of Nematoda has recently been reviewed by Poinar Jr. (2014b) and is for the most part restricted to relatively recent crown-group taxa, as are known from Early Cretaceous Lebanese amber (Poinar Jr. et al., 1994; Poinar Jr. and Buckley, 2006), Late Cretaceous Burmese amber (Grimaldi et al., 2002), and several species from Cenozoic ambers. Nematodes have shown an intimate relationship with their hosts for at least 40 million years (Poinar Jr., 2003b, 2014a).

The Early Devonian *Palaeonema phyticum* is identified as the earliest known nematode based on details of its buccal cavity (including teeth), the pharynx, and the ovaries (Poinar Jr. et al., 2008). *Palaeonema phyticum* dates an association with plants, the fossils being preserved in stomatal chambers, and are considered the first fossil herbivore (Dunlop and Garwood, 2018).

Sinusoidal horizontal burrows from the Early Ordovician were potentially produced by marine benthic nematodes (Baliński et al., 2013), but this is difficult to confirm.

NEMATOMORPHA

Among the most fascinating invertebrates are horsehair worms, hairworms, or gordian worms, close relatives of nematodes with both a parasitoid[13] life cycle and an arthropod as host. Nematomorphs include 5 described marine species that parasitize decapods and nearly 350 named limnoterrestrial species that infect mostly insects, as well as other terrestrial arthropods (Hanelt et al., 2005; Spiridonov and Schmatko, 2013). The adult worms of the terrestrial nematomorphs often emerge from the dying host in an aquatic environment, where they drive their host to a supposedly induced suicide (Thomas et al., 2002; Barquin et al., 2015). When in the aquatic environment, the host is often exposed to predators, and nematomorphs have often also managed to escape predation by wriggling out of the mouth, nose, or gills of the predator (fishes or frogs) that had consumed its host (Ponton et al., 2006). This escape was recorded from about 20% of the fish predators and 35% of frogs and can take up to 28 minutes. No other parasite is known to survive predation this way.

Reproductive adults are free living in water, where they mate as a knotted mass of multiple individuals, giving them their common name of gordian worms, as a reference to the Gordian knot—a legend of Phrygian Gordium associated with Alexander the Great and often used as a metaphor for an intractable problem solved easily by thinking outside the box ("cutting the Gordian knot"). This is so because adults are threadlike, ranging in size from 50 to 100 mm long and 1 to 3 mm in diameter, although some species are reported to measure up to 1 m in length. Another common name, horsehair worms, stems from the belief that they arose from the hair of horses' tails that fell into the water.

Numerous reports exist of patients passing worms per rectum and per urethra, or of worms being present in the vomit of humans and their pets, or even in the underwear of a woman (Hanelt et al., 2005). However, most of these cases seem to be instances of pseudoparasitsm, where the vertebrates most certainly ingested infected arthropods. Since nematomorphs are now known to survive the digestive tacts of their host predators (Ponton et al., 2006), it seems most likely that these cases are not true nematomorph infections.

SYSTEMATICS

About 360 species have been described in 18 genera; 87 species described since 1990. Besides the description of new taxa, several species were reinvestigated in recent

---

[13] A parasitoid is an organism that spends a significant part of its life history within a single host organism in a relationship where the host is ultimately killed.

years, in combination with a critical evaluation of the validity of genera, as many were not monophyletic. Phylogenetic relationships and synapomorphies among nematomorphs have been poorly studied above generic relationships. The group is divided into two main clades (often referred to as classes or orders).

Order Nectonematoidea

Nectonema (5 spp.)

Order Gordioidea

e.g., *Chordodes*, *Gordius*, *Paragordius*

NEMATOMORPHA: A SYNOPSIS

- Bilaterally symmetrical, unsegmented vermiform protostomes, with a long and thin body
- Primary body cavity completely surrounded by an extracellular matrix, not lined by mesoderm, sometimes reduced by parenchyme
- Well-developed fibrous cuticle with collagen (chitin reported in the larval cuticle), with one cuticular molt
- Body wall with longitudinal muscles only
- Gut reduced to various degrees in the adult
- Unique juvenile stage that parasitizes arthropods
- Ventral and dorsal epidermal cords housing the longitudinal nerve cords
- Without discrete excretory organs

Nectonematoids are marine, planktonic animals with a double row of natatory setae along both sides of the body, parasitizing decapod crustaceans, mostly crabs but also lobsters (Poinar Jr. and Brockerhoff, 2001; Schmidt-Rhaesa et al., 2012). Since their original finding, as free-living animals on the coast of Massachusetts (Verrill, 1879), a handful of species have been described, the latest being the parasitic phase of a New Zealand species (Poinar Jr. and Brockerhoff, 2001). The Norwegian *Nectonema munidae* has been studied in particular detail (Schmidt-Rhaesa, 1996a, b, 1997, 1998), most of the nectonematoids' anatomy referring to this species. The nonmarine species are much better studied, and adults lack the natatory setae. Both groups also differ in the internal pseudocoelomic cavity, which is small and fluid-filled in Nectonematoidea but largely filled with mesenchym in Gordioidea. *Nectonema* has a single gonad, while the gonads are paired in Gordioidea.

The cuticle, which can be either smooth or structured into so-called areoles, of taxonomic significance, is secreted by a cellular epidermis and lacks chitin in gordiids, but chitin has been detected in the larval cuticle of *Nectonema* (Neuhaus et al., 1996b; Neuhaus and Higgins, 2002). The epidermis is lined by a layer of longitudinal muscles, which are radially flattened, but circular muscles are lacking, as in nematodes (Schmidt-Rhaesa, 2013a). The structure of the muscle cells has been investigated in detail in both groups of nematomorphs, being of the coelomyarian type (Schmidt-Rhaesa, 1998). Interestingly, nematomorphs seem to have three regulatory muscle proteins—troponin, calponin, and caldesmon—the latter not

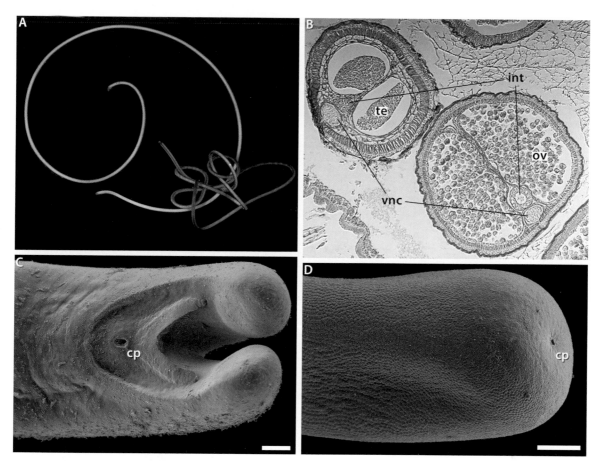

FIGURE 27.1. A, two adult individuals of *Gordius violaceus*; B, histological cross section through the middle body region of a male (upper left) and a female (lower right) *Paragordius varius* showing the ventral nerve cord (vnc), intestine (int), testis (te) and ovary (ov); C, posterior end of male *Gordius* sp. with the cloacal pore (cp), scale 100 μm; D, Posterior end of female *Beatogordius latastei* showing the cloacal pore (cp), scale 100 μm. Photo credits: B–D, Andreas Schimdt-Rhaesa.

known from any invertebrate genomes but identified in some invertebrate striated muscle cells (Meyer-Rochow and Royuela, 2018). While the lack of homologs for caldesmon may suggest unspecificity of the antibody, nematomorphs show that troponin and calponin can co-occur in the same type of muscle.

Intraepidermal nerves and a large, unpaired ventral nerve cord are located within a ventral epidermal cord in *Nectonema* and juvenile gordiids, but it becomes translocated to a subepidermal and even submuscular position in adult gordiids (Schmidt-Rhaesa and Henne, 2016). In *Nectonema* there is also a smaller nerve cord present inside a second dorsal epidermal cord (Schmidt-Rhaesa and Henne, 2016). Innervation of the musculature appears to come from the basiepidermal nerve net.

The cuticle differs considerably between the larva and the adult forms, as during the parasitic phase nutrients are absorbed through the integument, composed of physiologically active epidermal cells and the cuticle constituting a thin layer that is a minimal barrier for nutrient uptake. During development, the larval cuticle is replaced by a second cuticle, the adult cuticle, which is much thicker and multilayered, typical for the free-living phase, which has just a protective function (Schmidt-Rhaesa et al., 2005). The epidermis and the intestine decrease in size and cytological components during further development (Schmidt-Rhaesa, 2005), when the animal ceases to eat. A single molting event, when the larval cuticle is replaced by the adult one, has been documented in nematomorphs (Schmidt-Rhaesa, 2005).

The intestinal system of adult nematomorphs is reduced to a certain degree, and a mouth can be present or absent, depending on the species and perhaps on the developmental stage (Schmidt-Rhaesa, 2013a). The intestine is a thin tube in freshwater species and blindly ending in *Nectonema*. Because adults do not feed and nutrient absorption is restricted to juveniles, the short adult phase is probably just reproductive. When complete, the intestine joins the gonads in the posterior end, forming a cloaca. Gonads are tubelike and can become very voluminous, carrying a large quantity of gametes, as expected in a parasitic animal. The extension of the gonads is inversely correlated to the amount of parenchyme—a presumed storage tissue. The reproductive organs are not well understood in *Nectonema*.

The anterior nervous system of Nematomorpha is composed of a brain, a prominent ventral nerve cord, and a nerve plexus in the periphery. An additional dorsally situated longitudinal neurite bundle is only present in *Nectonema* species and has not been documented in Gordioidea. The shape of the brain in Gordioidea has been a contentious issue (Schmidt-Rhaesa, 1996a), probably due to fixation artifacts and difficulties, recently resolved by combining multiple imaging methodologies (Henne et al., 2017). The latter study shows a brain composed of a central neuropil and a ring-shaped structure with associated somata, unlike that of other cycloneuralians (Schmidt-Rhaesa and Henne, 2016; Henne et al., 2017). The unpaired ventral nerve cord emerges from the posteroventral part of the brain. Sensory structures are not clearly documented in nematomorphs, but integumental receptors may be present (Schmidt-Rhaesa et al., 2005).

LIFE CYCLE

The life cycle of the marine *Nectonema* species is not well understood. Larvae have been observed on very few occasions, and it is not known how they infect the final host or whether there are intermediate hosts. Because larvae have never been found inside any organism, it is unknown if they encyst.

The best known nematomorph life cycles are those of some gordioids (fig. 27.2), some of which are now maintained as laboratory cultures, including the American species *Paragordius varius* (see Hanelt and Janovy, 2004b), providing important insights into nematomorph biology. Adult gordioids are free-living in freshwater environments, while juveniles are obligate parasites in terrestrial arthropods with a paratenic aquatic host, for example, insects and molluscs (Hanelt and Janovy, 2003), requiring a transition from the aquatic to terrestrial environment. However, epibenthic aquatic larvae and their terrestrial definitive hosts do not overlap in habitat. It now seems that cysts play a major role in this habitat transition, as they have large infection rates, especially in snails (Hanelt et al., 2001), and cysts can not only survive metamorphosis in insects that transition from aquatic larvae to terrestrial adults (Hanelt and Janovy, 2004a) but even freezing (Bolek et al., 2013). It has also been shown that cysts formed in spurious paratenic hosts may eventually transfer to normal paratenic hosts (Hanelt and Janovy, 2004a).

Paratenic hosts in freshwater environments (including pancrustaceans, annelids, snails, and fishes) (Hanelt and Janovy, 2003; Schmidt-Rhaesa, 2013a) become infected by consuming mobile, nonswimming gordioid larvae that cross the gut wall and encyst. The paratenic hosts are later ingested by the definitive host, typically a predatory or omnivorous terrestrial arthropod. Species in the orders Orthoptera, Blattodea, Mantodea, and Coleoptera are the most frequent insect hosts, but they are also found in other insects as well as in myriapods and several groups of arachnids, including Araneae and probably Scorpiones (Schmidt-Rhaesa, 2013a; Spiridonov and Schmatko, 2013). Upon maturity, gordioids exit the definitive host and enter the aquatic environment, where they mate and lay eggs within a few days (Schmidt-Rhaesa et al., 2005). The transition from larva to adult requires a great deal of morphogenesis, which especially involves the body wall, during the transition

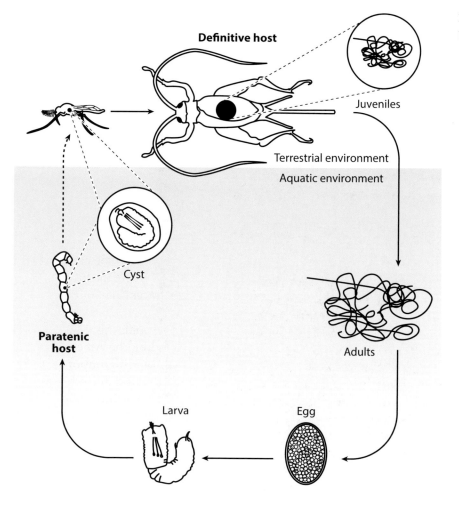

FIGURE 27.2. Life cycle of a gordiid. Based on multiple internet sources.

Definitive host

Juveniles

Terrestrial environment

Aquatic environment

Cyst

Paratenic host

Adults

Larva   Egg

from a feeding larva to a nonfeeding adult (Schmidt-Rhaesa, 2004; Henne et al., 2017).

Finding mates can be difficult in parasites that are released from the host into the external environment where the parasite can find itself trapped without a sexual partner. To solve this problem and guarantee reproduction, parasites in numerous phyla have evolved reproductive strategies, as predicted by the reproductive assurance hypothesis, such as hermaphroditism or parthenogenesis (Hanelt and Janovy, 1999, 2000; Bolek and Coggins, 2002). *Paragordius obamai* represents the only known nematomorph species to reproduce asexually, a strategy determined genetically that may have evolved as a means to assure reproduction (Hanelt et al., 2012). Nonetheless, in most sexual nematomorph species, sex ratios are often male-biased (Hanelt et al., 2012).

GENOMICS

No nematomorph nuclear genome is yet publicly available, but unpublished draft genomes have been sequenced for terrestrial species. Transcriptomic resources are likewise quite limited for nematomorphs (Dunn et al., 2008; Laumer et al., 2019). No nematomorph mitochondrial genome is yet available.

FOSSIL RECORD

While several Cambrian forms have been assigned to nematomorphs, these belong to Palaeoscolecida, a group of contentious systematic position in Ecdysozoa (see chapter 20). The oldest incontrovertible fossil is from Burmese Cretaceous amber, *Cretachordodes burmitis* (Poinar Jr. and Buckley, 2006; Schmidt-Rhaesa, 2013a). Younger amber fossils, associated with a cockroach host, are known from Dominican amber (Poinar Jr., 1999).

PANARTHROPODA

Should you find an animal with serially repeated, paired ventrolateral appendages—something that could happen from abyssal depths to the highest mountain tops—it is most probably a member of the clade Panarthropoda (sometimes referred to as Arthropoda, mostly in German textbooks). Tardigrades, onychophorans, and arthropods[14] (= euarthropods, according to Ortega-Hernández, 2016) constitute a clade that shares a suite of derived characters yet exhibits high disparity with respect to such basic morphological characters as segmentation, brain architecture, body cavities, and structure of the ventrolateral appendages.

The name Panarthropoda was introduced by Nielsen (1995), although a comparable clade had been recognized since the middle of the nineteenth century, mostly as either Arthropoda or Gnathopoda. In spite of this, the interrelationships and even the monophyly of Arthropoda were subjects of historical debate, as some scenarios placed Onychophora with Myriapoda and Hexapoda in the "phylum" Uniramia (Manton, 1972). Varied hypotheses have been put forward to describe relationships between members of Panarthropoda. Most recently this focuses on the relative positions of Tardigrada and Onychophora with respect to Arthropoda, whereas in the past it also included whether arthropods were monophyletic, diphyletic, or polyphyletic. Likewise, the membership of parasitic groups, such as Pentastomida, had been debated for centuries until molecular data corroborated evidence from sperm ultrastructure to suggest an affinity to branchiuran crustaceans (Abele et al., 1989). We thus follow here both the modern nomenclature and the composition of the three panarthropodan phyla (e.g., Minelli, 2009; Nielsen, 2012a; Brusca et al., 2016b).

Defining panarthropods morphologically is less straightforward than it might appear, as most characters typically used in textbooks are absent in one of the three phyla. They all have paired ventrolateral segmental appendages with terminal claws, but the nature of these appendages differs among them. Only arthropods have undergone an arthropodization process, with both segmental sclerites and

---

[14] Ortega-Hernández (2016) provides a historical review of the usage of the name Arthropoda in the early zoological literature. He argues that the name Arthropoda should be applied according to von Siebold's (1848) sense, in which Tardigrada were grouped with what was later named Euarthropoda. Later, Lankaster (1904) used Arthropoda to include Onychophora + Euarthropoda, which does not have priority but is more in line with current phylogenomic hypotheses. Given that (a) molecular data strongly contradict the monophyly of von Siebold's Arthropoda, (b) that the Code of Zoological Nomenclature does not rule above the family level, and thus there is no rule of priority, and (c) that the term Arthropoda has been nearly universally used in the sense adopted here in non-German texts, we follow the standard modern nomenclature and refer to the phylum Arthropoda excluding Onychophora and Tardigrada.

appendages sclerotized as podomeres separated by arthrodial membranes. In spite of this, at least onychophorans and arthropods share the same general patterns of gap gene expression along the proximodistal axis of the appendages (Janssen and Budd, 2010; Janssen et al., 2015); these data are not yet known for tardigrades.

Likewise, all three groups have a ganglionar supraesophageal brain, but that of tardigrades is composed of a single segment (Gross and Mayer, 2015); that of onychophorans is composed of two, the protocerebrum and deutocerebrum (Mayer et al., 2010b); and that of arthropods has three, protocerebrum, deutocerebrum, and tritocerebrum. The ventral nerve cords of these three groups also differ, with a paired ganglionated nerve cord in tardigrades and arthropods versus a lack of segmental ganglia in onychophorans. This is also concomitant with their external appearance, as onychophorans instead of external segments show an annulated cuticle. Nevertheless, the segment polarity protein engrailed is expressed in the posterior ectoderm of developing segments in each of the three panarthropod groups, suggesting that it plays a common role in establishing segmental boundaries (Gabriel and Goldstein, 2007). While segmented mesoderm and a mixocoel have also been proposed as synapomorphies for Panarthropoda (Nielsen, 2012a), these are not observed in tardigrades.

A longstanding debate about the correspondence between the panarthropod appendages (Scholtz, 2016) has been settled with the help of expression data from Hox genes and nervous system patterning genes, as well as neuronal tracing through development. It is, for example, now well established that a neuromere corresponding to the protocerebrum of arthropods innervates the eyes in all three groups, but among extant panarthropods only onychophorans have an appendage innervated by the protocerebrum, the antennae; it is inferred that the frontal appendages of many stem-group arthropods (such as Radiodonta) and other panarthropods may have been innervated by the protocerebrum (Ortega-Hernández et al., 2017). Fossilized traces of the brain and nerve tracts into the frontal appendage of the stem-group arthropod *Kerygmachela kierkegaardi* suggest a single-segmented (protocerebral) brain (Park et al., 2018). Expression of the head gap gene *orthodenticle* during early embryogenesis is restricted to this first head segment in all panarthropod groups (Smith et al., 2016).

The first leg of tardigrades is positionally homologous to the jaws of onychophorans, the cheliceres/chelifores of chelicerates, and the (first) antenna of mandibulates. However, the gene expression pattern of the Hox gene *labial*, which has a strong pattern in this second tardigrade body segment, differs from that of the other panarthropods, which starts in the third body segment, often together with *Hox3*, which is expressed in the third segment of all panarthropods, in this case also in tardigrades (Smith et al., 2016). Comparison of the expression domains of the patterning genes *six3*, *orthodenticle*, and *pax6* in the developing brain of the tardigrade *Hypsibius exemplaris* and those of arthropods (fig. 28.1) are most consistent with the tardigrade brain being composed of the protocerebrum alone (Smith et al., 2018), and this is most parsimoniously inferred to be the ancestral state for Panarthropoda as a whole (Ortega-Hernández et al., 2017; Park et al., 2018).

While Panarthropoda are recognized as monophyletic in most modern phylogenetic studies using morphology or molecules, phylogenomic analyses have re-

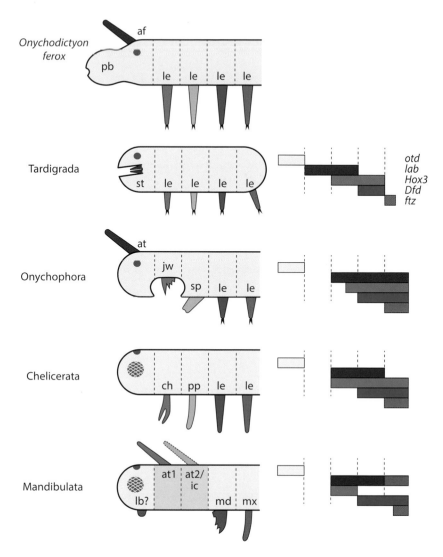

FIGURE 28.1. Anterior body segmentation of selected panarthropods. Illustrations at left based on Ou et al. (2012); Hox expression domains at right from Smith et al. (2016). Segment boundaries in left and right columns are aligned, with colors for Hox expression as follow: *otd*, light gray; *lab*, red; *Hox3*, green; *Dfd*, blue; *ftz*, dark gray. Other abbreviations: af, antenniform appendage; at, antenna; at1, first antenna; at2, second antenna; ch, chelicera; ic, intercalary segment; jw, jaw; lb, labrum; le, walking limbs/ lobopods; md, mandible; mx, maxilla; pb, proboscis; pp, pedipalp; sp, slime papilla; st, stylet.

peatedly identified a possible conflict with tardigrades. Tardigrades are sometimes believed to be the sister group of arthropods (see Yang et al., 2015 for a recent cladistic analysis), mostly due to similarities in the ganglionar nervous system (Mayer et al., 2013b; Gross and Mayer, 2015), as opposed to that of onychophorans. A tardigrade–arthropod grouping was formalized under the name Tactopoda, hypothesizing that a well-formed stepping gait is synapomorphic (Budd, 2001).

Tactopoda conflicts, however, with a series of phylogenomic analyses that have placed onychophorans closer to arthropods than are tardigrades and/or also cast some doubt on the membership of Tardigrada in Panarthropoda. Several analyses have found a relationship of Tardigrada to Nematoida[15] (Dunn et al., 2008; Hejnol

[15] This debate is often discussed as a nematode–tardigrade relationship, but this is not entirely precise, as several studies excluded nematomorphs. The monophyly of Nematoida is generally well supported, and thus we should refer to a nematoid–tardigrade relationship, although in a few studies

et al., 2009; Borner et al., 2014), while others suggest that this is due to a long-branch attraction artifact (Campbell et al., 2011; Rota-Stabelli et al., 2013). Many of these studies, however, relied on old ESTs, but newer analyses based on genome data or new transcriptomes continue to find the nematode–tardigrade clade (Laumer et al., 2015a; Yoshida et al., 2017). The debate is far from settled, as the same genomic data set that places Tardigrada with Nematoida suggests a relationship to the other Panarthropoda when only rare genomic changes in core developmental gene sets are analyzed (Yoshida et al., 2017).

As noted above, onychophorans are almost universally found to be the sister group of arthropods in modern molecular analyses, and several characters listed as panarthropod characters and thought to be reduced in tardigrades (whether their phylogenetic position was thought to be as either sister group of arthropods or of onychophorans) instead emerge as possible synapomorphies for the onychophoran–arthropod clade. These include the deutocerebral ganglion without expression of the Hox gene *labial*, the dorsal tubular heart with segmental paired ostia, and the segmental metanephridia. Tardigrades lack both a circulatory system and nephridia.

GENOMICS

The field of genomics was largely developed around the immense amounts of data available for panarthropods since the first draft genome of *Drosophila melanogaster* was published by Adams et al. (2000). Arthropod genomics is thus a healthy camp, with hundreds of genomes publicly available and many more underway (i5K Consortium, 2013).

However, less is known from the other two panarthropod groups. Genomics in tardigrades is emerging, due to interest in some of their biological properties, such as anhydrobiosis. The first draft genome of *Hypsibius exemplaris* was published (as *H. dujardini*), with a claim of extensive horizontal gene transfer (Boothby et al., 2015), and this was postulated to be correlated with them being such extremophiles. These results have, however, been dismissed as contamination artifacts (discussed in chapter 29). An additional genome is also available for the heterotardigrade *Ramazzottius varieornatus* (Hashimoto et al., 2016). No published onychophoran genome is yet available.

FOSSIL RECORD

Molecular dating predicts that panarthropods arose in the Precambrian. Various Ediacaran fossils with obvious differentiation along the anteroposterior axis and bodies composed of repeated units, such as *Spriggina floundersi* (fig. 1.4 H), *Marywadea ovata* (fig. 1.4 I), and *Parvancorina* species, have long been associated with arthropods (e.g., Peterson et al., 2008). However, these fossils lack appendages and are not demonstrably cephalized, and whether or not they possess bilateral

---

tardigrades nested within nematoids. Things become more complicated, as seen in the Ecdysozoa chapter, when we include Loricifera in the mix.

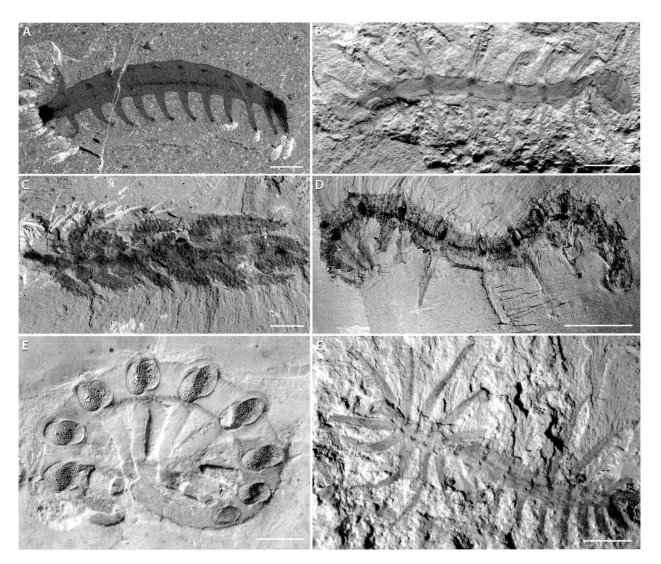

FIGURE 28.2. Cambrian lobopodians from the Burgess Shale (A) and Chengjiang (B–F). A, *Aysheaia pedunculata*, scale 5 mm; B, *Hallucigenia fortis*, scale 2.5 mm; C, *Diania cactiformis*, scale 5 mm; D, *Onychodictyon ferox*, scale 1 cm; E, *Microdictyon sinicum*, scale 2.5 mm; F, *Luolishania longicruris*, scale 2 mm. Photo credits: A, Jean-Bernard Caron; B, D, Derek Siveter; C, F, Xiaoya Ma; E, Lars Ramsköld.

symmetry (as opposed to so-called glide reflection symmetry, i.e., offset or zigzag units rather than segments) remains under debate. The reliable fossil record of Panarthropoda commences in the Fortunian Stage of the Cambrian, when the earliest trace fossils produced by arthropod-style locomotion appear (fig. 1.3 C). The Cambrian fossil record is numerically dominated by arthropods at the level of individual fossil specimens as well as species.

Although many Cambrian lobopodians (fig. 28.2) can be assigned to the stem groups of either Arthropoda or Onychophora (see those respective chapters), some phylogenetic analyses have placed particular lobopodians on the stem lineage of

Panarthropoda as a whole. This pertains, for example, to the Burgess Shale lobo-podian *Aysheaia pedunculata* (fig. 28.2 A) under some schemes of character weight-ing in a cladistic analysis by Smith and Ortega-Hernández (2014), whereas *Aysheaia* is instead resolved as a stem-group tardigrade in analyses by Caron and Aria (2017), who instead resolve such Cambrian lobopodians as luolishaniids (fig. 28.2 F) and hallucigeniids (fig. 28.2 B) as stem-group Panarthropoda. Considering such taxa as well the morphology of cycloneuralian outgroups and fossils placed on the stem lineages of the three panarthropod phyla, it is possible to predict that the ances-tral panarthropod was a macroscopic lobopodian with heteronomous body an-nulation, an anteriorly facing mouth with radial circumoral papillae, and paired dorsolateral epidermal structures (e.g., reticulate plates in *Microdictyon*: fig. 28.2 E) in segmental association with lobopodous limbs (Smith and Ortega-Hernández 2014).

TARDIGRADA

29

Among the most unusual of all invertebrates are water bears, the members of the phylum Tardigrada (fig. 29.1). Nearly 1,265 species of tardigrades have been described up to June 2018 (Degma and Guidetti, 2018), and between 10 and 30 new species have been added yearly during the past decade. Originally named *kleiner Wasserbär* (= little water bear) by German zoologist Johann August Ephraim Goeze in 1773 or *il tardigrado* by Italian biologist Lazzaro Spallanzani in 1776, tardigrades received phylum status in 1962 (Ramazzotti, 1962). "Water bears" are mostly microscopic, measuring between 50 µm and 1 mm, the latter in the giant species *Milnesium tardigradum*. Tardigrades inhabit marine, limnic, and terrestrial environments, and they are ubiquitous in both terrestrial and aquatic environments, thriving from the extreme cold soils of Antarctica to the abyss (Schmidt and Martínez Arbizu, 2015).

The capacity of tardigrades for undergoing cryptobiosis (both anhydrobiosis and cryobiosis) has fascinated zoologists as well as the general public, and much research now focuses on their amazing properties including resistance to radiation, heat and cold tolerance, and osmotic stress, as well as desiccation (Møbjerg et al., 2011; Heidemann et al., 2016; see a series of reviews in Schill, 2018), being able to withstand a change in temperature of over 400°C (Rahm, 1937b). Some of these responses to environmental variables result in seasonal cyclic changes in morphology and physiology, that is, cyclomorphosis (Kristensen, 1982; Møbjerg et al., 2007; Halberg et al., 2009b; Guidetti and Møbjerg, 2018). Tardigrades are also known for their capacity to survive in conditions found in outer space, and no group of invertebrates has more research done in space than tardigrades (e.g., Jönsson et al., 2008; Rebecchi et al., 2009; Persson et al., 2011; Rizzo et al., 2015). If an animal could escape the Earth's atmosphere and arrive on another planet, it would almost certainly have to be a tardigrade.

In addition to these ecological oddities, tardigrades are also morphologically bizarre, having been portrayed as "walking heads," as Hox expression domains reveal their five body segments to be the segmental homologues of the head of an arthropod plus a minute region of the terminal end of the body, having lost the equivalent of the thorax and most of the abdomen that would make an insect (Smith et al., 2016).

Phylogenetically, tardigrades have generally been associated with arthropods and onychophorans in a clade now known as Panarthropoda, but their relationship to these two other taxa has been in flux. While von Siebold (in von Siebold and Stannius, 1848) considered tardigrades part of the ingroup Arthropoda, most authors consider them separate taxa (irrespective of the rank assigned to them).

FIGURE 29.1. A, *Tanarctus bubulubus*, a buoyant marine arthrotardigrade from the Faroe Islands; B, the marine heterotardigrade *Echiniscoides sigismundi*, scale 50 μm; C, the eutardigrade *Dactylobiotus parthenogeneticus* molting its cuticle and leaving eggs inside the cuticle for protection; the hatchling shows the cuticular components of the stylet apparatus, pharyngeal apparatus, and claws. Photo credits: A–B, Reinhardt Møbjerg Kristensen; C, Noemí Guil.

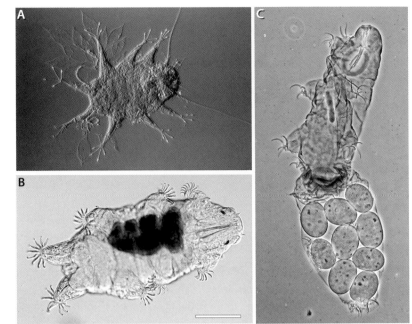

Traditionally, two prevalent hypotheses placed tardigrades as either sister group of arthropods—based mostly on the structure of the peripheral nervous system (e.g., Gross and Mayer, 2015)—or as sister group of onychophorans—based on their lobopodian body plan. Much recent research has thus focused on understanding their nervous system in comparison with those of arthropods and onychophorans. However, recent phylogenomic data contradict both of these hypotheses and instead place tardigrades either as sister group of onychophorans + arthropods (Campbell et al., 2011; Rota-Stabelli et al., 2013) or as related to nematoids (Hejnol et al., 2009; Borner et al., 2014; Laumer et al., 2015a) (see the discussion in previous chapters). Here we tentatively place Tardigrada as members of Panarthropoda, interpreting their paired segmental ventrolateral appendages as homologous, and follow current phylogenomic data recognizing a clade composed of Onychophora and Arthropoda (e.g., Laumer et al., 2019).

SYSTEMATICS

Molecular phylogenetics has contributed to understanding the systematic relationships of tardigrades to other ecdysozoans since the early days of the field (e.g., Garey et al., 1996a; Giribet et al., 1996). Likewise, molecular data have played a growing role in understanding tardigrade interrelationships (e.g., Jørgensen and Kristensen, 2004; Guidetti et al., 2009; Jørgensen et al., 2010; Guil and Giribet, 2012; Guil et al., 2013a; Guil et al., 2014; Fujimoto et al., 2017; see a review in Jørgensen et al., 2018), especially as phenotypic characters have been shown to be highly homoplastic (Guil et al., 2013b; but see Marchioro et al., 2013 for novel myoanatomical characters). Molecular data have also increasingly been applied to questions of species delimitation (e.g., Jørgensen et al., 2007; Guil and Giribet, 2009; Czechowski et al., 2012; Guil et al., 2013a; Faurby and Barber, 2015; Velasco-Castrillón et al., 2015; Cesari et al., 2016), revealing numerous species complexes, perhaps most notably in the case of *Hypsibius dujardini* and its commonly investigated laboratory strain, which is now named *H. exemplaris* (see Gąsiorek et al., 2018).

Traditionally divided into the three classes Heterotardigrada, Eutardigrada, and Mesotardigrada (Nelson, 2002), there are now doubts about the third class Mesotardigrada, which is based on a single species, *Thermozodium esakii*, from a hot spring near Nagasaki (Japan). The description of this species, by Gilbert Rahm (1937a, b), includes two drawings that do not seem to be accurate illustrations of a real specimen, and the type material as well as the type locality no longer exist, as they were both destroyed in an earthquake and no one has found any similar species there or in other regions (Suzuki et al., 2017). Mesotardigrada is now considered a nomen dubium (Grothman et al., 2017).

Heterotardigrada was traditionally comprised of two orders: Arthrotardigrada and Echiniscoidea. The traditional arthrotardigrades are marine species (with one exception), usually with a median cirrus on the head and telescopic legs bearing 4 to 6 toes with complex claws and/or sucking discs. Echiniscoids are primarily terrestrial with an armored cuticle with thickened plates, but there are some marine and freshwater unarmored species with telescopic legs bearing up to 13 claws. Arthrotardigrada is, however, paraphyletic with respect to Echiniscoidea (Jørgensen et al., 2010; Guil et al., 2013a; Fujimoto et al., 2017), and therefore here we consider Echiniscoidea to be a subclade of the order Arthrotardigrada, the only order of the class Heterotardigrada.

Eutardigrada includes the unarmored orders Apochela (terrestrial) and Parachela (primarily terrestrial and freshwater, with a few marine species); their legs terminate in claws without toes. Whether Eutardigrada is monophyletic or paraphyletic with respect to Heterotardigrada remains unresolved (Guil and Giribet, 2012), and the internal relationships of Parachela suggest the need of further superfamilies to accommodate the existing diversity (Marley et al., 2011; Guil et al., 2013b).

A recent estimate of species diversity has proposed that while about 95% of the limnoterrestrial species are currently known (>1,000 named), only about 19% of the marine ones are described (>180 named) (Bartels et al., 2016). This probably implies that many more marine lineages remain to be discovered.

    Class Heterotardigrada
       Order Arthrotardigrada
    Class Eutardigrada
       Order Apochela (= Milnesiidae)
       Order Parachela
          Eohypsibioidea
          Hypsibioidea
          Isohypsibioidea
          Macrobiotoidea

nervous system develops in an anterior-to-posterior gradient, beginning with the neural structures of the head. The brain develops as a dorsal, bilaterally symmetrical structure and contains a single central neuropil. The stomodeal nervous system develops separately and includes at least four separate, ringlike commissures. A circumbuccal nerve ring arises late in development and innervates the circumoral sensory field. The segmental trunk ganglia likewise arise from anterior to posterior and establish links with each other via individual pioneering axons. Each hemiganglion is associated with a number of peripheral nerves, including a pair of leg nerves and a branched, dorsolateral nerve.

The pattern of brain development (Gross and Mayer, 2015) as well as gene expression domains (Smith et al., 2016; Smith et al., 2017) thus support a single-segmented (unipartite) brain in tardigrades, unlike those of onychophorans (which are bipartite) or arthropods (tripartite), and contradicting earlier anatomical observations that suggested a tripartite brain (Dewel and Dewel, 1996; Persson et al., 2012; Persson et al., 2013; but see discussion in Møbjerg et al., 2018). Likewise, the tardigrade circumbuccal nerve ring cannot be homologized with the arthropod "circumoral" nerve ring, suggesting that this structure is unique to tardigrades (Gross and Mayer, 2015) and instead may be homologous to the cycloneuralian brain.

Most eutardigrades and echiniscoids have pigmented, cuplike eyes, found inside the lateral lobes of the brain rather than being adjacent to the body wall (Greven, 2007). Other sensory structures include the filamentous cirri or clava ("cephalic appendages") of marine tardigrades, probable chemoreceptors, each resembling the sensory setae of arthropods (Kristensen, 1981).

The developmental biology of tardigrades has recently been reviewed by Gross et al. (2015). Tardigrades lay a few eggs that can be deposited in the environment or into the old exuvium after ecdysis, corresponding to elaborate or smooth eggs, respectively (Gross et al., 2015). Early embryonic development has been studied in a few species, including classical studies (von Erlanger, 1895; von Wenck, 1914; Marcus, 1928a, 1929), although those results were somehow difficult to interpret. Eibye-Jacobsen (1996/97) studied the development of *Halobiotus crispae* and *Echiniscoides sigismundi* using light microscopy and transmission electron microscopy, but important aspects of development, such as the specific pattern of cleavage or the mode of mesoderm formation could not be determined. More recent studies have applied other techniques for cell lineage tracing, including 4D microscopy in *Thulinius stephaniae* (see Hejnol and Schnabel, 2005) and multiplane DIC recordings in *Hypsibius exemplaris* (see Gabriel et al., 2007). In the case of *T. stephaniae*, it has been shown that the embryos are highly regulative (indeterminate cleavage), but in the case of *H. exemplaris*, development is highly stereotyped. Whether or not this includes variation within the phylum remains unstudied. Likewise, some embryos seem to have synchronous cleavage during the first cleavages, while others are asynchronous from the onset. The blastula stage is a stereoblastula. While epiboly has been often cited as the mechanism for blastopore fate in tardigrades (Gabriel et al., 2007), it does not occur in *T. stephaniae* (see Hejnol and Schnabel, 2005). Mesoderm development continues to be contentious, but the presence of mesodermal pouches is accepted (Gross et al., 2015).

EXTREME EXTREMOPHILES

Tardigrades are well known for being able to push their physiological limits to extremes that most other organisms, including other extremophiles, cannot—these include abiotic factors, such as cold, heat, radiation, vacuum, and osmotic stress. Anhydrobiosis, the ability to withstand nearly complete desiccation, sometimes described as an ametabolic state (Wełnicz et al., 2011), has evolved in many limnoterrestrial animals, including nematodes and rotifers, but it is perhaps best understood in tardigrades, including some marine intertidal species (Jørgensen and Møbjerg, 2014). This constitutes a survival state under many of these extreme conditions. To enter the state of anhydrobiosis, many of these invertebrates are thought to require accumulation of compatible osmolytes, such as the nonreducing disaccharide trehalose to protect against dehydration damage (Hengherr et al., 2008). However, not all tardigrade species seem to accumulate trehalose during anhydrobiosis. Another important factor seems to be the accumulation of intracuticular lipids during desiccation, which act as a dehydration-dependent permeability barrier (Wright, 1989). In tardigrades and rotifers, the animals enter the so-called tun, a state that marks the entrance into anhydrobiosis, during which some species undergo 87% reduction of body volume (Halberg et al., 2013). During dehydration, tardigrades seem to undergo a series of anatomical changes in the contraction of their musculature, which mediates a structural reorganization vital for anhydrobiotic survival and is essential for resumption of life following rehydration (Halberg et al., 2013). Differential gene expression data suggest that de novo gene expression is required for successful transition to anhydrobiosis (Kondo et al., 2015). In addition, some species of tardigrades can enter into a diapause stage by forming a cyst with a cuticular covering that is abandoned once environmental conditions (often temperature) become favorable. It has also been documented that tardigrades can revive and successfully reproduce after being frozen for more than 30 years (Tsujimoto et al., 2016).

The loss of water can carry additional stresses, including oxidative stress, and tardigrades are able to deal with this by producing the pure superoxide anion radical, as opposed to what all other animals produce, which is either a mix of the superoxide anion and hydroxyl radicals, or just the latter (Savic et al., 2015). It is also interesting that some species are able to survive freezing in their active hydrated state, but this capacity is not phylogenetically correlated (Guidetti et al., 2011) and tolerance largely depends on the cooling rate (Hengherr et al., 2009; Hengherr and Schill, 2018).

Tardigrades are also well known to withstand elevated doses of many stressors, including ionizing radiation and UV radiation (reviewed by Jönsson et al., 2018). At least for UV radiation, their resistance is conferred both by high capacity for DNA damage repair and DNA protection in the anhydrobiotic state (Horikawa et al., 2013). While not as tolerant as the limnoterrestrial anhydrobiotic species, the marine heterotardigrades also withstand high doses of Gamma radiation (Jönsson et al., 2016). Interestingly, recent studies have shown that a tardigrade-unique DNA-associating protein termed "Damage suppressor," or "Dsup," suppresses X-ray–induced DNA damage by 40% in human cultured cells, improving radiotolerance

30

ONYCHOPHORA

Onychophora, its members commonly known as "velvet worms," or "peripatus," is the only entirely terrestrial animal phylum; that is, all species and life stages live on land—although they require environments highly saturated with water. This is why it is most common to find them walking about at night in temperate or tropical rainforests, while during the day one must look in the leaf litter, under stones, or inside extremely humid rotten logs. Their terrestrial mode of life requires tracheae for respiration and appendages adapted for sensing the environment, catching and chewing prey, and for locomotion.

The first onychophoran, *Peripatus juliformis*, from St. Vincent Island in the Caribbean, was described in 1826 as a "leg bearing slug" by the St. Vincent native Rev. Lansdown Guilding in his *Mollusca Caribbaeana* (Guilding, 1826). The group now comprises about 180 valid species (Oliveira et al., 2012), all restricted to the tropical belt, north and south of the Equator, and the temperate zones of the Austral continents. They are sometimes featured in popular science and folklore (Monge-Nájera and Morera-Brenes, 2015). During dry periods, velvet worms retire to protective burrows or other retreats where they can preserve moisture, while during wet periods they can be found actively hunting at night, inside damp fallen logs or in leaf litter in the regions where they live. They are therefore animals with low vagility and narrow ecological requirements (Murienne et al., 2014), highly structured genetically (McDonald and Daniels, 2012; Bull et al., 2013), and thus excellent subjects for biogeographic and conservation research (Hamer et al., 1997; Murienne et al., 2014; Myburgh and Daniels, 2015). Onychophorans live for several years, during which time periodic molting takes place, as often as every two weeks in some species.

The oldest known terrestrial fossil onychophoran is from the Carboniferous of Montceau-les-Mines, in France (Garwood et al., 2016), the group having changed little in their external morphology during the past 310 million years. A plethora of early Paleozoic lobopodians with annulated bodies and appendages (lobopods) are often associated with extant Onychophora, Tardigrada, or both (Ortega-Hernández, 2015).

The close relationship of onychophorans to arthropods has triggered a great deal of research in order to understand the earlier steps in the evolution of the largest animal phylum. The group has thus gone through a revival in the twenty-first century (Blaxter and Sunnucks, 2011), with novel research focusing on morphology and development of coelomic and nephridial structures and germ band formation (Mayer et al., 2004; Mayer et al., 2005; Mayer and Koch, 2005; Mayer, 2006a; Koch et al., 2014); structure of the brain and trunk nervous system and neurogenesis

(Eriksson et al., 2003; Strausfeld et al., 2006a; Strausfeld et al., 2006b; Mayer and Harzsch, 2007, 2008; Whitington, 2007; Peña-Contreras et al., 2008; Eriksson and Stollewerk, 2010a, b; Mayer et al., 2010b; Martin and Mayer, 2014; Martin et al., 2017); segment formation (Eriksson et al., 2010; Janssen et al., 2010; Mayer et al., 2010a); and visual pigments (Hering et al., 2012), as well as taxonomic, biogeographic, and biodiversity research (Murienne et al., 2014; Giribet et al., 2018).

SYSTEMATICS

Living onychophorans comprise two families, Peripatidae and Peripatopsidae (fig. 30.1). The former is circumtropical in distribution (with many species in the Neotropics, one in West Africa, and a few in southeast Asia and India), whereas the latter is circum-Austral (confined to the temperate Southern Hemisphere), with species in Chile, South Africa, New Guinea, Australia, and New Zealand.

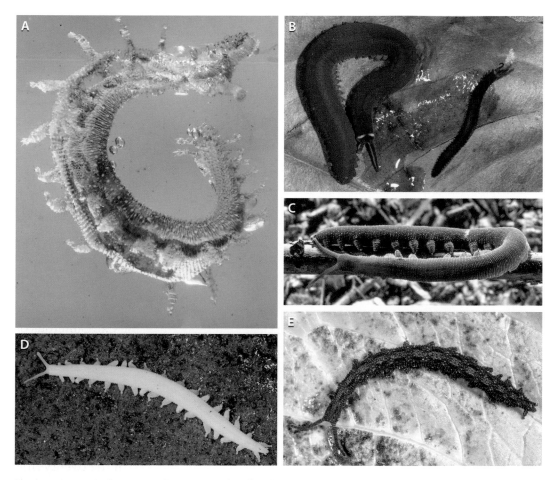

FIGURE 30.1. A, Cretaceous Burmese amber fossil, possibly of *Cretoperipatus burmiticus*; B, a female of the Neotropical placentotrophic viviparous *Macroperipatus torquatus* with newborn (Peripatidae); C, the only African Peripatidae, the viviparous *Mesoperipatus tholloni*; D, the cave-adapted matrotrophic *Peripatopsis alba* (Peripatopsidae); E, an oviparous species of *Ooperipatellus* from New Zealand (Peripatopsidae). Photo credit: A, Diying Huang.

The members of Peripatidae are characterized by having the genital opening between the penultimate leg pair; by having 19 to 43 leg pairs; by having a diastema on the inner blades of the jaws; and by body pigments being soluble in ethanol. A total of 74 species in 10 genera are currently recognized. The members of Peripatopsidae show the genital opening between the last leg pairs, have 13 to 29 leg pairs, and lack a diastema on the inner blades of the jaws; and their body pigments are not soluble in ethanol. A total of 106 species in 39 ill-defined genera are currently recognized.

Phylogenetic studies on onychophorans are limited to a few morphological and molecular data sets (Allwood et al., 2010; Oliveira et al., 2013), supporting the division of the group into two families and with good biogeographic structure (Murienne et al., 2014; Giribet et al., 2018). Recent Peripatidae divide into the Southeast Asian genus *Eoperipatus*, the West African monotypic *Mesoperipatus*, and a larger Neotropical clade that is sister group to the African species, but no molecular study has included the Indian *Typhloperipatus*, so its phylogenetic position remains untested. *Mesoperipatus* may constitute a group of cryptic species, instead of a monotypic genus (Costa and Giribet, 2016), and numerous cryptic species have been suggested among Peripatidae (e.g., Sampaio-Costa et al., 2009; Lacorte et al., 2011; Oliveira et al., 2011). The 99-My-old specimens from Burmese amber (Grimaldi et al., 2002) belong to Peripatidae (Oliveira et al., 2016). The Austral Peripatopsidae have more complex interrelationships, with certain biogeographical regions hosting more than one clade, as in Australia, New Zealand, and South Africa. Fine-level systematic work is clearly needed in these animals, as molecular data often indicate the lack of support for many previous cladistic morphological hypotheses and taxonomic groups and, in many cases, have shown cryptic diversity (Trewick, 2000; Oliveira et al., 2011; Daniels et al., 2013; Ruhberg and Daniels, 2013).

ONYCHOPHORA: A SYNOPSIS

- 180 recognized species; all terrestrial predators
- Adults ranging in length between 5 mm and 15 cm
- Bilaterally symmetrical worms, with ventrolateral pairs of legs (lobopods) corresponding to internal segmentation
- Cuticle containing α-chitin, externally annulated (not segmented externally); the cuticle is molted periodically
- Head with three pairs of appendages: antennae, jaws, and slime papillae that shoot glue; with one pair of simple rhabdomeric eyes (absent in some caverniculous species)
- Trunk with 13 to 43 pairs of annulated legs
- Body with circular, oblique, and longitudinal muscles
- Complete through-gut
- Subterminal genital opening
- Coelomic cavity largely reduced and merging to the primary cavity, forming a mixocoel; excretion via metanephridia
- Respiratory tracheae, which open to the outside by spiracles scattered over the body

- Brain formed by the fusion of two pairs of ganglia; CNS without trunk ganglia
- Direct development, which involves oviparity, or viviparity, the latter with or without a placenta

One of the most unique features of onychophorans bearing on their relationship to other metazoans is their central nervous system. It is ladderlike in structure but differs markedly from that of tardigrades and arthropods in several aspects (Eriksson and Stollewerk, 2010b). A large bilobed cerebral ganglion, probably formed by the fusion of the two anterior ganglia (protocerebrum and deutocerebrum), lies dorsal to the pharynx (Mayer et al., 2010b), supplying nerves to the antennae and eyes (protocerebrum), jaws (deutocerebrum), and lip papillae of the oral region (which receive nerves from both the protocerebrum and deutocerebrum) (Eriksson and Budd, 2001; Eriksson et al., 2003; Martin and Mayer, 2014; Mayer, 2016). The cerebral ganglion is followed by a pair of thick ventral nerve cords (fig. 30.2 A), without apparent segmental ganglia, connected by nonsegmental ring commissures, which are also connected to multiple longitudinal nerve tracts (Mayer and Harzsch, 2007).

Recent findings, based on neuroanatomical investigations, clearly revealed the presence of both segmental and nonsegmental features within the onychophoran ventral nervous system (Martin et al., 2017). This pattern shows a notable correspondence in position between structures such as musculature and peripheral nerves, thus leading to the assumption that the arrangement of the neurons in the onychophoran nervous system is dictated solely by the organization of the structures they supply (i.e., the lobopods), instead of being organized in functional units (ganglia) (Martin et al., 2017), as in arthropods and tardigrades.

Onychophoran appendages show different degrees of differentiation. The first pair of appendages is a pair of antennae, innervated by the protocerebrum, which are elongate and present a series of annulations similar to those of the body. The oral papillae house the opening of the slime gland, thought to have originated from modified metanephridia, and are responsible for secreting the glue that traps prey (Morera-Brenes and Monge-Nájera, 2010; Concha et al., 2015); they are innervated postcerebrally. The jaws are highly modified and have a highly cuticularized area with teeth; the presence or absence of a diastema is of taxonomic importance for distinguishing between the two recognized onychophoran families. The trunk appendages or lobopods are all similar in structure, bearing paired terminal claws and spinous pads (fig. 30.2 B). While tagmata in arthropods are generally defined by the staggered patterns of expression of Hox genes, in onychophorans all Hox genes are expressed until the posterior end of the embryo (Janssen et al., 2014).

Onychophoran limbs present regionalization with respect to expression of gap genes along the proximodistal axis—as in arthropods—even though they are unsegmented (Janssen et al., 2010; Janssen et al., 2015). However, onychophorans and arthropods differ in the dorsoventral limb gene expression, a pattern that seems less conserved among these two groups of panarthropods (Janssen et al., 2015).

The coelomic cavity of onychophorans and their associated excretory organs are also of general interest for understanding their relationship to arthropods. In onychophorans the coelom is restricted almost entirely to the gonadal cavities in the adult animal (Koch et al., 2014). In the Neotropical species *Epiperipatus biolleyi*, the fate of the embryonic coelomic cavities has been studied in detail (Mayer et al., 2004), providing evidence that embryonic coelomic cavities fuse with spaces of the primary body cavity (blastocoel). During embryogenesis, the somatic and splanchnic portions of the mesoderm separate and the former coelomic linings are transformed into mesenchymatic tissue. The resulting body cavity therefore represents a mixture of primary and secondary (coelomic) body cavities—that is, the "mixocoel." The hemocoel is also arthropod-like, being partitioned into sinuses, including a dorsal pericardial sinus.

A pair of nephridia lies in each leg-bearing body segment except the one possessing the genital opening, reinforcing the segmented nature of onychophorans. The nephridiopores are adjacent to the base of each leg (fig. 30.2 C), except in the fourth and fifth legs, whose nephridia open through distal nephridiopores on the transverse pads. The nephridial anlagen develop by reorganization of the lateral portion of the embryonic coelomic wall that initially gives rise to a ciliated canal (Mayer, 2006a). All other structural components, including the sacculus (fig. 30.2 B), merge after the nephridial anlage has been separated from the remaining mesodermal tissue (Mayer, 2006a). The nature of the excretory wastes is, however, not known.

The anteriormost nephridia are thought to have transformed into salivary glands and slime glands (fig. 30.2 A), in addition to the nephridial anlage in the antennal segment; the posterior nephridia seem to have developed into gonoducts in females.

Onychophorans are unique in their mode of development, as they may be oviparous or viviparous, with or without a placenta, and this is reflected in the great structural diversity of their ovaries (Mayer and Tait, 2009). Females of oviparous species (e.g., the members of the genus *Ooperipatus* from Australia) have an ovipositor and produce large, oval, yolky eggs with chitinous shells. Although some believe that this is the primitive onychophoran condition, oviparous species are rare among living onychophorans (fig. 30.1 E), and current phylogenetic evidence has shown that the oviparous clades are derived. Most living onychophorans are viviparous and have evolved highly specialized modes of development (Anderson, 1973). Most Old World viviparous species lack a placenta, whereas all New World viviparous species have a placental attachment to the oviduct wall (Anderson and Manton, 1972; Huebner and Lococo, 1994).

Velvet worms possess a tracheal system for respiration, which open by as many as 75 spiracles per segment (Gaffron, 1885), and oxygen-transport is mediated by an arthropod-type hemocyanin (Kusche et al., 2002), thus suggesting that the evolution of oxygen carriers preceded the divergence of Onychophora and Arthropoda. This is seen as an efficient circulatory system in a low-oxygen environment, but whether it evolved once in the common ancestor of onychophorans and arthropods or independently—since both groups have independent colonization of land—remains to be thoroughly tested.

Onychophorans are nocturnal, often secretive, animals, and their simple monochromatic eyes (Hering et al., 2012) have been homologized with the median ocelli of arthropods (Mayer, 2006b). The r-opsin (onychopsin) gene is expressed exclu-

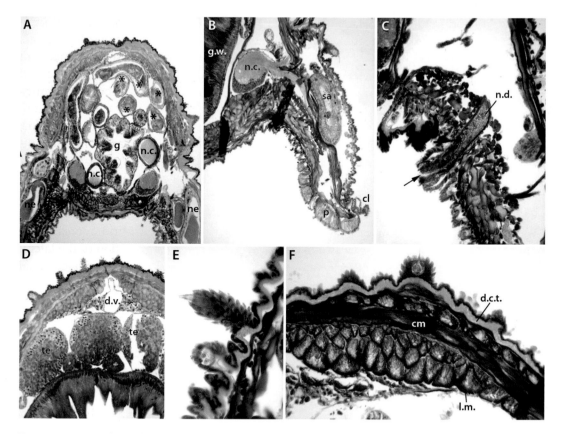

Figure 30.2. Histological sections through the body (A, D), legs (B, C), cuticle (E), and body wall (F) of the Australian peripatopsid *Euperipatoides kanangrensis*. Asterisks in A indicate slime glands; arrow in C indicates nephridiopore. Abbreviations: cl, claw; c.m., circular muscle; d.c.t., dermal connective tissue; d.v., dorsal vessel; g., gut; g.w., gut wall; l.m., longitudinal muscle; n.c., nerve cord; n.d., nephridial duct; ne, nephridia; p, pad; sa, sacculus; s.g., slime gland; te, testis.

sively in the photoreceptor cells of the eye, whereas c-opsins are expressed exclusively in the optic ganglion cells and the brain, suggesting that they do not have a role in vision (Beckmann et al., 2015). These findings thus suggest that onychopsin is involved in vision, whereas the onychophoran c-opsin may only have a photoreceptive, nonvisual function.

GENOMICS

An onychophoran genome (*Euperipatoides rowelli*) is publicly available (https://www.ncbi.nlm.nih.gov/bioproject/203089), although its annotation has not yet been produced. Illumina-based transcriptomes are available for a number of taxa (Fernández et al., 2014), and further transcriptomes are being applied to addressing the phylogenetic position of onychophorans with respect to the other panarthropod phyla and for developmental research (e.g., Franke et al., 2015).

Complete mitochondrial genomes are available for members of both Peripatidae and Peripatopsidae (Podsiadlowski et al., 2008; Braband et al., 2010a;

Braband et al., 2010b; Segovia et al., 2011). These have revealed that that onychophoran mitochondrial genomes have undergone major rearrangements. In addition, onychophorans show extreme tRNA editing (Segovia et al., 2011), which seems to have persisted through the long evolutionary history of the group.

FOSSIL RECORD

Various Cambrian lobopodians have been classified as stem-group Onychophora for many years (Ramsköld and Chen, 1998). The documentation of armored lobopodians from the Burgess Shale and Chengjiang has inspired several cladistic analyses aimed at placing various lobopodians in the stem lineages of each of the major panarthropod clades. One recent framework assigns a clade of lobopodians that have paired dorsoventral plates or spines on trunk segments (fig. 28.2 B, D–F) to the onychophoran stem group (Smith and Ortega-Hernández, 2014). A putative synapomorphy of *Hallucigenia* (taken as exemplary for the broader clade) and Onychophora is a shared mode of growth of the sclerites (claws and dorsoventral spines in *Hallucigenia*; claws and jaws in extant Onychophora; Oliveira and Mayer, 2013). This involves a stacking of constituent elements of the sclerites such that the outermost layer is shed in ecdysis and is replaced by the layer immediately beneath it (Smith and Ortega-Hernández, 2014). Under this scheme, Cambrian stem-group onychophorans include disparate morphologies and inferred ecologies, as exemplified by *Collinsium ciliosum* and allied species (Luolishaniidae), in which an anterior batch of trunk appendages is long, slender, and setose, and a posterior batch is stout and bears robust claws (Yang et al., 2015). These are inferred to have suspension feeding and anchoring functions, respectively.

Quantitative analyses of appendage tagmosis and morphospace occupation demonstrate that terrestrial crown-group onychophorans occupy a small portion of the morphospace represented by total-group Onychophora (Yang et al., 2015). An alternative phylogenetic hypothesis instead interprets the armored lobopodian clades of Luolishaniidae and Hallucigeniidae as a paraphyletic series in the panarthropod stem group rather than the onychophoran stem (Caron and Aria, 2017). In this framework, the oldest total-group onychophoran is *Antennacanthopodia gracilis*, from the early Cambrian of China (Ou et al., 2011), a species recovered in that position in previous analyses as well (Ma et al., 2014c; Yang et al., 2015).

The oldest terrestrial fossil velvet worm is *Antennipatus montceauensis* from the Carboniferous of Montceau-les-Mines (Garwood et al., 2016). Although it shows characteristics of modern onychophorans, it is not possible to determine whether it is in the onychophoran crown group because several diagnostic anatomical characters are not preserved. Tertiary fossils from Baltic and Dominican amber (Poinar Jr., 1996, 2000) are misidentifications, but the Cretaceous Burmese amber species *Cretoperipatus burmiticus* (fig. 30.1 A) is now well documented (Oliveira et al., 2016). Specimens of *Cretoperipatus* (fig. 30.1 A) show apomorphic characters of extant Onychophora such as leg pads, as well as specific details of the dermal papillae, the diastema in the jaws, and the position of the genital pad shared with Peripatidae.

ARTHROPODA

Arthropoda is by far the most species-rich animal phylum, outnumbering all others by an order of magnitude, constituting ca. 85% of the named animal species. Their extraordinary diversity is not confined to the Recent, as they have been the largest animal phylum since the early Cambrian, with some extinct lineages, like Trilobita, represented by more than 20,000 known species (Adrain, 2011). Insects alone consist of some 1.02 million described species (Zhang, 2011). In addition to numbers of species, arthropods constitute the largest proportion of the above-ground animal biomass in tropical rainforests, with 23.6 kg/ha, an abundance of 23.9 million individuals/ha, and a density on leaf surfaces of 280 individuals/m^2 leaf area—about 5 to 10 times higher than previously estimated (Dial et al., 2006).

Arthropods are also the dominant group of animals in the soil (e.g., Stork, 1996) and contribute more lineages to the marine meiofauna than any other animal phylum (Giere, 2009). They further constitute some of the largest marine blooms registered for any animal, as exemplified by krill (Atkinson et al., 2009)—with an estimated biomass of over 5 billion metric tons. Likewise, copepods are often the main component of the marine zooplankton, representing between 70 and 99% in terms of numbers and from 20 to 88% in terms of biomass for the <300-μm fraction (Thompson et al., 2013), and arthropods in general are the dominant animal group of cryptic benthic organisms in tropical reefs (Pearman et al., 2018). There is little doubt that arthropods are the animal lineage with the most impact and influence on a planetary scale.

Segmentation, on the one hand, and their rigid exoskeleton, on the other, are at the core of arthropod success. Cambrian fossils discussed below indicate that the primitive arthropod body plan was similar to that of onychophorans or the earlier developmental stages of many modern arthropods in having homonomous segmentation with serially similar limbs on each body segment. First, a specialization between head segments/appendages versus trunk segments/appendages probably provided a great evolutionary advantage to the first arthropods. Subsequently, the appearance of heteronomous segmentation, in which groups of segments were modified in concert to perform specialized functions, allowed for a much greater body regionalization. This process is known as "tagmosis." Typical arthropod tagmata include the prosoma and opisthosoma in chelicerates; cephalon, pereion, and pleon in malacostracans; head, thorax, and abdomen in insects; etc. Living arthropod diversity includes examples in which tagmosis is confined to the head and trunk, being clearly differentiated from each other, as in extant remipedes and myriapods. The functions of tagmata also vary in the different groups. While the anteriormost tagma usually serves the functions of feeding and sensing (there may be other tagmata also involved in sensory functions, such as in insects that hear with their legs), it also serves for locomotion in arachnids, while in insects it is the thorax that has a locomotory function.

Segmentation, body patterning, and appendage identity are well understood from developmental and genetic points of view, being favorite topics in the field of evolution. We know more about body patterning in arthropods than in any other animals, including humans. It is in arthropods where concepts such as modularity, axis formation, colinearity, or segmental identity have been developed and are best studied. The genetic and developmental model organism *Drosophila melanogaster* is perhaps the best-studied invertebrate species, forming the basis for much of modern genetics, including the discovery of Hox genes, which have long been known to play a fundamental role in specifying segmental identities along the anteroposterior body axis in arthropods (Lewis, 1978) and other animals. The interaction between homeotic proteins and the genes they regulate has been responsible for defining the different animal body plans and their distinct body parts (Carroll, 1995).

Heteronomous segmentation and their thick chitinous cuticle that is typical of the adult stages have also allowed arthropods to adapt to land in an unprecedented way compared to other animals, perhaps only paralleled by vertebrates, another group with a strong skeleton. The cuticle acts as an exoskeleton with two main functions in the terrestrial environment, providing structural support for locomotion (including flight in pterygote insects), and avoiding hydric stress by providing insulation to their interior milieu. The arthropod cuticle, and its constraints, form, and function, are discussed below. For now it suffices to stress that a rigid cuticle imposes important developmental constrains especially for growth, but also to the body plan design, requiring appendages with joints as well as intersegmental areas with arthrodial membranes to allow movement. The rigidity of their exoskeleton has other consequences for arthropod scholars: the abundance of characters for identifying and classifying them. It is therefore not a surprise that many theoretical contributions to systematics have been made by arthropod biologists.

Ecologically, arthropods span the spectrum from obligate carnivores to scavengers, detritivores, herbivores, to ecto- and endoparasites and parasitoids. Eusociality has evolved multiple times in disparate insect lineages (e.g., Blattodea, multiple instances in Hymenoptera), as well as in spiders, while others have developed subsociality. Eusociality in insects is often used as a model to compare with human societies, in which some individuals reduce their own reproductive potential to raise the offspring of others, and this has been justified with kin selection theory (Hamilton, 1962), based on the concept of inclusive fitness. This prevalent approach has been criticized, arguing that standard natural selection theory represents a simpler and superior approach (Nowak et al., 2010). The debate on eusociality in insects has always been heated!

Arthropods are also of immense importance to nonscientists in many ways. Arthropods in general, but especially crustaceans, are an important source of protein for humans, and many terrestrial groups have key ecosystem functioning properties, including pollination and nutrient recycling. Others, especially insects, can have a negative impact on agriculture as crop pests, act as vectors for infectious diseases such as malaria and dengue, or be main allergen agents, such as, for example, the dust mites in the genus *Dermatophagoides*. Finally, venomous arthropods abound in multiple clades (scorpions, pseudoscorpions and spiders among arachnids, centipedes, remipedes, and many insects), and these cannot only be potentially harmful—indeed, lethal—but culturally they are often dreaded or worshiped (e.g., spiders and scorpions).

SYSTEMATICS

While the definition of an arthropod has changed through history (see a detailed historical account in Ortega-Hernández, 2016), we follow here the most generalized usage of the term in non-Germanic textbooks; the latter often use Arthropoda as a synonym of what we call Panarthropoda. While it has been argued that the original name Arthropoda, proposed by von Siebold (1848) included Tardigrada (Ortega-Hernández, 2016), that composition differs considerably from the modern understanding of the group (see a discussion in chapter 28).

The deepest systematic divisions in Arthropoda have reached a reasonable level of consensus in the past decade after considerable debate since the mid-1990s (fig. 31.1). The multiple competing hypotheses have been discussed in many reviews, as summarized by Edgecombe (2010; see also Giribet and Edgecombe, 2012, 2013, 2019). One key controversy involved whether pycnogonids were sister group to all other euarthropods (the Cormogonida hypothesis) rather than being sister group to Euchelicerata (Giribet et al., 2001; Dunlop and Arango, 2005). This last hypothesis has been supported in most recent molecular analyses (e.g., Regier et al., 2010; Rota-Stabelli et al., 2013; Rehm et al., 2014) and is endorsed herein, the main morphological apomorphy being the deutocerebral appendage as a chelicera or chelifore.

Another recent debate involved whether Myriapoda is sister group to Chelicerata (the Myriochelata or Paradoxopoda hypotheses) rather than being sister group of a clade that unites Hexapoda with crustaceans, collectively named Pancrustacea or Tetraconata (the Mandibulata hypothesis). Mandibulata, thus, unites arthropods with the third cephalic appendage modified as a mandible. Support for Mandibulata has been found in most recent multigene or phylogenomic analyses (Regier et al., 2010; Rota-Stabelli et al., 2013), with Myriochelata/Paradoxopoda being interpreted as a result of saturated sites and deficient taxon sampling (Rota-Stabelli et al., 2011). Therefore, similarities in mode of neurogenesis between myriapods and chelicerates are now viewed as the plesiomorphic state for arthropods as a whole (Eriksson and Stollewerk, 2010a), rather than being apomorphic for Myriochelata.

A major systematic controversy in Euchelicerata is whether or not Arachnida is monophyletic. From the perspective of morphology this seems straightforward, the terrestrial arachnids sharing several compelling apomorphic characters to the exclusion of the primitively marine Xiphosura. That said, arachnid monophyly is contradicted with most molecular data sets, which tend to resolve Xiphosura within the arachnids (Sharma et al., 2014a), but this result may be impacted by long branch lengths affecting the position of several groups, notably the two main lineages of mites and ticks. However, a recent analysis of phylogenomic data conducting specific tests designed to counter systematic error supports a placement of Xiphosura within the arachnids, often as sister group to Ricinulei (Ballesteros and Sharma, 2019).

Euchelicerate ordinal relationships remain one of the open questions in arthropod phylogeny, though a clade that shares two pairs of book lungs—Arachnopulmonata—is consistently recovered (see a review in Giribet, 2018a). This unites scorpions, spiders (Araneae), and three extant orders grouped together as Pedipalpi: the Amblypygi, Uropygi, and Schizomida.

Myriapods and hexapods were long united in a terrestrial group named Atelocerata or Tracheata, to the exclusion of Crustacea. Today Atelocerata is seen to be supported only by a suite of morphological characters (Wägele and Kück, 2014), some of them linked to terrestrial habits, but is contradicted by a vast array of molecular analyses that relate hexapods to crustaceans (Pancrustacea or Tetraconata) rather than to myriapods, and Pancrustacea is also defended by a suite of neuroanatomical characters (Strausfeld and Andrew, 2011).

In this framework, Myriapoda is monophyletic (e.g., Fernández et al., 2016), but "Crustacea" is not. Crustaceans are a grade of primitively aquatic pancrustaceans, and, conversely, hexapods are terrestrial "crustaceans." The systematic framework for Pancrustacea outlined below is based on recurring patterns in well-sampled molecular phylogenetic analyses (Regier et al., 2010; von Reumont et al., 2012; Oakley et al., 2013; Misof et al., 2014; Schwentner et al., 2017; Schwentner et al., 2018), but it need be acknowledged that some of the groupings have few obvious morphological apomorphies (such as Allotriocarida, although potentially united by loss of the mandibular palp in ontogeny; Schwentner et al., 2017) or, indeed, are in conflict with morphological data (e.g., Multicrustacea).

Multilocus (Regier et al., 2010; Oakley et al., 2013) and phylogenomic (Misof et al., 2014) trees have emphatically settled on the monophyly of Hexapoda, after a few years in which mitogenomic phylogenies recovered Collembola with a separate crustacean sister group to other hexapods (Carapelli et al., 2007). Phylogenomic data resolve the relationships of primitively flightless hexapods with Ellipura (Collembola + Protura) as sister group to Diplura + Ectognatha (= Insecta). In this scheme, "Entognatha," which grouped Ellipura and Diplura as hexapods with the mouthparts enveloped by cranial folds, is paraphyletic (Misof et al., 2014; Schwentner et al., 2017). Diplura and Ectognatha had been recognized by morphologists as a clade, Cercophora, named for their shared terminal appendages, known as "cerci."

Insects, of course, constitute the basis for entire books and degree programs, so this chapter necessarily focuses on general aspects of Arthropoda rather than delving into the biology of insects in detail. The diversity of Insecta, especially the "Big Four" orders of Holometabola, is staggering, including some 116,800 living species of Hymenoptera, 386,500 species of Coleoptera, more than 157,300 of Lepidoptera, and over 155,400 of Diptera (Zhang, 2011). One family of beetles, the weevils, includes 51,000 described species, more than all other animal phyla apart from Mollusca and Chordata.

Subphylum Chelicerata
  Class Pycnogonida
  Euchelicerata
    Class Xiphosura
    Class Arachnida

Subphylum Mandibulata
  Myriapoda
    Class Chilopoda
    Progoneata
      Class Symphyla
      Dignatha
        Class Pauropoda
        Class Diplopoda
  Pancrustacea (= Tetraconata)
    Oligostraca
      Class Ostracoda
      Class Mystacocarida
      Icthyostraca
        Class Branchiura
        Class Pentastomida
    Altocrustacea
      Multicrustacea
        Class Malacostraca
        Class Copepoda
        Class Thecostraca
      Allotriocarida
        Class Cephalocarida
        Class Branchiopoda
        Labiocarida
          Class Remipedia
          Class Hexapoda
            Ellipura
              Order Collembola
              Order Protura
            Cercophora
              Order Diplura
              Insecta (= Ectognatha)
                Order Archaeognatha
                Dicondylia
                  Order Zygentoma
                  Pterygota

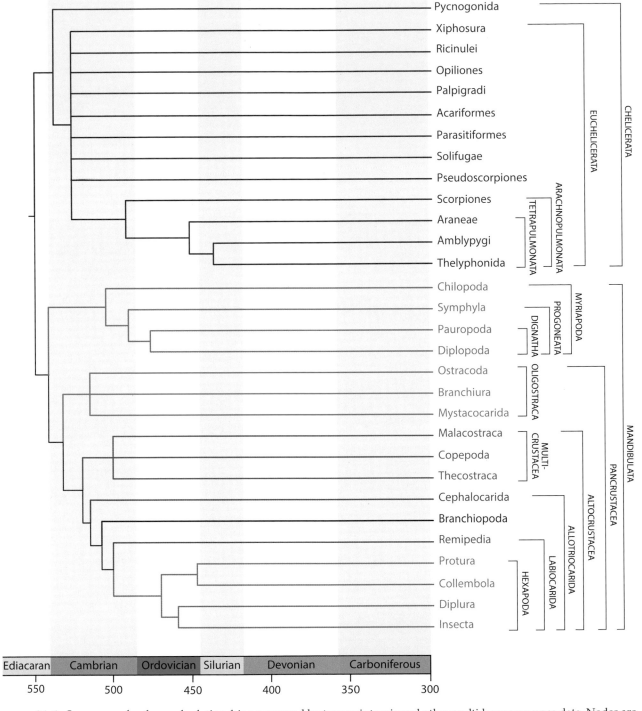

Figure 31.1. Summary of arthropod relationships recovered by transcriptomic and other multi-locus sequence data. Nodes are calibrated by molecular dating constrained by fossils. Modified from Edgecombe and Legg (2014) and Giribet and Edgecombe (2019).

ARTHROPODA: A SYNOPSIS

- Bilaterally symmetrical, primitively coelomate, segmented protostomes
- Coelom reduced to portions of the reproductive and excretory systems
- Body segments primitively associated with one pair of jointed appendages
- Body regionalized into two or more tagmata; in mandibulates a head, with sensory and feeding functions, and trunk, the latter often subdivided into a thorax (locomotory) and abdomen (digestive and respiratory functions); in chelicerates a prosoma, with feeding and locomotory functions, and an opisthosoma, with digestive and respiratory function
- Exoskeleton principally chitinous (typically with α-chitin), composed of sclerotized, articulated tergites dorsally; growth requires molting
- Appendages sclerotized, subdivided into stiffened podomeres separated by arthrodial membranes, with varied condyles and pivot joints at the articulations
- Compound eyes with rhabdomeric photoreceptors, modified to simple eyes in arachnids
- Brain tripartite, "rope ladder" nerve cord (sometimes largely condensed) composed of segmental ganglia joined by commissures and connectives
- Cardiovascular system in which hemolymph is pumped by a heart with dorsal segmental ostia and often with appendage "hearts" (pulsatile organs)
- Diverse respiratory systems, including gills in aquatic forms, book lungs in terrestrial arachnids, and a diversity of nonhomologous and noncentralized tracheal systems in insects, myriapods, and some arachnids
- Hemocyanin as the usual respiratory pigment
- All cells without cilia, except for sperm cells
- Development can be direct or indirect through a series of intermediate stages; animals can hatch with a final or a nonfinal number of segments

EXOSKELETON AND APPENDAGES

As in all ecdysozoans, the arthropod cuticle is a layered extracellular formation secreted by the epidermis. Each somite is encased by skeletal plates named "sclerites." Minimally, these consist of dorsal tergites, usually complemented by ventral sternites. The sides of the body, the pleuron, are typically membranous but in myriapods and hexapods include a variable number of sclerites known as "pleurites." At least some of these have been shown by gene expression to be derived from proximal segments of the leg that have been incorporated into the body wall (Coulcher et al., 2015). The inner surface of the sclerites usually bears apodemes, ridges or processes that serve as sites of muscle attachment. The sclerites are separated by areas of thin, flexible cuticle—arthrodial (or articular) membrane.

The outermost layer of the cuticle is called the "cement layer" or the "envelope" and is composed of natural lipids, wax esters, and proteins (Moussian, 2013). Immediately beneath the envelope is the epicuticle, which contains a waxy layer that confers a barrier to water loss in terrestrial arthropods and is mainly composed of proteins and lipids. The epicuticle is underlain by the exocuticle and endocuticle, collectively the procuticle, which is mostly constructed from layers of crystalline chitin (a long-chain polymer of N-acetylglucosamine, a derivative of glucose) and proteins. The epidermis contains unicellular glands, the ducts of which may extend to the cuticular surface.

Diagnostic of arthropods is sclerotization or tanning of the cuticle. This involves cross-linking of protein molecules with phenolic substances that confers a rigidity to the sclerites and also, usually, to the appendages. In various groups of arthropods, including various crustaceans and in most millipedes, mineralization makes for even greater rigidity. This typically involves calcium carbonate deposition in the procuticle, with calcite inserted between protein and fibers of chitin. Pigment, when present, is usually deposited in the exocuticle. Chitin in arthropod cuticle is in the form of α-chitin (e.g., Greven et al., 2016), referring to the antiparallel arrangement of the fibers in bundles of chitin, an exception to this being Pentastomida, the cuticle of which contains ß-chitin (Karuppaswamy, 1977).

Molting in arthropods is regulated by the steroid hormone ecdysone, dividing the life cycle into a series of instars between molts. Ecdysone is secreted by specialized glands in mandibulate arthropods, these being the prothoracic gland in insects and the X-organ in malacostracan crustaceans. While ecdysone is ubiquitous across Arthropoda, juvenile hormone plays an important role in insect postembryonic development, including regulation of molting and metamorphosis in holometabolous insects (Nijhout, 2013). Less is known about the hormonal mechanism of molting in arachnids or myriapods, although Chilopoda at least are known to employ the cerebral gland to regulate molting (reviewed by Rosenberg et al., 2011).

Numbers of postembryonic molts differ widely across Arthropoda as a whole (Minelli and Fusco, 2013), insects, for example, ranging from as few as just two to more than 30. Molt numbers can be fixed within species or subject to variability in response to such factors as diet, temperature, humidity, and population density.

As noted above, tagmosis is one of the most conspicuous features of the arthropod body plan. For some 20 years, the role of Hox genes in patterning segmental identities and with that the limits of tagmata have been appreciated (Hughes and Kaufman, 2002; Angelini and Kaufman, 2005). The prosoma of chelicerates for example shares broadly overlapping expression domains of the anterior six Hox genes (*labial, proboscipedia, Hox3, deformed, Sex combs reduced* and *Hox 6/fushi tarazu*), and the same is true of the opisthosoma, in which the posterior four Hox genes (*Antennapedia, Ultrabithorax, abdominal-A* and *Abdominal-B*) are expressed with overlap (Schwager et al., 2015). In mandibulate arthropods, which generally have differentiated gnathal appendages, the cephalic Hox genes have narrow expression domains, typically strongly expressed in just a single segment, which corresponds to a unique appendage morphology. The functional recruitment of trunk appendages into the head, in the form of maxillipeds (anterior trunk limbs used for feeding), is marked by expression of the typical cephalic Hox genes *Sex combs reduced* and/or *Antennapedia* in those segments, instead of the usual anterior trunk Hox gene *Ultrabithorax* (Averof and Patel, 1997).

In scorpions, all but one of the ten usual arthropod Hox genes are represented by two paralogues, and the expression boundaries of the duplicated four posterior Hox genes correspond to the distinctive division of the opisthosoma into two regions, the mesosoma and metasoma (Sharma et al., 2014b; Sharma et al., 2015). Arthropods inherited their complement of ten Hox genes from their most recent common ancestor with Onychophora, but have greater intersegmental variation; in onychophorans all Hox genes express to the posterior end of the embryo versus their greater degree of restriction in arthropods (Janssen et al., 2014). Recent reviews of segmentation and tagmosis include arthropods as a whole (Fusco and Minelli, 2013) and subgroups such as Chelicerata (Dunlop and Lamsdell, 2017).

The anatomical and ecological diversity of arthropods is to a large degree a consequence of their ability to differentiate the appendages along the anteroposterior axis of the body (fig. 31.2). Arthropods are the "Swiss Army knives" of the animal kingdom, with appendages of different tagmata or within a single tagma (notably the gnathal appendages) being specialized for different functions.

The developmental genetic basis for appendage structure has been revealed in various different arthropod models (Pechmann et al., 2010; Jockusch and Smith, 2015), usually based on comparison with *Drosophila melanogaster*. All arthropods pattern the proximodistal (P–D) axis of the appendages early in development by gap genes, which are expressed in broad domains in a conserved sequence along the P–D axis (Boxshall, 2013). The distal domain shows expression of *Distal-less*, and the proximal domain that in *D. melanogaster* corresponds to both coxa and trochanter shows expression of *extradenticle* and *homothorax*. Between these is the expression domain of the gap gene *dachshund*. Downstream of these gap genes, the position of joints is determined by the Notch signaling pathway, in particular the genes *Serrate*, *Delta*, and *fringe*. However, there is no one-to-one correlation between homologous pattern domains and homologous morphological domains among major groups, and the longstanding question of how the various podomeres that are given different names in different arthropod groups relate to each other is not answered by these data.

Appendages are characterized by sclerotized podomeres separated by an arthrodial membrane, the joints at which the procuticle is thin and flexible. The plane of movement at various joints is influenced by the number and positions of condyles, which act as fulcra. As well as their obvious functional importance, the arrangement of condyles between different podomeres can be a source of systematic characters, as, for example, in arachnids (Shultz, 2007).

The cell lineage of the thoracic appendages of the amphipod crustacean *Parhyale hawaiensis* has been reconstructed across several days of embryogenesis (Wolff et al., 2018), providing insights into limb morphogenesis, at the cellular scale, at an unprecedented level of detail. As in other studied malacostracans, each thoracic limb derives from cells from two neighboring parasegments (see below), each of which is initially represented by a single row of cells in the germ band. These data indicate that limb buds are derived from cells that are organized into compartments along the anteroposterior axis and the dorsoventral axis, a conserved feature shared with *Drosophila melanogaster*. These compartments intersect as the distal tip of the

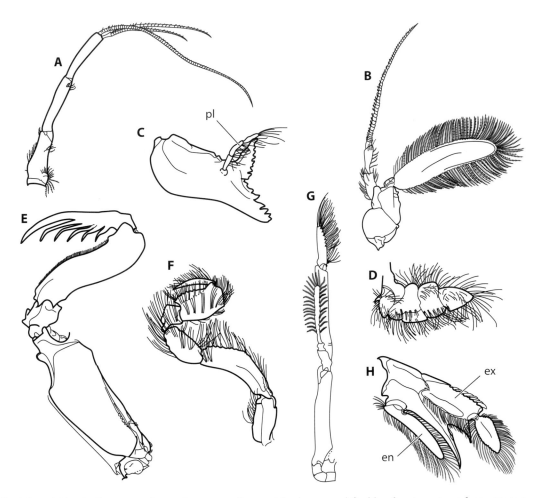

FIGURE 31.2. Serial variation in the appendages along an arthropod body, exemplified by the stomatopod crustacean *Oratosquilla nata* (not to scale): A, antennule (= first antenna); B, antenna (= second antenna); C, mandible; D, maxilla; E, first thoracopod; F, fifth thoracopod; G, eighth pereopod; H, uropods. Abbreviations: en, endopod; ex, exopod; pl, mandibular palp. Based on Tirmizi and Kazmi (1984).

developing limb. Elongation of the limb involves accelerated cell division in its distal part compared to that more proximally.

In extant arthropods the anteriormost appendage is that of the deutocerebral segment, the antenna in mandibulates (antennule in crustaceans) and the chelicera in chelicerates. The fact that this pair of appendages alone is anterior to the expression domain of the Hox cluster is one argument for recognizing a fundamental distinction between the antenna and all postantennal appendages (Boxshall, 2004). In this framework, the antenna is multisegmented and organized along a single P–D axis, whereas the postantennal appendages are at least primitively biramous. Ancestrally, the postantennal batch (including those of the head) was all structurally similar to each other, a state that is observed in fossil groups such as trilobites. In contrast to this, most arthropods display at least some degree of specialization of the cephalic appendages for feeding.

Biramous appendages have been subjected to different nomenclature for the constituent elements; we here follow the nomenclature of Boxshall (2004, 2013). The proximal limb stem that bears the rami is the protopodite. It commonly bears one or more spinose projections, gnathobases, on its inner margin. In crustaceans in particular the protopodite is subdivided into separate podomeres, the coxa and basis. In a biramous limb, the outer limb branch is the exopodite and the inner limb branch the endopodite. The latter is classically regarded as the homologue of the single limb branch of uniramous (or stenopodial) locomotory limbs of myriapods, hexapods, and arachnids, often called a "telopodite." The outer branch is inferred to have been lost as a consequence of terrestrialization. In various groups of crustaceans, epipodites are a third ramus, emanating from the coxal part of the protopodite, and mostly present on postcephalic limbs.

Comparison of uniramous and biramous limbs informed by clonal analysis in a single species, the amphipod *Orchestia cavimana*, showed that in this case uniramous limbs form, not by the loss of an exopodite, but rather by arresting a subdivision along the main limb axis that results in the biramous limbs (Wolff and Scholtz, 2008). However, this does not mean that all "biramous" limbs form the same way. Fossil taxa with two rami (such as trilobites) are classically described as having an exopodite and an endopodite, but unlike in *O. cavimana*, described above, the outer branch represents an outgrowth along a different axis from that of the inner branch. In such cases the more general term "exite" can be applied to the outer branch.

The identity of the anteriormost appendages and their innervation have long been debated, but are well understood now (fig. 31.3), largely based on the Hox expression domains noted above (chapter 28). Because several Sanger-based phylogenies at the time placed Pycnogonida as sister group to all other arthropods, a neuroanatomical study by Maxmen et al. (2005) suggested that the pycnogonid chelifores were innervated by the protocerebrum, but this was later shown to be an artifact and that the chelifores are innervated by the deutocerebrum (Brenneis et al., 2008). After resolving this issue, it is now accepted that chelifores are homologous to the chelicera of euchelicerates, as classically thought based on morphological similarities, to the antenna of myriapods, and to the first antenna of crustaceans (including the hexapod antenna) (Damen et al., 1998; Telford and Thomas, 1998). Following these correspondences, the pedipalp of chelicerates is homologous to the second antenna of most crustaceans (there is no tritocerebrally-associated appendage in the terrestrial mandibulates—myriapods and hexapods—which have a limbless intercalary segment). The first walking leg of chelicerates is thus innervated by the first postcerebral ganglion and corresponds to the mandible of mandibulates, and so on.

However, while homology of the anterior appendages is now well understood, there is no model to homologize posterior appendages other than by counting segments along the A–P axis. But are the appendages of the fifteenth trunk segment of a house centipede (their specialized terminal appendage pair) homologous to those of the fifteenth trunk segment of a scolopendromorph (with typically 21 or 23 trunk segments) or to the last trunk segment of the scolopendromorph? This question has received almost no attention in the literature.

In both, chelicerates and insects (studied from several species), the coeloms are closely associated with their visceral wall to the gut anlage, corresponding to a perivisceral coelom. They form by schizocoely within primarily apolar mesoderm. In chelicerates, the derivation of the nephridioducts from the embryonic coelom is confirmed, whereas in pancrustaceans they develop independently of coeloms; indeed, in arthropods as a whole a coelomic origin of sacculi is called into question (Koch et al., 2014).

A detailed review of the arthropod cardiovascular system was provided by Wirkner et al. (2013). The internal organs of arthropods bathe in hemolymph, the liquid that fills the body cavity, the hemocoel. The main organ that pumps the hemolymph is a heart, which may be accompanied by the movements of other organs, and an arterial system transfers the hemolymph to different parts of the body. Hemolymph flows from the hemocoel back into the heart or dorsal vessel through ostia, slit-shaped openings that are usually segmentally arranged. Septa divide the body cavity into sinuses through which hemolymph is channeled. These include the pericardial sinus, which encloses the dorsal vessel, and a perivisceral sinus that encloses other internal organs. Hemolymph further circulates between organs and muscles through narrow spaces called "lacunae." In addition to the heart, accessory pulsatile organs ("auxiliary hearts") supply body appendages with hemolymph in insects. These pulsatile organs are indispensable in the open circulatory system of insects, which mostly reduce the vascular system to the dorsal vessel, facilitating exchange of hemolymph to the antennae, long mouthparts, legs, wings, and abdominal appendages (Pass, 2000), and the accessory circulatory system becomes especially elaborate in the wings (Pass et al., 2015).

The cardiovascular system exhibits substantial differences between major groups. At one end of the spectrum, some small arthropods, such as Pauropoda, entirely lack circulatory organs and rely on diffusion and organ movement for transport, and at the other end such groups as horseshoe crabs have especially elaborately ramifying arterial and lacunar systems (Göpel and Wirkner, 2015). Crustaceans range from having only a short, saclike heart to the diverse arterial system of malacostracans, in which homologies of the cardiovascular system provide valuable systematic characters (Wirkner and Richter, 2010).

The exoskeleton's marked impermeability presents a challenge for gas exchange. Some small, lightly sclerotized arthropods, such as most pauropods, palpigrades, and some copepods, can oxygenate the tissues through the cuticle, and even at larger body sizes, as in pycnogonids, cuticular pores can facilitate oxygen diffusion (Lane et al., 2018), but these can affect the strength of the cuticle. However, the metabolic demands of most arthropods, especially terrestrial forms, require specialized gas-exchange organs. In most crustaceans, some kind of gills serves this role, typically represented by one of the rami of the appendages, often the epipods. Gills are usually folded in some manner to increase surface area.

Tracheae that open to spiracles are characteristic of hexapods, myriapods, and some orders of arachnids. In the latter, the spiracles are positioned on highly variable segments and fit to phylogeny suggests multiple origins. In hexapods and myriapods the spiracles are generally situated in the lateral part of the body, the pleuron, but they open dorsally, on the posteromedian part of the tergites, in scutigeromorph centipedes.

The book gills on opisthosomal segments 3–7 of the marine Xiphosura are widely accepted as the homologues of the book lungs of many terrestrial arachnids. In each case these lamellate respiratory organs originate from the posterior wall of the opisthosomal limb buds (Farley, 2010). In various araneomorph spider lineages, the second pair of book lungs may be missing or be replaced by tracheae.

Hemocyanin is the typical respiratory pigment in the hemolymph of arthropods, although hemoglobin is also known in various crustaceans. Many insects employ hexamerins in place of hemocyanin. Phylogenetic analysis of hemocyanin sequences has been useful for various questions in arthropod phylogeny, including corroborating a remipede/hexapod relationship (Ertas et al., 2009) and inferring the interrelationships of arachnid orders (Rehm et al., 2012).

NERVOUS SYSTEM

Arthropods have a dorsal, ganglionated brain composed of at least two preoral neuromeres, the proto- and deutocerebrum. A third neuromere, the tritocerebrum, is generally ascribed to the ancestral state of the arthropod brain/syncerebrum (Scholtz and Edgecombe, 2006; Strausfeld, 2012) though in various crustaceans it is topologically distinct, leading some to question whether a tripartite brain is general for arthropods as a whole (Richter et al., 2013). The tritocerebrum links the brain with the subesophageal ganglion (fig. 31.5) and with the stomatogastric nervous system, which supplies the foregut and the clypeolabral part of the head (Loesel et al., 2013).

In addition to receiving input from the eyes (see below), the arthropod protocerebrum has a system of unpaired midline neuropils, known as the "central body," connected with satellite neuropils (collectively, the central complex). In neopteran insects the central body is differentiated into three midline neuropils: the ellipsoid body, fan-shaped body, and superior arch (Loesel et al., 2002; Strausfeld, 2009). Satellite neuropils connected to these are the protocerebral bridge, its lateral bridge neuropils, and a pair of noduli. Some of these subunits, such as the fan-shaped body, protocerebral bridge, and lateral accessory lobes, are shared more widely with crustaceans, including malacostracans, remipedes, and branchiopods (Stegner et al., 2014).

In the context of a well-developed central complex as an apomorphic character of Pancrustacea, various midline and satellite neuropils (including the central body) need to be interpreted as lost in such diminutive crustaceans as cephalocarids (Stegner and Richter, 2011) and mystacocarids (Brenneis and Richter, 2010). The midline neuropil area in chelicerates is called the "arcuate body." Whereas the central body in mandibulates (including Chilopoda as a representative of Myriapoda) is embedded in the protocerebral matrix, the chelicerate arcuate body lies superficial to the protocerebrum (Loesel et al., 2002), and correspondences between the latter and a comparable arcuate midline neuropil in Onychophora (Strausfeld et al., 2006b) serve to polarize characters in arthropods. Across Arthropoda, the central body appears in crown-group arachnids are a transformation from compound eyes (Miether and Dunlop, 2016).

A basic distinction can also be made between the compound eye of xiphosuran chelicerates, which have corneal cones produced by thickening of the cornea, and the eyes of Mandibulata, in which the dioptric apparatus includes a crystalline cone. A conserved number of crystalline cone cells—an inferred ancestral number of four—is the basis for the name Tetraconata for the crustacean–hexapod (= Pancrustacea) clade. The number of retinular cells is also quite highly conserved in crustaceans and hexapods, the inferred ancestral state being eight such cells, in contrast to a larger, variable number in myriapods (Harzsch et al., 2007).

In most arthropods, the lateral compound eyes are complemented by various kinds of simple eyes, likewise innervated by the protocerebrum. These include the larval eyes of holometabolous insects, the three dorsal ocelli of most insects, and the naupliar eyes of crustaceans. Morphologically these range from simple pigment-cup ocelli (e.g., naupliar eyes) to structures capable of highly acute vision, such as the eyes of jumping spiders (Salticidae). The latter are diurnal hunters that use their anteromedian and anterolateral eyes to track prey; the anteromedian eyes are equipped with muscles that permit complex movements of the retina, and the retina is layered to facilitate enhanced color vision (Land and Nilsson, 2012). Visual organs can include as many as observed in horseshoe crabs, which have lateral compound eyes of more than 1,000 ommatidia, a pair of median (simple) ocelli, and three pairs of larval eyes (Battelle et al., 2016b), as well as photoreceptive organs on the telson that play a role in regulating the animal's circadian clock (Shuster et al., 2003).

Many of the best studied arthropods, such as the honeybee *Apis mellifera*, have trichromatic or tetrachromatic vision, associated with three or four visual opsins. However, a surprising diversity of visual opsins are being identified in other arthropods, including such taxa as horseshoe crabs (Battelle et al., 2016a), dragonflies (Futahashi et al., 2015), and branchiopod crustaceans, as exemplified by *Daphnia pulex*. The latter has 46 opsins, of which 27 are identified as visual (Colbourne et al., 2011), and has four-fold color vision, despite having compound eyes with only 22 ommatidia. The color vision of stomatopod crustaceans has attracted particular interest for the remarkable number of visual pigments that are concentrated in a band of enlarged ommatidia across the middle of the eye. Each of 12 tiers across this midband contains a different visual pigment, 8 of which are associated with the visible spectrum and 4 with the ultraviolet (Marshall et al., 2007). Stomatopods have the ability to discriminate circularly polarized light, which they produce to communicate with conspecifics (Gagnon et al., 2015).

The lateral eyes of arthropods are equipped with discrete, conserved neuropil areas that relay visual input to the protocerebrum. These are particularly elaborate in insects and decapod crustaceans, in which four optic neuropils are widely recognized—named the lamina, medulla, and the lobula complex (which consists of two parts, lobula and lobula plate) (fig. 31.5 C). These optic neuropils have a distinctive arrangement of crossed axons interconnecting them (Strausfeld, 2005). In contrast, branchiopods have two retinoptic neuropils connected by uncrossed axons (fig. 31.5 B). Two retinoptic neuropils likewise serve the compound eye of the centipede *Scutigera coleoptrata* (Sombke and Harzsch, 2015), and this arrangement is reconstructed as ancestral for arthropods.

Because arthropods are encased in their exoskeleton, epidermal nerve endings are separated from environmental stimuli, so external mechanoreceptors and chemoreceptors must penetrate the exoskeleton. The sense organs of arthropods include a diverse range of cuticular structures grouped as epidermal sensilla. These share the same basic components: sensory cells with dendritic processes and sheath cells, which are interspersed into the shaft of the sensillum (Müller et al., 2011). Mechanosensory sensilla are typically setae—movable, innervated cuticular processes set in sockets—and are especially common on the antennae (in mandibulates) and on the distal tarsomeres of other appendages in chelicerates. In contrast to simple setiform sensilla, mechanoreceptors can assume more complex forms, often being multi-innervated. Those used to detect air vibration in terrestrial arthropods are called "trichobothria."

The relative positions of the appendages are sensed by various kinds of proprioreceptors that function by transmitting information about neuronal deformation at articulations to the central nervous system. The slit sensilla of arachnids and the chordotonal organs of insects exemplify this kind of sense organs. Chemosensory sensilla, such as those used in olfaction, come in diverse shapes. Numerous kinds of compound sensilla are also found in different groups of arthropods, these being situated in depressed areas of the epidermis, with a pore at the external opening, such as Tömosváry's organs (also known as "temporal organs") in various groups of myriapods and in ellipuran hexapods (Collembola and Protura). In myriapods, these evidently play a role in carbon dioxide detection.

REPRODUCTION AND DEVELOPMENT

Although most arthropods are gonochoristic, hermaphroditism is known in some groups, such as remipedes, cephalocarids, and barnacles, and parthenogenesis (thelytoky) has originated in many groups, including populations of species that are otherwise gonochoristic (van der Kooi et al., 2017), as these populations tend to be characterized by broader ecological niches than their counterparts. Some large groups, such as insects, have direct insemination via copulation, whereas many other terrestrial lineages that have internal fertilization (within Arachnida and Myriapoda) transfer sperm via spermatophores. Deposition of a spermatophore is often associated with courtship rituals. Modified appendages used by the male for sperm transfer, such as the pedipalps of spiders and the gonopods of millipedes, are major sources of species-specific taxonomic characters. Sexual dimorphism can range from subtle to pronounced, examples of the latter including striking differences in size between females and males (e.g., in orb-weaving spiders) or differences in morphology and ecology, such as in strepsipteran insects in which the females are endoparasites of other insects and the adult males are short lived, nonfeeding winged forms.

Perhaps one of the most interesting questions in the evolution of sex in arthropods is the origin of haplodiploid systems in eusocial species. In haplodiploidy, males develop from unfertilized eggs and are haploid, whereas females develop from fertilized eggs and are diploid, as in the mictic phase of the monogonont rotifers (see chapter 39). This sex-determination system is sometimes called "arrhenotoky." Haplodiploidy is found in all members of Hymenoptera and Thysanoptera, as well as in some Hemiptera, Coleoptera, and some groups of mites. Another phenomenon is pseudo-arrhenotoky, or paternal genome elimination, in which males develop from fertilized eggs but where the paternal genome is heterochromatinized or lost in the somatic cells and not passed on to their offspring. This occurs in certain mites, Coleoptera and Hemiptera. A similar system is functional haplodiploidy, where, in males, the paternal genome is present in all tissues but is eliminated during spermatogenesis and not transmitted to the next generation. This system is found in some Hemiptera, Coleoptera, and Diptera. Hybrids of these systems also occur in a few groups, and all of these have been summarized by Normark (2003).

The literature on *Wolbachia* and sexual systems is growing exponentially since the discovery that experimental applications of antibiotics can convert thelytokous *Trichogramma* wasps back to arrhenotokous haplodiploidy (Stouthamer et al., 1990). *Wolbachia* is now well understood to be able to convert a lineage in which males are produced parthenogenetically to one in which females are produced parthenogenetically, by feminizing (through genome doubling) parthenogenetically produced sons (Normark, 2003).

With respect to reproductive structures, the gonopore may be single or paired, though taxon specific, and differs in position in different groups, as often reflected in classifications. Thus, in progoneate myriapods (the clade Progoneata) the gonopore opens behind the second pair of trunk legs, whereas their sister group, Chilopoda, is opisthogoneate, the gonopore situated at the posterior end of the trunk, as in hexapods. Euchelicerates have a fixed position of the gonopore on somite VIII, in the anterior part of the opisthosoma. Crustaceans display the most variability in gonopore position, though with conserved positions in some large groups, such as in Eumalacostraca, where it is located on thoracic segments 6 and 8 in females and males, respectively.

Sperm morphology and ultrastructure have played a significant role in the phylogenetics of all major groups of arthropods. Reviews of sperm structure and its systematic implications include those on chelicerates (Alberti, 1995), myriapods (Baccetti et al., 1979), and crustaceans (Jamieson, 1991) plus insects (Jamieson et al., 1999; Dallai et al., 2016). Flagellate sperm are regarded as primitive for arthropods as a whole—being the only cell type with cilia—and are known in select crustaceans (notably in remipedes) (Yager, 1989) and in the aquatic chelicerates, including pycnogonids (van Deurs, 1974) and xiphosurans. Some other major groups have highly modified aflagellate sperm, such as branchiopods and malacostracans, millipedes, and numerous groups of arachnids. Details of sperm ultrastructure have made key contributions to detecting sister-group relationships between pentastomid and branchiuran parasites (Wingstrand, 1972; Jamieson and Storch, 1992), and resolving the "entognathous" hexapods as paraphyletic because Diplura share apomorphies with Ectognatha (Dallai et al., 2011). They are also routinely used in cladistic analyses of chelicerate, myriapod, and hexapod relationships.

Arthropods show a variety of modes of cleavage (Scholtz and Wolff, 2013). These include superficial (or intralecithal) cleavage, in which cell membranes are lacking between the cleavage products, which migrate to the periphery of the egg, and total cleavage, in which blastomeres are separated by membranes, the blastomeres coming to surround a central cavity, the blastocoel. In numerous cases, cleavage involves traits of both of these modes, and the term "mixed cleavage" may pertain. Within major arthropod taxa, mode of cleavage typically varies between different subgroups.

For example, within Myriapoda, symphylans, pauropods, and some millipedes display total cleavage, whereas other millipedes and centipedes show superficial cleavage (Brena, 2015). Likewise in Chelicerata, xiphosurans have yolky eggs with superficial cleavage, and this is shared by several arachnid orders as well. Fit to phylogenetic trees suggests that instances of total cleavage in arachnids are probably derived from euchelicerate ancestors with superficial cleavage, whereas pycnogonids have total cleavage (Scholtz and Wolff, 2013). Although various arthropods had been described as having modified spiral cleavage (Anderson, 1973), no detailed correspondences are identified in studies of the same exemplar taxa, and there is no case to be made for spiral cleavage at the base of the Arthropoda (Scholtz, 1998). Crustaceans with total or mixed cleavage typically exhibit stereotypic early cleavage in which identifiable blastomeres have predictable subsequent fates, but this is not the case in hexapods, myriapods, or chelicerates (Ungerer and Scholtz, 2009).

Arthropod embryonic development is described in terms of a germ disc or germ band, a field of blastoderm cells on one side of the egg, the rest of which is composed of extra-embryonic tissue. Germ bands are often characterized as short or long, although, in fact, myriad intermediates are known (Chipman, 2015b). Short germ bands can be as extreme as consisting only of the developing head lobes, the subsequent segments being produced by a posterior growth zone. In the long germ band mode of development, segments are formed nearly simultaneously, as is best known for *Drosophila melanogaster*.

With respect to the generation of segments along the body axis, postembryonic development can be categorized as either anamorphic or epimorphic (Minelli and Fusco, 2013). Anamorphosis involves segment numbers increasing with successive molts, whereas epimorphosis refers to the complete adult segment number being attained in embryogenesis. In both cases, segments originate subterminally in a proliferative zone (but not in *D. melanogaster*; see below). Anamorphic development is classified into different modes according to whether or not molting and addition of segments continue for the entire life of the animal, without a fixed segment number (euanamorphosis), with a fixed segment number (teloanamorphosis), or whether an anamorphic phase is followed by an epimorphic phase, in which molts do not involve addition of segments (hemianamorphosis). This nomenclature was first used for millipedes (Enghoff et al., 1993) but applies to other arthropods as well.

Many arthropods are direct developers, and this mode is conserved across some large groups, such as arachnids. Crustaceans, being mainly aquatic, have numerous types of larvae, but a nauplius, in which the first and second antennae and mandibles are the appendicular complement, is widespread. Hexapods primitively have direct development, but insects include a diversity of developmental modes, including complex metamorphosis in the palaeopteran groups (Odonata and Ephemeroptera), with an aquatic larval form that looks drastically different than the terrestrial adults. The most diverse lineages within Insecta are either hemimetabolous (juveniles as feeding nymphs) or holometabolous (with larval and pupal stages).

As introduced earlier in this chapter, empirical and conceptual developments in evolutionary development biology have often been initiated based on arthropod models. *Drosophila melanogaster* was central to the discovery of the genes that pattern the anteroposterior axis of animal bodies (fig. 31.6), and in the case of segmented animals such as arthropods, ultimately specify segmental identity. In arthropods this involves the upstream to downstream sequence of gap genes, pair rule genes, segment polarity genes, and homeotic genes. While *Drosophila* is in many respects an atypical arthropod in that its segmental patterning is simultaneous along the body, the roles of segment polarity genes such as *engrailed* and *wingless* have been found to be conserved across the arthropods in establishing the initial boundaries of segments as parasegments (e.g., Damen, 2002 for arachnids). A contemporary account of the evo-devo of *Drosophila* is provided by Hartenstein and Chipman (2015). In the case of *Drosophila*, the genetic basis of most organ systems is well understood.

Other insect models have permitted comparison with *Drosophila*, with the aim of identifying conserved and variable molecular mechanisms that underpin similarities and differences in phenotypes. The red flour beetle *Tribolium castaneum* is especially widely studied, differing from *Drosophila* in having both a short germband embryo and larval appendages, and the milkweed bug *Oncopeltus fasciatus* is an attractive model system for a hemimetabolous insect (Chipman, 2015a). Over the past 20 years, gene expression data have accumulated for a range of chelicerates (reviewed by Schwager et al., 2015), myriapods (reviewed by Brena, 2015), and non-insect pancrustaceans.

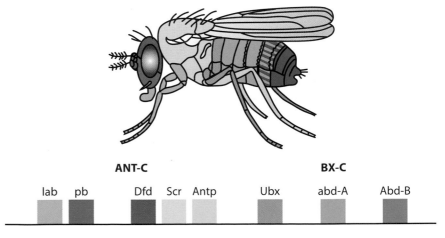

FIGURE 31.6. Hox gene expression correspondences in adult *Drosophila melanogaster*. Based on Wikipedia.

GENOMICS

Arthropoda includes the model organism par excellence, *Drosophila melanogaster*. The genome of *D. melanogaster* was thus the second available animal genome (Adams et al., 2000), soon after the release of the much more compact genome of *Caenorhabditis elegans*. Since the publication of that first arthropod genome, a steady number has been added every year, now totaling more than 420 genomes (as of May 2019), representing some 162 of the nearly 3,000 arthropod families (http://i5k.github.io/arthropod_genomes_at_ncbi). Several initiatives, including the 5,000 arthropod genomes initiative (i5K) community, continue to expand current genome sampling to unprecedented levels (i5K Consortium, 2013).

Chelicerate genomes now include multiple acarine species, especially those of commercial and medical importance—three members of Ixodida, five of Mesostigmata, nine of Sarcoptiformes, and one of Trombidiformes—as well as six spiders, two scorpions, a pseudoscorpion, and three of the four extant horseshoe crab species (Nossa et al., 2014; Kenny et al., 2016). Myriapod genomes consist of the centipede *Strigamia maritima* (Chipman et al., 2014) and the millipede *Trigoniulus corallinus* (Kenny et al., 2015). The former showed relatively little gene loss and shuffling compared to other available arthropod genomes, particularly compared to *Drosophila*, facilitating comparisons with non-arthropod bilaterians. Non-insect crustaceans include four species of Branchiopoda, eight Malacostraca, and seven Oligostraca, but genome-level data are missing for some classes of special phylogenetic interest, such as Remipedia and Cephalocarida.

As might be expected given their megadiversity, arthropod genomes are highly variable in size, displaying a twentyfold difference in genome size across the phylum, with much of this variation being witnessed even within groups such as mites (Pisani et al., 2013; see also www.genomesize.com). Recent studies have established that whole genome duplications have occurred multiple times in different arthropod clades, including in horseshoe crabs (Kenny et al., 2016), probably independently of them in the arachnopulmonate arachnids (Schwager et al., 2017), and numerous times in the evolution of hexapods (Li et al., 2018b).

Mitogenomics has a long history of study in arthropods. The mitochondrial genome typically includes the 13 coding genes, 2 ribosomal RNAs, and 20 tRNAs that are widespread across Bilateria (Boore, 1999; Pisani et al., 2013). Gene order rearrangements are common and have assisted with resolving some important phylogenetic questions, including the monophyly of Pancrustacea and the identification of pentastomids as oligostracan pancrustaceans (Lavrov et al., 2004). Mitochondrial gene order in arthropods is often described with reference to the inferred ancestral state in the horseshoe crab *Limulus polyphemus*. Mitogenomics is affected by considerable compositional heterogeneity in arthropods, notably AT enrichment, which is especially biased in pycnogonids, mites, and some insect lineages (Pisani et al., 2013).

FOSSIL RECORD

Arthropod groups with biomineralization of the exoskeleton are especially well represented in the fossil record; Ostracoda, for example, has more than 50,000 fossil species, and the wholly extinct Trilobita nearly 20,000 species, demonstrating that the enormous diversity of arthropods has been in effect throughout the Phanerozoic (Edgecombe and Legg, 2013).

The arthropod body fossil record commences at the base of Cambrian Series 2 (ca. 521Ma), when the first trilobites appear (fig. 31.7 G). Accounts of bivalved arthropods from earlier strata, in the Terreneuvian Series, are not dated with certainty. The body fossil record of arthropods is preceded by compelling evidence for their presence in the trace fossil record in sediments dated with confidence to the Fortunian, not far above the base of the Cambrian, although molecular dating often places the origin of arthropods in the Ediacaran (e.g., Rota-Stabelli et al., 2013; Schwentner et al., 2017). Terreneuvian sediments include locomotory and resting traces of arthropods such as *Rusophycus* (fig. 1.3 C), *Cruziana*, *Diplichnites*, and *Monomorphichnus* (Mángano and Buatois, 2014), at least some of which are likely to have been made by trilobitomorphs.

The biomineralized Cambrian fossil record dominated by trilobites is exuberantly supplemented in Burgess Shale-type biotas by a vast array of extinct groups that are known from details of their appendages and in many cases internal anatomy as well, particularly for the gut. In terms of either total number of fossil specimens or species diversity in such assemblages as the Chengjiang and Qingjiang biotas or the Burgess Shale, arthropods already outnumber all other animal phyla (Zhao et al., 2013; Caron et al., 2014; Fu et al., 2019). Burgess Shale-type biotas have been instrumental in permitting the sequence of branching events and character acquisition in the arthropod stem group to be inferred (Daley et al., 2018).

Although details of the arthropod stem group differ in various analyses, a number of patterns recur (Legg et al., 2013; Yang et al., 2015). The earliest branching stem-group arthropods are large-bodied lobopodians, as exemplified by such early Cambrian taxa as *Jianshanopodia decora* and *Megadictyon haikouensis* (Vannier et al., 2014). Apomorphic characters that identify them as total-group arthropods are enlarged, spinose, raptorial frontal appendages with presumed protocerebral innervation and segmental midgut diverticulae with a characteristic network of internal canals. These characters are retained by so-called gilled lobopodians, including *Kerygmachela kierkegaardi* (fig. 31.7 B) and *Pambdelurion whittingtoni* (fig. 31.7 C), which possess segmental trunk flaps and possibly lobopodial ventrolateral appendages. Radiodonta, including such familiar animals as *Anomalocaris* (fig. 31.7 A), are resolved crownward of these lobopodians in the arthropod stem group, possessing fully arthropodized frontal appendages and compound eyes (Paterson et al., 2011) but lacking trunk appendages.

Crownward of Radiodonta, the details of the arthropod stem group are particularly contentious, as several intensively studied taxa are variably regarded either as stem-group or crown-group Arthropoda. For example, fuxianhuiids (fig. 31.7 D), an early Cambrian group from China for which anatomical data even include remains of the nervous system (Ma et al., 2015b; Yang et al., 2016), are variably placed in the arthropod stem group (Yang et al., 2018) or in Mandibulata, in the arthropod crown (Aria and Caron, 2017). Likewise, megacheiran "great appendage" arthropods (fig. 31.7 E) are resolved either outside or inside the arthropod crown group—usually, in the latter position, as stem-group chelicerates. However, the

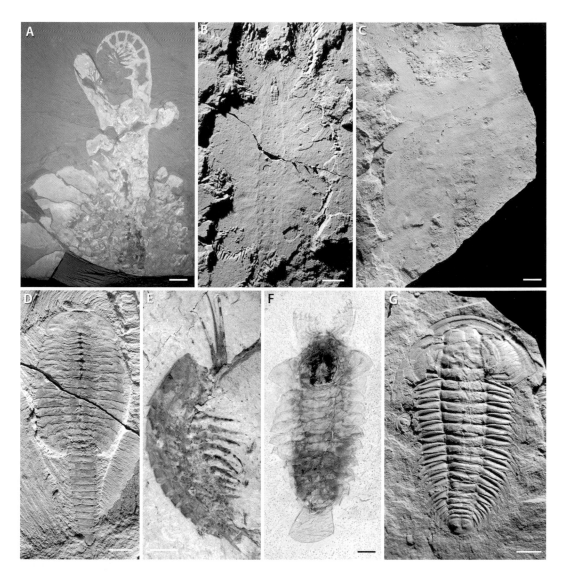

Figure 31.7. Cambrian total-group arthropods: A, *Anomalocaris canadensis*, scale 1 cm; B, *Kerygmachela kierkegaardi*, scale 1 cm; C, *Pambdelurion whittingtoni*, scale 1 cm; D, *Fuxianhuia protensa*, scale 5 mm; E, *Leanchoilia illecebrosa*, scale 1 mm; F, *Sanctacaris uncata*, scale 5 mm; G, *Redlichia takooensis*. Photo credits: A, Allison Daley; B, C, Jakob Vinther; D, E, Xiaoya Ma; F, David Legg; G, John Paterson.

monophyly of a group that includes all of these fossil taxa as well as the arthropod crown group is recognized by the Deuteropoda concept (Ortega-Hernández, 2016). This is based on the shared presence of a deutocerebral anteriormost head appendage. This organization contrasts with frontal appendages deeper in the arthropod stem group (in Radiodonta, for example), which are interpreted as protocerebral (Cong et al., 2014; Ortega-Hernández et al., 2017).

Chelicerata: The chelicerate fossil record was reviewed systematically by Dunlop (2010). The earliest chelicerates are Burgess Shale-type arthropods such as

Wisangocaris barbarahardyae and *Sanctacaris uncata* (fig. 31.7 F) (Jago et al., 2016). Paleozoic taxa with a horseshoe crab–like appearance, traditionally grouped as Synziphosurina, are now thought to be a paraphyletic assemblage that is stemward of all crown-group euchelicerates (Lamsdell, 2013). Eurypterids are one of the best known groups of fossil chelicerates, known from superb preservation of the appendages and even reproductive structures such as sperm-carrying organs (Kamenz et al., 2011) that reinforce their traditional position as sister group of arachnids.

Scorpions have traditionally been interpreted as sister group of all other arachnids, and supposed aquatic fossils used to underpin separate origins of terrestrial habits in crown-group scorpions and in other arachnids (a grouping that was called Lipoctena; Weygoldt and Paulus, 1979). Molecular data now provide support for scorpions being nested within Arachnida as closest relatives of Tetrapulmonata (Sharma et al., 2014a), a clade (including spiders) that bears two pairs of book lungs. The book lungs of living scorpions share characters consistent with homology with those of Tetrapulmonata (Scholtz and Kamenz, 2006). This homology has prompted a reinterpretation of some fossil scorpions that had previously been regarded as aquatic on either morphological or sedimentological grounds as potentially being terrestrial forms. The extinct (Silurian to Permian) arachnid order Trigonotarbida has been shown to bear book lungs on the same two segments as tetrapulmonates. The larger-bodied arachnid orders have fossil records that generally extend back to the Carboniferous (e.g., Amblypygi, Uropygi), and mites have fossils as old as the Early Devonian, represented by crown-group acariforms in the Rhynie Chert, Scotland. In addition to the trigonotarbids, a few other extinct groups of arachnids, such as Haptopoda, Phalangiotarbida, and Uraraneida, are classified as orders.

Myriapoda: Two of the four myriapod classes, Symphyla and Pauropoda, have a fossil record confined to amber, symphylans recently being documented from Late Cretaceous Burmese amber (Moritz and Wesener, 2018). The two more diverse classes, Diplopoda and Chilopoda have fossils as old as the Silurian (Shear and Edgecombe, 2010). No terrestrial or aquatic fossils have reliably been assigned to the myriapod stem group.

Penicillata (bristly millipedes) are sister group to all other millipedes, the latter (Chilognatha) united by the calcified cuticle (Blanke and Wesener, 2014). Penicillates occur in amber as old as the Early Cretaceous, from Lebanon, but most fossils belong to the more robust Chilognatha. Paleozoic millipedes are mostly assigned to extinct orders. The oldest (Silurian and Devonian) fossils as well as the large spiny millipedes of Carboniferous coal swamps, Euphoberiida, belong to the extinct group Archipolypoda. Arthropleurids, including the largest terrestrial arthropods of all time, are generally regarded as millipedes (Edgecombe, 2015). Mesozoic millipedes were very rare until most extant orders are found in Cretaceous (ca. 99 Ma) Burmese amber (Wesener & Moritz 2018).

Centipedes appear in the fossil record in the Late Silurian, represented by the extant order Scutigeromorpha (reviewed by Edgecombe, 2011). The Devonian includes an extinct order, Devonobiomorpha, represented by a single species from New York State. The Carboniferous has the first fossils of Scolopendromorpha and the Jurassic the oldest Geophilomorpha. No stem-group centipedes are known.

Current molecular dates for myriapods predict that the basal divergence between Chilopoda and Progoneata (Symphyla, Pauropoda, and Diplopoda) dates to the Cambrian, considerably older than the oldest body fossils (mid Silurian) or trace fossils (Late Ordovician) of myriapods (Lozano-Fernandez et al., 2016; Fernández et al., 2018). Inferred terrestrial habits (because the extant members are terrestrial and have terrestrial adaptations such as tracheae and Malpighian tubules) of the last common ancestor of myriapods are inconsistent with the absence of terrestrial flora in the Cambrian, although recent molecular dating of land plants predicts Cambrian origins (Morris et al., 2018). Trackways made by arthropods like euthycarcinoids on mid-Cambrian tidal flats are the oldest evidence of subaerial locomotion by arthropods (Collette et al., 2012), although these animals were more likely amphibious than fully terrestrial.

As for Mandibulata and Pancrustacea, a series of three-dimensionally preserved fossils, mostly from the late Cambrian Orsten of Sweden, have alternatively been identified as stem-group Crustacea (Haug et al., 2010) or as stem-group Mandibulata (Legg et al., 2013; Edgecombe, 2017). In either case, a group of superficially ostracod-like fossils known as Phosphatocopina are regarded as the closest extinct relatives of the crown group, sharing characters of the labrum and mouthparts with crustaceans. Interpreted as a paraphyletic grade, these fossils demonstrate stepwise evolutionary changes in the evolution of the gnathal appendages, including the mandible (Edgecombe, 2017).

Molecular evidence has provided strong support for remipedes (rather than other candidates proposed over the past decade or so, such as Branchiopoda or Malacostraca) being the closest relative of Hexapoda (Schwentner et al., 2017). The oldest definite hexapod is the Early Devonian collembolan *Rhyniella praecursor*. The coeval *Rhyniognatha hirsti*, known only from mandibles that have been interpreted as those of a pterygote insect by Engel and Grimaldi (2004), has recently been disputed as being an insect (Haug and Haug, 2017). Fossilized pterygote wings are recorded from the late Early Carboniferous onward, the earliest known fossils identified as orthopterid (Prokop et al., 2005). Hexapods are rare in the fossil record between the Early Devonian and the Late Carboniferous, an interval during which arachnids (and to a lesser extent myriapods) dominate the terrestrial arthropod fossil record that has been termed the "hexapod gap" (Schachat et al., 2018). Insects become abundant from the Late Carboniferous onward, coincident with direct evidence for the evolution of flight. Holometabolous insects are already known from the Middle Pennsylvanian (Nel et al., 2013).

32

SPIRALIA

In 1898, in a lecture at the Marine Biological Laboratory at Woods Hole, Massachusetts, developmental biologist Edmund B. Wilson proposed that certain similarities in cell lineages hint at an evolutionary relationship between groups since united as Spiralia (see Lambert, 2008). Since then it has become evident that the members of all phyla with spiral cleavage—a highly stereotyped egg cleavage pattern (e.g., Freeman and Lundelius, 1992) in which the 4d cell typically gives rise to the mesoderm of the adult individual (e.g., Henry and Martindale, 1998; Lambert, 2008; Hejnol, 2010)—constitute a clade of animals (Nielsen et al., 1996; Nielsen, 2012a).[16]

Because molecular data have shown that this clade also comprises animal groups with other types of developmental patterns, including the lophophorate phyla (Halanych et al., 1995), the name Lophotrochozoa has often been applied to Spiralia. We, however, follow a more precise nomenclature wherein Lophotrochozoa is restricted to a subclade of Spiralia (see chapter 44), following the original definition of the name Lophotrochozoa (Halanych et al., 1995), as discussed elsewhere (e.g., Giribet et al., 2009). The name Spiralia has been likewise preferred in recent invertebrate textbooks (e.g., Westheide and Rieger, 2007; Nielsen, 2012a; Brusca et al., 2016b). Furthermore, the term Lophotrochozoa refers to the presence of a lophophore or a trochophore larva, and neither character is found in Rouphozoa or Gnathifera, the two first offshoots of Spiralia.

Spiralia is a difficult clade to define based on morphology (Giribet, 2014c), and while spiral cleavage optimizes as a synapomorphy for this clade, it is homoplastic due to multiple losses. Nevertheless, spiral cleavage does not occur outside Spiralia. Current phylogenetic resolution of Spiralia suggests the presence of three major clades, Gnathifera, Rouphozoa and Lophotrochozoa (Struck et al., 2014; Laumer et al., 2015a; Kocot et al., 2017; Laumer et al., 2019). At least some members of each of these clades show spiral cleavage.

The phylogeny of Spiralia has been in flux, and some relationships remain unresolved, even when using large transcriptomic/genomic data sets (Laumer et al., 2015a; Kocot et al., 2017; Laumer et al., 2019; Marlétaz et al., 2019). Debate about the monophyly (Giribet et al., 2000) versus paraphyly (Struck et al., 2014) of Platyzoa (Cavalier-Smith, 1998) is now largely settled, this group constituting a grade composed of two clades, Gnathifera and Rouphozoa, the latter being the sister group of Lophotrochozoa (Struck et al., 2014; Laumer et al., 2015a; Laumer et al., 2019; but see Marlétaz et al., 2019). Lophotrochozoan internal relationships remain difficult to elucidate, especially due to the unstable position of Cycliophora in

[16] While Nielsen et al. (1996) recognized a clade that contained spiral cleavers as Spiralia, their clade differs in composition from our—including Nielsen, 2012a—current understanding of it.

relation to Entoprocta and the lophophorate taxa (Laumer et al., 2015a; Kocot et al., 2017). To complicate things even further, Orthonectida is now considered a highly modified parasitic member of Lophotrochozoa (Mikhailov et al., 2016), as is also another parasitic group, Dicyemida (Suzuki et al., 2010). The possibility that Chaetognatha belong with Gnathifera (Fröbius and Funch, 2017) is now supported by phylogenomic data (Laumer et al., 2019; Marlétaz et al., 2019) and has also been recovered by morphological phylogenetics (Vinther and Parry, 2019).

CELL LINEAGES AND FATES

Embryologists often refer to cell lineages, especially in those animals with highly stereotyped developmental modes. The cell lineage of spiralian embryos has been thoroughly studied and established since the late 1800s. Multiple techniques can be used to trace a cell lineage and establish a fate map. But not all embryos behave in the same manner, and in some cases cell fates are not established until later in development, if ever, while in other animals cell fates are already established during early cleavage, as early as the 2- or 4-cell stage (Gilbert and Raunio, 1997).

The study of a cell lineage can be done in different ways, including direct observation of developing embryos, but this becomes nearly impossible after a few cell divisions, and thus modern techniques include cell labeling or automated tracing through 4D microscopy (multifocal, time-lapse video recording system) (Schnabel et al., 1997). Experimental manipulation of embryos allows establishing whether they are determinative (cell fates are established early) or regulative (also called indeterminate; cell fates are established later and removal of early embryonic cells does not affect normal development).

An elaborate coding system for spiral cleavage was developed by pioneering embryologist Edmund B. Wilson for the annelid *Alitta succinea* (see Wilson, 1892) (fig. 32.1). Following Wilson's work, a simple account of the spiralian development

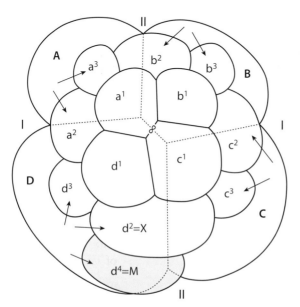

FIGURE 32.1. Cleavage map of *Alitta succinea* originally drawn by Wilson (1892), with the nomenclature used at the time.

follows (Gilbert and Raunio, 1997). Irrespective of some variation with respect to the point of fertilization and the establishment of the embryonic axes, the first cleavage already results in an asymmetry, with the two daughter cells being of different size (fig. 32.2 B). At the 4-cell stage, following the initial meridional divisions, the cells are named A, B, C, and D—the D cell is often slightly larger than the others, providing a point of reference for naming the cells—and are labeled clockwise (beginning with the left-most cell) when viewed from the animal pole (32.2 C–D). This stage is often referred to as the quartet of macromeres (Q).

The next cell division is more or less unequal, with the four cells near the animal pole being displaced in a dextrotropic fashion. The four smaller cells are called the first quartet of micromeres (1q cells) and are given individual codes of 1a, 1b, 1c, and 1d, and their macromere counterparts (1Q cells) are 1A, 1B, 1C, and 1D (i.e., A gives origin to the 1A and 1a cells) (fig. 32.2 E). In some instances the macromeres and micromeres are similar in size, but this nomenclature is used in describing all spiral embryos. The size discrepancy often depends upon the amount of yolk present at the vegetal pole in the original egg, which is often retained primarily in the larger macromeres. In some spiralian embryos (molluscs and annelids), vegetal cytoplasm is segregated into a cytoplasmic protrusion called a "polar lobe," which fuses with the D quadrant (fig. 32.2 B, D). The polar lobe has an important role in D quadrant specification and the subsequent development of the embryo of *Tritia obsoleta* (e.g., Clement, 1952; Lambert and Nagy, 2001; Lambert et al., 2016), but not in *Crepidula fornicata*, as shown by deletion experiments (Henry et al., 2006; Henry et al., 2017).

The division from 8 to 16 cells occurs levotropically and involves synchronous cleavage of each macromere and micromere. The macromeres (1Q) divide to produce a second quartet of micromeres ($2q = 2a$, 2b, 2c, 2d), and 4 daughter macromeres, whose prefix numeral is changed to "2." The first micromere quartet also divides and now comprises 8 cells, each of which is identifiable not only by the letter corresponding to its parent macromere but also by the addition of superscript numerals. For example, the 1a micromere divides to produce daughter cells $1a^1$ and $1a^2$ cells. The cell closest to the animal pole receives the superscript "1." Thus, the 16-cell stage includes the following cells:

Derivatives of the 1q: $1a^1$, $1b^1$, $1c^1$, $1d^1$, $1a^2$, $1b^2$, $1c^2$, $1d^2$
Derivatives of the 1Q: 2a, 2b, 2c, 2d, 2A, 2B, 2C, 2D

The next division (from 16 to 32 cells) again involves dextrotropic displacement, showing the typical alternating pattern of spiral development. The third micromere quartet (3q) is formed, the daughter macromeres are now given the prefix "3" (3Q), and all of the 12 existing micromeres divide. Superscripts are added to the derivatives of the first and second micromere quartets according to the rule of the position as stated above. Thus, the $1b^1$ cell divides to yield the $1b^{11}$ and $1b^{12}$ cells; the $1a^2$ cell divides to yield the $1a^{21}$ and $1a^{22}$ cells; and so on.

At the 64-cell stage, one derivative of the 3Q is 4d (the mesentoblast), a key cell that has been shown to give origin to the hindgut and mesoderm of the adult animal (Hejnol et al., 2007), and which has often been coded as a character used to define spiral cleavage in morphological data matrices (e.g., Nielsen et al., 1996).

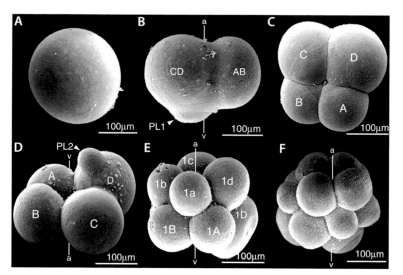

FIGURE 32.2. Scanning electron micrographs of early stages of development, from fertilized egg (A) to 12-cell stage (F) in the aplacophoran *Epimenia babai*. During the 2-cell stage (B), the first polar lobe (PL1) has protruded from the vegetal side as the two cells (AB/CD) divide meridionally. The polar lobe is then absorbed by the CD cell; C, animal view of the 4-cell embryo; D, vegetal view of the 4-cell embryo with the second polar lobe (PL2) protruding from the vegetal side of the D cell; E, lateral view of the 8-cell stage with labeled cells; F, lateral view of 8- to 16-cell stage. The animal–vegetal axis (a–v) is indicated. Photo credits: Akiko Okusu.

However, in the annelid *Capitella teleta* the mesoderm arises from 3c and 3d (Eisig, 1899; Meyer et al., 2010; Meyer and Seaver, 2010), instead of 4d.

The shifts between dextral and sinistral cleavages reflect the characteristic chirality of the spiralian mode of development. An alteration of this chirality has been shown to have an effect on the handedness of *Lymnaea* snail shells, which is determined by a single locus that functions maternally (Kuroda et al., 2009). There also seems to be a relation of chirality to the Nodal pathway (Grande and Patel, 2009).

This spiralian developmental pattern, first characterized for annelids by Wilson (1892), was also subsequently characterized in molluscs by another pioneering embryologist, Edwin Grant Conklin, who also established the first cell lineages of *Crepidula fornicata* and other gastropods by tracing the history of individual blastomeres through the entire development of the animal to the formation of adult organs (Conklin, 1897). Since then, many cell lineages have been established for a great diversity of spiralian embryos, most showing very similar patterns of early embryogenesis.

The fundamental embryonic axes are established by organizing centers that influence the fates of nearby cells. Among spiralians, an organizer (typically the 3D macromere) sets up the dorsoventral axis, which arises from one of the four basic cell quadrants during development (the dorsal, D quadrant) (Henry et al., 2017). In some species the D quadrant is specified conditionally, via cell–cell interactions, while in others it is specified autonomously, via asymmetric cell divisions (such as those involving the formation of polar lobes). The third quartet macromere (3D) typically serves as the spiralian organizer (e.g., Lambert et al., 2016); however, other cells born earlier or later in the D quadrant lineage can serve as the organizer, such as the 2d micromere in *Capitella teleta* (see Amiel et al., 2013) or the 4d micromere in the mollusc *Crepidula fornicata* (see Henry et al., 2017).

Spiral cleavage is thus a highly stereotyped mode of development with conserved blastomere fates and similarities in larval morphology across several phyla (e.g., van den Biggelaar et al., 1997; Henry and Martindale, 1998; Nielsen, 2004, 2005;

Halanych, 1991), although this marker was subsequently criticized for lack of resolution at this phylogenetic level (e.g., Halanych, 1991). Based on the discovery of the dicyemid *DoxC* gene, which encodes a spiralian peptide, dicyemids were proposed to be members of Spiralia (Kobayashi et al., 1999), and this position has been discussed based on phylogenetic analysis of more standard Sanger-based markers. *18S rRNA* data have suggested an uncertain position within Bilateria (Katayama et al., 1995), while others have suggested a sister-group relationship to nematodes (Pawlowski et al., 1996), based on the same marker. Analyses of *18S* and *28S rRNA* sequences suggested a close affinity of dicyemids (and orthonectids; see chapter 34) to annelids (Petrov et al., 2010), and ß-tubulin also suggests a rather uncertain animal affinity (Noto and Endoh, 2004). Another study using amino acid sequences of *innexin* suggested that dicyemids are sister group to the clade consisting of annelids and molluscs (Suzuki et al., 2010). Recent phylogenomic data, however, support a close relationship to Gastrotricha or to Rouphozoa more generally (Lu et al., 2017).

Morphological phylogeny of Dicyemida (Furuya et al., 2001; Furuya and Tsuneki, 2003), based on the cell lineages of the vermiform embryos, divide the group into three families. Recent application of molecular species delimitation has, however, suggested that co-occurring morphological species that were thought to be distinct are instead conspecific (Eshragh and Leander, 2015), suggesting that some criteria for defining morphospecies may require revision.

While some authors refer to the taxon as Rhombozoa, divided into the orders Dicyemida and Heterocyemida (Stunkard, 1982), others consider Rhombozoa a subtaxon of Dicyemida, generating a confusing terminology. Here we follow the nomenclature adopted by most current workers, who consider Dicyemida as the highest taxon and Rhombozoa as a synonym and divide Dicyemida into two or three families.

> Family Conocyemidae (2 species)
>
> Family Dicyemidae (> 120 species)
>
> Family Katharellidae (1 species)

◼ DICYEMIDA: A SYNOPSIS

- Simple vermiform animals with a large axial cell surrounded by 14 to 40 ciliated cells
- Axial cell contains a large polyploid nucleus and intracellular stem cells called "axoblasts"
- Without ECM, tissues or organs, symmetry, or gastrulation
- Metagenesis and spiral development of the infusoriform embryo

While most metazoans present an ECM, that is, a complex of proteoglycans, adhesive glycoproteins, and collagens of importance in morphogenesis and cell differentiation and central to gastrulation (Czaker, 2000), dicyemids lack any kind of typical metazoan ECM structures, including a basement membrane, and hence do not develop through gastrulation. Nonetheless, the ECM components fibronectin, laminin, and type IV collagen have all been shown by immunolabeling to be localized intracellularly (Czaker, 2000), therefore suggesting that the physical ECM has been lost secondarily.

Dicyemids display features that are convergent with those of protists, such as tubular mitochondrial cristae, the capacity for endocytosis from the outer surface, and the absence of collagenous tissue (Noto and Endoh, 2004), and this is why in the past dicyemids were often placed outside of animals.

There is little morphological differentiation among dicyemid cells, although gene expression data have shown regional specific expression patterns among somatic cells of vermiform stages and infusoriform larvae (Ogino et al., 2011), suggesting some level of differentiation among the dicyemid cells and thus more cellular complexity than previously anticipated.

Dicyemids have a complex metagenetic life cycle (one phase that reproduces sexually and one that reproduces asexually) consisting of two morphologically distinct stages (e.g., Lapan and Morowitz, 1975; Hochberg, 1990; Furuya and Tsuneki, 2003). The final vermiform stages, named "nematogen" and "rhombogen," derive from a vermiform embryo formed asexually from an agamete. The infusoriform embryo develops from a fertilized egg produced by the infusorigen, a hermaphroditic gonad that also forms from an agamete (Furuya and Tsuneki, 2007). Infusoriform larvae are unusual in that they consist of larger numbers of cells and are more complex in body organization than adults (Furuya et al., 2004), thus arguing for their secondary simplification due to their parasitic/commensal lifestyle.

In fact, the name "dicyemid" derives from the presence of two different types of embryos during their life cycle, and the switch between these two can be determined by the density of individuals in the renal sac of the host. A high population density in the renal sac may cause the shift from asexual to sexual reproduction (Lapan and Morowitz, 1975), the infusoriform embryos thus escaping from the renal sac in search of a new host, but it is not yet known how the infusoriform larva develops into a vermiform stage.

The vermiform stages consist of an axial cell and a single layer of 8 to 30 ciliated cells or peripheral cells, a number that is constant (dicyemids are eutelic) and species specific. At the anterior region of the animal, 4 to 10 peripheral cells with shorter and denser cilia than the posterior cells form the calotte, a structure that helps to attach to the renal tissue. While it was believed that different shapes of the calotte in co-occurring individuals represented different species (Furuya and Tsuneki, 2003), this does not seem to be the case (Eshragh and Leander, 2015).

Infusoriform embryos consist of 37 or 39 cells, more differentiated than in the vermiform stages. Internally there are 4 large cells or urn cells, each containing a germinal cell. At the anterior region of the infusoriform embryo is a pair of apical cells, each containing a refringent body composed of magnesium inositol hexaphosphate (Lapan, 1975). The external cells are mostly ciliated (Furuya and Tsuneki, 2003).

The development of the infusoriform embryo has been studied in detail for a number of species (Furuya et al., 1992, 1996; Furuya et al., 2001; Furuya and Tsuneki, 2003), and the early phases have been described as being spiral. Spiral development is, however, a highly stereotyped mode of development that is hardly comparable to the known infusoriform embryo development.

ciliary movement, only recently has it been widely recognized that orthonectids have muscles and a nervous system (Slyusarev and Starunov, 2016). This is also reflected in part in their genome, which presents a large repertoire of genes for the nervous system, although certain gene families and pathways have been lost. However, orthonectids' highly simplified genome still retains elements of the genetic toolkit of bilaterians.

The presence of muscular and nervous systems, involved in coordinating the movement of the free-swimming reproductive individuals was only recently recognized and characterized (Slyusarev and Manylov, 2001; Slyusarev, 2003; Slyusarev and Starunov, 2016), implying the existence of action potential generating ion channels, neurotransmitter receptors and electrical synapses, which have now been characterized in their genome (Mikhailov et al., 2016). However, orthonectids have lost the entire glutamate receptor family. The simplicity of the *I. linei* nervous system is associated not only with a decrease in the receptor diversity but also with molecular mechanisms responsible for the nervous system development, axon guidance, and synapse formation (Mikhailov et al., 2016).

CHAETOGNATHA

35

> The species of this genus are remarkable from the simplicity of their structure, the obscurity of their affinities, and from the abounding in infinite numbers over the intra-tropical and temperate areas. (Darwin, 1844: 1)

In his introduction to the genus *Sagitta*, Charles Darwin highlighted two salient features about this somehow ignored group of animals: their odd anatomy resulting in a lack of sensible phylogenetic hypotheses, and their ecological abundance in some marine environments. Arrow worms (fig. 35.1) are exclusively marine, voracious but small (most extant chaetognaths are only a few mm in size and planktonic, but there are some inshore benthic and deepwater species that may be up to 12 cm in length) animals with a streamlined elongate body with lateral fins and a post-anal tail, complex mouthparts, and lack of circular musculature. About 190 species are known today, and the group has been around since the Cambrian. Ecologically, chaetognaths are among the most abundant planktonic predators (Martini and Haddock, 2017), sometimes accounting for more than 10% of total zooplankton biomass, being outnumbered only by copepods, their major prey (Ball and Miller, 2006). Recent metabarcoding studies have shown that in some months almost 20% of the reads correspond to Chaetognatha in the Red Sea (Casas et al., 2017) and chaetognaths dominate among the holoplanktonic animals, whether in reads or in specimens examined visually (Lindeque et al., 2013). The debate about their affinities has gone on for decades, but novel genomic data may have found a stable position for one of the strangest free-living invertebrate groups. The group has recently been reviewed by Müller et al. (2018), who provide a state-of-the-art knowledge of chaetognaths.

SYSTEMATICS

The phylogenetic position of Chaetognatha in the animal kingdom has been convoluted since the time that Charles Darwin mentioned the obscurity of their affinities, a pessimistic view followed by Ghirardelli (1968) as well as many earlier authors, in part due to their unusual mode of embryonic development (Ghirardelli, 1995; Kapp, 2000). Because of their supposed trimeric body plan and to some interpretations of their embryogenesis, chaetognaths have traditionally been placed with deuterostomes (Hyman, 1959), although Hyman herself noted many anatomical similarities with certain "aschelminth" features and noted the difficulties in allying them with her other deuterostomes, as shown in the earliest

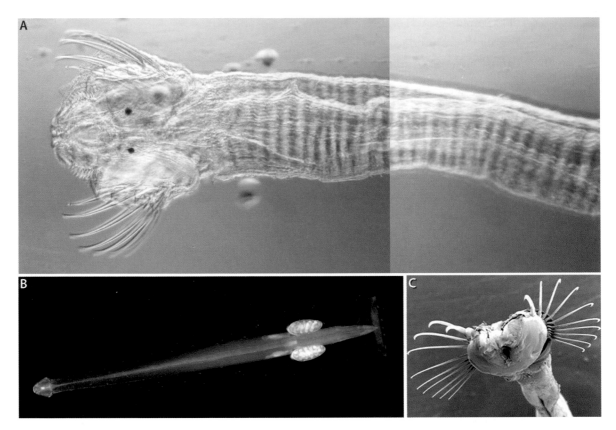

Figure 35.1. A, head and anterior end of a pelagic chaetognath from Panama; B, the deep-water bioluminescent species *Eukrohnia fowleri* with embryos in the lateral gelatinous pouches; C, scanning electron micrograph of the anterior end of *Caecosagitta macrocephala*, showing the grasping spines. Photo credits: B, C, Steven Haddock.

among animals, but she relied entirely on the classic studies on chaetognath development (e.g., Hertwig, 1880; Doncaster, 1902; John, 1933) without generating new data. More recent studies using cell-labeling techniques have shown certain characteristics reminiscent of spiralians (Shimotori and Goto, 2001), such as leiotropic second cleavage, also observed in earlier studies of chaetognath development where there is alternation of leiotropic and dexiotropic cleavage (Elpatiewsky, 1909).

It seems that the blastopore closes during embryogenesis and that the anus is usually formed later during development in the posterior end of the embryo (Hejnol and Martín-Durán, 2015), although without expressing brachyury in the hindgut (Takada et al., 2002). Further detailed studies of early embryogenesis of chaetognaths are needed to better understand some of these processes.

The central nervous system of chaetognaths has also received considerable attention, due to its particular configuration and its unique associated sensory organs (Bone and Goto, 1991; Shinn, 1997; Harzsch and Müller, 2007; Harzsch et al., 2009; Rieger et al., 2010; Rieger et al., 2011; Harzsch et al., 2016; Müller et al., 2018). The CNS (fig. 35.2) is characterized by a dorsal brain, a pair of vestibular ganglia each with an associated esophageal ganglion, a subesophageal ganglion, and a ventral nerve center. The brain innervates the eyes and the corona ciliata dorsally and connects with

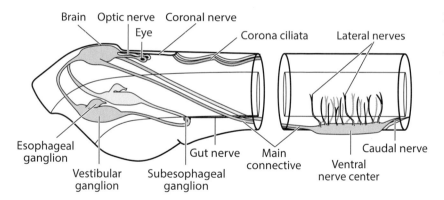

Brain Optic nerve Coronal nerve
Eye
Corona ciliata Lateral nerves
Esophageal ganglion
Vestibular ganglion Subesophageal ganglion Gut nerve Main connective Ventral nerve center Caudal nerve

Figure 35.2. Details of the nervous system of a chaetognath. Based on Harzsch et al. (2016).

the vestibular ganglia and with the subesophageal ganglion forming an anterior loop. It is also connected with the ventral nerve center through the paired main connective. From the ventral nerve center a paired caudal nerve and lateral nerves emerge. The ventral nerve center is absent in *Bathybelos typhlops*, which instead has a dorsal ganglion, also connected to the brain by a paired main connective (Bieri and Thuesen, 1990). Neurogenesis has been studied for the brain and the ventral nerve center of *Spadella cephaloptera* (Rieger et al., 2011; Perez et al., 2013).

A retrocerebral organ innervating the posterior end of the brain and positioned on the dorsal side of the head is composed of ciliary bodies and rootlets along the apical membranes of some microvilli-bearing epithelial cells (Harzsch et al., 2016); it is thus suspected to be a baroreceptor (Shinn, 1997).

Chaetognaths are equipped with three types of ciliary sense organs: the transversally oriented ciliary fence organs or "hair fans" (among other names; reviewed in Bone and Pulsford, 1978; Müller et al., 2014), a series of longitudinally (parallel to the anterior–posterior axis) oriented ciliary tuft organs, and a ciliary loop known as the "corona ciliata" (Müller et al., 2014). The first two have been examined in detail by Müller et al. (2014). The hair fans may be related to elaborate mating dances, as described in *Paraspadella gotoi* (Goto and Yoshida, 1985), as well as for detecting prey-produced vibrations (Horridge and Boulton, 1967; Feigenbaum, 1978). The hair fans are obliterated upon preservation in formalin (Feigenbaum, 1978), which is why they have not been described properly in many chaetognath species.

The corona ciliata is located on the dorsal side of the body and extends along the posterior region of the head ("neck"). It consists of numerous monociliary cells, forming a loop, and shows an extensive and intricate pattern of innervation (Bone and Goto, 1991; Shinn, 1997; Harzsch et al., 2016). The function of the corona ciliata is, however, unknown, but its microscopic architecture suggests that it is most likely a chemosensory organ (Harzsch et al., 2016). Other authors have attributed to it a function similar to that of the lateral line of fishes (Malakhov et al., 2005). The corona ciliata is subdivided into an internal canal filled with inflated electron-lucent microvilli and an external part carrying cilia (Malakhov et al., 2005).

Chaetognaths have a pair of functional eyes (fig. 35.1 A) of the ciliated photoreceptor class. These photoreceptors can be of different types, included inverted, with the photoreceptor cells pointing inward and embedded in a single large pigment cell, or everted with the photoreceptors pointing outward (Ball and Miller, 2006). The eyes have relatively few receptor units and lack lenses, so they apparently are not image-forming. However, two features that appear to be unique to chaetognath eyes are the conical body—for which some have suggested an optical role—and the unusual lamellar organization of the distal segment of at least some species (Goto et al., 1984).

Like virtually everything else, the muscles of chaetognaths are unique. The primary muscle (occupying ca. 80% of body wall volume) shows the characteristic structure of transversely striated muscles, as in insect asynchronous flight muscles. In addition, the body wall has a secondary muscle with a peculiar structure, displaying two sarcomere types (S1 and S2), which alternate along the myofibrils. S1 sarcomeres are similar to those in the slow striated fibers of many invertebrates. In contrast, S2 sarcomeres do not show a regular sarcomeric pattern, but instead exhibit parallel arrays of two filament types (Royuela et al., 2003).

Chaetognatha also display unusual neuromuscular innervations, with axonal varicosities that lack specialized junctions so that the presynapses are separated from the underlying muscles by a thick extracellular matrix through which the transmitter must pass (Harzsch et al., 2015).

The feeding behavior of chaetognaths has received some attention from an experimental point of view, as these animals are seen as voracious predators (Ball and Miller, 2006) that use their hair fans to sense vibrations to capture the prey (Horridge and Boulton, 1967). Two anatomical structures allow them to feed on large prey, the hood, which acts as a net surrounding the prey, and their impressive chitinous spines and teeth (fig. 35.1 A, C), interpreted by some authors to be homologous to the gnathiferan jaws (e.g., Fröbius and Funch, 2017). In addition to these physical structures, some chaetognaths are known to have tetrodoxin, an extremely potent sodium channel-blocking neurotoxin to paralyze their prey (Thuesen et al., 1988). The tetrodoxin seems to be produced by *Vibrio alginolyticus*, while chaetognaths may be a major agent to distribute the bacteria through the environment (Thuesen and Kogure, 1989).

Bioluminescence has been reported (e.g., in *Eukrohnia fowleri*, fig. 35.1 B, and *Caecosagitta macrocephala*, fig. 35.1 C), and the bioluminescent organs characterized in two species of chaetognaths (Thuesen et al., 2010), which are not closely related phylogenetically (Haddock et al., 2010).

Chaetognaths play an important role in the vertical carbon flux of the oceans, as illustrated for the Southern Ocean (Giesecke et al., 2009), through the downward transport of organic matter in the form of fecal pellets, as well as through vertical migrations through the life cycle, as in *Sagitta gazellae* (Ball and Miller, 2006).

Chaetognaths use their fins for gliding, until eventually they begin to sink. Sinking is, however, counteracted by active swimming with a series of short, rapid darts, followed by another glide cycle (Ball and Miller, 2006).

▬ GENOMICS

Genomic resources for chaetognaths are scarce and mostly rely on transcriptomic data (Matus et al., 2006; Marlétaz et al., 2008; Kocot et al., 2017; Laumer et al., 2019;

Marlétaz et al., 2019). Mitochondrial genomes have been characterized for multiple species of chaetognaths, showing a remarkably similar structure in most of the studied species (Shen et al., 2016), yet some major rearrangements have also been detected among species, as well as unusual levels of genetic variation, which may be due to large population sizes and/or an accelerated mutation rate of the mitochondrial genome (Marlétaz et al., 2017).

FOSSIL RECORD

Nearly complete body fossils of chaetognaths are known from Cambrian Stage 3 (see below), but the group is likely represented even earlier in the Cambrian by microfossils known as "protoconodonts," long recognized based on their structure and mode of growth as probably grasping spines of chaetognaths (Szaniawski, 2002). The recovery of partially articulated assemblages strengthens this interpretation (Vannier et al., 2007). The earliest protoconodonts are among the oldest known protostomes, being recovered from the Fortunian Stage of the Cambrian in rocks dated to ca. 535 Ma (Kouchinsky et al., 2012).

Protosagitta spinosa from the early Cambrian Chengjiang biota is certainly a chaetognath, its body differentiated into head, neck, trunk, and likely tail regions (Vannier et al., 2007; Shu et al., 2017). The head bears grasping spines arranged as in extant chaetognaths. Despite the obvious similarities to extant taxa in anatomy and presumably (pelagic) feeding ecology, it has not been established whether *P. spinosa* is a stem- or crown-group chaetognath. Another Chengjiang chaetognath, *Ankalodus sericus*, is known only from its feeding apparatus, which consists of several bundles of spines (Shu et al., 2017). Similar organization of the grasping spines is observed in a Burgess Shale chaetognath, *Capinatator praetermissus*, the body of which reaches nearly 10 cm exclusive of its fins (Briggs and Caron, 2017). After the Cambrian, the fossil record of chaetognaths is essentially barren until the Carboniferous, when grasping spines are again recovered (Doguzhaeva et al., 2002). The middle Cambrian Burgess Shale *Oesia disjuncta*, formerly interpreted as a chaetognath (Szaniawski, 2005), is now placed within Hemichordata (Nanglu et al., 2016) (see chapter 16).

36

GNATHIFERA

A clade of protostomes with a special type of cuticularized jaws, including Rotifera[17] (comprising also Acanthocephala) and Gnathostomulida, was recognized by Ahlrichs (1995a, b) but also discussed as a possibility by other authors (e.g., Reisinger, 1961; Sterrer et al., 1985). Gnathifera would later include what Ahlrichs called "new group A" (Ahlrichs, 1997), based on a new taxon of freshwater "aschelminth" from Greenland (Kristensen, 1995)—what would later become Micrognathozoa (Kristensen and Funch, 2000).

While early molecular phylogenetic analyses using a few markers failed to recover monophyly of Gnathifera (Giribet et al., 2004) or placed Cycliophora with Gnathifera (Giribet et al., 2000), most molecular analyses did support a sister-group relationship of Gnathostomulida and Rotifera when Micrognathozoa was not sampled (e.g., Witek et al., 2009; Struck et al., 2014; Wey-Fabrizius et al., 2014; Golombek et al., 2015; Kocot et al., 2017), as well as when including Micrognathozoa (Laumer et al., 2015a). This latter paper agrees with most earlier assessments of gnathiferan relationships, supporting a sister-group relationship of Micrognathozoa and Rotifera (Ahlrichs, 1997; Herlyn and Ehlers, 1997; Kristensen and Funch, 2000; Funch and Kristensen, 2002; Sørensen, 2003), at least in some of the most parsimonious reconstructions (Sørensen et al., 2000). This relationship seems also to be supported by a large array of neuroanatomical characters (Bekkouche and Worsaae, 2016a) and by the presence of a skeletal lamina in rotifers and in the dorsal epithelium of micrognathozoans (Ahlrichs, 1997; Kristensen and Funch, 2000; Nielsen, 2012a).

The synapomorphies of Gnathifera have been discussed in a series of papers (Ahlrichs, 1997; Sørensen, 2003; Bekkouche and Worsaae, 2016a) and include the ultrastructure of the jaws (Rieger and Tyler, 1995), formed by the rods that appear to be a basic brick in most of the jaw elements, with the jaws composed of rods with a lucent material and an electron-dense core; and the jaws with pincers caudally articulating into unpaired pedicles (Sørensen, 2003). In addition, the presence of monociliated cells in their epidermis is also present in some gnathiferan groups (Nielsen, 2012a).

Recently, an analysis of rotiferan Hox genes has identified similarities between Chaetognatha and Rotifera (but not examined in Gnathostomulida and Micrognathozoa) (Fröbius and Funch, 2017) and proposed the inclusion of Chaetognatha within Gnathifera. In addition to the molecular putative synapomorphies (e.g., inclusion of the *MedPost* gene, previously only known for chaetognaths), it has been

[17] After the recognition that Acanthocephala renders Rotifera paraphyletic, a number of authors adopted the term Syndermata for this clade, while others preferred to include Acanthocephala within Rotifera. Syndermata and Rotifera are thus equivalent here.

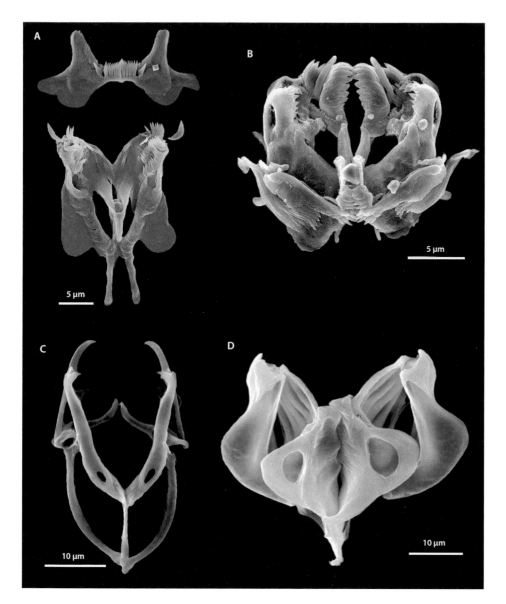

Figure 36.1. Representative jaws of the different gnathiferan groups: A, basal plate and jaws of the gnathostomulid *Gnathostomula armata*; B, the micrognathozoan *Limnognathia maerski*; C, the monogonont rotifer *Encentrum astridae*; D, the monogonont rotifer *Brachionus calyciflorus*. Photo credits: Martin Vinther Sørensen.

proposed that the high chitin content of the spines and teeth and the structure of the chitinous cuticle of the chaetognath head could be homologous to the chitinous parts and membranes of the pharynx in gnathiferans. A chaetognath–gnathiferan clade was recovered in phylogenomic analyses that include numerous new chaetognath transcriptomes (Laumer et al., 2019; Marlétaz et al., 2019), as well as in a phylogenetic analysis of morphological characters including fossils by Vinther and Parry (2019), although with Chaetognatha nested within a paraphyletic Gnathifera.

▬ MICROGNATHOZOA: A SYNOPSIS

- Triploblastic, bilateral, nonsegmented acoelomate; without a circulatory system or special gas exchange structures
- Body divided into three main regions: a head, an accordion-like thorax, and an abdomen
- Mouth opening anteroventrally; gut incomplete (dorsal anus opens temporarily)
- Pharyngeal apparatus consisting of a complex jaw apparatus with four sets of jawlike elements and sets of striated muscles
- Epidermis with supporting dorsal and lateral plates (intracellular matrix), not syncytial
- Ventral ciliation consisting of preoral ciliary field and paired cyliophores around mouth and along midline of thorax and abdomen
- Excretion via three pairs of protonephridia with monociliated terminal cells
- Sensory organs include stiff monociliated cells supported by microvilli and nonciliated internal eyes (phaosomes)
- Posterior end with ciliated pad and one pair of glands
- Males unknown; probably parthenogenetic

Among the most conspicuous characters of this microscopic animal is the pharyngeal apparatus (fig. 37.1), up to 30 μm wide, but with a complexity not seen in any other animal, including numerous hard jaw elements with intricate musculature and a buccal ganglion. The hard elements of the complex jaw apparatus (fig. 36.1 B) have received direct attention in both *Limnognathia maerski* (Sørensen, 2003) and an unnamed species of *Limnognathia* (De Smet, 2002). The musculature and innervation were studied later (Bekkouche et al., 2014; Bekkouche and Worsaae, 2016a). Some of the most conspicuous jaw elements include the large paired fibularium, the main jaws, the ventral jaws, and the dorsal jaws. The pharyngeal musculature includes a unique major ventral muscle plate, formed by eight to ten longitudinal cross striated muscle fibers, supporting the entire jaw apparatus, as well as other sets of striated muscles. These complex jaws are used for feeding upon bacteria on the surfaces of mosses and sediments.

One of the most characteristic behaviors of micrognathozoans is their slow spiral motion when moving freely in the water column or at the bottom of a petri dish, a movement that seems to be mediated by the trunk ciliophores, also used for crawling or gliding, this caused by the movement of the rows of ciliophores in unison. A typical escape mechanism involves contraction of trunk muscles. The body wall musculature has been studied in detail by Bekkouche et al. (2014) and consists of seven main pairs of longitudinal muscles and thirteen pairs of oblique dorsoventral muscles in the thoracic and abdominal regions. Additional minor muscles and the pharyngeal apparatus musculature are also found. The integument has an intracellular lamina that is considered homologous to the intraskeletal lamina of rotifers (see chapter 39).

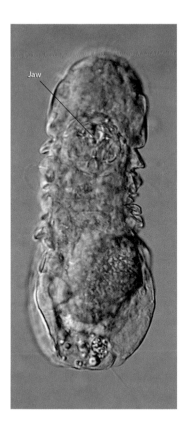

FIGURE 37.1. Live specimen of *Limnognathia maerski* with the visible jaw apparatus in the anterior third of the body (top). Photo credit: Reinhardt Møbjerg Kristensen.

Micrognathozoa have three pairs of relatively simple protonephridia with a single collecting tube on each side of the animal (R. M. Kristensen, pers. comm.). While two pairs were reported in the original description of *L. maerski* and in several subsequent publications (e.g., Bekkouche and Worsaae, 2016a), a recent account reported three pairs, two in the accordionlike thorax and one pair extending into the abdomen (Worsaae and Kristensen, 2016). The terminal cells are monociliated, similar to those of gnathostomulids and unlike the multiciliated terminal cells of rotifers.

The nervous system of *Limnognathia maerski* is relatively simple, consisting of an anterior, slightly bilobed, large dorsal brain, paired sub-esophageal ganglia, two trunk commissures, two pairs of ventral longitudinal nerves and peripheral nerves (Bekkouche and Worsaae, 2016a). In addition, the pharyngeal ganglion has two apical tufts of ciliary receptors that control the movement of the buccal apparatus, a trait shared by all Gnathifera (Bekkouche and Worsaae, 2016a). A series of peripheral nerves extends from the ventral cords, connecting to sensory cilia, most of which are monociliated collar receptors, whereas others are more complex, involving multiple sensory cells. These sensory structures are known, from anterior to posterior, as apicalia, frontalia, lateralia, dorsalia, and caudalia. In addition, the anterior end of the animal bears a pair of lateral hyaline vesicles called "phaosomes" that may be unpigmented inner microvillar eyes.

two gnathiferan clades and chaetognaths (Laumer et al., 2019; Marlétaz et al., 2019) is now found in large phylogenomic analyses. In most data sets that did not sample Micrognathozoa, gnathostomulids were found to be sister group to Rotifera (Struck et al., 2014; Golombek et al., 2015), consistent with the Gnathifera concept.

The phylogenetic system of gnathostomulids was largely established by Sterrer (1972), who divided gnathostomulids into two orders, according to the presence (Bursovaginoidea) or absence (Filospermoidea) of a female bursa–vagina system. He further divided Bursovaginoidea into the suborders Scleroperalia (with a cuticularized bursa) and Conophoralia (with a soft bursa and giant sperm) (fig. 38.1). However, in his proposed phylogenetic tree, Sterrer (1972) placed Conophoralia within Scleroperalia, making this suborder paraphyletic. A largely similar result, although with a different rearrangement of the scleroperalian families, was found in a morphological cladistic analysis of 55 morphological characters (Sørensen, 2002b). Later, a combined analysis of an expanded morphological matrix and a handful of molecular markers corroborated the sister-group relationship of Filospermoidea and Bursovaginoidea, as well as the reciprocal monophyly of Scleroperalia and Conophoralia (Sørensen et al., 2006), and this study largely provides the bases for the modern classification of the group. A recent taxonomic account of the phylum was provided by Sterrer and Sørensen (2015). From the 102 species in 26 genera currently recognized (Sterrer and Sørensen, 2015 and subsequent updates), 72 have been authored or co-authored by Austrian/Bermudian zoologist Wolfgang Sterrer.

> Order Filospermoidea
>> Family Haplognathiidae (10 spp. in the genus *Haplognathia*)
>> Family Pterognathiidae (18 spp. in 2 genera)
> Order Bursovaginoidea
>> Suborder Scleroperalia
>> Family Agntahiellidae (3 spp. in 2 genera)
>> Family Clausognathiidae (monotypic)
>> Family Mesognathariidae (6 spp. in 3 genera)
>> Family Rastrognathiidae (monotypic)
>> Family Paucidentulidae (monotypic)
>> Family Onychognathiidae (9 spp. in 5 genera)
>> Family Problognathiidae (monotypic)
>> Family Gnathostomulidae (24 spp. in 5 genera)
>> Suborder Conophoralia
>> Family Austrognathiidae (25 spp. in 3 genera)

While Sterrer (1972) predicted more than a thousand gnathostomulid species, this estimate has been revised more recently to "a few hundreds," given the large number of supposedly cosmopolitan species (Sterrer and Sørensen, 2015). Nonetheless, no population genetic analysis has yet been conducted on any of these supposed

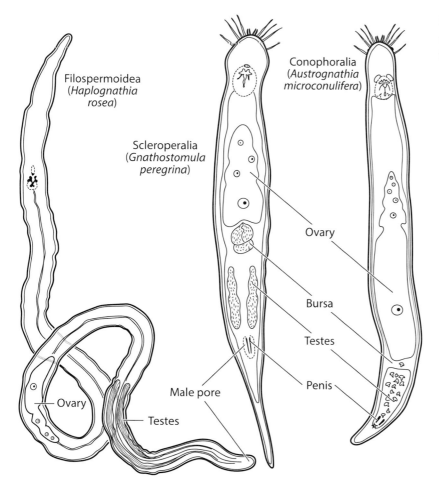

FIGURE 38.1. Habitus of the three main morphologies of gnathostomulids. Based on Sterrer (1972).

Filospermoidea
(*Haplognathia rosea*)

Scleroperalia
(*Gnathostomula peregrina*)

Conophoralia
(*Austrognathia microconulifera*)

Ovary

Bursa

Testes

Penis

Male pore

Ovary

Testes

Ovary

species with widespread distributions, and thus the true magnitude of Gnathostomulida remains to be tested.

GNATHOSTOMULIDA: A SYNOPSIS

- Triploblastic, acoelomate, bilaterally symmetrical unsegmented animals
- Body divided into head and trunk, in some cases with a small tail
- Completely ciliated monolayered epidermis, consisting of monociliated cells
- Through-gut, although the anus can be underdeveloped or closed
- Complex jaws and pharyngeal apparatus
- Without a circulatory system or special structures for gas exchange
- Serially arranged paired protonephridia consisting of monociliated terminal cells
- Direct developers, possibly with spiral cleavage
- Exclusively marine, interstitial, often confined to sulfide-rich sands
- Hermaphroditic

GENOMICS

No published data are available for nuclear genomes of gnathostomulids, and just a few ESTs (Dunn et al., 2008) and transcriptomes (Struck et al., 2014; Laumer et al., 2015a; Laumer et al., 2019) have been generated for the purpose of testing their phylogenetic position. Nearly complete mitochondrial genomes are available for two species (Golombek et al., 2015).

FOSSIL RECORD

No fossils have been attributed to gnathostomulids, although there has been discussion about similarities between the gnathostomulid jaws and some microconodonts (Durden et al., 1969; Ochietti and Cailleux, 1969), as well as structures interpreted as jaws in the Burgess Shale species *Amiskwia sagittiformis* (Caron and Cheung, 2019).

ROTIFERA

Rotifera or "wheel animals" account for over 2,030 named species (Segers, 2007) of microscopic, generally free-living organisms (although many are sessile) and another 1,200 or so species of acanthocephalans (Smales, 2015), obligate intestinal parasites of vertebrates that can reach up to 60 cm in length. Rotifers (also referred to as Syndermata) are often divided into four main clades with very different biologies. Free-living rotifers (Bdelloidea and Monogononta) are microscopic animals, mostly much smaller than 1 mm, generally characterized by the eponymous rotatory or wheel organ (the corona) and by the gnathiferan buccal apparatus, in this case formed by a muscular pharynx called a "mastax" and the "trophi" (or jaws; fig. 36.1 C–D). Among the most notable rotifers are bdelloids, referred to as "sleeping beauties" (Ricci, 2016) and "evolutionary scandals" (Maynard Smith, 1986), due to their two most salient characteristics: an extreme anhydrobiotic capacity and the apparent lack of sex (they are obligate parthenogens)—which paradoxically has made them a model to study the evolution of sex. They exist in virtually any moist or wet (including terrestrial) habitats and have also been found—along with platyhelminths, annelids, and arthropods—at 1.4 km soil depths in paleometeoric fissure water (Borgonie et al., 2015).

Monogononts are facultative cyclical parthenogens, with dwarf males, and are free-living in marine and freshwater environments. Some species are sessile as adults and about 25 of these species form aggregations sometimes referred to as "colonies" (Wallace, 1987; Fontaneto and De Smet, 2015), some aggregations reaching 3,500 individuals (Fontaneto and De Smet, 2015). Aggregating species are known in Gnesiotrocha (Flosculariidae and Conochilidae; although the validity of these families is currently disputed; see Meksuwan et al., 2015)—but the bdelloid *Philodina megalotrocha* exhibits some form of gregarism.

Seisonideans are exclusively sexual, highly modified, elongated rotifers that live epizoically on small crustaceans of the genus *Nebalia*. Acanthocephalans, or "thorny-headed worms," once considered their own phylum, are a diverse group of endoparasites using pancrustaceans as intermediate hosts and vertebrates as final hosts (Semmler Le, 2016).

In addition to their ability to undergo anhydrobiosis, bdelloid rotifers are well known for their resistance to ionizing radiation, both phenomena often being connected to their eutelic condition and to the ability to repair damaged DNA. Likewise, these phenomena are now linked to the presence of large proportions of exogenous DNA in some of the species that are able to desiccate. Anhydrobiosis has also been explained in light of their ridding themselves of deadly fungal parasites through complete desiccation, followed by dispersal by wind to establish new populations in the absence of the parasite (Wilson and Sherman, 2010).

A few free-living rotifer species are a common food source in aquaculture (Hagiwara et al., 1997), as they can grow in large quantities and thus are said to live at low Reynolds numbers.[19] Due to their abundance, they have been used as model organisms for ecotoxicological studies (Snell and Janssen, 1995). Most of the experimental work has therefore been conducted on some of these economically and ecologically important species, especially the *Brachionus plicatilis* species complex. Free-living forms can be found in any habitat where water is available, even if these are ephemeral and subjected to heat stress (Smith et al., 2012), as they are able to produce dormant stages that resist desiccation, although these are different in Bdelloidea and Monogononta. No resistance forms are found in the strictly marine Seisonidea. Their ability to withstand desiccation as well as freezing tolerance (Gilbert, 1974) and other extreme conditions have been studied in detail. These properties are often related to aspects of the bdelloids' supposed asexual condition (Pouchkina-Stantcheva et al., 2007).

Rotifers have also been studied as models for aging (King, 1969; Austad, 2009; Snell et al., 2012), and other aspects they share with other extremophiles, including ionizing radiation (Gladyshev and Meselson, 2008; Krisko et al., 2012). As in tardigrades, many of these aspects are now receiving detailed molecular attention in terms of genes and pathways related to DNA damage and repair. The ecological aspects of the parasitism and interactions of the acanthocephalans with their hosts and the environment have been summarized by Kennedy (2006).

SYSTEMATICS

Once considered two separate animal phyla, the parasitic acanthocephalans are now understood as a subclade of Rotifera, which includes also Seisonidea, a clade of crustacean ectoparasites. This clade has also been named Syndermata, in reference to the syncytial epidermis found in all the members of the phylum, to distinguish them from the "old" Rotifera, which excluded Acanthocephala. In fact, some modern treatises still treat the monophyletic Acanthocephala independently from the paraphyletic Rotifera (Schmidt-Rhaesa, 2015), but we prefer to use Rotifera as the higher taxon name (= Syndermata) instead of using it as a paraphyletic entity or even getting rid of it. This relationship between acanthocephalans and rotifers has been recognized for a long time based on morphological evidence (e.g., Clément, 1980; Lorenzen, 1985; Ahlrichs, 1995a, 1997; Rieger and Tyler, 1995; Wallace et al., 1996; Garey et al., 1998), including sperm ultrastructure (Ferraguti and Melone, 1999), although some studies have endorsed a relationship of acanthocephalans to the annelid clade Myzostomida, based on sperm ultrastructure and spermiogenesis (Mattei and Marchand, 1987,

---

[19] In fluid mechanics, the Reynolds number (Re) is a dimensionless number that gives a measure of the ratio of inertial forces to viscous forces for given flow conditions. The Reynolds number is an important parameter that describes whether flow conditions lead to laminar or turbulent flow.

1988; but see Funch et al., 2005). Molecular evidence (e.g., Winnepenninckx et al., 1995b; Garey et al., 1996b; Garey et al., 1998; Giribet and Ribera, 1998; Near et al., 1998; García-Varela et al., 2000; Mark Welch, 2000; García-Varela et al., 2002; Near, 2002; Herlyn et al., 2003; Giribet et al., 2004; García-Varela and Nadler, 2006; Sørensen and Giribet, 2006), including mitogenomics (Gazi et al., 2016; Sielaff et al., 2016) and phylogenomics (e.g., Hausdorf et al., 2007; Wey-Fabrizius et al., 2014; Laumer et al., 2015a), has been conclusive with respect to the inclusion of Acanthocephala within Rotifera.

The interrelationships of Rotifera or some of their subclades have been investigated using explicit morphological data matrices (Melone et al., 1998; Monks, 2001; Sørensen, 2002a), but these studies did not consider the new concept of Rotifera (they either focused on Acanthocephala or on the free-living forms). An analysis combining morphology and Sanger-based DNA sequence data recovered the traditional tree, with Acanthocephala as sister group to the remaining Rotifera when analyzing only morphological data, but placed Acanthocephala with Seisonidea and Bdelloidea with molecules or a combination of molecules and morphology, dividing the clade into Hemirotifera (= Hemirotatoria) and Monogononta (= Eurotatoria) (Sørensen and Giribet, 2006). This topology is largely supported by subsequent phylogenomic data sets (Wey-Fabrizius et al., 2014; Laumer et al., 2015a), although in one case Seisonidea appears as sister group to all other rotifers instead of just Acanthocephala (Laumer et al., 2015a), with Bdelloidea being the sister group of Acanthocephala—constituting a clade named Lemniscea, supported by the presence of lemnisci (see below). This result is also supported in mitogenomic analyses (Sielaff et al., 2016).

Bdelloidea comprises ca. 460 species in three families, with a rather conserved body plan, and are typically used as the model to describe the anatomy of rotifers. Seisonidea comprises only four species in the genera *Seison* and *Paraseison* (Leasi et al., 2011), epizoic symbionts on the members of the leptostracan genus *Nebalia*. Acanthocephalans constitute a diverse group with ca. 1,200 named species. Their internal phylogeny has received considerable attention due to the interest in understanding these vertebrate parasites. Old classification systems divided Acanthocephala into Palaeoacanthocephala, Eoacanthocephala, and Archiacanthocephala, with the class Polyacanthocephala proposed more recently (Amin, 1987), and this classification system is supported by some molecular analyses (García-Varela et al., 2002), but contradicted in others (Gazi et al., 2016). Whether Palaeoacanthocephala is mono- or diphyletic is debatable (Herlyn et al., 2003), but its monophyly is supported in mitogenomic analyses (Gazi et al., 2016). We therefore follow a division of Acanthocephala into the three traditional clades.

Monogononta, with ca. 1570 described species (Segers, 2007), is the most diverse rotifer group and also displays the greatest morphological disparity (Sørensen and Giribet, 2006). Monogononts are mostly free-swimming, but some are sessile. They divide into Gnesiotrocha (with the orders Collothecaceae and Flosculariaceae) and Pseudotrocha (with the single order Ploima). These clades were supported by the combined analysis of Sørensen and Giribet (2006).

Class Hemirotatoria (= Hemirotifera)
    Subclass Bdelloidea
    Subclass Acanthocephala
        Archiacanthocephala
        Palaeoacanthocephala
        Eoacanthocephala
    Subclass Seisonidea (= Seisonacea)
Class Eurotatoria
    Subclass Monogononta
        Gnesiotrocha
            Order Collothecaceae
            Order Flosculariaceae
        Pseudotrocha
            Order Ploima

ROTIFERA: A SYNOPSIS

- Pseudocoelomate, nonsegmented bilaterians
- Protonephridia
- Eutely; lack of mitosis upon completion of embryonic development
- Syncytial epidermis with intracellular lamina
- Lack of cuticle
- Feeding apparatus with muscular mastax and trophi (= jaw)
- Two corona of cilia (rotatory or wheel organ)
- Toes with pedal glands
- Cryptobiosis (in adult bdelloids or in embryos of monogononts)

Most morphological knowledge of the free-living rotifers is based on the females, as males are unknown for bdelloids and they are smaller and of simpler organization in monogononts. These females show a great deal of variation across groups, but they are all bilaterally symmetrical and with a clear distinction of the dorsal and ventral sides. The body is generally divided into head with a corona of cilia, trunk, and toes (or foot) with pedal glands, located behind the cloacal opening, which is dorsal. The body often appears "segmented" due to foldings of the epidermis, but this is generally not considered segmentation (some authors refer to this as pseudosegmentation). Generally, the neck appears as a pseudosegment separating head and trunk. The head and toes can retract into the trunk. The anatomy of the parasitic elongated Seisonidea includes numerous modifications, including the absence of toes (see a review in Ahlrichs and Riemann, 2018).

The head bears the characteristic ciliary corona used both for locomotion and/or food collection, but some sessile species lack the typical corona, as is the case of Collothecaceae (Monogononta, Gnesiotrocha), which includes species with indirect development and adults with an infundibulum, instead of the corona. The head also bears the mouth opening and a diversity of sense organs, including eyespots, tentacles, and antennae. Additional head structures can be found, including a rostrum used for the characteristic "leechlike" movement of bdelloids. Large anterolateral setae that facilitate saltation through the water column as a means of escape from predators can also be present (Hochberg and Ablak Gurbuz, 2007; Hochberg et al., 2017b). The head region includes usually—in addition to the anterior portion of the digestive system, including the mastax, and the brain—a glandular retrocerebral organ, which opens ventral to the rostrum in monogononts or in the rostrum (or proboscis) in bdelloids. This glandular organ is comprised of paired subcerebral glands, an unpaired retrocerebral sac, and ducts leading to the coronal surface, but its function remains poorly understood (Wallace, 2002). The retrocerebral organ has also been described in males (Riemann and Kieneke, 2008).

Most rotifers are planktonic solitary species, but several gnesiotrochan species are sessile and produce tubular sheaths around their bodies, first secreted as an unsegmented tube by juveniles. The tubes have a variable morphology and are produced by different glands, adding segments as they grow. Tubes consist of an inner mucuslike layer that extends from the base to the foot region and an external layer that is secreted by the so-called cement cells, which is instead a modified region of the syncytial integument that forms a beltlike gland around the animal (Yang and Hochberg, 2018). Externally, the tube consists of a series of thickened rings and elongated girdles, both of which are somewhat fibrous in appearance.

The body of adult acanthocephalans consists of a large, sausage-shaped trunk or metasoma with a more slender anterior end, or presoma (also called "praesoma"), consisting of an armed proboscis with hooks and a hookless neck (Taraschewski, 2015). The presoma occupies more than 50% of the young acanthocephalan's body, while in the gravid females it is just a small anterior appendage of a huge metasoma. The presoma, when not attached to the host, can be everted and retracted into the metasoma, mediated both by the contractions of the internal muscles of the proboscis and the trunk and by the flow of liquid inside the tegument and the lemnisci, a pair of tube-shaped prolongations of the presomal tegument projecting into the body cavity. The presence of lemnisci in bdelloids has been used to justify a group composed of bdelloids and acanthocephalans, but the homology of these structures in the two clades remains to be further investigated (Garey et al., 1998). Some have also argued that the presoma may be homologous to the proboscis of bdelloids (Lorenzen, 1985), but this has been rejected by others (Clément, 1993).

All acanthocephalans are gutless, obligatory endoparasites and typically measure 1–2 cm in length, although a few larger species can reach up to 50 cm, as in some species in the genus *Macracanthorhynchus*. Internally there is a ligament cord, to which the gonads attach.

In life, the proboscis or the entire presoma becomes surrounded or encapsulated by tissue of the intestinal wall of the host, the metasoma remaining inside the gut lumen. Those species with long necks can penetrate deep into the gut wall or even project into the peritoneal cavity. Some can go further, digging through the gut and completely entering the peritoneal cavity, waiting for the next host, a common practice if the first vertebrate is a paratenic host (Taraschewski, 2015). While short-necked species can change the point of attachment frequently, the long-necked ones tend to remain anchored to a single spot.

FIGURE 39.1. Schematic drawing of the stratified epidermis of *Seison annulatus*. The blue surface represents the cytoplasm, which includes the two sublayers of the internal layer. Modified from Ahlrichs (1997).

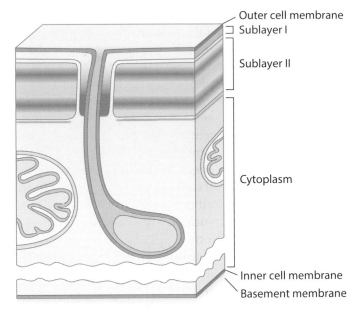

Outer cell membrane
Sublayer I
Sublayer II
Cytoplasm
Inner cell membrane
Basement membrane

Despite their major differences in body plan, a few characters unite the parasitic acanthocephalans to the free-living forms, as summarized in Lorenzen (1985) and Ahlrichs (1997), among others. In most males of Monogononta the intestine is reduced to a cord of tissue without a lumen to which the testis attaches. The ligament cord of Acanthocephala, to which the gonads attach, has been interpreted also as a reduced intestine (von Haffner, 1950) that would be homologous to the intestinal cord of monogononts, but this requires further study. The syncytial organization of the epidermis (fig. 39.1), with intraepidermal skeletal condensations and infoldings of the apical plasma membrane, is another characteristic of all the members of the phylum (Albrecht et al., 1997). The outer epidermal cell membrane intrusions with bulbs and an anterior insertion of the flagellum on sperm cells have also been discussed in this context by Ahlrichs (1997).

The body cavity of rotifers is considered a pseudocoel, as it is lined by extracellular matrix and not by an epithelium. It is a spacious body cavity, apparently lacking fibrils or microfilaments of collagen (Clément and Wurdak, 1991; Clément, 1993; Fontaneto and De Smet, 2015), but with free amoeboid cells or amoebocytes, at least in some species, which form a highly dynamic, three-dimensional network of filopodia. The pseudocoel acts as a hydrostatic skeleton, whose ionic composition and volume are regulated by the protonephridia.

The protonephridia of the free-living rotifers have been studied using light microscopy first (e.g., Pontin, 1964), but mostly by transmission electron microscopy (e.g., Braun et al., 1966; Mattern and Daniel, 1966; Warner, 1969; Schramm, 1978; Clément and Wurdak, 1991; Ahlrichs, 1993a, b; Riemann and Ahlrichs, 2010). Ahlrichs (1993a) provided the first description of the complete protonephridial system of a rotifer, which was known to show considerable variation across species (e.g., Remane, 1933). It consists of two protonephridia lying ventrolaterally in the pseudocoelomic cavity, formed by three to four multinucleated cells with multiple terminal organs of flame bulbs, attached to the collecting tubules or capillary canals, which in turn connect to the main canal. The two main canals discharge together in a single contractile urinary bladder (in monogononts) or in a contractile cloaca.

The flame bulbs are somehow similar to those of Platyhelminthes (chapter 43), which are the textbook example of a flame bulb. They are hollow, conical, and laterally compressed, the lumen draining into the capillary canals of the syncytium, the distal end being closed by a protoplasmic cap. Variation in the flame bulb morphology and number (2 to 100 per side) is observed among the different groups. While the protonephridial system is not well studied in the male monogononts, in some but not all cases it is reduced, with fewer flame bulbs.

Among acanthocephalans, protonephridia have been reported so far only in Oligacanthorhynchida (a subclade of Archiacanthocephala) (Dunagan and Miller, 1986b), while in the other groups it is assumed that the crypts of the outer membrane enlarging the surface of the integument are involved in excretion and osmoregulation (Taraschewski, 2015). These protonephridia have been studied with scanning electron microscopy and transmission electron microscopy (Dunagan and Miller, 1985, 1986b, 1991; Krapf and Dunagan, 1987), but few have paid attention to them recently. Two or three nephridial designs have been described, one of a dendritic type, organized as branches of a tree in which each final branch terminates in a ciliated bulb, and a capsular (or sac) type, in which all ciliated bulbs empty directly into a common chamber (see reviews in Dunagan and Miller, 1986b, 1991). A rudimentary type, consisting of a single cell and no ducts to the outside, has also been described (von Haffner, 1942), but it is unclear whether this is an excretory organ or not. The first two types are syncytial with three nuclei located in the capsule or stem wall and none in the flame bulbs. Both types consist of two clusters of flame bulbs that empty separately into an expandable excretory bladder, which in turn empties into ducts of the reproductive system, constituting a urogenital system that empties to the outside through a gonopore located at the tip of the penis in males and the vagina in females (Dunagan and Miller, 1986b).

The digestive system consists of a through tube, with a mouth, buccal tube, pharynx with mastax, esophagus, stomach, intestine, cloaca, and cloacal aperture or anus. While this system is straight, in some species the anus is displaced anteriorly, becoming U-shaped. In the members of the family Habrotrochidae, the stomach is syncytial and without a lumen. The intestinal system of the male may be fully developed, as in *Rhinoglena frontalis*, which feeds as a male (Melone, 2001), to completely absent, with all sorts of intermediate forms, some with a well-developed intestinal system but without trophi. In some cases, the digestive tract is reduced to a strand without a lumen that supports the testis. The digestive system is completely reduced in acanthocephalans, which have no mouth, and uptake nutrients that have been digested by the host through their body surface.

The trophi comprises a set of hard, cuticularized extracellular elements or rods (Rieger and Tyler, 1995) that are added through ontogeny (De Smet and Segers, 2017). The different elements receive specific names, as do the many types of trophi (Fontaneto and De Smet, 2015). These diverse structures are used for rotiferan taxonomy, as they are rich in characters (Sørensen, 2002a). The musculature associated with the trophi can be complex and is composed of a series of paired and un-paired muscles (e.g., Sørensen et al., 2003; Wilts et al., 2010; Wulfken et al., 2010). While in most species studied there was no growth of the trophi after hatching, *Asplanchna priodonta* demonstrates posthatching growth (Fontaneto and Melone, 2005).

The characteristic corona (the alternative names "wheel organ," or "rotatory organ," refer to the optical effect when the cilia are moving) of most nonparasitic rotifers comprises a ciliated area, and the buccal field around the, usually, ventrally located mouth. The buccal field is evenly ciliated, composed of short cilia, and extends apically to form a circumapical band, which delimits an unciliated apical field. The cilia at the anterior margin of the circumapical band are strong and form the preoral trochus. The cilia at the posterior margin are finer and form the post-oral cingulum. The trochus and the cingulum delimit a finely ciliated groove. Additional nomenclature applies to these bands, depending on whether or not they consist of a single row of cilia. In many species the marginal cilia of the circum-apical band are elongated and often located on a pair of lateral, earlike projections, the auricles, which may help with swimming. The different types of corona are summarized in many rotiferan treatises, most recently in Fontaneto and De Smet (2015).

Nielsen (e.g., 2012a) has argued that the wheel organ is homologous to the trocho-phore bands and thus interprets rotifers as being neotenic descendants of a hypo-thetical protostomian ancestor, the so-called gastroneuron—this idea dating back to Hatschek (1878). This hypothesis is hardly supported by developmental data, and lack of detailed cell lineages in rotifers prevents addressing this formally. It is also contradicted by phylogenetic information, as rotifers are most closely allied to other groups without a trochophore larva.

Cell constancy and eutely have been addressed in a diversity of species and or-gans of rotifers (summarized in Gilbert, 1989). While the number of nuclei has been determined to be constant in some species (e.g., 959 in *Epiphanes senta*), other species show considerable variability in nuclear number in certain tissues. Still, it seems that there is a total absence of cell division after completing development (Clément and Wurdak, 1991). But cell constancy and eutely are not necessarily the same. While some eutelic animals, like *Caenorhabditis elegans*, have a fixed number of somatic cells, they continue to divide their germline. This is, however, not the case of rotifers, which seem to also stop cell division even in the germline (Pagani et al., 1993).

The muscular system of rotifers (fig. 39.2) has received enormous attention since immunostaining and confocal laser microscopy became widely used, with virtually every new species, as well as many well-known species, being described at this level, with dozens of papers published since the pioneering work of Hoch-berg and Litvaitis (2000a) and Kotikova et al. (2001). A recent summary has been provided by Fontaneto and De Smet (2015). The somatic muscular system consists of two major groups of muscles (either thin filaments or arranged in separate bands), a system of outer circular bands and a system of inner, paired, and bilat-erally symmetrical longitudinal bands. The muscles insert on the skeletal syncytial integument. Somatic muscles of bdelloids are smooth or obliquely striated, whereas in monogononts they are primarily cross-striated (Clément, 1993).

The circular muscles of the trunk comprise generally between three and seven separate muscles that can constitute closed rings, open ventrally or ventrally and dorsally, or split into dorsolateral and ventrolateral bands. The development often depends on whether the species are soft bodied (more complete rings) or whether their bodies are stiffened by integumentary plates (with lorica), in which case the musculature is reduced. In species with moveable spines or appendages, the circular muscles of the trunk are absent and may be transformed into the muscles responsible for moving the appendages (e.g., Hochberg and Ablak Gurbuz, 2007, 2008). The inner longitudinal retractor muscles are present in the head, trunk, and toes, some spanning the length of the animal. They allow bending of the body as well as retracting the head and foot into the trunk. In Seisonidea the overall muscular arrangement consists of segmentally organized longitudinal fibres that extend the length of the body and are surrounded by incomplete circular bands (Leasi et al., 2012).

In contrast, the musculature of Acanthocephala has received less attention and recent work has focused on light microscopy and dissection (e.g., Herlyn and Ehlers, 2001; Herlyn, 2002). It is the longitudinal musculature that allows the presoma to withdraw within the metasoma. The muscular system of the metasoma consists of subtegumental circular and longitudinal muscles, which have a tubular nature (forming the so-called lacunar system) and contain a gelatinous substance rich in glycogen granules, mitochondria, and fibers (Taraschewski, 2000). Early work on the lacunar system used corrosive casting to elucidate its three-dimensionality, concluding that it probably served as a hydrostatic skeleton (Miller and Dunagan, 1985b). Inside the metasoma of female worms, a fibrous septum containing muscles fibers (the ligament sacs) separates the body cavity longitudinally into two compartments. In males the ligament strand keeps the testis in position.

NERVOUS SYSTEM

As in the case of the muscular system, study of the nervous system of rotifers (fig. 39.2) has benefited from the recent immunohistochemical developments and recent reviews (Fontaneto and De Smet, 2015; Hochberg, 2016; Semmler Le, 2016), but these are lacking in Seisonidea. Earlier work on nervous systems was largely summarized by Bullock and Horridge (1965), supplemented by the transmission electron microscopy work by Clément and Wurdak (1991). Most recent works employ immunohistochemistry (e.g., Kotikova et al., 2005; Hochberg, 2007b, 2009, 2016; Kotikova and Raikova, 2008; Leasi et al., 2009; Hochberg and Lilley, 2010). The main structure of the central nervous system is the brain, an oblong or spindle-shaped mass that can be quite large. It is located behind the corona, dorsal to the mastax, and is easily visualized through the body wall. The brain can be surrounded by epithelial or muscular cells and has a constant number of cells (e.g., 183 in *Epiphanes senta* and 249 in *Synchaeta tavina*) (Fontaneto and De Smet, 2015).

A pair of lateral or ventrolateral nerve cords, composed of a bundle of axons, emanates from the brain and extends caudally, fusing in a posterior caudal or

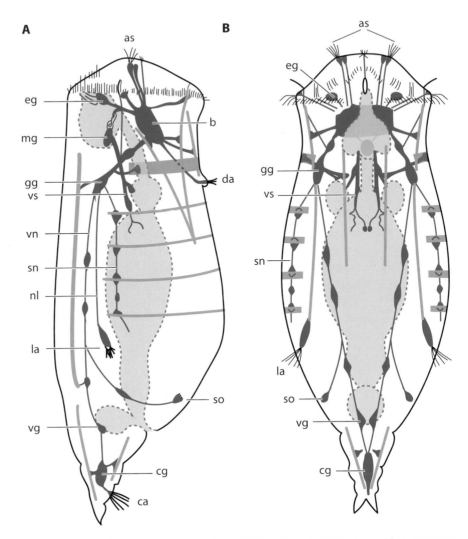

Figure 39.2. Schematic of a rotifer nervous and muscle systems in lateral (A) and ventral (B) views. Blue, nervous system; yellow, intestinal system, bladder, and pedal glands; pink, muscles. Abbreviations: as, apical sense organs; b, brain; ca, caudal antenna; cg, caudal ganglion; da, dorsal antenna; eg, epipharyngeal ganglion; gg, geniculate ganglion; la, lateral antenna; mg, mastax ganglion; nl, nerve to lateral antenna; sn, scalar nerve; so, supra-anal sense organ; vn, ventral nerve; vg, vesicular ganglion; vs, visceral nerve. Modified from Remane (1933).

pedal ganglion. The cords may be connected by one or more commissures. In addition, the cords may have branching lateral nerves at secondary ganglia. Whether there is any correspondence between secondary ganglia and the observed external pseudosegmentation has not been addressed in the literature. The coronal region with its sensory structures is innervated by a pair of neurites emanating from the brain, which may form a neuronal ring; another pair of neurites innervates the sensory antennae (Hochberg and Lilley, 2010). A comparison of different rotifer nervous systems can be found in Kotikova (1998). The brain also innervates the mastax and salivary glands. The caudal ganglion innervates the foot.

In males the nervous system appears well developed, except for the mastax ganglion, which is usually absent in most species, as they have a reduced nonfunctional mastax.

Rotifers have a diversity of sensory organs—including mechanoreceptors, photoreceptors, and chemoreceptors—or sensory complexes combining multiple functions (Clément and Wurdak, 1991; Fontaneto and De Smet, 2015). These include sensory bristles, papillae, chemoreceptive pores and sensory pits, among others, associated with the corona. There may also be prominent dorsal, lateral, and caudal antennae (fig. 39.2). The dorsal antenna is found in most species and is generally located medially on the head or neck, but it can be on the trunk. This is a paired structure in the embryo, fusing into an unpaired organ in most species. In bdelloids it forms a long tube, telescopically retractable by paired muscle cells, and has three pairs of nerve endings. Paired lateral antennae are present in monogononts, generally located laterally and symmetrically in the posterior half of the trunk, but all sorts of displacements, asymmetries, and even partial or total fusion are known. Each antenna contains a single sensory neuron. A single, rarely paired, small caudal antenna, consisting of a ciliary pit or a tuft of cilia, is generally found dorsally on the distal foot pseudosegment. Other sensory organs (papillae or sensory pits) may be located in the same region instead of the caudal antenna.

Rotifer photoreceptors have been discussed thoroughly (Clément, 1980; Clément and Wurdak, 1984). Red or black (often depending on diet) eyes, eyespots, or ocelli (different nomenclatures have been used) are generally found in monogononts and bdelloids, but these may disappear in the adult stage of sessile species. Eyes are generally found on the apical field, rostrum, lateral sides of the corona, or on the brain, and may have different structure. The cerebral eyes (similar to those of Platyhelminthes) are single or fused paired eyes; the other types of eyes are usually paired. A photoreceptive function has also been suggested for the phaosome of the bdelloid *Philodina roseola*—the phaosome is an unpaired structure, consisting of sensory nerve endings in a spherical bulb containing fan-shaped, flattened cilia. It is located at the base of the rostrum, near the openings of the retorecerebral apparatus (Clément, 1980; Clément and Wurdak, 1991).

Most of what we know about the nervous system of acanthocephalans derived from the work of American parasitologists T. T. Dunagan and D. M. Miller (summarized in Miller and Dunagan, 1985a; Dunagan and Miller, 1991). Most subsequent work has focused on the presoma (e.g., Dunagan and Miller, 1987; Dunagan and Bozzola, 1989, 1992a, b; Herlyn et al., 2001), especially of the archiacanthocephalan *Macracanthorhynchus hirudinaceus*, a pig parasite. No nervous system or sense organs are known for the acanthor, and little information is available for other developmental stages (Hehn et al., 2001). In adult acanthocephalans, the nervous system consists of a cerebral ganglion inside the proboscis, comprised of 50 to 109 cells (information compiled in Miller and Dunagan, 1985a), and a set of ganglia in the male, for copulation and insemination. Two lateral posterior longitudinal main nerves in the metasoma communicate with the ganglia. Additional nerves emanate from the cerebral ganglion (six major nerves in *M. hirudinaceus*). Two lateral sense organs and a single or paired apical sense organ can be found in different groups (Dunagan and Miller, 1983, 1986a).

▬ REPRODUCTION AND DEVELOPMENT

Embryonic development has received relatively little attention and has focused almost exclusively on monogononts, with a few key studies dating back to the mid–twentieth century (de Beauchamp, 1956; Pray, 1965; Lechner, 1966), but mostly neglected since then (see Boschetti et al., 2005). The current state of the field has been summarized by Hejnol (2015b). While rotifer development has often been referred to as "modified spiral" cleavage, there is no indication of a spiral pattern, and partial fate map information is only available for the monogonont *Asplanchna girodi* (see Lechner, 1966), and it is insufficient to interpret homologies between spiral cleavage and the rotifer mode, as no information about the origin of mesoderm is known in rotifers. Germ cell specification has been studied at the molecular level in *Brachionus plicatilis* (Smith et al., 2010).

In general, cleavage (fig. 39.3) is holoblastic and unequal, with a transverse furrow in the first division, a typical 16-cell stage, and early gastrulation by epiboly of the large D blastomere. As described by Hejnol (2015b), the embryo gives off one polar body before the onset of cleavage, and the position of the polar body marks the animal pole of the embryo. The first cell division is unequal, dividing the egg into a large CD blastomere, whose descendants will give rise to the germovitellarium and other internal tissues, while the smaller AB blastomere will contribute mainly to the ectoderm. The AB blastomere divides equally into blastomeres A and B, while CD divides unequally into the smaller C blastomere and the larger D blastomere, the size of blastomere C being similar to those of A and B. This 4-cell stage looks similar to the 4-cell stages of unequal cleaving spiralians, which also display a larger blastomere (D) and three smaller blastomeres. The polar body is located in the center of the four blastomeres, where all blastomeres are in contact with each other. Interestingly, during the subsequent cell divisions, bdelloids and monogononts differ in the further spatial interrelationships of the blastomeres and polar body. While in monogononts the polar body is located first on the tip of the small blastomeres that are formed until the 16-cell stage (e.g., Lechner, 1966), the polar body in the bdelloid species is located close to the large blastomeres that will gastrulate.

Even less has been done recently with embryonic development of acanthocephalans since it was last reviewed more than three decades ago (Schmidt, 1985), and the data mostly rely on studies conducted at the end of the nineteenth century and that of Meyer (1928).

The development of the embryo that gives rise to the acanthor (see below) has recently been summarized by Hejnol (2015b). The fertilized egg is oval in shape and has two polar bodies that mark the future anterior end of the animal. The first cell division is equatorial to the longitudinal egg axis and gives rise to the animal blastomere AB (in contact with the polar bodies) and the slightly larger vegetal blastomere CD. The two blastomeres divide equally in the next cell division to produce the animal blastomere B3, the median blastomeres A3 and C3, and the vegetal blastomere D3. The subsequent cell divisions are unequal. Blastomere D3 is the first to divide unequally into micromere D4.1 and macromere D4.2. The other

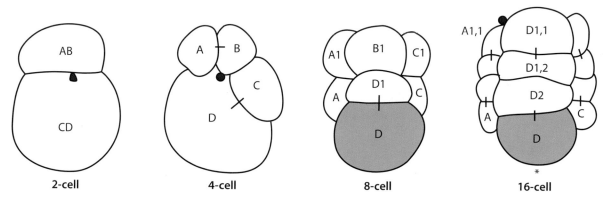

FIGURE 39.3. *Asplanchna* cleavage up to the 16-cell stage. Polar body in red; blastomere D, which will form the germovitellarium, in brown. Based on Hejnol (2015b).

macromeres follow a similar pattern in that they give off smaller micromeres toward the animal pole.

In different species the embryo becomes syncytial at different times, ranging from the 4-cell to the 36-cell stage. A clear fate map has not been determined. Up to the 25-cell stage of *Gigantorhynchus gigas*, only macromeres divide, the micromeres remaining arrested (Meyer, 1928). The quartet of macromeres thus forms "mother cells" of all micromeres before the micromeres start to divide again in the 25-cell stage. During further development, the micromeres condense and become internalized to form a "central nuclear mass," which has been interpreted as a form of gastrulation. These condensed nuclei will be part of all the internal tissues except the epidermis. The embryo then begins to form up to four egg layers of different density.

Development in Rotifera is often referred to as direct, that is, the egg giving rise to a juvenile that resembles the adult, but this is a simplification of a more complex situation. Rotifers and acanthocephalans can have complex life cycles—with mictic and amictic phases, each giving rise to embryos—and acanthocephalans have different developmental stages. Even in the free-living rotifers, a few species of Gnesiotrocha have a short-lived "larva" (it is similar to the adults of other monogononts) that undergoes a dramatic metamorphosis upon settlement (Kutikova, 1995), replacing the "larval" corona with the adult infundibulum—a highly unusual cup-shaped head that appears to develop from the anterior foregut (Hochberg and Hochberg, 2015, 2017). Indirect development is maximized in Acanthocephala.

Acanthocephalans are gonochoristic and may show sexual dimorphism in size, longevity, and pathogenicity. The fully embryonated eggs, containing an acanthor, are discharged by a gravid female. After ingestion by an intermediate parathenic arthropod host, the acanthor hatches and penetrates the body cavity, where it continues developing. Inside the intermediate host it goes through an acanthella stage and becomes the cystacanth, the final stage within the paratenic host. In some cases, a second paratenic host, usually a fish, is required if the final host is a marine mammal. The host–parasite relationship could be one of strict specificity, while some species can use a multitude of parathenic hosts.

As adults, acanthocephalans inhabit the intestinal tract of vertebrates. Transmission of adult parasites (termed postcyclic transmission) is possible through consumption of infected hosts (e.g., Skuballa et al., 2010).

The acanthor (see Albrecht et al., 1997) is surrounded by four eggshells (embryonic envelopes) and is composed of three syncytia: a frontal syncytium, a central syncytium, and an epidermal syncytium. The central syncytium shows a mass of condensed nuclei and twelve decondensed nuclei and gives rise to ten longitudinal muscle bands and two oblique retractor muscles. A single decondensed nucleus can be assigned to each of the twelve muscular systems. The epidermal syncytium embeds the other two syncytia and forms the wrinkled epidermis, which shows an extracellular glycocalyx and intrasyncytial condensations. Prominent recurved hooks, which mark the anterior end of the acanthor, and body spines are intraepidermal differentiations. Partly branched tubular infoldings of the epidermal plasma membrane of the acanthor represent precursors of the pore ducts typical of the adult epidermis.

Upon hatching, the spindle-shaped acanthor immediately increases in size and begins penetrating into the wall of the mesenteron of the intermediate host. The acanthor may then cease activities for hours or even days, or it may proceed through the wall of the intestine (Schmidt, 1985). Development continues with a series of reorganizations that form the primordia for the proboscis apparatus, the brain, as well as the muscular and reproductive elements of the body. At this stage some nuclei stop dividing, fixing the final number of cells of the animal. As development continues, the worm acquires all of the structures of the adult and is capable of infecting the definitive host. This stage is the so-called cystacanth, which is not encysted.

Contrary to the exclusively parthenogenetic bdelloids, many monogonont rotifers follow a complex cycle (fig. 39.4) composed of a parthenogenetic amictic phase (diploid eggs) that gives rise to amictic females and a mictic phase (haploid eggs) that includes sexual reproduction (giving origin to females if fertilized) and the presence of haploid dwarf males resulting from absence of fertilization (Nogrady et al., 1993; Melone and Ferraguti, 1999). Monogononts—unlike bdelloids, which can enter a dormant stage at any point in their life cycle—can only enter dormancy as a so-called resting egg, a fertilized mictic egg (Rao and Sarma, 1985; Boschetti et al., 2011). When a male fertilizes a haploid egg, a diploid zygote originates and is encased in a composite shell made of two or three layers, then starts cleavage until it arrests development and becomes dormant. This "resting egg" is a diapausing embryo that will remain dormant for variable periods, from days to years, resuming its development after being activated by external (e.g., illumination, temperature, salinity) and/or internal stimuli (Boschetti et al., 2011).

Evo-devo work has yet to exploit the rotifer system and virtually nothing has been done on acanthocephalans. Fröbius and Funch (2017) recently compared the Hox complement of rotifers and reported the expression of five Hox genes during embryogenesis of the rotifer *Brachionus manjavacas*. While these genes define different functional components of the nervous system, they did not show the usual

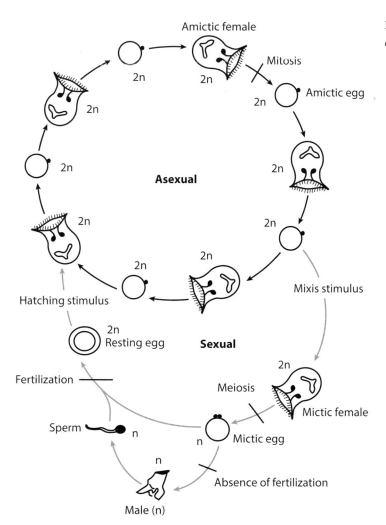

Figure 39.4. Life cycle of the monogonont rotifer *Brachionus plicatilis*. Based on Hoff and Snell (1987).

bilaterian staggered expression along the anteroposterior axis. The difference in homeobox composition with respect to Rouphozoa + Lophotrochozoa on the one hand and presence of motifs that are uniquely shared with Chaetognatha on the other (Fröbius and Funch, 2017) may contribute to resolving the phylogenetic position of Chaetognatha with respect to Gnathifera, but further data are needed for Gnathostomulida and Micrognathozoa

GENOMICS

Rotiferan genomes have been of general interest due to the apparent lack of sex in bdelloids and the presence of ancestral tetraploidy (Mark Welch et al., 2008; Hur et al., 2009), or more recently as a means to study cryptobiosis (Nowell et al., 2018). Full genomes have been sequenced for *Adineta vaga* (Flot et al., 2013), *A. ricciae* (Nowell et al., 2018), *Rotaria magnacalcarata* (Eyres et al., 2015; Nowell et al., 2018), *R. macrura* (Nowell et al., 2018), and *Brachionus calyciflorus* (Kim et al., 2018). A large fraction of exogenous DNA was identified in most cases (e.g., 8% for *A. vaga*). Horizontal gene transfer from environmental DNA has been shown to be the main cause of this exogenous DNA, with estimates of an average of 12.8 gains versus 2.0 losses of foreign genes per million years (Eyres et al., 2015). Additional unassembled unpublished genomes are available in GenBank, including *Brachionus manjavacas*.

An explanation for the presence of exogenous DNA in the bdelloid genomes hypothesizes that they incorporate foreign DNA when they repair their chromosomes following double-strand breaks (DSBs) upon rehydration (Debortoli et al., 2016), but this has been questioned by Nowell et al. (2018), who showed that many colinearity breaks are the result of short-read assembly. The expectation that the ability to survive desiccation must have a profound influence on bdelloids' genome structure (Gladyshev and Arkhipova, 2010) is now reconsidered. Furthermore, the same mechanism seems to operate in their extreme resistance to ionizing radiation, explained by the presence of an unusually effective system of antioxidant protection for cellular constituents, including those required for DSB repair (Krisko et al., 2012). This mechanism thus allows bdelloids to recover and continue reproducing after desiccation or after doses of IR causing hundreds of DSBs per nucleus.

One of the supposed most extraordinary aspects of bdelloids, a clade known from fossils as far back as the Paleogene (Poinar Jr. and Ricci, 1992; Waggoner and Poinar Jr., 1993) and, based on molecular dating, estimated to be even older, is the absence of sex (Ricci, 2016). Despite much observation of field and laboratory populations and except for one heavily qualified account of having twice seen a male among many females in a sample from a Danish lake (Wesenberg-Lund, 1930), neither males, hermaphrodites, mating, nor meiosis have ever been reported within the class (Birky, 2010). The fact that bdelloids could have evolved without genetic exchange is, however, hard to accept, and the hypothesis that bdelloid males, perhaps rare or invisible, do exist has persisted for years (Wesenberg-Lund, 1930). Even if males existed, by being extremely rare, their effect on the global populations would be effectively nonexistent (Birky, 2010).

A solution to the mystery of bdelloid supposed lack of sex has recently been proposed. A study in *Macrotrachela quadricornifera* has revealed a striking pattern of allele sharing consistent with sexual reproduction and with meiosis (Signorovitch et al., 2015), even if of an atypical sort, in which segregation occurs without requiring homologous chromosome pairs, as in evening primroses (Golczyk et al., 2014). This *Oenothera*-like meiosis is, however, not universally accepted (Flot et al., 2016), and some still consider the question of bdelloid recombination an open question (Nowell et al., 2018).

Transcriptomic resources are available for a relatively large number of species (Hausdorf et al., 2007; Wey-Fabrizius et al., 2014; Eyres et al., 2015). The recent development of RNAi for rotifers (Snell et al., 2011) represents, from a developmental point of view, a step toward functional genomics research in this understudied group of animals.

Rotifer mitochondrial genomes show some unusual properties. The monogonont species in the genus *Brachionus* are unique in having their mitochondrial genome organized into two circular chromosomes of unequal copy number (Suga et al., 2008; Hwang et al., 2014; Kim et al., 2017), something that has not been reported for any other organism, including other rotifers/acanthocephalans (e.g., Min and Park, 2009; Gazi et al., 2016).

FOSSIL RECORD

The fossil record of rotifers is sparse and, so far, restricted to the Cenozoic. Most of the record was summarized by Valentine (2004). Fossils have been reported from nonmarine Eocene deposits in South Australia (Southcott and Lange, 1971), from Miocene amber from the Dominican Republic (Poinar Jr. and Ricci, 1992; Waggoner and Poinar Jr., 1993; note that these deposits were then dated as being older, i.e., Eocene), from Holocene peat deposits in northern Ontario, Canada (Warner and Chengalath, 1988), and Holocene deposits of an Antarctic continental lake (Swadling et al., 2001). No fossils are known for Acanthocephala or Seisonidea.

40 PLATYTROCHOZOA

After the dismantlement of Platyzoa, it is now generally considered that Rouphozoa constitutes the sister group of Lophotrochozoa (Struck et al., 2014; Laumer et al., 2015a; Laumer et al., 2019; but see Marlétaz et al., 2019), and together they constitute a clade named Platytrochozoa (= Lophotrochozoa in Marlétaz et al., 2019). No morphological synapomorphies were proposed when establishing this taxon (Struck et al., 2014), but members of both Rouphozoa (i.e., Platyhelminthes) and Lophotrochozoa (Mollusca, Annelida, Nemertea, Entoprocta) have spiral development, a character that has also been suggested for Gnathostomulida (Riedl, 1969), although to some this feature remains questionable. If the absence of spiral development were confirmed for Gnathostomulida, Spiralia would need to be restricted to Rouphozoa + Lophotrochozoa, and thus Platytrochozoa would be a synonym of Spiralia, but under the current definition, this taxon is helpful for referring to the sister group of Gnathifera.

Platytrochozoa is thus a large clade of protostomes with an array of disparate morphologies and an early origin in the chronicle of animal life. While Gnathifera show a relatively conserved body plan, the same cannot be said for Platytrochozoa, which have probably diverged drastically despite the conserved and stereotypic spiral cleavage present in many of its members. Although speculative, it would not be that surprising to explain the morphological disparity of the group by, first, a divergence in body patterning after early developmental conservatism, followed by a relaxation of the early developmental pattern, as seen in the nonpolyclad Platyhelminthes, Gastrotricha, and many clades of Trochozoa, including cephalopod Mollusca or the lophophorate phyla. In fact, these observations and ideas have led some to propose that phyla may be defined as a collection of species whose gene expression at the mid-developmental transition is both highly conserved among them, yet divergent relative to other species (Levin et al., 2016), although this has proven highly controversial (Hejnol and Dunn, 2016). Nonetheless, it would be interesting to understand how certain fundamental body-plan elements (like the presence of blind versus through-guts or the presence or absence of primary and secondary body cavities) are drastically more variable in some clades than in others, and Platytrochozoa could be seen as perhaps the most extreme animal clade in which these anatomical transitions have occurred multiple times.

ROUPHOZOA

After a turbulent phylogenetic history, several recent phylogenomic studies have found that Platyhelminthes (flatworms, tapeworms, flukes, and their relatives) and Gastrotricha form a clade named Rouphozoa (Struck et al., 2014; Laumer et al., 2015a; Lu et al., 2017; Laumer et al., 2019), although a recent study has disputed this clade (Marlétaz et al., 2019). Platyhelminthes had been allied traditionally to a diversity of animal groups or placed as a basal bilaterian clade due to their acoelomate body plan and lack of an anus, while gastrotrichs were often grouped with other "aschelminths" due to their cuticle and nervous system of the cycloneuralian type.

Cavalier-Smith (1998) introduced the name Platyzoa for a clade diagnosed as "ciliated non-segmented acoelomates or pseudocoelomates lacking vascular system; gut (when present) straight, with or without anus," which included two phyla, Platyhelminthes and Acanthognatha, the latter consisting of Trochata (= Rotifera–Acanthocephala; he recognized the possible paraphyly of Rotifera with respect to Acanthocephala) and Monokonta, a clade that included Gnathostomulida and Gastrotricha. Platyzoa was subsequently endorsed by some molecular phylogenetic analyses (e.g., Giribet et al., 2000; Dunn et al., 2008; Hejnol et al., 2009), but later shown to be a systematic artifact (Struck et al., 2014; Laumer et al., 2015a; see also Kocot et al., 2017) and paraphyletic, consisting of two groups: Gnathifera and its sister clade, consisting of Rouphozoa and Lophotrochozoa.

The name Rouphozoa was introduced by Struck et al. (2014) as a derivation of the Greek word *rouphao*, for "ingesting by sucking," referring to the preferred feeding mode of platyhelminths and gastrotrichs, but this character is not uncommon among metazoans with strong sucking pharynxes, and the plesiomorphic state within Platyhelminthes is a pharynx symplex (found in Catenulida and Macrostomorpha), engulfing instead of sucking the prey, and thus the name does not really reflect a synapomorphy of the group. Few valid synapomorphies are thus found for these sister taxa, as they both seem to have symplesiomorphic traits for Spiralia, such as lack of coeloms, complete or nearly complete body ciliation, and protonephridia. The clade is, however, strongly supported in recent phylogenomic analyses (Laumer et al., 2015a; Lu et al., 2017; Laumer et al., 2019; but see Marlétaz et al., 2019), and a character that could be used to define it is the presence of the duogland organs (Tyler and Rieger, 1980), a putative synapomorphy of Rouphozoa, even though this was once considered a striking case of convergence (Tyler, 1988). However, similar organs also exist in other groups, including annelids and nematodes; other types of adhesive organs are also present in many other invertebrates, including byssal threads of bivalves, pedal adhesion in gastropods, and tube feet in echinoderms, among others (Tyler, 1988), but homology with the duo-gland system is highly unlikely.

FOSSIL RECORD

Apart from indirect evidence for parasitic platyhelminths and fossil evidence for free-living platyhelminths (see chapter 43), a very meager fossil record is known for members of Rouphozoa. The fossilization potential for both main groups is low, despite the presence of a relatively thick cuticle in gastrotrichs (Maas, 2013) and jaws in some platyhelminth groups.

GASTROTRICHA

Gastrotricha (from the Greek *gasteros* and *trichos*, the "hairy bellies") is a cosmopolitan phylum of microscopic aquatic, limnic, and semiterrestrial invertebrates comprising about 840 described species, divided into the two orders Macrodasyida and Chaetonotida (Dal Zotto et al., 2018). Macrodasyida includes taxa living mostly interstitially in marine sandy bottoms—although a freshwater lineage derives from within this marine group (Todaro et al., 2012)—while Chaetonotida comprises species found from marine to limnic environments. Much of the knowledge on this group is owed to the work of German zoologist Adolf Remane and US zoologist William Hummon, but the first monograph of the group is that of Zelinka (1890). Gastrotrich biology is perhaps at its peak with the combination of molecular techniques, confocal laser microscopy, and a broad interest in meiofaunal diversity, with some extremely active hubs in Italy, Sweden, Brazil, and the United States.

Despite this recent revival, led in part by the molecular revolution (see Hochberg et al., 2017a), gastrotrichs remain a somewhat neglected group. Nonetheless, gastrotrich researchers have pioneered the use of virtual vouchering techniques, especially online videos. The Gastrotricha World Portal (www.gastrotricha.unimore .it) serves as a repository of these and many other resources, including bibliographic tools (Hochberg et al., 2017a).

Gastrotrichs are generally smaller than 1 mm long (down to 75 μm), but they can reach up to 3.5 mm in the species of the giant *Megadasys* (Kånneby et al., 2012). They superficially resemble flatworms but are easily distinguished by their adhesive tubes at the end of the body and the cilia restricted to the ventral side of the animal. Due to their small size, gastrotrich anatomy has required the use of transmission electron microscopy (e.g., Ruppert, 1991a), but more recently considerable advances have been made with the combination of immunostaining and confocal laser microscopy. A state-of-the-art account of the group has recently been provided by Kieneke and Schmidt-Rhaesa (2015).

Gastrotrichs are important components of all sorts of aquatic habitats, where they can become locally abundant. A report estimates densities of up to 2.6 million individuals per m^2 in freshwater lakes in Poland (Nesteruk, 1996) and 0.67 million *Chaetonotus* individuals per m^2 in a freshwater Galapagos crater, in which a species of gastrotrich was the dominant meiofaunal species (Muschiol and Traunspurger, 2009). While densities are also high in some marine environments, these are not as high as in fresh waters. However, the distribution in marine environments is very large, as gastrotrichs are known from the subtidal to the deep sea (Kieneke, 2010), including the Kuril–Kamchatka Trench, below 5,700 m (Schmidt and Martínez Arbizu, 2015), where they are part of the mesopsammon. Planktonic forms have evolved multiple times from benthic marine groups (Kånneby and Todaro, 2015). In limnic environments gastrotrichs can be either benthic (psammic) or live among vegetation (epiphytic), some species being able to swim. They show high seasonality in presence and abundance (e.g., Nesteruk, 2017), but their patterns of distribution are poorly known (Balsamo et al., 2008).

▬ SYSTEMATICS

The systematic position of Gastrotricha has been in flux and is discussed in part in chapter 41. They were once considered closely related to various meiofaunal protostome groups, such as rotifers in the clade Trochelminthes (Hyman, 1951b); kinorhynchs in Nematorhyncha (Bütschli, 1876); nematodes (Ruppert, 1982); or gnathostomulids in Monokonta (Cavalier-Smith, 1998) or in Neotrichozoa (Zrzavý et al., 1998). The latter relationship has been a favorite for some authors, due to the presence of monociliated epithelia in Gnathostomulida and at least some Gastrotricha (Rieger, 1976)—yet monociliated epithelia have also been discussed as being plesiormorphic for animals. Other organ systems have driven the relationships with the other phyla.

The internal phylogeny of Gastrotricha has been tackled using morphological cladistic analyses (Travis, 1983; Hochberg and Litvaitis, 2000b, 2001b; Kieneke et al., 2008b), including detailed sperm ultrastructure analyses (Marotta et al., 2005; Kieneke and Schmidt-Rhaesa, 2015), myoanatomy (Hochberg and Litvaitis, 2001d; Kieneke and Schmidt-Rhaesa, 2015), and neuroanatomy (Rothe et al., 2011). A detailed discussion of competing hypotheses can be found in Kieneke et al. (2008b), but in general, it is accepted that the group divides into Macrodasyida and Chaetonotida, the latter dividing into Paucitubulatina and the monogeneric Multitubulatina (*Neodasys*). Early molecular analyses explored the feasibility of the *18S rRNA* marker for gastrotrich phylogenetics (Wirz et al., 1999; Manylov et al., 2004) and soon included meaningful taxon sampling (Todaro et al., 2003; Todaro et al., 2006; Petrov et al., 2007; Todaro et al., 2012). These studies lacked resolution at deep levels and some failed to find monophyly of Paucitubulatina, or even Gastrotricha. A subsequent analysis was able to recover monophyly of the two main gastrotrich clades, with the exception of *Neodasys*, which continued to nest within Macrodasyida (Paps and Riutort, 2012)—as also suggested by the strap shape, the presence of multiple adhesive tubes distributed along the body, musculature, and the serially repeated protonephridia. Monophyly of Chaetonotida (Paucitubulatina + Multitubulatina), was, however, found in a combined morphological and *18S rRNA* analysis (Zrzavý, 2003), which found paraphyly of Macrodasyida with respect to Chaetonotida, as well as in a recent detailed analysis of Chaetonotida (Bekkouche and Worsaae, 2016b), though the latter study used only three Macrodasyida species as outgroups.

Another series of studies incorporated data from a second nuclear ribosomal RNA marker (*28S rRNA*) and the mitochondrial *cytochrome* c *oxidase subunit I*, resulting in more robust hypotheses but mostly focused on Chaetonotida (Kånneby et al., 2013; Bekkouche and Worsaae, 2016b), and also highlighting the non-monophyly of the large family Chaetonotidae, which includes Dasydytidae and Neogosseidae (Kånneby et al., 2013; Kånneby and Todaro, 2015; Bekkouche and Worsaae, 2016b). In these analyses, most genera are non-monophyletic and genera containing both marine and freshwater species never form clades, instead grouping with other species according to habitat, highlighting the problems of the current classification based on cuticular structures. These markers were also used to answer shallower phylogenetic relationships within specific subclades (Todaro et al., 2011; Kånneby et al., 2012).

Despite the large number of samples studied, and the number of cladistic analyses of anatomical data, the current classification system of gastrotrichs shows conflict between molecules and morphology, at least based on traditional characters. It seems unlikely that either Macrodasyida or Chaetonotida are monophyletic, but it is probable that the chaetonotid subclade Paucitubulatina may be monophyletic. In the absence of robust molecular hypotheses based on large numbers of genes, we use here the current gastrotrich classification but stressing that this continues to be as provisional as ever and will likely require drastic reassignment of genera and families, and that will require reaccommodating *Neodasys* within Macrodasyida.

Order Macrodasyida
    Family Cephalodasyidae
    Family Dactylopodolidae
    Family Hummondasyidae
    Family Lepidodasyidae
    Family Macrodasyidae
    Family Planodasyidae
    Family Redudasyidae
    Family Thaumastodermatidae
    Family Turbanellidae
    Family Xenodasyidae

Order Chaetonotida
  Suborder Multitubulatina
    Family Neodasyidae
  Suborder Paucitubulatina
    Family Muselliferidae
    Family Xenotrichulidae
    Family Chaetonotidae (incl. Dasydytidae, Neogosseidae)
    Family Proichthydidae
    Family Dichaeturidae

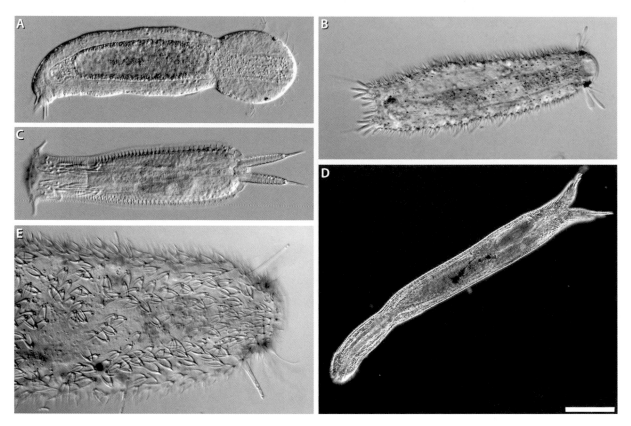

FIGURE 42.1. A, *Dactylopodola baltica* (Macrodasyida); B, *Thaumastoderma heideri* (Macrodasyida); C, *Xenotrichula velox* (Chaetonotida, Paucitubulatina); D, *Diuronotus aspetos* (Chaetonotida, Paucitubulatina), scale 100 μm; E, detail of the anterior end of *Thaumastoderma heideri* showing the body covered in scales. Photo credits: Martin Vinther Sørensen.

In the current system, Macrodasyida (fig. 42.1 A–B) includes strap-shaped species, all but two marine or estuarine, which are oviparous hermaphrodites with cross-fertilization. Paucitubulatina (fig. 42.1 C–D), with bowling-pin-shaped species, includes a few hermaphroditic marine species, but two thirds are limnic. The limnic species have been considered mostly parthenogenetic, but now they are understood as combining parthenogenesis followed by simultaneous hermaphroditism in a rather unusual life cycle (Weiss, 2001).

GASTROTRICHA: A SYNOPSIS

- Triploblastic, bilateral, unsegmented acoelomates
- Paired extensions in the posterior end, carrying one to many pairs of adhesive tubes
- Cuticle well developed, often forming plates and spines; outer cuticle multilayered
- External ciliation restricted to ventral surface; with monociliate or multiciliate epidermal cells; external cilia covered by outer layers of cuticle

- Gut complete, with a terminal mouth and with a ventral posterior anus (in most species); with a muscular pharynx
- Unique muscular system, with discrete longitudinal and circular muscle strands and helicoidal muscles around the gut
- One pair or serially repeated pairs of protonephridia
- Without special circulatory or gas exchange structures
- Hermaphroditic, but in some species hermaphroditism is preceded by parthenogenesis
- Direct development

Gastrotrichs are generally microscopic animals. The body shape of Paucitubulatina is generally bowling-pin shaped, with a distinct head region (sometimes with cephalic structures, like "tentacles"), neck, and a more or less bulbous trunk, but these regions are less differentiated in macrodasyids, which often have the body about the same diameter throughout their length. *Neodasys*, despite being classified as a Chaetonotida, nests within Macrodasyida and is strap shaped.

At the posterior end, gastrotrichs often present paired extensions called "feet" or "furca," the latter being the larger ones of Paucitubulatina. The feet carry one pair (in Paucitubulatina) or more (in Macrodasyida) of adhesive tubes; some swimming chaetonotids lack feet, and the members of *Diuronotus* bear two pairs of adhesive tubes, while *Dichaetura* may have one or two pairs of adhesive tubes and one pair of solid spines. Some species may have a tail (e.g., species of *Urodasys*). The adhesive tubes have a secretory function and are accompanied by a cilium. They are found along the body in Macrodasyida and *Neodasys*, in which they receive different names depending on their location (see Kieneke and Schmidt-Rhaesa, 2015).

The mouth opens in a terminal or subterminal position, in some species leading to a funnel- or barrel-shaped buccal cavity that may be very broad and occupy the anterior end of the animal. The body ciliation and dorsal cuticular structures—including spines, scales, or both—are often used for taxonomy in the group. The locomotory cilia are restricted to the ventral side of the animal and have specific patterns of distribution, forming bands or patches, again, of taxonomic importance. In some species, cilia are tightly packed, forming cirri. In others, isolated stiff cilia, of probable sensory function, are found at the anterior end of the animals.

The epidermal cilia are always completely enclosed by the outer membrane layer of the cuticle, a feature unique to the phylum Gastrotricha. Monociliated epidermal cells have been reported in a number of species (Rieger, 1976; Ruppert, 1991a), but this is not found in all gastrotrichs.

The epidermis and the pharynx are covered by a multilayered cuticle, consisting of the endocuticle and the epicuticle (also called "exocuticle"). It is this epicuticle that covers the entire body, including the sensory and locomotory cilia (Rieger and Rieger, 1977; Ruppert, 1991a), and in some cases both cuticular layers surround each cilium (Hochberg, 2008). The cuticle can be smooth or sculptured, and its thickness varies across species. The epicuticle is composed of a varying number of layers (1 to 25), each usually being trilaminate (Neuhaus et al., 1996b). The endocuticle is granular or fibrous in structure and can be subdivided. In Paucitubulatina, there are local thickenings of the cuticle in the head region that form cuticlular plates. When cuticular structures, such as spines or scales, are present, these are formed by endocuticle.

The gastrotrich cuticle does not contain chitin (Neuhaus et al., 1996b) and is not molted, a major difference with the cuticle of Ecdysozoa (see chapter 20), yet their similarity was used to suggest a sister-group relationship of Gastrotricha to Ecdysozoa (Schmidt-Rhaesa et al., 1998). That relationship is not supported by current phylogenetic data, which instead suggest that either a similar cuticle was ancestral for Protostomia or, more likely, that it evolved twice, once in the last common ancestor of Ecdysozoa and once in Gastrotricha. A common origin, a hypothesis put forward by Schmidt-Rhaesa (2002), would suggest a multiplication of the epicuticle in gastrotrichs, whereas ecdysozoans added the chitinous exocuticle (see also Kieneke and Schmidt-Rhaesa, 2015).

The epidermis of macrodasyids often contains an arrangement of so-called epidermal glands (see Teuchert, 1977), each consisting of a single flask-shaped glandulocyte, with an apical pore that releases the secretion. Their role is, however, not well understood (Kieneke and Schmidt-Rhaesa, 2015). The adhesive tubes are other conspicuous glandular structures. They are tubelike extensions of the body cuticle containing two types of glandulocytes, morphologically similar to viscid and releasing gland cells of Platyhelminthes and so are called a "duo-gland" system, considered functionally comparable to the duo-gland organs of Platyhelminthes (Tyler and Rieger, 1980). The species of *Neodasys*, however, have a single gland and was hypothesized to be the ancestral type within gastrotrichs (Tyler et al., 1980), at a time when a relationship between gastrotrichs and platyhelminths was not supported. With our current understanding of the position of *Neodasys* as a derived macrodasyid (Todaro et al., 2003; Todaro et al., 2006; Petrov et al., 2007; Paps and Riutort, 2012), and in light of the Rouphozoa hypothesis, the homology between the duo-gland organs of Gastrotricha and Platyhelminthes is favored here (contra Tyler, 1988).

Despite earlier discussions about the possible presence of intercellular lacunae in the gastrotrich body plan (Ruppert, 1991b), their body has been interpreted as acoelomate (Rieger et al., 1974; Kieneke and Schmidt-Rhaesa, 2015). Therefore, they lack a circulatory system or respiratory structures. Mesoderm-derived so-called Y-cells have been reported (Teuchert, 1977; Travis, 1983). These cells are arranged in a longitudinal row lateral to the pharynx and the intestine, and they are in close relation to the longitudinal musculature in the posterior of the animal. While an antagonistic role of the Y-cells with respect to the musculature has been proposed, their function is not well understood. Hemoglobin-containing cells, in a similar position (and probably homologous) have been documented in *Neodasys* by Ruppert and Travis (1983). Red-pigmented Y-cells have also been documented in other species (Hochberg et al., 2014).

A post-anal chordoid organ—an organ consisting of probably muscle-derived, disc-shaped cells, containing myofibrils that span transversally through the cytoplasm—has been reported in multiple species of Macrodasyida (Schöpfer-Sterrer, 1969; Rieger et al., 1974; Kieneke and Schmidt-Rhaesa, 2015). These cells are similar in ultrastructure to the notochord of cephalochordates (Rieger et al., 1974) and to the chordoid organ of the chordoid larva of cycliophorans (Funch, 1996).

The excretory system of gastrotrichs consists of one or more pairs of protonephridia, and these were compared in Kieneke et al. (2007). The freshwater Paucitubulatina typically have one pair of protonephridia (Kieneke et al., 2008a; Kieneke and Hochberg, 2012), while serially arranged protonephridia (a minimum of two pairs) are typically found in Macrodasyida (Teuchert, 1973; Neuhaus, 1987), but a recent study has reported two pairs of protonephridia in marine Paucitubulatina (Bekkouche and Worsaae, 2016b). *Neodasys* species have three pairs of protonephridia, resembling Macrodasyida (Maas, 2013). The protonephridia of chaetonotids consist of a bicellular terminal organ with a so-called composite filter, a voluminous and aciliar canal cell with a convoluted lumen, and a nephridiopore cell (Kieneke and Schmidt-Rhaesa, 2015). Protonephridial units in *Neodasys* and Macrodasyida are tricellular, with monociliated terminal and canal cells, but without cilia in the nephridiopore cell (Kieneke and Schmidt-Rhaesa, 2015). The latter study provides a detailed hypothesis of the evolution of protonephridia in gastrotrichs.

Gastrotrichs feed on detritus, bacteria and algae. Their digestive system consists of a through-gut, with a terminal or slightly subterminal mouth opening followed by a sometimes enlarged buccal cavity (which can be funnel shaped), a large myoepithelial pharynx, a midgut, and ending in a ventral anus. The anus is a simple pore at the posterior of the animal, although it may be absent in species of *Urodasys*. Paucitubulatina species have a short, cuticularized hindgut. This through-gut in gastrotrichs contrasts with the blind gut of their likely sister group, Platyhelminthes.

The pharynx is highly muscularized, a typical Y-shaped pharynx with monociliated, myoglanduloepithelial cells surrounding the lumen, similar to that of nematodes, and thus a relationship to nematodes was proposed (Ruppert, 1982); similarities to the pharynxes of tardigrades and bryozoans were also suggested. It is now understood that this type of pharynx evolved independently in multiple animal lineages (Nielsen, 2013). The orientation of the pharyngeal lumen (Y-shaped or inverted Y) has been used as a fundamental character to distinguish between Chaetonotida and Macrodasyida, but given the phylogenetic position of *Neodasys* within Macrodasyida, it seems that the Y-shaped pharynx of Paucitubulatina and *Neodasys* may be convergent. The pharyngeal cuticle is similar to the body cuticle, but often with more epicuticular layers than the body cuticle (Ruppert, 1991a).

A unique character of the gastrotrich digestive system is the pair of pharyngeal pores that connect the lumen with the exterior and that are found in Macrodasyida (except in the members of the genus *Lepidodasys*). The position of the pores changes in different species, and the opening and closing is controlled by a muscular sphincter. Although not entirely understood, excess water from sucking the food may be eliminated through the pores (Ruppert, 1991a).

The digestive system shows some variation, especially with respect to the mouth and structure of the pharynx (see Kieneke and Schmidt-Rhaesa, 2015). In some paucitubulatines the mouth opening appears to be reinforced by a strong cuticle, the mouth ring, which supports other cuticular structures, including spines, sometimes forming what is called a "buccal cage" or "mouth basket." Other modifications, often with concentrations of cilia, also exist. Toothlike structures, movable by musculature, may also be present in the buccal tube of some species.

Based on light microscopy, Remane (1936) provided detailed comparative information of the musculature across gastrotrich species. More recently, the muscular system of gastrotrichs has been studied using a combination of transmission electron microscopy (e.g., Rieger et al., 1974; Teuchert, 1977; summarized in Ruppert, 1991a) and confocal laser scanning microscopy along with immunostaining (Hochberg and Litvaitis, 2001a, c, d, e, 2003a; Hochberg, 2005; Leasi et al., 2006; Kieneke and Schmidt-Rhaesa, 2015; Bekkouche and Worsaae, 2016b). The state-of-the-art gastrotrich muscle systems is summarized in Kieneke and Schmidt-Rhaesa (2015) and Bekkouche and Worsaae (2016b), who provide broad and detailed comparisons across taxonomic groups. Musculature has also become a key character suite to reconstruct the phylogeny of Gastrotricha (Hochberg and Litvaitis, 2001d; Kieneke and Schmidt-Rhaesa, 2015; Bekkouche and Worsaae, 2016b).

The body wall musculature consists of a system of separate strands of longitudinal and circular muscles, known since the earlier studies of gastrotrich anatomy. The introduction of confocal laser scanning microscopy in gastrotrichs, pioneered by Rick Hochberg and Marian Litvaitis, discovered helicoidally arranged muscles that partially enwrap the pharynx and midgut, these representing a synapomorphy of gastrotrichs (Hochberg and Litvaitis, 2001c). Additional muscles, including dorsoventral muscles, are found in species of Paucitubulatina (Bekkouche and Worsaae, 2016b). The pharyngeal musculature is dense and consists of radial pharyngeal muscles and at least in some groups includes longitudinal muscles restricted to the pharyngeal region. In addition, there is also musculature associated with the adhesive glands and the genital structures.

The nervous system of gastrotrichs has received attention in the context of the Cycloneuralia hypothesis and has often been used to postulate a relationship to members of Ecdysozoa. This relationship has been based on the idea of a circumpharyngeal neuropil (or an orthogon) proposed by earlier transmission electron microscopy studies (Teuchert, 1977; Wiedermann, 1995) but not supported by current interpretations (Schmidt-Rhaesa and Rothe, 2014). In order to address this key hypothesis about the evolution of the gastrotrich brains, many recent papers have examined the nervous system of selected species using immunohistochemistry (e.g., Joffe and Wikgren, 1995; Hochberg and Litvaitis, 2003b; Hochberg, 2007a; Rothe and Schmidt-Rhaesa, 2008, 2009; Hochberg and Atherton, 2011; Rothe et al., 2011; Schmidt-Rhaesa and Rothe, 2014, 2016; Bekkouche and Worsaae, 2016b). Up-to-date reviews have recently become available (Rothe et al., 2011; Kieneke and Schmidt-Rhaesa, 2015; Bekkouche and Worsaae, 2016b; Schmidt-Rhaesa and Rothe, 2016).

The brain (fig. 42.2) is bilaterally symmetrical, "dumb-bell-like" (Hochberg, 2007a), and located in a dorsal to dorsolateral position, with bridgelike dorsal and ventral commissures; a pair of longitudinal cords runs lateroventrally and connects the two caudal ganglia (Rothe et al., 2011). Very fine ventral commissures can be found connecting the longitudinal cords. The dorsal commissure of the brain is a bridge of neurons with somata on both sides of this bridge, as identified early by Zelinka (1890) and Remane (1936) and confirmed by subsequent immunohistochemical studies (see a discussion in Kieneke and Schmidt-Rhaesa, 2015). The nerve cords can have a basiepithelial and intraepidermal position, as is the case of *Turbanella cornuta* (e.g., Ruppert, 1991a), but they are subepidermal in other species,

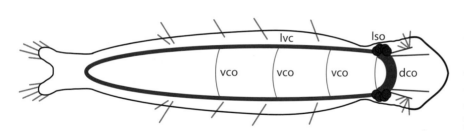

FIGURE 42.2. Schematic representation of the nervous system of a gastrotrich. The brain consists of a dorsal commissure (dco) with lateral somata (lso). The somatogastric nervous system consists of a pair of lateroventral nerve cords (lvc), connected by fine ventral commissures (vco). Based on Schmidt-Rhaesa and Rothe (2016).

including *Neodasys* (Kieneke and Schmidt-Rhaesa, 2015). The nature of the intraepidermal nervous system has played a role in discussions of the relationships between xenoturbellids and hemichordates (Reisinger, 1960) but may be more labile than previously thought.

Red- or black-pigmented eyespots are recognizable with light microscopy in a few species of gastrotrichs, as are palps of varying sizes and shapes. Additional sensory organs, although not well understood, are the so-called piston pit or pestle organ, a pair of lateral depressions in the head region with cilia. All documented gastrotrich sensory organs consist of ciliated receptor cells, with the possible exception of rhabdomeric photoreceptors mentioned by Gagné (1980: p. 122) as his "own unpublished observation." Gastrotrichs have two types of ciliary head sensory organs with sheath and receptor cells (Teuchert, 1976), which are very similar to the receptors of Kinorhyncha (Neuhaus, 1997). These are innervated by the brain, from which a number of neurites extend into the anterior end and innervate the anterior and posterior head sensory structures (Liesenjohann et al., 2006). Additional pharyngeal nervous structures (two pairs of sensory cilia) have been reported by Bekkouche and Worsaae (2016b). The latter study also describes a unique canal system associated with the pharyngeal musculature with associated nerves, formed of multiple pharyngeal canals (Bekkouche and Worsaae, 2016b). The canal system, running along the anterior–posterior axis of the animal (not related to the pharyngeal pores described above), may be unique to *Diuronotus* or Muselliferidae, but its function remains unknown.

■ REPRODUCTION AND DEVELOPMENT

The fascinating reproductive biology of gastrotrichs, including different modes of hermaphroditism (sequential and simultaneous) and reciprocal sperm transfer, at least in *Macrodasys*, has been nicely summarized by Kieneke and Schmidt-Rhaesa (2015). In the case of *Macrodasys*, the gonads are paired, hermaphroditic, and polarized into male and female regions, with male and female gametes being produced simultaneously and displaying internal fertilization with reciprocal cross-insemination. Sperm transfer, however, is indirect, when the autosperm is transferred externally from the simple male pores to the caudal organ and copulatory tube, while the animals are coupled. The caudal organ then functions as a penis, conducting the autosperm into the partner's seminal receptacle (Ruppert, 1978).

In addition to hermaphroditism, a few Macrodasyida species are parthenogenetic, with only female gonads and lacking accessory reproductive organs. The life cycle of the mostly freshwater Paucitubulatina can get more complex and combine an apomictic parthenogenetic phase with an amphimictic hermaphroditic phase (Balsamo, 1992).

Gastrotrichs are generally oviparous, oviposition sometimes occurring through rupture of the body wall, as in some species of catenulid Platyhelminthes and Gnathostomulida. One species has been documented as being ovoviviparous (Hummon and Hummon, 1983). They are all direct developers.

Cleavage in gastrotrichs has received little attention, and relatively few studies exist (de Beauchamp, 1929; Swedmark, 1955), the most prominent being those of *Lepidodermella squamata* by Sacks (1955) and the partial cell lineage of *Turbanella cornuta* by Teuchert (1968). The first cleavage is total and results in two cells of equal size, named AB and CD. The second cleavage plane is perpendicular to the first and shifts the blastomeres 90 degrees with respect to the first ones—what is called "rhomboid" (Teuchert, 1968) or "tetrahedral" (Malakhov, 1994). At this stage there are already major differences between Paucitubulatina and Macrodasyida in terms of synchrony and size. The asynchrony in macrodayids leads to an intermediate three-cell stage. Here the division of blastomere D produces a dorsal and a ventral cell, called "E," which gives origin to the mesoderm. It is in the eight-cell stage that the blastomeres detach from each other in the central part to form a small blastocoel in *T. cornuta*. Gastrulation occurs by migration of the two cells derived from cell E, which have remained arrested during the sixth and seventh cleavage. By this time the embryo is bilaterally symmetrical. The archenteron remains a ventral pit. The embryo then stretches and bends and closes the ventral furrow by growth of the lateral ectodermal cells (Teuchert, 1968; Kieneke and Schmidt-Rhaesa, 2015).

Gastrotrich sperm morphology has received considerable attention in a comparative framework (e.g., Ferraguti and Balsamo, 1995; Fregni, 1998; Marotta et al., 2005; Kieneke and Schmidt-Rhaesa, 2015), as it includes huge variation in the shape and size of the nucleus as well as in other characters, including presence/absence of mitochondria and a flagellum.

Gastrotrichs are often cited as an example of eutely, which implies a fixed number of somatic cells with lack of mitosis, but this is difficult to assert in a group where complete cell lineages do not exist. Furthermore, the fact that gastrotrichs have the capacity of wound healing and regeneration of large body parts (Manylov, 1995), and to repair their body wall after rupture for oviposition, leaves little room for the possibility of eutely.

GENOMICS

Gastrotrich genomics is in its infancy, but all the studied species have small genomes, the smallest being 50 Mb and the largest 616 Mb, with most ranging between 100 and 250 Mb (Balsamo and Manicardi, 1995). The first ESTs of a gastrotrich were those of *Turbanella ambronensis* in the study of Dunn et al. (2008). Struck et al.

(2014) included Illumina-based transcriptomes of four species, and three additional Illumina-based transcriptomes were added by Laumer et al. (2015a) and Lu et al. (2017). The first mitochondrial genome was published recently, being one of the last animal phyla for which mitochondrial genomes were largely unavailable (Golombek et al., 2015).

FOSSIL RECORD

Despite being cuticularized, gastrotrichs currently have no known fossil record (Maas, 2013).

43

PLATYHELMINTHES

Platyhelminthes[20] (the "flatworms") constitutes a cosmopolitan phylum of aquatic and terrestrial invertebrates known from about 29,300 (Zhang, 2011; based on estimates from S. Tyler) to 30,000 described species (Caira and Littlewood, 2013). These include both free-living and parasitic worms, three-quarters of the diversity belonging to a clade of mostly vertebrate parasitic forms (flukes, tapeworms, and their relatives). Flatworms have become one of the most important models for studying regeneration (mostly on planarians or triclads but also on macrostomorphs), and the flukes and tapeworms are among the most studied parasites. An important number of species impact on human health, parasitizing 200 million people in 74 countries (Chitsulo et al., 2004), as is the case for the five species of human blood flukes in the genus *Schistosoma*, the agent of schistosomiasis or bilharzia. In addition, platyhelminths have played a major role in understanding animal evolution, especially given their acoelomate condition, completely ciliated epidermis, and blind digestive system with a generally ventral mouth—features that made them a candidate for the "most basal" bilaterians. It is for this reason that many books and science documentaries persist in referring to the platyhelminths as the first animals to incorporate a third embryonic layer (the mesoderm), the first excretory organs, or the first steps toward cephalization, their generally ventral mouth and blind gut supposed to signal an "intermediate" evolutionary position. They are thus often referred to as "basal bilaterians" or "basal protostomes"—see our discussion about "basal" in the introductory chapter. Paradoxically, their complex reproductive organs were cited as an oddity of parallel origins of complexity. Instead, the core of Platyhelminthes is now known to be a derived group of protostomes, deeply nested within Spiralia, yet some former members of the phylum, Acoelomorpha, still retain this key evolutionary position—as sister group to all other bilaterians—and display many of these supposedly primitive traits (see chapter 12).

From an ecological perspective, free-living Platyhelminthes (fig. 43.1) are key predators in marine (with most lineages being meiofaunal), limnic, and terrestrial environments, where geoplanids are an important component of the rainforest ecosystems (Álvarez-Presas et al., 2014), as evidenced by the charismatic genus *Obama* from the Brazilian Atlantic forest (e.g., Carbayo et al., 2016). Other groups may also thrive on land, including rhabdocoels (Van Steenkiste et al., 2010). Free-living platyhelminths can be extremely abundant in the marine meiofauna but also as macrofauna, and although not abundant, polyclads have been found down to the hadal zone, in the Kuril–Kamchatka and Peru–Chile Trenches (Jamieson, 2015).

[20] This group has been referred to as Platyelmia, Platyelminthes, or Plathelminthes by different authors.

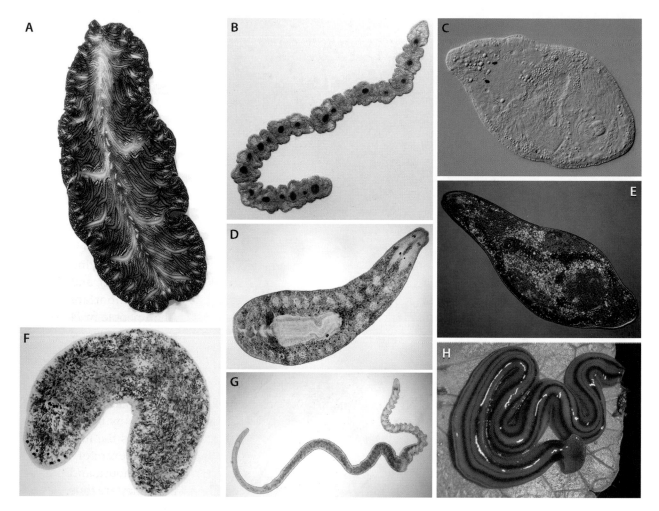

FIGURE 43.1. Diversity of platyhelminths. A, the Polycladida *Pseudobiceros bedfordi* from the Great Barrier Reef, Australia; B, the freshwater Catenulida *Catenula lemnae*; C, an unidentified interstitial Rhabdocoela (Solenopharyngidae); D, an undescribed genus of marine Tricladida (Maricola) from a marine interstitial habitat in Bocas del Toro, Panama; E, a marine interstitial Rhabdocoela (genus *Kytorhynchus*) from Sardinia; F, an interstitial Polycladida with dinoflagellate symbionts from Bocas del Toro (genus *Chromoplana*); G, a juvenile of the Proseriata genus *Polystyliphora*, with 20 accessory penis stylets visible; H, an invasive terrestrial flatworm in the genus *Bipalium*, from Panama. Photo credits: B–G, Christopher Laumer.

One catenulid and one rhabdocoel have also been found, along with rotifers, annelids, and arthropods, at 1.4 km soil depths in paleometeoric fissure water (Borgonie et al., 2015). While the parasitic tapeworms can measure up to 8.8 m in length (the longest being a specimen preserved in the Meguro Parasitological Museum, Tokyo), free-living species range in size from less than 1 mm to up to 1 m, the latter in the terrestrial hammerhead flatworms of the genus *Bipalium* (Kawakatsu et al., 1982), a common invasive genus around the world (Justine et al., 2018).

Flatworms lack mitosis in all differentiated cells (Ehlers, 1985a) but have totipotent stem cells, the neoblasts, that move freely throughout the body and continuously replace somatic tissues. Neoblasts, which have received much interest in recent years as a candidate for a stem cell system (Shibata et al., 2010; Baguñà, 2012; Reddien, 2013), are partly responsible for the high regenerative ability of flatworms and are discussed at length below.

SYSTEMATICS

The question of monophyly of Platyhelminthes has been debated for quite some time (Smith et al., 1986), and numerous groups, including xenoturbellids, gnathostomulids, and even nemerteans and myzostomes have been classified as platyhelminths at one time or another after Linnaeus's Vermes was dismantled. Likewise, the sister-group relationship and exact phylogenetic position of Platyhelminthes have been a matter of heated debate. On the one hand, many authors may have pictured some "Platyhelminthes" as sister group to all other bilaterians; for example, Hyman (1940) placed "Primitive acoel flatworms" at the base of Bilateria but Platyhelminthes + Nemertea in a basal trichotomy of Protostomia, composed of two other lineages, "Aschelminthes" and "Trochophore" (which included arthropods, annelids, molluscs, and other Schizocoela). More explicit phylogenetic statements have placed Platyhelminthes + Gnathostomulida (in the clade Platyhelminthomorpha) as the sister group to all other bilaterians (Ax, 1985), or Platyhelminthes + Nemertea nested within protostomes (Nielsen et al., 1996), to mention just two radically different positions. The phylogenetic position of Platyhelminthes is discussed in more detail in chapter 41 (Rouphozoa).

Traditionally, Platyhelmintes was split into three classes—Turbellaria, Trematoda, and Cestoda (e.g., Hyman, 1951a)—or more recently four, with the addition of Monogenea to the other three (e.g., Brusca and Brusca, 1990), and these systems continued in use until quite recently, at least in the English language literature (e.g., Brusca and Brusca, 2003; Pechenik, 2010). Instead, the Germanic literature has long recognized that Turbellaria is a paraphyletic assemblage and that the parasitic "classes" form a clade deeply nested within "Turbellaria" (e.g., Ax, 1984, 1996; Ehlers, 1984, 1985a, b; Smith et al., 1986; Westheide and Rieger, 1996). One aspect that has changed since these earlier studies of platyhelminth phylogeny is the placement of Acoelomorpha (see chapter 12), and thus many of the earlier discussions about whether Acoelomorpha or Catenulida represented the sister group to all other Platyhelminthes (or all other "Turbellaria") (e.g., Karling, 1974) is ignored here. Platyhelminthes is thus now considered to include two main taxa, Catenulida and Rhabditophora.

Molecular phylogenetics arrived early to Platyhelminthes (e.g., Baverstock et al., 1991; Riutort et al., 1992; Riutort et al., 1993; Joffe et al., 1995). The first comprehensive molecular phylogenetic study used complete *18S rRNA* sequence data for a diverse set of Platyhelminthes (Katayama et al., 1996), and found that acoels were sister group to all other platyhelminths, but the taxonomic samples lacked nonplatyhelminth bilaterians. Inter-ordinal relationships showed low support, but Catenulida was nested deeply within Rhabditophora. Carranza et al. (1997) first showed that Platyhelminthes, as understood at the time, was not monophyletic, as Acoela and Catenulida branched outside the core of Platyhelminthes (Nemertodermatida were inside Platyhelminthes, but this sequence was later found to be chimeric). This study supported the clades Rhabditophora and Neodermata as monophyletic. A follow-up paper using the same marker placed Catenulida within Platyhelminthes and resolved the position of acoels as sister group to Nephrozoa (Ruiz-Trillo et al., 1999), leaving Nemertodermatida also outside Nephrozoa with novel data on *myosin heavy chain type II* (Ruiz-Trillo et al., 2002). Another tour-de-force was the publication of the first total evidence analysis of morphology and *18S rRNA* for a large number of Platyhelminthes (Littlewood et al., 1999), confirming the placement of acoels outside Nephrozoa, and providing well-supported resolution to many internal nodes. These included the sister-group relationship of Catenulida and Rhabditophora, and many groups that are now well understood (e.g., Macrostomorpha, Adiaphanida, Neodermata), but others were not sufficiently resolved. Several additional studies focused on *18S rRNA* (Campos et al., 1998; Jondelius, 1998; Littlewood and Olson, 2001; Willems et al., 2006). Many of the early contributions to Platyhelminthes phylogeny were compiled in Littlewood and Bray (2001). From then, most subsequent molecular work focused on specific subclades, until Laumer and Giribet (2014) reevaluated platyhelminth phylogeny using four markers with the aim to better understand the origins of ectolecithality from an endolecithal ancestor (see below). A follow-up study used *18S rRNA* and *28S rRNA* to reevaluate relationships of Platyhelminthes, with emphasis on Adiaphanida (Laumer and Giribet, 2017).

A major breakthrough was the use of transcriptomic data, which resulted in a series of highly congruent studies on the relationships among the major platyhelminth lineages (Egger et al., 2015; Laumer et al., 2015a; Laumer et al., 2015b; International Helminth Genomes Consortium, 2019), which supported most major clades outlined below, and resolved a long quest in Platyhelminthes studies (e.g., Justine, 1991; Jondelius, 1992a, b; Watson and Rohde, 1993; Watson et al., 1993), the sister-group relationship of the limnoterrestrial species *Bothrioplana semperi* (the only species in Bothrioplanida) to Neodermata—the largest clade of vertebrate parasites. We adopt here the classification system proposed by Brusca et al. (2016b), which largely draws on recent phylogenomic work on platyhelminth phylogeny (fig. 43.2).

Finally, considerable attention has been devoted to the interrelationships of the main neodermatan clades and the cercomeromorph hypothesis—whether Cestoda is sister group to Monogenea instead of to Trematoda (Lockyer et al., 2003; Park et al., 2007; Laumer et al., 2015b)—, as well as to those of its main clades (Olson et al., 2001; Olson and Littlewood, 2002; Cribb et al., 2003; Olson et al., 2003), with dozens of papers focusing on the phylogeny of specific parasitic groups (see International Helminth Genomes Consortium, 2019 for a hypothesis based on full genomes). Some of this work has focused on the coevolution between hosts and parasites (e.g., Olson et al., 2010; Caira et al., 2014).

Subphylum Catenulidea
    Class Catenulida
Subphylum Rhabditophora
    Infraphylum Macrostomorpha
        Order Haplopharyngida
        Order Macrostomida

Infraphylum Trepaxonemata

  Superclass Amplimatricata

    Order Polycladida

    Order Prorhynchida

  Superclass Gnosonesimora

    Order Gnosonesimida

  Superclass Euneoophora

    Class Rhabdocoela

      Order Dalytyphloplanida

      Order Kalyptorhynchia

    Class Proseriata

      Order Unguiphora

      Order Lithophora

    Class Acentrosomata

      Subclass Adiaphanida

        Order Tricladida

        Order Prolecithophora

        Order Fecampiida

      Subclass Bothrioneodermata

        Infraclass Bothrioplanata

          Order Bothrioplanida

        Infraclass Neodermata

          Trematoda

          Aspidogastrea (aspidogastrean flukes)

          Digenea (digenean flukes)

          Cercomeromorpha

            Monogenea (monogenean flukes)

            Cestoda (tapeworms and relatives)

▬ PLATYHELMINTHES: A SYNOPSIS

- Triploblastic, acoelomate, bilaterally symmetrical worms
- Body cylindrical or in the larger species, flattened dorsoventrally
- Incomplete gut (without anus; but an anal pore is present in *Haplopharynx*); gut absent in some parasitic forms (Cestoda and in the fecampiid *Notentera ivanovi*)
- Cephalized, with a central nervous system comprising an anterior cerebral ganglion and longitudinal nerve cords connected by transverse commissures (ladderlike nervous system)
- Multiciliated epidermal cells; few cilia in Catenulida

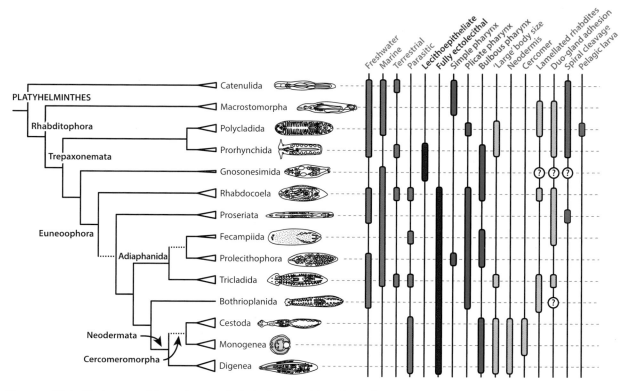

Figure 43.2. Phylogeny of the major platyhelminth taxa with ecological, anatomical and developmental characteristics mapped on the tree. Based on Laumer et al. (2015b), https://doi.org/10.7554/eLife.05503.010.

- Protonephridia as excretory/osmoregulatory structures
- Absence of circulatory or discrete gas exchange structures
- Mucus produced by rhabdoids/rhabdites (absent in most Catenulida, although some have rhabdoids)
- Generally hermaphroditic, with complex reproductive systems
- Biflagellate sperm (aflagellate in Catenulida and Macrostomida)
- Direct or indirect development; larval development in Polycladida (Götte's and Müller's larvae); complex life cycles with many intermediate stages in the parasitic forms
- Spiral cleavage and 4d-derived mesoderm present in Polyclada and probably Catenulida

The typical acoelomate body plan of Platyhelminthes has been used as the textbook example of this condition, where no internal body cavity (other than the gut) exists in a triploblastic animal. As discussed above, this made Platyhelminthes the center of attention for the evolution of bilateral animals, or at least protostomes, a phylogenetic position their relatively simple body plan has not really earned. Yet being acoelomate—lacking a circulatory system and respiratory structures—has dire consequences for their body plan. Platyhelminths need to maintain a surface-to-volume ratio that allows for most of their body functions. They achieve this by becoming extremely flattened, especially in the larger species, and by increasing

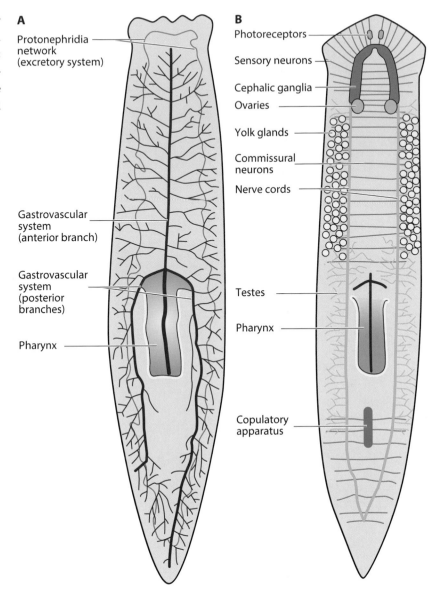

FIGURE 43.3. Anatomy of two freshwater planarians, illustrating major anatomical features in triclads. A, *Dendrocoelum lacteum* with the gastrovascular and excretory systems, and B, an idealized *Schmidtea* with the reproductive and nervous systems. Based on Newmark and Sánchez-Alvarado (2002).

A

Protonephridia network (excretory system)

Gastrovascular system (anterior branch)

Gastrovascular system (posterior branches)

Pharynx

B

Photoreceptors

Sensory neurons

Cephalic ganglia

Ovaries

Yolk glands

Commissural neurons

Nerve cords

Testes

Pharynx

Copulatory apparatus

the volume of the digestive system (fig. 43.3), which becomes extremely ramified in Polycladida. The largest Platyhelminthes are obviously the parasitic species, as is also the case in Rotifera (chapter 39).

Platyhelminthes are a diverse group that shows great disparity in body plans (fig. 43.1), cuticular structures, digestive systems, and lifestyles. This is reflected in the many exceptions to the general descriptions often provided in textbooks, which tend to refer to the classical freshwater planarians (Tricladida). We concentrate here on some of the aspects that we find important without attempting to describe each morphology, yet we try to provide an outline of the disparity of the group. The body shape can thus be cylindrical (generally in small species), extremely flattened dorsoventrally, with an oval to ribbonlike shape, but many other shapes exist. The head can be more or less defined and bear the sense organs, but it may or may not bear the opening of the blind digestive system or mouth, which is often ventral and not necessarily anterior—in some cases, as in the cylindrostomid prolecitophorans, the mouth can be posterior—although prorhynchids have a terminal anterior mouth. The anterior part of most neodermatans has attachment organs, such as hooks and suckers, or the scolex of tapeworms.

While Platyhelminthes are the archetype of unsegmented acoelomates, some tapeworms are conspicuously "segmented"; some eucestodes are monozoic, lacking internal proglottization and external segmentation (Caryophyllidea), others are polyzoic but externally unsegmented, and all other eucestodes display classic proglottization. Many textbooks have traditionally equated segmentation with metamerism, but this definition is not generally followed in some of the current literature (Minelli and Fusco, 2004; Minelli, 2017), as at least some non-metameric ecdysozoans are clearly segmented (Tardigrada, Kinorhyncha).

The typical proglotticized adult tapeworm has its body divided into a scolex, a short neck, and a strobila composed of proglottids (up to 4,500), which are continually produced by the neck, the oldest being those farther from the neck. Each proglottid contains a hermaphroditic reproductive tract, and when mature it ruptures, allowing the self-fertilized eggs to leave the host, either passively in the feces or actively in some cases. These proglottids are considered segments by some, but of an independent origin from those in other animals, as the mode of addition of segments is proximal (at the neck region) instead of distal (Koziol et al., 2016; Minelli, 2016). Interestingly, this is not very different from the stolonization process of syllid annelids, which continue to add stolons between the end of the adult body and the anterior portion of the older stolon, the most distant stolons being the oldest ones. However, in these tapeworms there is no septum between proglottids, and the sister group to all other eucestodes is nonsegmented (Li et al., 2017c).

The body wall of the free-living species is composed of epidermis, basement membrane, and the smooth musculature, often arranged in circular, diagonal, and longitudinal layers (from external to internal). The area between the body wall and the internal organs is usually filled with mesenchyme, which includes a variety of cells, muscle fibers, and connective tissue.

The epidermis is often ciliated and can be cellular or syncytial, with gland cells and nerve endings. Ciliation in catenulids is sparser than in rhabditophorans, and this characteristic has been used to explain the evolution from monociliated to multiciliated epithelia, catenulids representing sort of an intermediate step (Ax, 1984, 1996). Cilia are responsible for the gliding movement of free-living flatworms, even if some larger species may also use their muscles for turning or swimming. The two first offshoots of the Platyhelminthes tree, Catenulida and Macrostomorpha, lack a basement membrane (Rieger, 1981).

Gland cells of different types are found in the epidermis around the body, and these can be used to avoid desiccation, to facilitate movement, or to capture prey. Some well-studied rhabditophoran-only glands are the duo-gland adhesion systems that provide both granules for temporary attachment and granules for breaking the attachment. Multiple duo-glands are found in an animal. In the case of *Macrostomum lignano,* about 130 of these adhesive organs are located in a horseshoe-shaped arc along the ventral side of a tail plate, allowing the animals to adhere to and release from a substrate multiple times within a second. They are also often found in the attaching structures of ectocommensal species, such as the horseshoe crab ectocommensals in the genus *Bdelloura* (Tricladida), and various Temnocephalida. This reversible adhesion relies on adhesive organs comprised of three cell types: an adhesive gland cell, a releasing gland cell, and an anchor cell, which is a modified epidermal cell responsible for structural support, and the structural integrity of the anchor cell is essential for the adhesion process (Lengerer et al., 2014). Similar or homologous organs are also found in Gastrotricha (Tyler and Rieger, 1980; see discussion in chapter 41) and Annelida (Martin, 1978).

In most free-living forms, epidermal and subepidermal glands produce a wide range of solid secretions called rhabdoids, rod-shaped cellular inclusions that produce copious amounts of mucus when moved to the surface of the epithelium. This mucus helps with ciliary gliding and probably with avoiding desiccation and predation. When the rhabdoids have a cortical layer (cortex) composed of concentrically arranged lamella, and a medulla that varies from homogeneous to granular, they are called "rhabdites" or "lamellate rhabdites" (Ax, 1996). Lamellate rhabdites are the character that defines Rhabditophora, although they are conspicuously absent in Neodermata and in a few other groups. Similar cellular bodies are also found in Nemertea, Annelida, Gastrotricha, and Acoelomorpha.

Some flatworms acquire nematocysts from their prey, which are stored in cells on the dorsal surface in more than 30 species belonging to Catenulida, Macrostomorpha, Polycladida, Rhabdocoela, Proseriata, and Prolecithophora (reviewed in Karling, 1966; Goodheart and Bely, 2017). Acid secretions, probably of a defensive nature, have been documented in some polyclads (Thompson, 1965). In some marine species the outer epidermis may be armed with numerous calcareous spicules, which may have a structural role and form inclusions or a true spicular skeleton. These are now known from a relatively large number of species (Rieger and Sterrer, 1975a, b).

The external covering, the tegument of the members of Neodermata, is highly modified. It consists of a syncytium of fused nonciliated cytoplasmic extensions of large cells whose cell bodies (with the nuclei) lie in the mesenchyme. This tegument is important for the uptake and circulation of nutrients and exchange of gases and nitrogenous wastes. In Cestoda, which have no digestive system, the uptake of nutrients occurs solely through the tegument, the surface of which is increased by microfolds called "microtriches." These microtriches may interdigitate with the intestinal microvilli of the host. Developmentally, the earlier stages have a ciliated epidermis that is shed to develop the new syncytial tegument, also called "neodermis," hence the name Neodermata.

Platyhelminthes are typically carnivorous predators or scavengers that may feed on biofilms, but a few are herbivorous. Some species of Prolecithophora and Rhabdocoela present kleptoplasty (Douglas, 2007), the alga transferring ca. 30 to 40% of fixed carbon to the animal. Land planarians in the genus *Bipalium* are well known predators of earthworms (Fiore et al., 2004), and due to their invasive nature, have an ecological impact on soil biology, as does the so-called New Zealand flatworm, *Arthurdendyus triangulatus*, which has become another major pest in Europe. The presence of tetrodotoxin helps them subdue large prey items, such as earthworms (Stokes et al., 2014). Likewise, the invasive "New Guinea flatworm," *Platydemus manokwari*, is a predator that specializes in land snails, which it tracks through their mucus secretion (Iwai et al., 2010; Justine et al., 2015). Symbiosis and parasitism are also frequent among planarians.

The flatworms' digestive system includes the mouth, pharynx, and the blind intestine or digestive cavity, typically a blind gut that can be more or less ramified but that can be reduced in some parasitic forms. The position of the mouth is variable, ventral or terminal, and can be located in the posterior or anterior halves of the animal. The pharynx is ectodermal in origin and can be highly muscularized, and it bears pharyngeal glands that produce mucus and, in some species, digestive enzymes. Catenulids and macrostomorphs have a simple pharynx (or pharynx simplex), which is essentially a short tube connecting the mouth and the intestine. The pharynx of most other free-living flatworms is more complex and can be protruded, and food can be partially digested externally, as in triclads.

The gut has to distribute nutrients throughout the body and thus needs to increase its surface in the larger species. Triclads have three main branches of the gut (fig. 43.3 A)—one anterior and two posterior and parallel—and attain one of the largest sizes of the free-living platyhelminths with the terrestrial planarians, but the dendrocoelid genus *Baikalobia* also attains monstrous sizes. Polyclads have multiple branches of the digestive system, to reach all parts of the large body of these species, some reaching 9 cm in diameter or more. Digestion is often extracellular, inside the gut, until the partially digested material is phagocytized by the single layer of intestinal cell lining and digestion is continued intracellularly. Even though the digestive system is generally blind, an anal pore has been described for the macrostomorph *Haplopharynx rostratus* (see Karling, 1965), and several polyclad species possess multiple pores at the end of the branches of the digestive tract on their dorsal side (Lang, 1884; Kato, 1943), probably associated with their body size (Ehlers, 1985a).

The digestive cavity is highly reduced in flukes, the esophagus leading to a pair of intestinal caeca or a single caecum. Cestodes lack any traces of the mouth or digestive system.

Flatworms, with a few exceptions, have protonephridia mostly for osmoregulation, but also for excretion. Proof of this is in the absence of protonephridia in some small marine catenulids or in the presence of more nephridial units in freshwater than in marine species. These are of the flame bulb type (as in Rotifera) and can be single (in some catenulids) or occur in pairs. The flame bulbs of most species have a filtration apparatus constructed as a weir, defined as a structure consisting of ribs (rods) which are outgrowths of the terminal cell, or the terminal as well as the proximal canal cells, the ribs connected by a "membrane" of extracellular substance through which filtration occurs (Rohde, 2001). Some protonephridia, however, lack a weir altogether (*Haplopharynx rostratus*), and excretion/osmoregulation is apparently by exocytosis (Rohde and Watson, 1998). In some macrostomorphs there are ribs but no "membrane" (Brüggemann, 1986).

The flame bulbs are typically connected to a network of collecting tubules that open to the exterior through one or multiple nephridiopores, and in the case of flukes, they may have a storage area or bladder that connects with a single posterior nephridiopore (in digenean flukes), or a pair of anterior nephridiopores (in monogenean flukes). In tapeworms the collecting tubes connect along proglottids, opening to a bladder in the terminal proglottid. Once the terminal proglottid has been lost, they open directly to the exterior.

The mechanism that controls protonephridial regeneration in planarians is understood at an unprecedented level (Rink et al., 2011; Scimone et al., 2011) and is controlled by many genes also involved in excretory systems, not necessarily protonephridia, in other metazoans.

Protonephridia have been used as a source of characters to understand platyhelminth evolution (e.g., Ehlers, 1985b; Ehlers and Sopott-Ehlers, 1986; Rohde, 1991), including an explicit parsimony analysis of protonephridial ultrastructure (Rohde, 2001). Details of the ultrastructure of the main types of platyhelminth protonephridia are provided in these studies (see also Rieger et al., 1991).

NERVOUS SYSTEM

The nervous system of free-living Platyhelminthes has recently been reviewed by Hartenstein (2016), and has as well for Neodermata (Biserova, 2016), and our knowledge of the model species *Dugesia japonica*, *Schmidtea mediterranea*, and *Macrostomum lignano* is now finely detailed, especially during regeneration (Cebrià, 2007). As for many other anatomical characteristics, the nervous system shows considerable variation across groups. A common pattern among the different species is a more or less compact anterior brain with the nerve cell bodies arranged at the periphery of a central neuropil of nerve cell processes and synapses; and a so-called orthogon (Reisinger, 1925) comprising the central nervous system of the trunk.

The orthogon consists of paired nerve cords (ventral, lateral, dorsal), although generally the ventral pair is more developed, and in some cases, including *Schmidtea meditarranea*, only the ventral pair remains, which also displays a ganglionar pattern, similar to that of segmented animals (fig. 43.3 B). These nerve cords are interconnected by transverse fiber bundles (commissures). Additionally, the pharynx shows some variation, in triclads with two internal concentric nerve rings (Baguñà and Ballester, 1978) that are directly connected to the brain or to the nerve cords. A peripheral subepithelial nervous system may also be present, which extends to the reproductive organs and adhesive suckers, when present. As in the case of the muscle arms of nematodes, echinoderms, cephalochordates, and kinorhynchs, long central processes of the muscle cells have been described as invading the nerve cords (Chien and Koopowitz, 1972; Hartenstein, 2016).

Most free-living flatworms have cup-shaped ocelli, formed by small groups of rhabdomeric photoreceptor cells flanked by pigment cells, and as in gastrotrichs, they can be cerebral or epidermal eyes. Additional intracerebral photoreceptors consisting of a single rhabdomeric sensory cell and two-cup or mantle cells have been described in *Microstomum spiculifer* (see Sopott-Ehlers, 2000). The number and position of the eyes, whether they form clusters or whether they are located on the pseudotentacles, have been traditionally assigned taxonomic importance, especially in polyclads, but they are now considered highly homoplastic (Bahia et al., 2017).

Additional photosensory organs can be found in *Stenostomum leucops* (Palmberg and Reuter, 1992). Ciliary pits, probably of a chemosensory function, are found in several platyhelminth lineages. Statocysts are also found in the brain of several species mostly of catenulids and proseriates and in the rhabdocoel genus *Lurus* (Ehlers, 1991). In proseriates they are used as a character to divide between Unguiphora (without a statocyst) and Lithophora (with a statocyst) (Laumer et al., 2014). In *Monocelis* species, the statocyst is located dorsal to the fused pair of eyes. Interestingly, the statocyst does not seem to regenerate in proseriates (Girstmair et al., 2014).

REPRODUCTION, DEVELOPMENT, AND REGENERATION

Platyhelminths possess a unique totipotent stem cell system, the neoblasts, which can comprise about 20% or more of the cells in the adult animal (Baguñà et al., 1990). Neoblasts are supposed to be competent for the renewal of all cell types, including germ cells, during postembryonic development and regeneration (Rieger et al., 1999) as well as after feeding, and platyhelminths thus lack mitosis in all differentiated cells (Ehlers, 1992c; Littlewood et al., 1999). Only neoblasts are responsible for the maintenance of a healthy population of differentiated somatic cells. This was elegantly demonstrated by Baguñà et al. (1989), who introduced purified neoblasts into irradiated planarian hosts, which resumed mitotic activity and blastema formation, while introduction of differentiated healthy cells never did so. This experiment was later repeated with a subpopulation called "clonogenic neoblasts," with identical results (Wagner et al., 2011), confirming that even a single cell can be sufficient to regenerate an entire animal, a clear signal of totipotency. The power of these neoblasts is what confers on flatworms their phenomenal regeneration powers, distributed across the phylum (see Cebrià et al., 2015), even when compared to other powerful regenerators such as sea stars (Echinodermata, Asteroidea).

The ability to specifically label the regenerative stem cells, combined with the use of double-stranded RNA to inhibit gene expression in planarians, reignited interest in flatworms as an experimental model for studying regeneration and the control of stem cell proliferation (Ladurner et al., 2000; Newmark and Sánchez Alvarado, 2000). Numerous recent articles have revised the role of neoblasts in regenerating planarians (Sánchez Alvarado, 2003; Handberg-Thorsager et al., 2008), but until recently we have not understood what molecular mechanisms allow for regeneration of the correct missing body parts. The identification of position-control genes and the role of muscle in harboring positional information to neoblasts are changing our understanding of planarian regeneration (Reddien, 2018). A historical account on neoblasts can be found in Baguñà (2012) and a detailed model of the cellular and molecular basis for planarian regeneration in Reddien (2018). Cells similar to planarian neoblasts are also found in schistosomes (Collins et al., 2013; Wang et al., 2013).

It is well known that planarians can be cut into pieces—up to 279! (Morgan, 1898)—and each piece can regenerate into a complete organism within a few weeks (Handberg-Thorsager et al., 2008). Cells at the location of the wound site proliferate to form a blastema that will differentiate into new tissues and regenerate the missing parts of the piece of the cut planarian. In fact, most planarians can regenerate heads at anterior-facing wounds and tails at posterior-facing wounds throughout the body. New tissues can grow due to pluripotent stem cells that have the ability to create all the various cell types.

Other classical experiments have shown that a longitudinal cut along the midline of the head can produce a two-headed planarian (among many other such examples). How a small fragment of planarian maintains an A–P axis and is able to regenerate the head in the anterior region and the tail in the posterior region is now well understood; it involves Wnt signaling through β-catenin, which controls posterior identity during body plan formation (Petersen and Reddien, 2009a, b). This mechanism, common for establishing polarity in other metazoans, is nearly identical to that found in acoels (Srivastava et al., 2014). Alteration of these gradients can result in planarians that regenerate two heads at each end of the body or alternatively with two tails.

Two-headed forms have also been observed in *Dugesia japonica* that have traveled to space. In this experiment, the heads and tails of a series of specimens were amputated and then, without further manipulation, placed in the International Space Station for five weeks. After space-exposed worms were returned to the laboratory, one specimen in 15 regenerated into a double-headed phenotype—normally an extremely rare event—and this phenotype persisted in subsequent amputations (Morokuma et al., 2017). The biology behind this result remains a mistery. In addition, it has been shown that in some dendrocoelid with regeneration-deficient powers, Wnt signaling has been aberrantly modified, but a rescue of its normal downregulated function results in the reestablishment of the regenerative ability (Sikes and Newmark, 2013), linking regeneration to maintenance of A–P polarity.

De-growth to sizes smaller than new-born planarians has been registered in starving triclads (see the discussion in Baguñà et al., 1990), a phenomenon rarely seen in other animals. It has also been shown that after starvation the first tissues to be resorbed are reproductive. This de-growth is reversible if food becomes available. Another curiosity is the supposed memory transfer through cannibalism in planarians studied by American biologist James V. McConnell in the 1950s and 1960s (McConnell, 1962). McConnell reported that when planarians conditioned to respond to a stimulus were ground up and fed to conspecifics, the recipients learned to respond to the stimulus faster than a control group did, providing evidence for what he believed to be a chemical basis for memory. His results were disputed when cannibalized untrained planarians produced the same results (Hartry et al., 1964). Interestingly, new research has shown that decapitated trained planarians are able to retain long-term memory (Shomrat and Levin, 2013), which was interpreted as a demonstration that these factors for memory must be stored in neural circuits outside the brain. More recent experiments with the mollusc *Aplysia californica* have shown that memory can be transferred through RNA and thus has an epigenetic basis (Bédécarrats et al., 2018) (see chapter 46, Mollusca).

The ability to regenerate is also put to use for asexual reproduction, which is common among limnic and terrestrial species. Asexual reproduction by paratomy (transverse fission) is common among different clades of Platyhelminthes (Rieger, 1986). In some species of Catenulida and Macrostomorpha, incomplete fission may produce chains of individuals that remain attached until they mature. Postpharyngeal architomy (when specimens split in half behind the pharynx) or fragmenting are also found in some species of triclads. Asexual reproduction is also important in the gonochoristic flukes when they cannot find mates.

Sexual reproduction in free-living flatworms has attracted attention for many reasons, the ecology and evolutionary consequences of different reproductive alternatives having recently been reviewed (Ramm, 2017). According to this author, flatworms depict five alternative strategies: (i) being free-living or parasitic (parasitism has evolved in multiple lineages); (ii) reproducing sexually or asexually; (iii) being gonochoristic or hermaphroditic; and for hermaphroditic species, (iv) either outcrossing or selfing, and (v) balancing investment into the male versus the female sex function (sex allocation of resources).

Considering this variation in sexuality and lifestyles, flatworms are mostly hermaphroditic, with complex male and female reproductive tracts. The male system can have single (e.g., Macrostomorpha), paired (e.g., Rhabdocoela), or multiple (e.g., Polycladida) testes. These have more or less complex reproductive systems, including sperm ducts and a seminal vesicle for storing sperm, often with associated prostate glands. A copulatory organ may exist, and these are in some cases transformed into hunting structures, as in the genus *Prorhynchus*, with an elongate needlelike stylet, which serves a dual function in predation, with the stylet being used to deliver secretions of unknown composition into prey prior to engulfment by the pharynx (Laumer, 2014).

The female system is more variable than the male one, and it has been the focus of many evolutionary studies about the utilization of yolk (see Martín-Durán and Egger, 2012). In archoophoran flatworms (Catenulida, Macrostomorpha, Polycladida), the ova are endolecithal (those that do not produce distinct yolk-bearing cells and instead have what is called a "germovitellarium" that produces both, eggs and yolk). In the neoophoran condition, the ova are ectolecithal. Ectolecithality may have a single origin in flatworms (with a loss in Polycladida) or two independent origins (in Prorhynchida and in the clade uniting *Gnosonesima* + Euneoophora). Irrespectively, endolecithality provides spatial partitioning of the products of oogenesis into almost or entirely yolkless oocytes and separate, specialized yolk-bearing cells, or vitellocytes (Gremigni, 1983; Laumer and Giribet, 2014). The germarium is separate from the yolk gland (vitellarium), which produces the yolk that is deposited on the yolk-free ova inside the eggshell. Unlike in other metazoans, vitellocytes are largely or entirely responsible for the synthesis and storage of yolk and the formation of a protective shell by marginal granules and shell gland secretions (Laumer and Giribet, 2014).

In both types of oogenesis, the eggs are moved via an oviduct to the female atrium, which may also have special chambers for sperm storage (the copulatory bursa and seminal receptacle). Accessory cement glands for the production of eggshells and egg cases may also be present.

Male and female reproductive tracts often have separate openings, the male pore often being anterior. In a few cases the male atrium opens inside the mouth, and

Some of the most important digenean flukes are schistosomes, the agents of schistosomiasis (fig. 43.4). The three main species infecting humans are *Schistosoma haematobium*, *S. japonicum*, and *S. mansoni*. Two other species, more localized geographically, are *S. mekongi* and *S. intercalatum*. In addition, other species of schistosomes, which parasitize birds and mammals, can cause cercarial dermatitis in humans. Adult schistosomes are gonochoristic and sexually dimorphic, the females being much larger than the males. A male and a female become paired before reproductive maturation of the female and remain so for the rest of their lives. In fact, the completion of female growth and the maintenance of reproductive function are dependent on this close association (Popiel, 1986). Under favorable conditions the eggs hatch and release miracidia, which swim and penetrate specific snail intermediate hosts (such as those in the genera *Bulinus*, *Oncomelania*, *Biomphalaria*, and *Neotricula*).

The life stages in the snail include two generations of sporocysts, asexually amplifying themselves within the sporocyst, and the production of large numbers of cercariae, given the small chance that a given cercaria will find a final host. Upon release from the snail, the infective cercariae swim by pulling with their muscular tail, penetrate the skin of the human host (this occurring in infected freshwaters), and shed their forked tail, becoming schistosomulae. This infection can be controlled by a tetracyclic alkaloid produced by *Rotaria rotatoria*, a rotifer that can be found growing on the intermediate host shell (Gao et al., 2019). The schistosomulae migrate through several tissues and stages to the vascular system. Adult worms reside in the mesenteric vessels in various locations or in the veins (in the case of *S. haematobium*). The females, 7 to 20 mm in length, deposit eggs in the small venules of the portal and perivesical systems and are moved progressively toward the lumen of the intestine (in *S. mansoni* and *S. japonicum*) and of the bladder and ureters (in *S. haematobium*), and thus are eliminated with feces or urine, respectively, until they return to freshwater.

Until recently it had been difficult to understand how an animal can have multiple body plans (up to five different stages) and alternate sexual (in the mammalian host) and asexual (in the intermediate host) reproduction. A recent study has figured part of this out by studying larval stem cells (Wang et al., 2018). In the case of *S. mansoni*, subsets of larvally derived stem cells seem to be the source of adult stem cells and the germline, revealing that stem cell heterogeneity may be driving the propagation of the schistosome life cycle.

Digenean fluke cycles are well characterized in a few cases, especially those that impact livestock, such as the liver fluke, *Fasciola hepatica*, or a hybrid species that affects humans. Genomes are rapidly becoming available for many of these pathogens, including the Chinese liver fluke *Chlonorchis sinensis*, which causes chlonorchiasis. This species uses a snail in the genus *Parafossarulus* as first intermediate host and freshwater fish as a secondary host, encysting in the muscle tissue. Some parasitic flatworms are notorious for having a larval phase able to change the behavior or physiology of the intermediate hosts to favor transmission (Lafferty, 1999); this has been termed *parasite increased trophic transmission*. PITT is well known from a few digeneans, including *Euhaplorchis californiensis*, which is known to modify

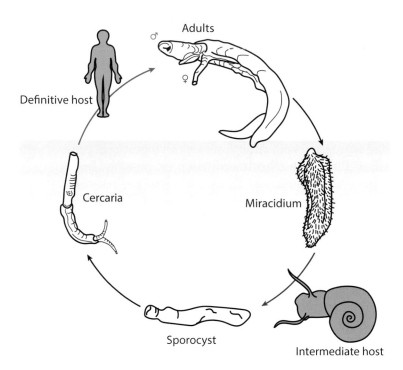

Definitive host

Adults

♂

♀

Cercaria

Miracidium

Sporocyst

Intermediate host

Figure 43.4. Life cycle of *Schistosoma mansoni.* Based on Gao et al. (2019).

the behavior of killifishes, its secondary host, so as to be exposed to its shorebird predators.

The Lancet liver fluke *Dicrocoelium dendriticum* parasitizes the liver of a cow or another grazing mammal. Snails are primary hosts, while ants act as secondary hosts, some of the larvae infecting the brain and changing the behavior of the ant, which climbs up a blade of grass at night waiting to be eaten. Curiously, if not eaten by dawn, the flukes release their control and the ant continues its normal life until the next night. But perhaps one of the most striking PITT cases is that of *Leucochloridium paradoxum*, or the green-banded broodsac flatworm, which uses snails in the genus *Succinea* as the intermediate host. The flatworms' sporocysts infect the host snails' tentacles, which become enlarged, colorful, and pulsatile and take on the appearance of caterpillars; the snails' behavior is changed as well: they now shun the hidden, shaded places they normally prefer, thereby exposing themselves to birds—who happen to be the flatworms' final hosts (Wesołowska and Wesołowski, 2014).

Tapeworms or cestodes represent an extreme example in the evolution of parasitism in Platyhelminthes, with their complete loss of a gut and a segmented body plan. As adults they are almost exclusively enteric parasites of vertebrates, with complex life cycles involving ontogenetically distinct larval stages that first develop in arthropod hosts, although variation in everything from their basic body architecture to their host associations is found among the estimated 6,000 species of tapeworms (Olson et al., 2012). Although tapeworm infections are not as prevalent as those of *Schistosoma* or *Fasciola*, and have been used for weight loss in some cultures, cestode diseases remain a significant threat to our health and agriculture, fluke genome revealed a lack of certain lipid and fatty acid synthesis genes that make them dependent on the host (Berriman et al., 2009; Wang et al., 2011). This genome also has an unusual structure, with many micro-exon genes (from 6 to 36 bases, in multiples of 3), where these exons contain 75% of the coding sequence. Schistosome genomes have undergone significant protein–domain-loss events but contain a detailed molecular repertoire to allow the pathogen to locate and penetrate the host, then nourish and interact with the environment within its host (Zhou et al., 2009). Further genomic adaptations, despite synteny across many neodermatan genes, include loss of genes and pathways that are ubiquitous in other animals, including homeobox families (Tsai et al., 2013; Hahn et al., 2014), but are enriched in specialized pathways for detoxification, usage of host nutrients, and families of known antigens that may redirect host immune response (Tsai et al., 2013; Zheng et al., 2013). Neodermatans have also lost the *piwi* and *vasa* genes, which are considered essential for animal development (Hahn et al., 2014).

In addition, the macrostomid *Macrostomum lignano* (Wudarski et al., 2017) and the planarian *Schmidtea mediterranea* (Grohme et al., 2018) have been sequenced due to their known regenerative properties. *Macrostomum lignano* has become a model for studying karyological instabilities and possible speciation via whole genome duplication (Zadesenets et al., 2016; Zadesenets et al., 2017a; Zadesenets et al., 2017b). Transgenesis has been developed in this species for the first time in free-living Platyhelmintes, opening new doors to experimental work in this early-divergent lineage of flatworms. As in other model organisms, including *Drosophila melanogaster* and *Caenorhabditis elegans*, the *S. mediterranea* genome lacks a number of highly conserved genes found in many other animals. In this case, it lacks critical components of the mitotic spindle assembly checkpoint. The genome of *S. mediterranea* is highly polymorphic and repetitive, harboring a novel class of giant retroelements (Grohme et al., 2018). It also displays stable heterozygosity (lack of recombination), possibly due to an inversion, making *S. mediterranea* a novel genetic system in which to dissect complex inheritance mechanisms (Guo et al., 2016).

Mitochondrial genomes in flatworms have a codon usage different from other invertebrates, with Rhabditophora (but not Catenulida) having the codon AAA coding for the amino acid asparagine, rather than the usual lysine, and AUA for isoleucine, rather than the usual methionine (Telford et al., 2000). Both start codons and stop codons have also been reported as more variable than in other invertebrates (Le et al., 2003; Wey-Fabrizius et al., 2013).

Dozens of mitochondrial genomes have been available for Platyhelminthes since the early days of mitogenomics, focusing both on parasitic (e.g., von Nickisch-Rosenegk et al., 2001; Le et al., 2003) and free-living species (Ruiz-Trillo et al., 2004; Sakai and Sakaizumi, 2012). Most were from Neodermata (Solà et al., 2015) until very recently, even though mitochondrial genes are routinely used for phylogenetic inference of platyhelminth subclades (e.g., Aguado et al., 2017). Interestingly, with the advent of RNAseq, a large number of complete mitochondrial genomes are being assembled (e.g., Aguado et al., 2016; Bachmann et al., 2016), including the first catenulid species, *Stenostomum leucops* and *S. sthenum*, which differ greatly from rhabditophorans (Egger et al., 2017; Rosa et al., 2017). In general, mitochondrial gene order is poorly conserved among flatworms.

FOSSIL RECORD

Various body fossils have been attributed to Platyhelminthes, but most lack diagnostic characters. The fossil record of free-living forms is poorer than for parasites, although the latter is sometimes indirect (e.g., blisters in shells). Reliable early indirect evidence includes hook circlets on Devonian placoderm and acanthodian fishes that compare with the hooks of parasitic monogeneans (reviewed by De Baets et al., 2015). A single specimen preserved in Eocene Baltic amber named *Palaeosoma balticus* has been identified as a member of the suborder Typhloplanoida (Poinar Jr., 2003a), and eggs attributed to cestodes are known from coprolites as old as the Permian (Dentzien-Dias et al., 2013), setting a constraint on when complex life cycles evolved. Casts of body fossils at the ends of trails from Middle Triassic tidal flat muds are indentified as free-living platyheminths (Knaust, 2010), in part because some specimens have calcareous spicules associated with the inferred body wall of the worm, as in many instances of extant free-living flatworms (Rieger and Sterrer, 1975a).

44 LOPHOTROCHOZOA

One of the great mysteries of animal evolution is the sudden appearance in the Fortunian Stage and diversification in Cambrian Stage 2 of a series of sclerites—either organic or biomineralized, and usually disarticulated—followed by the appearance in Cambrian Stage 3 of animal fossils represented by more readily interpreted carbonaceous compressions of entire bodies. These sclerites, called "small shelly fossils," tend to be associated with the shells, spicules, and setae of such extant groups as molluscs, brachiopods, annelids, and some of their stem group members (such as halkieriids, wiwaxiids, hyoliths, and tommotiids).

Traditionally, animals with sclerites/shells/setae were considered as isolated, phylogenetically unrelated groups, although many of them had been aggregated based on a concept called Coeloscleritophora (Bengtson and Mizzarzhevsky, 1981), implying that "coeloscterites" have been lost multiple times (Bengston, 2006). This draws upon a suite of these Cambrian sclerites having a similar "hollow" structure involving thin walls that surround a cavity with a restricted opening to the outside on the basal surface (Porter, 2008). This grade of sclerite organization is shared by taxa that, on other evidence, have disparate systematic affinities, such as chancelloriids (discussed as likely Porifera in chapter 4) and halkieriids (discussed as aculiferan molluscs in chapter 46). When mineralized (usually by aragonite), the mineralization of "coeloscterites" is thought to have an organic template. Phylogenomic analyses (e.g., Dunn et al., 2008) suggested that groups with chitinous setae (such as annelids and brachiopods) and those with spicules and/or calcium carbonate sclerites (such as molluscs) share a common ancestor that could have evolved the ability to secrete these types of extracellular structures. A hypothesis of homology recognizes both as epidermal formations whose secretory cells develop into a cup or a follicle with microvilli at its base (Giribet et al., 2009).

Interestingly, this clade of animals primarily with spicules and/or sclerites also includes animals without any type of hard parts, such as nemerteans, phoronids, and entoprocts, and the transition from making sclerites/spicules/shells to not making anything has not been studied. One recent argument, however, speaks in favor of homology of the hard parts, as it has been proposed that there has been co-option of Hox genes for making shells, spicules, and chaetae (Schiemann et al., 2017). More specifically, the latter study showed expression of Hox genes in two brachiopod species in the chaetae and shell fields—as brachiopods have both shells and chaetae. The shared expression of Hox genes, together with *Arx*, *Zic*, and Notch pathway components are thus observed in chaetae and shell fields in brachiopods, molluscs, and annelids, and this has been interpreted as providing molecular evidence supporting the conservation of the molecular basis for these lophotrocho-zoan characters (Gazave et al., 2017; Schiemann et al., 2017).

This proposal fits neatly with the solid phylogenetic evidence for the monophyly of Lophotrochozoa. More work is needed to understand the diversity of mecha-nisms for forming calcified supportive and protective structures—which are also known to have evolved in other animals, including many cnidarian skeletons, vertebrate skeletons, echinoderm stereoms, and even in some marine flatworms (Rieger and Sterrer, 1975a)—but recent comparative work on molluscan secre-tomes has shown high variability in the genes being used for secreting shells even in closely related species (Kocot et al., 2016).

The name Lophotrochozoa has been used for many different clades of proto-stomes. Here we use the term Lophotrochozoa in its original intention: "defined as the last common ancestor of the three traditional lophophorate taxa, the mol-lusks, and the annelids, and all of the descendants of that common ancestor" (Ha-lanych et al., 1995: 1642). The term was subsequently extended to include other members of the clade Spiralia as well as Chaetognatha (Aguinaldo et al., 1997), a notation followed especially by evolutionary developmental biologists. This exten-sion seems unjustified for a node-defined taxon, especially when an alternative name, Spiralia, has long been in place for that clade. We thus restrict Lophotrocho-zoa to the non-gnathiferan, non-rouphozoan spiralians (see Giribet et al., 2007; Struck et al., 2014; Laumer et al., 2015a)—the clade that constitutes the sister group of Rouphozoa. A recent phylogenomic study by Marlétaz et al. (2019), however, found Platyhelminthes and Gastrotricha nested within the non-gnathiferan spira-lians, and thus they apply the term Lophotrochozoa to our Platytrochozoa (= Rouphozoa + Lophotrochozoa).

The compound name of Lophotrochozoa implies that this clade includes the members of Lophophorata (i.e., Brachiopoda, Bryozoa, and Phoronida), and the for-mer Trochozoa (i.e., Annelida, Mollusca), plus other groups with a "cryptic trocho-phore" larva (Nemertea), plus Entoprocta and perhaps Cycliophora. The position of the latter two groups, probably closely related, is unstable, the presence of Cy-cliophora often conditional on the position of Entoprocta outside of Lophotrochoza (when Cycliophora are included) or with Bryozoa (when Cycliophora are excluded) (Laumer et al., 2015a).

While there is now some support for lophophorate monophyly (e.g., Nesnidal et al., 2014), that group is often nested within the trochozoans, and thus the clade name Lophotrochozoa seems appropriate (an alternative followed by others is to use the name Trochozoa for this clade). While the lophophore has been discussed as a valid anatomical structure (e.g., if entoprocts were sister group to bryozoans, there would be no need to require a lophophore to be derived from a coelomic com-partment), similar coelomic structures are found among pterobranch hemichor-dates, to the point that it has been suggested that the tentaculated arms of ptero-branchs are homologous with the lophophores of brachiopods, phoronids, and bryozoans (Halanych, 1993). This was soon rejected by Halanych (1996a), who later considered these structures to be convergent based on the phylogenetic position of the lophophorates (within Protostomia) and the pterobranchs (within Deutero-stomia). Lophophore anatomy is discussed in more detail in chapter 49.

Another hypothesis that has been discussed in the recent literature is the putative clade Lacunifera, which groups Entoprocta with Mollusca, based on certain similarities of the creeping larva of entoprocts with adult molluscs (Haszprunar and Wanninger, 2008; Merkel et al., 2015). These include a similar tetraneurous nervous system and a complex serotonin-expressing apical organ (Merkel et al., 2015). This is also based on interpretations of the body cavity of Entoprocta to be a lacunar circulatory system, similar to that of molluscs (e.g., Salvini-Plawen and Bartolomaeus, 1995; Ax, 1999; Haszprunar and Wanninger, 2008). This topic is discussed further in chapter 51 (Entoprocta).

TROCHOPHORE LARVAE

Homology and definition of trochophore larvae are disputed, but it is generally agreed that the cell lineage of the prototroch—a preoral belt of specialized ciliary cells derived from the trochoblast cells—is nearly identical across trochozoan phyla (Nielsen, 2004). The prototroch cilia usually beat posteriorly, and together with the postoral metatroch and a circumoral ciliated field they form a complex character, the so-called downstream-collecting system (Nielsen, 1985; Jenner, 2004). In a typical spiralian embryo, the trochoblasts differentiate early in development into ciliated cells and give rise to the prototroch, and thus emphasis has been placed on understanding the cell lineage of the prototroch (Damen and Dictus, 1994b), which is now regarded as the main character defining a trochophore (Rouse, 1999b) (fig. 44.1).

For example, in the gastropod *Patella vulgata*, trochoblasts that form the prototroch are of different clonal origin and the four quadrants of the embryo have an unequal contribution to the prototroch (Damen and Dictus, 1994b). After initial ciliation some trochoblasts become deciliated, and for some cells the choice between a larval and an adult cell fate is conditionally specified. Cell-lineage analysis demonstrates that the various autonomously and conditionally specified trochoblasts are organized according to the dorsoventral axis of the embryo.

The cell lineage of the other ciliary bands is not necessarily homologous between members of different phyla (Rouse, 1999b; Henry et al., 2007), and it is now known that differences exist for the cell lineage of the metatroch even within closely related gastropod species (Hejnol et al., 2007; Gharbiah et al., 2013).

It is likely that a planktotrophic trochophore was the ancestral larval type (Nielsen, 2005), at least in Lophotrochozoa, but probably not for the whole clade Spiralia. Even in some palaeonemerteans, a peroral band of cells with arrested cleavage has been postulated to be homologous to the prototroch based on position, cell lineage, and fate, even if these cells are not differentially ciliated (Maslakova et al., 2004b).

Central to the presence of a trochophore larva is the trochaea theory promoted by Danish zoologist Claus Nielsen, which proposes a series of adaptive modifications of an ancestral ring of compound cilia at the blastoporal rim of a modified gastraea, the trochaea, leading to the prototroch, metatroch, and telotroch of the trochophora larvae (e.g., Nielsen, 1985, 2012a, b, 2018a; Nielsen et al., 2018). Due to its supposed central role in understanding protostome (or at least lophotrochozoan) evolution, many authors have interpreted other larval forms as derived trocho-

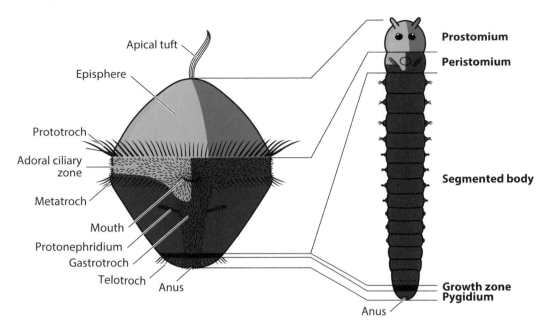

Figure 44.1. Interpretation of the contribution of the four quadrants to larval and adult animals, as exemplified by a polychaete and its trochophore. Based on Nielsen (2004).

phores. This quest for trochophores has sometimes gone too far, and it is unclear whether the mere presence of certain ciliary bands, especially without information about their cell lineage, suffices for postulating homology of trochophore larvae (see a detailed discussion of the different proposals in Jenner, 2004).

It has, for example, been asked whether a new type of nemertean nonfeeding trochophore-like pilidium is a trochophore (Maslakova and von Dassow, 2012) and whether the chordoid larva of Cycliophora is another type of trochophore (Funch, 1996); the latter has also been asked of entoproct larvae (Nielsen, 1967a). Or, finally, is the rotiferan wheel organ homologous with the ciliary bands of trochophore larvae? (Nielsen, 1995, 2001, 2012a). Certainly, further developmental data will be required for all these groups before more definitive answers can be provided, but it seems that environmental pressure in the aquatic environment would be prone to convergence in developing fields of ciliated cells, and it is most likely that most of these structures are not homologous with the prototroch of trochophore larvae.

FOSSIL RECORD

As discussed above, the fossil record of lophotrochozoans is rich in early Cambrian disarticulated sclerites and shells, these providing evidence for the many stem and crown groups that are discussed in subsequent chapters. A case has been made for a Precambrian putative lophotrochozoan. *Namacalathus hermanastes*, a goblet-shaped, stalked, apparently calcareous fossil from the terminal Ediacaran (approx. 550–541 Mya) of Namibia (fig. 1.4 F), was originally interpreted as a possible cnidarian-like animal (Grotzinger et al., 2000). It was later considered a putative lophotrochozoan, suggested to be allied to Lophophorata due to its sessile lifestyle and calcareous foliated structure with a possible inner organic-rich layer (Zhuravlev et al., 2015), although extensive recrystallization restricts the histological comparisons that have been made with brachiopods and bryozoans (Landing et al., 2018). While the presence of Ediacaran lophotrochozoans would reconcile some of the large differences existing between the metazoan fossil record and current estimates of early animal divergences (dos Reis et al., 2015), lophotrochozoan affinities for *Namacalathus* are not readily reconciled with the typical hexaradial symmetry of the cup, which is more comparable to Cnidaria (Cunningham et al., 2017) or to the putative stem-group ctenophores in the genus *Siphusauctum* (Zhao et al., 2019).

CYCLIOPHORA

While we often have the impression that all body plans have already been discovered, cycliophorans remind us, not once, but twice, how some of the small invertebrates are often neglected. Originally described as ectocommensals on the common Norway lobster or scampi (Funch and Kristensen, 1995), *Nephrops norvegicus*, identifying the first known cycliophoran, *Symbion pandora*, exposed how little we know about the commensals of a species that is commercially exploited in Europe. Soon enough, cycliophorans were reported from Mediterranean *Homarus gammarus* (Nedved, 2004)—although this is now known to be another undescribed species. The story repeated itself, after discovering a second species of cycliophoran in the American lobster (Obst et al., 2006), *Homarus americanus*, a species of even more economic impact. Three additional undescribed cryptic species have now been reported after genetic analyses, two from *Homarus americanus* and one from the European *Homarus gammarus* (Obst et al., 2005; Baker et al., 2007). Cycliophorans are thus a phylum of just a few microscopic marine species with a unique body plan, so far restricted to commensalism in nephropid lobsters (or their symbiotic/epizoic copepods; see Neves et al., 2014a) in the North Atlantic and the Mediterranean Sea. Cycliophorans have not been found in other than nephropid pancrustaceans in their known range, but it would not be surprising if they inhabited other nephropid species elsewhere.

The story of the discovery of cycliophorans is one of frustration combined with a stroke of luck. Co-discoverer Peter Funch, then a graduate student at the Natural History Museum of Denmark, in Copenhagen, was working on rotifers at the time, when he realized that all his samples, after a summer of arduous work in Ghent, Belgium, were contaminated by rotifers from the tap water system used to rinse his meiofaunal samples from Greenland (Funch et al., 1996). Upon his return to Copenhagen, he went to see one of his mentors, Prof. Claus Nielsen, to vent his frustration over an apparently wasted summer. Out of utter generosity, Prof. Nielsen handed him a sample of "a strange rotifer" from the mouthparts of *Nephrops norvegicus*, which ended up launching Peter's career (see Funch and Neves, 2018, for a personal account).

Among the most curious aspects of cycliophorans is their extremely complicated life cycle, with numerous, more or less connected, stages. Some of these stages and the main alternating asexual and sexual cycles are discussed below. Here it suffices to say that by growing on the cuticle of an ecdysozoan, the sessile feeding stage needs a quick mechanism to recolonize the molting lobster.

▰ SYSTEMATICS

Cycliophorans are so unique that few characters can be homologized to any other animals, to the point that budding—a unique mode of reproduction in noncolonial animals but common in colonial ones—was used as the main character to suggest a relationship to Entoprocta and Bryozoa in their original description (Funch and Kristensen, 1995). A relationship to Entoprocta was also suggested based on cladistic analyses of morphological data of Zrzavý et al. (1998) and Sørensen et al. (2000). However, the position of this clade among protostomes was not satisfactorily resolved.

Early analyses of *18S rRNA* sequence data suggested a relationship of Cycliophora to Rotifera (Winnepenninckx et al., 1998; Giribet et al., 2000), but, again, support for this position was deficient, and a subsequent analysis combining molecular and morphological data suggested a clade composed of Cycliophora, Rotifera, and Myzostomida that was named Prosomastigozoa ("forward-flagellar animals") by Zrzavý et al. (2001). No information is, however, available for the mode of sperm movement in cycliophorans, and this clade is now interpreted as artificial. The possible relationship of cycliophorans to rotifers was also suggested by myoanatomical studies (Wanninger, 2005). Yet another feature, the distinctive longitudinal rod or chordoid organ of the chordoid larva, consists of vacuolized cells with circular myofilaments, comparable to a similar structure in gastrotrichs (Funch, 1996). Further molecular studies, using additional data, however, rescued the entoproct–cycliophoran relationships suggested in earlier literature on cycliophorans (Passamaneck and Halanych, 2006; Paps et al., 2009b).

Phylogenomic analyses using first ESTs (Hejnol et al., 2009; Nesnidal et al., 2013b), and later Illumina-based trancriptomes (Laumer et al., 2015a; Kocot et al., 2017; Laumer et al., 2019), also corroborated the entoproct–cycliophoran group but had difficulties placing it among other spiralians. Hejnol et al. (2009) found support for Polyzoa, a clade including these two phyla and Bryozoa, the name previously introduced by Cavalier-Smith (1998). However, poor gene sampling in the transcriptomes of all these taxa rendered the monophyly of Polyzoa, as well as the exact position of Entoprocta–Cycliophora, unsettled or contradicted (Nesnidal et al., 2013b). Polyzoa, however, has been adopted in some recent textbooks (Nielsen, 2012a), the resemblance between the cyclic degeneration of the gut and buccal funnel of cycliophorans on the one hand, and the cyclic renewal of the polypides of bryozoans on the other, having been pointed out. We, however, do not subscribe to the proposed homology between the ring of compound cilia and an archaeotroch proposed by Nielsen (2012a) or the proposed homology between the chordoid larva and the trochophore larva proposed by Funch (1996). A relationship of Cycliophora–Entoprocta to the remaining lophophorates is supported in the new wave of densely sampled phylotranscriptomic analyses.

The phylum remains monogeneric to date and contains just two named species. A phylogeny of the known species was proposed by Baker and Giribet (2007), based on four Sanger markers. In that study, the two European samples were more closely related than the samples living on *Homarus* on both sides of the Atlantic. The population structure of these species, including the detection of three cryptic species

Figure 45.1. Life habitus of a feeding stage of *Symbion americanus*. Photo credit: Peter Funch.

on the American lobster, was investigated by Obst et al. (2005) and Baker et al. (2007). While no characters have been proposed to distinguish between these species, some of the anatomical variation described (e.g., the number of longitudinal muscle fibers in the trunk of the feeding individuals) may be able to define species, as the number of these fibers appears to be fixed in *S. pandora* (Neves et al., 2009a, 2010a).

> Class Eucycliophora
> > Order Symbiida
> > > Family Symbiidae
> > > > Genus *Symbion*
> > > > > *Symbion americanus*
> > > > > *Symbion pandora*

CYCLIOPHORANS: A SYNOPSIS

- Triploblastic bilateral unsegmented noncolonial acoelomates
- Complex life cycle alternating asexual and sexual phases with sessile and many free-swimming individuals
- Feeding individuals with a ciliated buccal funnel, oval trunk, posterior adhesive disc, and a U-shaped gut

- Feeding stages attached to setae on the mouthparts of nephropid lobsters
- Layered cuticle characterized by a polygonal sculptured surface
- Without circulatory or respiratory organs
- One pair of protonephridia in the chordoid larva (not in any of the other stages)
- Unique dwarf male
- Egg cleavage deficiently known; holoblastic with four micromeres and four macromeres at the eight-cell stage

Among the most striking characteristics of cycliophorans is their unique life cycle (fig. 45.2), originally reconstructed as consisting of an asexual phase with a feeding stage attached to the lobster from which individuals continuously bud from inside through a mechanism described as replacement of buccal funnel (Funch and Kristensen, 1995). It is believed that each replacement leaves a scar in the cuticle, as clearly shown in *Symbion americanus* (see Obst et al., 2006). During the asexual cycle, a small Pandora larva (ca. 170 μm long) may escape from the feeding stage to re-colonize the same host. This larva possesses a buccal funnel inside its body and develops into a new feeding stage as soon as it settles onto the recently molted lobster. This can be done in response to external stimuli, such as molting, and it has been suggested that it is the contact with hemolymph that may trigger the release of the Pandora larva. Because cycliophorans are often observed in a petri dish in the laboratory, the Pandora larva probably emerges as a consequence of stress generated after severing the lobster's mouthparts.

The sexual phase of the life cycle is more complex and has been revised a number of times (e.g., Funch and Kristensen, 1999; Neves et al., 2012). The feeding stage was originally thought to give origin to either a Prometheus larva (i.e., a male-producing larval stage) or a female through budding. The Prometheus larva escapes the progenitor individual, but what happens after then is not clear until it begins with a settled female individual. Nonetheless, many aspects of this cycle are now better understood, as it has been shown to display key differences at the level of gene expression between the sexual and asexual phases (Neves et al., 2017).

The Prometheus larva (originally interpreted as a dwarf male) is 120 μm long. Once it settles on a feeding individual, it develops one to three 40-μm males inside its body (Obst and Funch, 2006). Females carrying a single oocyte settle on the same host individual and are impregnated by a dwarf male in a process that is poorly understood (Neves, 2016b) but that may be mediated by the male penis. Males are also unique in having a very small number of somatic cells (ca. 50) in adults (Neves et al., 2009b), a reduced number of cells with respect to the young males, which have ca. 200 cells (Neves and Reichert, 2015). The adult male is thus smaller than the young males by a process of nuclear loss; muscle and epidermal cells of the mature male lack nuclei. These males thus constitute one of the smallest known metazoans. The impregnated female then migrates from the maternal body to the host cuticle, where it settles into a sheltered area of the mouthparts, encysts, and produces an embryo that develops into a chordoid larva. The chordoid larva hatches from the cyst, settling on a new host, where it develops into a feeding stage, initiating the asexual cycle.

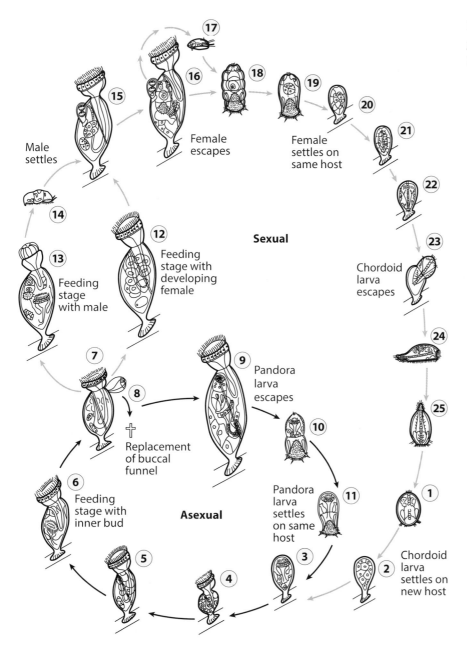

FIGURE 45.2. Life cycle of *Symbion pandora*. Modified from Funch and Kristensen (1995).

The sessile feeding stages are the most conspicuous cyclophoran individuals, easily observed in nearly all fresh individuals of *Nephrops* and *Homarus*, but often not in highly stressed animals kept in tanks under suboptimal conditions. The feeding individuals measure between 350 and 500 μm (*S. americanus* being larger than *S. pandora*), and the body divides into an anterior densely ciliated buccal funnel, an oval trunk, and a posterior peduncle with an adhesive disc to attach to the host's setae. The body is covered by a cuticle, which leaves growth rings. The feeding stage is equipped with a U-shaped gut, the anus opening at the base of the buccal funnel. There is no coelomic body cavity, the small body being packed with large mesenchymal cells. Food is driven into the digestive cavity by ciliary action and by the action of contractile myoepithelial cells that are disposed circularly and below the ciliated ring, like a large short tube with a wider base and top; a complex sphincter is found near the anus. The stomach consists of large glandular cells. The buccal funnel contains two longitudinal muscles, and two to eight longitudinal muscle fibers are found along the distal end of the trunk (six in *S. americanus* and two to eight in *S. pandora*). No protonephridia have been observed in the feeding stage. The nervous system of the feeding stage remains poorly understood, and even the presence or absence of a brain is controversial (see Neves, 2016a).

The Pandora larva, the Prometheus larva, and the female share a similar external morphology, with an anteroventral ciliated field, a posterior ciliated tuft, and four bundles of long cilia in an anterior position, which constitute a sensory organ. The Prometheus larva of the American species has two posterior "toes." The chordoid larva differs considerably, with its ventral chordoid organ; a different ciliation pattern, including a pair of dorsal ciliated organs; lack of a digestive system; and by the presence of a pair of protonephridia (Funch, 1996; Wanninger, 2005; Neves et al., 2010b).

Little is known about the early cleavage of the fertilized egg, and observations have only been made through the female's body. Embryos up to an eight-cell stage have been observed and consist of four macromeres and four micromeres (Neves et al., 2012). Cleavage appears to be holoblastic. The asexual development of the dwarf male inside the Prometheus larva is better understood (Neves and Reichert, 2015), showing the reduction of nucleated cells discussed earlier.

Due to their uncertain phylogenetic position, abundance, and tractable size, extensive myo- and neuroanatomical work has been conducted on the different life stages of cycliophorans, mostly on the swimming ones (Wanninger, 2005; Neves et al., 2009a; Neves et al., 2009b; Neves et al., 2010a; Neves et al., 2010b; Neves et al., 2012; Neves and Reichert, 2015). The musculature of the swimming stages includes ventral and dorsal longitudinal muscles as well as dorsoventral muscles. The architecture of the nervous system of all free-living stages was recently summarized by Neves (2016a). In general, all free-swimming stages possess a dorsal brain composed of a pair of lateral clusters of perikarya interconnected by a commissural neuropil and two distinct ventral longitudinal neurites (four in the chordoid larva) (Neves, 2016b). There is no apical organ in any of the cycliophoran larval stages.

We are just beginning to grasp aspects of the ecology of cycliophorans. Feeding individuals feed in a more or less passive way, when the lobster feeds (Funch et al., 2008), and thus their distribution on the lobster mouthparts tries to optimize the host food resources (Obst and Funch, 2006). Because some individuals of the American lobster can host up to three species of cycliophorans (Baker et al., 2007), it is possible that these take on different positions along the mouthparts, with some sort of ecological segregation, though this remains to be tested.

An intriguing paper described a possible novel host–symbiont relationship of cycliophorans to harpacticoid copepods collected from the mouthparts of a European lobster (Neves et al., 2014a), making this the first case not strictly associated with the lobster, but an indirect association.

GENOMICS

Genomic resources are limited for cycliophorans, but an EST library is available for *Symbion pandora* (Hejnol et al., 2009) and Illumina-based transcriptomes are available for *S. americanus* (see Laumer et al., 2015a) and *S. pandora* (see Neves et al., 2017). The latter study further characterizes the transcriptomes of different life stages as well as sexual and asexual feeding stages to show that there is considerable differential gene expression among them. The asexual stages differentially express genes predominantly related to RNA processing and splicing as well as protein folding, suggesting a high degree of regulation at the transcriptional and post-transcriptional levels. The sexual stages express genes related to signal transduction and neurotransmission. This study (Neves et al., 2017) thus shows how important gene expression and regulation may be in organisms with alternating life cycles.

diversity of medicinal applications through human history, first documented from ancient Greece and recently looked into for possible surgical applications (Li et al., 2017a). The conotoxins of the cone snails (Conidae) are a model for both medical (e.g., Becker and Terlau, 2008) and evolutionary research (e.g., Chang and Duda, 2012) due to the fast evolution and expansion of conotoxin genes.

Finally, the fossil record of shelled molluscs is so rich that this phylum has been the focus of many ecological, paleobiological, and biogeographical studies, especially driven by the unparalleled *Paleobiology Database*, in which Mollusca features as the most prominent fossil resource. Key literature related to major ecological questions, such as latitudinal geographic gradients and their evolutionary dynamics, uses molluscs as examples (e.g., Roy et al., 1998; Jablonski et al., 2006). Molluscs served as the basis for the escalation hypothesis, which postulates that the interactions between organisms drove the evolution of protective structures such as shells (Vermeij, 1987). Likewise, molluscs have been featured as models to study other biogeographical phenomena, including bathymetric diversity gradients (Rex et al., 2005).

■■■ SYSTEMATICS

While the term Mollusca was coined by Linnaeus, he classified molluscs within his Class Vermes, some in Intestina (i.e., the shipworm *Teredo*), the shell-less ones as Mollusca (i.e., *Limax, Doris, Sepia*; and which also included many other soft-bodied animals, including cnidarians, echinoderms, priapulans, or annelids), and the great majority in his Testaria (the shelled species, which also included barnacles) (Linnaeus, 1758). What we now understand by the term "mollusc" has little similarity to Linnaeus's definition.

Molluscan systematics has been a fruitful enterprise, with a series of influential papers focusing on the relationships among the molluscan classes. Many others have centered on the interrelationships of each of its classes down to species relationships. These papers number in the hundreds, so we will not attempt to give a detailed account here. Instead, we refer the readers to some key morphological papers dealing with relationships among classes or within the molluscan classes (e.g., Salvini-Plawen and Steiner, 1996; Ponder and Lindberg, 1997; Waller, 1998; Reynolds and Okusu, 1999; Lindgren et al., 2004; Simone, 2011; Bieler et al., 2014); to key phylogenomic papers (Kocot et al., 2011; Smith et al., 2011; Kocot et al., 2013; Zapata et al., 2014; González et al., 2015; Lemer et al., 2016; Tanner et al., 2017; Pabst and Kocot, 2018; Cunha and Giribet, 2019; Lemer et al., 2019; Kocot et al., 2019); and to a synthetic treatment of these hypotheses (Sigwart and Lindberg, 2015). Likewise, the diversity of Mollusca and scope of the literature limits us from including detailed classifications of each class here, but we adopt a classification system derived from recent extensive work on the different classes, often combining morphology and molecules.

Monoplacophorans were common in the early Paleozoic and were thought to have become extinct during the Devonian period, until the discovery of living specimens during the Danish *Galathea* expedition of the mid-1950s, from abyssal depths off Costa Rica (Lemche, 1957). They were immediately heralded as a "living fossil" (see Lindberg, 2009). The group is now represented by more than 30 species from a late Mesozoic radiation (Kano et al., 2012; Haszprunar and Ruthen-steiner, 2013). Monoplacophorans played an important role in the debate about the origins of molluscs because of their serially repeated anatomical structures, including gills and nephridia (Lemche and Wingstrand, 1959, 1987), the basis for the metameric theory of molluscan origins.

A major systematic debate on mollusc phylogeny has centered on the monophyly (Scheltema, 1993) or paraphyly (Salvini-Plawen, 1980; Haszprunar, 2000) of Aculifera—the molluscs with spicules on their epidermis (Caudofoveata and Solenogastres) or on the girdle (Polyplacophora)—while nearly all morphological phylogenies agreed on the monophyly of Conchifera (the classes with a discrete shell). The paraphyly of Aculifera implied that Polyplacophora was sister group to Conchifera, constituting the clade Testaria (see a discussion of these hypotheses in Haszprunar, 2000; Giribet, 2014a; Vinther, 2014; Wanninger and Wollesen, 2019). Clear implications of these two hypotheses have to do with whether molluscs were primitively vermiform animals that acquired their shell secondarily or whether they were shelled animals primitively and one lineage, Aplacophora, became secondarily vermiform and lost its shell (Ivanov, 1996; Scherholz et al., 2013).

This debate seemed to settle after a diversity of molecular data and fossils converged on the monophyly of Aculifera (e.g., Kocot et al., 2011; Smith et al., 2011; Vinther et al., 2012; Osca et al., 2014; Vinther, 2014; Kocot et al., 2019), but only one of these included data on Monoplacophora (Smith et al., 2011). Nevertheless, and despite the initial proposal being based on a partially contaminated sequence of a monoplacophoran (Giribet et al., 2006), several Sanger-based studies and mitogenomic analyses have continued to find monophyly of Serialia, a clade composed of Polyplacophora and Monoplacophora (Wilson et al., 2010; Stöger et al., 2013; Stöger et al., 2016) that supports the notion of molluscs being primitively metameric (Götting, 1980). Nonetheless, the exact phylogenetic position of Monoplacophora continues to be debated, the phylogenomic resolution favoring a sister-group relationship to Cephalopoda (Smith et al., 2011). Clearly, further work is needed with respect to the phylogenetic position of Monoplacophora.

Recent analyses support a clade composed of Scaphopoda, Gastropoda, and Bivalvia. Kocot et al. (2011) coined the term Pleistomollusca for a supposed clade joining Gastropoda and Bivalvia, but the relationships among the three classes remain poorly understood (Kocot et al., 2011; Smith et al., 2011; Vinther et al., 2012; Osca et al., 2014; Cunha and Giribet, 2019). Although not as bleak as seen by Sigwart and Lindberg (2015), a few important aspects of the molluscan interclass relationships remain poorly resolved, chiefly the precise phylogenetic positions of Scaphopoda and Monoplacophora, which, based on transcriptomic/genomic data since 2011 (Kocot et al., 2011; Smith et al., 2011), surprisingly remain unexplored. A sister-group relationship of Scaphopoda and Cephalopoda has been proposed based on neuroanatomical data (Sumner-Rooney et al., 2015), but this relationship has not been supported by any molecular analysis so far.

The literature focusing on internal relationships for the different molluscan classes or subclades is immense and can be classified into morphological, molecular, or a combination of both. Among the molecular approaches, most have focused on PCR-amplified markers sequenced with Sanger technologies, mitogenomics,

FIGURE 46.1. Proposed phylogenetic tree of the main molluscan lineages.

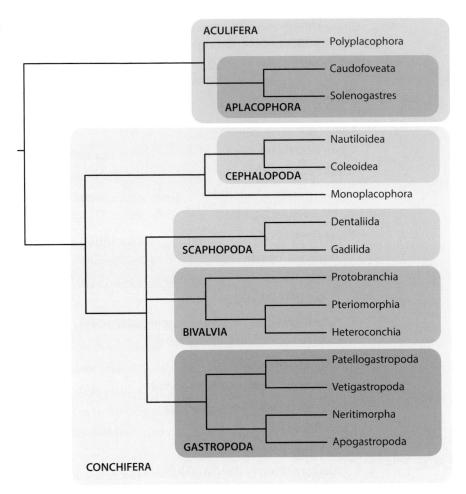

- Visceral mass concentrating all the internal organs
- Body covered by thick epidermal–cuticular sheet of skin, the mantle, dorsal to the visceral mass, which forms a cavity (the mantle cavity) in which are housed the ctenidia, osphradia, nephridiopores, gonopores, and anus
- Mantle with cells that secrete calcareous epidermal spicules, plates, or the shell
- Heart lies in the pericardial chamber and is composed of separate ventricle(s) and atrium/atria
- Buccal region with odontophore and radula (absent in bivalves)
- Complete gut, with marked regional specialization, including large digestive caeca
- Respiratory system of gills or ctenidia (many marine molluscs), secondary gills (nudibranchs) or lungs (pulmonate gastropods)
- Metanephridia (protonephridia in some larvae); in some nephridia can be large and complex (named "kidneys")
- Cleavage spiral; discoidal in cephalopods
- Development indirect or direct; when indirect, generally with a trochophore larva, followed by a veliger larva in gastropods and bivalves

Molluscs are too diverse to be treated exhaustively in a single chapter of a textbook like this, and numerous synthetic works have been published through the years. Some of the most utilized treatises are the 10-volume series *The Mollusca*, published in the 1980s, and the two-part series of *Mollusca: The Southern Synthesis* (Beesley et al., 1998), as well as key volumes on anatomy, evolution, and phylogeny of Mollusca (e.g., Hyman, 1967; Taylor, 1993; Ponder and Lindberg, 2008b), in addition to the many invertebrate series to which we constantly refer in this book, including the *Microscopic Anatomy of Invertebrates* and the mollusc volumes of the *Reproductive Biology of Invertebrates* and *Reproduction of Marine Invertebrates*. These account for thousands of pages synthesizing our knowledge on molluscs. The intention of this chapter is thus to provide a broad synthesis of the phylum, but without attempting to be as exhaustive as in most other chapters.

The molluscan body plan shows great disparity (fig. 46.2), and therefore many textbooks tend to use what has been known as a "hypothetical ancestral mollusc," or "HAM." Such "archetypes" or "hypothetical ancestors" are sometimes used for pedagogic reasons, yet others have made of this a research program and a matter of heated debate. As discussed by Lindberg and Ghiselin (2003), there are so many inceptions of HAM, and they do not come "clearly labeled with warnings about the harm that they might do if mistaken for real organisms." We will, therefore, not discuss such an ancestor or attempt to give another view of a molluscan ancestor. But in spite of this, multiple trends in molluscan evolution are worth considering here.

Such trends require hypothetical character reconstructions to understand complex processes such as torsion, which is a fundamental process in gastropod evolution. Many extinct and extant molluscs are bilaterally symmetrical, with the mouth located at the anterior end and the anus at the posterior. Chitons, monoplacophorans, and aplacophorans closely conform to this body plan. However, many cephalopods have a coiled planispiral shell, still bilaterally symmetrical, or, in the mostly extant shell-less forms, the anus is facing anteriorly and does not open at the end of the body. This gets more complicated in gastropods, as a second process, torsion, by which the visceral mass rotates 180 degrees relative to the head/foot region during ontogenesis (Wanninger et al., 2000), is superimposed upon the shell coiling. The process of torsion causes the shell to coil toward the left or toward the right—sometimes within the same species. Although shell coiling is not necessarily present in the adult form of some limpetlike gastropods, torsion is seen in their internal anatomy and in their development (Wanninger et al., 2000), as the anus is located on the right side of the animal, opening anteriorly in the dextral species.

The direction in which a gastropod shell coils, whether dextral (right-handed) or sinistral (left-handed), originates in early development (Schilthuizen and Davison, 2005). It is highly biased toward dextral (more than 90% of species), but sinistral forms are known from mutant individuals of dextral species, mixed populations (e.g., *Amphidromus perversus*), entire species (e.g., *Neptunea contraria*), or even larger clades. In addition to the coiling of the shell, the body can be independently dextrally or sinistrally coiled, with four possible combinations, the most common being a dextral animal with a dextrally coiled shell, or "dextral orthostrophic" (Frýda, 2012). "Sinistral orthostrophic" is the mirror image and the second most frequent form, but "dextral hyperstrophic" (dextral animal with a

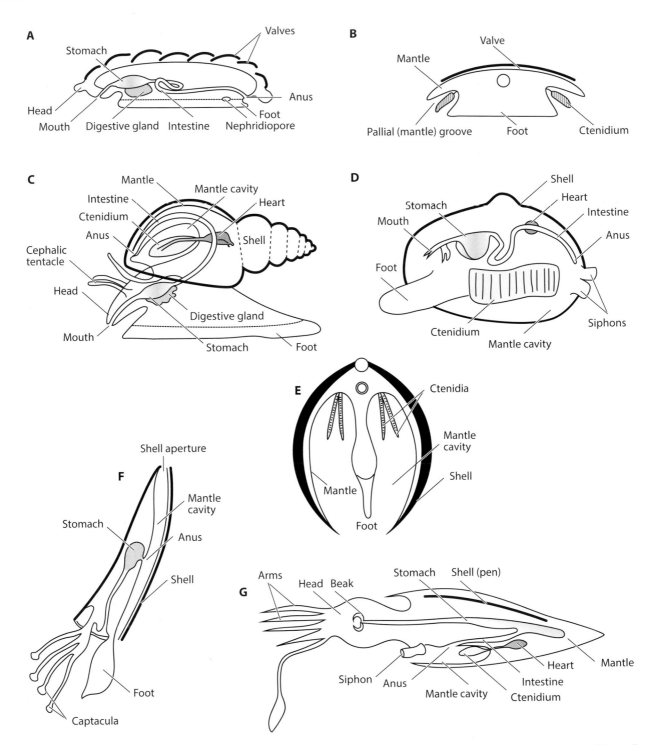

Figure 46.2. Body plan of selected molluscs. A, a chiton (Polyplacophora) in midsagittal view; B, cross-section of a chiton; C, side view of a snail (Gastropoda); D, virtual dissection of a clam (Bivalvia); E, cross-section of a bivalve; F, sagittal section of a tusk shell (Scaphopoda); G, midsagittal view of a squid (Cephalopoda). Based on Brusca and Brusca (1990).

sinistrically coiled shell) and "sinistral hyperstrophic" (its mirror image) also exist in nature. In addition, a condition called "heterostrophy" is found in multiple clades of gastropods where the protoconch (the gastropod larval shell) and the teleoconch (the adult shell) appear to be coiled in opposite directions (e.g., Cylichnidae, Pyramidellidae, Architectonicidae).

Torsion has deep implications for the anatomy of a snail, and through combination with the process of anopedal flexure (also in scaphopods), the gut becomes U-shaped (fig. 46.2 C). As a consequence, we often refer to some organs with the terms "pretorsional" and "posttorsional." For example, the right ctenidium of bivalves is supposed to be the equivalent of the left adult ctenidium of gastropods, that is, the posttorsional left ctenidium. In fact, some of the posttorsional right organs, such as the gonad, are reduced or disappear in adult gastropods. The posttorsional right ctenidium is absent in many gastropods and reduced in others (e.g., some vetigastropods, such as Fissurellidae and Haliotidae). This obviously affects many other symmetrical structures of gastropods, such as the nephridia, and especially their nervous system (see below) and sensory organs such as the osphradia (Page, 2006).

Torsion has been documented during embryonic development in all gastropods, and many remain torted as adults (their twisted, figure-eight nervous system defines a condition called "streptoneury"). However, a clade of gastropods has become secondarily detorted, the visceral nerves being untwisted, in a condition known as "euthyneury." Detorsion can be complete in some nudibranchs or partial, as in many panpulmonates.

The evolutionary implications of the process of torsion have long been discussed in the literature, going back to Garstang's famous verse "The Ballad of the Veliger," and are nicely summarized in Brusca et al. (2016a). While torsion has been often interpreted as a tradeoff (as it implies the loss of some organs), gastropods have become the most successful group of molluscs. For comparison, the coiled-shell cephalopods only inhabit the last chamber of the shell. This has important implications, as they have to carry a large shell that is of little value to the animal. Instead, most gastropods are able to withdraw entirely within their shell, yet this requires that all relevant organs, while internally residing within the visceral mass, need to open near the shell aperture. The conical spiral of gastropods may be more effective than a tightly packed planispiral shell, especially for reaching large sizes; planispiral shells also exist in extant gastropods, but often in tiny species (e.g., Omalogyridae), and are known from the early gastropod fossil record. However, huge planispiral fossil cephalopods were once diverse in the oceans.

Understanding snail chirality from genetic, developmental, and ecological perspectives are growing fields of research (e.g., Shibazaki et al., 2004; Grande and Patel, 2009). For example, it has been shown that sinistral *Satsuma* snails (Camaenidae) survive predation by *Pareas iwasakii* snakes and that stylommatophoran snail speciation by reversal of coil has accelerated in the geographic range of pareid snakes (Hoso et al., 2010). At least in some cases chirality is determined by a single genetic locus with delayed inheritance; that is, the genotype of the offspring is dictated by the genotype of the individual acting as a mother (Schilthuizen and Davison, 2005).

The process of torsion can be controlled by the expression of just two genes, *nodal* and *Pitx*, and the side of the embryo where they are expressed is related to shell chirality (Grande and Patel, 2009). Another study has shown that in *Lymnaea stagnalis*, the chiral blastomere arrangement at the 8-cell stage (but not the 2- or 4-cell stage) determines the left–right asymmetry throughout the developmental program, and this is the mechanism that acts upstream of the *nodal* signaling pathway (Kuroda et al., 2009).

Patterning of the body along its anteroposterior axis involves striking differences in the deployment of Hox genes between studied Aculifera and Conchifera (reviewed by Wanninger and Wollesen, 2019). Polyplacophora have a staggered expression of Hox genes along the A–P axis (Fritsch et al., 2015; Fritsch et al., 2016), conforming to the colinear arrangement that is typical of Bilateria as a whole. In contrast, until recently it has been thought that Hox genes are expressed in a noncolinear fashion in conchiferans. The gastropod *Gibbula varia* exhibits structure-specific expression of the Hox genes, with expression in, for example, the apical organ, the foot, the shell field, and the prototroch (Samadi and Steiner, 2010). Likewise in Cephalopoda, Hox genes are not obviously employed in A–P patterning but are expressed in particular organ systems, including the gills, arms, and funnel. But more recently a study on the scaphopod *Antalis entalis* has revealed that the mid-stage trochophore exhibits a near-to-staggered expression of all nine identified Hox genes (Wollesen et al., 2018), and a reevaluation of published data on other conchiferans suggest that a staggered expression may be found in some early developmental stages.

▬ THE INTEGUMENT, MANTLE, AND SHELL

Before exploring the different molluscan groups, a couple of other characters are worth discussing. Molluscs have a thick epidermis with the ability to secrete calcium carbonate spicules, shell plates, or shells. Modifications of these shells are common, and gastropods have lost the shell in multiple lineages. The molluscan shell is discussed at length below.

The molluscan body wall typically consists of a cuticle (when present), epidermis, and muscles (fig. 46.3 A). The cuticle is composed of sclerotized proteins, such as conchin, and in the case of aplacophorans, it also contains the polysaccharide chitin, which is present in several other structures in molluscs (Peters, 1972). The epidermis is usually a single layer of ciliated, highly secretory cuboidal or columnar cells. Additional secretory gland cells are abundant in outer surfaces, especially the sole of the foot. Other secretory cells are abundant on the dorsal mantle surface, the so-called shell glands, which produce the calcium carbonate sclerites or shells. Shell glands form through invagination of a shell field in the gastrula of conchiferans, then go through evagination again to form a new shell field or remain internal in species with internal shells (Kniprath, 1981). The muscle layers are typically circular and diagonal, and the inner muscles are longitudinal. Muscles associated with the foot (pedal retractors) or adductor muscles of bivalves can be prominent.

The mantle and mantle cavity are unique structures of molluscs, and they have, in part, been discussed as a reason for their evolutionary success. The mantle is a sheetlike tissue that primarily covers the dorsal part of the animal that secretes and is in contact with the shell in those molluscs with an external shell. The edges or "skirt" of the mantle form a double fold that encloses a water space between them and the foot, called the "mantle (or pallial) cavity." This space, where water circulates via ciliary or muscular action, houses the ctenidia, osphradia, anus, nephridiopores, and gonopores. The mantle cavity of chitons and monoplacophorans is considered to be the ancestral condition; it is bilaterally symmetrical, the water circulating from front to back, oxygenating the multiple pairs of ctenidia and clearing gametes and nitrogenous wastes from the gonopores and nephridiopores. This mantle cavity is highly modified in gastropods due to the process of torsion, the mantle cavity concentrated on the right side of the animal. The pulmonary cavity or lung of pulmonate gastropods is supposedly homologous to this mantle cavity (Ruthensteiner, 1997). The mantle cavity of aplacophorans is highly reduced, but that of bivalves is much enlarged, basically filling the shell between the thin mantle and the foot, hosting a pair of hypertrophied ctenidia (in Autobranchia; Protobranchia have primitive ctenidia, similar to those of other groups of molluscs). In autobranch bivalves, the mantle edges may present different degrees of fusion, forming siphons (inhalant and exhalant) and other mantle structures.

Cephalopods are highly modified; in species without an external shell (most extant species), the "body" is indeed the mantle. Water is drawn into the mantle cavity—which hosts the ctenidia, the anal opening as well as the reproductive and excretory pores—by muscular action of the mantle and is expelled through a special funnel-shaped organ called the "siphon," which is used as a means for locomotion through jet propulsion.

The shells of many gastropods are smooth and even colorful internally, where in contact with the mantle, while being rough on the outside. Many gastropods, including cowries and olive snails, have shiny and colorful shells also externally, the reason being that the mantle covers the shell when the animal is active. When the animal is disturbed or resting, the whole mantle withdraws within the shell through a narrow aperture, exposing the beautiful shell.

Spicules or shells are formed by either of two polymorphs of calcium carbonate, aragonite or calcite. Their formation is at the interface with the environment and involves controlled deposition of calcium carbonate within a framework of macromolecules that are secreted from the dorsal mantle epithelium (Kocot et al., 2016). In aplacophorans, aragonitic spicules (sclerites or scales) form extracellularly in the epidermis and are embedded in the cuticle. A typical spicule begins as a solid tip, continues to an open-ended hollow spicule, and finally becomes a closed-ended hollow spicule (Okusu, 2002), although solid spicules are also common. These spicules are secreted by serially arranged glandular cells (spiculoblasts) located on the dorsal mantle in Caudofoveata (Nielsen et al., 2007) or in a noniterated fashion in the case of Solenogastres (Todt and Wanninger, 2010). Chitons are significant because they show both shell plates (similar to the shells of conchiferans) and small calcified girdle, or perinotum, elements on an elongate body (somewhat reminiscent of aplacophoran sclerites). A homology has been proposed between the cuticle/associated girdle mantle epithelium of chitons and the periostracum/inner side of the outer mantle fold of bivalves (Checa et al., 2017).

Understanding shell formation requires understanding multiple events, including the initial deposition of the larval shell, and its continuous growth (and repair) through the life of the animal, a process that is just beginning to be understood and that relies on specific asymmetries of the mantle edge. It is this mantle edge that controls the different processes of shell deposition and calcification, as characterized by gene expression (Herlitze et al., 2018). Progress is being made from a molecular point of view, with focus on the mantle secretome as a key to understanding the formation of the diversity of molluscan shells (Jackson et al., 2006; Marin et al., 2012; McDougall and Degnan, 2018). This formation of the shell implies the deposition of a diversity of proteins (many of which have quickly evolved and constitute expanded gene families of repeated low-complexity motifs), as well as the process of biomineralization (Furuhashi et al., 2009; McDougall et al., 2013; McDougall and Degnan, 2018).

Confusion exists between the terms "shell field" and "shell gland," which are applied by different authors (Eyster and Morse, 1984). Generally, the first larval shell (protoconch I in gastropods or prodissoconch I in bivalves) is formed by a single or discrete shell field—circular in gastropods, saddlelike in bivalves, as two dorsally hinged shells form (Eyster and Morse, 1984). The subsequent larval shell (protoconch II and prodissoconch II) and the adult shell (teleoconch), which continues to grow indefinitely, are secreted from cells located in shell glands in folds along the mantle margin, first as the periostracum, which then biomineralizes (Kniprath, 1981; Checa, 2000; Checa et al., 2014). The periostracum is largely composed of conchin, similar to that of the cuticle, and sits up on top of the shell. The calcium carbonate layers have different types of crystal that precipitate on a matrix of conchin, including prismatic, spherulitic, laminar, and crossed lamellar.

The number of layers and types of minerals of the shells (fig. 46.3 B) can be group specific, but they also vary in different parts of the shells and have been extensively used for taxonomic purposes (e.g., Taylor et al., 1969; Carter, 1990). Nacre is a specific type of laminar (tablet) layer found in the iridescent parts of the shells of some gastropods (vetigastropods), cephalopods and bivalves (pteriomorphians), while in most monoplacophorans "nacre" is a similar structure consisting of foliated aragonite (Checa et al., 2009), also found in several Cambrian molluscs (bivalves and others), and with an origin independent of modern nacre (Vendrasco et al., 2011). Nacre tablet thickness has been proposed as a paleothermometer to estimate temperature in shallow-water paleoenvironments (Gilbert et al., 2017).

Understanding the mechanisms that contribute to shell coloration is still in its infancy, despite the enormous variation displayed by molluscs and appreciated by humans. Most of what is currently known has been summarized by Williams (2017). The main shell pigments found in Mollusca comprise carotenoids, melanin, and tetrapyrroles, including porphyrins and bile pigments. Some of these colors are heritable, while others depend on the diet. Colors may have a strong phylogenetic signal in some clades.

The shells of the different molluscan classes may have different structure and composition. The shell plates of chitons, for example, are perforated by vertical canals for the aesthetes—extensions of the nervous system that serve a variety of sensory functions (see below). In bivalves, the two valves are typically hinged dorsally by a system of teeth and sockets and an organic elastic proteinaceous ligament, with a powerful adductor musculature that controls the closing of the valves. The mantle edge, siphons, and muscles leave characteristic impressions in the inner valves known as "pallial line," "siphonal line," and "adductor muscle scars," among other less conspicuous muscle impressions.

Fossil cephalopods have external shells, but among the extant species, only nautiloids have an external, chambered, planispiral shell, with septae separating the different chambers, the animal inhabiting the broadest (terminal) one. The septae are traversed by a tissue connection called the "siphuncle," which helps regulate buoyancy by controlling the amount of gas and liquid in each chamber. This system, however, limits the depth at which *Nautilus* can live, as its shell implodes below ca. 800 m (Kanie et al., 1980), and thus promotes speciation across islands separated by deep waters (Combosch et al., 2017). Another chambered cephalopod shell exists in *Spirula spirula*, but it is internalized, as is the straight shell of cuttlefishes and the completely decalcified pen of squids. A shell is mostly lacking in octopods, although a shell rudiment is present. The evolution and reduction of the shell in cephalopods have long fascinated zoologists (see Kröger et al., 2011).

Perhaps even more spectacular is the evolution of the gastropod shell and its numerous variations from conical to spiral, from tight spirals to loose ones, and finally the numerous losses in multiple lineages, especially within Heterobranchia. Whether the earliest gastropods had a limpetlike shell, like that of patellogastropods and many vetigastropods (limpetlike shells have also originated in many other lineages, including highly derived heterobranchs), or a spiral one, growing along a vertical axis, is a matter of debate. The gastropod shell is highly variable in size, shape, presence or absence of siphonal canals, and ornamentation, color, and so forth. In some groups the shells cement to the substrate, as in vermetids; in others the shell is internalized, vestigial, or lost altogether, although it is always present in the larval stage and may then be discarded or resorbed. Many shell-less gastropods, especially heterobranchs, use additional defense mechanisms, including mimetic and aposematic coloration, kleptocnidia (Goodheart and Bely, 2017), acidic secretions, or the acquisition and/or de novo synthesis of numerous secondary metabolites (Wägele et al., 2006; Cimino and Ghiselin, 2009). The field of chemical ecology has used nudibranchs and other heterobranchs as a main subject for study (e.g., Avila et al., 2018).

THE MOLLUSCAN FOOT AND LOCOMOTION

The foot is another hallmark of molluscs, with a primarily locomotory function, but in some groups it also serves to delimit the lower end of the mantle cavity. Highly muscular, it forms the base of the animal; it contacts and adheres to the substrate in chitons, monoplacophorans, and gastropods, especially in the limpetlike forms, in which it has a creeping ventral ciliated sole with numerous mucus

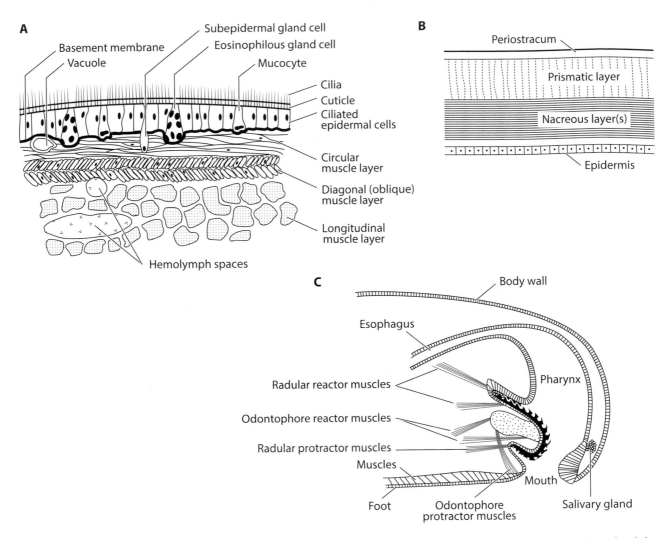

A
Basement membrane
Vacuole
Subepidermal gland cell
Eosinophilous gland cell
Mucocyte
Cilia
Cuticle
Ciliated epidermal cells
Circular muscle layer
Diagonal (oblique) muscle layer
Longitudinal muscle layer
Hemolymph spaces

B
Periostracum
Prismatic layer
Nacreous layer(s)
Epidermis

C
Body wall
Esophagus
Pharynx
Radular reactor muscles
Odontophore reactor muscles
Radular protractor muscles
Muscles
Foot
Odontophore protractor muscles
Mouth
Salivary gland

FIGURE 46.3. Idealized schematic cross sections through A, the integument of a mollusc; B, the shell; C, the mouth and radular apparatus. Based on Brusca and Brusca (1990).

gland cells; indeed, special pedal glands for secreting mucus are found in many gastropods. The foot is retractable into the shell in the nonlimpet gastropods, which still display its flat creeping sole, modified, in scaphopods and bivalves, for digging and reduced or entirely lost in aplacophorans, in which it does not have a locomotory function. The foot of cephalopods is likewise highly modified.

The foot of gastropods is characterized by the presence of an operculum, a hard, disc-shaped structure on the dorsal surface of the foot and primarily designed to close the shell aperture (Checa and Jiménez-Jiménez, 1998). It is attached to its dorsal epithelium at a region called the opercular disc. Opercula are also found in other animal groups (e.g., Annelida, Bryozoa), including other molluscs (ammonites and nautiloids), but these protective structures are of different origins. An operculum is found in the larva of nearly all gastropods—except in pleurobranchomorphs—and often in the adult stage. In the adults, an operculum is absent in patellogastropods and other limpetlike lineages; in many heterobranchs, including most pulmonates; and in different lineages of vetigastropods and caeno-gastropods, as well as in the terrestrial neritimorphs. The operculum can be pro-teinaceous (made of conchiolin) or calcareous. In one deep-sea species, *Chrysomal-lon squamiferum* (Peltospiridae), inhabiting hydrothermal vents and dubbed the "scaly-foot gastropod" (Van Dover et al., 2001; Warén et al., 2003), the foot is armored at the sides with iron-mineralized sclerites; the shell also contains an iron sulfide layer on top of the periostracum. It is the only metazoan that uses iron sulfide as part of its skeleton (Chen et al., 2015).

The foot musculature consists of a set of pedal retractors, which attach to the shell and mantle, and a series of smaller muscles that allow the sole to contract and expand, to control movement. In many bivalves, the foot is laterally compressed and highly modified for digging into the sediment, with powerful pedal retrac-tors and a pair of protractors, using a series of movements for digging, anchoring, and pulling the shell down. In a few bivalves in the family Galeommatidae, the foot is used for crawling, as in gastropods.

Many bivalves have adopted a sessile life form by either cementing the shell to the substrate or by using the special anchoring threads called the "byssus." The process of cementation has originated in multiple lineages of bivalves (Yonge, 1979), and many cementing bivalves use byssal threads when they are juveniles to an-chor to the substrate. The byssus is found mostly in adult pteriomorphians (and in the heterodont *Dreissena*), but it is also found in the juveniles of many heteroconchs. The byssus is a nonliving biomaterial that consists of a bundle of collagenous threads attached distally to hard surfaces and fused proximally through a stem to byssal retractor muscles at the base of the foot (Waite and Broomell, 2012). The bys-sal threads (colloquially called the "beard" in mussels) are made one at a time in the ventral groove of the foot in a matter of minutes, and are secreted as a quickly polymerizing liquid by the byssus gland, located on the foot. They have received considerable attention from a biomaterials perspective (Waite and Broomell, 2012; Liu et al., 2015a), due both to their strength and to their ability to quickly polymer-ize in a wet environment, which may confer possible clinical implications, espe-cially for suturing wet wounds (Lee et al., 2011). Recently, it has been realized that byssus performance is affected by lower pHs, and thus ocean acidification could have a negative impact on byssally attached bivalves (O'Donnell et al., 2013).

Swimming has evolved multiple times in molluscs, including some bivalves able to move by flapping their valves, and in the caenogastropod heteropods the foot is modified for swimming and the shell is greatly reduced and decalcified. Many het-erobranchs such as nudibranchs are able to swim long distances by undulating their bodies, while anaspidean sea hares use their parapodial lobes and pteropod sea angels live exclusively in the water column. But swimming has been perfected in the cephalopods, and particularly in the shell-less forms, through the jet-propulsion mechanism described above. This swimming involves structures and musculature of the mantle and also of the foot, which has been transformed evo-lutionarily into prehensile arms and a funnel tube, the latter representing the pos-terior end of the foot (Lee et al., 2003).

▰ DIGESTIVE SYSTEM, RADULAE, AND GILLS

The digestive system of molluscs consists of a through-gut, generally with an anterior mouth and a posterior anus. It is more or less elaborate, depending on their varied dietary preferences, and many of their structures, like their unique radular apparatus and the bivalve stomach generally enveloped by the digestive gland, contain large numbers of phylogenetically informative characters. The hepatopancreas, or midgut gland, provides the functions of the mammalian liver and pancreas, including the production of digestive enzymes and absorption of digested food.

In the case of bivalves, the stomach has been used as a key character set to define major phylogenetic groups (e.g., Purchon, 1960), although reevaluation of these characters did not prove highly informative (Bieler et al., 2014). Likewise, the feeding apparatus of gastropods has played an important role in their phylogeny and systematics (e.g., Taylor et al., 1993). Finally, the gills of autobranch bivalves are also fundamental feeding structures, discussed below, and were the basis of another traditional classification system for bivalves (Stasek, 1963).

The anterior portion of the digestive tract of most molluscs (fig. 46.3 C) bears a muscular structure in the anterior portion of the pharynx, called the "odontophore," which serves as a support for one of the hallmarks of molluscs, the radula. This radular apparatus is present in virtually all molluscs, with the exception of bivalves, which are classically interpreted as "lacking a head." The radula can contain a few radular plates, as in some aplacophorans and a variety of Cambrian fossils, while in other molluscs the radular teeth form a series of transverse rows, up to 50 or so, with many teeth per row.

The radula, formed of chitin, grows by adding new rows of teeth, through special cells called "odontoblasts," as the older, most-distal ones are worn out. The radula of chitons is hardened with magnetite and other iron oxides (Joester and Brooker, 2016), goethite in the case of patellogastropods (Barber et al., 2015). The radula clearly reflects the type of diet of its bearer, with herbivorous species often having many small teeth, or, in sacoglossan algal-feeders, are even monoseriate to pierce each algal cell individually, while carnivores tend to have fewer and more heterogeneous teeth per row. In some cases, radulae have adaptations for boring into the shells of other molluscs, as in Muricidae and Naticidae, but the most highly modified radula is that of Conidae, here transformed into a hollow, harpoonlike structure used for injecting their potent venom into prey. A thorough and updated discussion of the radulae across molluscan taxa is provided by Brusca et al. (2016a) and Joester and Brooker (2016).

Jaws are also present in many molluscs, and they are highly modified in cephalopods to form the beak. Cephalopods use a multitude of mechanisms to capture, crush, and tear their large prey, combining their prehensile arms with suckers and hooks, their beak, and the radula.

In bivalves, feeding occurs by different mechanisms, the structures involved not necessarily related to the digestive system. For example, protobranchs use labial palps to draw food particles from the sediment into the anterior portion of the digestive system, while the autobranchs use their hypertrophied gills for filter feeding, drawing food particles into the mantle cavity, where they are sorted and directed toward the anterior part of the digestive system by ciliary action. A few, mostly deep-sea bivalves—although there are some also inhabiting relatively shallow waters—have transformed their gills into a muscular septum that allows them to suck up small prey or to ingest small gastropods and other benthic animals. Filter feeding also occurs in other sessile molluscs, such as vermetids, tethydid nudibranchs with hypertrophied mouth velums (something similar is also seen in predatory tunicates), thecosomes secreting a mucus web to intercept phytoplankton, and some deep-sea cephalopods use their webbed arms or filament-like arms to trap "marine snow."

Finally, many molluscs, and especially bivalves, have developed intimate symbioses with other organisms, including photosynthetic zooxanthellate algae (in the giant clams of the genus *Tridacna* and some other relatives in the family Cardiidae) (see Li et al., 2018a) and several chemoautotrophic bacteria (mostly sulphide-oxidizing but also methanotrophic). The latter symbioses have evolved as a widespread nutritional strategy in multiple lineages of protobranchs, pteriomorphians, and heterodont bivalves from shallow waters to the deep sea (Oliver and Taylor, 2012; Distel et al., 2017), as well as in some hydrothermal-vent gastropod lineages (Suzuki et al., 2006). The so-called solar-powered sacoglossans are capable of obtaining and cultivating chloroplast organelles incorporated in their own tissues. Nonfeeding related symbioses with bioluminescent bacteria have also been well characterized in cephalopods (Sepiolidae) (Nishiguchi et al., 2004).

Molluscs typically have gills (ctenidia) for respiration, but some lineages lack ctenidia altogether, while others have secondarily derived gills. The supposedly primitive ctenidia consist of a longitudinal axis projecting from the mantle cavity with, typically, a series of alternating ciliary filaments along the axis (these are opposed and not alternating in some protobranch bivalves). The cilia are thus responsible for drawing water into the mantle cavity. Gills are serially repeated in monoplacophorans and chitons—two pairs are present in nautiloids and a pair in most other molluscs—but they are absent in scaphopods and neomeniomorph aplacophorans. In bivalves the gills evolved from primitive ctenidia with solely a respiratory function, to hyperthrophied organs for filter feeding, correlated with the loss of the entire head and feeding apparatus, as discussed earlier.

As described above, in gastropods one ctenidium is often lost as a consequence of torsion. In heterobranchs, the typical ctenidia are absent, but other respiratory structures appear, including gills, which are often found in the same position as the primitive ctenidium. In nudibranchs and sacoglossans, the cerata—dorsal and lateral outgrowths of the body surface—can aid in gas exchange, and in some nudibranch taxa, secondary elaborate gills appear surrounding the anus. In many terrestrial gastropods, gas exchange occurs within the mantle cavity in a highly vascularized region known as a "lung." Lunglike structures seem to have evolved multiple times within heterobranchs.

Gills have played an important role in the taxonomy of bivalves. Protobranchs have a primitive gill not used for feeding, but in autobranchs the gills have different configurations in terms of plication and interconnection of the gill filaments, which define different lineages. The gills are substituted for a muscular septum in the so-called septibranch anomalodematans, which have become secondarily carnivorous.

BODY CAVITIES AND EXCRETION

The nature of the molluscan coelom has been a matter of lively discussion, especially in relation to its phylogenetic affinities to annelids versus other phyla. As in entoprocts, the primary body cavity is composed of numerous lacunae (Salvini-Plawen and Bartolomaeus, 1995; Schmidt-Rhaesa, 2007), some of which are canalized to form the dorsal heart. The dorsal heart is surrounded by the pericard—a small coelomic compartment that forms by schizocoely (Moor, 1983; Salvini-Plawen and Bartolomaeus, 1995). The pericard is also associated with another coelomic derivative, the renopericardial duct, which can be paired or single (Bartolomaeus, 1996/97; Fahrner and Haszprunar, 2002). The system is often referred to as the "renopericardial complex." The closed circulatory system of cephalopods, with endothelially lined blood vessels and heart, could also be considered a coelomic cavity.

Protonephridia have been reported from the larvae of chitons, gastropods, and bivalves, typically one pair, but multiple terminal cells are found in some freshwater snails. In adult molluscs metanephridia are present, opening into the pericard. The sites of filtration are podocytes in the pericard epithelium. In all groups except Aplacophora, the pericardioduct is specialized for reabsorption and active secretion. This structure is known as a "kidney" (Schmidt-Rhaesa, 2007).

Rhogocytes, also termed "pore cells," are the site of hemocyanin or hemoglobin biosynthesis in gastropods, occurring as solitary or clustered cells in the connective tissue and may have a function related to excretion and various metabolic processes (Stewart et al., 2014; Kokkinopoulou et al., 2015). These free-floating cells have been proposed as a synapomorphy of Mollusca (Haszprunar, 1996b; Wanninger and Wollesen, 2015).

NERVOUS SYSTEM

The molluscan nervous system has recently been reviewed for Aculifera, Monoplacophora, and Scaphopoda (Sigwart and Sumner-Rooney, 2016); Bivalvia (Wanninger, 2016b); Cephalopoda (Wild et al., 2015; Wollesen, 2016); and Gastropoda (Voronezhskaya and Croll, 2016), and we refer to these state-of-the-art summaries for further details. Likewise, an excellent summary has recently been provided by Brusca et al. (2016a). The central nervous system depicts characters shared with many other protostomes but with special adaptations in the different molluscan classes, especially with respect to the series of ganglia and nerve cords that innervate the different body regions, including the head region, foot, and the viscera. An enormous diversity is found in the distribution and anatomy of the anterior ganglia (Faller et al., 2012); ganglia are less developed in aculiferans than in other molluscan classes, and they attain maximal condensation in the highly complex cephalopod brain. The nervous system of sea hares in the genus *Aplysia* and of the squid *Doryteuthis pealeii* have received special attention due to their giant neurons and ganglia (see below).

In gastropods the nervous system is modified as a consequence of the process of torsion, and the anterior region includes cerebral, pedal, and pleural ganglia and the twisted longitudinal nerve cords with visceral ganglia, as well as one supaesophageal and one subesophageal ganglion. In most molluscs the cerebral ganglion is often connected to the buccal ganglia by a buccal commissure or by a cerebrobuccal nerve ring. There are also subradular ganglia connected by subradular connectives. In aculiferans, which supposedly have a plesiomorphic nervous system, there is a pair of longitudinal pedal (ventral) nerve cords and a pair of visceral (lateral) nerve cords with numerous commissures between the different longitudinal cords (Shigeno et al., 2007; Todt et al., 2008; Redl et al., 2014). In bivalves there is fusion of ganglia, ending up with a pair of anterior cerebropleural ganglia, a pair of pedal ganglia, and a pair of posterior visceral ganglia. The cerebropleural ganglia are connected by a dorsal commissure over the esophagus.

The nervous system of Cephalopoda is much more elaborate, with a massive brain encased in a cartilaginous capsule that is penetrated by the esophagus and large optic lobe, with numerous optic nerves innervating the large eyeball (Wollesen et al., 2009). A large anterior brachial nerve innervates each arm, other anterior nerves innervate the buccal region, and posterior nerves innervate the body. These giant nerves are exceptionally well studied in squids, which combine first-, second- and third-order giant nerves, those of *Doryteuthis pealeii* having been the subject of much research in nerve physiology due to their fast axonal transport and to the fact that may exceed 500 μm in diameter and can be easily dissected (Song et al., 2016).

Mollusc sense organs are numerous and include a diversity of sensory tentacles, statocysts, photoreceptors (including complex eyes), and unique organs called "osphradia." Many gastropods have cephalic tentacles that may bear eyes, but many other types of cephalic, epipodial, and mantle tentacles can be found within Gastropoda. Gastropod eyes range from pigment pits (as in limpets) to pinhole eyes (as in abalone) or, in many lineages (e.g., periwinkles), image-forming lensed eyes. Many nudibranchs bear a pair of anterior rhinophores of chemosensory function on the dorsal part. In Solenogastres, cirri (bundles of cilia) and the papillae from the vestibulum have sensory function.

Osphradia are patches of sensory epithelium, a chemosensory organ located on or near the gills or on the mantle wall. They function as chemoreceptors and perhaps also monitor the amount of sediment in the inhalant current. However, little is known about the biology of osphradia, and their anatomy differs markedly throughout the phylum (Lindberg and Sigwart, 2015); those of aplacophorans and *Nautilus* are not considered homologous. In gastropods with two gills, an osphradium is present on the supporting membrane of each gill; in the gastropods that possess a single gill, there is only one osphradium, and it lies on the gill membrane or on the mantle cavity wall anterior and dorsal to the attachment of the gill itself. Osphradia are reduced or absent in gastropods that have lost both gills, that possess a highly reduced mantle cavity, or that have taken up a strictly pelagic existence. Osphradia are best developed in benthic predators and scavengers, such as neogastropods.

Scaphopods lack eyes, tentacles, and osphradia but have captacula, structures used for feeding that may have a tactile function. Bivalves have a diversity of sensory organs often located on the mantle edge, including tentacles and eyespots. The tentacles can be located on the siphonal apertures or along the mantle. Photoreceptors and eyes of different types have evolved multiple times in the headless bivalves (Morton, 2008), including compound eyes in ark clams (Nilsson, 1994) and mirror optics in the many eyes along the mantle edge in scallops (Speiser et al., 2011b). Chitons also lack statocysts, tentacles, and cephalic eyes but have the specialized aesthetes or "shell eyes," as well as the so-called adanal sensory structures. The aesthetes are branches of the nervous system that serve a variety of sensory functions. In chiton species within Schizochitonidae and in two subfamilies of Chitonidae, a number of the aesthetes are capped with an ocellus that includes an aragonitic lens and is able to form images (Speiser et al., 2011a; Li et al., 2015a). In addition, what has been termed the "Schwabe organ" is a pigmented sensory organ found on the ventral surface of lepidopleuridan chitons (Sigwart et al., 2014), which represents the adult expression of the chiton larval eye, being retained and elaborated in adult lepidopleuridans, and thus of a photoreceptive nature (Sumner-Rooney and Sigwart, 2015). Several other epidermal structures and receptor types are present in molluscs, including, for example, the adoral sense organs of protobranch bivalves (Schaefer, 2000).

Cephalopod sense organs are highly developed, the camera-type coleoid eyes having been portrayed as an example of evolutionary convergence with those of vertebrates, and—as in vertebrates—with a cornea, iris, lens, pupil, and retina, but this retina is of the direct type, unlike in vertebrates. Nautiloids have simpler eyes (lacking a lens and using pinhole optics), osphradia, and a statocyst that is less complex than that of coleoids, which lack osphradia but possess numerous chemosensory and tactile cells on their arms (especially in the suckers). The visual system of cephalopods is especially important because it has triggered the evolution of their striking color display, which is elegantly controlled by the nervous system. The integument is full of chromatophores, pigment cells that can be individually expanded or contracted by muscular action, causing the pigment to be exposed (when expanded) or concentrated into an inconspicuous dot (when contracted).

With this mechanism cephalopods are able to change color as well as texture, also through muscular action of the skin cells, almost instantaneously. These changes are not only used for camouflage but also for courtship and aggression, as in the dramatic flashes of blue color displayed by the highly venomous Indo-Pacific blue-ringed octopuses in the genus *Hapalochlaena*. In addition, many deep-sea squids are bioluminescent, this having evolved multiple times and in different organs, including the mantle and eyes. The bioluminescence can be intrinsic or extrinsic, using symbiotic bacteria in complex light organs that mimic the behavior of an eye. This is particularly well studied in the shallow-water bobtail squids (Sepiolidae), which use the ventral light organ for counterillumination (Jones and Nishiguchi, 2004) and control the symbiont growth by venting excess bacteria daily. In the deep-sea species, bioluminescence has other roles, including communication and prey attraction.

Most coleoids have an ink sac located near the intestine. The ink is used to deter predators and for conspecific communication when it is released through the anus and mantle cavity. The ink is rich in mucus and melanin pigment, and in the odd bobtail squid *Heteroteuthis dispar* a bioluminescent mucus from glands near the ink sac is secreted with the ink. The relationship of ink to human life dates back millennia; it has been widely used in Mediterranean and Asian cooking (think of black pasta or black rice), writing, and making dyes. More recently, cephalopod ink has had a role in drug discovery due to its antimicrobial, anticancer, and many other cosmetic and medical applications (Derby, 2014).

Research on the physiological basis of memory storage in neurons largely based on the model sea hare *Aplysia californica* was recognized in the award of the 2000 Nobel Prize in Physiology and Medicine to Austrian–American neuroscientist Eric Kandel. Recent neurobiological work using the same model sea hare has tackled the old question of memory transfer and whether long-term memory is mediated by synaptic strength or epigenetic modifications. A series of recent experiments by Bédécarrats et al. (2018) has shown that the memory for long-term sensitization to external stimuli can be successfully transferred by injecting RNA from sensitized into naive animals. Furthermore, a similar neuronal behavior to that produced in sensitization can be reproduced in vitro by exposing neurons to RNA from trained *A. californica*. These experiments add to the previous work on memory transfer in Platyhelminthes (see chapter 43).

▬ REPRODUCTION AND DEVELOPMENT

Many molluscan lineages are gonochoristic, but Solenogastres, some Monoplacophora and most Heterobranchia are hermaphroditic. Gonochoristic molluscs typically have a pair of gonads that discharge their gametes into the water environment, fertilization being external and development indirect. Numerous exceptions to all these aspects are found, including the evolution of hermaphroditism and of internal fertilization in many lineages, but especially in the nonmarine environments. Likewise, many internal fertilizers become brooders, including cephalopods, which may display different degrees of parental care. The whole spectrum of reproductive modes can be found in several clades of molluscs, and in some cases sex is determined by a combination of genetics and the environment (Santerre et al., 2013). As a consequence of torsion, gastropods have a single gonad, which in hermaphroditic species can be an ovotestis; others have separate male and female gonads or are protandric. The reproductive tracts of gastropods become really complex, with numerous ducts, glands, and copulatory and storage structures; they are therefore an important source of taxonomic characters. Due to the enormous divergence in reproductive systems, we direct the reader to more specific studies on mollusc reproductive organs.

Interesting noncanonical reproductive systems are found in some gregarious species, such as the slipper shell *Crepidula fornicata*, where individuals stack on top of each other, females at the bottom and males on top. Microsatellite analyses of this system have determined that paternity is mostly from the largest male attached to the female (Proestou et al., 2008). In this system detached males can eventually become female. Many terrestrial pulmonates are simultaneous hermaphrodites and display complex mating behaviors, as in the famous banana slug, *Ariolimax dolichophallus*, with complex courtship following slime threads, reciprocal penetration, and, sometimes, reciprocal apophallation (Leonard et al., 2002).

The most complex mating behaviors are, however, found in cephalopods. Cephalopods are gonochoristic, with a single gonad in the posterior region of the visceral mass. Males produce a spermatophore, stored in a large reservoir called "Needham's sac" and released into the mantle cavity. In females, an oviducal gland at the end of the oviduct secretes the protective membrane for the eggs, complemented by the secretions of the nidamental glands. In some cases, mating behavior can be preceded by vivid color displays. Male coleoids use a modified arm, called a "hectocotylus," as an intromittent organ to transfer the spermatophores into the oviducal opening of females. In nautiloids, four small arms form an organ, called a "spadix," for sperm transfer. Eggs are then laid individually or in clutches, some species dying soon after depositing their eggs. Among the most bizarre reproductive modes in cephalopods is that of the paper nautilus, *Argonauta* species, in which the female uses two arms to sculpt a "shell-looking" calcareous house in which the eggs are laid, while the dwarf male may share this brood chamber with the embryos.

The early egg development of molluscs has recently been reviewed by Wanninger and Wollesen (2015), who provide details of the best-known models. It is best characterized in gastropods (see reviews in Collier, 1997; Wanninger and Wollesen, 2015), scaphopods (Lacaze-Duthiers, 1858), and chitons (Henry et al., 2004), and to a lesser extent in bivalves (Zardus and Morse, 1998) and aplacophorans (Okusu, 2002) (fig. 32.2), conforming to a typical spiral pattern.

Fertilization depicts some unique features in chitons, which have elaborate egg hulls and elongate sperm heads, characters that strongly correlate and that show a clear phylogenetic signal (Okusu et al., 2003; Buckland-Nicks, 2008). Other oddities of chiton fertilization are that the sperm mitochondria and centrioles appear not to penetrate the egg surface (Buckland-Nicks, cited in Wanninger and Wollesen, 2015). In most studied molluscs, meiosis is only completed after fertilization, when the polar bodies are shed. This is the animal pole of the embryo, and the point of entry of the sperm cell determines the orientation of the first cleavage furrow. Early cleavage is usually total (holoblastic) but may be equal (e.g., in *Turbo cornutus*) or unequal (in *Tritia obsoleta*). Polar lobes may appear as early as the first cleavage (fig. 32.2 B), a stage termed the "trefoil" stage, known from scaphopods, aplacophorans, and some gastropods. Polar lobes have an important role in morphogenesis, as demonstrated by the classical experiments involving their removal (Clement, 1968). The typical spiral cleavage is visible at this stage.

Detailed cell lineages have been established for a number of species, most prominently *Crepidula fornicata*, for which a classical fate map was available (Conklin, 1897) and has recently been refined (Hejnol et al., 2007). No information is available about the cell genealogies of solenogastres, caudofoveates, monoplacophorans, or cephalopods, but the fate of the prototroch is well known in a number of additional molluscs, including chitons, scaphopods, and gastropods (Damen and Dictus,

1994a). The fate of the metatroch is also known for *Tritia obsoleta*, but in this case it differs from that of *C. fornicata* (and from that of annelids), casting doubt on the homology of the metatroch across spiralians (Gharbiah et al., 2013). *Tritia* has also been used as a model to understand determinative development, especially with respect to eye formation. The first quartet micromeres 1a and 1c normally develop the eyes in *T. obsoleta*, and these do not develop if 1a and 1c are removed at the eight-cell stage. However, in a classical experiment on regulative eye development, Sweet (1998) showed that eyes can form if the precursors of 1a and 1c are removed at the two- or four- cell stage, demonstrating that proximity to the D macromere is important for determining whether or not a micromere will go on to develop an eye.

In *C. fornicata* the establishment of a single D quadrant appears to rely on a combination of both autonomous (via inheritance of the polar lobe) and conditional mechanisms (involving induction via the progeny of the first quartet micromeres). D quadrant identity in *C. fornicata* is established between the fifth and sixth cleavage stages, as in other spiralians that use conditional specification. Subsequently, following the next cell cycle, organizer activity takes place soon after the birth of the 4d micromere. Therefore, unlike in other spiralians that use conditional mechanisms of specification, in this mollusc species the specification of the D quadrant and the activity of the dorsoventral organizer are temporally and spatially uncoupled (Henry et al., 2017). This is not the case in the development of *Tritia obsoleta*, in which the polar lobe has an important role in both D quadrant specification and the subsequent development of the embryo (e.g., Clement, 1952; Lambert and Nagy, 2001; Lambert et al., 2016).

It has been widely discussed that a "molluscan cross"—a specific pattern of embryonic development during the 64-cell stage—is typically found in molluscs (as well as in sipunculan annelids and entoprocts) (Merkel et al., 2012), but the significance of this character has been questioned due to the lack of an explicit framework of cell identity across spiralian embryos and their supposed "crosses" (Jenner, 2003).

The mode of gastrulation depends on egg size and yolk content. In species with small eggs, gastrulation occurs by invagination on the vegetal pole and subsequent formation of a fluid-filled coeloblastula. Species with large, yolky eggs usually exhibit epiboly and a massive stereoblastula (Wanninger and Wollesen, 2015). The blastopore may develop into the mouth or may close. Mesoderm derives from two embryonic sources, ectomesoderm (in gastropods mostly formed by the third quartet micromeres with minor contributions from second quartet derivatives) and endomesoderm (derived from progenies of the 4d mesentoblast, which divides and forms two mesodermal bands, right and left) (Wanninger and Wollesen, 2015).

Indirect development is found in many molluscan lineages, with a free-swimming typical trochophore bearing an apical ciliated tuft and a prototroch. In some groups (e.g., chitons), the trochophore is the only larval stage, but it has important modifications with respect to the typical trochophore (Henry et al., 2004). Likewise, the trochophore of Caudofoveata has, in addition to the prototroch and a telotroch, serially arranged glandular cells (spiculoblasts) at the dorsal mantle and a ventral suture as a possible site of a former foot sole (Nielsen et al., 2007). In many

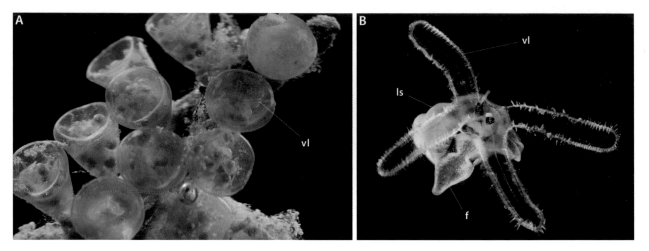

FIGURE 46.4. A, Gastropod egg capsules attached to a piece of dead coral, with multiple larvae per capsule; B, Veliger larva of a gastropod. Abbreviations: f, foot; ls, larval shell (protoconch); vl, velar lobe.

bivalves and gastropods, the trochophore transforms into a second larval type called a "veliger," which has a shell, a foot, and a swimming structure, the velum (fig. 46.4 B)—two large, ciliated lobes that develop from the prototroch, which can be used not only for swimming but also for feeding, and then they are subdivided in multiple lobes. The larval shell of gastropods is called a "protoconch" and that of bivalves a "prodissoconch," generally with two different stages, distinguishing a lecithotrophic and a planktotrophic larval phase (although in some cases the larva remains entirely lecithotrophic); these are called "PI" and "PII."

In gastropods veligers also have an operculum, although this is lost in the adults of many species. It is during this stage that the process of torsion occurs. The veliger continues to develop, and eyes and other adult structures form, until settlement, when the larva transforms into a juvenile. Brooding is common in some species, which may release veliger larvae instead of trochophores. In most gastropods (except in patellogastropods and vetigastropods), the trochophore larva is largely suppressed and a veliger hatches from the egg, encapsulated into a maternally generated egg capsule (fig. 46.4 A)—a characteristic structure found in many members of this clade of gastropods named Angiogastropoda due to this encapsulation (Cunha and Giribet, 2019). In other cases, development is direct, including in most pulmonates and many caenogastropods, the veliger stage happening within the egg case.

Freshwater bivalves of the clade Unionida ("naiads") have parasitic larvae that are brooded in the mother's ctenidia until they are expelled and attach to the gills of fishes, which transport them upstream. Glochidia, haustoria, and lasidia are some of these typical ectoparasitic larvae that can develop within epithelial cysts formed by the host fish (Wächtler et al., 2001).

Other so-called test, or pericalymma, larvae are found in protobranch bivalves and Solenogastres (Zardus and Morse, 1998), which lack a veliger stage. In the bivalve pericalymma, the test entirely covers the developing larva, including the shell, and is cast off during metamorphosis when the juvenile protobranch commences its benthic life.

Cephalopod development is very different from that of other molluscs, not retaining traces of spiral development (Boletzky, 1989). They produce embryos with large amounts of yolk and always show direct development. Cleavage is discoidal, the early embryo consisting of a monolayered disc of blastomeres (termed a blastodisc) on the animal pole, with the first cleavage furrow determining the longitudinal axis of the embryo, and the second and third cleavages being perpendicular to the first one (see Wanninger and Wollesen, 2015). Cephalopod embryos develop entirely within the egg case, as they slowly consume the large yolk sac by "ingesting" its content through the mouth, which opens directly into the yolk sac.

GENOMICS

Molluscan genomics had a slow start, compared to that of other large animal clades, but has recently picked up. Genomes of 27 molluscs are publicly available (as of May 2019), many originally driven by the aquaculture and biomedical industries (Takeuchi, 2017). The first molluscan genome was that of the gastropod *Lottia gigantea*, as a model to better understand animal evolution (Simakov et al., 2013). Since then, a plethora of gastropod genomes are now publicly available, including those of the pulmonate freshwater snail *Radix auricularia*, used as a model to investigate climate change; the biomedically important *Biomphalaria glabrata*, an intermediate host for schistosomes; the commercially important *Haliotis discus hannai* (Adema et al., 2017; Nam et al., 2017; Schell et al., 2017); and the invasive golden apple snail *Pomacea canaliculata* and three other ampullariids that show expanded gene families related to environmental sensing and cellulose digestion, which may have facilitated them becoming invasive pests (Liu et al., 2018; Sun et al., 2019). A few additional unpublished gastropod draft genomes are available in GenBank, including neurotoxin models *Colubraria reticulata* and *Conus tribblei*, the neurobiology model *Aplysia californica*, and the common freshwater snails *Lymnaea stagnalis* and the sinistral *Physella acuta*.

In addition, many bivalves of commercial importance as food, for the pearl industry, important hosts of symbionts, or invasive species have had their genomes sequenced. These include the pteriomorphians *Crassostrea gigas*, *Saccostrea glomerata*, *Pinctada fucata*, and *P. f. martensii*, *Mytilus galloprovincialis*, *Modiolus philippinarum*, *Bathymodiolus platifrons*, *Patinopecten yessoensis*, and *Argopecten irradians* (Zhang et al., 2012; Takeuchi et al., 2012; Murgarella et al., 2016; Takeuchi et al., 2016; Du et al., 2017a; Du et al., 2017b; Sun et al., 2017; Wang et al., 2017a; Powell et al., 2018), including the invasive freshwater species *Limnoperna fortunei* (Uliano-Silva et al., 2018); and the heterodont *Ruditapes philippinarum* (Mun et al., 2017). Unpublished genomes available in GenBank include a chromosome-level genome for the Akoya pearl oyster *Pinctada imbricata*, the freshwater palaeoheterodont mussle *Venustaconcha ellipsiformis*; and a few heterodonts, two highly invasive species, *Dreissena polymorpha,* and *Corbicula fluminea*, and the shipworm *Bankia setacea*.

In addition, the genome of three cephalopods, including *Octopus bimaculoides* (Albertin et al., 2015), the commercially important common octopus (*Octopus vulgaris*) (Zarrella et al., 2019), and the developmental bobtail squid model *Euprymna vulgaris* investigated for its light organ (Belcaid et al., 2019). These molluscan genomes have been studied for a diversity of reasons, including biomineralization and the expanded protein families that are fundamental for forming shells and nacre (Du et al., 2017b).

RNA editing, a posttranscriptional process that provides sequence variation, is frequently used in coleoid cephalopods to diversify proteins, especially those associated with neural function (Liscovitch-Brauer et al., 2017), probably explaining the complex behaviors of these cephalopods. Mollusc genomics is thus a great resource for understanding a diversity of biological, ecological, and behavioral phenomena in a group of great disparity and complexity, as well as a potential source for food and some of the most incredible biomaterials, such as byssus, nacre, or iron-rich radulae.

Mitogenomics has been thoroughly explored for molluscan phylogenetics (see a recent review in Stöger and Schrödl, 2013), with hundreds of complete mitogenomes currently available, most of them for Gastropoda and Bivalvia. In the case of bivalves, mitogenomics often produces phylogenetic trees that are at odds with those obtained based on morphology or nuclear genomes (e.g., Doucet-Beaupré et al., 2010; Plazzi et al., 2013; Plazzi et al., 2016), as the rearrangement rate on the mitochondrial genome is too high to be used as a phylogenetic marker and several genes have unusually high rates of evolution (Plazzi et al., 2016). Bivalve mitochondrial genomes are also notorious for presenting the rare phenomenon of doubly uniparental inheritance (DUI), where males receive mtDNA from both parents and transmit their paternal mtDNA to their sons while females receive mtDNA only from their mother (Hoeh et al., 1996), a phenomenon first described in *Mytilus* (see Zouros et al., 1994) and now known in most major bivalve clades (Theologidis et al., 2008).

The role of DUI and its evolutionary implications, having two mitochondrial genomes evolving independently but with the possibility of recombination (Tsaousis et al., 2005), has remained unclear, but it has been postulated that it may be correlated with sex determination. The retention of the paternal copy in male bivalves is probably due to a modification of the ubiquitination mechanism and DNA methylation could be involved in the maintenance of DUI in bivalves (Capt et al., 2018). Mitogenomics of gastropods also conflicts with the deep splits found by other methods (Grande et al., 2008), as they have many rearrangements when compared to other animal groups, but mitogenomes have been useful in resolving the phylogeny of shallower clades of gastropods (e.g., Osca et al., 2015; Uribe et al., 2016).

FOSSIL RECORD

The molluscan fossil record is extraordinarily rich and has been used to document planetary events since the Terreneuvian, the earliest Cambrian. The earliest molluscs are represented by small shelly fossils (e.g., Parkhaev, 2008, 2017) from the upper part of the Fortunian in Siberia, China, Mongolia, and Iran. Although the systematic affinities of some of these are subject to debate, they include "sachitids" (sclerites of such taxa as *Halkieria* and *Siphogonuchites*), which are mostly regarded as allied to Aculifera (Vinther, 2014, 2015), caplike shells identified as Monoplaco-

FIGURE 46.5. Early Cambrian putative conchiferan molluscs from South Australia: A, *Aldanella* cf. *golubevi,* scale 200 µm; B, *Pelagiella subangulata,* scale 200 µm; C, *Emargimantus angulatus,* scale 200 µm; D, *Ilsanella enallaxa,* scale 200 µm; E, F, *Helcionella histosia,* scale 100 µm; G, *Watsonella crosbyi,* scale 100 µm; H, *Anabarella australis,* scale 200 µm; I, J, *Pojetaia runnegari,* scale 150 µm. Photo credits: Glenn A. Brock.

pora (such as *Purella*), and likely Gastropoda (e.g., *Latouchella*). The distinction between monoplacophoran versus gastropod identities for cap-shaped, univalved early Cambrian shells is especially difficult, although in some cases protoconch morphology permits identification (Parkhaev, 2017).

A measure of the challenge in assigning some small shelly fossils to molluscan clades or even elsewhere within Spiralia is shown by some conical shells with serial muscle impressions that had been attributed to Monoplacophora having been reidentified as brachiopods (Dzik, 2010), and even such coiled forms as *Aldanella* (fig. 46.5 A) that have been allied to gastropods have a mode of growth more comparable to the enigmatic hyoliths (Dzik and Mazurek, 2013). Some early Cambrian molluscs, such as *Pelagiella* (fig. 46.5 B), preserve asymmetrical muscle scars that have been regarded as indicating a degree of torsion and thus membership in the gastropod total-group (Runnegar, 1981). Well-preserved material of *Pelagiella* shows that the aragonitic shell was composed of multiple layers with distinct lamellar microstructures (Li et al., 2017b), and paired chaetal bundles are known to project from the aperture in this taxon (Thomas, 2018). The expectation from small shelly fossils that early molluscs were millimetric in size is in some cases contradicted by the discovery that typically "small shellies" are larval shells. This is demonstrated by a "helcionellid" (a grade of early conchiferans: fig. 46.5 C–F) representing just the apex of a 2-cm limpetlike mollusc (Martí Mus et al., 2008).

Whereas the split between Conchifera and Aculifera (i.e., crown group Mollusca) is confidently constrained by Fortunian small shelly fossils, the possibility that the Ediacaran fossil *Kimberella quadrata* (fig. 46.6A–B) could be a mollusc or mollusclike bilaterian has been explored numerous times since that idea was first formalized by Fedonkin and Waggoner (1997). Its inferred dorsal surface is resistant and has been regarded as a non-biomineralized shell (Fedonkin et al., 2007b; Vinther, 2015), whereas other students of the same fossils have interpreted this tuberculate surface as mineralized sclerites (Ivantsov, 2010). The softer ventral side has been described as a sole and compared to a molluscan foot. The body is certainly zoned laterally, this zonation viewed as representing the foot and mantle separated by a groove in the mollusc model. Apparent internal structures have been interpreted as a pharynx with paired pouches (Vinther, 2015).

Trace fossils of the genus *Kimberichnus* are associated with *Kimberella* body fossils in both South Australia and the White Sea, Russia (Gehling et al., 2014). These are arcuate sets of ridges, often with a fan-shaped arrangement, interpreted as scratch marks that depict systematic excavation of a microbial mat (fig. 46.6 B). The apparent anterior end of *Kimberella* occasionally shows a proboscis-like extension that evidently bore denticles that produced the scratch marks (Fedonkin et al., 2007b). Some workers have interpreted the style of mat excavation as similar to radular feeding in Mollusca, though it has been noted that the feeding of *Kimberella/Kimberichnus* indicates rearward movement (Gehling et al., 2014), the reverse of that in molluscs (Parkhaev, 2017).

A few taxa originally described from the Burgess Shale, some of them formerly of uncertain or debated systematic position, have been documented by new collections that have supported placement in the mollusc stem group. *Wiwaxia corrugata* is the most morphologically complete and best known of these (fig. 46.6 D)—the genus now includes additional recently described species from several Cambrian Konservat-Lagerstätten (reviewed by Yang et al., 2014). Its molluscan affinities are best shown by its radula, which had been interpreted as such by Scheltema et al. (2003) before its more detailed description (Smith, 2012b) overturned previous comparisons with Annelida based on the microvillar organization of its sclerites (Butterfield, 1990). The *Wiwaxia* radula consists of two or three rows of teeth, each row with a single axial tooth flanked on each side by 8 to 16 curved lateral teeth (fig. 46.6 E). *Wiwaxia* sclerites are found as isolated (but recognizable) small carbonaceous fossils in Cambrian shales from many parts of the world (Slater et al., 2017).

Odontogriphus omalus from the Burgess Shale (fig. 46.6 C) was recognized as a "naked" (shell-less), dorsoventrally flattened mollusc when its radula was identi-

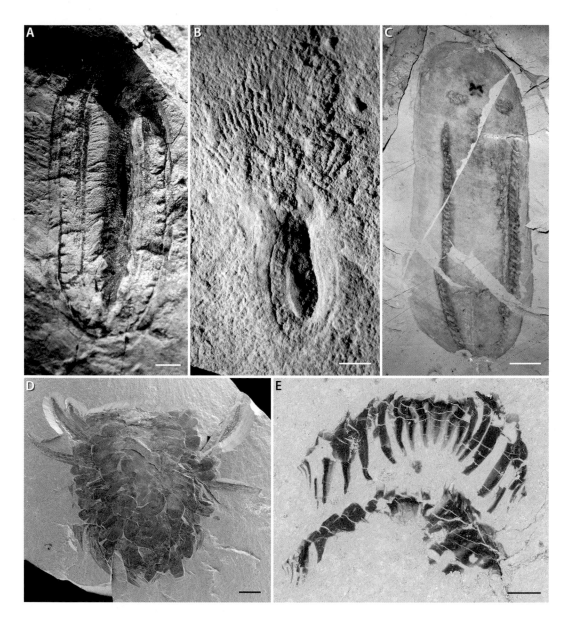

FIGURE 46.6. Ediacaran problematica and Cambrian stem-group Mollusca. A–B, *Kimberella quadarata*, scales 5 mm; B, with *Kimberichnus* feeding trace; C, *Odontogriphus omalus*, scale 1 cm; D, E, *Wiwaxia corrugata*; D, complete specimen, scale 5 mm; E, radula, scale 100 μm. Photo credits: A–B, Jakob Vinther; C, Jean-Bernard Caron; D–E, Martin Smith.

fied and the close similarity of its radula to that of *Wiwaxia* was documented (Caron et al., 2006; Smith, 2012b). Recent phylogenetic analyses recover *Odontogriphus* and *Wiwaxia* as a clade of stem-group Mollusca (Vinther et al., 2017).

Sachitid sclerites include some of the earliest molluscan fossils. The scleritome of the halkieriid sachitid *Halkieria evangelista* was revealed by articulated specimens from Sirius Passet, Greenland (Conway Morris and Peel, 1995). These show a single dorsal shell plate at the anterior and posterior ends of the animal (fig. 46.7 A)

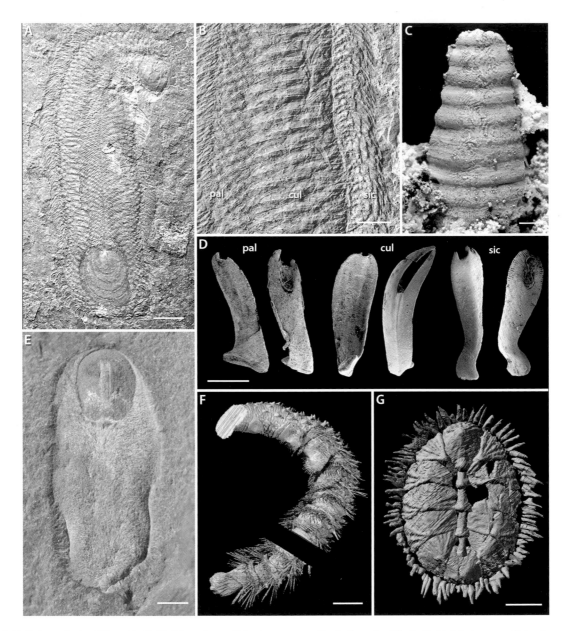

FIGURE 46.7. Paleozoic Aculifera (A, B, D–G) and possible stem-group Cephalopoda (C). A, B, *Halkieria evangelista*; A, articulated scleritome, scale 5 mm; B, palmate (pal), cultrate (cul) and siculate (sic) sclerites, scale 2 mm; C, *Tannuella elinorae*, scale 2 mm; D, *Australohalkieria superstes*, sclerites labeled as in B, scale 200 μm for palmate and siculate sclerites, 300 μm for cultrates; E, *Calvapilosa kroegeri*, scale 2 mm; F, *Acaenoplax hayae*, scale 2.5 mm; G, *Protobalanus spinicoronatus*, scale 2.5 mm. Photo credits: A, B, E, G, Jakob Vinther; C, Glenn Brock; D, Susannah Porter; F, Mark Sutton.

and a zonation of sclerite types across the body (fig. 46.7 B, D) comparable to that of chitons, a pattern used to ally the group with Aculifera (Vinther, 2014, 2015). The sclerites of halkieriids appear to have been aragonitic (Porter, 2004; Vinther, 2009). The discovery of *Calvapilosa kroegeri*, an Early Ordovician sachitid with a single anterior shell plate (fig. 46.7 E), strengthens the molluscan affinities of sachitids because the fossils preserve a ribbonlike, polystichous radula (Vinther et al., 2017).

Orthrozanclus, a genus with two species originally described from the Burgess Shale and subsequently from Chengjiang (Conway Morris and Caron, 2007; Zhao et al., 2017), resembles *Halkieria* in its sclerite zonation but has a single anterior shell plate. Its sclerites do not appear to have been biomineralized, but it has been argued that the relief of the shell plate suggests it was mineralized (Zhao et al., 2017). *Orthrozanclus*, *Calvapilosa*, and *Halkieria* are resolved in phylogenetic analyses as a grade of stem-group Aculifera (Vinther et al., 2017), suggesting that "Sachitida" is paraphyletic. Arguments have been made that *Wiwaxia* and *Halkieria/Orthrozanclus* are distantly allied, the latter pair's sclerite zonation (fig. 46.7 B) being compared to camellan stem-group brachiozoans (Zhao et al., 2017). We consider the molluscan interpretation to better account for the radula of *Calvapilosa*, the correspondences in sclerite zonation between *Halkieria* and Aculifera, and similarities between the complex internal canal system in halkieriid sclerites and the aesthete system of aculiferans (Vinther, 2009).

Whereas "sachitids" bear one or two dorsal shell plates, fossils inform on the more crownward evolution of Aculifera. Eight shell plates are reconstructed for the last common ancestor of extant aculiferans, based on similarities between chitons and aplacophorans. Neontological evidence from the latter comes from the alternation between transverse spicule bands and seven fields devoid of spicules in the postlarva of a solenogaster (Scheltema and Ivanov, 2002) and transverse bands of papillae at the sites of aragonitic spicule formation in the larva of a caudofoveate (Nielsen et al., 2007). The Silurian *Acaenoplax hayae* (Sutton et al., 2004) and *Kulindroplax perissokomos* (Sutton et al., 2012) are aculiferans with seven dorsal shell plates (fig. 46.7 F).

Kulindroplax is especially informative for its combination of an aplacophoran-like spiculate body and polyplacophoran-like valves, particularly resembling those of fossil paleoloricate "chitons," which may be aplacophorans rather than polyplacophorans (Vendrasco et al., 2004; Sigwart and Sutton, 2007; Sutton et al., 2012). An extinct Middle and Late Paleozoic group known as "multiplacophorans" have as many as 17 shell plates arranged in 7 transverse rows (Vendrasco et al., 2004; Vinther et al., 2012) (fig. 46.7 G). Although unquestionably allied to chitons, whether they are stem- or crown-group Polyplacophora has been debated (reviewed by Vinther, 2014).

Bivalves are confidently identified in the early Cambrian, although from younger strata than the first aculiferans, monoplacophorans, and gastropods. The first bivalves are *Fordilla troyensis* and *Pojetaia runnegari*, plus a few additional species in these genera, which occur in roughly coeval rocks from the Tommotian, Cambrian Stage 2 (Pojeta Jr., 2000; Vendrasco et al., 2011). A review of the Cambrian fossil record reaffirmed the status of *Fordilla* and *Pojetaia* (fig. 46.5 I–J) as bivalves based on their muscle scar organization and their hinge and ligament structure (Elicki and Gursu, 2009). Both *Fordilla* and *Pojetaia* had a laminar inner shell microstructure reminiscent of the foliated aragonite of modern monoplacophorans (Checa et al., 2009). This microstructure is shared by the Cambrian molluscs *Anabarella* (fig. 46.5 H) and *Watsonella* (fig. 46.5 G), providing evidence that they are likely close to the ancestry of bivalves (Vendrasco et al., 2011). Intriguingly, none of the Cambrian molluscs unambiguously display nacre, even though appropriate preservation is known, and it has been hypothesized that nacre likely evolved independently in different conchiferan lineages in response to predation pressure during the Great Ordovician Biodiversification Event (Vendrasco et al., 2011).

Among extinct high-level molluscan groups, Rostroconchia has been ranked as high as class level (Pojeta Jr. et al., 1972) and is often related to bivalves (Ponder and Lindberg, 2008a) or alternatively as a paraphyletic grade of stem-group scaphopods (Vinther, 2014). Rostroconchs have left and right valves like Bivalvia but are not really symmetrical, as they affix to the substrate by the left valve. The group has a temporal range from at least the late Cambrian to the Permian, with some earlier Cambrian taxa allied to this lineage in some interpretations.

Each of the two main rostroconch groups, Ribeirioida and Conocardioida, has been posited to be more closely related to crown-group scaphopods by different workers, with more recent evidence from the protoconch form decanting toward conocardioids as sister group of living scaphopods (Peel, 2006; Vinther, 2015). Various tubular shells dating to as early as the Ordovician have been embroiled in controversy over whether they represent scaphopods or non-molluscan groups (such as annelids), but scaphopods identifiable as Dentaliida range from the Early Carboniferous onward (Peel, 2006; Parkhaev, 2017).

The ca. 180 genera and ca. 1,000 species of extant cephalopods are joined by some 4,000 extinct genera, nearly half of which are Paleozoic nautiloids (Kröger et al., 2011). Cephalopods are first convincingly represented in the late Cambrian in the form of chambered phragmocones of the genus *Plectronoceras* (Chen and Teichert, 1983), which demonstrably possess a siphuncle, although geologically earlier chambered shells of Cambrian helcionellid molluscs (Brock and Paterson, 2004) such as *Tannuella* (46.7 C) have been suggested as possible stem-group Cephalopoda (Kröger et al., 2011). *Plectronoceras* and other early cephalopods were apparently oriented with the body situated ventrally and the shell directed upward. The proposal that the Burgess Shale animal *Nectocaris pteryx* is a soft-bodied cephalopod (Smith and Caron, 2010) has been solidly critiqued (Kröger et al., 2011), principally for lack of general characters of Mollusca and for showing similarities to derived coleoids rather than to more generalized early nautiloids.

The more diverse of the two extant cephalopod clades, Coleoidea, has its earliest, well-corroborated stem-group representatives in the Devonian, bactritids and ammonoids, which are united by the form of their protoconchs, among other characters. The latter include ammonites, perhaps the most iconic invertebrate fossils of the Mesozoic. Pelagic habits that allowed for widespread geographic distributions and relatively high evolutionary turnover made ammonoids important for biozonation of Mesozoic marine strata. The straight-shelled (orthoconic) bactritids contrast with the slightly curved (cyrtoconic) shells of early ammonoids. Ammonoids suffered major extinctions in the Late Devonian, end-Permian, and end-Triassic mass extinction events, but they recovered and reradiated after each of these before finally succumbing to the end-Cretaceous mass extinction. The belemnites, a common Jurassic–Cretaceous coleoid group, also vanished in the end-Cretaceous extinction, along with much of the diversity within Nautiloidea.

Although various Early Paleozoic nautiloids have been implicated in the origins of the lineage leading to the extant *Nautilus* and *Allonautilus* on the one hand and ammonoids and coleoids on the other, implying a split in the cephalopod crown group as early as the Ordovician, molecular dating suggests this divergence may be no earlier than the mid Paleozoic, that is, Silurian (reviewed by Kröger et al., 2011). The five living orders of Coleoidea are supplemented by six extinct orders (Nishiguchi and Mapes, 2008). "Nautiloidea" in the broad sense is certainly paraphyletic with respect to Coleoidea, with various extinct Paleozoic "nautiloid" orders likely being stem-group Cephalopoda. The straight-shelled Orthocerida, a group that first appeared near the base of the Ordovician, are often regarded as the "nautiloids" that are most closely related to Coleoidea, that is, stem-group coleoids (Nishiguchi and Mapes, 2008).

ANNELIDA

Annelids include all types of so-called "true worms" or "segmented worms": the mainly marine polychaetes, the terrestrial earthworms (and their many marine and freshwater oligochaete counterparts), the leeches, and many other worms that at some point in time had received phylum status. These are Pogonophora and Vestimentifera (both now in the family Siboglinidae), Echiura, Sipuncula,[21] Diurodrilida, *Lobatocerebrum*, Myzostomida, and even possibly the parasitic Orthonectida (here treated in chapter 34). On the other hand, other animals once considered annelids now receive their own phylum status (e.g., Gnathostomulida). Defining an annelid morphologically is thus a challenging enterprise due to the enormous differences in body plan among the members of this diverse and disparate clade, and thus we follow a strictly phylogenetic definition of the phylum. Anatomical characters traditionally used to identify Annelida, including their typically conspicuous segmentation and presence of chaetae, are not found in a number of its constituents, so below, while we discuss major characters and developments in annelid biology and systematics, we caution the reader on the many exceptions for all of these.

Despite their chaotic systematics, Annelida includes about 23,000 valid species (G. W. Rouse, pers. comm.) of ubiquitous invertebrates that have been able to adapt to all sorts of marine (benthic and pelagic), limnic, and terrestrial environments (fig. 47.1). In the ocean they are well known for their numerous adaptations to the meiobenthos (many of the so-called archiannelids now constituting three main clades), and to truly extreme environments. These include the tube worms (pogonophorans and vestimentiferans) that inhabit hydrothermal vents and hydrocarbon seeps, or their closest relatives and more recently discovered extremophiles dubbed "bone devouring worms," or *Osedax* (Rouse et al., 2004), now known from whale falls in oceans around the globe. These lifestyles are possible due to evolution of intimate symbioses with other organisms (see below).

Clitellate annelids are one of the conquerors of limnic and terrestrial environments, with ca. 9,000 species inhabiting the continental landmasses. In the oceans, annelids can have sedentary or errant lifestyles in benthic or (less commonly) pelagic environments, some species spending their entire life in the water column, a condition that has evolved multiple times even within small clades of annelids (Osborn and Rouse, 2008). Many of these have also developed bioluminescence (Verdes and Gruber, 2017). Some sedentary species secrete tubes, which can be

---

[21] As with Priapulida, we prefer the vernacular term "sipunculan" over "sipunculid" to distinguish between species belonging to Sipuncula from those belonging to Sipunculidae.

organic or inorganic (e.g., of calcium carbonate). They are also found from the intertidal to the deepest parts of the oceans, with many species inhabiting depths greater than 6,000 m (Kirkegaard, 1956; Wolff, 1960; Saiz et al., 2016). Attaching to black smokers, the Pompeii worm *Alvinella pompeiana* has been found to thrive at sustained temperatures between 45 to 60°C and possibly up to 105°C for short periods (Grime and Pierce, 2012), making this species the most heat-tolerant animal known after particular extremophilic tardigrades. Annelids also include other extremophiles, including the hydrothermal tube worms (Siboglinidae), *Riftia pachyptila*, also one of the fastest-growing invertebrates (Lutz et al., 1994), due to their intimate symbiosis with chemoautotrophic bacteria (Cavanaugh et al., 1981).

Feeding ecology is extremely varied and includes carnivory (predation as well as scavenging), filter feeding, deposit feeding, symbiotic relationships with chemoautotrophic and heterotrophic bacteria (sometimes with a reduction of the gut), and parasitism on invertebrates and vertebrates. Many of these feeding modes have obvious consequences on the worm's ground plan, growth rate, and so on. Suspension feeding is typical of tube-dwelling annelids, most familiar of which are fan worms and feather-duster worms (Sabellidae and Serpulidae), which trap particles using currents generated by a crown of branching tentacles called "radioles."

Some earthworms, like the Mekong giant earthworm (*Amynthas mekongianus*) from Laos, can reach up to 2.9 m in length (Grime and Pierce, 2012), similar to that of the largest "polychaete," the infamous bobbit worm (*Eunice aphroditois*)—although sometimes it is reported to be up to 6 m long (Glasby et al., 2000). A length of two to three meters is also reported for a few marine species, including *Glycera dibranchiata*, used as bait in Canada, and many of the beach worm species of *Australonuphis*, occurring on surf beaches on the east coast of Australia. The smallest annelids are meiofaunal forms, smaller than 1 mm in length. Segments, when present, can range from six or fewer to over a thousand (Glasby et al., 2000).

Many terrestrial earthworms have important roles for soil ecology, including two nearly ubiquitous, globally distributed earthworms, *Lumbricus terrestris* and *Aporrectodea caliginosa*, which can be abundant and often compete for habitat. Their activity is beneficial because it enhances soil nutrient cycling through the rapid incorporation of detritus into mineral soils and because their secretions enhance the activity of other beneficial soil microorganisms (Bhadauria and Saxena, 2010)

Medicinal leeches have been in use since 200 BC and are still used, although for quite different purposes, mostly related to avoiding coagulation in severed fingers and other surgical procedures, due to their anticoagulant properties. Annelid venoms are also a growing area of applied research and inquiry (von Reumont et al., 2014a), venom glands often being associated with the buccal apparatus of predatory species, as is the case of the bloodworms in the genus *Glycera* (von Reumont et al., 2014b). An array of toxins is also known from transcriptomic approaches in fireworms (Amphinomidae) (Verdes et al., 2018), despite the lack of known venom glands and the recent confirmation that their calcareous chaetae do not act as needles for injecting venom (Tilic et al., 2017).

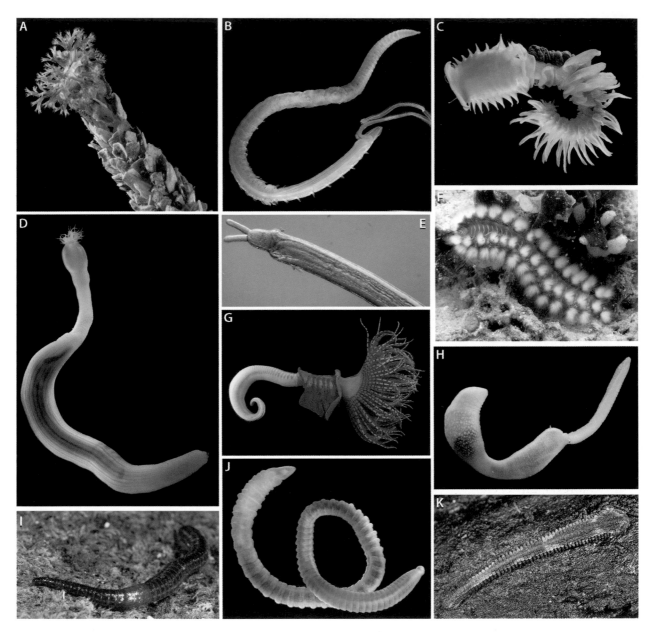

Figure 47.1. Diversity of annelids. A, *Owenia* sp. (Oweniidae); B, *Magelona johnstoni* (Magelonidae); C, *Chaetopterus* sp. (Chaetopteridae); D, the sipunculan *Siphonosoma cumanense* (Sipunculidae); E, the unsegmented *Polygordius* sp. (Polygordiidae); F, the fireworm *Haermodice carunculata* (Amphinomidae); G, *Protula* sp. (Serpulidae); H, the echiuran *Thalassema* sp. (Thalassematidae); I, the earthworm *Megascolides maoricus* (Megascolecidae); J, the developmental model *Capitella teleta* (Capitellidae); K, the leech *Mesobdella gemmata*, one of the few terrestrial leeches in South America. Photo credits: A, J, Fred Pleijel; B, © Hans Hillewaert, CC BY-SA 4.0, https://commons.wikimedia.org/w/index.php?curid=395854; E, Katrine Worsaae; F, Juan Moles.

SYSTEMATICS

Membership in the phylum Annelida has been a controversial matter, but in recent years, with the advent of molecular phylogenetics, annelids have incrementally incorporated animal groups that were once considered their own phyla, including, more or less chronologically, Echiura, Pogonophora, and Vestimentifera (McHugh, 1997), as well as Sipuncula (Struck et al., 2007). Echiura and Sipuncula have long been treated as separate animal phyla, but the presence of chaetae in echiurans had suggested an annelid affinity, while this was not the case for sipunculans. The development of the CNS of echiurans shows remnants of segmentation (Hessling, 2002; Hessling and Westheide, 2002), as does that of sipunculans (Kristof et al., 2008; Kristof and Maiorova, 2016), even though as adults the latter present a single non-ganglionated ventral nerve cord. Pogonophorans were once considered members of their own phylum of deuterostome animals (Ivanov, 1955), a position that was based upon misinterpretation of the anatomy of incomplete animals, but subsequent work kept them with the deuterostomes (Hyman, 1959). Likewise, Vestimentifera were elevated to phylum rank by Jones (1985). Since then, these animals were treated in most textbooks as their own phylum of protostomes for about two decades.

Likewise, since *Diurodrilus* may lack segmentation and other typical annelid characters, it was once proposed to be a member of the clade Spiralia with uncertain affinities (Worsaae and Rouse, 2008), but recent phylogenomic analyses have shown that it nests within annelids (Laumer et al., 2015a). Another mysterious group, Lobatocerebridae, has been considered as an annelid by most authors (Rieger, 1980), but despite their similarities to other meiofaunal annelids, such as Dinophilidae and Protodrilidae, having been pointed out, they also attracted comparisons with Gnathostomulida and catenulid Platyhelminthes (Kerbl et al., 2015). Ultimately, however, phylogenomic analyses confirmed their annelid affinities (Laumer et al., 2015a). In another case, the membership of Myzostomida in Annelida has long been disputed, but multiple sources of molecular data now support an affinity to segmented worms (Bleidorn et al., 2007; Bleidorn et al., 2009; Helm et al., 2012; Andrade et al., 2015).[22] Finally, other groups, including Orthonectida, have been suggested to be highly modified annelids (Schiffer et al., 2018), but this is not universally endorsed (Mikhailov et al., 2016; Lu et al., 2017) and may require further study.

In addition to the fluid membership to the phylum Annelida, internal systematics has been a matter of heated debate and quite frustrating to resolve using morphology or PCR-based molecular approaches. Traditional annelid classifications, originally derived from de Quatrefages (1865), divided "polychaetes" into two orders, Errantia and Sedentaria (e.g., Fauvel, 1923, 1927), a system that was later abandoned. Cladistic analysis underpinned an influential morphology-based classification (Rouse and Fauchald, 1997) in which the polychaetes were divided into three main taxa, Scolecida, Canalipalpata, and Aciculata.

Canalipalpata and Aciculata share the presence of cephalic palps, and on that basis were grouped as Palpata. Aciculata, with numerous synapomorphies, has

---

[22] The history of the placement of Myzostomida has been elegantly reviewed by Bleidorn et al. (2014).

largely withstood molecular phylogenetic analyses, whereas Canalipalpata was united based on only a few apomorphic characters and has consistently been resolved as non-monophyletic in molecular studies (e.g., Struck et al., 2011; Kvist and Siddall, 2013; Weigert et al., 2014; Andrade et al., 2015; Laumer et al., 2015a; Helm et al., 2018), as recently summarized by Weigert and Bleidorn (2016). Scolecida were united mostly by absence characters, and the monophyly of the group is doubtful. Although the Rouse and Fauchald (1997) classification recognized a formal taxon Polychaeta in opposition to Clitellata (earthworms and leeches), molecular data have consistently and convincingly placed Clitellata within a grade of polychaetes, and Polychaeta was abandoned. We thus refer to "polychaetes" or "nonclitellate annelids" in an informal sense to acknowledge that the constituent members are a paraphyletic group.

Phylogenomic analyses form the basis for current understanding of annelid phylogeny and systematics (fig. 47.2). A recurring theme has been the monophyly of Pleistoannelida, a clade proposed by Struck (2011) to group Errantia and Sedentaria (the two major groups of de Quatrefages, somewhat redefined), and probably Myzostomida (Andrade et al., 2015). Nearly all analyses coincide in the presence of a grade of taxa excluded from Pleistoannelida that includes Palaeoannelida (Oweniidae + Magelonidae), Chaetopteridae, and Amphinomida + Sipuncula. More recently, two other taxa have been united with Chaetopterida, forming the clade Chaetopteriformia: Apistobranchidae and Psammodrilidae (Helm et al., 2018). The internal relationships of Errantia and Sedentaria remain less understood.

Clitellates constitute more than one-third of the described species of annelids, having diversified enormously in the continental environments, and also include numerous species of medical importance, most prominently the medicinal leeches (Hirudinida). In spite of their ubiquity, the long quest for the clitellate sister group (e.g., Purschke et al., 2000; Purschke, 2003; Purschke et al., 2014) still defies annelid systematists. Finally, the highly modified siboglinids (including pogonophorans, vestimentiferans, and *Osedax*), the sponge symbiont *Spinther*, the echinoderm (and anthozoan) parasites of the clade Myzostomida, and the meiofaunal Dinophilidae have been particularly difficult to place phylogenetically (Andrade et al., 2015).

For Sipuncula, we follow the revised system proposed by Kawauchi et al. (2012) and refined by the phylogenomic analysis of Lemer et al. (2015). The traditional taxonomy of Echiura (now the family Echiuridae) has not been supported by recent phylogenetic analyses based on molecular data (Goto et al., 2013), wherein the former orders Xenopneusta and Heteromyota were nested within the order Echiuroinea, and thus all higher structure of this clade is abandoned for the time being. Instead, that study supported two main groups distinguished by the presence/absence of marked sexual dimorphism involving dwarf males and the paired/nonpaired configuration of the gonoducts (genital sacs) (Goto et al., 2013).

A revised classification system, based largely on recent phylogenomic analyses (Struck et al., 2011; Kvist and Siddall, 2013; Weigert et al., 2014; Andrade et al., 2015; Laumer et al., 2015a; Helm et al., 2018), was recently summarized by Weigert and Bleidorn (2016) and Rouse (2016) and is currently followed by many workers. We follow the latter (and the number of species listed) here, while recognizing the higher clades Palaeoannelida and Pleistoannelida, proposed by Struck (2011) and Weigert and Bleidorn (2016), respectively:

Palaeoannelida
 Oweniidae (< 50 spp.)
 Magelonidae (70 spp.)
Chaetopteriformia
 Apistobranchidae (6 spp.)
 Psammodrilidae (8 spp.)
 Chaetopteridae (70 spp.)
Unnamed clade
 Amphinomida (> 200 spp.)
 Sipuncula (150 spp.)
  Sipunculidae
  Golfingiidae
  Siphonosomatidae
  Antillesomatidae
  Phascolosomatidae
  Aspidosiphonidae
Pleistoannelida
 Errantia
  Protodrilida (> 60 spp.)
  Aciculata
   Eunicida (> 1,000 spp.)
   Phyllodocida (> 4,600 spp.)
 Sedentaria
  Orbiniidae (175 spp.)
  Cirratuliformia (560 spp.)
  Opheliidae (153 spp.)
  Spionida (650 spp.)
  Sabellida (860 spp.)
  Siboglinidae (164 spp.)
  Spintheridae (12 spp.)
  Dinophilidae (16 spp.)
  Maldanomorpha (303 spp.)
  Terebelliformia (870 spp.)
  Capitellida (370 spp.)
   Capitellidae
   Echiuridae

Clitellata (ca. 6,000 spp.)
    Capilloventridae
    Naididae
    Crassiclitellata
    Enchytraeidae
    Lumbriculidae
    Hirudinoidea
        Acanthobdellida
        Branchiobdellida
        Hirudinida
Incertae sedis: Myzostomida (156 spp.)

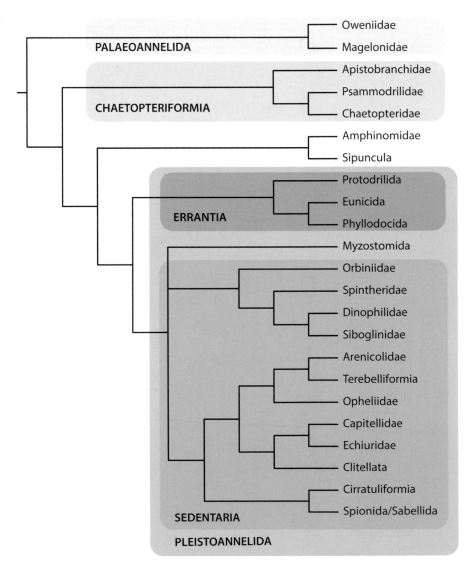

Figure 47.2. Proposed phylogeny of Annelida based on published phylogenomic analyses.

PALAEOANNELIDA

Oweniidae
Magelonidae

CHAETOPTERIFORMIA

Apistobranchidae
Psammodrilidae
Chaetopteridae

Amphinomidae
Sipuncula

ERRANTIA

Protodrilida
Eunicida
Phyllodocida

Myzostomida

Orbiniidae
Spintheridae
Dinophilidae
Siboglinidae

Arenicolidae
Terebelliformia
Opheliidae
Capitellidae
Echiuridae
Clitellata
Cirratuliformia
Spionida/Sabellida

SEDENTARIA

PLEISTOANNELIDA

▬ ANNELIDA: A SYNOPSIS

- Bilaterally symmetrical, mostly coelomate or exceptionally acoelomate worms
- Segmented (some lineages are unsegmented, including Sipuncula, Echiura, *Lobatocerebrum*, *Diurodrilus*, *Polygordius*); segments are added caudally (except in Myzostomida)
- Cleavage typically spiral, with a trochophore larva (lost in many lineages, including Clitellata)
- Complete through-gut, lost in some extremophiles; with a U-shaped gut in Sipuncula
- Closed circulatory system (absent in many meiofaunal groups) and a diversity of respiratory pigments
- Protonephridia (in the trochophore larva and in the adult of some species) or more typically with metanephridia in the adults
- Typically, with a well-developed nervous system with a dorsal cerebral ganglion, circumenteric connective, and a paired ganglionated ventral nerve cord (modified in many lineages)
- Head typically composed of a presegmental prostomium and peristomium (only in the segmented species)
- Commonly with head appendages (e.g., palps and antennae), and segmental parapodia
- With lateral, segmentally arranged epidermal chaetae (absent in Sipuncula and *Lobatocerebrum* and in a few other taxa)
- With a diversity of feeding ecologies
- Gonochoristic (many marine species) or hermaphroditic (clitellates); development indirect with trochophore larva, or direct (in clitellates)

The complexity of annelid body plan disparity requires discussion of so much variation for each anatomical, ecological, behavioral, and physiological trait that it would take much more space than it is allocated in this book—and much more collective knowledge. We encourage the reader to seek additional sources, a favorite of ours being Rouse (2016), together with recent state-of-the-art reviews of development (Bleidorn et al., 2015) and the nervous system and associated structures (Helm and Bleidorn, 2016; Kristof and Maiorova, 2016; Purschke, 2016). Below we summarize some of the most important aspects of the anatomy and biology of annelids.

In the majority of annelids, the body is distinctly segmented and subdivided into four regions (fig. 44.1). The presegmental prostomium derives from the anterior part of the trochophore larva (the episphere). The mouth is situated in the first segment, the peristomium, which derives from the prototroch ciliary band and buccal region of the larva. The prostomium and peristomium comprise the usual annelid head, but one or more body segments may also be fused to the head. Body segments are serially repeated, with segmentation variably being homonomous or heteronomous. The latter is observed in most tube-dwelling forms (fig. 47.1 C, G). The terminal part of the body is the pygidium. Like the prostomium, the pygidium has been classically viewed as nonsegmental, but this is no longer the case (see the discussion below). Obviously, segmentation (or its absence) plays a major role in understanding many of the features of annelids.

In the overwhelming majority of annelids, a coelom is developed from the larval metatrochophore stage onward (for a review, see Koch et al., 2014). An ancestral state can be reconstructed for the annelid coelom as segmentally arranged paired cavities separated by median mesenteries and anterior and posterior septa, but the latter are lacking in either particular regions or throughout the body in several subgroups (including Echiura and Sipuncula). Coelomic fluid is moved by cilia or by body wall contraction. A peritoneal lining of the coelomic cavities is observed to be incomplete or even absent in polychaetes (Koch et al., 2014). The coelom is more commonly lined by myoepithelia, and a myoepithelium precedes any other kind of lining in the course of development. The mode of coelomogenesis differs between pre- and postmetamorphic stages, with strictly segmental coelomic cavities being confined to the latter. This difference has been noted to correspond to differences in Hox gene expression (Novikova et al., 2013).

Sipunculans typically have two coeloms, a spacious trunk coelom and an additional tentacular coelom, and possess, in addition to amoebocytes and erythrocytes, the characteristic multicellular ciliated urns (Schmidt-Rhaesa, 2007). These cells appear to be phagocytic in function. The coelom of echiurans is similar to the large coelom of sipunculans and also possess urns. *Lobatocerebrum* is acoelomate (Rieger, 1980), as are many other interstitial annelids (Rieger and Purschke, 2005). Myzostomes have a coelom reduced mostly to branched spaces in which the gametes mature. In leeches, having a solid body that makes it almost impossible to squeeze, the coelom is reduced to small channels.

Nephridia are predominantly metanephridia in adult annelids, though they are frequently modified from protonephridia in larvae. The first nephridial structures to form in most annelids are a pair of protonephridia, known as "head kidneys," anteriorly in the trochophore. Typically, the nephridia are serially organized as one pair per segment—although the nephridiopore may lie on the segment behind the nephrostome, especially in Clitellata—but reductions from this pattern can be as extreme as in several groups of Sedentaria, in which only a single large pair of metanephridia is present. A comprehensive account of nephridial ultrastructure and nephriogenesis is provided by Koch et al. (2014). Various meiofaunal lineages, as well as the parasitic myzostomes, have segmental protonephridia, but segmental metanephridia are more common and are typical of macrofaunal species. Which of the two kinds of nephridia form is dependent on the structure of the proximal cells of the nephridial anlage, whether or not they form a ciliated funnel (and develop as metanephridia) or develop as a protonephridium with solenocytes (Bartolomaeus and Quast, 2005). Metanephridia also function as gonoducts for the release of gametes that are temporarily housed in the coelomic cavity.

While not all annelids are segmented, all segmented spiralians are annelids, the only exceptions being the segmented nemertean *Annulonemertes minusculus* and tapeworms. Segments are normally delimited by body rings or annuli externally and by the subdivision of the coelom by intersegmental septa internally. Segments are generated by a posterior growth zone anterior to the pygidium, which contains the anus. Contrary to most thinking that the pygidium is postsegmental, the model species *Platynereis dumerilii* exhibits complex neural and muscular features, as well as a spacious coelom-like cavity, that are suggestive of a segmental nature (Starunov et al., 2015).

The cellular basis for segment formation is especially well characterized in the large embryos of direct-developing clitellates. In leeches, segmentation is linked to the subdivision of cells known as "teloblasts," stem cells that are responsible for sequential segment addition. These give rise to daughter cells, called "blast cells," via a cycle of unequal divisions. However, the pattern of segment formation is highly variable in annelids (Balavoine, 2014) and does not even follow an anterior–posterior axis in myzostomes, at least for the segmentally associated external structures (Kato, 1952). The process of segmentation is well understood in *Capitella teleta* (fig. 47.1 J), in which an initial group of segments appears simultaneously and then the rest are added sequentially from a growth zone (Irvine and Seaver, 2006).

Parapodia are a typical component of annelid segmentation but are absent in clitellates, and various lineages within Sedentaria, and of course in the nonsegmented sipunculans and echiurans. The parapodia are most often biramous, consisting of a dorsal notopodium and a ventral neuropodium. Each ramus usually bears a cluster of chaetae. The structure and function of parapodia can be highly variable, and those of different body regions may be specialized for particular functions, for example, respiration or locomotion. Especially elaborate regionalization of the body, including the parapodia, is seen in *Chaetopterus* (fig. 47.1 C), which filter-feeds in U-shaped burrows. A series of small anterior parapodia are followed by three hypertrophied notopodial fans on segments 14 to 16, used to generate a flow of water through the tube.

Chaetae are classified into varied types based on their structure (reviewed by Merz and Woodin, 2006). They may be similar or different on the noto- and neuropodium. Capillary chaetae, which are tapering and unjointed, are widely distributed across polychaetes, whereas compound (jointed) chaetae have a more restricted systematic distribution. The distal part of a chaeta may have a single hook (e.g., falcate chaetae) or a dentate hook, or the chaeta may form a series of teeth (uncini). Chaetae internalized within the parapodia, known as "aciculae," serve as a "skeletal" support for the parapodium. The uniramous parapodia of myzostomes, being equipped with aciculae, provide one of the arguments for their identity as annelids. Chaetae are chitinous, although in the families Amphinomidae and Euphrosinidae they are impregnated with calcium carbonate. In Hirudinoidea (leeches and their relatives; fig. 47.1 K), chaetae are confined to a few anterior segments in one subgroup, Acanthobdellida.

Chaetae in annelids arise from chaetoblasts, the apical microvilli of single cells. These chaetoblasts are virtually indistinguishable from those of brachiopods (Lüter, 2000b), which likewise form in ectodermal invaginations (chaetal follicles). The definitive form of a chaeta is influenced by pulses of microvilli formation and cellular-level interactions between the chaetoblast and the adjoining follicle cells (Tilic and Bartolomaeus, 2016). Interestingly, Notch, a key signaling pathway involved in neurogenesis in other bilaterians, plays no role in neurogenesis in the polychaete *Platynereis dumerilii* but instead is co-opted for the process of chaetogenesis (Gazave et al., 2017).

The circulatory system can include a middorsal and midventral longitudinal vessel for the anterior and posterior transport of the blood, respectively, supplemented by segmental vessels that supply the parapodia, nephridia, gut, and body wall muscles. As is so often the case in Annelida, examples of drastic reduction of the circulatory organs occur, and are not always restricted, to lineages of small body size. Specialized pumping structures ("hearts") are commonly lacking but can be prominent in tube-dwelling forms.

The most common respiratory pigment is hemoglobin, although other pigments occur in particular groups. For example, hemerythrins, previously known only in Magelonidae and in Sipuncula, have been found to be quite widespread taxonomically (Costa-Paiva et al., 2017).

Like many organ systems, the musculature of annelids is variable. It has been inferred that an outer layer of circular muscle and several (four?) longitudinal muscle bands are ancestral for annelids (Tzetlin and Filippova, 2005), typically with prominent dorsal and ventral longitudinal bands. However, circular musculature is lacking in some groups, in which case other muscle groups (e.g., oblique, parapodial, and diagonal or dorsoventral muscles) may instead be present (Purschke and Müller, 2006). Clitellates are unique among Annelida in possessing a complete layer of longitudinal muscle rather than having it organized in bands. Annelid muscle fibers are generally double obliquely striated. In sipunculans, in addition to the body wall musculature there are prominent head/introvert retractor muscles, a set of four being the primitive state within the group (Schulze and Rice, 2009). The circular and sometimes the longitudinal body wall musculature are split into bands that in the species with smooth body musculature transform into a smooth sheath through development.

Muscle arms (sarconeural junctions or muscle tails), as typically known from nematodes but also reported in other invertebrates (Chien and Koopowitz, 1972) and reaching to the central nervous system, have been described in some annelids, including syllids (Wissocq, 1970) and lumbricid earthworms (Mill and Knapp, 1970).

Annelids provide a typical example of animals that combine coelom-based and muscular-based movement. During burrowing in nereids, the hydrostatic skeleton and body wall musculature produce the peristaltic movement, and the parapodia and chaetae the anchoring mechanism, for advancing in one direction. When the longitudinal musculature of a segment contracts, the segment becomes wider, and vice versa. Swimming is also possible through metachronal movement, similar to that used for crawling. Many interstitial annelids, however, lack coeloms and instead have bodies completely covered with cilia, therefore moving mostly by ciliary movement. As for leeches, these lack large coelomic spaces and thus use anterior and posterior suckers, which ordinarily serve to attach to prey, as anchoring mechanisms for crawling, done by attaching the anterior sucker, detaching the posterior sucker and moving it near to the anterior one, then detaching the anterior sucker and stretching the body to advance. Swimming is also possible by the leeches quickly undulating their bodies.

The gut is differentiated into a foregut, midgut, and hindgut; the foregut and hindgut are ectodermal, and the midgut is endodermal. The foregut includes the pharynx, and in some cases this is everted as a proboscis. Jaws, most characteristic of aciculate polychaetes, are derived from the cuticular lining of the foregut. While jaw replacement has been described in annelids (Paxton, 2005), and this mechanism has been equated to molting in ecdysozoans, it need be noted that this property is specific to Onuphidae rather than being a general feature of Annelida as a whole. Clitellates (notably earthworms) may have the posterior part of the foregut elaborated into a crop and gizzard. The midgut is dominated by the intestine, which may be a simple tube or be variably folded or have outgrowths (caeca). Extracellular digestion occurs mostly in the midgut lumen. In leeches the midgut is enlarged and associated with prominent caeca. Intestinal caeca are also abundant in *Aphrodita*. In sipunculans, the digestive tract is U-shaped, the mouth located at the tip of the introvert and the anus located dorsally and anteriorly. The gut is, furthermore, highly coiled around the ascending intestine, the bottom of the intestine being attached to the body wall by a spindle muscle. This digestive system is unique among all animals, even when compared to Cambrian fossil sipunculans (see below). Coiled but not U-shaped intestines can be common in annelids and other animals.

The regionalization of the gut into its ectodermal and endodermal parts has been studied from the perspective of gene expression in a few annelids. The fore-, mid- and hindgut depict some conserved expression patterns for the transcription factor FoxA and genes of the GATA family in *Capitella teleta* (see Boyle and Seaver, 2008), *Chaetopterus variopedatus*, and the sipunculan *Themiste lageniformis* (see Boyle and Seaver, 2010), despite differences in developmental and feeding modes.

Various annelid lineages have independently established symbioses with numerous types of bacteria, the most renowned of these being that of the giant tube worms around hydrothermal vents (Cavanaugh et al., 1981; Cavanaugh, 1983). Siboglinids lack a functional digestive tract and instead derive their energy from symbiotic chemoautotrophic gamma-Proteobacteria stored in an organ known as the "trophosome." In the case of *Osedax*, the trophosome houses heterotrophic (instead of chemoautotrophic) gamma-Proteobacteria (Rouse, 2016).

These symbiotic associations range from a few genera within largely aposymbiotic groups in Tubificidae or Glossiphoniidae to clades in which all species live in symbiosis, such as in Siboglinidae and Alvinellidae (Bright, 2005). The ties between partners range from rather loose and occasional associations (as in some mud-dwelling tubificids) to regular ectosymbiosis (alvinellids feed on the vent bacteria that grow on their external surfaces) to obligatory extra- or intracellular (gutless oligochaetes, glossiphoniid leeches, siboglinid tube worms). Bacterial associations span a range from beaches to the deep sea, from rotting wood at the sea bottom to freshwater ponds. In the sea, chemoautotrophy in sulfidic environments is common, but trophic specializations such as digestion of wood or blood also drive mutually beneficial associations with bacteria (Bright, 2005). There are 80 species of gutless marine clitellates in the family Naididae inhabiting anaerobic, sulfide-rich coral-sand sediments, which also host subcuticular symbiotic bacteria.

Leeches are well known for their blood-feeding habit, which nonetheless has been lost multiple times in the evolution of the group (Siddall et al., 2011). The saliva released when a leech bites contains active ingredients with anti-inflammatory, thrombolytic, and blood- and lymph-circulation enhancing properties, and although a specific analgesic substance is yet to be identified (Koeppen et al., 2014), leech bites experienced by humans often go unnoticed. Because of this property, leeches are currently being used to treat conditions in which pain is a major symptom. The salivary glands also produce a series of anticoagulant proteins (Kvist et al., 2017).

NERVOUS SYSTEM

A comprehensive account of the annelid nervous system (Purschke, 2016) is complemented by descriptions of the highly autapomorphic nervous systems of Sipuncula (Kristof and Maiorova, 2016) and Myzostomida (Helm and Bleidorn, 2016), and neurogenesis has been explored in detail in *Platynereis dumerilii* (Starunov et al., 2017). The traditional view of the ventral nerve cord being "rope ladder" in structure, with paired segmental ganglia joined by connectives and commissures, was challenged by Müller (2006), who documented considerable deviation from that pattern; indeed, a "rope ladder" nerve cord is in fact present in few annelids (fig. 47.3). In many annelids the ventral nerve cord has an unpaired appearance and the ganglia are not very distinct, or as many as four longitudinal nerve cords may be developed (as in amphinomids). The number of connectives per segment varies from as many as five to as few as one, the latter generally being seen as a result of fusion. When ganglia are present, lateral nerves extending from them commonly bear a pedal ganglion that in turn is the source of nerves into the parapodia. The nervous system is variably intraepidermal/basiepidermal or subepidermal. Debate surrounds the plesiomorphic state for the phylum as a whole, although the basiepidermal condition in palaeoannelids such as Megalonidae and Oweniidae (and not just small "archiannelids") increases the likelihood that this is ancestral for Annelida (Purschke, 2016). The central nervous system is supplemented by a stomatogastric nervous system that serves the pharynx and anterior part of the gut.

The annelid brain is primitively associated with the prostomium, but it is displaced posteriorly in clitellates (as far back as the third body segment) as well as in most of the non-clitellates that likewise have a reduced prostomium. In the basal grade of Annelida, the brain is not clearly divided into distinct regions, as is likewise the case in the relatively simple brains of Sedentaria. In errant annelids, in contrast, the brain is regionalized into units designated as fore-, mid-, and hindbrain. These regions typically innervate the palps, the eyes and antennae, and the nuchal organs, respectively. In myzostomes the brain is a bilobed neuropil located in the muscular sucking pharynx (Helm and Bleidorn, 2016), and the pharynx is innervated in a similar manner to Errantia that likewise have a muscular pharynx. In some burrowing forms that lack head appendages, notably Echiura and Oweniidae, the brain assumes the form of a single dorsal "medullary" commissure, probably a derived state (Rimskaya-Korsakova et al., 2016). The brain is most conspicuously segmental in leeches, being a complex of six ganglia, associated with segmentation of the subesophageal ganglion and a ventral nerve cord that is likewise markedly "rope ladder" in its organization (Purschke et al., 1993).

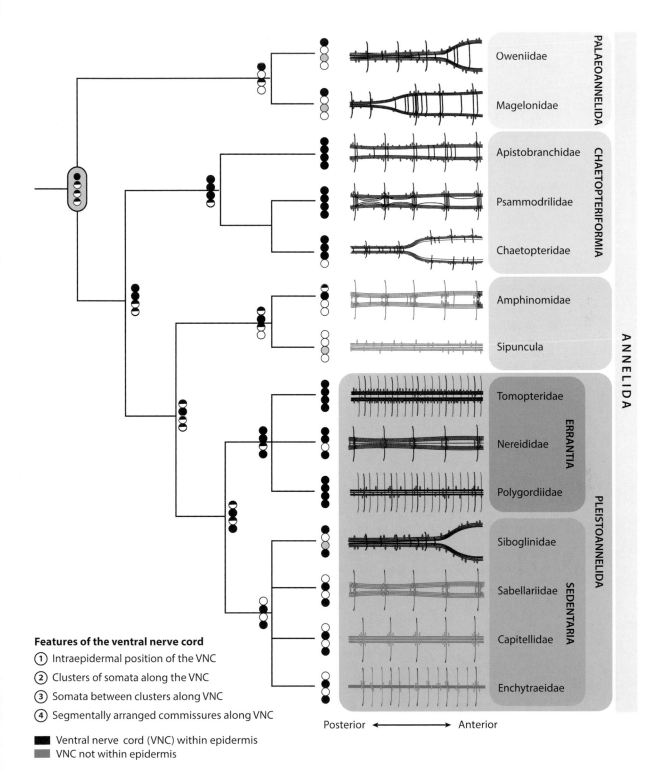

Features of the ventral nerve cord

① Intraepidermal position of the VNC

② Clusters of somata along the VNC

③ Somata between clusters along VNC

④ Segmentally arranged commissures along VNC

▉ Ventral nerve cord (VNC) within epidermis

▉ VNC not within epidermis

Posterior ⟵⟶ Anterior

FIGURE 47.3. Variation in structure of the ventral nerve cord in Annelida mapped on a phylogenetic tree and with parsimonious optimization of selected characters of the CNS (for more characters and additional details see original source in Helm et al., 2018). Character reconstruction as follows: absent (white circle), present (black circle), inapplicable (gray circle) and equivocal reconstruction (black/white circle). The four circles are ordered 1 to 4 from top to bottom.

Mushroom bodies ("corpora pedunculata") have a scattered systematic distribution in annelids (reviewed by Heuer et al., 2010). While best developed within Errantia, they are known from Sipuncula and some other "basal" lineages.

A peculiarity of the annelid ventral nerve cord is the so-called giant fiber system (reviewed by Purschke, 2016). This involves fibers an order of magnitude thicker than regular axons, and in some cases they are even among the largest nerves known from any animal. They mediate such rapid escape responses as contraction of longitudinal muscle and parapodial movements. Giant fibers have a broad systematic distribution in annelids, including many families of polychaetes as well as oligochaetes, though they are absent in early-derived annelid lineages such as oweniids and chaetopterids.

Annelids include a broad range of sensory structures. Chemoreceptive cells are concentrated on the head appendages (the palps, antennae, and tentacular cirri). At least in the case of the palps, a common pattern of innervation underpins a hypothesis of homology (Orrhage and Müller, 2005). Ciliated bands or pits situated between the two rami of the parapodia, most common in Sedentaria, are known as "lateral organs" (reviewed by Purschke, 2016), although their function is not well understood. Nuchal organs are found in most non-clitellate annelids, evidently playing a role in chemosensation. They have a conserved position at the posterior edge of the prostomium and are usually paired and densely ciliated (Purschke, 1997). Nuchal organs range from simple slits to an outgrowth of the prostomium, the caruncle of Amphinomidae, or the sipunculan nuchal–cerebral organ complex (Purschke et al., 1997).

Photoreceptor cells are found in most annelids (Purschke et al., 2006). Those that allow discrimination of the direction of light are most commonly pigment-cup ocelli. The familiar evo-devo model *Platynereis dumerilii*, for example, has a quite typical arrangement of paired pigment-cup eyes on each side of its head. These have the usual protostome rhabdomeric photoreceptors and r-opsins. The cerebral eyes of *P. dumerilii*, however, have ciliary photoreceptors and express c-opsin (Arendt et al., 2004). The cerebral eyes of errant polychaetes can have lenses, and they have some capacity for image formation. Within Sedentaria, feather-duster worms (Sabellidae) include an example of relatively recent evolution of compound eyes within Annelida (Nilsson, 1994; Bok et al., 2016). These tube-dwelling worms may have up to four pairs of cerebral ocelli, each with a single photoreceptor and few pigment cells, as well as segmental ocelli on the thorax and abdomen and sometimes on the pygidium. These ocelli are complemented by eyes on the crown of radioles—these radioles being the structures sabellids use both for filter-feeding and for respiration. In some sabellids, ocelli or single or paired compound eyes are scattered along all radioles, whereas in others, such as *Megalomma*, small compound eyes are at the tips of all radioles; in other species, two radioles are modified to accommodate enlarged compound eyes at their tips. These can have hundreds of facets.

Radiolar eyes are of additional interest because, as noted above for *Platynereis*, they provide an example within Protostomia of ciliary input to the sensory membrane of the photoreceptors, an exception to the general dichotomy of rhabdomeric versus ciliary photoreception in protostomes and deuterostomes, respectively. The phototransduction cascade of a species of *Megalomma* corroborates a ciliary contribution to radiolar eyes in that the opsins are most closely related to those associated with vertebrate photoreceptors (Bok et al., 2017). Radiolar eyes, also including compound eyes, are present elsewhere in Sabellida within the family Serpulidae.

Photoreception in Annelida also involves structures known as "phaeosomes," vacuoles into which a photoreceptive process projects. These are the typical light-receptive organs in clitellates, in which they are microvillar rather than ciliary (Döring et al., 2013). On a related topic, bioluminescence has evolved multiple times in annelids, appearing in at least 13 lineages (Verdes and Gruber, 2017).

REPRODUCTION AND DEVELOPMENT

Most annelids have a trochophore larva (exceptions include direct developing clitellates). The prototroch encircles the larva just in front of the mouth, dividing the trochophore into the episphere (anteriorly) and hyposphere (posteriorly) (fig. 44.1). The prototroch is composed of compound cilia and originates from trochoblasts of the first two quartets of micromeres. An apical ciliary tuft is usually present as well. Additional ciliary bands are variably present in different annelids (reviewed by Rouse, 1999b; and Bleidorn et al., 2015). These include a perianal telotroch (absent in many planktotrophic larvae), a midventral neurotroch, and a metatroch posterior to the mouth (absent in lecithotrophic larvae). The telotroch delimits the posterior growth zone. Specialized kinds of trochophore larvae occur in particular groups, such as the bell-shaped mitraria of Oweniidae, in which the ciliary bands are monociliated and the larva undergoes an abrupt and drastic change in body shape at metamorphosis (Smart and Von Dassow, 2009). In spite of this, however, the precursors of the adult nervous system are identifiable in the mitraria (Helm et al., 2016). The nervous system of trochophores typically includes a prototroch nerve, a metatroch nerve, a telotroch nerve, and an apical organ (Purschke, 2016).

Annelid reproduction and development have recently been reviewed by Bleidorn et al. (2015). As in every other aspect, they show a variety of reproductive strategies, including sexual as well as a diversity of asexual reproduction modes and regeneration (Kostyuchenko et al., 2016). For sexual reproduction, different types of free spawning, brooding, and encapsulation of embryos in cocoons can be distinguished, and all types involve either planktotrophic or lecithotrophic developmental stages (Wilson, 1991). In general, species with planktotrophic versus lecithotrophic development can be distinguished by the latter having larger, yolkier eggs (Irvine and Seaver, 2006). Poecilogeny (the presence of multiple modes of development), with lecitothrophic and planktotrophic larvae occurring in the same species, has been reported for some annelid species (Levin, 1984). Many limnoterrestrial species are direct developers.

Syllids constitute an especially striking group in terms of reproductive diversity. The primitive mode of sexual reproduction is characterized by epitoky (as epigamy and stolonization, respectively), swarming, and external fertilization. Stolonization is the process by which a stolon—a structure resembling the adult but containing only gametes—forms, generally at the posterior of the animal, but stolons can also form on the parapodia, as in *Ramysillis,* or on other parts of the body (Aguado et al., 2015). In meiofaunal species many other modes have evolved, including external brooding, direct sperm transfer, internal fertilization, viviparity, parthenogenesis, and simultaneous as well as successive hermaphroditism (Franke, 1999). Studies of stolonization using comparative transcriptomics and gene expression data show a hormonal control (Aguado et al., 2015; Álvarez-Campos et al., 2019). During the breeding season, as the stolons mature they detach from the adult and the gametes are released into the water column. The process is synchronous for each species and has been reported to be under environmental and endogenous control, probably via endocrine regulation. Indeed, several neurohormones, such as methylfarnesoate, dopamine, and serotonin, are differentially expressed during stolonization and have been seen as potential candidates to trigger the stolon formation, the correct maturation of gametes, and the release of stolons from the adult when gametogenesis completes. Furthermore, this process seems to be under circadian control (Álvarez-Campos et al., 2019).

Sexual dimorphism involving dwarf males is found in some echiurans and *Osedax*, as well as in a few other species of annelids, especially parasites, and in some cases the male and female had not been recognized as conspecific until recently (e.g., Vortsepneva et al., 2008). In the two former cases, dwarf males are environmentally determined when a larva settles on an already established female and generally becomes male (Jaccarini et al., 1983; Worsaae and Rouse, 2010). While echiurans generally have a one-to-one sex ratio, in the case of *Osedax* each female has a "harem" of males living along her tube, sometimes more than a hundred per female (Rouse et al., 2004; Rouse et al., 2008). Other *Osedax*, however, seem to have reverted to a state in which males and females are similar (Rouse et al., 2015). The case of the males of *Osedax* seems to be a striking example of paedomorphosis, as they retain morphological traits typical of siboglinid trochophore larvae, including a ciliary band that appears to be a putative prototroch and opisthosomal chaetae (Rouse et al., 2004). This and other heterochronies have been a favorite topic of annelid workers, especially when explaining the diversity of the many meiofaunal groups that seem to have lost many of the "typical" annelid adult characters (Westheide, 1987; Struck et al., 2002; Worsaae et al., 2018).

Sexual reproduction in clitellates involves the eponymous clitellum, a region of glandular tissue that partially or totally encircles the worm's body, on or near the gonopores. During mating, the hermaphroditic worms align their respective pores for the reciprocal sperm transfer into the other individual. The clitellum is also involved in producing the cocoon, which, as it is produced and moved along the body of the animal, first receives eggs from the female gonopores and then the sperm stored in the seminal receptacles.

Annelid sperm ultrastructure is exceptionally well known (Jamieson and Rouse, 1989; Ferraguti, 1999; Rouse, 2005), but specific evolutionary trends are difficult to infer, due to the diversity of reproductive strategies within the group.

Embryonic development is typically spiral (see chapter 32), and in fact the coding system for spiral cleavage was developed based on the annelid *Alitta succinea* by Wilson (1892). Cleavage is typically holoblastic, giving origin to a coeloblastula or in some cases to a stereoblastula. Gastrulation is by invagination, epiboly, or a combination of both.

Although annelid cell lineages have been known for over a century (Wilson, 1892; Eisig, 1899; Nelson, 1904), detailed fate maps were lacking due to the difficulty of microinjection. Most of what we know about annelid development is as a result of work on just a few species: *Helobdella triserialis* (e.g., Weisblat and Shankland, 1985); *Platynereis dumerilii*, for which a detailed staging ontology has been established (e.g., Fischer et al., 2010); and *Capitella teleta* (see Seaver, 2016). A detailed fate map has recently been established for *C. teleta* (see Meyer et al., 2010). In this case, it has been shown that annelids differ considerably from other spiralians in many major developmental features. Contrary to most other spiral developers, there are four to seven distinct origins of mesoderm, all ectomesodermal. In addition, the left and right mesodermal bands arise from 3d and 3c, respectively, whereas 4d generates a small number of trunk muscle cells, the primordial germ cells, and the anus. The establishment of mesoderm from 3c and 3d rather than from 4d confirms what had already been established by Eisig (1899) but not accepted by most subsequent workers (see a historical account in Seaver, 2016).

Unlike all other known animals, *C. teleta* can regenerate the germline following removal of the germline progenitors in the early embryo, and different mechanisms operate in the larval and adult phases: a regulative event in the early stage larva and a stem cell transition event after metamorphosis, when the animals are capable of substantial body regeneration (Dannenberg and Seaver, 2018).

Organizing activity (see chapter 32), despite being a trademark of spiralian development, has not been well characterized in annelids until recently, with the 2d micromere being found to be the organizer in *Capitella teleta* (Amiel et al., 2013). It is this 2d somatoblast that generates the trunk ectoderm (Meyer et al., 2010; Meyer and Seaver, 2010). Leech embryos lack an organizer.

Although polychaetes are supposed to have nonregulative embryos, recent studies have shown some regulative capacity (Yamaguchi et al., 2016). In *C. teleta* the larval eyes arise from 1a and 1c, while generally their individual deletion produces a one-eyed larva. However, sometimes this deletion results in a two-eyed larva, indicating some form of regulation. If 1b is deleted in addition to the other individual cells, a one-eyed larva invariably results, so 1b helps compensate.

Hox gene expression data are available for a handful of annelids, but are best documented in the development of *Capitella teleta* (see Fröbius et al., 2008). Only the anterior Hox genes (*labial*, *proboscipedia*, *Hox3*) are expressed in the larval ectoderm before the appearance of segments. Each of the nine thoracic segments has a unique Hox expression boundary. Spatial and temporal colinearity of Hox genes seen in *C. teleta* is shared by *Chaeopterus*, but not by the leech *Helobdella robusta* (see Bleidorn et al., 2015).

▨ GENOMICS

Despite the relatively early sequencing of the genome of *Capitella teleta* (although published much later), annelid genomics lags behind comparable groups in terms of abundance and diversity. Currently, this and the genomes of the leech *Helobdella robusta* and the earthworm *Eisenia fetida* are the only published annelid genomes (Simakov et al., 2013; Zwarycz et al., 2016). In addition, the genomes of *Hydroides elegans* (Serpulidae) and the oligochaete *Amynthas corticis* (Megascolecidae) are available in NCBI. In contrast, and due especially to interest in the phylogenetic relationships of annelids, the number of available transcriptomes is considerable (see references above).

Annelid comparative genomics remains in its infancy, and only recently have there been some comparative analyses among the three published genomes. The numerous homeobox gene expansions and losses, exemplified by a phenomenal level of homeobox gain in *Eisenia fetida* compared to the allied *Helobdella robusta*, have been interpreted as an explanation for the enormous body plan diversity of the group (Zwarycz et al., 2016). Genomic studies of the earlier lineages would be fundamental to understand the degree of novelty within this phylum.

Since the publication of the first annelid mitochondrial genome (Boore and Brown, 1995), that of the earthworm *Lumbricus terrestris*, numerous species have been sequenced (e.g., Bleidorn et al., 2007; Mwinyi et al., 2009; Li et al., 2015b). While the gene order has been supposed to be conserved across annelids, it is now clear that a conserved gene order mostly applies just to Pleistoannelida but that the earlier branches have very different mitogenome orders (Weigert et al., 2016).

Rare genomic changes in the form of novel microRNAs have been used to corroborate a relationship between Myzostomida (sampled from *Myzostoma cirriferum*) and annelids (Helm et al., 2012).

FOSSIL RECORD

Various Proterozoic fossils have been claimed to be annelids, but none are supported by convincing apomorphies of Annelida or any of its subgroups. The oldest well-corroborated annelids are from the early Cambrian (Stage 3) Chengjiang (Liu et al., 2015b; Han et al., 2019) and Sirius Passet Konservat-Lagerstätten (Conway Morris and Peel, 2008). The latter includes one taxon with pygidial cirri (Vinther et al., 2011), a derived character of the annelid crown group that is otherwise unknown in Cambrian annelids. The Burgess Shale includes several species of annelids, some abundant and permitting detailed anatomical reconstruction (Conway Morris, 1979). Phylogenetic analyses resolve most Cambrian taxa as members of the annelid stem group and contribute to a picture of the earliest annelids as relatively large-bodied polychaetes with well-developed parapodia and paired palps, and the head poorly delineated from the trunk (Eibye-Jacobsen, 2004; Parry et al., 2015; Parry et al., 2016). An abundant, well-preserved stem-group annelid from the Burgess Shale, *Kootenayscolex barbarensis*, bears neurochaetae on its peristomium, underpinning a hypothesis that the absence of peristomial chaetae in crown-group annelids is a secondary loss (Nanglu and Caron, 2018).

The fossil record of annelids is numerically dominated by their jaws, a fossil type known as "scolecodonts," which first appears in the latest Cambrian and radiated in the Ordovician (Hints and Eriksson, 2007). Scolecodonts are attributed to Aciculata and, particularly, Eunicida, the jawed extant clades. Paleozoic scolecodonts are classified in numerous extinct families, suggesting that the taxonomic composition of Eunicida has changed markedly. The arrangement of the multi-element jaw apparatus of eunicids as dorsal mandibles and ventral maxillae permits an apparatus-based approach to be applied to the systematics of scolecodonts, integrating extinct and extant taxa more closely than had been the case previously (Paxton, 2009; Paxton and Eriksson, 2012; reviewed by Parry et al., 2019).

The Paleozoic fossil record indicates some significant extinct groups, such as machaeridians. Disarticulated calcitic shell plates of this group are common fossils from the Early Ordovician to the Permian, and even access to articulated specimens, which demonstrate that the shell plates are arranged in inner and outer series, left the systematic position uncertain until the biomineralized scleritome was found in association with soft parts (Vinther et al., 2008). Machaeridians have segmental parapodia bearing chaetae and are accordingly recognized as annelids. More precisely, the alternating attachment sites of the shell plates have been homologized with the alternating dorsal cirri of scaleworms (Vinther et al., 2008), which would place machaeridians in the annelid crown group.

Other Paleozoic fossil annelids include instances of whole-body, soft-part preservation from most geological periods. Particularly informative examples are *Kenostrychus clementsi*, an aciculate polychaete from the Silurian of the United Kingdom (Sutton et al., 2001), at least four taxa from the Lower Devonian Hunsrück Slate of Germany (Briggs and Bartels, 2010), the pyritized phyllodocidan *Arkonips topororum* from the Middle Devonian of Canada (Farrell and Briggs, 2007), and representatives assigned to several extant families from the Upper Carboniferous Mazon Creek, Illinois (Thompson, 1979). The first appearance of various extant annelid groups is reviewed by Parry et al. (2014: table 1).

Sipunculan fossils are known from the lower Cambrian of Chengjiang (Huang et al., 2004b), and they show a rather modern body plan with the exception of the coiling of the gut, present in all extant sipunculan worms but not in the fossils. Additional trace fossils have been suggested as having been produced by sipunculans by different authors, and these are reviewed in Cutler (1994).

Another significant contribution to the annelid fossil records is provided by tubes, particularly calcified ones. These provide constraints on the likely timing of origin of families classified within Canalipalpata, such as Sabellidae and especially Serpulidae (Ippolitov et al., 2014). Various Ediacaran and Paleozoic tubes previously attributed to serpulids have been rejected, but well-corroborated records are known from the Triassic onward, and they have been relatively common fossils since the Jurassic. Fossilized worm tubes in hydrothermal vents are known from at least the Silurian, and can be attributed to Siboglinidae from the Jurassic onward (Georgieva et al., 2019). Borings assigned to the siboglinid *Osedax* have been identified from bones of plesiosaurs and marine turtles from the Cretaceous, providing insight into these worms' feeding habits before the radiation of whales (Danise and Higgs, 2015).

Clitellate fossils are geologically young, and their fossil record has been relegated to a few suggested body fossil impressions and cocoons. The first possible body fossil is from the Middle Ordovician Trenton Limestone of Quebec (Conway Morris et al., 1982). Cocoons attributed to leeches first appear in the Triassic (Bomfleur et al., 2012), and more recently cocoons have been described from the lower Eocene of Antarctica (McLoughlin et al., 2016). The latter include examples in which spermatozoa are preserved with a helical organization comparable to that of the acrosome of the extant crayfish worm *Branchiobdella* (see Bomfleur et al., 2015).

48 NEMERTEA

Nemertes, a marine nymph of the Mediterranean, was one of 50 daughters of Nereus and Doris; she was considered the Nereis of "unerring" counsel, wisest of the sisters. Nemerteans, commonly known as "ribbon worms," are, however, often ignored in studies of marine diversity, especially when compared to molluscs and annelids, despite being a major group of marine predators and scavengers and one of the few animal groups to have colonized freshwater and terrestrial environments. With nearly 1,300 described species (Gibson, 1995; Kajihara et al., 2008), ribbon worms constitute the eleventh largest animal phylum, yet are considered as a minor group by some, perhaps due to their often-cryptic nature and poor representation in zoological collections due to their soft bodies, which hinder preservation. Nemerteans are a major and voracious predator group in places like Antarctica and include what is considered by most the longest animal ever recorded, *Lineus longissimus*, which can measure more than 30 m in length (McIntosh, 1873–1874; Gittenberger and Schipper, 2008). Many species are, however, small or even microscopic.

Of all the animal phyla, nemerteans are unique in presenting a protonephridial excretory system similar to that of some acoelomates while possessing coeloms, as evidenced by their rhynchocoel and their closed circulatory system (Turbeville, 1986). The rhynchodeal apparatus, a unique characteristic of nemerteans, is used for capturing prey, often using nemertean-produced toxins (Whelan et al., 2014) to paralyze them, but it may also be used for defense, as often observed when nemerteans are handled. Previous claims about the presence of bacterial-derived tetrodotoxin-like substances (Carroll et al., 2003) have been questioned (Strand et al., 2016). It is surprising how many videos about nemertean predation are available on the internet, fascinating the observers, who often do not know what type of animals they are watching. Perhaps the most epic is Sir David Attenborough's BBC video of the giant Antarctic *Parborlasia corrugatus* devouring a dead seal pup (www.youtube.com/watch?v=HG17TsgV_qI).

Nemerteans (also known as Rhynchocoela) thus encompass animals with many odd characteristics, somehow resembling flatworms, but some attaining giant sizes, with a circulatory system (and coelomically derived body cavities). They are also often beautifully colored, can maintain commensal relationships with many other invertebrates, and can even display parental care, secreting a cocoon in which they guard the eggs for long periods of time, as in the Antarctic *Antarctonemertes* (Taboada et al., 2013). Their developmental modes are also varied, as are their ecological niches with benthic infaunal and epifaunal species, as well as pelagic ones.

SYSTEMATICS

The phylogenetic position of Nemertea within metazoans has been in flux, but two main threads have been followed. Some authors have proposed a relationship to Platyhelminthes due to their overall similarities, especially in the excretory system and compact body wall, and that grouping was named Parenchymia by Nielsen (1995). However, the supposed similarities in the integument of these two groups of protostomes have been interpreted as probably plesiomorphic (Norenburg, 1985). Another thread places nemerteans with other coelomate spiralians, as the sister group of Trochozoa (Ax, 1995), a proposal that more or less captures the idea posed by Turbeville and Ruppert (1985) of nemerteans representing a stepping-stone between aceolomates and coelomates.

Molecular data have further corroborated the closer relationship to molluscs and annelids than to platyhelminths (e.g., Turbeville et al., 1992; Winnepenninckx et al., 1995a), and later as members of the clade Lophotrochozoa. Their more precise position within Lophotrochozoa remains contentious; a clade named Kryptrochozoa has been proposed as the least inclusive clade containing the Brachiopoda and Nemertea (Giribet et al., 2009), but others have contested that clade (Nesnidal et al., 2013b; Kocot et al., 2017). Nonetheless, the inclusion of Nemertea within Lophotrochozoa is now supported by nearly all phylogenomic analyses (e.g., Dunn et al., 2008; Struck and Fisse, 2008; Hejnol et al., 2009; Kocot et al., 2017; Laumer et al., 2019), including those based on novel genomic data (Luo et al., 2017), which also support Kryptrochozoa. But more recently a phylogenomic analysis has found support for Parenchymia (Marlétaz et al., 2019), a result that would need to be corroborated in subsequent analyses. Almost anecdotally, a handful of authors proposed a close relationship of nemerteans to chordates and interpreted nemerteans as chordates' ancestors, in what it is known as "Jensen's hypothesis" (see a discussion in Sundberg et al., 1998).

Nemerteans are typically unsegmented, and thus the discovery of an interstitial "segmented" nemertean, *Annulonemertes minusculus*, led Berg (1985) to postulate alternative scenarios of nemertean evolution in which segmentation played a key role. Subsequent molecular analyses showed *A. minusculus* to be a derived hoplonemertean (Sundberg and Strand, 2007), thus inferring an acquisition of segmentation not homologous to that of other segmented lophotrochozoans.

The classification of nemerteans is based on the traditional system of Coe (1943), now refined with molecular phylogenetic analyses. An account of the history of nemertean classification is given in Brusca et al. (2016b) and largely follows Stiasny-Wijnhoff (1936), who accepted as classes Schultze (1851)'s division of nemerteans into the classes Anopla and Enopla. Anopla, now recognized as paraphyletic, was divided into Palaeonemertea and Heteronemertea, and Enopla into Hoplonemertea and Bdellonemertea, but *Bdellonemertes* now appears nested within Hoplonemertea. The latter was further subdivided into Monostilifera and Polystilifera. A consensus in the nemertean community recently dismissed Anopla and Enopla (Strand et al., 2018).

The internal phylogeny of nemerteans has received substantial attention, including analyses of morphological data matrices (e.g., Sundberg, 1990; Sundberg and Hylbom, 1994; Härlin and Sundberg, 1995; Crandall, 2001; Maslakova and

Norenburg, 2001; Schwartz and Norenburg, 2001), but morphological results were often not satisfactory, and most trees resulted in large polytomies due to poor description of morphology and high levels of homoplasy (Schwartz and Norenburg, 2001). Nemertean phylogenetics thus switched to molecular data, with numerous published papers focusing on specific branches of the nemertean tree. Several studies tackled nemertean phylogeny globally, first using a handful of Sanger-based markers (e.g., Sundberg et al., 2001; Thollesson and Norenburg, 2003; Andrade et al., 2012; Kvist et al., 2014; Kvist et al., 2015), and later using phylotranscriptomics (Andrade et al., 2014).

Nearly all published molecular analyses support monophyly of Pilidiophora, including Hubrechtidae—formerly considered a member of Palaeonemertea—plus Heteronemertea. Pilidiophora is characterized by the eponymous pilidium larva (Maslakova, 2010b) and by a specific bilateral arrangement of the proboscis musculature (Chernyshev et al., 2013). The sister group of Pilidiophora is Hoplonemertea, both constituting a clade referred to as Neonemertea. Sanger-based analyses have often found paraphyly or low support for Palaeonemertea (barring Hubrechtidae), but its monophyly is strongly supported in the larger phylogenomic analyses. Internal resolution of the major clades in Palaeonemertea and Hoplonemertea is largely consistent across data sets, and with the exception of the former Bdellonemertea, largely agrees with the traditional classifications. However, many families and genera of nemerteans are not monophyletic and may require extensive revision. Because such rampant non-monophyly may be derived from poor character selection used in nemertean taxonomy, new standards are emerging with respect to the morphological and anatomical traits that should be described in nemerteans (Sundberg et al., 2016). Traditionally, the principal anatomical features used to distinguish between these clades included proboscis armature, mouth location relative to the position of the cerebral ganglion, gut shape, layering of the body wall muscles, position of the longitudinal nerve cords and larval features.

> Class Palaeonemertea
> Class Pilidiophora
> > Order Heteronemertea
> Class Hoplonemertea
> > Order Polystilifera
> > > Pelagica
> > > Reptantia
> > Order Monostilifera
> > > Cratenemertea
> > > Distromatonemertea

There are nearly 1,300 described species of nemerteans that have been accounted for by Gibson (1995) and a subsequent update (Kajihara et al., 2008). Numerous regional checklists are also available. Because nemerteans are abundant in the marine meiofauna, the existence of numerous cryptic species (Leasi and Norenburg,

2014) has the potential to increase the known diversity substantially. However, although there seems to be strong evidence for cross-continental distributions of terrestrial and limnic species (Moore et al., 2001; Koroleva et al., 2014), cryptic diversity, at least among terrestrial forms, has also been reported (Mateos and Giribet, 2008).

A recent population genomics study has shown that the fissiparous species *Lineus pseudolacteus* is a triploid hybrid between Atlantic populations of *L. sanguineus* and *L. lacteus* (Ament-Velasquez et al., 2016), opening the door to understanding some of the recalcitrant problems in nemertean taxonomy.

NEMERTEA: A SYNOPSIS

- Triploblastic, coelomate, bilaterally symmetrical unsegmented animals[23]
- Body divided into head, trunk, and tail, the latter sometimes filiform
- Completely multiciliated epithelium, and two or three layers of body wall muscles arranged in various ways
- Unique proboscis apparatus, opening anteriorly and lying dorsal to the gut and surrounded by a coelomic hydrostatic chamber called the "rhynchocoel"
- Closed circulatory system; sometimes with hemoglobin as respiratory pigment
- Complete digestive tract, with a through-gut and a posterior anus
- Excretory system consisting of a network of protonephridia (lacking in a few deep-sea species)
- Bilobed cerebral ganglion that surrounds the proboscis apparatus (not the gut), and two or more longitudinal nerve cords connected by transverse commissures
- Marine, limnic, or terrestrial predators
- Mostly gonochoristic; early development typically spiralian; development direct or indirect; when indirect, with a number of different larvae, including the pilidium
- Asexual reproduction by fragmentation not uncommon

The body wall of nemerteans comprises a ciliated epidermis, the dermis, the muscle layers surrounding the gut and other internal organs, and a mesenchyme. The pseudostratified epidermis contains multiciliated epidermal cells, gland cells and granule-containing basal or interstitial cells resting on a well-developed basement membrane. Both circular and longitudinal muscles are present within the epidermis, woven between the cells of this layer (Turbeville and Ruppert, 1983), and the muscles concentrate on the anterior part of the animal for peristaltic burrowing. In palaeonemerteans the dermis is extremely thin or composed of only a homogeneous gel-like layer, while in heteronemerteans it is typically quite thick and densely fibrous and usually includes a variety of gland cells. The subepidermal layers of circular and longitudinal muscles are well developed, and their organization varies among taxa and may occur in either a two- or three-layered plan. Internal to

---

[23] With the exception of *Annulonemertes minusculus* discussed above.

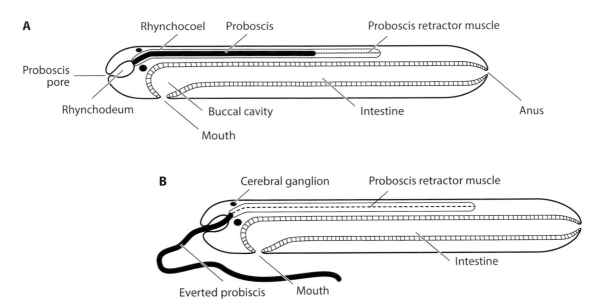

FIGURE 48.1. Schematic of a nemertean with the proboscis retracted (A) and everted (B). Based on Brusca and Brusca (1990).

the muscle layers is a dense, more or less solid mesenchyme, although in some nemerteans the thick muscle layers nearly obliterate this inner mass. The mesenchyme includes a gel matrix and often a variety of loose cells, fibers, and dorso-ventrally oriented muscles.

Perhaps the most salient synapomorphy of Nemertea is its unique proboscis apparatus, a structure that gives the phylum its other name, Rhynchocoela (from the Greek *rhynchos*, "snout"; *coel*, "cavity"), present in all known nemertean species except in *Arhynchonemertes axi* (see Riser, 1988).

The proboscis apparatus is a complex arrangement of tubes, muscles, and hydraulic systems; the lumen is called the "rhynchodeum" (fig. 48.1). The proboscis itself is an elongate, eversible blind tube and is either associated with the foregut (in hoplonemerteans) or opens through a separate proboscis pore (in the other clades). The proboscis has been studied in great detail in a species of Hubrechtidae (Chernyshev et al., 2013), and the musculature has been evaluated across a large portion of the nemertean diversity, showing phylogenetic structure (Chernyshev, 2015). The glandular epithelium consists of four types of gland cells (e.g., Junoy et al., 2000), which are not associated with any glandular system. A nervous plexus lies in the basal part of the glandular epithelium and includes multiple irregularly anastomosing nerve trunks. The proboscis musculature includes four layers: endothelial circular, inner diagonal, longitudinal, and outer diagonal; inner and outer diagonal muscles consist of noncrossing fibers. The endothelium consists of apically situated support cells with rudimentary cilia and subapical myocytes (Chernyshev et al., 2013).

The proboscis may be branching in some species (Sun, 2006) and can be regionally specialized and bear stylets in various arrangements in most members of Hoplonemertea. Each calcified stylet is composed of a central organic matrix surrounded by an inorganic cortex composed of calcium phosphate. The stylets are formed within large, uninuclear epithelial cells called "styletocytes" (Stricker, 1984). The processes of styletogenesis and stylet replacement (during growth and after losing stylets during prey capture) have been studied in a few species of nemerteans.

The proboscis is surrounded by a closed, fluid-filled, coelomic space called the "rhynchocoel," which is surrounded by additional muscle layers. The inner blind end of the proboscis is connected to the posterior wall of the rhynchocoel by a proboscis retractor muscle. In a few taxa, eversion and retraction are accomplished hydrostatically, without a retractor muscle. Eversion of the proboscis is accomplished by contraction of the muscles around the rhynchocoel; this increases the hydrostatic pressure within the rhynchocoel itself, squeezing on the proboscis and causing its eversion. The everted proboscis moves with the muscles in its wall; the proboscis is retracted back inside the body by the coincidental relaxation of the muscles around the rhynchocoel and contraction of the proboscis retractor muscle. The retracted proboscis may extend nearly to the posterior end of the worm, and usually only a portion of it is extended during eversion. It is not uncommon that the proboscis is released under stress, since the animals have the ability to regenerate.

The development of the closed circulatory system of nemerteans was studied early and constituted the basis for considering nemerteans as coelomates (Turbeville and Ruppert, 1985; Turbeville, 1986, 1991; Bartolomaeus, 1988). There is a good deal of variation in the architecture of nemertean circulatory systems, and these could probably be explored in more detail for systematic purposes. The simplest arrangement occurs in certain palaeonemerteans, in which a single pair of longitudinal vessels extends the length of the body, connecting anteriorly by a cephalic lacuna and posteriorly by an anal lacuna. Others may include transverse vessels, enlargement and compartmentalization of the lacunar spaces, and the addition of a middorsal vessel. The walls of the blood vessels are only slightly contractile, and general body movements generate most of the blood flow, which is often nondirectional, currents often reversing directions.

The blood is a colorless fluid in which various cells are suspended, at least some of which contain hemoglobin. Although conclusive evidence is lacking, the circulatory system appears to be involved in the transport of nutrients, gases, neurosecretions, and excretory products (Giribet, 2016c). Gas exchange is epidermal, without any specialized structures.

The excretory system of most nemerteans consists of two (Bartolomaeus, 1985) to thousands of flame bulb protonephridia (also called "flame cells"), similar to those found in free-living flatworms (and rotifers), one of the main reasons why nemerteans were often associated with platyhelminths. Deep-sea pelagic hoplonemerteans seem to lack protonephridia altogether. The terminal cells may be found in the connective tissue or are more or less intimately associated with the circular system (Jespersen, 1987), sometimes without connection between the blood vessels and the protonephridia (Bartolomaeus, 1988). The flame bulbs of terrestrial nemerteans may exhibit a remarkable structure, being large binucleate structures with riblike skeletal supports (Jespersen and Lützen, 1989).

Great variation in nephridial structure is found among marine species. In the simplest case, a single pair of flame bulbs leads to two nephridioducts, each with its own laterally placed nephridiopore. More complex conditions include rows of single flame bulbs or clusters of flame bulbs with multiple ducts. In some species the walls of the nephridioducts are syncytial and lead to hundreds or even thousands of pores on the epidermis. The most elaborate conditions occur in certain terrestrial nemerteans where ca. 70,000 clusters of flame bulbs (six to eight in each cluster) lead to as many surface pores. In some heteronemerteans (e.g., *Baseodiscus*), the excretory system discharges into the foregut.

There is some morphological and experimental evidence that the protonephridia also play an important role in osmoregulation, especially in freshwater and terrestrial species, in which the excretory system is most elaborate.

Some members of Heteronemertea and Hoplonemertea have invaded freshwater environments and must control changes in salinity. A few are terrestrial, although restricted to moist shady habitats and nocturnal life to avoid desiccation. In addition, the terrestrial species tend to secrete a mucus coat that reduces water loss. Likewise, intertidal nemerteans remain in moist areas during low tide periods to prevent osmotic stress. A marine meiofaunal species from North Carolina (fig. 55.1 B) lives in sediments at about 1 m depth above water level (pers. observation), probably relying on the water that fills the interstices of sand by capillarity.

The architecture of the nervous system of several species of nemerteans has recently been investigated with high anatomical resolution (Beckers et al., 2011; Beckers et al., 2013; Beckers, 2015) and was nicely summarized by Beckers and von Döhren (2016). It shows considerable variation among the major nemertean taxa.

The central (or medullated) nervous system typically consists of a ring-shaped brain and lateral branching medullary cords, as well as several longitudinal nerves, nerve plexuses, and sensory structures. A system of muscle fibers associated with the brain and lateral nerve cords is present in all major groups of hoplonemerteans (Petrov and Zaitseva, 2012).

The brain is bilaterally symmetrical, forming a ring around the rhynchocoel, with dorsal and ventral connections of the two brain halves; the ventral commissural tract is more prominent than the dorsal one. The members of Pilidiophora possess a brain that expands caudally and is associated with a conspicuous sensory organ, the cerebral organ (Beckers, 2015). Although this organ is also found in hoplonemerteans and certain palaeonemertean species, only in Pilidiophora is it directly connected to the dorsal lobe of the brain and terminating in a layer of neurons close to the blood vessel (see below). This particular morphology is now interpreted as an apomorphic character for Pilidiophora (Beckers, 2015). The paired, laterally located medullary cords are continuous with the posterior face of the ventral brain lobes, extending the entire length of the animal, and unite in the anal commissural tract dorsal to the intestine. The location of the medullary cords inside the body tissue varies in a taxon-specific manner. Major longitudinal nerves projecting from the brain comprise cephalic nerves, paired oesophageal nerves and proboscis nerves. Additional longitudinal nerves may be present in different taxa.

Sense organs of different types occur in different species, and these comprise eyes, cerebral organs, frontal organs, lateral organs, dorsal epidermal pits, statocysts, groups of so-called sensory cirri, and lateral organs (Turbeville et al., 1992). Frontal organs, eyes, and cerebral organs are among the most common sensory organs and are located in the head region. Epidermal sensory cells, distributed over the entire length of the body, are found in all species.

Pigmented eyes, situated in the head, mainly anterior to the brain, are the most conspicuous sense organs. While eyes are virtually absent in palaeonemerteans, they are common in neonemerteans, where they vary in number from a single pair to many. They are subepidermal, pigment-cup ocelli containing numerous rhabdomeric photoreceptor cells, sometimes with additional lenslike structures (von Döhren and Bartolomaeus, 2007). The optical nerves are connected to the frontal part of the brain in Hoplonemertea.

Much of the sensory input important to nemerteans is chemosensory. Ribbon worms are very sensitive to dissolved chemicals in their environment and employ this sensitivity in food location, substratum testing, general water analysis, and probably mate location (Giribet, 2016c). Probably, too, all nemerteans respond to contact with chemical stimuli, and many are capable of distance chemoreception of materials in solution. At least two different nemertean structures may be implicated in the initiation of chemotactic responses: paired cerebral organs and frontal glands (= cephalic glands).

The paired cerebral organs situated in the head region show the most pronounced differences among nemertean sense organs. They are present in some Palaeonemertea, Pilidiophora, and in most species of Hoplonemertea but are absent in cephalothricids, *Carinoma*, and Pelagica. Cerebral organs comprise a ciliated epidermal canal of differing, taxon-specific dimensions. The opening to the exterior of the canal is via small lateral pits, close behind the brain region in *Tubulanus*, *Carinina*, and the heteronemerteans *Apatronemertes*, *Baseodiscus*, and *Valencinia*, although the pits are dislocated ventrally in the latter; via simple lateral or ventral cephalic grooves in front of the brain in most monostiliferous Hoplonemertea; via transverse grooves with secondary slits in the lateral margins of the head behind the brain in polystiliferous Hoplonemertea and the monostiliferous hoplonemertean *Nipponnemertes*; or they open to the posterior end of lateral horizontal cephalic slits in lineid Heteronemertea (Beckers and von Döhren, 2016).

Cephalic slits are furrows of variable depth that occur laterally on the heads of many ribbon worms. These furrows are lined with a ciliated sensory epithelium supplied with nerves from the cerebral ganglion. Water circulates through the cephalic slits and over this presumably chemosensory epithelial lining. In Heteronemertea the cerebral organs are intimately connected to the inferior part of the dorsal brain lobe, giving the impression that they are a part of the brain. The cerebral organs are reported to be in close proximity to the blood vessels in some Hoplonemertea and in all Pilidiophora, in which the posterior end of the cerebral organ projects into large blood lacunae, separated from it only by the endothelium and its extracellular matrix.

Frontal organs are located at the anterior tip of the head in most hetero- and hoplonemerteans but are absent in Palaeonemertea, Hubrechtida, and Pelagica. They either comprise a single protrusible epidermal pit (in Hoplonemertea and basally branching Heteronemertea) or three ciliated canals arranged in the pattern of an upright triangle projecting into the body in Heteronemertea. In species where there is one pit, it is usually associated with extensive cephalic glands discharging into it. Whether the frontal organ is homologous to the larval apical organ of other spiralians requires further testing.

Most ribbon worms are active predators on small invertebrates, but some have been documented (videos available on the internet) capturing live fish or even cephalopods. Some are scavengers, feeding on all sorts of decaying animal matter, including large vertebrates, and yet others are said to feed on vegetable matter. There is evidence to suggest that species of the commensal genus *Malacobdella*, which inhabit the mantle cavity of bivalve molluscs, feed largely on phytoplankton captured from their host's feeding and gas exchange currents. Some species are capable of tracking prey over long distances, and this is probably done by chemotactic responses. Ribbon worms that hunt and track can recognize the trails left by potential prey, as they fire their proboscis along the trail ahead of them to capture food.

The behavior involved in the capture and ingestion of live prey is different from that associated with scavenging on dead material. In predation the proboscis is employed both in capturing prey and in moving it to the mouth for ingestion. The proboscis is everted and wrapped around the victim, which is physically subdued by the proboscis and may be killed by its toxic secretions. Nemerteans with an armed proboscis (Hoplonemertea) actually use the stylets to pierce the prey's body to introduce the toxin. In other cases, it has been recorded that large prey are first attacked with the proboscis to inject venom, and then tracked for ingestion at a later stage.

Once captured, the prey is drawn to the mouth by retraction of the proboscis and usually swallowed whole. The mouth is expanded and pressed against the food, and swallowing is accomplished by peristaltic action of the body wall muscles aided by ciliary currents in the anterior region of the gut. Scavenging usually does not involve the proboscis; the food is directly ingested by muscular action of the body wall and foregut. In some predatory hoplonemerteans (those in which the lumen of the proboscis is connected with the anterior gut lumen), the foregut itself may be everted for feeding on animals too large to be swallowed whole. In such cases, fluids and soft tissues are generally sucked out of the prey's body. Favorite food items of nemerteans include polychaete and oligochaete worms, small crustaceans, or molluscs, many of these having been detected in PCR assays.

Species of the hoplonemertean genus *Carcinonemertes* are ectoparasites on brachyuran crabs—they are egg predators that spend their entire life on the crabs—with reports of up to 99% infestation rates of *Carcinonemertes errans* on the commercially important Pacific Dungeness crab (*Cancer magister*) (Dunn and Young, 2013). Others are known to be commensals on other marine invertebrates, including ascidians, where they can complete their life cycle but do not seem to feed upon the host (Junoy et al., 2010).

REPRODUCTION AND DEVELOPMENT

Most nemerteans show remarkable powers of regeneration (see a recent review in Bely et al., 2014), as shown in a seminal series of papers on regeneration by Coe (1929, 1930, 1932, 1934a, b). The posterior portions of the body, which may often be lost, can be regenerated by most nemerteans, but head regeneration is only possible in a few lineages and originated independently at least four times within the phylum (Zattara et al., 2019). The proboscis, which can be lost during prey capture or during stressful situations, can also regenerate. In addition, nemerteans can undergo growth and degrowth, with reports of animals being maintained in a starved condition for over a year shrinking in size but otherwise appearing to be normal (Dawydoff, 1928). Some species of *Lineus* can reproduce asexually by fissiparity, a process sometimes referred to as "fragmentation."

Most ribbon worms are gonochoristic, although protandric and even simultaneous hermaphrodites are known. The gonads are generally specialized patches of mesenchymal tissue arranged serially along each side of the intestine, alternating with the midgut diverticula. In most nemerteans the development of gonads occurs along nearly the entire length of the body, but in a few species the gonads are more restricted, often toward the anterior end. The gonads begin to enlarge and hollow just prior to the onset of the breeding season. Specialized cells in the walls of the rudimentary ovaries and testes proliferate eggs and sperm into the lumina of the gonadal sacs. In females, additional cells produce yolk.

With the proliferation of gametes, the gonadal sacs expand to almost fill the area between the gut and the body wall. When the animals are nearly ready to spawn, mating behavior is initiated and the worms become increasingly active. The presence of a ripe conspecific can stimulate the release of gametes from other mature individuals. During mating activities, multiple worms may join into a mucus-covered mating mass. The gametes are released through temporary pores or by rupturing the body wall through contraction of the body wall muscles or of special mesenchymal muscles surrounding the gonads (Giribet, 2016c).

Fertilization is often external, either in the water column or in the gelatinous mass of mucus produced during mating. In some cases, egg cases can form, and part or all of the embryonic development occurs within them. Internal fertilization with a protrusible penis occurs in certain species; some use suckers to clasp the female. In some cases the sperm is released into the mucus surrounding the mating worms and then move into the ovaries of the female; once fertilized, the eggs are usually deposited in egg capsules, where they develop, although some Antarctic species can produce cocoons that are brooded by the mother (Taboada et al., 2013). Some terrestrial species are ovoviviparous with direct development, with the embryos retained within the body of the female. Ovoviviparity is also known in a few deep-sea pelagic forms.

Egg development is well understood in *Cerebratulus lacteus* (e.g., Wilson, 1900; Martindale and Henry, 1995; Henry and Martindale, 1996, 1998), and early development through the gastrula stage is similar in most of the species studied to date. Cleavage is holoblastic and spiral, producing either three (*Tubulanus*) or, more typically, four quartets of micromeres. Gastrulation is usually by invagination of the

49

LOPHOPHORATA

Lophophorata (or Tentaculata) was a staple of zoology textbooks for decades (e.g., Hyman, 1959), an association between the coelomate groups Brachiopoda, Phoronida, and Bryozoa based on the shared presence of a lophophore—a mesosomal tentacle crown with an upstream-collecting ciliary band (Nielsen, 2002)—although the mesosomal nature of the bryozoan lophophore has long been rejected. The group was recognized for a long time as a clade of coelomate protostomes, named Molluscoidea by Grobben (1908), but this came into question in many early molecular phylogenetic analyses. In some analyses, brachiopods and phoronids were grouped together, but bryozoans were placed elsewhere in the Protostomia (Halanych et al., 1995; Halanych, 1996a), commonly allying with entoprocts (Helmkampf et al., 2008b)—as strongly advocated by Nielsen (1971, 1995). In other studies, a lack of resolution in the then employed markers based on Sanger sequencing (e.g., Zrzavý et al., 1998; Giribet et al., 2000; Passamaneck and Halanych, 2006; Helmkampf et al., 2008a) left lophophorates largely unresolved. However, recent phylogenomic studies have provided some support for the monophyly of Lophophorata (Nesnidal et al., 2013b; Laumer et al., 2015a; Kocot et al., 2017; Laumer et al., 2019; Marlétaz et al., 2019), a clade that we adopt here, although with caution.

Reanalysis of the literature and comparison between Brachiopoda and Bryozoa allows the hypothesis that a protocoel is lacking in all Lophophorata and that merely two unpaired coelomic cavities, one tentacle and one trunk coelom, can be assumed for the ground pattern of this taxon (Bartolomaeus, 2001; Gruhl et al., 2005), although this is not universally accepted, as some authors still advocate a trimeric body plan (e.g., Temereva et al., 2015).

It is also worth noting in this context that while Lophophorata is now generally supported (e.g., Brusca et al., 2016b), two other groups of animals are often allied to the three phyla listed above: Entoprocta and Cycliophora. Indeed, in the original report of Cycliophora by Funch and Kristensen (1995), a relation of the new animals to bryozoans and entoprocts was postulated, based mostly on their ability for budding. Bryozoa (also called Ectoprocta) and Entoprocta (also called Kamptozoa) have had a long and convoluted history, with many early workers considering entoprocts as part of Bryozoa, sometimes in a group named Polyzoa (see a historical account in Nielsen, 2012a), a clade that was "resurrected" for Bryozoa, Entoprocta, and Cycliophora by Cavalier-Smith (1998), who placed Entoprocta and Cycliophora in his concept of Kamptozoa (different from that of most other authors). Monophyly of Polyzoa was subsequently found in molecular phylogenomic analyses (Hejnol et al., 2009; Laumer et al., 2015a; Kocot et al., 2017). Interestingly, some phylogenomic analyses recover a group of Bryozoa + Entoprocta (Hausdorf et al.,

2007; Struck et al., 2014), but this clade often disappears when Cycliophora is added (Hejnol et al., 2009; Nesnidal et al., 2013b; Kocot et al., 2017; Laumer et al., 2019).

Perhaps most striking is the study of Laumer et al. (2015a), which explored the effect of including Cycliophora in the data set and highlighted some of the inconsistencies found in many of the other phylogenomic analyses including these taxa. When Cycliophora is included, Entoprocta and Bryozoa tend to position themselves near Cycliophora, outside of Lophotrochozoa; but when Cycliophora is excluded, or in the most complete data set, Lophophorata is monophyletic (with or without Entoprocta included in the analyses). Given the longstanding history of grouping Entoprocta with Bryozoa, the strong evidence for Brachiopoda and Phoronida being protostomes, and the growing evidence for grouping these taxa as the clade Lophophorata, we treat these four phyla in this section. Cycliophora are dealt with above (chapter 45), although their affinities remain poorly understood. In fact, one of the most recent phylogenomic analyses dealing with Lophophorata did not include Cycliophora, but placed Entoprocta with Mollusca (Marlétaz et al., 2019).

A corollary of this new scenario is the homology of the tentacular crown of all members of Lophophorata. Traditionally, entoprocts were not considered members of Lophophorata because of their recognized membership in Protostomia (largely due to the presence of spiral cleavage), while lophophorates were assigned to Deuterostomia—at least in the second half of the twentieth century. Another difference has to do with the coelomic nature of the lophophore, which is supposed to be developmentally derived from the mesocoel in the coelomate taxa (at least in brachiopods and phoronids) (Nielsen, 2002) but not in bryozoans nor in entoprocts, since the latter do not have a coelom.

Innervation of the lophophore, specifically a homology between a set of the main nerves in representatives of each of the three phyla, suggests monophyly of Lophophorata (Temereva and Tsitrin, 2015; Temereva, 2017b), but it is difficult to homologize these to the tentacular nerves of Entoprocta, which may be the sister group of Bryozoa.

FOSSIL RECORD

Brachiopoda and Bryozoa are both major groups in the fossil record, as recounted in their respective chapters. The potential stem groups of Brachiozoa and its two constituent phyla, Brachiopoda and Phoronida, have been a topic of much recent research activity (Skovsted et al., 2008; Skovsted et al., 2009b; Holmer et al., 2011; Skovsted et al., 2011; Murdock et al., 2014; Zhang et al., 2014), summarized in chapter 52. Recent paleontological scenarios for the origin and early radiation of lophophorates have concentrated on brachiopods and phoronids and have mostly excluded bryozoans from consideration.

A recent addition to candidates for membership in Lophophorata are hyoliths. These are common Paleozoic fossils with an operculum, a conical, aragonitic shell and, in some taxa, a pair of lateral spines called "helens." Hyolitha has most often been regarded as *incertae sedis*, related to molluscs or assigned to their own phylum. Recent examination of large numbers of specimens from the middle Cambrian *Haplophrentis carinatus* from the Burgess Shale and Spence Shale Konservat-Lagerstätten

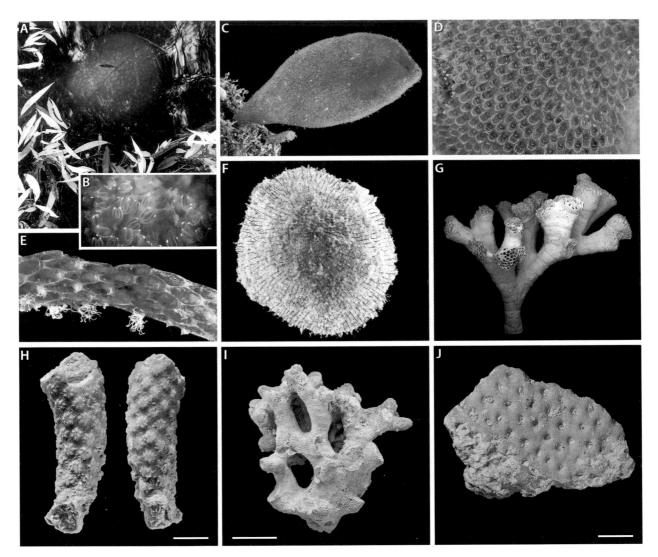

Figure 50.1. A–B, large spherical colony of the limnic phylactolaemate *Pectinatella magnifica* (Pectinatellidae); C, a ctenostome of the genus *Flustrellidra* (Flustrellidridae); D, an encrusting species of the cheilostome *Steginoporella* sp. (Steginoporellidae); E, mid-branch part of a colony with autozooids of the Antarctic cheilostome *Carbasea curva* (Flustridae); F, a free-living bryozoan in the family Selenariidae; G, scanning electron micrograph of a colony of the cyclostome *Fasciculipora ramosa* (Frondiporidae); H, the Ordovician Cystoporata *Constellaria* sp. (Constellariidae), scale 5 mm; I, the Jurassic Cyclostomata *Ceriocava corymbosa* (Cavidae), scale 10 mm; J, the Ordovician Cryptostomata *Stictoporellina* sp. (Stictoporellidae), scale 10 mm. Photo credits: A–B, Kira Treibergs; G, Blanca Figuerola; H–J, Paul Taylor.

While most bryozoans are colonial and sessile, some notable exceptions exist. The members of the cheilostome family Cupuladriidae form conical, free-living colonies on soft bottoms, unattached to the substrate (Dick et al., 2003), and are able to walk on their vibracula (a modified zooid; see below). Likewise, the phylactolaemate *Cristatella mucedo* can creep along twigs in lakes and ponds. The genus *Monobryozoon*, which includes two species from the Northeast Atlantic (Gray, 1971), is primarily a solitary, free-living meiofaunal bryozoan, although it can reproduce asexually by budding. It is also notable that in the phylactolaemate *Stephanella hina*, decolonization—the separation of individual zooids from their colony—does not seem to have an effect on the individuals (Schwaha et al., 2016).

Among the sessile bryozoans, many grow on hard substrates, seagrass blades, algae, and blades of kelp, but some prefer animal substrates. For example, in Antarctica, bryozoans are commonly found growing on pycnogonids, which are a major predator of the same bryozoans when they grow on hard substrates, and it has been postulated that carapaces of pycnogonids act as refugia for the bryozoans from competition for space on hard substrata, ice scour, and even predation by their host (Key et al., 2013). A pelagic *Alcyonidium* has also been reported for Antarctica, forming a floating hollow, spherical colony (Peck et al., 1995). Bathymetrically, bryozoans are found from the intertidal down to 8,300 m in the South Pacific Ocean's Kermadec Trench (Hayward, 1981).

The faunal compositions of bryozoans in the southern hemisphere may be useful for reconstructing the breakup of Gondwana (Figuerola et al., 2017) and as models for studying the closing of the isthmus of Panama (Dick et al., 2003). Fossil bryozoans have also played a key role in the study of such evolutionary theories as punctuated equilibrium (Cheetham, 1986).

From an anthropocentric point of view, several freshwater bryozoans are hosts for the parasite that causes proliferative kidney disease in salmonid fish—the endoparasitic myxozoan *Tetracapsuloides bryosalmonae*—the disease having substantial impacts on aquaculture and wild fisheries (Hartikainen and Okamura, 2015). Due to the fact that many bryozoan species are typically associated with human activity, including commercial and recreational vessels, several marine bryozoans have now been studied in the context of biological invasions (e.g., Johnson et al., 2012; Loxton et al., 2017). Bryozoans are also important in the study of chemical ecology, both as adults (e.g., Blackman and Walls, 1995; Figuerola et al., 2013) or as larvae (Lopanik et al., 2004), even though at least some of the secondary metabolites are produced by bacterial symbionts.

▬ SYSTEMATICS

The phylogenetic affinities of bryozoans have been contradictory, but most workers have recognized three main taxa, often given the class rank: Phylactolaemata, Stenolaemata, and Gymnolaemata. These were recovered as monophyletic in the first comprehensive molecular analysis of bryozoans using nuclear ribosomal RNAs and mitochondrial genes (Fuchs et al., 2009) and corroborated in subsequent analyses (Waeschenbach et al., 2012b). However, the traditional division of Gymnolaemata into Ctenostomata and Cheilostomata was not supported, as Ctenostomata appears paraphyletic with respect to Cheilostomata, and neither Anasca or Ascophora (subgroups of Cheilostomata) are monophyletic (Waeschenbach et al., 2012b).

Although many alternative classifications are currently in place (see Nielsen, 2018b), we follow the one derived from the phylogenetic analyses of Waeschenbach et al. (2012b) here (see also Westheide and Rieger, 2007). This classification system is congruent with the (nonnumerical) cladistic analysis of Carle and Ruppert (1983) but contrasts with the parsimony analysis of Cuffey and Blake (1991), who found Phylactolaemata to be the sister group of Gymnolaemata. Some classifications place Stenolaemata (or Cyclostomata) as the sister group of Eurystomata (= Ctenostomata + Cheilostomata) sensu Marcus (1938), as a subclade of Gymnolaemata (e.g., Ax, 2003; Nielsen, 2016a), but this system is not supported by molecular studies. A cladistic analysis including extinct and extant groups (Todd, 2000) has suggested that Ctenostomata is a paraphyletic group where both Stenolaemata and Gymnostomata nest. This is rooted in the idea that the unmineralized ctenostomes gave rise to the skeletonized stenolaemates and cheilostomes, a relationship contradicted by all molecular analyses to date.

Phylactolaemata comprises some 70+ species that occur exclusively in limnic environments, commonly referred to as freshwater bryozoans. They are characterized by their chitinous or gelatinous cystids, monomorphic colonies, and generally cylindrical zooids. Like many freshwater species of invertebrates, they can produce asexual resting bodies, known as "statoblasts," from which new colonies will form. Their family-level phylogenetic relationships have been studied using nuclear ribosomal RNA genes and/or mitochondrial genes (Wood and Lore, 2005; Hartikainen et al., 2013). Freshwater bryozoans are also found among the cruciform stoloniferan Ctenostomata (Smith et al., 2003).

The sister-group relationship between the mostly marine Stenolaemata and Gymnolaemata is supported by a number of morphological synapomorphies, including (1) polypide recycling (see below), (2) lack of a basal lamina in the funiculus, (3) presence of parietal muscles, (4) lack of an epistome, (5) anal budding direction, and (6) complete body wall separating the zooids (Taylor and Waeschenbach, 2015).

Stenolaemata (with the only extant group Cyclostomata) are exclusively marine calcified bryozoans, while Gymnolaemata includes both the uncalcified Ctenostomata and the calcified, generally polymorphic Cheilostomata. Because of their calcification, the systematics of cyclostomes has been based mostly on skeletal characters, but those have been shown to be highly homoplastic (Waeschenbach et al., 2009; Taylor et al., 2015). The systematics of the uncalcified ctenostomes also seems to be at odds with recent molecular results, resulting in the synonymy of such common genera as *Bowerbankia* and *Vesicularia* with *Amathia* (Waeschenbach et al., 2015).

The most comprehensive molecular estimate to date found Ctenostomata to be paraphyletic to the inclusion of a monophyletic Cheilostomata (Waeschenbach et al., 2012b), although further taxon and gene sampling would be desirable to provide a better supported bryozoan phylogeny and to test specific hypotheses of cheilostome polyphyly (Jebram, 1992) or that some living ctenostomes may be more closely related to stenolaemates than to the cheilostomes (Todd, 2000).

Molecular species-level systematics has been studied in a variety of taxa, including *Membranipora membranacea*, one of the most widespread species of bryozoan, showing an antitropical distribution and inhabiting all major ocean basins (Schwaninger, 2008). Cryptic speciation is rampant, even among common well-known species (e.g., Gómez et al., 2007; Nikulina et al., 2007; Waeschenbach et al., 2012a; Fehlauer-Ale et al., 2014).

Class Phylactolaemata

Class Stenolaemata

    Order Cyclostomata

    Order Trepostomata †

    Order Cystoporata †

    Order Cryptostomata †

    Order Fenestrata †

Class Gymnolaemata

  Subclass Eurystomata

    Order Ctenostomata

    Order Cheilostomata

▨ BRYOZOA: A SYNOPSIS

- Bilaterally symmetrical colonial coelomates
- Skeleton usually mineralized, typically calcitic; or with chitin
- Coelom connecting three body regions, the epistome (only present in some phylactolaemates), the lophophore, and the trunk
- Zooids within a colony produced by asexual budding; often polymorphic, divided into autozooids and heterozooids
- Circular or U-shaped lophophore
- U-shaped through-gut, the anus close to the mouth, opening outside the lophophore
- Without circulatory or excretory structures; without respiratory pigments
- Mostly benthic, mostly sessile; in marine and limnic environments

Understanding the anatomy of bryozoans entails disentangling the individual zo-oids. Colony shape is a different matter and is based on growth patterns that are usually species specific, and that may change during the life of the colony.

Variation in bryozoan zooids has recently been reviewed and can be divided into three categories: astogenetic, ontogenetic, and polymorphic (Schack et al., 2018). Astogenetic variation encompasses the differences in shape and size among the ancestrula, zooids in the zone of astogenetic change, and zooids within the zone of astogenetic repetition. Ontogenetic variation refers to changes as a zooid develops. Both astogenetic and ontogenetic variation are continuous. This is in contrast to polymorphic variation, which is discontinuous and displays abrupt changes in shape, size, and other characteristics of the individual zooids.

A recent classification of the different types of zooids was given by Schack et al. (2018), and we follow it here, but some variation occurs between authors. Like au-tozooids, the typical lophophore-bearing zooids, polymorphs also exhibit ontoge-netic variation. There are two main categories of bryozoan polymorphism: auto-zooidal polymorphs, which retain the lophophore and include reproductive individuals (not all may be able to feed) and heterozooids, which do not have a ten-tacular crown and are not able to feed. Heterozooids include avicularia, vibracula, kenozooids, and cyclostome gonozooids. In avicularia, which possess a highly modified cystid and musculature, the tentacle crown is reduced to a vestige. Vibracula have a flagellum-like operculum used for cleaning the colony or for locomotion in the nonsessile species. In kenozooids (reduced anchoring individuals) the polypide is completely absent.

Avicularia (fig. 50.2 C) have an enlarged operculum that articulates at the base against the rostrum and forms a "jaw" able to capture small possible enemies. However, despite general belief, there is little evidence that vibracula and avicularia are used for defense of the colony (Winston, 1984), although avicularia have been documented to capture syllid polychaetes (Winston, 1986). Vibracula are adventitious avicularia with extreme morphology, in which the rostrum is extremely elongated and their hinge structure allows it to rotate over the surface of the colony (unlike avicularia, which can only swing in one plane). In addition, the colony may have a series of protective structures, including spines, which may be derived kenozooids, and body wall projections. Rhizoids are elongated kenozooids that extend to the substratum or other areas of the colony, providing attachment to the substrate and colony support, especially in flexible erect colonies. Reproductive polymorphs, including ovicells, are discussed below.

Cheilostome zooecium morphology is highly variable, and due to their calcification, these cystids are often used for taxonomic purposes. Most species have box-shaped zooecia, and some areas may be uncalcified, conferring some movement to the zooids. We follow the terminology of Nielsen (2016a) here. The outer surface of the box that bears the orifice for the lophophore is called the frontal surface. The frontal surface has a partly uncalcified flexible frontal membrane. Contraction of parietal protractor muscles pulls the frontal membrane inwards, increases coelomic pressure, and projects the polypide out. An operculum offers protection when the polypide is retracted, but the frontal membrane is a weak point for predation, so protective spines (in cribrimorphs), a frontal shield or cryptocyst, or an ascus (in ascophorans) offer different degrees of protection, but these structures have evolved multiple times, as evidenced by the non-monophyly of ascophoran bryozoans (Waeschenbach et al., 2012b).

In addition to the ontogeny of individual zooids, bryozoans undergo polypide recycling, whereby polypides periodically degenerate to produce a brown body and are replaced by a new polypide within the same zooid. In some cases up to five polypide cycles can be observed, and the polypide activity is linked to environmental conditions (Barnes and Clarke, 1998). This mechanism resembles the cycliophoran recycling of the feeding individuals (Funch and Kristensen, 1995).

Unlike in some other colonial animals, at least for the autozooids, bryozoan zooids are clearly demarcated by the elements of the polypide, though the interconnections between these autozooids are different across bryozoan groups. In phylactolaemates, which lack septa, the coelom is continuous among zooids, and the funiculus (a tubular tissue cord) extends from the bottom of the U-shaped gut to the body wall. All other bryozoans lack this contiguous coelom, as the zooids are separated. In some gymnolaemates that bud off stolons, a stolonal funiculus passes across a pore in each stolonal septum, connecting the funiculus of each zooid. In the non-stoloniferous gymnolaemates the cystid walls may have pores penetrated by tissue plugs that connect the funiculus of adjacent zooids. Finally, in cyclostomes there are interzooidal pores that allow communication of coelomic fluid, but there are no funicular connections.

The coelomic system of bryozoans is organized in two ways, but a tripartite nature has been rejected even for those species with an epistome, since the coelomic cavities of the epistome, lophophore, and the trunk are connected (Gruhl et al., 2009; Schwaha et al., 2011). In phylactolaemates there is typically an epistome that contains the preoral coelomic cavity, whereas gymnolaemates, but also some phylactolaemates, lack an epistome and, consequently, the preoral coelom (Gruhl et al., 2009; Temereva, 2017c). The preoral coelomic cavity of phylactolaemates, when present, is confluent with the trunk coelom and is lined by peritoneal and myoepithelial cells. The lophophore coelom extends into the tentacles and is connected to the trunk coelom by two weakly ciliated coelomic ducts on either side of the rectum. The lophophore coelom passes the epistome coelom on its anterior side. This region has traditionally been called the "forked canal" and hypothesized to represent the site of excretion (Gruhl et al., 2009).

The bryozoan feeding apparatus includes the lophophore, a circular or U-shaped supportive structure with its own coelom (see above) and musculature (Tamberg and Shunatova, 2017). The lophophore bears a single row of ciliated tentacles that lead toward the mouth, which is located in the center. From there, the digestive system leads to a U-shaped gut that opens to the anus, outside of the lophophore. Bryozoans feed by generating water currents with their ciliated tentacles, collecting small particles—in the size range of nanoplankton and microplankton—from the flow (Tamberg and Shunatova, 2016).

There is no circulatory system in bryozoans, but given their small size and distances between zooids, distribution of metabolites within zooids is by diffusion, and between zooids it is facilitated by the common coelom in phylactolaemates and by the cystid pores of cyclostomes or by the funicular cords in the other groups (Nielsen, 2016a). Likewise, there are no specific organs for gas exchange, nor are there respiratory pigments.

Excretion is poorly understood in bryozoans. Metanephridia are known from adult phoronids and brachiopods (e.g., Bartolomaeus, 1989; Lüter, 1995), but no discrete excretory organ is known from bryozoans. Some authors have proposed that the forked canal of the coelom could act as an excretory organ (see a discussion in Gruhl et al., 2009; Schwaha et al., 2011), as it exhibits a dense ciliation, as in metanephridia, but there is no evidence for an excretory function or an excretory pore (earlier reports of an excretory pore have not been confirmed).

The polypide musculature is highly reduced in most groups due to the sessile lifestyle. In phylactolaemates sheets of circular and longitudinal muscles are present between the epidermis and the peritoneum. In the other groups, the musculature forms groups of separate muscles, including the protractors that attach to the frontal membrane of gymnolaemates and the tentacular retractor and protractor muscles. A recent study has compared fifteen species of ctenostomes, showing a certain degree of conservation in a series of diverse colony forms. However, several myoanatomical features such as the cardiac sphincter, basal cystid muscles, tentacle sheath muscles, and apertural muscle arrangement vary across taxa (Schwaha and Wanninger, 2018). As an example of the evolution of functional innovation through vestigialization, the much more developed musculature of the avicularia seems to have evolved from the autozooid (Carter et al., 2011).

The structure of the nervous system of bryozoans has been thoroughly revised by Gruhl and Schwaha (2016). The adult nervous system consists of a cerebral ganglion located at the base of the lophophore, with nerve cords that extend from the brain to innervate the ring of tentacles. These nerve cords also embrace the pharynx, forming a circumpharyngeal nerve ring. In phylactolaemates, in which the lophophore is more elaborate, the tentacles are located in the arms of the horseshoe-shaped lophophore, with radial nerves and ganglion horns (Gruhl and Bartolomaeus, 2008; Schwaha et al., 2011). Additional neurite bundles emerging from the lateral and basal parts of the cerebral ganglion innervate the epidermis, intestine, musculature, and coelomic epithelia (Gruhl and Schwaha, 2016). Interzooidal communication is probably mediated by visceral nerves innervating part of the digestive tract and the tentacle sheath nerves that on their distalmost part reach into the body wall and consequently probably act in interzooidal communication (Bobin, 1977; Schwaha and Wanninger, 2015).

The serotonin-lir nervous system of bryozoans has recently been studied by Schwaha and Wanninger (2015), showing a consistent pattern across clades. Phylactolaemates and gymnolaemates both have a "serotonergic gap." As for sensory organs, bryozoans have few, the only known receptors being tactile cells on the tentacles and on the sensory papillae of the avicularia and ciliated sensory cell on the introvert of *Rhamphostomella ovata* (see Shunatova and Nielsen, 2002).

The larval nervous system is best studied for the gymnolaemate coronate and cyphonautes larvae, beginning with the pioneering work of Woollacott and Zimmer (1972) and followed by a series of studies using transmission electron microscopy (e.g., Hughes and Woollacott, 1978, 1980; Stricker et al., 1988; Zimmer and Woollacott, 1989a, b, 1993). More recently a diversity of larvae has been studied by immunostaining and confocal laser microscopy (Wanninger et al., 2005; Santagata, 2008a, b; Gruhl, 2009, 2010; Nielsen and Worsaae, 2010; Schwaha et al., 2015; Gruhl and Schwaha, 2016). In the studied larvae, all nervous structures are transitory, being completely resorbed during metamorphosis.

Bryozoan larvae lack a distinct brain, and different names have been applied to their nerve nodule, which is located near the pyriform organ. The pyriform organ is formed of ciliated sensory cells, often found in a larger structure called the apical disc. There is also an apical organ, a neuromuscular strand and the corona. Phototaxis can play a key role in larval emergence from the brood chamber as well as in settlement and seems to invert (from positive to negative) as the larva becomes competent (Wendt and Woollacott, 1999). Photoreceptors formed by a single sensory cell as the functional unit have been characterized in at least a few larval species (see a review in Hughes and Woollacott, 1980).

The best treatment of bryozoan reproductive biology is still that of Reed (1991). As in other colonial animals, asexual reproduction by budding plays a fundamental role in colony growth. Budding involves only elements of the body wall, but the exact method is taxon-specific and may form first the cystid (e.g., in cheilostomes) or the polypide (in phylactolaemates). In addition to budding, phylactolaemates produce encapsulated dormant structures called "statoblasts," a form of asexual buds formed in large numbers during adverse conditions and protected by a pair of chitinous walls (Francis, 2001). The statoblasts function as reproductive, survival, and dispersal agents, and can withstand freezing, desiccation, and other stresses. Because the statoblast valves are made of sclerotized chitin, they preserve well when buried in lake sediments, and along with other microfossil remains they provide information on past aquatic environments (Francis, 2001). While statoblast morphology (including floatoblasts and sessoblasts) may be species specific, it does not seem to have a strong phylogenetic signal, with many shapes and features being convergent (Hirose et al., 2011). Colony fragmentation is also important for asexual propagation, as in many other colonial animals.

One of the most interesting aspects of bryozoan biology is their diversity of sexual reproductive strategies. Bryozoan colonies are hermaphroditic, but in general individual zooids have separate sexes, sometimes with marked sexual dimorphism, especially in some cyclostomes. Because of this hermaphroditism and the prevalence of internal fertilization, multiple authors have discussed the idea of self-fertilization (see a historical review in Ostrovsky, 2008). More recent experiments have shown that selfing is possible in these hermaphroditic colonies, but the viability of the larvae is drastically reduced in the first generation and nonexisting in the second generation (Johnson, 2010). Hermaphroditic zooids also exist in some species, as in *Membranipora membranacea*, which is also a broadcast spawner (Temkin, 1994). Genetic research has also demonstrated that a colony in *Bugulina stolonifera* is composed of multiple founding individuals, with larval settlement contributing to maximizing outcrossing, at least in a largely inbreeding population at Woods Hole (Johnson and Woollacott, 2010), where more than 90% of the attached individuals could not have derived from the colony on which they were attached.

The diversity of strategies for sexual reproduction has recently been reviewed by Ostrovsky (2013). He identified five major reproductive patterns depending on the type of oogenesis (oligolecithal versus macrolecithal), ovary structure, and type of embryonic incubation (nonplacental versus placental). In the placental mode, a diversity of matrotrophic strategies exist with the different types of brood chambers and have evolved in all major bryozoan lineages (Ostrovsky et al., 2009). Likewise, brood chambers (ovicells) seem to have evolved independently on multiple occasions in cheilostomes (Ostrovsky, 2013).

The ovaries are diffuse and arise from the peritoneum of the cystid wall. The testes usually develop from the funiculus, the sperm being released into the coelom, and from there exiting the polypide through the tips of the abfrontal tentacles, but it is unclear how they enter the female zooid to fertilize the eggs. A few species of cheilostomes and ctenostomes release the fertilized eggs into the water column, but, as discussed earlier, most species, including phylactolaemates, brood their embryos. Numerous brooding structures and mechanisms have been described in the different species of bryozoans, perhaps the most intriguing one being that of cheilostomes, which form ovicells, each of these consisting of a calcified double-walled ooecium enclosing a brooding cavity. In many species, the developing female zooids induce the developing distal zooid (male or female) to produce an ovicell by a specialization of the proximal part of the frontal wall (Nielsen, 2016a). The ovicell is thus not a zooid, but a specialized region of the zooid and evolves from the ooecium. The ooecium is a body-wall outgrowth formed generally from the distal daughter zooid. Spines provided by the distal daughter evolve into ooecia through fusion or reduction in the number of spines, flattening, loss of basal articulation, and relocation of spine bases. The egg-producing zooid then transfers the egg to the ovicell and closes it with a body-wall plug (ooecial vesicle) or autozooidal operculum, or both (Schack et al., 2018).

Cyclostome reproduction involves a rare mode of sexual reproduction in which a single zygote produces multiple offspring—on the order of one to many hundreds of larvae from each zygote—called "polyembryony" (Zimmer, 1997; Hughes et al., 2005; Pemberton et al., 2007). This elevated number of broods is held in large embryo chambers called "gonozoids," produced by the hypertrophy of single female zooids or by the fusion of multiple chambers.

Early cleavage has been studied in just a few species of bryozoans (e.g., Barrois, 1877; Calvet, 1900; Marcus, 1938; Corrêa, 1948), displaying a unique stereotypic cleavage pattern with a biradial arrangement of the blastomeres that is widely conserved within the group (Vellutini et al., 2017). Most of this knowledge has been summarized in a series of reviews (Reed, 1991; Zimmer, 1997; Nielsen, 2012a; Santagata, 2015b). Recent developmental work has focused on the common and widespread cheilostome *Membranipora membranacea* (Stricker et al., 1988; Nielsen and Worsaae, 2010; Santagata, 2015b), for which the cell lineage, MAPK signaling, and the expression of 16 developmental genes have been characterized (Vellutini et al., 2017). Interestingly, this work showed that the molecular identity and the fates of early bryozoan blastomeres are similar to the putative homologous blastomeres in spiral-cleaving embryos, concluding that bryozoans have retained traits of spiral development, despite the evolution of a novel cleavage geometry.

Two main developmental types are found in bryozoans: polyembryony with very simple, lecithotrophic larvae is found in Cyclostomata, and normal embryology with planktotrophic or lecithotrophic larvae is found in Ctenostomata and Cheilostomata. The characteristic planktotrophic, pyramidal-shaped cyphonautes larva is found in representatives of both ctenostome and cheilostome groups and can live in the plankton for months. It is characterized by the ring of cilia around the margin of the shell, a U-shaped ciliated ridge and a complete gut. Ctenostomata and Cheilostomata can also have a diversity of spheroid-shaped nonfeeding larvae, the most common of which is the coronate larva, almost completely ciliated. An intermediate pyramidal-shaped nonfeeding larva, sometime called pseudocyphonautes, can also be found. Additional terminology exists for other types of larvae that do not fit these types, including the vesiculariform larva (Zimmer and Woollacott, 1993). Likewise, the short-lived dispersal phase of phylactolaemates is considered to lack some characteristics of larvae, yet it is often referred to as "larva" (e.g., Zimmer and Woollacott, 1977a; Nielsen, 2016a). The ecology of the bryozoan larvae has been discussed in detail elsewhere (see Nielsen, 1971; Zimmer and Woollacott, 1977b).

Metamorphosis from larva to ancestrula has been studied in a few species (e.g., Woollacott and Zimmer, 1971, 1978; Zimmer and Woollacott, 1977a; Lyke et al., 1983). The settling cyphonautes everts the adhesive (or metasomal) sac found inside the larva, which spreads out in the substratum and secretes the basal wall of the ancestrula. The shells open widely and detach while the pallial epithelium secretes

FIGURE 50.2. A, larva of the cheilostomate *Bugulina stolonifera*, scale 50 μm; B, ancestrula of *B. stolonifera*, scale 200 μm; C, polymorphic ovicelled (ov) colony of *Bugula minima* with zooids and avicularia (av) (from Winston and Woollacott, 2008), scale 250 μm. Photo credits: Robert Woollacott.

the frontal wall. Variation of this type of metamosphosis exists in other larvae, including the *Bugulina stolonifera* larva (fig. 50.2 A), one of the best-studied larvae, where in the ancestrula (fig. 50.2 B) the body wall epidermis derives exclusively from the metasomal sac and the aboral epithelium contributes only to the tentacle sheath (Woollacott and Zimmer, 1978).

GENOMICS

Bryozoan genomics lags behind most other invertebrate phyla with similar diversity. ESTs are available for a handful of species (Dunn et al., 2008; Nesnidal et al., 2013b; Laumer et al., 2019), and the few transcriptomic resources are limited to a few species generated in the broader context of inferring animal phylogeny (Laumer et al., 2015a; Laumer et al., 2019) or for developmental studies, as is the case of *Bugula neritina* (see Wong et al., 2012; Wong et al., 2014) and *Bugulina stolonifera* (Treibergs, 2019).

Mitochondrial genomes are also limited to three species representing Phylactolaemata, Stenolaemata, and Gymnolaemata (Waeschenbach et al., 2006; Jang and Hwang, 2009; Gim et al., 2018), but little comparative work is available.

FOSSIL RECORD

There are no unequivocal bryozoans of Cambrian age (Taylor et al., 2013), although various candidates have been proposed through the decades. Most recently, *Pywackia baileyi*, a rod-shaped fossil with a covering of polygonal calyces, from the late Cambrian of Mexico, has been proposed to be the first bryozoan (Landing et al., 2010). It has alternatively been interpreted as an octocoral (Taylor et al., 2013), and this countered by Landing et al. (2015). The lack of bryozoans in Cambrian Burgess Shale-type Konservat-Lagerstätten may be attributed to the thin body wall cuticle of non-biomineralized bryozoans, as are inferred to have been in existence by then, and the relatively offshore, fine siliclastic bottoms that dominate this style of preservation, unfavorable to bryozoan suspension feeding (Taylor and Waeschenbach, 2015).

Unambiguous bryozoans are nonetheless found in the Early Ordovician, the oldest of which are stenolaemates from the Tremadocian of China (Ma et al., 2015a; Taylor and Waeschenbach, 2015). Stenolaemata is represented today only by the order Cyclostomata, which comprises less than 10% of living bryozoan species (versus more than 80% being in the gymnolaemate order Cheilostomata) (Bock and Gordon, 2013). Although stenolaemates are thus a fairly minor group now, they were the dominant Paleozoic bryozoans, with cyclostomes being joined by six extinct orders that are grouped together as a monophyletic Palaeostomata (Ma et al., 2014a) (fig. 50.1 H, J). They are united by being "free-walled," lacking calcified exterior walls between the zooids. Five of these six extinct orders first appeared in the Lower and Middle Ordovician, a time of major radiation of this class. This is marked not only by the first appearance of high-ranking groups but also by prolific speciation and ecological radiation (Taylor and Ernst, 2004). Stenolaemates are in fact among the most abundant macrofossils in some Paleozoic fossil sites.

The oldest gymnolaemates also first appear in the Ordovician, represented by borings produced by ctenostomes into calcareous substrates. Stenolaemate diversity was diminished in the major Paleozoic mass extinctions (end Ordovician, Late Devonian, end Permian). The latter was especially severe, with Triassic bryozoans being rare, known from just a few genera. Cyclostomes, however, recovered to become the most diverse group of Jurassic (fig. 50.1 I) and, especially, Cretaceous Bryozoa.

Although modern bryozoans are mostly cheilostomes, the first appearance of this order is not recorded until the Late Jurassic, and there seems little question that cheilostomes have a Mesozoic origin. Nonetheless, they diversified rapidly in the Cretaceous to overtake Cyclostomata by the latter part of that period. Cheilostomes represent one of multiple events of bryozoans evolving a calcareous skeleton, inferred to have formed de novo from a soft-bodied ancestor. Molecular evidence for ctenostomes being paraphyletic with respect to cheilostomes (Waeschenbach et al., 2012b) suggests that some fossil ctenostomes are stem-group Cheilostomata. The period of rapid diversification of cheilostomes has been associated with the appearance of larval brooding, indicated in the fossil record by the presence of ovicells; brooded larvae being non-planktotrophic might facilitate allopatric speciation compared to nonbrooded larvae that can feed in the plankton for months, although molecular phylogenies predict that brooding has convergent origins within cheilostomes (Taylor and Waeschenbach, 2015). The success of cheilostomes has also been linked to the presence of avicularia and other zooids that are often linked to the ability to deter micropredators. Avicularia appear in the Albian Stage of the Early Cretaceous, when the cheilostome diversification accelerated (Cheetham et al., 2006).

Although Phylactolaemata is sister group of the other two bryozoan classes, their non-biomineralized skeleton confers a relatively young and incomplete fossil record. Their best fossil evidence is in the form of the chitinized statoblasts, ranging back to the Permian (Vinogradov, 1996).

Entoprocta—their name referring to the "inside" position of the anus; sometimes also called Endoprocta or Kamptozoa—is a group of minute, sessile, solitary or colonial animals with a crown of feeding tentacles superficially resembling a hydroid. From the 180 or so described species (Borisanova and Potanina, 2016), all but two are marine or live in brackish waters (Nielsen, 2016b). The two known limnic species, the widespread *Urnatella gracilis* and the recently described Thai species *Loxosomatoides sirindhornae*, must therefore have invaded freshwater environments twice independently, as they are in different families (Wood, 2005; Schwaha et al., 2010). Entoprocts are often found growing on other invertebrates or encrusting algae or hard substrates, generally at shallow depths not exceeding 200 m, although recently they have been recorded from depths of up to 5,220 m (Borisanova et al., 2015).

Despite their small size and being relatively poorly known, interest in entoproct development dates back to the nineteenth century (Harmer, 1885). Entoprocts were characterized by Marcus (1939) as having spiral cleavage, originally contrasting with the idea that they are related to bryozoans and other lophophorates. Their life cycle has supported a sister-group relationship to bryozoans (Nielsen, 1971) or to bryozoans and cycliophorans (Funch and Kristensen, 1995; Nielsen, 2012a), but their foot morphology has been used as an argument for a sister-group relationship with Mollusca (Haszprunar and Wanninger, 2008; Merkel et al., 2015). The reality is that the exact phylogenetic position of entoprocts remains difficult to elucidate even when using large amounts of genomic data (e.g., Hejnol et al., 2009; Laumer et al., 2015a; Kocot et al., 2017; Laumer et al., 2019). Until very recently molecular data strongly contradicted a sister-group relationship to Mollusca, as predicted by the Lacunifera hypothesis, but analyses by Marlétaz et al. (2019) do indeed recover a mollusc–entoproct clade, something that requires further testing.

SYSTEMATICS

The current classification system of Entoprocta was established by Emschermann (1972b), who divided the phylum into the clades Solitaria (with Loxosomatidae) and Coloniales, which in turn divided into Astolonata (the then new family Loxokalypodidae) and Stolonata (Pedicellinidae and Barentsiidae). Certain characters, including the foot, have been suggested to bear relevant phylogenetic information (Iseto and Hirose, 2010), but few explicit morphological hypotheses have been proposed. The four-family system is thus followed by most authors (Nielsen, 2016b). Cuticular pores and hollow spines, in addition to muscles, are often used in taxonomy of entoprocts.

Molecular phylogenetic analyses of a handful of Sanger-based markers have supported a main division between Solitaria and Coloniales, but within Solitaria the genus *Loxosomella* is paraphyletic with respect to *Loxomitra* (Fuchs et al., 2010; Borisanova et al., 2015) and *Loxosoma* (K. Kocot, pers. comm.); monophyly of either Pedicellinidae and/or Barentsiidae is unsupported (Fuchs et al., 2010; Borisanova et al., 2015). *Loxokalypus socialis* (the only accepted species in the family Loxokaly-podidae) and the limnic species are so far unsampled in molecular phylogenies, and thus their positions, especially that of *Urnatella*, remain a mystery. We thus use here a conservative system while recognizing that loxosomatid genera, and especially the Coloniales families, may require extensive revision once the existing paradigm is properly tested phylogenetically.

Class Solitaria

   Family Loxosomatidae

Class Coloniales

   Family Loxokalypodidae

   Family Barentsiidae

   Family Pedicellinidae

ENTOPROCTA: A SYNOPSIS

- Triploblastic, bilateral, unsegmented pseudocoelomates
- Sessile, solitary or colonial animals, with a stalk bearing a cup-shaped body (calyx) including the internal organs, with a crown of tentacles
- U-shaped through-gut, the mouth and anus opening inside the crown of tentacles
- A pair of protonephridia; increased number in *Urnatella*
- Mostly hermaphroditic
- Spiral cleavage
- Indirect development, either via swimming- or creeping-type larva
- All but two species marine

The entoproct body plan consists of a cup-shaped body, the calyx, on top of a stalk, the calyx bearing a crown of ciliated tentacles, forming a closed horseshoe or almost a circle, which delimit the atrium or vestibule, where the gonoducts and protonephridia open. The cilia draw food to the mouth, located at the end of a buccal funnel, coalescing with the tentacles. The anus is situated at the end of an anal tubercle, in the atrium (hence the name Entoprocta); the gut is therefore U-shaped. In the solitary forms the stalk attaches to the substrate, sometimes bearing a complex foot (in loxosomatids). In Coloniales the stalk attaches to the branched stolons or to an enlarged basal plate.

The body wall and adult musculature have recently been studied using immunostaining (Wanninger, 2004; Fuchs et al., 2006; Schwaha et al., 2010; Borisanova et al., 2012) to complement earlier studies based on anatomical sectioning (Reger, 1969). These studies have shown that the muscular system of Coloniales is

less developed than that of loxosomatids, and the continuous stalk–calyx musculature typical of loxosomatids is often divided into separate muscles by a cuticular
diaphragm. Some colonial species present a sphincter below the stalk–calyx diaphragm (Schwaha et al., 2010). All entoprocts have a tentacle musculature with
outer main muscles and thinner inner muscles (Nielsen and Rostgaard, 1976;
Schwaha et al., 2010). Loxosomatids have two atrial ring muscles, but a single thick
muscle is found in the colonial *Loxosomatoides sirindhornae* (Schwaha et al., 2010).

The digestive tract muscles include esophageal ring muscles, an intestinal
sphincter and an anal sphincter, but other muscles, rectal retractors and rectum
musculature, are not present in the investigated colonial species. The calyx musculature, its development, and the stalk musculature, including the presence or absence of oblique muscles, is often used in taxonomy of the group but requires
further study. The typical behavior of stalk bending at the muscular urn-shaped
segments is what has given them the alternative name Kamptozoa, meaning "bending animal."

The myoanatomy of the "creeping larva" has also received recent attention
(Merkel et al., 2015), as the foot with a creeping sole has been used as a putative
synapomorphy to unite Entoprocta with Mollusca (Haszprunar, 1996a; Ax, 1999).
The muscle systems include numerous entoproct-specific muscles and some muscles that have been suggested to be likely creeping-type, larva-specific structures,
similar to those of molluscs. The entoproct-specific muscles include frontal organ
retractors, several other muscle fibers originating from the frontal organ, and longitudinal prototroch muscles. The muscles that resemble those of molluscs include
paired sets of dorsoventral muscles that intercross ventrally above the foot sole and
a paired enrolling muscle that is distinct from the musculature of the body wall
(Merkel et al., 2015). The myoanatomy of the swimming-type larva, although still
complex, is much simpler than that of the creeping-type larva (Fuchs and Wanninger, 2008).

While some authors maintain that there is no persistent body cavity in entoprocts, the area between the gut and body being filled with mesenchyme (Nielsen,
2016b), others provide evidence for a primary body cavity separated from all surrounding tissues by a basal lamina, which underlies the ectodermal and endodermal epithelia and covers the excretory system, gonads, muscular cells, and nerves
that cross the body cavity (Borisanova et al., 2014). They thus consider the body cavity of Entoprocta to be a hemocoel, as previously described by Bartolomaeus
(1993). This hemocoel has also been interpreted as a lacunar circulatory system,
similar to that of molluscs (e.g., Salvini-Plawen and Bartolomaeus, 1995; Ax, 1999;
Haszprunar and Wanninger, 2008), leading to the Lacunifera hypothesis, but others
reject this homology (Borisanova et al., 2014), especially since no phylogenetic analysis of molecular data had found support for Lacunifera, until the recent publication of Marlétaz et al. (2019).

Adult entoprocts usually possess a single pair of protonephridia in the calyx,
within the hemocoel and between the stomach and esophagus. These open into
the atrium by separate nephropores in loxosomatids or a single pore in pedicelli-
nids and barentsiids (Franke, 1993; Nielsen and Jespersen, 1997). The freshwater
Urnatella has long been known to have complex protonephridia (Emschermann,

FIGURE 51.1. Metamorphosis of three types of entoproct larva. Originals courtesy of Claus Nielsen.

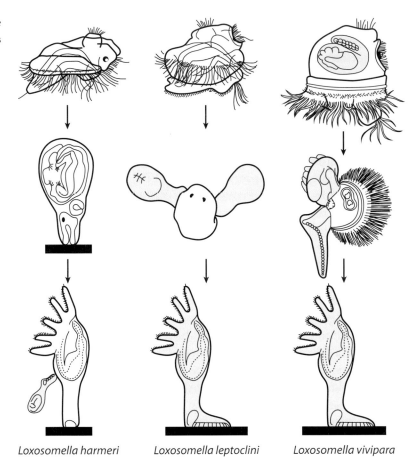

Loxosomella harmeri   *Loxosomella leptoclini*   *Loxosomella vivipara*

With a few exceptions, most entoprocts have a short larval life after breaking free from the egg envelope before settlement and metamorphosis—irrespective of whether they were lecitotrophic or planktotrophic. Metamorphosis (fig. 51.1) varies greatly in the different species, as beautifully summarized by Nielsen (1971), but little is known about the histological processes that occur during settlement and metamorphosis and nothing about the fate of the larval neural or muscular components and the transition to the adult individual.

Some species of *Loxosomella* settle with a frontal organ (fig. 51.1 left), retaining the larval gut to adulthood, while other loxosomatids develop internal or external buds from the larval episphere while the larval body disintegrates (fig. 51.1 center). In the colonial species, the larva settles by specific cells located above the retracted prototroch, undergoing unequal growth of the body mass and rotating the gut so that the atrial surface points away from the substrate (Nielsen, 2016b).

▨ GENOMICS

No genome data are available for any member of Entoprocta. Transcriptomic resources are available for a species of *Pedicellina* (from Hejnol et al., 2009), *Loxosoma pectinaricola*, and *Barentsia gracilis* (from Halanych and Kocot, 2014), but these are rather limited. Mitochondrial genomes are available for two species of Loxosomatidae (Yokobori et al., 2008).

FOSSIL RECORD

A tentaculate, stalked goblet-shaped organism from the early Cambrian Chengjiang biota, *Cotyledion tylodes*, has been redescribed as a stem-group entoproct (Zhang et al., 2013). It is a sessile form with the body differentiated into a calyx, stalk, and holdfast, either solitary or gregariously attached to skeletal parts of other organisms. The calyx and stalk, U-shaped gut, and mouth and anus surrounded by a crown of flexible (and possibly retractable) tentacles are the basis for aligning *Cotyledion* with Entoprocta, but it differs from crown-group entoprocts in having the calyx and stalk covered with elongate, ovoid sclerites and in its macroscopic size. The composition of the sclerites (i.e., whether they were biomineralized) is uncertain.

Crown-group entoprocts are known as fossils only from an occurrence of the extant genus *Barentsia* in the Late Jurassic of the United Kingdom, preserved as a colony overgrown by an oyster (Todd and Taylor, 1992).

Figure 52.1. Early Cambrian Brachiozoa. A–J represent tommotiids: A, *Eccentrotheca helenia*, scale 500 µm; C–E, *Paterimitra* sp., scales 250 µm; F–G, *Micrina etheridgei*, scales 250 µm; H–I, *Tannuolina* sp., scales 500 µm (H), 250 µm (I); J, the camenellan *Dailyatia ajax*, scale 200 µm; K, *Yuganotheca elegans*, scale 5 mm. Photo credits: A–I, Christian Skovsted; J, Glenn A. Brock; K, Derek Siveter.

PHORONIDA

Phoronids were first discovered in the autumn of 1845, after J. Müller recovered, from the Helgoland Sea, a new animal form that he later named *Actinotrocha branchiata*, thinking that it was the adult of a new animal (Müller, 1846). The first adult phoronids, *Phoronis hippocrepia* and *P. ovalis*, were not described until a decade later by Wright (1856), who considered them to "possess characters common to the Polyzoa [. . .], the Tunicata [. . .], and the Annelida, in which last class they probably ought to take their place." *Phoronis* was named after "one of the surnames of Isis," the Egyptian Goddess.

It was Schneider (1862) who described the metamorphosis of *Actinotrocha* into an adult "worm," one he thought was related to sipunculans. Apparently, the prolific Russian author Aleksandr O. Kowalevsky (1867) clarified the connection between *Actinotrocha* and *Phoronis* (see Emig et al., 2005), but a wrong citation has been used in the literature.[24] One of the earliest and most complete studies of the life cycle of *Phoronis hippocrepia* from the Atlantic North American coast was published more than a century ago (Brooks and Cowles, 1911).

Because of the dual nature of the discovery of this phylum, there have been multiple rulings of the International Code of Zoological Nomenclature about name priorities, and for a long time, a dual naming system was used (see the details in Emig et al., 2005) until very recently.

Phoronids are a group of exclusively marine benthic animals, ranging in size from 1.5 to 45 cm in length, when extended (Emig et al., 2005). There are currently 15 accepted species (Temereva and Chichvarkhin, 2017; Temereva and Neldyudov, 2017), many of which are supposedly widespread. They are found in all oceans except around Antarctica, mostly in shallow depths, but they can make it down to 400 to 600 m. Phoronids secrete a chitinous tube that attaches to hard substrates or to the tubes of other animals, like ceriantharians, or else they inhabit soft sediments, where they can retreat rapidly. This habitat difference seems correlated with anatomical characters of phylogenetic significance (Temereva and Neldyudov, 2017).

SYSTEMATICS

Phoronids are divided into two genera, *Phoronis* and *Phoronopsis*, and the relationships of their species were first examined using morphological characters by Emig (1985), who found *Phoronopsis* to be nested within a paraphyletic *Phoronis*—a result

---

[24] Emig et al. (2005) incorrectly cite the following reference: Kowalevsky, A. 1867. Anatomie und Entwicklung von Phoronis. Mémoires de l'Académie Impériale de Saint-Pétersbourg 10(15), 1–148; but article 10(15) is another Kowalevsky reference on ascidians.

FIGURE 53.1. Original drawing of *Phoronis hippocrepia* by Wright (1856).

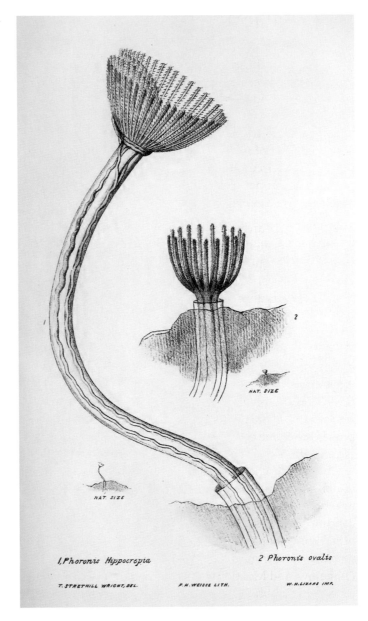

also supported by the morphological and molecular data of subsequent analyses (Santagata and Cohen, 2009; Hirose et al., 2014; Temereva and Neldyudov, 2017). It now seems clear that *Phoronis* is paraphyletic, while *Phoronopsis*, characterized by a folding in the trunk wall forming a collar under the lophophore, is monophyletic (Temereva and Neldyudov, 2017). *Phoronis ovalis* is now widely recognized as the sister species to all other phoronids and thus has been studied in detail, as it may have certain plesiomorphic character states that help connect phoronids with other lophophorate groups (Temereva, 2017b). The phylum remains with a single family and the two genera discussed above, but the generic designations obviously need revision.

▬ PHORONIDA: A SYNOPSIS

- Bilaterally symmetrical, triploblastic, vermiform lophophorates
- Secrete a chitinous tube they inhabit
- Two or three (disputed) coelomic cavities in the adult
- Body divided into flap-like epistome, lophophore-bearing collar, and trunk
- U-shaped through-gut, the anus located near the mouth
- One pair of metanephridia in the trunk
- Closed circulatory system
- Transient peritoneal gonads; can be gonochoristic or hermaphroditic
- Indirect development though an actinotroch larva

Deuterostomes are characterized by a trimeric coelomic compartmentalization consisting of an anterior protocoel, followed by a mesocoel and a posterior metacoel. Because the lophophorates have traditionally been placed with Deuterostomia (at least for most of the twentieth century), it has often been said that this trimeric organization is also found in phoronids, with the epistome containing a protocoel, the lophophore containing the mesocoel, and the trunk containing a metacoel. The interpretation of some of these cavities has long been debated (Cowles, 1904). A detailed ultrastructural study of the actinotroch larva and adult of *Phoronis muelleri* did not show an anterior cavity lined by epithelium (Bartolomaeus, 2001). Instead, what appears to be a cavity inside the larval episphere on the light microscopic level is an enlarged subepidermal extracellular matrix with an amorphous filling, into which several muscle cells are embedded (Bartolomaeus, 2001). Larvae, thus, possess only one coelomic cavity, the large trunk coelom that is adopted in the adult organization.

The second coelomic cavity of the adult, the lophophore coelom, develops as a double layer of epithelialized mesodermal cells at the base of the adult tentacle buds and becomes filled with fluid during metamorphosis (Bartolomaeus, 2001). This condition has also been described for *P. ovalis*, supposedly the sister group to all other known phoronids (Gruhl et al., 2005). However, subsequent work on *Phoronopsis harmeri* maintains that a coelomic cavity is present also in the epistome (Temereva and Malakhov, 2011), although it is difficult to discern the origin of such a cavity. The nature of the trimeric coelom of phoronids thus continues to puzzle biologists.

The coelomic cavity (and the chitinous tube) provide support to the animal, which has a weak muscular system, although it can quickly retreat into the tube, sometimes buried deep into the sediment. Anecdotally, we once tried to collect an undescribed species of sand phoronid by scuba, and while we were able to quickly cut a piece of the lophophore with scissors, we were unable to retrieve a single specimen by "surprising it" or by coring the sand with a tube longer than 50 cm.

The evolution of the asymmetry of the circulatory system of phoronids has recently been discussed by Temereva and Malakhov (2003).

One of the most important characteristics of phoronids is their large lophophore, used for feeding. It derives from the metacoel during development, constituting the metasome, and is formed by parallel ridges curved in an oval, horseshoe shape or, in the larger species, forming two symmetrical spirals; the tentacles show differences between these alternative lophophore shapes (Temereva and Malakhov, 2009b). The lophophore has been studied in detail using transmission electron microscopy and immunostaining techniques, with emphasis on the coelom, musculature, and nervous system (Herrmann, 1997; Temereva and Malakhov, 2009b; Temereva, 2017b). The lophophores of some species may contain accessory reproductive glands (Zimmer, 1967).

The chitinous tube is another trademark of phoronids. It is secreted by gland cells located in the epidermis, being sticky at first and then acquiring a parchment-like consistency, to which sand grains and other inorganic particles are often attached, especially in the soft-bottom species.

Phoronids have a pair of metanephridia in the trunk, which derive from the larval protonephridia (Bartolomaeus, 1989; Temereva and Malakhov, 2006). The adult metanephridia have a characteristic double nephrostome, the large and small nephrostomes; these open into the metacoel and share a nephridioduct. Each nephridiopore, which opens near the anus, is also used for releasing the gametes.

The adult nervous system of phoronids has been studied in detail and has received special attention in recent studies (see Santagata, 2002; Temereva and Malakhov, 2009a; Temereva, 2012, 2017b). A group of basiepidermal ciliated neuronal cell bodies concentrated between the mouth and the anus (Fernández et al., 1996), and of uncertain origin (although often said to be dorsal), appears to be the adult "brain" (Santagata, 2015c) and is connected to a collar nerve ring with unciliated neuronal cells along its length, at the base of the tentacles (Temereva and Malakhov, 2009a). Sporadically distributed neuronal cells and fibers are found throughout the surface of the trunk epithelium. A subset of these cells and fibers are serotonergic (Santagata, 2002), but the most centralized neuronal structure is the giant nerve fiber embedded in the anterior portion of the trunk epithelium (Fernández et al., 1996; Temereva and Malakhov, 2009a). The number (one or two), position, and shape of the giant nerve fibers are important taxonomic characters for phoronids (Emig, 1974).

The reproductive biology of phoronids has recently been summarized by Santagata (2015c) and Temereva (2018). All known phoronids have benthic adults and planktotrophic larvae, with the exception of *Phoronis ovalis*, which has a creeping lecithotrophic larva. Development can occur in the water column or by brooding embryos to an early larval or competent larval stage on specialized nidamental glands at the base of the lophophore. Two species incubate embryos inside the parental tube (Zimmer, 1991). Viviparity was recently reported in *P. embryolabi*, which incubates embryos in the trunk coelom to produce feeding actinotroch larvae (Temereva and Chichvarkhin, 2017).

Reproductive patterns in phoronids are connected with the mode of oogenesis, with small eggs being broadcast while larger eggs are often brooded (Temereva, 2018). An exception to the rule is *P. embryolabi*, with numerous small eggs (60 µm) that are incubated in the parental coelom (Temereva, 2018).

The embryonic development of *Phoronis vancouverensis* has been studied using cell-labeling techniques (Freeman and Martindale, 2002) and 4D microscopy (Pennerstorfer and Scholtz, 2012). Despite general belief that phoronids have a somehow modified radial cleavage pattern, a recent study applying 4D microscopy has shown that many cell divisions post–third quartet are oblique and with an alternation of dextral and sinistral cell division, concluding that the cleavage pattern of *Phoronis muelleri* displays several characters consistent with the pattern of spiral cleavage (Pennerstorfer and Scholtz, 2012). The early cleavage program does not generate cells of unique identity, cell fates being established at later developmental time points (Freeman and Martindale, 2002). These authors also showed that mesodermal cells form at ectodermal–endodermal boundaries from both germ layers, which may be consistent with reports proposing a dual origin of the coelom, with the anterior coelom/s originating by immigration, while the metacoel originates by enterocoely (Malakhov and Temereva, 1999; Temereva and Malakhov, 2007). The postembryonic development from larva to adult has been studied by a few authors (Silén, 1954).

The phoronid actinotroch larva has received perhaps more attention than the adult, with numerous recent studies focusing on its development, coelom formation, and nephridial origin. The larval intestine originates by ingression of posterior ectoderm (Freeman and Martindale, 2002) and has been followed through postembryonic development (Temereva, 2010). Other recent studies have used immunohistochemistry to study larval myoanatomy (Santagata, 2002; Santagata and Zimmer, 2002) and neuroanatomy (Hay-Schmidt, 1989, 1990; Santagata and Zimmer, 2002; Temereva, 2012, 2017a).

The feeding ecology and the directionality of particle collecting by the cilia of adult and larval phoronids have been characterized (Riisgård, 2002; Temereva and Malakhov, 2010). A row of sensory stiff laterofrontal cilia is found between the band of water-pumping lateral cilia and the band of particle-transporting frontal cilia (Riisgård, 2002). These cilia are also found in bryozoans, brachiopods, and pterobranchs.

GENOMICS

The complete nuclear genome of *Phoronis australis* has been sequenced (Luo et al., 2017) and comprises ca. 20,500 genes and high heterozygosity. The genome has eight Hox genes in one Hox cluster and three ParaHox genes. *Scr* and *Antp* were not found in the *P. australis* genome. Given that these two Hox genes are expressed in the shell-forming epithelium in brachiopods (Schiemann et al., 2017), their absence in the phoronid lineage may be related to the lack of a shell (Luo et al., 2017).

Mitochondrial genomes are available for *Phoronis psammophila* and *Phoronopsis harmeri*, differing only in the relative position of *atp6* and some *tRNAs*, and are highly conserved with respect to chiton molluscs (Helfenbein and Boore, 2004; Podsiadlowski et al., 2014).

FOSSIL RECORD

The lower Cambrian Chengjiang *Iotuba chengjiangensis* (and its likely synonym *Eophoronis chengjiangensis*) has been interpreted as a phoronid (Chen and Zhou, 1997; Chen, 2004; Hou et al., 2017), based on a supposedly U-shaped gut and anterior tentacles resembling a lophophore, but this affinity as well as the morphological interpretations have subsequently been questioned (Huang et al., 2004c; see the discussion in Ma et al., 2010). The gut has been regarded alternatively as looped because of curvature of the trunk (Huang et al., 2004c). *Iotuba* may instead be allied to the scalidophoran *Louisella pedunculata* (see Conway Morris, 2006) or some other priapulans, its affinities remaining problematic (Hou et al., 2017). Possible trace fossils in the form of burrows attributed to phoronids (Fenton and Fenton, 1934; Joysey, 1959) seem circumstantial, as discussed in earlier reviews of the fossil record of phoronids (Budd and Jensen, 2000; Valentine, 2004). Some problematic solitary Paleozoic fossils (such as tentaculitids and cornulitids) have been compared to phoronids, as have modular biomineralized forms such as hederelloid "bryozoans" (Taylor et al., 2010; Landing et al., 2018).

BRACHIOPODA

Brachiopods, or "lamp-shells," are regarded as something of an enigmatic group in modern seas, nowadays restricted to just a few hundred living species (443 extant species accounted for by Zhang, 2011), mostly confined to relatively deep waters or sciophilous environments. However, they are among the most intensively studied and familiar groups of invertebrates in the fossil record, with a known diversity of ca. 12,000 extinct species classified as some 6,000 genera spanning the early Cambrian to the Recent (Carlson, 2016). Because of their abundance and ubiquity in the fossil record, brachiopods are often used to document large-scale paleoecological changes (e.g., Mii et al., 2001). The five extant orders of brachiopods are supplemented by some 20 extinct orders, most of them based exclusively on "hard part" characters. The rhynchonelliform brachiopods were among the most abundant animals during the Paleozoic, drastically declining after that period. Possible outcompetition by bivalves has been evoked to account for this decline (Thayer, 1985), but metabolic rates do not suggest this replacement, as the metabolic space has always been broader in bivalves than in brachiopods, even during the Paleozoic (Payne et al., 2014).

Brachiopods are sessile, exclusively marine lophophorate animals, which prefer colder waters, with the exception of the tropical linguliforms. Three characteristics of brachiopods make them stand out among related phyla. Like the other lophophorates, they have a lophophore, but in this case with an internal skeletal element, a variably elaborate lophophore support. Like bivalve molluscs, brachiopods have a shell formed by two valves, often of different polymorphs of calcium carbonate, but they can also have calcium phosphate (in Lingulida). Their valves, instead of being hinged dorsally, as in bivalves (as well as in other bivalved molluscs and arthropods), are jointed such that one valve is dorsal and the other ventral, and these are hinged posteriorly. Finally, they share with annelids the presence of chaetae (often called "setae" in brachiopods) with the exact same composition and ultrastructure (Lüter, 2000b).

SYSTEMATICS

First named by Dumeril in 1806 within the "Mollusques," "Brachiopodes" have been classified many ways by emphasis being placed on different morphological characters, such features as lophophore form, the arrangement of the pedicle relative to the shell valves, and articulations between the valves. The distinction between inarticulate and articulate brachiopods had a long history in classification, with Inarticulata and Articulata being recognized and corroborated by molecular data. The former includes the extant Linguliformea and Craniiformea, which lack articulatory structures between the shell valves but have very different lifestyles, and the latter includes the extant Rhynchonelliformea. The articulate brachiopods include the three living orders Rhynchonellida, Thecideida, and Terebratulida, all of which have well-developed articulation structures in the form of ventral teeth and dorsal sockets.

Monophyly of Inarticulata has been endorsed in some cladistic analyses based on morphology (Carlson, 1995; Williams et al., 1996), whereas others instead group Craniiformea and Rhynchonelliformea in a putative clade Calciata, named based on calcite shell mineralogy (Holmer et al., 1995), versus a phosphatic shell in Linguliformea. Nearly all molecular analyses have found monophyly of both Articulata and Inarticulata (Cohen and Gawthrop, 1996; Cohen et al., 1998; Cohen, 2000; Cohen and Weydmann, 2005; Sperling et al., 2011), including when using transcriptomic data (Kocot et al., 2017; Laumer et al., 2019; Marlétaz et al., 2019).

While some authors have proposed that Phoronida is a subtaxon of Brachiopoda based on DNA sequence data of rRNAs (Cohen and Gawthrop, 1996; Cohen, 2000; Cohen and Weydmann, 2005), current evidence does not support this hypothesis (see chapter 52). Brachiopod monophyly has consistently been recovered in morphology-based phylogenetic analyses (Carlson, 1995; Holmer et al., 1995; Williams et al., 1996). Analyses based on nuclear housekeeping and ribosomal genes likewise recover monophyly of Brachiopoda, and restricting the analysis to slow-evolving sites strengthened support for brachiopod monophyly (Sperling et al., 2011), as do all recent phylogenomic analyses (Nesnidal et al., 2013b; Laumer et al., 2015a; Kocot et al., 2017; Laumer et al., 2019; Marlétaz et al., 2019). The best supported internal relationship in the same analyses was Lingulida and Craniida (= the extant Inarticulata) to the exclusion of Rhynchonelliformea (= Articulata).

The classification system of brachiopods has changed extensively through the years (Carlson, 1995; Williams et al., 1996), and we follow some of the aspects of the most recent revision by Carlson (2016), which includes the vast extinct diversity of the group, but reject the inclusion of Phoronida within Brachiopoda and retain Articulata and Inarticulata, due to the broad use in the recent literature and their monophyletic status (fig. 54.1).

Inarticulata
    Subphylum Linguliformea
        Class Lingulata
            Order Lingulida (25 extant spp.)
            Order Siphonotretida †
            Order Acrotretida †
        Class Paterinata †
            Order Paterinida †
    Subphylum Craniiformea
        Class Craniata
            Order Craniida (11 extant spp.)

Order Craniopsida †

Order Trimerellida †

Articulata

Subphylum Rhynchonelliformea

Class Obolellata †

Class Chileata †

Class Kutorginata †

Class Strophomenata †

Class Rhynchonellata

Order Pentamerida †

Order Rhynchonellida (ca. 50 extant species)

Order Atrypida †

Order Terebratulida (ca. 300 extant species)

Order Athyridida †

Order Thecideida (23 extant species)

Order Spiriferida †

Order Spiriferinida †

BRACHIOPODA: A SYNOPSIS

- Triploblastic, bilaterally symmetric, trimeric, coelomate lophophorates
- Body enclosed between two shells (valves); one dorsal and one ventral
- Often sessile, attached to the substratum by a fleshy pedicle, or cemented
- Valves lined (and produced) by mantle lobes formed by outgrowths of the body wall and creating a water-filled mantle cavity
- Body composed of an epistome, lophophore, viscerae, and mantle
- Lophophore circular to variably coiled; with or without internal skeletal support structures
- With setae, homologous to annelid chaetae
- Gut generally U-shaped; anus present (Inarticulata) or absent (Articulata)
- With one or two pairs of metanephridia
- Circulatory system rudimentary and open; restricted to distribution of nutrients
- Without specialized respiratory structures; oxygen transport through coelomic fluid
- Most are gonochoristic and undergo mixed or indirect life histories, with lobate larvae
- Gametes develop from transient gonadal tissue on peritoneum of metacoel
- Cleavage holoblastic, radial, and nearly equal; coeloblastulae usually gastrulate by invagination; blastopore closes and mouth and anus form secondarily

authors have, however, postulated that the shared and specific expression of Hox genes, together with *Arx*, *Zic*, and Notch pathway components in setae/chaetae and shell fields in brachiopods, molluscs, and annelids provide molecular support for the conservation of the molecular basis for these lophotrochozoan characters (Schiemann et al., 2017). The specificity of such homologies probably requires further study in additional species.

The mantle bears some analogies to that of bivalves, being attached to the dorsal and ventral valves and forming a water-filled mantle cavity. The lophophore and the viscera are located within the mantle cavity, as are the gills, viscerae, and foot in bivalves. The mantle edge often bears chitinous setae that may act as a sieve to prevent the entrance of large particles. These setae are of nearly identical structure as the chaetae of annelids (Storch and Welsch, 1972; Lüter, 2000b). Brachiopods have two types of setae, the adult setae (also found in larval Lingulida) and the larval setae of non-lingulid brachiopods. The larval setae, built by a chaetoblast, are distally surrounded by a single invaginated epidermal cell. The adult setae are built by a chaetoblast within a setal follicle, and several follicle cells are in contact with the setal surface and are involved in chaetogenesis (Lüter, 2000b). The mechanism of forming the adult setae is nearly indistinguishable from the mode of chaetogenesis in annelids.

The brachiopod lophophore (fig. 54.2) is a pair of tentacle-bearing arms that extend anteriorly into the mantle cavity, being evident when the valves open. Its shape varies between different species, ranging from circular or U-shaped to highly coiled (the spirolophe condition). The brachiopod lophophore, unlike those of the other lophophorate groups, is contained within the valves and is essentially immovable, especially so when it contains support elements. In most brachiopods, the lophophore is supported hydrostatically by the coelomic pressure, but in some derived articulate species (fossil and extant) there exist a series of mineralized supports: the brachidium (crura, spiralia, and loops in the fossil literature); brachial ridges on the interior of the dorsal valve; and brachiophores (Carlson, 2016). The lophophore and tentacles are profusely innervated, although different groups of brachiopods show alternative innervation patterns (Temereva and Kuzmina, 2017).

In order to feed, the valves must be opened so the lophophore can perform its filter-feeding, and the mechanism of valve operation differs among species (Lüter, 2016b). In articulates, two pairs of muscles are responsible for valve movements: the diductors open the valves while the adductors close them. Inarticulates, on the other hand, lack diductor muscles (and hinges), and the valves open by contracting the posterior adductor muscles and relaxation of the anterior muscle groups, while using the central adductors to close the valves.

The food particles are then drawn by ciliary action into the mouth, which is sometimes followed by a pharynx, which leads into an esophagus, the stomach, digestive diverticula, and the pylorus (James, 1997). The digestive tube is generally U-shaped (not in Craniidae). The brachiopod digestive system was classified by McCammon (1981) into three groups: (a) the infaunal inarticulates (Lingulida), with long intestines and an anal opening; (b) the epifaunal inarticulates (Craniida), with a pouchlike intestine, opening into an anus; and (c) the articulate spe-

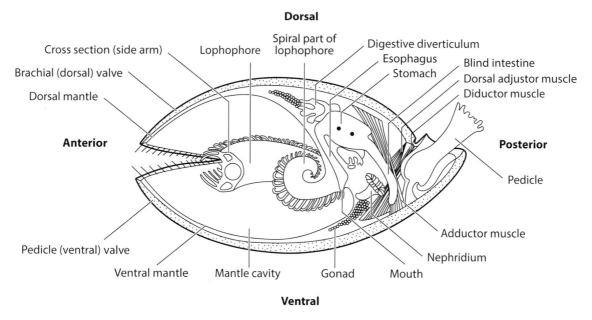

FIGURE 54.2. Generalized articulate rhynchonelliform brachiopod. Based on Brusca and Brusca (1990).

cies, with a blind intestine. When present, the anus opens anteriorly on the right side of the body, but in *Novocrania anomala* it opens posteriorly.

The pedicle emerges from the posterior of the animal, through the ventral valve (fig. 54.2), and may bear papillae or extensions that help to adhere to the substratum. The pedicle of Lingulida has intrinsic musculature, but that of the articulates lacks muscles. Because their structure and development are quite different, some authors are of the opinion that the pedicles evolved independently in the two groups (Carlson, 2016), although this seems implausible.

Brachiopods have a complex open circulatory system that includes a main central heart and several accessory hearts (Kuzmina et al., 2016) with a dual function, propulsion of blood in the ramified vessels and sinuses, and ultrafiltration of liquid that moves from the blood system to the perivisceral coelom (Kuzmina and Malakhov, 2014). The circulatory system is thus probably restricted to nutrient distribution (Lüter, 2016b) and not to oxygen transport, which seems to be mediated by certain hemerythrin-containing coelomocytes. No respiratory organs or specialized respiratory structures have been observed in brachiopods.

The coelomic cavities of brachiopods have received some attention (Lüter, 2000a; Kuzmina et al., 2006; Temereva et al., 2015; Temereva, 2017c) due to their role in the discussion of the deuterostome versus protostome debate. Some authors have argued for a bipartite coelom (with mesocoel and metacoel) in brachiopods (Lüter, 1996, 2000a), but an epistome preoral coelom (protocoel) was recently demonstrated in *Lingula anatina* (Temereva et al., 2015) and subsequently confirmed (Temereva, 2017c), the lophophore containing two main compartments: the preoral coelom (protocoel) and the lophophoral coelom (mesocoel). The epistome seems to be solid in rhynchonelliforms.

Epitheliomuscular/myoepithelial cells, largely corresponding to those of cnidarians (see chapter 8), have been found in a diversity of brachiopods, including the coelomic mesenteries (Storch and Welsch, 1974, 1976), the periesophageal coelom (Kuzmina and Malakhov, 2011), in the accessory hearts and main heart (Kuzmina and Malakhov, 2014; Kuzmina et al., 2016), and in the lophophoral coelom (Temereva, 2017c). In addition to the coelomic linings, all epithelia are monolayered and all ciliated cells are monociliate (Nielsen, 2012a).

Brachiopods have one or two pairs of metanephridia—two pairs in most rhynchonellids, except in *Cryptopora boettgeri* (see Helmcke, 1940)—disposed at each side of the esophagus, connected by a ileoparietal band (a tissue "bridge") (Lüter, 1995; James, 1997). The second pair of metanephridia of rhynchonellids is supported by a gastroparietal band. The structure of the metanephridia is similar in all known species, with the nephrostome opening to the metacoel and the nephridioducts opening via nephridiopores into the mantle cavity (Lüter, 2016b). Metanephridia also function as gonoducts, discharging the gametes from the coelom into the mantle cavity.

Older accounts postulated that brachiopods excrete ammonia mostly by diffusion via the tissue surface, mostly of the mantle and the lophophore (James et al., 1992), and that metanephridia may not play a role in excretion (James, 1997), as in the terebratellid *Calloria inconspicua* the metanephridia remain closed until the animal becomes sexually mature, during the second year of life (Percival, 1944). This idea is contradicted by more recent observations of the ectodermally derived nephridial duct internalizing liquid droplets from the nephridial lumen (C. Lüter, pers. comm.) and by the description of several podocyte-like cell arrangements in different places of the brachiopod body. In addition, the ectodermally derived canal of the metanephridium of *Terebratalia transversa* already starts to invaginate during the larval stage (Santagata, 2011: fig. 4 A). During or immediately after metamorphosis, these ducts become connected to the outwardly growing metanephridial funnels, which are derived from coelomic epithelium, that is, mesoderm (C. Lüter, pers. comm.). From the very beginning of the juvenile's filter-feeding lifestyle, these metanephridia are functioning excretory organs. Therefore, it is now clear that in addition to acting as gonoducts, the nephridia also discharge phagocytic coelomocytes that accumulate metabolic waste (Lüter, 2016b).

The brachiopod central nervous system has recently been summarized by Lüter (2016a), and a comprehensive study of the larval nervous system was provided by Santagata (2011). Other recent works have focused on the innervation of the lophophore (Temereva and Tsitrin, 2015; Temereva and Kuzmina, 2017). The nervous system is somewhat reduced, with a dorsal ganglion connected by a circumenteric nerve ring to a ventral ganglion. From these ganglia and the nerve ring, a series of nerves emerge and extend to the muscles, mantle (especially the mantle edge), setae, and lophophore.

Little is known about sense organs in brachiopods. In linguliforms the swimming juvenile has a median tentacle with primary receptor cells at its tip, which directly project into a circumoesophageal nerve ring (Hay-Schmidt, 1992; Lüter, 1996), but its function remains unknown. A pair of statocysts has been reported in the late larvae and adults of disciniids (e.g., Chuang, 1968). Eyes have also been reported in some rhynchonelliforms (see Lüter, 2016a) and linguliform (e.g., Chuang, 1968) larvae. Neuropeptide-mediated behavioral responses have been registered in larvae (Thiel et al., 2017).

REPRODUCTION AND DEVELOPMENT

The gonads are formed from the peritoneum during the reproductive season and extend into the mantle canals. Brachiopods are generally free spawners, but some species retain the embryos in the lophophore for some time. Craniiformea and Rhynchonelliformea have lecitotrophic larvae, whereas Linguliformea have planktotrophic larvae.

The embryology of a few brachiopod species has been studied (e.g., Conklin, 1902; Long and Stricker, 1991; Nielsen, 2005; Lüter, 2007) and recently reviewed by Nielsen (2012a) and Santagata (2015a). Early cleavage is typically radial and holoblastic, but quite a lot of variability has been reported (Freeman, 2003). Early development fate maps are available for a few brachiopod species, specially by the dedicated work of American developmental biologist Gary Freeman (1993, 1995, 1999, 2000, 2003).

Developmental work on brachiopods has focused on several aspects of their unusual body plan. They have been a key taxon for understanding major transitions in developmental patterns in bilaterians, including protostomy versus deuterostomy, as both forms of development occur within the group: protostomy in *Terebratalia transversa* and deuterostomy in *Novocrania anomala* (see Martín-Durán et al., 2016a). In the case of brachiopods, it has been shown that the switch between protostomy and deuterostomy depends on the differential activity of Wnt signaling, together with the timing and location of mesoderm formation. Segment-polarity genes have also been studied in brachiopods, and they are involved in nonsegmental roles in body patterning, including shell formation (Vellutini and Hejnol, 2016; Schiemann et al., 2017). Interestingly, *Terebratalia transversa*, unlike most other known animals, exhibits a spatial noncolinear Hox gene expression, both in the pre- and postmetamorphic stages (Schiemann et al., 2017; Gąsiorowski and Hejnol, 2019).

Sperm ultrastructure has played an important role in phylogenetic inference of many animal groups, but data on brachiopods are restricted to three inarticulates (*Lingula anatina*, *Novocrania anomala*, and *Discinisca tenuis*) (Afzelius and Ferraguti, 1978a; Hodgson and Reunov, 1994) and two articulates (*Terebratulina retusa* and *Kraussina rubra* (Afzelius and Ferraguti, 1978a; Hodgson and Reunov, 1994) and have not been analyzed cladistically.

GENOMICS

The complete nuclear genome of the inarticulate *Lingula anatina* totals 425 Mb and has a gene number (ca. 34,000) higher than its closest relatives, by extensive expansion of gene families, especially in its unique biomineralization (Luo et al., 2015a).

For example, molluscs and brachiopods share biomineralization genes that encode for enzymes such as chitin synthase (CHS) and bone morphogenetic protein (BMP) signaling. However, they show several domain combinations to produce lineage-specific shell matrix collagens, alanine-rich fibers, and novel shell matrix proteins (SMPs) (Luo et al., 2015a).

Mitochondrial genomes are available for one inarticulate (Luo et al., 2015b) and a few articulate (Stechmann and Schlegel, 1999; Noguchi et al., 2000; Helfenbein et al., 2001; Karagozlu et al., 2017) species.

FOSSIL RECORD

As discussed in Lophophorata, brachiopod origins have been addressed in the broader context of a brachiopod–phoronid clade, and the sclerite-bearing tommotiids are the strongest candidates for stem-group brachiopods. As discussed under Brachiozoa (chapter 52), tommotiid sclerites and brachiopod shells share several fine structural similarities (Balthasar et al., 2009). Tannuolinid tommotiids such as *Tannuolina* (see Skovsted et al., 2014) and *Micrina* (see Holmer et al., 2002) have especially been compared in detail with Cambrian and Recent brachiopods, and their perforate sclerites include the shell-penetrating pores that housed setae, as in crown-group brachiopods (Skovsted et al., 2014). Scleritome organization differs between different tannuolinids, *Micrina* effectively having two valves like crown-group brachiopods (fig. 52.1 F–G) but *Tannuolina* having additional sclerites (fig. 52.1 H–I).

The geologically oldest crown-group brachiopods are paterinids (fig. 54.3 B, C, E), a group of Linguliformea confined to the Cambrian and Ordovician. They first appear at the base of Cambrian Stage 2 in Siberia. The calcareous-shelled Inarticulata, Craniiformea, have the extant order Craniida first appearing near the base of the Ordovician, but include two extinct orders in the Paleozoic, one of which (Trimerellida) had an original aragonitic shell (Balthasar et al., 2011), in contrast to high magnesium calcite in craniids.

The biomineralized shell of brachiopods permits considerable information on soft anatomy to be inferred via, for example, the lophophore support, impressions of muscles, and the mantle canals. In addition, soft parts such as the pedicle are preserved in some fossil brachiopods, including forms as far back as the early Cambrian. Exceptionally preserved linguliforms are especially well known from the Cambrian Chengjiang biota (fig. 54.3 A) and provide evidence for the U-shaped gut and anteriorly placed anus, as well as the pedicle growing from the ventral mantle (Zhang and Holmer, 2013), and in some cases even the preservation of the soft-tissue of the lophophore (Zhang et al., 2004). Soft anatomy, including exceptionally preserved setae, is also known for the common Chengjiang species *Heliomedusa orienta* (fig. 54.3 F), underpinning its transfer from Craniopsida to either Discinoidea (Lingulida) (Chen et al., 2007a) or a position in the brachiopod stem group (Zhang and Holmer, 2013). Setae are likewise preserved in some Burgess Shale brachiopods (fig. 54.3 D, E). Soft anatomical preservation in Cambrian brachiopods was reviewed by Holmer and Caron (2006). Soft tissue preservation is rare for rhychonelliforms, but a Silurian example includes a schizolophe lophophore and the pedicle (Sutton et al., 2005).

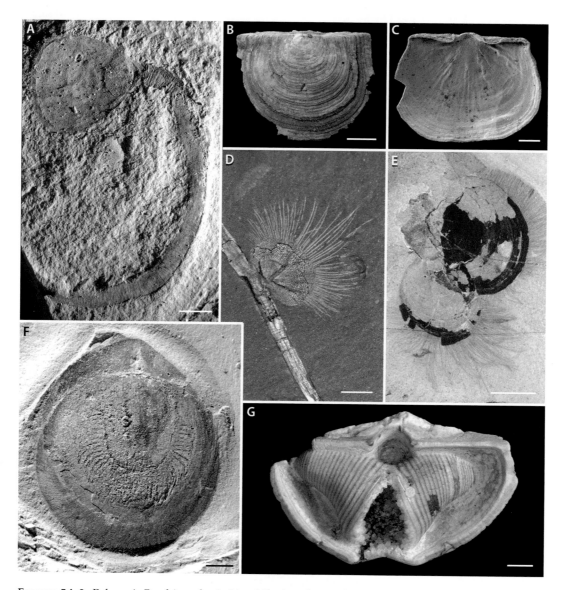

FIGURE 54.3. Paleozoic Brachiopoda: A, *Lingulella chengjiangensis*, with pedicle preserved, scale 2.5 mm; B, C, the paterinid *Askepasma toddense*, scales 1 mm; D, *Micromitra burgessensis*, with setae preserved, attached to the cnidarian? *Tubulella flagellum*, scale 3 mm; E, *Paterina zenobia*, with setae preserved, scale 5 mm; F, *Heliomedusa orienta*, with soft part preservation, scale 2 mm; G, *Spirifer striata*, with spiral lophophore support prepared, scale 1 cm. Photo credits: A, F, Derek Siveter; B, C, Glenn A. Brock; D, E, Tim Topper.

Brachiopods were a fairly low diversity group in the Cambrian, although they already occupied much of their full spectrum of lifestyles by the early Cambrian (Topper et al., 2015). In terms of diversification, however, their prolific radiation in the Ordovician, and sustained throughout the Paleozoic, is one of the hallmarks of the Great Ordovician Biodiversification Event. Although rhynchonelliforms are represented by several orders in the Cambrian (fig. 54.1), they underwent a steady diversification in the Early Ordovician, then more than doubled generic diversity to over 200 genera in the Darriwilian Stage of the Middle Ordovician (Harper et al., 2017). The earliest (Cambrian) rhynchonelliforms belong to extinct orders that had effective hinges between the valves, but either lacked teeth or other such articulatory structures that are known from later members (collectively Neoarticulata, which includes the three extant orders) or had teeth of uncertain homology.

Most of rhychonelliform diversity can be sorted into two major clades, rhynchonellates and strophomentates (Williams et al., 1996), that had split from each other by the Cambrian. The former includes the three living articulate orders, whereas the strophomenates are nearly exclusively Paleozoic. Brachiopods were largely decimated in the end Permian mass extinction. Although the Triassic and Jurassic include the last records of a few typically Paleozoic orders (Productida, Spiriferinida, Athyridida), Mesozoic and Cenozoic brachiopods otherwise consist exclusively of the extant orders. Terebratulida, a group that first appeared near the base of the Devonian but was a minor component of mid and Late Paleozoic faunas, has dominated brachiopod diversity throughout the Mesozoic and Cenozoic. The end Permian extinction marked a replacement of diverse groups that lacked pedicles as adults by those that are attached by a pedicle to hard substrates (Carlson, 2016).

Within the radiation of articulate brachiopods, innovations in lophophore supports provide a major basis for identifying groups. The earliest fossil brachiopods lacked any such mineralized support, the lophophore being supported hydrostatically, and this condition is common in the Cambrian and Ordovician. After the end Ordovician mass extinction, diversity of Silurian and Devonian articulate brachiopods was dominated by taxa that had crura or spiralia (fig. 54.3 G) as lophophore supports, whereas Late Paleozoic shells often have brachial ridges (Carlson, 2016). Mineralized lophophore supports allow brachiopods to be classified according to lophophore types. The spiral loops known as the "spirolophe" condition of the lophophore are reconstructed as the basal state for brachiopods as a whole (Carlson, 2016) and are especially common in the Paleozoic.

PROBLEMATICA

In this book we have discussed many animal body plans and have tried to understand the phylogenetic position and other aspects of most large animal clades. But we have always been fascinated by the so-called Problematica, which are often the last chapters/sections in other textbooks. Westheide and Rieger (2007) discuss Myxozoa and Mesozoa in their concluding chapter—the latter including classical "mesozoans" (Dicyemida and Orthonectida) and other groups, such as Chaetognatha and Xenoturbellida. More often discussed are groups like *Buddenbrockia* and the remaining Myxozoa, Dicyemida and Orthonectida, or curious animals like *Salinella salve* or *Diurodrilus* (Nielsen, 2012a). Inspiring us to undertake studies on rare animals is—among others discussing these "Problematica"—Haszprunar et al. (1991), who discussed *Trichoplax adhaerens* (Placozoa), *Buddenbrockia plumatellae* (Cnidaria), *Salinella salve* (Incertae sedis; fig. 55.1 A), Mesozoa, *Xenoturbella bocki* (Xenacoelomorpha), *Lobatocerebrum psammicola*, and *Plactosphaera pelagica* (Hemichordata). All but one of these taxa are discussed in other sections of this book.

The phylogenetic position of some of the taxa listed above is now generally well understood (for example, Myxozoa is a subclade of Cnidaria; *Lobatocerebrum* and *Diurodrilus* are derived annelids), whereas others are still unstable. Nonetheless, it is clear that Chaetognatha, Dicyemida, and Orthonectida are spiralians, chaetognaths now recognized as close relatives of Gnathifera, and that Xenoturbellida forms a clade with Acoelomorpha, most plausibly as the sister group of Nephrozoa.

The enigmatic *Jennaria pulchra*, originally allied with annelids (Rieger, 1991a, b), continues to be untested. It was found in humid sand above the water level in North Carolina, but when searching in that exact site we could only find an interstitial

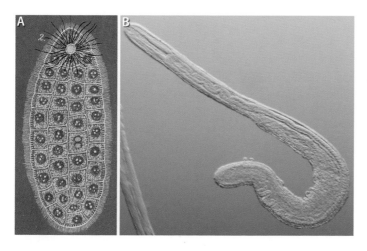

FIGURE 55.1. A, original illustration of *Salinella salve*; B, a North Carolina cephalotricid nemertean, which inhabits the same sediments from which *Jennaria pulchra* was described. Photo credits: B, Katrine Worsaae.

cephalotricid nemertean (fig. 55.1 B). More effort should be spent trying to find this mysterious animal.

Finally, we know about *Salinella salve* as much as we did at the turn of the nineteenth century. Classified with Mesozoa by several authors (Hyman, 1940; Westheide and Rieger, 1996), some have recognized a possible affinity to protists and suggested the need for a separate phylum (Hyman, 1940), for which Monoblastozoa was erected (Blackwelder, 1963). *Salinella salve* was described by Frenzel (1892) from a water culture obtained from a salt pan in Córdoba, Argentina, but it has never been found again, despite multiple attempts to re-collect it. These include a documented recent attempt in 2012 by German malacologist Michael Schrödl (see Dunning, 2012), a renowned specialist in mesopsammic molluscs (e.g., Jörger et al., 2014). Schrödl discovered that Johannes Frenzel had not collected the soil samples from which he cultured *Salinella* himself, and finding it difficult to obtain reliable soil samples, was not able to recover any *Salinella* specimens. This mysterious organism, although based on a detailed original description by a serious scholar and meticulous artist (e.g., fig. 55.1 A), is questioned by most. A detailed account of *Salinella salve*, which is supposed to have a through-gut and reproduce asexually by fission, is given by Brusca et al. (2016b), who have continued to feature this animal despite their skeptical views on its existence (Brusca and Brusca, 1990, 2003), as do other authors (Minelli, 2009).

BIBLIOGRAPHY

Abad, P.; Gouzy, J.; Aury, J. M.; Castagnone-Sereno, P.; Danchin, E. G. J.; Deleury, E.; Perfus-Barbeoch, L.; Anthouard, V.; Artiguenave, F.; Blok, V. C.; et al. 2008. Genome sequence of the metazoan plant-parasitic nematode *Meloidogyne incognita*. Nature Biotechnology 26, 909–915.

Abele, L. G.; Kim, W.; Felgenhauer, B. E. 1989. Molecular evidence for inclusion of the phylum Pentastomida in the Crustacea. Molecular Biology and Evolution 6, 685–691.

Achatz, J. G.; Chiodin, M.; Salvenmoser, W.; Tyler, S.; Martinez, P. 2013. The Acoela: on their kind and kinships, especially with nemertodermatids and xenoturbellids (Bilateria incertae sedis). Organisms Diversity and Evolution 13, 267–286.

Achatz, J. G.; Martinez, P. 2012. The nervous system of *Isodiametra pulchra* (Acoela) with a discussion on the neuroanatomy of the Xenacoelomorpha and its evolutionary implications. Frontiers in Zoology 9, 27.

Adams, E. D. M.; Goss, G. G.; Leys, S. P. 2010. Freshwater sponges have functional; sealing epithelia with high transepithelial resistance and negative transepithelial potential. PLoS One 5, e15040.

Adams, M. D.; Celniker, S. E.; Holt, R. A.; Evans, C. A.; Gocayne, J. D.; Amanatides, P. G.; Scherer, S. E.; Li, P. W.; Hoskins, R. A.; Galle, R. F.; et al. 2000. The genome sequence of *Drosophila melanogaster*. Science 287, 2185–2195.

Addamo, A. M.; Vertino, A.; Stolarski, J.; García-Jiménez, R.; Taviani, M.; Machordom, A. 2016. Merging scleractinian genera: the overwhelming genetic similarity between solitary *Desmophyllum* and colonial *Lophelia*. BMC Evolutionary Biology 16, 149.

Adell, T.; Martín-Durán, J. M.; Saló, E.; Cebrià, F. 2015. Platyhelminthes. In: Wanninger, A. (ed.), Evolutionary Developmental Biology of Invertebrates 2: Lophotrochozoa (Spiralia). Springer, Vienna, pp. 21–40.

Adema, C. M.; Hillier, L. W.; Jones, C. S.; Loker, E. S.; Knight, M.; Minx, P.; Oliveira, G.; Raghavan, N.; Shedlock, A.; do Amaral, L. R.; et al. 2017. Whole genome analysis of a schistosomiasis-transmitting freshwater snail. Nature Communications 8, 15451.

Adrain, J. M. 2011. Class Trilobita Walch, 1771. In: Zhang, Z.-Q. (ed.), Animal biodiversity: An outline of higher-level classification and survey of taxonomic richness. Magnolia Press, Auckland, pp. 104–109.

Adrianov, A. V.; Maiorova, A. S. 2015. *Pycnophyes abyssorum* sp. n. (Kinorhyncha: Homalorhagida), the deepest kinorhynch species described so far. Deep-Sea Research. II: Topical Studies in Oceanography 111, 49–59.

Adrianov, A. V.; Malakhov, V. V. 1995a. The phylogeny and classification of the class Kinorhyncha. Zoosystematica Rossica 4, 23–44.

Adrianov, A. V.; Malakhov, V. V. 1995b. The phylogeny, classification and zoogeography of the class Priapulida. 1. Phylogeny and classification. Zoosystematica Rossica 4, 219–238.

Adrianov, A. V.; Malakhov, V. V. 2001a. Symmetry of priapulids (Priapulida). 1. Symmetry of adults. Journal of Morphology 247, 99–110.

Adrianov, A. V., Malakhov, V. V. 2001b. Symmetry of priapulids (Priapulida). 2. Symmetry of larvae. Journal of Morphology 247, 111–121.

Afzelius, B. A.; Ferraguti, M. 1978a. Fine structure of brachiopod spermatozoa. Journal of Ultrastructure Research 63, 308–315.

Afzelius, B. A.; Ferraguti, M. 1978b. The spermatozoon of *Priapulus caudatus* Lamarck. Journal of Submicroscopic Cytology 10, 71–79.

Aguado, M. T.; Glasby, C. J.; Schroeder, P. C.; Weigert, A.; Bleidorn, C. 2015. The making of a branching annelid: an analysis of complete mitochondrial genome and ribosomal data of *Ramisyllis multicaudata*. Scientific Reports 5, 12072.

Aguado, M. T.; Grande, C.; Gerth, M.; Bleidorn, C.; Noreña, C. 2016. Characterization of the complete mitochondrial genomes from Polycladida (Platyhelminthes) using next-generation sequencing. Gene 575, 199–205.

Aguado, M. T.; Noreña, C.; Alcaraz, L.; Marquina, D.; Brusa, F.; Damborenea, C.; Almon, B.; Bleidorn, C.; Grande, C. 2017. Phylogeny of Polycladida (Platyhelminthes) based on mtDNA data. Organisms Diversity and Evolution 17, 767–778.

Aguinaldo, A. M. A.; Turbeville, J. M.; Lindford, L. S.; Rivera, M. C.; Garey, J. R.; Raff, R. A.; Lake, J. A. 1997. Evidence for a clade of nematodes, arthropods and other moulting animals. Nature 387, 489–493.

Ahlrichs, W. H. 1993a. On the protonephridial system of the brackish-water rotifer *Proales reinhardti* (Rotifera, Monogononta). Microfauna Marina 8, 39–53.

Ahlrichs, W. H. 1993b. Ultrastructure of the protonephridia of *Seison annulatus* (Rotifera). Zoomorphology 113, 245–251.

Ahlrichs, W. H. 1995a. *Seison annulatus* and *Seison nebaliae*—Ultrastruktur und Phylogenie. Verhandlungen der Deutschen zoologischen Gesellschaft 88, 155.

Ahlrichs, W. H. 1995b. Zur Ultrastruktur und Phylogenie von *Seison nebaliae* Grube, 1859, und *Seison annulatus* Claus, 1876—Hypothesen zu phylogenetischen Verwandtschaftsverhältnissen der Bilateria. Cuvillier, Göttingen.

Ahlrichs, W. H. 1997. Epidermal ultrastructure of *Seison nebaliae* and *Seison annulatus*, and a comparison of epidermal

Ax, P. 2001. Das System der Metazoa. III. Ein Lehrbuch der phylogenetischen Systematik. Spektrum Akademischer Verlag, Stuttgart.

Ax, P. 2003. Multicellular animals: Order in Nature—System made by man, vol. 3. Springer, Berlin.

Ax, P.; Dörjes, J. 1966. *Oligochoreus limnophilus* nov. spec., ein kaspisches Faunenelement als erster Süsswasservertreter der Turbellaria Acoela in Flüssen Mitteleuropas. Internationale Revue der gesamten Hydrobiologie und Hydrographie 57, 15–44.

Azevedo, R. B. R.; Cunha, A.; Emmons, S. W.; Leroi, A. M. 2000. The demise of the Platonic worm. Nematology 2, 71–79.

Babcock, L. E.; Grunow, A. M.; Sadowski, G. R.; Leslie, S. A. 2005. *Corumbella*, an Ediacaran-grade organism from the Late Neoproterozoic of Brazil. Palaeogeography, Palaeoclimatology, Palaeoecology 220, 7–18.

Babonis, L. S.; DeBiasse, M. B.; Francis, W. R.; Christianson, L. M.; Moss, A. G.; Haddock, S. H. D.; Martindale, M. Q.; Ryan, J. F. 2018. Integrating embryonic development and evolutionary history to characterize tentacle-specific cell types in a ctenophore. Molecular Biology and Evolution 35, 2940–2956.

Babonis, L. S.; Martindale, M. Q. 2017. *PaxA*, but not *PaxC*, is required for cnidocyte development in the sea anemone *Nematostella vectensis*. EvoDevo 8, 14.

Baccetti, B.; Burrini, A. G.; Dallai, R.; Pallini, V. 1979. Recent work in myriapod spermatology. (The spermatozoon of Arthropoda 31.). In: Camatini, M. (ed.), Myriapod biology. Academic Press, London, pp. 97–104.

Baccetti, B.; Rosati, F. 1968. The fine structure of the Polian vesicles of holothurians. Zeitschrift für Zellforschung und Mikroskopische Anatomie 90, 148–160.

Bachmann, L.; Fromm, B.; Patella de Azambuja, L.; Boeger, W. A. 2016. The mitochondrial genome of the egg-laying flatworm *Aglaiogyrodactylus forficulatus* (Platyhelminthes: Monogenoidea). Parasites and Vectors 9, 285.

Baeumler, N.; Haszprunar, G.; Ruthensteiner, B. 2012. Development of the excretory system in a polyplacophoran mollusc: stages in metanephridial system development. Frontiers in Zoology 9, 23.

Baguñà, J. 2012. The planarian neoblast: the rambling history of its origin and some current black boxes. International Journal of Developmental Biology 56, 19–37.

Baguñà, J.; Ballester, R. 1978. The nervous system in planarians: Peripheral and gastrodermal plexuses, pharynx innervation, and relationship between central nervous system structure and acoelomate organization. Journal of Morphology 155, 237–252.

Baguñà, J.; Martinez, P.; Paps, J.; Riutort, M. 2008. Back in time: a new systematic proposal for the Bilateria. Philosophical Transactions of the Royal Society B: Biological Sciences 363, 1481–1491.

Baguñà, J.; Riutort, M. 2004a. The dawn of bilaterian animals: the case of acoelomorph flatworms. BioEssays 26, 1046–1057.

Baguñà, J.; Riutort, M. 2004b. Molecular phylogeny of the Platyhelminthes. Canadian Journal of Zoology 82, 168–193.

Baguñà, J.; Romero, R.; Saló, E.; Collet, J.; Auladell, C.; Ribas, M.; Riutort, M.; García-Fernàndez, J.; Burgaya, F.; Bueno, D. 1990. Growth, degrowth and regeneration as developmental phenomena in adult freshwater planarians. In: Marthy, H.-J. (ed.), Experimental Embryology in Aquatic Plants and Animals. Plenum Press, New York, pp. 129–162.

Baguñà, J.; Saló, E.; Auladell, C. 1989. Regeneration and pattern formation in planarians. III. Evidence that neoblasts are totipotent stem cells and the source of blastema cells. Development 107, 77–86.

Bahia, J.; Padula, V.; Schrödl, M. 2017. Polycladida phylogeny and evolution: integrating evidence from 28S rDNA and morphology. Organisms Diversity and Evolution 17, 653–678.

Bahrami, A. K.; Zhang, Y. 2013. When females produce sperm: Genetics of *C. elegans* hermaphrodite reproductive choice. Genes Genomes Genetics 3, 1851–1859.

Bai, X.; Adams, B. J.; Ciche, T. A.; Clifton, S.; Gaugler, R.; Kim, K.-s.; Spieth, J.; Sternberg, P. W.; Wilson, R. K.; Grewal, P. S. 2013. A lover and a fighter: the genome sequence of an entomopathogenic nematode *Heterorhabditis bacteriophora*. PLoS One 8, e69618.

Bailly, X.; Laguerre, L.; Correc, G.; Dupont, S.; Kurth, T.; Pfannkuchen, A.; Entzeroth, R.; Probert, I.; Vinogradov, S.; Lechauve, C.; et al. 2014. The chimerical and multifaceted marine acoel *Symsagittifera roscoffensis*: from photosymbiosis to brain regeneration. Frontiers in Microbiology 5, 498.

Baird, A. H.; Bhagooli, R.; Ralph, P. J.; Takahashi, S. 2009. Coral bleaching: the role of the host. Trends in Ecology and Evolution 24, 16–20.

Baker, A. N.; Rowe, F. W. E.; Clark, H. E. S. 1986. A new class of Echinodermata from New Zealand. Nature 321, 862–864.

Baker, J. M.; Funch, P.; Giribet, G. 2007. Cryptic speciation in the recently discovered American cycliophoran *Symbion americanus*; genetic structure and population expansion. Marine Biology 151, 2183–2193.

Baker, J. M.; Giribet, G. 2007. A molecular phylogenetic approach to the phylum Cycliophora provides further evidence for cryptic speciation in *Symbion americanus*. Zoologica Scripta 36, 353–359.

Balavoine, G. 2014. Segment formation in Annelids: patterns, processes and evolution. International Journal of Developmental Biology 58, 469–483.

Baliński, A.; Sun, Y. 2017. Early Ordovician black corals from China. Bulletin of Geosciences 92, 1–12.

Baliński, A.; Sun, Y.; Dzik, J. 2013. Traces of marine nematodes from 470 million years old Early Ordovician rocks in China. Nematology 15, 567–574.

Ball, E. E.; Miller, D. J. 2006. Phylogeny: The continuing classificatory conundrum of chaetognaths. Current Biology 16, R593–596.

Ballesteros, J. A.; Sharma, P. P. 2019. A critical appraisal of the placement of Xiphosura (Chelicerata) with account of known sources of phylogenetic error. Systematic Biology 68, https://doi.org/10.1093/sysbio/syz011.

Balsamo, M. 1992. Hermaphroditism and parthenogenesis in lower Bilateria: Gnathostomulida and Gastrotricha. Collana U.Z.I. Selected Symposia and Monographs 6, 309–327.

Balsamo, M.; d'Hondt, J.-L.; Kisielewski, J.; Pierboni, L. 2008. Global diversity of gastrotrichs (Gastrotricha) in fresh waters. Hydrobiologia 595, 85–91.

Balsamo, M.; Manicardi, G. C. 1995. Nuclear-DNA content in Gastrotricha. Experientia 51, 356–359.

Balser, E. J.; Ruppert, E. E. 1990. Structure, ultrastructure, and function of the preoral heart kidney in *Saccoglossus kowalevskii* (Hemichordata, Enteropneusta) including new data on the stomochord. Acta Zoologica 71, 235–249.

Balthasar, U.; Butterfield, N. J. 2009. Early Cambrian "soft-shelled" brachiopods as possible stem-group phoronids. Acta Palaeontologica Polonica 54, 307–314.

Balthasar, U.; Cusack, M.; Faryma, L.; Chung, P.; Holmer, L. E.; Jin, J.; Percival, I. G.; Popov, L. E. 2011. Relic aragonite from Ordovician–Silurian brachiopods: Implications for the evolution of calcification. Geology 39, 967–970.

Balthasar, U.; Skovsted, C. B.; Holmer, L. E.; Brock, G. A. 2009. Homologous skeletal secretion in tommotiids and brachiopods. Geology 37, 1143–1146.

Barbeitos, M. S.; Romano, S. L.; Lasker, H. R. 2010. Repeated loss of coloniality and symbiosis in scleractinian corals. Proceedings of the National Academy of Sciences of the USA 107, 11877–11882.

Barber, A. H.; Lu, D.; Pugno, N. M. 2015. Extreme strength observed in limpet teeth. Journal of the Royal Society, Interface 12, 20141326.

Barker, G. M. (ed.) 2001. The Biology of Terrestrial Molluscs. CABI Publishing, Wallingford.

Barnes, D. K. A.; Clarke, A. 1998. Seasonality of polypide recycling and sexual reproduction in some erect Antarctic bryozoans. Marine Biology 131, 647–658.

Barquin, A.; McGehee, B.; Sedam, R. T.; Gordy, W. L.; Hanelt, B.; Wise de Valdez, M. R. 2015. Calling behavior of male *Acheta domesticus* crickets infected with *Paragordius varius* (Nematomorpha: Gordiida). Journal of Parasitology 101, 393–397.

Barrois, J. 1877. Recherches sur l'embryologie des Bryozoaires. Six-Horemans, Lille.

Barshis, D. J.; Ladner, J. T.; Oliver, T. A.; Seneca, F. O.; Traylor-Knowles, N.; Palumbi, S. R. 2013. Genomic basis for coral resilience to climate change. Proceedings of the National Academy of Sciences of the USA 110, 1387–1392.

Bartels, P. J.; Apodaca, J. J.; Mora, C.; Nelson, D. R. 2016. A global biodiversity estimate of a poorly known taxon: phylum Tardigrada. Zoological Journal of the Linnean Society 178, 730–736.

Bartolomaeus, T. 1985. Ultrastructure and development of the protonephridia of *Lineus viridis* (Nemertini). Microfauna Marina 2, 61–83.

Bartolomaeus, T. 1988. No direct contact betwen the excretory system and the circulatory system in *Prostomatella arenicola* Friedrich (Hoplonemertini). Hydrobiologia 156, 175–181.

Bartolomaeus, T. 1989. Ultrastructure and relationship between protonephridia and metanephridia in *Phoronis muelleri* (Phoronida). Zoomorphology 109, 113–122.

Bartolomaeus, T. 1993. Die Leibeshöhlenverhältnisse und Nephridialorgane der Bilateria—Ultrastruktur, Entwicklung und Evolution. Georg, Göttingen, pp. 293.

Bartolomaeus, T. 1996/97. Ultrastructure of the renopericardial complex of the interstitial gastropod *Philinoglossa helgolandica* Hertling, 1932 (Mollusca: Opisthobranchia). Zoologischer Anzeiger 235, 165–176.

Bartolomaeus, T. 2001. Ultrastructure and formation of the body cavity lining in *Phoronis muelleri* (Phoronida, Lophophorata). Zoomorphology 120, 135–148.

Bartolomaeus, T.; Ax, P. 1992. Protonephridia and metanephridia—their relation within the Bilateria. Zeitschrift für zoologische Systematik und Evolutionsforschung 30, 21–45.

Bartolomaeus, T.; Quast, B. 2005. Structure and development of nephridia in Annelida and related taxa. Hydrobiologia 535/536, 139–165.

Bartolomaeus, T.; Quast, B.; Koch, M. 2009. Nephridial development and body cavity formation in *Artemia salina* (Crustacea: Branchiopoda): no evidence for any transitory coelom. Zoomorphology 128, 247–262.

Battelle, B.-A.; Ryan, J. F.; Kempler, K. E.; Saraf, S. R.; Marten, C. E.; Warren, W. C.; Minx, P. J.; Montague, M. J.; Green, P. J.; Schmidt, S. A.; et al. 2016a. Opsin repertoire and expression patterns in horseshoe crabs: Evidence from the genome of *Limulus polyphemus* (Arthropoda: Chelicerata). Genome Biology and Evolution 8, 1571–1589.

Battelle, B. A.; Sombke, A.; Harzsch, S. 2016b. Xiphosura. In: Schmidt-Rhaesa, A.; Harzsch, S.; Purschke, G. (eds.), Structure and Evolution of Invertebrate Nervous Systems. Oxford University Press; Oxford; pp. 428–442.

Baughman; K. W.; McDougall, C.; Cummins, S. F.; Hall, M.; Degnan, B. M.; Satoh, N.; Shoguchi, E. 2014. Genomic organization of Hox and ParaHox clusters in the echinoderm, *Acanthaster planci*. Genesis 52, 952–958.

Baumgarten, S.; Simakov, O.; Esherick, L. Y.; Liew, Y. J.; Lehnert, E. M.; Michell, C. T.; Li, Y.; Hambleton, E. A.; Guse, A.; Oates, M. E.; et al. 2015. The genome of *Aiptasia*, a sea anemone model for coral symbiosis. Proceedings of the National Academy of Sciences of the USA 112, 11893–11898.

Baumiller, T. K.; Salamon, M. A.; Gorzelak, P.; Mooi, R.; Messing, C. G.; Gahn, F. J. 2010. Post-Paleozoic crinoid radiation in response to benthic predation preceded the Mesozoic marine revolution. Proceedings of the National Academy of Sciences of the USA 107, 5893–5896.

Baverstock, P. R.; Fielke, R.; Johnson, A. M.; Bray, R. A.; Beveridge, I. 1991. Conflicting phylogenetic hypotheses for the parasitic platyhelminths tested by partial sequencing of 18S ribosomal RNA. International Journal for Parasitology 21, 329–339.

Bayer, C.; Heindl, N. R.; Rinke, C.; Lücker, S.; Ott, J. A.; Bulgheresi, S. 2009. Molecular characterization of the symbionts associated with marine nematodes of the genus *Robbea*. Environmental Microbiology Reports 1, 136–144.

Bayha, K. M.; Chang, M. H.; Mariani, C. L.; Richardson, J. L.; Edwards, D. L.; DeBoer, T. S.; Moseley, C.; Aksoy, E.; Decker, M. B.; Gaffney, P. M.; et al. 2014. Worldwide phylogeography of the invasive ctenophore *Mnemiopsis leidyi* (Ctenophora) based on nuclear and mitochondrial DNA data. Biological Invasions 17, 827–850.

(Nematomorpha) after exposure to freezing. Journal of Parasitology 99, 397–402.

Boletzky, S. v. 1989. Recent studies on spawning, embryonic-development, and hatching in the Cephalopoda. Advances in Marine Biology 25, 85–115.

Bomfleur, B.; Kerp, H.; Taylor, T. N.; Moestrup, Ø.; Taylor, E. L. 2012. Triassic leech cocoon from Antarctica contains fossil bell animal. Proceedings of the National Academy of Sciences of the USA 109, 20971–20974.

Bomfleur, B.; Mörs, T.; Ferraguti, M.; Reguero, M. A.; McLoughlin, S. 2015. Fossilized spermatozoa preserved in a 50-Myr-old annelid cocoon from Antarctica. Biology Letters 11, 20150431.

Bone, Q.; Goto, T. 1991. The nervous system. The Biology of Chaetognaths. Oxford University Press, Oxford, pp. 18–31.

Bone, Q.; Pulsford, A. 1978. The arrangement of ciliated sensory cells in Spadella (Chaetognatha). Journal of the Marine Biological Association of the United Kingdom 58, 565–570.

Boore, J. L. 1999. Animal mitochondrial genomes. Nucleic Acids Research 27, 1767–1780.

Boore, J. L.; Brown, W. M. 1995. Complete sequence of the mitochondrial DNA of the annelid worm Lumbricus terrestris. Genetics 141, 305–319.

Boothby, T. C.; Tenlen, J. R.; Smith, F. W.; Wang, J. R.; Patanella, K. A.; Osborne Nishimura, E.; Tintori, S. C.; Li, Q.; Jones, C. D.; Yandell, M.; et al. 2015. Evidence for extensive horizontal gene transfer from the draft genome of a tardigrade. Proceedings of the National Academy of Sciences of the USA 112, 15976–15981.

Borchiellini, C.; Manuel, M.; Alivon, E.; Boury-Esnault, N.; Vacelet, J.; Le Parco, Y. 2001. Sponge paraphyly and the origin of Metazoa. Journal of Evolutionary Biology 14; 171–179.

Borgonie, G.; Linage-Alvarez, B.; Ojo, A. O.; Mundle, S. O.; Freese, L. B.; Van Rooyen, C.; Kuloyo, O.; Albertyn, J.; Pohl, C.; Cason, E. D.; et al. 2015. Eukaryotic opportunists dominate the deep-subsurface biosphere in South Africa. Nature Communications 6, 8952.

Borisanova, A. O.; Chernyshev, A. V.; Malakhov, V. V. 2012. Myoanatomy of Barentsia discreta (Busk, 1886) (Kamptozoa: Coloniales). Biologiya Morya (Vladivostok) 38, 22–34.

Borisanova, A. O.; Chernyshev, A. V.; Malakhov, V. V. 2014. The structure of the body cavity Pedicellina cernua (Pallas, 1775) and Barentsia discreta (Busk, 1886) (Kamptozoa, Coloniales). Russian Journal of Marine Biology 40, 426–439.

Borisanova, A. O.; Chernyshev, A. V.; Neretina, T. V.; Stupnikova, A. N. 2015. Description and phylogenetic position of the first abyssal solitary kamptozoan species from the Kuril–Kamchatka Trench area: Loxosomella profundorum sp. nov. (Kamptozoa: Loxosomatidae). Deep-Sea Research. II: Topical Studies in Oceanography 111, 351–356.

Borisanova, A. O.; Potanina, D. M. 2016. A new species of Coriella, Coriella chernyshevi n. sp (Entoprocta, Barentsiidae), with comments on the genera Coriella and Pedicellinopsis. Zootaxa 4184, 376–382.

Borner, J.; Rehm, P.; Schill, R. O.; Ebersberger, I.; Burmester, T. 2014. A transcriptome approach to ecdysozoan phylogeny. Molecular Phylogenetics and Evolution 80, 79–87.

Børve, A.; Hejnol, A. 2014. Development and juvenile anatomy of the nemertodermatid Meara stichopi (Bock) Westblad 1949 (Acoelomorpha). Frontiers in Zoology 11, 50.

Boschetti, C.; Leasi, F.; Ricci, C. 2011. Developmental stages in diapausing eggs: an investigation across monogonont rotifer species. Hydrobiologia 662, 149–155.

Boschetti, C.; Ricci, C.; Sotgia, C.; Fascio, U. 2005. The development of a bdelloid egg: a contribution after 100 years. Hydrobiologia 546, 323–331.

Botting, J. P.; Butterfield, N. J. 2005. Reconstructing early sponge relationships by using the Burgess Shale fossil Eiffelia globosa, Walcott. Proceedings of the National Academy of Sciences of the USA 102, 1554–1559.

Botting, J. P.; Cárdenas, P.; Peel, J. S. 2015. A crown-group demosponge from the Early Cambrian Sirius Passet biota, North Greenland. Palaeontology 58, 35–43.

Botting, J. P.; Muir, L. A. 2018. Early sponge evolution: A review and phylogenetic framework. Palaeoworld 27, 1–29.

Botting, J. P.; Muir, L. A.; Van Roy, P.; Bates, D.; Upton, C. 2012. Diverse Middle Ordovician palaeoscolecidan worms from the Builth-Llandrindod Inlier of central Wales. Palaeontology 55, 501–528.

Botting, J. P.; Zhang, Y.; Muir, L. A. 2017. Discovery of missing link between demosponges and hexactinellids confirms palaeontological model of sponge evolution. Scientific Reports 7, 5286.

Bottjer, D. J.; Davidson, E. H.; Peterson, K. J.; Cameron, R. A. 2006. Paleogenomics of echinoderms. Science 314, 956–960.

Bouchet, P.; Bary, S.; Héros, V.; Marani, G. 2016. How many species of molluscs are there in the world's oceans, and who is going to describe them? Mémoires du Muséum national d'histoire naturelle 208, 9–24.

Bourlat, S. J.; Juliusdittir, T.; Lowe, C. J.; Freeman, R.; Aronowicz, J.; Kirschner, M.; Lander, E. S.; Thorndyke, M.; Nakano, H.; Kohn, A. B.; et al. 2006. Deuterostome phylogeny reveals monophyletic chordates and the new phylum Xenoturbellida. Nature 444, 85–88.

Bourlat, S. J.; Nakano, H.; Åkerman, M.; Telford, M. J.; Thorndyke, M. C.; Obst, M. 2008. Feeding ecology of Xenoturbella bocki (phylum Xenoturbellida) revealed by genetic barcoding. Molecular Ecology Resources 8, 18–22.

Bourlat, S. J.; Nielsen, C.; Lockyer, A. E.; Littlewood, D. T.; Telford, M. J. 2003. Xenoturbella is a deuterostome that eats molluscs. Nature 424, 925–928.

Bourlat, S. J.; Rota-Stabelli, O.; Lanfear, R.; Telford, M. J. 2009. The mitochondrial genome structure of Xenoturbella bocki (phylum Xenoturbellida) is ancestral within the deuterostomes. BMC Evolutionary Biology 9, 107.

Boxshall, G. 2013. Arthropod limbs and their development. In: Minelli, A.; Boxshall, G.; Fusco, G. (eds.), Arthropod Biology and Evolution: Molecules, Development, Morphology. Springer, Heidelberg, pp. 241–267.

Boxshall, G. A. 2004. The evolution of arthropod limbs. Biological Reviews 79, 253–300.

Boyer, B. C. 1971. Regulative development in a spiralian embryo as shown by cell deletion experiments on the acoel, Childia. Journal of Experimental Zoology 176, 97–105.

Boyer, B. C.; Henry, J. Q.; Martindale, M. Q. 1996. Dual origins of mesoderm in a basal spiralian: cell lineage analyses in the polyclad turbellarian *Hoploplana inquilina*. Developmental Biology 179, 329–338.

Boyer, B. C.; Henry, J. Q.; Martindale, M. Q. 1998. The cell lineage of a polyclad turbellarian embryo reveals close similarity to coelomate spiralians. Developmental Biology 204, 111–123.

Boyle, M. J.; Seaver, E. C. 2008. Developmental expression of *foxA* and *gata* genes during gut formation in the polychaete annelid *Capitella* sp. I. Evolution and Development 10, 89–105.

Boyle, M. J.; Seaver, E. C. 2010. Expression of FoxA and GATA transcription factors correlates with regionalized gut development in two lophotrochozoan marine worms: *Chaetopterus* (Annelida) and *Themiste lageniformis* (Sipuncula). EvoDevo 1, 2.

Braasch, I.; Gehrke, A. R.; Smith, J. J.; Kawasaki, K.; Manousaki, T.; Pasquier, J.; Amores, A.; Desvignes, T.; Batzel, P.; Catchen, J.; et al. 2016. The spotted gar genome illuminates vertebrate evolution and facilitates human-teleost comparisons. Nature Genetics 48, 427–437.

Braband, A.; Cameron, S. L.; Podsiadlowski, L.; Daniels, S. R.; Mayer, G. 2010a. The mitochondrial genome of the onychophoran *Opisthopatus cinctipes* (Peripatopsidae) reflects the ancestral mitochondrial gene arrangement of Panarthropoda and Ecdysozoa. Molecular Phylogenetics and Evolution 57, 285–292.

Braband, A.; Podsiadlowski, L.; Cameron, S. L.; Daniels, S.; Mayer, G. 2010b. Extensive duplication events account for multiple control regions and pseudo-genes in the mitochondrial genome of the velvet worm *Metaperipatus inae* (Onychophora, Peripatopsidae). Molecular Phylogenetics and Evolution 57, 293–300.

Braden, B. P.; Taketa, D. A.; Pierce, J. D.; Kassmer, S.; Lewis, D. D.; De Tomaso, A. W. 2014. Vascular regeneration in a basal chordate is due to the presence of immobile, bifunctional cells. PLoS One 9, e95460.

Brasier, M. 2009. Darwin's Lost World: The Hidden History of Animal Life. Oxford University Press, Oxford.

Bråte, J.; Adamski, M.; Neumann, R. S.; Shalchian-Tabrizi, K.; Adamska, M. 2015. Regulatory RNA at the root of animals: dynamic expression of developmental lincRNAs in the calcisponge *Sycon ciliatum*. Proceedings of the Royal Society B: Biological Sciences 282, 20151746.

Brauchle, M.; Bilican, A.; Eyer, C.; Bailly, X.; Martínez, P.; Ladurner, P.; Bruggmann, R.; Sprecher, S. G. 2018. Xenacoelomorpha survey reveals that all 11 animal homeobox gene classes were present in the first bilaterians. Genome Biology and Evolution 10, 2205–2217.

Braun, G.; Kümmel, G.; Mangos, J. A. 1966. Studies on ultrastructure and function of a primitive excretory organ protonephridium of rotifer Asplanchna priodonta. Pflugers Archiv für die Gesamte Physiologie des Menschen und der Tiere 289, 141–154.

Braun, K.; Kaul-Strehlow, S.; Ullrich-Luter, E.; Stach, T. 2015. Structure and ultrastructure of eyes of tornaria larvae of *Glossobalanus marginatus*. Organisms Diversity and Evolution 15; 423–428.

Brena, C. 2015. Myriapoda. In: Wanninger, A. (ed.), Evolutionary Developmental Biology of Invertebrates 3: Ec-

dysozoa I: Non-Tetraconata. Springer, Vienna, pp. 141–189.

Brenneis, G.; Richter, S. 2010. Architecture of the nervous system in Mystacocarida (Arthropoda, Crustacea)—An immunohistochemical study and 3D reconstruction. Journal of Morphology 271, 169–189.

Brenneis, G.; Ungerer, P.; Scholtz, G. 2008. The chelifores of sea spiders (Arthropoda, Pycnogonida) are the appendages of the deutocerebral segment. Evolution and Development 10, 717–724.

Brenner, S. 1973. The genetics of behaviour. British Medical Bulletin 29, 269–271.

Brenner, S. 1974. The genetics of *Caenorhabditis elegans*. Genetics 77, 71–94.

Bresslau, E. 1909. Die Entwicklung der Acoelen. Verhandlungen der Deutschen Zoologischen Gesellschaft 19, 314–324.

Bridge, D.; Cunningham, C. W.; Schierwater, B.; DeSalle, R.; Buss, L. W. 1992. Class-level relationships in the phylum Cnidaria: evidence from mitochondrial genome structure. Proceedings of the National Academy of Sciences of the USA 89, 8750–8753.

Briggs, D. E. G. 2015. The Cambrian explosion. Current Biology 25, R864–R868.

Briggs, D. E. G.; Bartels, C. 2010. Annelids from the Lower Devonian Hunsrück Slate (Lower Emsian, Rhenish Massif, Germany). Palaeontology 53, 215–232.

Briggs, D. E. G.; Caron, J.-B. 2017. A large Cambrian chaetognath with supernumerary grasping spines. Current Biology 27, 2536–2543.

Briggs, D. E. G.; Kear, A. J. 1994. Decay of *Branchiostoma*: implications for soft-tissue preservation in conodonts and other primitive chordates. Lethaia 26, 275–287.

Briggs, D. E. G.; Kear, A. J.; Baas, M.; De Leeuw, J. W.; Rigby, S. 1995. Decay and composition of the hemichordate *Rhabdopleura*: Implications for the taphonomy of graptolites. Lethaia 28, 15–23.

Bright, M. 2005. Microbial symbiosis in Annelida. Symbiosis 38, 1–45.

Brock, G. A.; Paterson, J. R. 2004. A new species of *Tannuella* (Helcionellida, Mollusca) from the Early Cambrian of South Australia. Memoirs of the Association of Australasian Palaeontologists 30, 133–143.

Bronstein, O.; Kroh, A. 2019. The first mitochondrial genome of the model echinoid *Lytechinus variegatus* and insights into Odontophoran phylogenetics. Genomics 111, 710–718.

Brooke, N. M.; Garcia-Fernàndez, J.; Holland, P. W. 1998. The ParaHox gene cluster is an evolutionary sister of the Hox gene cluster. Nature 392, 920–922.

Brooks, W. K.; Cowles, R. P. 1911. *Phoronis architecta*: its life history, anatomy and breeding habits. Memoirs of the National Academy of Sciences 10, 71–148.

Broun, M.; Bode, H. R. 2002. Characterization of the head organizer in hydra. Development 129, 875–884.

Brown, R. 1983. Spermatophore transfer and subsequent sperm development in a homalorhagid kinorhynch. Zoologica Scripta 12, 257–266.

Brüggemann, J. 1986. Feinstruktur der Protonephridien von *Paromaostomum proceracauda* (Plathelminthes, Macrostomida). Zoomorphology 106, 147–154.

Brunanská, M.; Fagerholm, H.-P.; Moravec, F. 2007. Structure of the pharynx in the adult nematode *Anguillicoloides crassus* (Nematoda: Rhabditida). Journal of Parasitology 93, 1017–1028.

Brunet, T.; Bouclet, A.; Ahmadi, P.; Mitrossilis, D.; Driquez, B.; Brunet, A. C.; Henry, L.; Serman, F.; Bealle, G.; Menager, C.; et al. 2013. Evolutionary conservation of early mesoderm specification by mechanotransduction in Bilateria. Nature Communications 4, 2821.

Brusca, R. C.; Brusca, G. J. 1990. Invertebrates. Sinauer, Sunderland, MA.

Brusca, R. C.; Brusca, G. J. 2003. Invertebrates. 2nd ed. Sinauer, Sunderland, MA.

Brusca, R. C.; Lindberg, D. R.; Ponder, W. F. 2016a. Phylum Mollusca. In: Brusca, R. C.; Moore, W.; Shuster, S. M. (eds.), Invertebrates, 3rd ed. Sinauer, Sunderland, MA, pp. 453–530.

Brusca, R. C.; Moore, W.; Shuster, S. M. 2016b. Invertebrates, 3rd ed. Sinauer, Sunderland, MA.

Buckland-Nicks, J. 2008. Fertilization biology and the evolution of chitons. American Malacological Bulletin 25, 97–111.

Budd, G. E. 1998. The morphology and phylogenetic significance of *Kerygmachela kierkegaardi* Budd (Buen Formation, Lower Cambrian, N Greenland). Transactions of the Royal Society of Edinburgh: Earth Sciences 89, 249–290.

Budd, G. E. 2001. Tardigrades as 'stem-group arthropods': The evidence from the Cambrian fauna. Zoologischer Anzeiger 240, 265–279.

Budd, G. E.; Jensen, S. 2000. A critical reappraisal of the fossil record of the bilaterian phyla. Biological Reviews 75, 253–295.

Bull, J. K.; Sands, C. J.; Garrick, R. C.; Gardner, M. G.; Tait, N. N.; Briscoe, D. A.; Rowell, D. M.; Sunnucks, P. 2013. Environmental complexity and biodiversity: the multi-layered evolutionary history of a log-dwelling velvet worm in montane temperate Australia. PLoS One 8, e84559.

Bullivant, J. S. 1968. The method of feeding of lophophorates (Bryozoa, Phoronida, Brachiopoda). New Zealand Journal of Marine and Freshwater Research 2, 135–146.

Bullock, T. H.; Horridge, G. A. 1965. Structure and Function in the Nervous Systems of Invertebrates. W. H. Freeman, San Francisco/London.

Burke, M.; Scholl, E. H.; Bird, D. M.; Schaff, J. E.; Colman, S. D.; Crowell, R.; Diener, S.; Gordon, O.; Graham, S.; Wang, X. G.; et al. 2015. The plant parasite *Pratylenchus coffeae* carries a minimal nematode genome. Nematology 17, 621–637.

Burkhardt, P.; Stegmann, C. M.; Cooper, B.; Kloepper, T. H.; Imig, C.; Varoqueaux, F.; Wahl, M. C.; Fasshauer, D. 2011. Primordial neurosecretory apparatus identified in the choanoflagellate *Monosiga brevicollis*. Proceedings of the National Academy of Sciences of the USA 108, 15264–15269.

Butler, A. D.; Cunningham, J. A.; Budd, G. E.; Donoghue, P. C. J. 2015. Experimental taphonomy of *Artemia* reveals the role of endogenous microbes in mediating decay and fossilization. Proceedings of the Royal Society B: Biological Sciences 282.

Butler, P. G.; Richardson, C. A.; Scourse, J. D.; Wanamaker, A. D.; Shammon, T. M.; Bennell, J. D. 2010. Marine climate in the Irish Sea: analysis of a 489-year marine master chronology derived from growth increments in the shell of the clam *Arctica islandica*. Quaternary Science Reviews 29, 1614–1632.

Bütschli, O. 1876. Untersuchungen über freilebende Nematoden und die Gattung *Chaetonotus*. Zeitschrift für wissenschaftliche Zoologie 26, 363–413.

Bütschli, O. 1910. Vorlesungen über vergleichende Anatomie. Lfg 1: Einleitung. Vergleichende Anatomie der Protozoen. Integument und Skelett der Metazoen. W. Engelmann, Leipzig.

Butterfield, N. J. 1990. A reassessment of the enigmatic Burgess Shale fossil *Wiwaxia corrugata* (Matthew) and its relationship to the polychaete *Canadia spinosa* Walcott. Paleobiology 16, 287–303.

Byrne, M. 1985. Evisceration behavior and the seasonal incidence of evisceration in the holothurian *Eupentacta quinquesemita* (Selenka). Ophelia 24, 75–90.

Byrne, M. 1994. Ophiuroidea. In: Harrison, F. W.; Chia, F.-S. (eds.), Microscopic Anatomy of Invertebrates, vol. 14: Echinodermata. Wiley-Liss, New York, pp. 247–343.

Byrne, M. 2001. The morphology of autotomy structures in the sea cucumber *Eupentacta quinquesemita* before and during evisceration. Journal of Experimental Biology 204, 849–863.

Byrne, M.; Martinez, P.; Morris, V. 2016. Evolution of a pentameral body plan was not linked to translocation of anterior Hox genes: the echinoderm HOX cluster revisited. Evolution and Development 18, 137–143.

Byrne, M.; O'Hara, T. D. (eds.). 2017. Australian Echinoderms: Biology, Ecology and Evolution. CSIRO Publishing and ABRS, Melbourne/Canberra.

C. elegans Sequencing Consortium. 1998. Genome sequence of the nematode *C. elegans*: A platform for investigating biology. Science 282, 2012–2018.

Caira, J. N.; Jensen, K.; Waeschenbach, A.; Olson, P. D.; Littlewood, D. T. J. 2014. Orders out of chaos—molecular phylogenetics reveals the complexity of shark and stingray tapeworm relationships. International Journal for Parasitology 44, 55–73.

Caira, J. N.; Littlewood, D. T. J. 2013. Worms, Platyhelminthes. In: Levin, S. A. (ed.), Encyclopedia of Biodiversity, 2nd ed., vol. 7. Academic Press, Waltham, pp. 437–469.

Calderón-Urrea, A.; Vanholme, B.; Vangestel, S.; Kane, S. M.; Bahaji, A.; Pha, K.; Garcia, M.; Snider, A.; Gheysen, G. 2016. Early development of the root-knot nematode *Meloidogyne incognita*. BMC Developmental Biology 16, 10.

Calvet, L. 1900. Contributions a l'histoire naturelle des Bryozoaires ectoproctes marins. Coulet et fils, Montpellier.

Cameron, C. B. 2005. A phylogeny of the hemichordates based on morphological characters. Canadian Journal of Zoology 83, 196–215.

Cameron, C. B.; Bishop, C. D. 2012. Biomineral ultrastructure, elemental constitution and genomic analysis of biomineralization-related proteins in hemichordates. Proceedings of the Royal Society B: Biological Sciences 279, 3041–3048.

Cameron, R. A.; Kudtarkar, P.; Gordon, S. M.; Worley, K. C.; Gibbs, R. A. 2015. Do echinoderm genomes measure up? Marine Genomics 22, 1–9.

Campbell, L. I.; Rota-Stabelli, O.; Edgecombe, G. D.; Marchioro, T.; Longhorn, S. J.; Telford, M. J.; Philippe, H.; Rebecchi, L.; Peterson, K. J.; Pisani, D. 2011. MicroRNAs and phylogenomics resolve the relationships of Tardigrada and suggest that velvet worms are the sister group of Arthropoda. Proceedings of the National Academy of Sciences of the USA 108, 15920–15924.

Campbell, R. D. 1974. Cnidaria. In: Giese, A. C.; Pearse, J. S. (eds.), Reproduction of Marine Invertebrates, vol. 1: Acoelomate and Pseudocoelomate Metazoans. Academic Press, London, pp. 133–199.

Campos, A.; Cummings, M. P.; Reyes, J. L.; Laclette, J. P. 1998. Phylogenetic relationships of Platyhelminthes based on 18S ribosomal gene sequences. Molecular Phylogenetics and Evolution 10, 1–10.

Candia Carnevali, M. D. 2006. Regeneration in echinoderms: repair, regrowth, cloning. Invertebrate Survival Journal 3, 64–76.

Candia Carnevali, M. D.; Wilkie, I. C.; Lucca, E.; Andrietti, F.; Melone, G. 1993. The Aristotle's lantern of the sea-urchin Stylocidaris affinis (Echinoida, Cidaridae): functional morphology of the musculo-skeletal system. Zoomorphology 113, 173–189.

Cañestro, C.; Bassham, S.; Postlethwait, J. 2005. Development of the central nervous system in the larvacean Oikopleura dioica and the evolution of the chordate brain. Developmental Biology 285, 298–315.

Cañestro, C.; Bassham, S.; Postlethwait, J. H. 2008. Evolution of the thyroid: anterior–posterior regionalization of the Oikopleura endostyle revealed by Otx, Pax2/5/8, and Hox1 expression. Developmental Dynamics 237, 1490–1499.

Cannon, J. T.; Kocot, K. M.; Waits, D. S.; Weese, D. A.; Swalla, B. J.; Santos, S. R.; Halanych, K. M. 2014. Phylogenomic resolution of the hemichordate and echinoderm clade. Current Biology 24; 2827–2832.

Cannon, J. T.; Rychel, A. L.; Eccleston, H.; Halanych, K. M.; Swalla, B. J. 2009. Molecular phylfrogeny of Hemichordata, with updated status of deep-sea enteropneusts. Molecular Phylogenetics and Evolution 52, 17–24.

Cannon, J. T.; Swalla, B. J.; Halanych, K. M. 2013. Hemichordate molecular phylogeny reveals a novel cold-water clade of harrimaniid acorn worms. Biological Bulletin 225, 194–204.

Capt, C.; Renaut, S.; Ghiselli, F.; Milani, L.; Johnson, N. A.; Sietman, B. E.; Stewart, D. T.; Breton, S. 2018. Deciphering the link between doubly uniparental inheritance of mtDNA and sex determination in bivalves: Clues from comparative transcriptomics. Genome Biology and Evolution 10, 577–590.

Carapelli, A.; Lió, P.; Nardi, F.; van der Wath, E.; Frati, F. 2007. Phylogenetic analysis of mitochondrial protein coding genes confirms the reciprocal paraphyly of Hexapoda and Crustacea. BMC Evolutionary Biology 7, suppl. 2, S8.

Carbayo, F.; Francoy, T. M.; Giribet, G. 2016. Non-destructive imaging to describe a new species of Obama land planarian (Platyhelminthes, Tricladida). Zoologica Scripta 45, 566–578.

Carle, K. J.; Ruppert, E. E. 1983. Comparative ultrastructure of the bryozoan funiculus: A blood-vessel homologue. Zeitschrift für zoologische Systematik und Evolutionsforschung 21, 181–193.

Carlos, A. A.; Baillie, B. K.; Kawachi; M.; Maruyama, T. 1999. Phylogenetic position of Symbiodinium (Dinophyceae) isolates from tridacnids (Bivalvia), cardiids (Bivalvia), a sponge (Porifera), a soft coral (Anthozoa), and a free-living strain. Journal of Phycology 35, 1054–1062.

Carlson, S. J. 1995. Phylogenetic relationships among extant brachiopods. Cladistics 11, 131–197.

Carlson, S. J. 2016. The evolution of Brachiopoda. Annual Review of Earth and Planetary Sciences 44, 409–438.

Caron, J.-B.; Aria, C. 2017. Cambrian suspension-feeding lobopodians and the early radiation of panarthropods. BMC Evolutionary Biology 17, 29.

Caron, J.-B.; Conway Morris, S.; Shu, D. 2010. Tentaculate fossils from the Cambrian of Canada (British Columbia) and China (Yunnan) interpreted as primitive deuterostomes. PLoS One 5, e9586.

Caron, J.-B.; Cheung, B. 2019. Amiskwia is a large Cambrian gnathiferan with complex gnathostomulid-like jaws. Communications Biology 2, 164.

Caron, J.-B.; Gaines, R. R.; Aria, C.; Mángano, M. G.; Streng, M. 2014. A new phyllopod bed-like assemblage from the Burgess Shale of the Canadian Rockies. Nature Communications 5, 3210.

Caron, J.-B.; Morris, S. C.; Cameron, C. B. 2013. Tubicolous enteropneusts from the Cambrian Period. Nature 495, 503–506.

Caron, J.-B.; Scheltema, A.; Schander, C.; Rudkin, D. 2006. A soft-bodied mollusc with radula from the Middle Cambrian Burgess Shale. Nature 442, 159–163.

Carranza, S.; Baguñà, J.; Riutort, M. 1997. Are the Platyhelminthes a monophyletic primitive group? An assessment using 18S rDNA sequences. Molecular Biology and Evolution 14, 485–497.

Carré, C.; Carré, D. 1989a. Haeckelia bimaculata sp. nov., une nouvelle espèce méditerréenne de cténophore (Cydippida Haeckeliidae) pourvue de cnidocystes et de pseudocolloblastes. Comptes Rendus de L'Académie des Sciences Serie III. Sciences de la Vie 308, 321–327.

Carré, D.; Carré, C. 1989b. Acquisition de cnidocystes et différenciation de pseudocolloblastes chez les larves et les adultes de deux cténophores du genre Haeckelia Carus, 1863. Canadian Journal of Zoology 67, 2169–2179.

Carré, D.; Carré, C.; Mills, C. E. 1989. Novel cnidocysts of Narcomedusae and a medusivorous ctenophore, and confirmation of kleptocnidism. Tissue and Cell 21, 723–734.

Carroll, S.; McEvoy, E. G.; Gibson, R. 2003. The production of tetrodotoxin-like substances by nemertean worms in conjunction with bacteria. Journal of Experimental Marine Biology and Ecology 288, 51–63.

Carroll, S. B. 1995. Homeotic genes and the evolution of arthropods and chordates. Nature 376, 479–485.

Carter, J. G. 1990. Shell microstructural data for the Bivalvia. In: Carter, J. G. (ed.), Skeletal Biomineralization: Patterns,

Processes and Evolutionary Trends, vol. 1. Van Nostrand Reinhold, New York, pp. 297–411.

Carter, M. C.; Lidgard, S.; Gordon, D. P.; Gardner, J. P. A. 2011. Functional innovation through vestigialization in a modular marine invertebrate. Biological Journal of the Linnean Society 104, 63–74.

Cartwright, P.; Halgedahl, S. L.; Hendricks, J. R.; Jarrard, R. D.; Marques, A. C.; Collins, A. G.; Lieberman, B. S. 2007. Exceptionally preserved jellyfishes from the Middle Cambrian. PLoS One 2, e1121.

Casanova, J. P. 1985. Description de l'appareil génital primitif du genre Heterokrohnia et nouvelle classification des chaetognathes. Comptes rendus de l'Académie des Sciences Paris 301, 397–402.

Casas, L.; Pearman, J. K.; Irigoien, X. 2017. Metabarcoding reveals seasonal and temperature-dependent succession of zooplankton communities in the Red Sea. Frontiers in Marine Science 4, 241.

Castellanos-Martinez, S.; Aguirre-Macedo, M. L.; Furuya, H. 2016. Two new species of dicyemid mesozoans (Dicyemida: Dicyemidae) from Octopus maya Voss & Solis-Ramirez (Octopodidae) off Yucatan, Mexico. Systematic Parasitology 93, 551–564.

Castresana, J.; Feldmaier-Fuchs, G.; Yokobori, S.; Satoh, N.; Pääbo, S. 1998. The mitochondrial genome of the hemichordate Balanoglossus carnosus and the evolution of deuterostome mitochondria. Genetics 150, 1115–1123.

Castro, A.; Becerra, M.; Manso, M. J.; Anadón, R. 2015. Neuronal organization of the brain in the adult amphioxus (Branchiostoma lanceolatum): A study with acetylated tubulin immunohistochemistry. Journal of Comparative Neurology 523, 2211–2232.

Cavalcanti, F. F.; Klautau, M. 2011. Solenoid: a new aquiferous system to Porifera. Zoomorphology 130, 255–260.

Cavalier-Smith, T. 1998. A revised six-kingdom system of life. Biological Reviews 73, 203–266.

Cavalier-Smith, T.; Allsopp, M. T. E. P.; Chao, E. E.; Boury-Esnault, N.; Vacelet, J. 1996. Sponge phylogeny, animal monophyly, and the origin of the nervous system: 18S rRNA evidence. Canadian Journal of Zoology 74, 2031–2045.

Cavanaugh, C. M. 1983. Symbiotic chemoautotrophic bacteria in marine invertebrates from sulphide-rich habitats. Nature 302, 58–61.

Cavanaugh, C. M.; Gardiner, S. L.; Jones, M. L.; Jannasch, H. W.; Waterbury, J. B. 1981. Prokaryotic cells in the hydrothermal vent tube worm Riftia pachyptila Jones: Possible chemoautotrophic symbionts. Science 213, 340–342.

Cea, G.; Gaitán-Espitia, J. D.; Cárdenas, L. 2015. Complete mitogenome of the edible sea urchin Loxechinus albus: genetic structure and comparative genomics within Echinozoa. Molecular Biology Reports 42, 1081–1089.

Cebrià, F. 2007. Regenerating the central nervous system: how easy for planarians! Development Genes and Evolution 217, 733–748.

Cebrià, F.; Saló, E.; Adell, T. 2015. Regeneration and growth as modes of adult development: The Platyhelminthes as a case study. In: Wanninger, A. (ed.), Evolutionary Developmental Biology of Invertebrates 2: Lophotrochozoa (Spiralia). Springer, pp. 41–78.

Cesari, M.; McInnes, S. J.; Bertolani, R.; Rebecchi, L.; Guidetti, R. 2016. Genetic diversity and biogeography of the south polar water bear Acutuncus antarcticus (Eutardigrada : Hypsibiidae)—evidence that it is a truly pan-Antarctic species. Invertebrate Systematics 30, 635–649.

Chang, D.; Duda, T. F., Jr. 2012. Extensive and continuous duplication facilitates rapid evolution and diversification of gene families. Molecular Biology and Evolution 29, 2019–2029.

Chang, E. S.; Neuhof, M.; Rubinstein, N. D.; Diamant, A.; Philippe, H.; Huchon, D.; Cartwright, P. 2015. Genomic insights into the evolutionary origin of Myxozoa within Cnidaria. Proceedings of the National Academy of Sciences of the USA 112, 14912–14917.

Chapman, J. A.; Kirkness, E. F.; Simakov, O.; Hampson, S. E.; Mitros, T.; Weinmaier, T.; Rattei, T.; Balasubramanian, P. G.; Borman, J.; Busam, D.; et al. 2010. The dynamic genome of Hydra. Nature 464, 592–596.

Checa, A. 2000. A new model for periostracum and shell formation in Unionidae (Bivalvia, Mollusca). Tissue and Cell 32, 405–416.

Checa, A. G.; Jiménez-Jiménez, A. P. 1998. Constructional morphology, origin, and evolution of the gastropod operculum. Paleobiology 24, 109–132.

Checa, A. G.; Ramírez-Rico, J.; González-Segura, A.; Sánchez-Navas, A. 2009. Nacre and false nacre (foliated aragonite) in extant monoplacophorans (= Tryblidiida: Mollusca). Naturwissenschaften 96, 111–122.

Checa, A. G.; Salas, C.; Harper, E. M.; Bueno-Pérez, J. d. D. 2014. Early stage biomineralization in the periostracum of the 'living fossil' bivalve Neotrigonia. PLoS One 9, e90033.

Checa, A. G.; Vendrasco, M. J.; Salas, C. 2017. Cuticle of Polyplacophora: structure, secretion, and homology with the periostracum of conchiferans. Marine Biology 164, 64.

Cheetham, A. H. 1986. Tempo of evolution in a Neogene bryozoan: rates of morphologic change within and across species boundaries. Paleobiology 12, 190–202.

Cheetham, A. H.; Sanner, J.; Taylor, P. D.; Ostrovsky, A. N. 2006. Morphological differentiation of avicularia and the proliferation of species in mid-Cretaceous Wilbertopora Cheetham, 1954 (Bryozoa: Cheilostomata). Journal of Paleontology 80, 49–71.

Chen, C.; Linse, K.; Copley, J. T.; Rogers, A. D. 2015. The 'scaly-foot gastropod': a new genus and species of hydrothermal vent-endemic gastropod (Neomphalina: Peltospiridae) from the Indian Ocean. Journal of Molluscan Studies 81, 322–334.

Chen, H.-X.; Sun, S.-C.; Sundberg, P.; Ren, W.-C.; Norenburg, J. L. 2012. A comparative study of nemertean complete mitochondrial genomes, including two new ones for Nectonemertes cf. mirabilis and Zygeupolia rubens, may elucidate the fundamental pattern for the phylum Nemertea. BMC Genomics 13, 139.

Chen, H.-X.; Sundberg, P.; Norenburg, J. L.; Sun, S.-C. 2009a. The complete mitochondrial genome of Cephalothrix simula (Iwata) (Nemertea: Palaeonemertea). Gene 442, 8–17.

Chen, J.; Zhou, G. 1997. Biology of the Chengjiang fauna. Bulletin of the National Museum of Natural Science (Taichung) 10, 11–105.

Chen, J.-Y. 2004. The dawn of the animal world [in Chinese]. Jiangsu Science and Technical Press, Nanjing.

Chen, J.-Y. 2008. Early crest animals and the insight they provide into the evolutionary origin of craniates. Genesis 46, 623–639.

Chen, J.-Y.; Bottjer, D. J.; Davidson, E. H.; Dornbos, S. Q.; Gao, X.; Yang, Y.-H.; Li, C.-W.; Li, G.; Wang, X.-Q.; Xian, D.-C.; et al. 2006. Phosphatized polar lobe-forming embryos from the Precambrian of southwest China. Science 312, 1644–1646.

Chen, J.-Y.; Dzik, J.; Edgecombe, G. D.; Ramsköld, L.; Zhou, G.-Q. 1995. A possible Early Cambrian chordate. Nature 377, 720–722.

Chen, J.-Y.; Huang, D.-Y.; Chuang, S.-H. 2007a. Reinterpretation of the Lower Cambrian brachiopod Heliomedusa orienta Sun and Hou, 1987a as a discinid. Journal of Paleontology 81, 38–47.

Chen, J.-Y.; Huang, D.-Y.; Peng, Q.-Q.; Chi, H.-M.; Wang, X.-Q.; Feng, M. 2003. The first tunicate from the Early Cambrian of South China. Proceedings of the National Academy of Sciences of the USA 100, 8314–8318.

Chen, J.-Y.; Teichert, C. 1983. Cambrian Cephalopods. Geology 11, 647–650.

Chen, J. Y.; Bottjer, D. J.; Li, G.; Hadfield, M. G.; Gao, F.; Cameron, A. R.; Zhang, C. Y.; Xian, D. C.; Tafforeau, P.; Liao, X.; et al. 2009b. Complex embryos displaying bilaterian characters from Precambrian Doushantuo phosphate deposits, Weng'an, Guizhou, China. Proceedings of the National Academy of Sciences of the USA 106, 19056–19060.

Chen, J. Y.; Oliveri, P.; Li, C. W.; Zhou, G. Q.; Gao, F.; Hagadorn, J. W.; Peterson, K. J.; Davidson, E. H. 2000. Precambrian animal diversity: putative phosphatized embryos from the Doushantuo formation of China. Proceedings of the National Academy of Sciences of the USA 97, 4457–4462.

Chen, J. Y.; Schopf, J. W.; Bottjer, D. J.; Zhang, C. Y.; Kudryavtsev, A. B.; Tripathi, A. B.; Wang, X. Q.; Yang, Y. H.; Gao, X.; Yang, Y. 2007b. Raman spectra of a lower Cambrian ctenophore embryo from southwestern Shaanxi, China. Proceedings of the National Academy of Sciences of the USA 104, 6289–6292.

Chen, Z.; Chen, X.; Zhou, C.; Yuan, X.; Xiao, S. 2018. Late Ediacaran trackways produced by bilaterian animals with paired appendages. Science Advances 4, eaao6691.

Chen, Z.; Zhou, C.; Meyer, M.; Xiang, K.; Schiffbauer, J. D.; Yuan, X.; Xiao, S. 2013. Trace fossil evidence for Ediacaran bilaterian animals with complex behaviors. Precambrian Research 224, 690–701.

Chernyshev, A. V. 2015. CLSM Analysis of the phalloidin-stained muscle system of the nemertean proboscis and rhynchocoel. Zoological Science 32, 547–560.

Chernyshev, A. V.; Magarlamov, T. Y.; Turbeville, J. M. 2013. Morphology of the proboscis of Hubrechtella juliae (Nemertea, Pilidiophora): Implications for pilidiophoran monophyly. Journal of Morphology 274, 1397–1414.

Chia, F. S.; Burke, R. D. 1978. Echinoderm metamorphosis. In: Chia, F. S.; Rice, M. E. (eds.), Settlement and Metamorphosis of Marine Invertebrates. Elsevier, North Holland, NY, pp. 219–234.

Chien, P.; Koopowitz, H. 1972. The ultrastructure of neuromuscular systems in Notoplana acticola, a free-living polyclad flatworm. Zeitschrift für Zellforschung und Mikroskopische Anatomie 133, 277–288.

Chiodin, M.; Achatz, J. G.; Wanninger, A.; Martinez, P. 2011. Molecular architecture of muscles in an acoel and its evolutionary implications. Journal of Experimental Zoology, pt. B: Molecular Development and Evolution 316, 427–439.

Chiodin, M.; Borve, A.; Berezikov, E.; Ladurner, P.; Martinez, P.; Hejnol, A. 2013. Mesodermal gene expression in the acoel Isodiametra pulchra indicates a low number of mesodermal cell types and the endomesodermal origin of the gonads. PLoS One 8, e55499.

Chipman, A. D. 2015a. An embryological perspective on the early arthropod fossil record. BMC Evolutionary Biology 15, 285.

Chipman, A. D. 2015b. Hexapoda: Comparative aspects of early development. In: Wanninger, A. (ed.), Evolutionary Developmental Biology of Invertebrates 5. Ecdysozoa III: Hexapoda. Springer, Vienna, pp. 93–110.

Chipman, A. D.; Ferrier, D. E.; Brena, C.; Qu, J.; Hughes, D. S.; Schroder, R.; Torres-Oliva, M.; Znassi, N.; Jiang, H.; Almeida, F. C.; et al. 2014. The first myriapod genome sequence reveals conservative arthropod gene content and genome organisation in the centipede Strigamia maritima. PLoS Biology 12, e1002005.

Chitsulo, L.; Loverde, R.; Engels, D.; Barakat, R.; Colley, D.; Cioli, D.; Engels, D.; Feldmeier, H.; Loverde, P.; Olds, G. R.; et al. 2004. Schistosomiasis. Nature Reviews Microbiology 2, 12–13.

Chitwood, B. G.; Chitwood, M. D. 1933. The characters of a Protonematode. Journal of Parasitology 20, 1–130.

Chourrout, D.; Delsuc, F.; Chourrout, P.; Edvardsen, R. B.; Rentzsch, F.; Renfer, E.; Jensen, M. F.; Zhu, B.; de Jong, P.; Steele; R. E.; et al. 2006. Minimal ProtoHox cluster inferred from bilaterian and cnidarian Hox complements. Nature 442, 684–687.

Chuang, S.-H. 1968. The larvae of a discinid (Inarticulata, Brachiopoda). Biological Bulletin 135, 263–272.

Chun, C. 1880. Die Ctenophoren des Golfes von Neapel und der angrenzenden Meeres-Abschnitte. Wilhelm Engelmann, Leipzig.

Cimino, G.; Fontana, A.; Gavagnin, M. 1999. Marine opisthobranch molluscs: Chemistry and ecology in sacoglossans and dorids. Current Organic Chemistry 3, 327–372.

Cimino, G.; Ghiselin, M. T. 2009. Chemical defense and the evolution of opisthobranch gastropods. Proceedings of the California Academy of Sciences 60, 175–422.

Clark, A. G.; Eisen, M. B.; Smith, D. R.; Bergman, C. M.; Oliver, B.; Markow, T. A.; Kaufman, T. C.; Kellis, M.; Gelbart, W.; Iyer, V. N.; et al. 2007. Evolution of genes and genomes on the Drosophila phylogeny. Nature 450, 203–218.

Clausen, S.; Hou, X.-G.; Bergström, J.; Franzén, C. 2010. The absence of echinoderms from the Lower Cambrian Chengjiang fauna of China: Palaeoecological and

Coulcher, J. F.; Edgecombe, G. D.; Telford, M. J. 2015. Molecular developmental evidence for a subcoxal origin of pleurites in insects and identity of the subcoxa in the gnathal appendages. Scientific Reports 5, 15757.

Courtney, R.; Sachlikidis, N.; Jones, R.; Seymour, J. 2015. Prey capture ecology of the cubozoan *Carukia barnesi*. PLoS One 10, e0124256.

Cowles, R. P. 1904. Origin and fate of the body-cavities and the nephridia of the *Actinotrocha*. The Annals and Magazine of Natural History: Zoology, Botany, and Geology 14, 69–78.

Crandall, F. B. 2001. A cladistic view of the Monostilifera (Hoplonemertea) with interwoven rhynchocoel musculature: a preliminary assessment. Hydrobiologia 456, 87–110.

Crawford, B. J.; Campbell, S. S. 1993. The microvilli and hyaline layer of embryonic asteroid epithelial collar cells: a sensory structure to determine the position of locomotory cilia. The Anatomical Record 236, 697–709.

Creer, S.; Fonseca, V. G.; Porazinska, D. L.; Giblin-Davis, R. M.; Sung, W.; Power, D. M.; Packer, M.; Carvalho, G. R.; Blaxter, M. L.; Lambshead, P. J. D.; et al. 2010. Ultrasequencing of the meiofaunal biosphere: practice, pitfalls and promises. Molecular Ecology 19, 4–20.

Cribb, T. H.; Bray, R. A.; Olson, P. D.; Littlewood, D. T. J. 2003. Life cycle evolution in the Digenea: a new perspective from phylogeny. Advances in Parasitology 54, 197–254.

Crosby, C. H.; Bailey, J. V. 2018. Experimental precipitation of apatite pseudofossils resembling fossil embryos. Geobiology 16, 80–87.

Cross, J. H. 1996a. Enteric nematodes of humans. In: Baron, S. (ed.), Medical Microbiology, Galveston.

Cross, J. H. 1996b. Filarial nematodes. In: Baron, S. (ed.), Medical Microbiology, Galveston.

Crowe, J. H.; Newell, I. M.; Thomson, W. W. 1970. *Echiniscus viridis* (Tardigrada) fine structure of the cuticle. Transactions of the American Microscopical Society 89, 316–325.

Cuervo-González, R. 2017. *Rhodope placozophagus* (Heterobranchia) a new species of turbellarian-like Gastropoda that preys on placozoans. Zoologischer Anzeiger 270, 43–48.

Cuffey, R. J.; Blake, D. B. 1991. Cladistic analysis of the phylum Bryozoa. Bulletin de la Société des Sciences Naturelles de l'Ouest de la France Memoire Hors de Série 1, 97–108.

Cunha, A.; Azevedo, R. B. R.; Emmons, S. W.; Leroi, A. M. 1999. Variable cell number in nematodes. Nature 402, 253.

Cunha, T. J.; Giribet, G. 2019. Resolving deep gastropod relationships. Proceedings of the Royal Society B: Biological Sciences.

Cunningham, J. A.; Liu, A. G.; Bengtson, S.; Donoghue, P. C. J. 2017. The origin of animals: Can molecular clocks and the fossil record be reconciled? BioEssays 39, 1–12.

Cunningham, J. A.; Thomas, C.-W.; Bengtson, S.; Kearns, S. L.; Xiao, S.; Marone, F.; Stampanoni, M.; Donoghue, P. C. J. 2012. Distinguishing geology from biology in the Ediacaran Doushantuo biota relaxes constraints on the timing of the origin of bilaterians. Proceedings of the Royal Society B: Biological Sciences 279, 2369–2376.

Cunningham, J. A.; Vargas; K.; Pengju, L.; Belivanova, V.; Marone, F.; Martínez-Pérez, C.; Guizar-Sicairos, M.; Holler, M.; Bengtson, S.; Donoghue, P. C. J. 2015. Critical appraisal of tubular putative eumetazoans from the Ediacaran Weng'an Doushantuo biota. Proceedings of the Royal Society B: Biological Sciences 282, 20151169.

Cutler, E. B. 1994. The Sipuncula. Their systematics, biology, and evolution. Cornell University Press, Ithaca.

Cwiklinski, K.; Dalton, J. P.; Dufresne, P. J.; La Course, J.; Williams, D. J.; Hodgkinson, J.; Paterson, S. 2015. The *Fasciola hepatica* genome: gene duplication and polymorphism reveals adaptation to the host environment and the capacity for rapid evolution. Genome Biology 16, 71.

Czaker, R. 2000. Extracellular Matrix (ECM) components in a very primitive multicellular animal, the dicyemid mesozoan *Kantharella antarctica*. The Anatomical Record 259, 52–59.

Czaker, R.; Janssen, H. H. 1998. Outer extracellular matrix (ECM) in the dicyemid mesozoan *Kantharella antarctica*. Journal of Submicroscopic Cytology and Pathology 30, 349–353.

Czechowski, P.; Sands, C. J.; Adams, B. J.; D'Haese, C. A.; Gibson, J. A. E.; McInnes, S. J.; Stevens, M. I. 2012. Antarctic Tardigrada: a first step in understanding molecular operational taxonomic units (MOTUs) and biogeography of cryptic meiofauna. Invertebrate Systematics 26, 526–538.

Dal Zotto, M.; Leasi, F.; Todaro, M. A. 2018. A new species of Turbanellidae (Gastrotricha, Macrodasyida) from Jamaica, with a key to species of *Paraturbanella*. ZooKeys, 105–119.

Daley, A. C.; Antcliffe, J. B.; Drage, H. B.; Pates, S. 2018. Early fossil record of Euarthropoda and the Cambrian Explosion. Proceedings of the National Academy of Sciences of the USA 115, 5323–5331.

Dallai, R.; Gottardo, M.; Beutel, R. G. 2016. Structure and evolution of insect sperm: New interpretations in the age of phylogenomics. Annual Review of Entomology 61, 1–23.

Dallai, R.; Mercati, D.; Carapelli, A.; Nardi, F.; Machida, R.; Sekiya, K.; Frati, F. 2011. Sperm accessory microtubules suggest the placement of Diplura as the sister-group of Insecta s.s. Arthropod Structure and Development 40, 77–92.

Daly, M.; Brugler, M. R.; Cartwright, P.; Collins, A. G.; Dawson, M. N.; Fautin, D. G.; France, S. C.; McFadden, C. S.; Opresko, D. M.; Rodríguez, E.; et al. 2007. The phylum Cnidaria: A review of phylogenetic patterns and diversity 300 years after Linnaeus. Zootaxa 1668, 127–182.

Damen, P.; Dictus, W. J. A. G. 1994a. Cell lineage of the prototroch of *Patella vulgata* (Gastropoda, Mollusca). Developmental Biology 162, 364–383.

Damen, P.; Dictus, W. J. A. G. 1994b. Cell-lineage analysis of the prototroch of the gastropod mollusc *Patella vulgata* shows conditional specification of some trochoblasts. Roux's Archives of Developmental Biology 203, 187–198.

Damen, W. G. M. 2002. Parasegmental organization of the spider embryo implies that the parasegment is an evolutionary conserved entity in arthropod embryogenesis. Development 129, 1239–1250.

Damen, W. G. M.; Hausdorf, M.; Seyfarth, E. A.; Tautz, D. 1998. A conserved mode of head segmentation in arthropods revealed by the expression pattern of hox genes in a spider. Proceedings of the National Academy of Sciences of the USA 95, 10665–10670.

Daniels, S. R.; McDonald; D. E., Picker, M. D. 2013. Evolutionary insight into the *Peripatopsis balfouri* sensu lato species complex (Onychophora: Peripatopsidae) reveals novel lineages and zoogeographic patterning. Zoologica Scripta 42, 656–674.

Danise, S.; Higgs, N. D. 2015. Bone-eating *Osedax* worms lived on Mesozoic marine reptile deadfalls. Biology Letters 11, 20150072.

Dannenberg, L. C.; Seaver, E. C. 2018. Regeneration of the germline in the annelid *Capitella teleta*. Developmental Biology 440, 74–87.

Danovaro, R.; Dell'anno, A.; Pusceddu, A.; Gambi, C.; Heiner, I.; Kristensen, R. M. 2010. The first Metazoa living in permanently anoxic conditions. BMC Biology 8, 30.

Danovaro, R.; Gambi, C.; Dell'Anno, A.; Corinaldesi, C.; Pusceddu, A.; Neves, R. C.; Kristensen, R. M. 2016. The challenge of proving the existence of metazoan life in permanently anoxic deep-sea sediments. BMC Biology 14, 43.

Danovaro, R.; Gambi, C.; Della Croce, N. 2002. Meiofauna hotspot in the Atacama Trench, eastern South Pacific Ocean. Deep-Sea Research. I. Oceanographic Research Papers 49, 843–857.

Darras, S.; Fritzenwanker, J. H.; Uhlinger, K. R.; Farrelly, E.; Pani, A. M.; Hurley, I. A.; Norris, R. P.; Osovitz, M.; Terasaki, M.; Wu, M.; et al. 2018. Anteroposterior axis patterning by early canonical Wnt signaling during hemichordate development. PLoS Biology 16, e2003698.

Darwin, C. R. 1844. Observations on the structure and propagation of the genus *Sagitta*. Annals and Magazine of Natural History, Ser.1 13, 1–6.

David, B.; Lefebvre, B.; Mooi, R.; Parsley, R. 2000. Are homalozoans echinoderms? An answer from the extraxial-axial theory. Paleobiology 26, 529–555.

Davidson, E. H. 1989. Lineage-specific gene expression and the regulative capacities of the sea urchin embryo: a proposed mechanism. Development 105, 421–445.

Davidson, E. H.; Rast, J. P.; Oliveri, P.; Ransick, A.; Calestani, C.; Yuh, C.-H.; Minokawa, T.; Amore, G.; Hinman, V.; Arenas-Mena, C.; et al. 2002. A genomic regulatory network for development. Science 295, 1669–1678.

Dawydoff, C. 1928. Sur la réversibilité des processus du développement. Les phases extrêmes de la réduction des Némertes. Comptes rendus Académie des Sciences Paris 186, 911–913.

De Baets, K.; Dentzien-Dias, P.; Upeniece, I.; Verneau, O.; Donoghue, P. C. J. 2015. Constraining the deep origin of parasitic flatworms and host-interactions with fossil evidence. Advances in Parasitology 90, 93–135.

de Beauchamp, P. 1929. Le développement des Gastrotriches. Bulletin de la Société Zoologique de France 54, 549–558.

de Beauchamp, P. 1956. Le développement de *Ploesoma hudsoni* (Imhof) et l'origine des feuillets chez les Rotiferes. Bulletin de la Société Zoologique de France 81, 374–383.

de Goeij, J. M.; van Oevelen, D.; Vermeij, M. J. A.; Osinga, R.; Middelburg, J. J.; de Gocij, A. F. P. M.; Admiraal, W. 2013. Surviving in a marine desert: The sponge loop retains resources within coral reefs. Science 342, 108–110.

De Ley, P.; Blaxter, M. 2004. A new system for Nematoda: combining morphological characters with molecular trees, and translating clades into ranks and taxa. Nematology Monographs and Perspectives 2, 633–653.

de Mendoza, A.; Ruiz-Trillo, I. 2011. The mysterious evolutionary origin for the GNE gene and the root of Bilateria. Molecular Biology and Evolution 28; 2987–2991.

de Mendoza, A.; Suga, H.; Permanyer, J.; Irimia, M.; Ruiz-Trillo, I. 2015. Complex transcriptional regulation and independent evolution of fungal-like traits in a relative of animals. eLife 4, e08904.

de Quatrefages, A. 1847. Etudes sur les types inférieurs de l'embranchement des annelés. Mémoire sur l'echiure de Gaertner (*Echiurus Gaertnerii* NOB). Annales des sciences naturelles zoologie 7, 307–343.

de Quatrefages, A. 1865. Histoire naturelle des Annelés marins et d'eau douce. Annélides et Géphyriens. Libraries Encyclopédique de Roret, Paris.

de Rosa, R. 2001. Molecular data indicate the protostome affinity of brachiopods. Systematic Biology 50, 848–859.

De Smet, W. H. 2002. A new record of *Limnognathia maerski* Kristensen & Funch, 2000 (Micrognathozoa) from the subantarctic Crozet Islands, with redescription of the trophi. Journal of Zoology 258, 381–393.

De Smet, W. H.; Segers, H. 2017. Ontogeny of the jaws of monogonont rotifers: the malleate trophi of *Rhinoglena* and *Proalides* (Ploima, Epiphanidae). Invertebrate Biology 136, 422–440.

Debortoli, N.; Li, X.; Eyres, I.; Fontaneto, D.; Hespeels, B.; Tang, C. Q.; Flot, J. F.; Van Doninck, K. 2016. Genetic exchange among bdelloid rotifers is more likely due to horizontal gene transfer than to meiotic sex. Current Biology 26, 723–732.

Decraemer, W.; Coomans, A.; Baldwin, J. 2014. Morphology of Nematoda. In: Schmidt-Rhaesa, A. (ed.), Handbook of Zoology: Gastrotricha, Cycloneuralia and Gnathifera, vol. 2: Nematoda. De Gruyter, Berlin/Boston, pp. 1–59.

Degma, P.; Guidetti, R. 2018. Tardigrade taxa. In: Schill, R. O. (ed.), Water Bears: The Biology of Tardigrades. Zoological Monographs, vol. 2. Springer Nature Switzerland, Cham, pp. 371–409.

Degnan, B. M.; Adamska, M.; Richards, G. S.; Larroux, C.; Leininger, S.; Bergum, B.; Calcino, A.; Taylor, K.; Nakanishi, N.; Degman, S. M. 2015. Porifera. In: Wanninger, A. (ed.), Evolutionary Developmental Biology of Invertebrates 1: Introduction, Non-Bilateria, Acoelomorpha, Xenoturbellida, Chaetognatha. Springer, Vienna, pp. 65–106.

Dehal, P.; Satou, Y.; Campbell, R. K.; Chapman, J.; Degnan, B.; De Tomaso, A.; Davidson, B.; Di Gregorio, A.; Gelpke, M.; Goodstein, D. M.; et al. 2002. The draft genome of *Ciona intestinalis*: Insights into chordate and vertebrate origins. Science 298, 2157–2167.

Deheyn, D.; Watson, N. A.; Jangoux, M. 1998. Symbioses in *Amphipholis squamata* (Echinodermata, Ophiuroidea, Amphiuridae): geographical variation of infestation and effect

G. D.; et al. 2008. Broad phylogenomic sampling improves resolution of the animal tree of life. Nature 452, 745–749.

Dunn, C. W.; Leys, S. P.; Haddock, S. H. 2015. The hidden biology of sponges and ctenophores. Trends in Ecology and Evolution 30, 282–291.

Dunn, C. W.; Ryan, J. F. 2015. The evolution of animal genomes. Current Opinion in Genetics and Development 35, 25–32.

Dunn, F. S.; Liu, A. G.; Donoghue, P. C. J. 2018. Ediacaran developmental biology. Biological Reviews 93, 914–932.

Dunn, P. H.; Young, C. M. 2013. Finding refuge: The estuarine distribution of the nemertean egg predator *Carcinonemertes errans* on the Dungeness crab, *Cancer magister*. Estuarine, Coastal and Shelf Science 135, 201–208.

Dunning, H. 2012. Gone missing, circa 1892. The Scientist 26, 84–84.

Dupre, C.; Yuste, R. 2017. Non-overlapping neural networks in *Hydra vulgaris*. Current Biology 27, 1085–1097.

Durden, C. J.; Rodgers, J.; Yochelson, E. L.; Riedl, R. J. 1969. Gnathostomulida: Is there a fossil record? Science 164, 855–856.

Dylus, D. V.; Czarkwiani, A.; Blowes, L. M.; Elphick, M. R.; Oliveri, P. 2018. Developmental transcriptomics of the brittle star *Amphiura filiformis* reveals gene regulatory network rewiring in echinoderm larval skeleton evolution. Genome Biology 19, 26.

Dzik, J. 2010. Brachiopod identity of the alleged monoplacophoran ancestors of cephalopods. Malacologia 52, 97–113.

Dzik, J.; Mazurek, D. 2013. Affinities of the alleged earliest Cambrian gastropod *Aldanella*. Canadian Journal of Zoology 91, 914–923.

Eaves, A. A.; Palmer, A. R. 2003. Reproduction: widespread cloning in echinoderm larvae. Nature 425, 146.

Edgecombe, G. D. 2009. Palaeontological and molecular evidence linking arthropods, onychophorans, and other Ecdysozoa. Evolution: Education and Outreach 2, 178–190.

Edgecombe, G. D. 2010. Arthropod phylogeny: An overview from the perspectives of morphology, molecular data and the fossil record. Arthropod Structure and Development 39, 74–87.

Edgecombe, G. D. 2011. Chilopoda—The fossil history. In: Minelli, A. (ed.), Treatise on Zoology—Anatomy, Taxonomy, Biology. The Myriapoda, vol. 1. Brill, Leiden/Boston, pp. 355–361.

Edgecombe, G. D. 2015. Diplopoda—Fossils. In: Minelli, A. (ed.), Treatise on Zoology—Anatomy, Taxonomy, Biology. The Myriapoda, vol. 2. Brill, Leiden/Boston, pp. 337–351.

Edgecombe, G. D. 2017. Inferring arthropod phylogeny: Fossils and their interaction with other data sources. Integrative and Comparative Biology 57, 467–476.

Edgecombe, G. D.; Giribet, G.; Dunn, C. W.; Hejnol, A.; Kristensen, R. M.; Neves, R. C.; Rouse, G. W.; Worsaae, K.; Sørensen, M. V. 2011. Higher-level metazoan relationships: recent progress and remaining questions. Organisms, Diversity and Evolution 11, 151–172.

Edgecombe, G. D.; Legg, D. A. 2013. The Arthropod Fossil Record. In: Minelli, A., Boxshall, G.; Fusco, G. (eds.), Arthropod Biology and Evolution—Molecules, Development, Morphology. Springer, Heidelberg, pp. 393–415.

Edgecombe, G. D.; Ma, X. Y.; Strausfeld, N. J. 2015. Unlocking the early fossil record of the arthropod central nervous system. Philosophical Transactions of the Royal Society B: Biological Sciences 370, 20150038.

Eeckhaut, I.; Flammang, P.; LoBue, C.; Jangoux, M. 1997. Functional morphology of the tentacles and tentilla of *Coeloplana bannworthi* (Ctenophora, Platyctenida), an ectosymbiont of *Diadema setosum* (Echinodermata, Echinoida). Zoomorphology 117, 165–174.

Eernisse, D. J.; Albert, J. S.; Anderson, F. E. 1992. Annelida and Arthropoda are not sister taxa: A phylogenetic analysis of spiralian metazoan morphology. Systematic Biology 41, 305–330.

Egger, B.; Bachmann, L.; Fromm, B. 2017. *Atp8* is in the ground pattern of flatworm mitochondrial genomes. BMC Genomics 18, 414.

Egger, B.; Lapraz, F.; Tomiczek, B.; Müller, S.; Dessimoz, C.; Girstmair, J.; Skunca, N.; Rawlinson, K. A.; Cameron, C. B.; Beli, E.; et al. 2015. A transcriptomic-phylogenomic analysis of the evolutionary relationships of flatworms. Current Biology 25, 1347–1353.

Ehlers, U. 1984. Phylogenetisches System der Plathelminthes. Verhandlungen des naturwissenschaftlichen Vereins in Hamburg, 291–294.

Ehlers, U. 1985a. Das Phylogenetische System der Plathelminthes. Gustav Fischer, Stuttgart.

Ehlers, U. 1985b. Phylogenetic relationships within the Platyhelminthes. In: Conway Morris, S.; George, J. D.; Gibson, R.; Platt, H. M. (eds.), The origins and relationships of lower invertebrates. Systematics Association/Clarendon Press, Oxford, pp. 141–158.

Ehlers, U. 1991. Comparative morphology of statocysts in the Platyhelminthes and the Xenoturbellida. Hydrobiologia 227, 263–271.

Ehlers, U. 1992a. Dermonephridia—modified epidermal cells with a probable excretory function in *Paratomella rubra* (Acoela, Plathelminthes). Microfauna Marina 7, 253–264.

Ehlers, U. 1992b. Frontal glandular and sensory structures in *Nemertoderma* (Nemertodermatida) and *Paratomella* (Acoela): ultrastructure and phylogenetic implications for the monophyly of the Euplathelminthes (Plathelminthes). Zoomorphology 112, 227–236.

Ehlers, U. 1992c. No mitosis of differentiated epidermal cells in *Rhynchoscolex simplex* Leidy, 1851 (Catenulida). Microfauna Marina 7, 311–321.

Ehlers, U. 1993. Ultrastructure of the spermatozoa of *Halammohydra schulzei* (Cnidaria, Hydrozoa): the significance of acrosomal structures for the systematization of the Eumetazoa. Microfauna Marina 8, 115–130.

Ehlers, U. 1994. Absence of a pseudocoel or pseudocoelom in *Anoplostoma vivipara* (Nematodes). Microfauna Marina 9, 345–350.

Ehlers, U. 1997. Ultrastructure of the statocysts in the apodous sea cucumber *Leptosynapta inhaerens* (Holothuroidea, Echinodermata). Acta Zoologica 78, 61–68.

Ehlers, U.; Sopott-Ehlers, B. 1986. Vergleichende Ultrastruktur von Protonephridien: ein Beitrag zur Stammesgeschichte der Plathelminthen. Verhandlungen der Deutschen zoologischen Gesellschaft 79, 168–169.

Ehlers, U.; Sopott-Ehlers, B. 1997. Ultrastructure of the sub-epidermal musculature of *Xenoturbella bocki*, the adelpho-taxon of the Bilateria. Zoomorphology 117, 71–79.

Eibye-Jacobsen, D. 2004. A reevaluation of *Wiwaxia* and the polychaetes of the Burgess Shale. Lethaia 37, 317–335.

Eibye-Jacobsen, J. 1996/97. New observations on the embryology of the Tardigrada. Zoologischer Anzeiger 235, 201–216.

Eisig, H. 1899. Zur Entwicklungsgeschichte der Capitelliden. Mittheilungen Aus der Zoologischen Station zu Neapel 13, 1–292.

Eitel, M.; Francis, W. R.; Varoqueaux, F.; Daraspe, J.; Osigus, H.-J.; Krebs, S.; Vargas, S.; Blum, H.; Williams, G. A.; Schierwater, B.; et al. 2018. Comparative genomics and the nature of placozoan species. PLoS Biology 16, e2005359.

Eitel, M.; Guidi, L.; Hadrys, H.; Balsamo, M.; Schierwater, B. 2011. New insights into placozoan sexual reproduction and development. PLoS One 6, e19639.

Eitel, M.; Osigus, H.-J.; DeSalle, R.; Schierwater, B. 2013. Global diversity of the Placozoa. PLoS One 8, e57131.

Eitel, M.; Schierwater, B. 2010. The phylogeography of the Placozoa suggests a taxon-rich phylum in tropical and subtropical waters. Molecular Ecology 19, 2315–2327.

Elicki, O.; Gursu, S. 2009. First record of *Pojetaia runnegari* Jell, 1980 and *Fordilla* Barrande, 1881 from the Middle East (Taurus Mountains, Turkey) and critical review of Cambrian bivalves. Palaeontologische Zeitschrift 83, 267–291.

Elpatiewsky, W. 1909. Die Urgeschlechtszellen bildung bei *Sagitta*. Anatomischer Anzeiger 35, (226–239).

Emig, C. C. 1974. The systematics and evolution of the phylum Phoronida. Zeitschrift für zoologische Systematik und Evolutionsforschung 12, 128–151.

Emig, C. C. 1985. Phylogenetic systematics in Phoronida (Lophophorata). Zeitschrift für zoologische Systematik und Evolutionsforschung 23, 184–193.

Emig, C. C.; Roldán, C.; Viéitez, J. M. 2005. Phoronida. In: Ramos Sánchez, M. Á. (ed.), Fauna Ibérica, vol. 27: Lophophorata: Phoronida, Brachiopoda. Museo Nacional de Ciencias Naturales, Consejo Superior de Investigaciones Científicas, Madrid, pp. 21–56.

Emschermann, P. 1965. Das Protonephridiensystem von *Urnatella gracilis* Leidy (Kamptozoa) Bau, Entwicklung und Funktion. Zeitschrift für Morphologie und Oekologie der Tiere 55, 859–914.

Emschermann, P. 1972a. Cuticular pores and spines in the Pedicellinidae and Barentsiidae (Entoprocta), their relationship, ultrastructure, and suggested function, and their phylogenetic evidence. Sarsia 51, 7–16.

Emschermann, P. 1972b. *Loxokalypus socialis* gen. et sp. nov. (Kamptozoa, Loxokalypodidae fam. nov.), ein neuer Kamptozoentyp aus dem nördlichen Pazifischen Ozean. Ein Vorschlag zur Neufassung der Kamptozoensystematik. Marine Biology 12, 237–254.

Engel, M. S.; Grimaldi, D. A. 2004. New light shed on the oldest insect. Nature 427, 627–630.

Enghoff, H.; Dohle, W.; Blower, J. G. 1993. Anamorphosis in millipedes (Diplopoda)—the present state of knowledge with some developmental and phylogenetic considerations. Zoological Journal of the Linnean Society 109, 103–234.

Entchev, E. V.; Kurzchalia, T. V. 2005. Requirement of sterols in the life cycle of the nematode *Caenorhabditis elegans*. Seminars in Cell and Developmental Biology 16, 175–182.

Ereskovsky, A. 2010. The comparative embryology of sponges. Springer, Dordrecht.

Eriksson, B. J.; Budd, G. E. 2001. Onychophoran cephalic nerves and their bearing on our understanding of head segmentation and stem-group evolution of Arthropoda. Arthropod Structure and Development 29, 197–209.

Eriksson, B. J.; Stollewerk, A. 2010a. Expression patterns of neural genes in *Euperipatoides kanangrensis* suggest divergent evolution of onychophoran and euarthropod neurogenesis. Proceedings of the National Academy of Sciences of the USA 107, 22576–22581.

Eriksson, B. J.; Stollewerk, A. 2010b. The morphological and molecular processes of onychophoran brain development show unique features that are neither comparable to insects nor to chelicerates. Arthropod Structure and Development 39, 478–490.

Eriksson, B. J.; Tait, N. N.; Budd, G. E. 2003. Head development in the onychophoran *Euperipatoides kanangrensis* with particular reference to the central nervous system. Journal of Morphology 255, 1–23.

Eriksson, B. J.; Tait, N. N.; Budd, G. E.; Janssen, R.; Akam, M. 2010. Head patterning and Hox gene expression in an onychophoran and its implications for the arthropod head problem. Development Genes and Evolution 220, 117–122.

Erkut, C.; Vasilj, A.; Boland, S.; Habermann, B.; Shevchenko, A.; Kurzchalia, T. V. 2013. Molecular strategies of the *Caenorhabditis elegans* dauer larva to survive extreme desiccation. PLoS One 8, e82473.

Ernst, A.; Königshof, P. 2008. The role of bryozoans in fossil reefs—an example from the Middle Devonian of the Western Sahara. Facies 54, 613–620.

Erpenbeck, D.; Voigt, O.; Adamski, M.; Adamska, M.; Hooper, J. N. A.; Wörheide, G.; Degnan, B. M. 2007. Mitochondrial diversity of early-branching Metazoa is revealed by the complete mt genome of a haplosclerid demosponge. Molecular Biology and Evolution 24, 19–22.

Erpenbeck, D.; Wörheide, G. 2007. On the molecular phylogeny of sponges (Porifera). Zootaxa 1668, 107–126.

Ertas, B.; von Reumont, B. M.; Wägele, J. W.; Misof, B.; Burmester, T. 2009. Hemocyanin suggests a close relationship of Remipedia and Hexapoda. Molecular Biology and Evolution 26, 2711–2718.

Erwin, D. H.; Valentine, J. W. 2013. The Cambrian Explosion: The Construction of Animal Biodiversity. Roberts, Greenwood Village, CO.

Eshragh, R.; Leander, B. S. 2015. Molecular contributions to species boundaries in dicyemid parasites from eastern Pacific cephalopods. Marine Biology Research 11, 414–422.

Eszterbauer, E.; Atkinson, S.; Diamant, A.; Morris, D. W.; El-Matbouli, M.; Hartikainen, H. 2015. Myxozoan life cycles: practical approaches and insights. In: Okamura, B.; Gruhl, A.; Bartholomew, J. L. (eds.), Myxozoan Evolution, Ecology and Development. Springer, Cham, pp. 175–198.

Foth, B. J.; Tsai, I. J.; Reid, A. J.; Bancroft, A. J.; Nichol, S.; Tracey, A.; Holroyd, N.; Cotton, J. A.; Stanley, E. J.; Zarowiecki, M.; et al. 2014. Whipworm genome and dual-species transcriptome analyses provide molecular insights into an intimate host-parasite interaction. Nature Genetics 46, 693–700.

Francis, D. R. 2001. Bryozoan statoblasts. In: Smol, J. S.; Birks, H. J. B.; Last, W. M. (eds.), Developments in Paleoenvironmental Research. Tracking Environmental Change Using Lake Sediments, vol. 4: Zoological Indicators. Kluwer Academic, Dordrecht, pp. 105–123.

Francis, L. 1976. Social organization within clones of the sea anemone *Anthopleura elegantissima*. Biological Bulletin 150, 361–376.

Frank, U.; Plickert, G.; Müller, W. A. 2009. Cnidarian interstitial cells: The dawn of stem cell research. In: Rinkevich, B.; Matranga, V. (eds.), Stem Cells in Marine Organisms. Springer, Dordrecht, pp. 33–59.

Franke, F. A.; Schumann, I.; Hering, L.; Mayer, G. 2015. Phylogenetic analysis and expression patterns of Pax genes in the onychophoran *Euperipatoides rowelli* reveal a novel bilaterian Pax subfamily. Evolution and Development 17, 3–20.

Franke, H. D. 1999. Reproduction of the Syllidae (Annelida: Polychaeta). Hydrobiologia 402, 39–55.

Franke, M. 1993. Ultrastructure of the protonephridia in *Loxosomella fauveli*, *Barentsia matsuhimana* and *Pedicellina cernua*. Implications for the protonephridia in the ground pattern of the Entoprocta (Kamptozoa). Microfauna Marina 8, 7–38.

Franzén, Å. 1996. Ultrastructure of spermatozoa and spermiogenesis in the hydrozoan *Cordylophora caspia* with comments on structure and evolution of the sperm in the Cnidaria and the Porifera. Invertebrate Reproduction and Development 29, 19–26.

Franzén, Ä. 1955. Comparative morphological investigations into the spermiogenesis among Mollusca. Zoologiska Bidrag fran Uppsala 30, 399–456.

Franzén, Å.; Afzelius, B. A. 1987. The ciliated epidermis of *Xenoturbella bocki* (Platyhelminthes, Xenoturbellida) with some phylogenetic considerations. Zoologica Scripta 16, 9–17.

Freeman, G. 1980. The role of cleavage in the establishment of the anterior-posterior axis of the hydrozoan embryo. In: Tardent, P.; Tardent, R. (eds.), Developmental and Cellular Biology of Coelenterates. Proceedings of the 4th International Coelenterate Conference, Interlaken, Switzerland September 1979. Elsevier/Biomedical Press, Amsterdam, New York/Oxford, pp. 97–108.

Freeman, G. 1993. Regional specification during embryogenesis in the articulate brachiopod *Terebratalia*. Developmental Biology 160, 196–213.

Freeman, G. 1995. Regional specification during embryogenesis in the inarticulate brachiopod *Glottidia*. Developmental Biology 172, 15–36.

Freeman, G. 1999. Regional specification during embryogenesis in the inarticulate brachiopod *Discinisca*. Developmental Biology 209, 321–339.

Freeman, G. 2000. Regional specification during embryogenesis in the craniiform brachiopod *Crania anomala*. Developmental Biology 227, 219–238.

Freeman, G. 2003. Regional specification during embryogenesis in Rhynchonelliform brachiopods. Developmental Biology 261, 268–287.

Freeman, G.; Lundelius, J. W. 1992. Evolutionary implications of the mode of D quadrant specification in coelomates with spiral cleavage. Journal of Evolutionary Biology 5, 205–247.

Freeman, G.; Martindale, M. Q. 2002. The origin of mesoderm in phoronids. Developmental Biology 252, 301–311.

Fregni, E. 1998. The spermatozoa of macrodasyid gastrotrichs: observations by scanning electron microscopy. Invertebrate Reproduction and Development 34, 1–11.

Frenzel, J. 1892. Untersuchungen über die mikroskopische Fauna Argentiniens. Salinella salve nov. gen. nov. spec. Ein vielzelliges, inturorienartiges Tier (Mesozoon). Archiv für Naturgeschichte 58, 66–96.

Frickhinger, K. A. 1999. Hemichordata acorn worms. The fossils of Solnhofen. A documentation of fauna and flora of the Solnhofen Formation. 2: New specimens, new details, new results with an introduction to the geology of the Solnhofen Formation. Goldschneck-Verlag, Korb, pp. 76–79.

Fritsch, M.; Wollesen, T.; de Oliveira, A. L.; Wanninger, A. 2015. Unexpected co-linearity of Hox gene expression in an aculiferan mollusk. BMC Evolutionary Biology 15, 151.

Fritsch, M.; Wollesen, T.; Wanninger, A. 2016. Hox and Para-Hox gene expression in early body plan patterning of polyplacophoran mollusks. Journal of Experimental Zoology (Mol. Dev. Evol.) 326B, 89–104.

Fritzenwanker, J. H.; Genikhovich, G.; Kraus, Y.; Technau, U. 2007. Early development and axis specification in the sea anemone *Nematostella vectensis*. Developmental Biology 310, 264–279.

Fritzsch, G.; Bohme, M. U.; Thorndyke, M.; Nakano, H.; Israelsson, O.; Stach, T.; Schlegel, M.; Hankeln, T.; Stadler, P. F. 2008. PCR survey of *Xenoturbella bocki* Hox genes. Journal of Experimental Zoology, pt. B: Molecular and Developmental Evolution 310B, 278–284.

Fröbius, A. C.; Funch, P. 2017. Rotiferan *Hox* genes give new insights into the evolution of metazoan bodyplans. Nature Communications 8, 9.

Fröbius, A. C.; Matus, D. Q.; Seaver, E. C. 2008. Genomic organization and expression demonstrate spatial and temporal *Hox* gene colinearity in the lophotrochozoan *Capitella* sp. I. PLoS One 3, e4004.

Frýda, J. 2012. Phylogeny of Palaeozoic gastropods inferred from their ontogeny. In: Talent, J. A. (ed.), Earth and Life. Springer Science+Business Media B.V., pp. 395–435.

Fryer, G.; Stanley, G. D. 2004. A Silurian porpitoid hydrozoan from Cumbria, England, and a note on porpitoid relationships. Palaeontology 47, 1109–1119.

Fu, D.; Tong, G.; Dai, T.; Liu, W.; Yang, Y.; Zhang, Y.; Cui, L.; Li, L.; Yun, H.; Wu, Y.; Sun, A.; Liu, C.; Pei, W.; Gaines, R. R.; Zhang, X. 2019. The Qingjiang biota–a Burgess Shale-type Konservat Lagerstätte from the early Cambrian of South China. Science 363, 1338–1342.

Fuchs, J.; Bright, M.; Funch, P.; Wanninger, A. 2006. Immunocytochemistry of the neuromuscular systems of *Loxosomella vivipara* and *L. parguerensis* (Entoprocta: Loxosomatidae). Journal of Morphology 267, 866–883.

Fuchs, J.; Iseto, T.; Hirose, M.; Sundberg, P.; Obst, M. 2010. The first internal molecular phylogeny of the animal phylum Entoprocta (Kamptozoa). Molecular Phylogenetics and Evolution 56, 370–379.

Fuchs, J.; Obst, M.; Sundberg, P. 2009. The first comprehensive molecular phylogeny of Bryozoa (Ectoprocta) based on combined analyses of nuclear and mitochondrial genes. Molecular Phylogenetics and Evolution 52, 225–233.

Fuchs, J.; Wanninger, A. 2008. Reconstruction of the neuromuscular system of the swimming-type larva of *Loxosomella atkinsae* (Entoprocta) as inferred by fluorescence labelling and confocal microscopy. Organisms, Diversity and Evolution 8, 325–335.

Fujimoto, S.; Jørgensen, A.; Hansen, J. G. 2017. A molecular approach to arthrotardigrade phylogeny (Heterotardigrada, Tardigrada). Zoologica Scripta 46, 496–505.

Fujita, T.; Ohta, S. 1990. Size structure of dense populations of the brittle star *Ophiura sarsii* (Ophiuroidea: Echinodermata) in the bathyal zone around Japan. Marine Ecology Progress Series 64, 113–122.

Fukami, H.; Chen, C. A.; Budd, A. F.; Collins, A.; Wallace, C.; Chuang, Y. Y.; Chen, C.; Dai, C. F.; Iwao, K.; Sheppard, C.; et al. 2008. Mitochondrial and nuclear genes suggest that stony corals are monophyletic but most families of stony corals are not (Order Scleractinia, Class Anthozoa, Phylum Cnidaria). PLoS One 3, e3222.

Funch, P. 1996. The chordoid larva of *Symbion pandora* (Cycliophora) is a modified trochophore. Journal of Morphology 230, 231–263.

Funch, P.; Kristensen, R. M. 1995. Cycliophora is a new phylum with affinities to Entoprocta and Ectoprocta. Nature 378, 711–714.

Funch, P.; Kristensen, R. M. 1999. Cycliophora. In: Knobil, E.; Neill, J. D. (eds.), Encyclopedia of Reproduction. Academic Press, San Diego, pp. 800–808.

Funch, P.; Kristensen, R. M. 2002. Coda: the Micrognathozoa—a new class or phylum of freshwater meiofauna? In: Rundle, S. D.; Robertson, A. L.; Schmid-Araya, J. M. (eds.), Freshwater meiofauna: biology and ecology. Backhuys, Leiden, pp. 337–348.

Funch, P.; Neves, R. 2018. Cycliophora. In: Schmidt-Rhaesa, A. (ed.) Handbook of Zoology: Miscellaneous Invertebrates. De Gruyter, Berlin/Boston, pp. 87–110.

Funch, P.; Segers, H.; Dumont, H. J. 1996. Rotifera in tap water in G[h]ent, Belgium. Biologisch jaarboek Dodonaea 63, 53–57.

Funch, P.; Sørensen, M.; Obst, M. 2005. On the phylogenetic position of Rotifera—have we come any further? Hydrobiologia 546, 1–18.

Funch, P.; Thor, P.; Obst, M. 2008. Symbiotic relations and feeding biology of *Symbion pandora* (Cycliophora) and *Triticella flava* (Bryozoa). Vie et Milieu 58, 185–188.

Furuhashi, T.; Schwarzinger, C.; Miksik, I.; Smrz, M.; Beran, A. 2009. Molluscan shell evolution with review of shell calcification hypothesis. Comparative Biochemistry and Physiology B: Biochemistry and Molecular Biology 154, 351–371.

Furuya, H. 1999. Fourteen new species of dicyemid mesozoans from six Japanese cephalopods, with comments on host specificity. Species Diversity 4, 257–319.

Furuya, H.; Hochberg, F. G.; Tsuneki, K. 2001. Developmental patterns and cell lineages of vermiform embryos in dicyemid mesozoans. Biological Bulletin 201, 405–416.

Furuya, H.; Hochberg, F. G.; Tsuneki, K. 2004. Cell number and cellular composition in infusoriform larvae of dicyemid mesozoans (phylum Dicyemida). Zoological Science 21, 877–889.

Furuya, H.; Hochberg, F. G.; Tsuneki, K. 2007. Cell number and cellular composition in vermiform larvae of dicyemid mesozoans. Journal of Zoology 272, 284–298.

Furuya, H.; Tsuneki, K. 2003. Biology of dicyemid mesozoans. Zoological Science 20, 519–532.

Furuya, H.; Tsuneki, K. 2007. Developmental patterns of the hermaphroditic gonad in dicyemid mesozoans (Phylum Dicyemida). Invertebrate Biology 126, 295–306.

Furuya, H.; Tsuneki, K.; Koshida, Y. 1992. Development of the infusoriform embryo of *Dicyema japonicum* (Mesozoa: Dicyemidae). Biological Bulletin 183, 248–257.

Furuya, H.; Tsuneki, K.; Koshida, Y. 1996. The cell lineages of two types of embryo and a hermaphroditic gonad in dicyemid mesozoans. Development Growth and Differentiation 38, 453–463.

Fusco, G.; Minelli, A. 2013. Arthropod segmentation and tagmosis. In: Minelli, A.; Boxshall, G.; Fusco, G. (eds.), Arthropod Biology and Evolution: Molecules, Development, Morphology. Springer, Heidelberg, pp. 197–221.

Futahashi, R.; Kawahara-Miki, R.; Kinoshita, M.; Yoshitake, K.; Yajima, S.; Arikawa, K.; Fukatsu, T. 2015. Extraordinary diversity of visual opsin genes in dragonflies. Proceedings of the National Academy of Sciences of the USA 112, E1247–E1256.

G10KCOS. 2009. Genome 10K: a proposal to obtain whole-genome sequence for 10,000 vertebrate species. Journal of Heredity 100, 659–674.

Gabriel, W. N.; Goldstein, B. 2007. Segmental expression of Pax3/7 and Engrailed homologs in tardigrade development. Development Genes and Evolution 217, 421–433.

Gabriel, W. N.; McNuff, R.; Patel, S. K.; Gregory, T. R.; Jeck, W. R.; Jones, C. D.; Goldstein, B. 2007. The tardigrade *Hypsibius dujardini*, a new model for studying the evolution of development. Developmental Biology 312, 545–559.

Gad, G. 2009. A clearly identifiable postlarva in the life cycle of a new species of *Pliciloricus* (Loricifera) from the deep sea of the Angola Basin. Zootaxa 2096, 50–81.

Gaffron, E. 1885. Beiträge zur Anatomie und Histologie von *Peripatus*. I und II. Thiel. Zoologische Beiträge 1, 33–60, 145–163.

Gagné, G. D. 1980. Ultrastructure of the sensory palps of *Tetranchyroderma papii* (Gastrotricha, Macrodasyida). Zoomorphologie 95, 115–125.

Gagnon, Y. L.; Templin, R. M.; How, M. J.; Marshall, N. J. 2015. Circularly polarized light as a communication signal in mantis shrimps. Current Biology 25, 3074–3078.

Gahan, J. M.; Bradshaw, B.; Flici, H.; Frank, U. 2016. The interstitial stem cells in *Hydractinia* and their role in regeneration. Current Opinion in Genetics and Development 40, 65–73.

Gaidos, E.; Dubuc, T.; Dunford, M.; McAndrew, P.; Padilla-Gamiño, J.; Studer, B.; Weersing, K.; Stanley, S. 2007. The

Precambrian emergence of animal life: a geobiological perspective. Geobiology 5, 351–373.

Gaines, R. R. 2014. Burgess Shale-type preservation and its distribution in space and time. In: Laflamme, M.; Schiffbauer, J. D.; Darroch, S. A. F. (eds.), Reading and Writing of the Fossil Record: Preservational Pathways to Exceptional Fossilization. Paleontological Society Papers, vol. 20, pp. 123–146.

Gaiti, F.; Jindrich, K.; Fernandez-Valverde, S. L.; Roper, K. E.; Degnan, B. M.; Tanurdžić, M. 2017. Landscape of histone modifications in a sponge reveals the origin of animal cis-regulatory complexity. eLife 6, e22194.

Gao, J.; Yang, N.; Lewis, F. A.; Yau, P.; Collins, J. J.; Sweedler, J. V.; Newmark, P. A. 2019. A rotifer-derived paralytic compound prevents transmission of schistosomiasis to a mammalian host. bioRxiv https://doi.org/10.1101/426999.

García-Arrarás, J. E.; Greenberg, M. J. 2001. Visceral regeneration in holothurians. Microscopy Research and Technique 55, 438–451.

García-Bellido, D. C.; Lee, M. S. Y.; Edgecombe, G. D.; Jago, J. B.; Gehling, J. G.; Paterson, J. R. 2014. A new vetulicolian from Australia and its bearing on the chordate affinities of an enigmatic Cambrian group. BMC Evolutionary Biology 14, 214.

Garcia-Fernàndez, J.; Benito-Gutiérrez, E. 2009. It's a long way from amphioxus: descendants of the earliest chordate. BioEssays 31, 665–675.

Garcia-Fernàndez, J.; Holland, P. W. 1994. Archetypal organization of the amphioxus Hox gene cluster. Nature 370, 563–566.

García-Varela, M.; Cummings, M. P.; Pérez-Ponce de León, G.; Gardner, S. L.; Laclette, J. P. 2002. Phylogenetic analysis based on 18S ribosomal RNA gene sequences supports the existence of class Polyacanthocephala (Acanthocephala). Molecular Phylogenetics and Evolution 23, 288–292.

García-Varela, M.; Nadler, S. A. 2006. Phylogenetic relationships among Syndermata inferred from nuclear and mitochondrial gene sequences. Molecular Phylogenetics and Evolution 40, 61–72.

García-Varela, M.; Pérez-Ponce de León, G.; de la Torre, P.; Cummings, M. P.; Sarma, S. S. S.; Laclette, J. P. 2000. Phylogenetic relationships of Acanthocephala based on analysis of 18S ribosomal RNA gene sequences. Journal of Molecular Evolution 50, 532–540.

Garey, J. R.; Krotec, M.; Nelson, D. R.; Brooks, J. 1996a. Molecular analysis supports a tardigrade-arthropod association. Invertebrate Biology 115, 79–88.

Garey, J. R.; Near, T. J.; Nonnemacher, M. R.; Nadler, S. A. 1996b. Molecular evidence for Acanthocephala as a subtaxon of Rotifera. Journal of Molecular Evolution 43, 287–292.

Garey, J. R.; Schmidt-Rhaesa, A.; Near, T. J.; Nadler, S. A. 1998. The evolutionary relationships of rotifers and acanthocephalans. Hydrobiologia 387/388, 83–91.

Garm, A. 2017. Sensory biology of starfish—with emphasis on recent discoveries in their visual ecology. Integrative and Comparative Biology 57, 1082–1092.

Garm, A.; Ekström, P.; Boudes, M.; Nilsson, D. E. 2006. Rhopalia are integrated parts of the central nervous system in box jellyfish. Cell and Tissue Research 325, 333–343.

Garm, A.; Nilsson, D.-E. 2014. Visual navigation in starfish: first evidence for the use of vision and eyes in starfish. Proceedings of the Royal Society B: Biological Sciences 281, 20133011.

Garstang, W. 1894. Preliminary note on a new theory of the phylogeny of the Chordata. Zoologischer Anzeiger 17, 122–125.

Garstang, W. 1928. The morphology of the Tunicata and its bearing on the phylogeny of the Chordata. Quarterly Journal of Microscopical Sciences 72, 51–187.

Garwood, R. J.; Edgecombe, G. D.; Charbonnier, S.; Chabard, D.; Sotty, D.; Giribet, G. 2016. Carboniferous Onychophora from Montceau-les-Mines, France, and onychophoran terrestrialization. Invertebrate Biology 135, 179–190.

Gąsiorek, P.; Stec, D.; Morek, W.; Michalczyk, Ł. 2018. An integrative redescription of Hypsibius dujardini (Doyère, 1840), the nominal taxon for Hypsibioidea (Tardigrada: Eutardigrada). Zootaxa 4415, 45–75.

Gąsiorowski, L.; Bekkouche, N.; Sørensen, M. V.; Kristensen, R. M.; Sterrer, W.; Worsaae, K. 2017a. New insights on the musculature of filospermoid Gnathostomulida. Zoomorphology 136, 413–424.

Gąsiorowski, L.; Bekkouche, N.; Worsaae, K. 2017b. Morphology and evolution of the nervous system in Gnathostomulida (Gnathifera, Spiralia). Organisms Diversity and Evolution 17, 447–475.

Gąsiorowski, L.; Hejnol, A. 2019. Hox gene expression in postmetamorphic juveniles of the brachiopod Terebratalia transversa. EvoDevo 10, 1.

Gasmi, S.; Neve, G.; Pech, N.; Tekaya, S.; Gilles, A.; Perez, Y. 2014. Evolutionary history of Chaetognatha inferred from molecular and morphological data: a case study for body plan simplification. Frontiers in Zoology 11, 84.

Gavelis, G. S.; Wakeman, K. C.; Tillmann, U.; Ripken, C.; Mitarai, S.; Herranz, M.; Özbek, S.; Holstein, T.; Keeling, P. J.; Leander, B. S. 2017. Microbial arms race: Ballistic "nematocysts" in dinoflagellates represent a new extreme in organelle complexity. Science Advances 3, e1602552.

Gazave, E.; Lapébie, P.; Ereskovsky, A. V.; Vacelet, J.; Renard, E.; Cárdenas, P.; Borchiellini, C. 2012. No longer Demospongiae: Homoscleromorpha formal nomination as a fourth class of Porifera. Hydrobiologia 687, 3–10.

Gazave, E.; Lemaître, Q. I.; Balavoine, G. 2017. The Notch pathway in the annelid Platynereis: insights into chaetogenesis and neurogenesis processes. Open Biology 7, 160242.

Gazi, M.; Kim, J.; García-Varela, M.; Park, C.; Littlewood, D. T. J.; Park, J.-K. 2016. Mitogenomic phylogeny of Acanthocephala reveals novel Class relationships. Zoologica Scripta 45, 437–454.

Gee, H. 2016. A home for Xenoturbella. Nature 530, 43.

Gehling, J. G.; Runnegar, B. N.; Droser, M. L. 2014. Scratch traces of large Ediacara bilaterian animals. Journal of Paleontology 88, 284–298.

Gehrke, A. R.; Neverett, E.; Luo, Y.-J.; Brandt, A.; Ricci, L.; Hulett, R. E.; Gompers, A.; Ruby, J. G.; Rokhsar, D. S.; Reddien, P. W.; et al. 2019. Acoel genome reveals the regulatory landscape of whole-body regeneration. Science 363, eaau6173.

Georgieva, M. N.; Little, C. T. S.; Watson, J. S.; Sephton, M. A.; Ball, A. D.; Glover, A. G. 2019. Identification of fossil worm tubes from Phanerozoic hydrothermal vents and cold seeps. Journal of Systematic Palaeontology 17, 287–329.

Geyer, L. B.; Palumbi, S. R. 2003. Reproductive character displacement and the genetics of gamete recognition in tropical sea urchins. Evolution 57, 1049–1060.

Gharbiah, M.; Nakamoto, A.; Nagy, L. M. 2013. Analysis of ciliary band formation in the mollusc Ilyanassa obsoleta. Development Genes and Evolution 223, 225–235.

Ghedin, E.; Wang, S. L.; Spiro, D.; Caler, E.; Zhao, Q.; Crabtree, J.; Allen, J. E.; Delcher, A. L.; Guiliano, D. B.; Miranda-Saavedra, D.; et al. 2007. Draft genome of the filarial nematode parasite Brugia malayi. Science 317, 1756–1760.

Ghirardelli, E. 1968. Some aspects of the biology of the chaetognaths. Advances in Marine Biology 6, 271–375.

Ghirardelli, E. 1995. Chaetognaths: two unsolved problems: the coelom and their affinities. In: Lanzavecchia, G.; Valvassori, R.; Candia Carnevali, M. D. (eds.), Body cavities: function and phylogeny. Selected Symposia and Monographs U.Z.I., Mucchi, Modena, pp. 167–185.

Giard, A. 1877. Sur les Orthonectida, classe nouvelle d'animaux parasites des Échinodermes et des Turbellariés. Comptes rendus Academie des Sciences Paris 85, 812–814.

Gibson, R. 1995. Nemertean genera and species of the world: an annotated checklist of original names and description citations, synonyms, current taxonomic status, habitats and recorded zoogeographic distribution. Journal of Natural History 29, 271–562.

Giere, O. 2009. Meiobenthology: The Microscopic Motile Fauna of Aquatic Sediments. Springer, Berlin.

Gierer, A.; Berking, S.; Bode, H.; David, C. N.; Flick, K.; Hansmann, G.; Schaller, H.; Trenkner, E. 1972. Regeneration of Hydra from reaggregated cells. Nature 239, 98–101.

Giese, A. C.; Pearse, J. S.; Pearse, V. B. 1991. Reproduction of Marine Invertebrate, vol. VI: Echinoderms and Lophophorates. Boxwood Press, Pacific Grove, CA.

Giesecke, R.; González, H. E.; Bathmann, U. 2009. The role of the chaetognath Sagitta gazellae in the vertical carbon flux of the Southern Ocean. Polar Biology 33, 293–304.

Gilbert, J. J. 1974. Dormancy in rotifers. Transactions of the American Microscopical Society 93, 490–513.

Gilbert, J. J. 1989. Rotifera. In: Adiyodi, K. G.; Adiyodi, R. G. (eds.), Reproductive Biology of Invertebrates. Wiley, Chichester, pp. 179–199.

Gilbert, P. U. P. A.; Bergmann, K. D.; Myers, C. E.; Marcus, M. A.; DeVol, R. T.; Sun, C.-Y.; Blonsky, A. Z.; Zhao, J.; Karan, E. A.; Tamre, E.; et al. 2017. Nacre tablet thickness records formation temperature in modern and fossil shells. Earth and Planetary Science Letters 460, 281–292.

Gilbert, S. F.; Raunio, A. M. 1997. Embryology: Constructing the Organism. Sinauer, Sunderland, MA.

Gim, J.-S.; Ko, E.-J.; Kim, H.-G.; Kim, Y.-M.; Hong, S.; Kim, H.-W.; Gim, J.-A.; Joo, G.-J.; Jo, H. 2018. Complete mitochondrial genome of the freshwater bryozoan Pectinatella

magnifica (Phylactolaemata: Plumatellida) assembled from next-generation sequencing data. Mitochondrial DNA, pt. B: Resources 3, 373–374.

Giribet, G. 2003. Molecules, development and fossils in the study of metazoan evolution; Articulata versus Ecdysozoa revisited. Zoology 106, 303–326.

Giribet, G. 2004. ¿Articulata o Ecdysozoa?: una revisión crítica sobre la posición de los artrópodos en el reino animal. In: Llorente Bousquets, J. E.; Morrone, J. J.; Yáñez Ordóñez, O.; Vargas Fernández, I. (eds.), Biodiversidad, Taxonomía y Biogeografía de Artrópodos de México: Hacia una síntesis de su conocimiento, vol. IV. Facultad de Ciencias, UNAM, México, pp. 45–62.

Giribet, G. 2010. A new dimension in combining data? The use of morphology and phylogenomic data in metazoan systematics. Acta Zoologica 91, 11–19.

Giribet, G. 2014a. On Aculifera: A review of hypotheses in tribute to Christopher Schander. Journal of Natural History 48, 2739–2749.

Giribet, G. 2014b. Protostomes: The Greatest Animal Diversity. In: Vargas, P.; Zardoya, R. (eds.), The Tree of Life. Sinauer, Sunderland, MA, pp. 229–233.

Giribet, G. 2014c. Spiralians: Animals with Spiral Cleavage and their Relatives. In: Vargas, P.; Zardoya, R. (eds.), The Tree of Life. Sinauer, Sunderland, MA, pp. 235–240.

Giribet, G. 2016a. Genomics and the Animal Tree of Life: Conflicts and future prospects. Zoologica Scripta 45, 14–21.

Giribet, G. 2016b. New animal phylogeny: Future challenges for animal phylogeny in the age of phylogenomics. Organisms Diversity and Evolution 16, 419–426.

Giribet, G. 2016c. Phylum Nemertea. The ribbon worms. In: Brusca, R. C.; Moore, W.; Shuster, S. M. (eds.), Invertebrates, 3rd ed. Sinauer, Sunderland, MA, pp. 433–452.

Giribet, G. 2018a. Current views on chelicerate phylogeny—A tribute to Peter Weygoldt. Zoologischer Anzeiger 273, 7–13.

Giribet, G. 2018b. Phylogenomics resolves the evolutionary chronicle of our squirting closest relatives. BMC Biology 16, 49.

Giribet, G.; Buckman-Young, R. S.; Sampaio Costa, C.; Baker, C. M.; Benavides, L. R.; Branstetter, M. G.; Daniels, S. R.; Pinto-da-Rocha, R. 2018. The 'Peripatos' in Eurogondwana?—Lack of evidence that south-east Asian onychophorans walked through Europe. Invertebrate Systematics 32, 842–865.

Giribet, G.; Carranza, S.; Baguñà, J.; Riutort, M.; Ribera, C. 1996. First molecular evidence for the existence of a Tardigrada + Arthropoda clade. Molecular Biology and Evolution 13, 76–84.

Giribet, G.; Distel, D. L.; Polz, M.; Sterrer, W.; Wheeler, W. C. 2000. Triploblastic relationships with emphasis on the acoelomates and the position of Gnathostomulida, Cycliophora, Plathelminthes, and Chaetognatha: A combined approach of 18S rDNA sequences and morphology. Systematic Biology 49, 539–562.

Giribet, G.; Dunn, C. W.; Edgecombe, G. D.; Hejnol, A.; Martindale, M. Q.; Rouse, G. W. 2009. Assembling the spiralian tree of life. In: Telford, M. J.; Littlewood, D. T. (eds.),

Hanelt, B.; Janovy, J. 2004a. Life cycle and paratenesis of American gordiids (Nematomorpha: Gordiida). Journal of Parasitology 90, 240–244.

Hanelt, B.; Janovy, J. 2004b. Untying a Gordian knot: the domestication and laboratory maintenance of a Gordian worm, *Paragordius varius* (Nematomorpha: Gordiida). Journal of Natural History 38, 939–950.

Hanelt, B.; Janovy, J., Jr. 2003. Spanning the gap: experimental determination of paratenic host specificity of horsehair worms (Nematomorpha: Gordiida). Invertebrate Biology 122, 12–18.

Hanelt, B.; Thomas, F.; Schmidt-Rhaesa, A. 2005. Biology of the phylum Nematomorpha. Advances in Parasitology 59, 243–305.

Hanelt, B.; Van Schyndel, D.; Adema, C. M.; Lewis, L. A.; Loker, E. S. 1996. The phylogenetic position of *Rhopalura ophiocomae* (Orthonectida) based on 18S ribosomal DNA sequence analysis. Molecular Biology and Evolution 13, 1187–1191.

Harii, S.; Yasuda, N.; Rodriguez-Lanetty, M.; Irie, T.; Hidaka, M. 2009. Onset of symbiosis and distribution patterns of symbiotic dinoflagellates in the larvae of scleractinian corals. Marine Biology 156, 1203–1212.

Härlin, M. S.; Sundberg, P. 1995. Cladistic analysis of the eureptantic nemerteans (Nemertea: Hoplonemertea). Invertebrate Taxonomy 9, 1211–1229.

Harmer, S. F. 1885. On the structure and development of *Loxosoma*. Quarterly Journal of Microscopical Science 25, 261–337.

Harper, D. A. T.; Popov, L. E.; Holmer, L. E.; Smith, A. 2017. Brachiopods: origin and early history. Palaeontology 60, 609–631.

Harris, J. R.; Markl, J. 1999. Keyhole limpet hemocyanin (KLH): a biomedical review. Micron 30, 597–623.

Harrison, F. W.; De Vos, L. 1991. Porifera. In: Harrison, F. W. (ed.), Placozoa, Porifera, Cnidaria, and Ctenophora. Wiley-Liss, New York, pp. 29–89.

Harrison, P. L. 2011. Sexual reproduction of scleractinian corals. Coral Reefs: An Ecosystem in Transition. Springer Science+Business Media, pp. 59–85.

Harrison, P. L.; C., B. R.; Bull, G. D.; Oliver, J. K.; Wallace, C. C.; Willis, B. L. 1984. Mass spawning in tropical reef corals. Science 223, 1186–1189.

Harrison, P. L.; Wallace, C. C. 1990. Reproduction, dispersal and recruitment of scleractinian corals. In: Dubinsky, Z. (ed.), Ecosystems of the World: Coral Reefs. Elsevier, Amsterdam, pp. 133–207.

Hart, M. W.; Miller, R. L.; Madin, L. P. 1994. Form and feeding mechanism of a living *Planctosphaera pelagica* (Phylum Hemichordata). Marine Biology 120, 521–533.

Hartenstein, V. 2016. Platyhelminthes (excluding Neodermata). In: Schmidt-Rhaesa, A.; Harzsch, S.; Purschke, G. (eds.), Structure and Evolution of Invertebrate Nervous Systems. Oxford University Press, Oxford, pp. 74–92.

Hartenstein, V.; Chipman, A. D. 2015. Hexapoda: A *Drosophila*'s view of development. In: Wanninger, A. (ed.), Evolutionary Developmental Biology of Invertebrates 5. Ecdysozoa III: Hexapoda. Springer, Vienna, pp. 1–91.

Hartikainen, H.; Okamura, B. 2015. Ecology and evolution of malacosporean-bryozoan interactions. In: Okamura, B.; Gruhl, A.; Bartholomew, J. L. (eds.), Myxozoan Evolution, Ecology and Development. Springer, Cham, pp. 201–216.

Hartikainen, H.; Waeschenbach, A.; Wöss, E.; Wood, T.; Okamura, B. 2013. Divergence and species discrimination in freshwater bryozoans (Bryozoa: Phylactolaemata). Zoological Journal of the Linnean Society 168, 61–80.

Hartry, A. L.; Morton, W. D.; Keithlee, P. 1964. Planaria: Memory transfer through cannibalism reexamined. Science 146, 274–275.

Harvey, T. H. P. 2010. Carbonaceous preservation of Cambrian hexactinellid sponge spicules. Biology Letters 6, 834–837.

Harvey, T. H. P.; Butterfield, N. J. 2017. Exceptionally preserved Cambrian loriciferans and the early animal invasion of the meiobenthos. Nature Ecology and Evolution 1, 0022.

Harvey, T. H. P.; Dong, X. P.; Donoghue, P. C. J. 2010. Are palaeoscolecids ancestral ecdysozoans? Evolution and Development 12, 177–200.

Harvey, T. H. P.; Ortega-Hernández, J.; Lin, J.-P.; Zhao, Y. L.; Butterfield, N. J. 2012. Burgess Shale-type microfossils from the middle Cambrian Kaili Formation, Guizhou Province, China. Acta Palaeontologica Polonica 57, 423–436.

Harzsch, S. 2004. Phylogenetic comparison of serotonin-immunoreactive neurons in representatives of the Chilopoda, Diplopoda, and Chelicerata: implications for arthropod relationships. Journal of Morphology 259, 198–213.

Harzsch, S.; Melzer, R. R.; Müller, C. H. G. 2007. Mechanisms of eye development and evolution of the arthropod visual system: The lateral eyes of Myriapoda are not modified insect ommatidia. Organisms Diversity and Evolution 7, 20–32.

Harzsch, S.; Müller, C. H. G. 2007. A new look at the ventral nerve centre of *Sagitta*: implications for the phylogenetic position of Chaetognatha (arrow worms) and the evolution of the bilaterian nervous system. Frontiers in Zoology 4, 14.

Harzsch, S.; Müller, C. H. G.; Perez, Y. 2015. Chaetognatha. In: Wanninger, A. (ed.), Evolutionary Developmental Biology of Invertebrates 1: Introduction, Non-Bilateria, Acoelomorpha, Xenoturbellida, Chaetognatha. Springer, Vienna, pp. 215–240.

Harzsch, S.; Müller, C. H. G.; Rieger, V.; Perez, Y.; Sintoni, S.; Sardet, C.; Hansson, B. 2009. Fine structure of the ventral nerve centre and interspecific identification of individual neurons in the enigmatic Chaetognatha. Zoomorphology 128, 53–73.

Harzsch, S.; Perez, Y.; Müller, C. H. G. 2016. Chaetognatha. In: Schmidt-Rhaesa, A.; Harzsch, S.; Purschke, G. (eds.), Structure and Evolution of Invertebrate Nervous Systems. Oxford University Press, Oxford, pp. 652–664.

Hashimoto, T.; Horikawa, D. D.; Saito, Y.; Kuwahara, H.; Kozuka-Hata, H.; Shin, I. T.; Minakuchi, Y.; Ohishi, K.; Motoyama, A.; Aizu, T.; et al. 2016. Extremotolerant tardigrade genome and improved radiotolerance of human cultured cells by tardigrade-unique protein. Nature Communications 7, 12808.

Hashimoto, T.; Kunieda, T. 2017. DNA protection protein, a novel mechanism of radiation tolerance: Lessons from tardigrades. Life 7, 26.

Hasse, C.; Rebscher, N.; Reiher, W.; Sobjinski, K.; Moerschel, E.; Beck, L.; Tessmar-Raible, K.; Arendt, D.; Hassel, M. 2010. Three consecutive generations of nephridia occur during development of *Platynereis dumerilii* (Annelida, Polychaeta). Developmental Dynamics 239, 1967–1976.

Haszprunar, G. 1996a. The Mollusca: coelomate turbellarians or mesenchymate annelids? In: Taylor, J. D. (ed.), Origin and evolutionary radiation of the Mollusca. Oxford University Press, Oxford, pp. 1–28.

Haszprunar, G. 1996b. The molluscan rhogocyte (pore-cell, Blasenzelle, cellule nucale), and its significance for ideas on nephridial evolution. Journal of Molluscan Studies 62, 185–211.

Haszprunar, G. 1996c. Plathelminthes and Plathelminthomorpha—paraphyletic taxa. Journal of Zoological Systematics and Evolutionary Research 34, 41–48.

Haszprunar, G. 2000. Is the Aplacophora monophyletic? A cladistic point of view. American Malacological Bulletin 15, 115–130.

Haszprunar, G.; Rieger, R. M.; Schuchert, P. 1991. Extant "Problematica" within or near the Metazoa. In: Simonetta, A. M.; Conway Morris, S. (eds.), The early evolution of Metazoa and the significance of problematic taxa. Cambridge University Press, Cambridge, pp. 99–105.

Haszprunar, G.; Ruthensteiner, B. 2013. Monoplacophora (Tryblidia)—Some unanswered questions. American Malacological Bulletin 31, 189–194.

Haszprunar, G.; Vogler, C.; Wörheide, G. 2017. Persistent gaps of knowledge for naming and distinguishing multiple species of crown-of-thorns-seastar in the *Acanthaster planci* Species Complex. Diversity 9, 22.

Haszprunar, G.; Wanninger, A. 2008. On the fine structure of the creeping larva of *Loxosomella murmanica*: Additional evidence for a clade of Kamptozoa (Entoprocta) and Mollusca. Acta Zoologica 89, 137–148.

Hatschek, B. 1878. Studien über die Entwicklungsgeschichte der Anneliden. Ein Beitrag zur Morphologie der Bilaterien. Arbeiten aus dem Zoologischen Institute der Universität Wien und der Zoologischen Station in Triest 1, 277–404.

Hatschek, B. 1888. Lehrbuch der Zoologie. Eine morphologische Übersicht des Thierreiches zur Einführung in das Studium dieser Wissenschaft. Lief 1. Gustav Fischer, Jena.

Haug, C.; Haug, J. T. 2017. The presumed oldest flying insect: more likely a myriapod? PeerJ 5, e3402.

Haug, J. T.; Waloszek, D.; Haug, C.; Maas, A. 2010. High-level phylogenetic analysis using developmental sequences: The Cambrian †*Martinssonia elongata*, †*Musacaris gerdgeyeri* gen. et sp. nov. and their position in early crustacean evolution. Arthropod Structure and Development 39, 154–173.

Hausdorf, B.; Helmkampf, M.; Meyer, A.; Witek, A.; Herlyn, H.; Bruchhaus, I.; Hankeln, T.; Struck, T. H.; Lieb, B. 2007. Spiralian phylogenomics supports the resurrection of

Bryozoa comprising Ectoprocta and Entoprocta. Molecular Biology and Evolution 24, 2723–2729.

Hay-Schmidt, A. 1989. The nervous system of the actinotroch larva of *Phoronis muelleri* (Phoronida). Zoomorphology 108, 333–351.

Hay-Schmidt, A. 1990. Catecholamine-containing, serotonin-like, and FMRFamide-like immunoreactive neurons and processes in the nervous system of the early actinotroch larva of *Phoronis vancouverensis* (Phoronida): distribution and development. Canadian Journal of Zoology 68, 1525–1536.

Hay-Schmidt, A. 1992. Ultrastructure and immunocytochemistry of the nervous system of the larvae of *Lingula anatina* and *Glottidia* sp. (Brachiopoda). Zoomorphology 112, 189–205.

Hay-Schmidt, A. 2000. The evolution of the serotonergic nervous system. Proceedings of the Royal Society of London B: Biological Sciences 267, 1071–1079.

Hayward, P. J. 1981. The Cheilostomata (Bryozoa) of the deep sea. Galathea Report 15, 21–68.

Hehn, N.; Ehlers; U., Herlyn, H. 2001. Ultrastructure of the acanthella of *Paratenuisentis ambiguus* (Acanthocephala). Parasitology Research 87, 467–471.

Heidemann, N. W. T.; Smith, D. K.; Hygum, T. L.; Stapane, L.; Clausen, L. K. B.; Jørgensen, A.; Hélix-Nielsen, C.; Møbjerg, N. 2016. Osmotic stress tolerance in semi-terrestrial tardigrades. Zoological Journal of the Linnean Society 178, 912–918.

Heiner Bang-Berthelsen, I.; Schmidt-Rhaesa, A.; Kristensen, R. M. 2013. Loricifera. In: Schmidt-Rhaesa; A. (ed.), Handbook of Zoology. Gastrotricha, Cycloneuralia and Gnathifera, vol. 1: Nematomorpha, Priapulida, Kinorhyncha, Loricifera. De Gruyter, Berlin/Boston, pp. 349–371.

Heiner, I. 2008. *Rugiloricus bacatus* sp. nov. (Loricifera—Pliciloricidae) and a ghost-larva with paedogenetic reproduction. Systematics and Biodiversity 6, 225–247.

Heiner, I.; Boesgaard, T. M.; Kristensen, R. M. 2009. First time discovery of Loricifera from Australian waters and marine caves. Marine Biology Research 5, 529–546.

Heiner, I.; Kristensen, R. M. 2008. *Urnaloricus gadi* nov. gen. et nov. sp. (Loricifera, Urnaloricidae nov. fam.), an aberrant Loricifera with a viviparous pedogenetic life cycle. Journal of Morphology 270, 129–153.

Heinzeller, T.; Welsch, U. 2001. The echinoderm nervous system and its phylogenetic interpretation. In: Roth, G.; Wullimann, M. (eds.), Brain Evolution and Cognition. Wiley/Spektrum, New York, pp. 41–75.

Hejnol, A. 2010. A twist in time—The evolution of spiral cleavage in the light of animal phylogeny. Integrative and Comparative Biology 50, 695–706.

Hejnol, A. 2015a. Acoelomorpha and Xenoturbellida. In: Wanninger, A. (ed.), Evolutionary Developmental Biology of Invertebrates 1: Introduction, Non-Bilateria, Acoelomorpha, Xenoturbellida, Chaetognatha. Springer, Vienna, pp. 203–214.

Hejnol, A. 2015b. Gnathifera. In: Wanninger, A. (ed.), Evolutionary Developmental Biology of Invertebrates 2: Lophotrochozoa (Spiralia). Springer, Vienna, pp. 1–12.

Higgins, R. P.; Storch, V. 1991. Evidence for direct development in *Meiopriapulus fijiensis*. Transactions of the American Microscopical Society 110, 37–46.

Hillier, L. W.; Coulson, A.; Murray, J. I.; Bao, Z.; Sulston, J. E.; Waterston, R. H. 2005. Genomics in *C. elegans*: So many genes, such a little worm. Genome Research 15, 1651–1660.

Hilton, W. A. 1922. Nervous systems and sense organs IX. The Bryozoa. Journal of Entomology and Zoology 14, 76–79.

Hilton, W. A. 1928. IX. The Bryozoa. The nervous system and sense organs, vol. 1: The more primitive animals including Annulata and beginning with some small groups of Arthropoda. Department of Zoology, Pomona College, Claremont, California, pp. 71–79.

Hinman, V.; Cary, G. 2017. The evolution of gene regulation. eLife 6, e.27291.

Hints, O.; Eriksson, M. E. 2007. Diversification and biogeography of scolecodont-bearing polychaetes in the Ordovician. Palaeogeography Palaeoclimatology Palaeoecology 245, 95–114.

Hirose, E.; Hirose, M. 2007. Body colors and algal distribution in the acoel flatworm *Convolutriloba longifissura*: histology and ultrastructure. Zoological Science 24, 1241–1246.

Hirose, M.; Dick, M. H.; Mawatari, S. F. 2011. Are plumatellid statoblasts in freshwater bryozoans phylogenetically informative? Zoological Science 28, 318–326.

Hirose, M.; Fukiage, R.; Katoh, T.; Kajihara, H. 2014. Description and molecular phylogeny of a new species of *Phoronis* (Phoronida) from Japan, with a redescription of topotypes of *P. ijimai* Oka, 1897. ZooKeys, 1–31.

Hirose, M.; Yamamoto, H.; Nonaka, M. 2008. Metamorphosis and acquisition of symbiotic algae in planula larvae and primary polyps of *Acropora* spp. Coral Reefs 27, 247–254.

Hoberg, E. P.; Alkire, N. L.; de Queiroz, A.; Jones, A. 2001. Out of Africa: origins of the *Taenia* tapeworms in humans. Proceedings of the Royal Society B: Biological Sciences 268, 781–787.

Hochberg, A.; Hochberg, R. 2015. Serotonin immunoreactivity in the nervous system of the free-swimming larvae and sessile adult females of *Stephanoceros fimbriatus* (Rotifera: Gnesiotrocha). Invertebrate Biology 134, 261–270.

Hochberg, A.; Hochberg, R. 2017. Musculature of the sessile rotifer *Stephanoceros fimbriatus* (Rotifera: Gnesiotrocha: Collothecaceae) with details on larval metamorphosis and development of the infundibulum. Zoologischer Anzeiger 268, 84–95.

Hochberg, F. G. 1990. Diseases caused by protists and mesozoans. In: Kinne, O. (ed.), Diseases of Marine Animals, vol. 3. Biologische Anstalt Helgoland, pp. 47–202.

Hochberg, F. G.; Hofrichter, R. 2005. Dicyemida (Rhombozoa, "Mesozoa"). In: Hofrichter, R. (ed.), El mar Mediterráneo. Fauna, flora, ecología. II/1: Guía sistemática y de identificación. Procariotas, protistas, hongos, algas, animales (hasta Nemertea). Ediciones Omega, Barcelona, pp. 388–393.

Hochberg, R. 2005. Musculature of the primitive gastrotrich *Neodasys* (Chaetonotida): functional adaptations to the interstitial environment and phylogenetic significance. Marine Biology 146, 315–323.

Hochberg, R. 2007a. Comparative immunohistochemistry of the cerebral ganglion in Gastrotricha: an analysis of FMRFamide-like immunoreactivity in *Neodasys cirritus* (Chaetonotida), *Xenodasys riedli* and *Turbanella* cf. *hyalina* (Macrodasyida). Zoomorphology 126, 245–264.

Hochberg, R. 2007b. Topology of the nervous system of *Notommata copeus* (Rotifera: Monogononta) revealed with anti-FMRFamide, -SCPb, and -serotonin (5-HT) immunohistochemistry. Invertebrate Biology 126, 247–256.

Hochberg, R. 2008. Ultrastructure of feathered triancres in the Thaumastodermatidae and the description of a new species of Tetranchyroderma (Gastrotricha : Macrodasyida) from Australia. Journal of the Marine Biological Association of the United Kingdom 88, 729–737.

Hochberg, R. 2009. Three-dimensonal reconstruction and neural map of the serotonergic brain of *Asplanchna brightwellii* Rotifera, Monogononta). Journal of Morphology 270, 430–441.

Hochberg, R. 2016. Rotifera. In: Schmidt-Rhaesa, A.; Harzsch, S.; Purschke, G. (eds.), Structure and Evolution of Invertebrate Nervous Systems. Oxford University Press, Oxford, pp. 122–131.

Hochberg, R.; Ablak Gurbuz, O. 2007. Functional morphology of somatic muscles and anterolateral setae in *Filinia novaezealandiae* Shiel and Sanoamuang, 1993 (Rotifera). Zoologischer Anzeiger 246, 11–22.

Hochberg, R.; Ablak Gurbuz, O. 2008. Comparative morphology of the somatic musculature in species of *Hexarthra* and *Polyarthra* (Rotifera, Monogononta): Its function in appendage movement and escape behavior. Zoologischer Anzeiger 247, 233–248.

Hochberg, R.; Atherton, S. 2011. A new species of *Lepidodasys* (Gastrotricha, Macrodasyida) from Panama with a description of its peptidergic nervous system using CLSM, anti-FMRFamide and anti-SCPB. Zoologischer Anzeiger 250, 111–122.

Hochberg, R.; Atherton, S.; Kieneke, A. 2014. Marine Gastrotricha of Little Cayman Island with the description of one new species and an initial assessment of meiofaunal diversity. Marine Biodiversity 44, 89–113.

Hochberg, R.; Lilley, G. 2010. Neuromuscular organization of the freshwater colonial rotifer, *Sinantherina socialis*, and its implications for understanding the evolution of coloniality in Rotifera. Zoomorphology 129, 153–162.

Hochberg, R.; Litvaitis, M. K. 2000a. Functional morphology of the muscles in *Philodina* sp. (Rotifera: Bdelloidea). Hydrobiologia 432, 57–64.

Hochberg, R.; Litvaitis, M. K. 2000b. Phylogeny of Gastrotricha: a morphology-based framework of gastrotrich relationships. Biological Bulletin 198, 299–305.

Hochberg, R.; Litvaitis, M. K. 2001a. Functional morphology of muscles in *Tetranchyroderma papii* (Gastrotricha). Zoomorphology 121, 37–43.

Hochberg, R.; Litvaitis, M. K. 2001b. Macrodasyida (Gastrotricha): A cladistic analysis of morphology. Invertebrate Biology 120, 124–135.

Hochberg, R.; Litvaitis, M. K. 2001c. A muscular double-helix in Gastrotricha. Zoologischer Anzeiger 240, 59–66.

Hochberg, R.; Litvaitis, M. K. 2001d. The muscular system of *Dactylopodola baltica* and other macrodasyidan gastrotrichs in a functional and phylogenetic perspective. Zoologica Scripta 30, 325–336.

Hochberg, R.; Litvaitis, M. K. 2001e. The musculature of *Draculiciteria tessalata* (Chaetonotida, Paucitubulatina): implications for the evolution of dorsoventral muscles in Gastrotricha. Hydrobiologia 452, 155–161.

Hochberg, R.; Litvaitis, M. K. 2003a. Organization of muscles in Chaetonotida Paucitubulatina (Gastrotricha). Meiofauna Marina 12, 47–58.

Hochberg, R.; Litvaitis, M. K. 2003b. Ultrastructural and immunocytochemical observations of the nervous systems of three macrodasyidan gastrotrichs. Acta Zoologica 84, 171–178.

Hochberg, R.; Todaro, M. A.; Araujo, T. Q.; Atherton, S.; Balsamo, M.; Chang, C. Y.; Di Dimenico, M.; Garraffoni, A. R.; Guidi, L.; Känneby, T.; et al. 2017a. A tribute to William Hummon—Gastrotrich biologist *extraordinaire*. Proceedings of the Biological Society of Washington 130, 113–119.

Hochberg, R.; Yang, H.; Moore, J. 2017b. The ultrastructure of escape organs: setose arms and cross-striated muscles in *Hexarthra mira* (Rotifera: Gnesiotrocha: Flosculariaceae). Zoomorphology 136, 159–173.

Hodgson, A. N.; Reunov, A. A. 1994. Ultrastructure of the spermatozoon and spermatogenesis of the brachiopods *Discinisca tenuis* (Inarticulata) and *Kraussina rubra* (Articulata). Invertebrate Reproduction and Development 25, 23–31.

Hoeh, W. R.; Stewart, D. T.; Sutherland, B. W.; Zouros, E. 1996. Cytochrome c oxidase sequences comparisons suggest an unusually high rate of mitochondrial DNA evolution in *Mytilus* (Mollusca: Bivalvia). Molecular Biology and Evolution 13, 418–421.

Hoekstra, L. A.; Moroz, L. L.; Heyland, A. 2012. Novel insights into the echinoderm nervous system from histaminergic and FMRFaminergic-like cells in the sea cucumber *Leptosynapta clarki*. PLoS One 7, e44220.

Hoekzema, R. S.; Brasier, M. D.; Dunn, F. S.; Liu, A. G. 2017. Quantitative study of developmental biology confirms *Dickinsonia* as a metazoan. Proceedings of the Royal Society B: Biological Sciences 284, 20171348.

Hoff, F. H.; Snell, T. W. 1987. Plankton culture manual. Florida Aqua Farms, Dade City.

Hoffmeyer, T. T.; Burkhardt, P. 2016. Choanoflagellate models—*Monosiga brevicollis* and *Salpingoeca rosetta*. Current Opinion in Genetics and Development 39, 42–47.

Holland, J. W.; Okamura, B.; Hartikainen, H.; Secombes, C. J. 2011. A novel minicollagen gene links cnidarians and myxozoans. Proceedings of the Royal Society B: Biological Sciences 278, 546–553.

Holland, L. Z. 2015. Cephalochordata. In: Wanninger, A. (ed.), Evolutionary Developmental Biology of Invertebrates 6: Deuterostomia. Springer, Vienna, pp. 91–134.

Holland, L. Z.; Holland, N. D. 2007. A revised fate map for amphioxus and the evolution of axial patterning in chordates. Integrative and Comparative Biology 47, 360–372.

Holland, L. Z.; Laudet, V.; Schubert, M. 2004. The chordate amphioxus: an emerging model organism for developmental biology. Cellular and Molecular Life Sciences 61, 2290–2308.

Holland, N. D. 2003. Early central nervous system evolution: An era of skin brains? Nature Reviews Neuroscience 4, 617–627.

Holland, N. D.; Chen, J. 2001. Origin and early evolution of the vertebrates: new insights from advances in molecular biology, anatomy, and palaeontology. BioEssays 23, 142–151.

Holland, N. D.; Clague, D. A.; Gordon, D. P.; Gebruk, A.; Pawson, D. L.; Vecchione, M. 2005. 'Lophenteropneust' hypothesis refuted by collection and photos of new deep-sea hemichordates. Nature 434, 374–376.

Holland, N. D.; Jones, W. J.; Ellena, J.; Ruhl, H. A.; Smith, K. L. 2009. A new deep-sea species of epibenthic acorn worm (Hemichordata, Enteropneusta). Zoosystema 31, 333–346.

Holland, P. W. H. 2013. Evolution of homeobox genes. WIREs Developmental Biology 2, 31–45.

Holmer, L. E.; Caron, J. B. 2006. A spinose stem group brachiopod with pedicle from the Middle Cambrian Burgess Shale. Acta Zoologica 87, 273–290.

Holmer, L. E.; Popov, L. E.; Bassett, M. G.; Laurie, J. 1995. Phylogenetic analysis and ordinal classification of the Brachiopoda. Palaeontology 38, 713–741.

Holmer, L. E.; Skovsted, C. B.; Larsson, C.; Brock, G. A.; Zhang, Z. 2011. First record of a bivalved larval shell in Early Cambrian tommotiids and its phylogenetic significance. Palaeontology 54, 235–239.

Holmer, L. E.; Skovsted, C. B.; Williams, A. 2002. A stem group brachiopod from the Lower Cambrian: support for a *Micrina* (halkieriid) ancestry. Palaeontology 45, 875–882.

Holterman, M.; Karssen, G.; van den Elsen, S.; van Megen, H.; Bakker, J.; Helder, J. 2009. Small rubunit rDNA-based phylogeny of the Tylenchida sheds light on relationships among some high-impact plant-parasitic nematodes and the evolution of plant feeding. Phytopathology 99, 227–235.

Holterman, M.; van der Wurff, A.; van den Elsen, S.; van Megen, H.; Bongers, T.; Holovachov, O.; Bakker, J.; Helder, J. 2006. Phylum-wide analysis of SSU rDNA reveals deep phylogenetic relationships among nematodes and accelerated evolution toward crown clades. Molecular Biology and Evolution 23, 1792–1800.

Hooge, M. D.; Tyler, S. 1999. Body-wall musculature of *Praeconvoluta tornuva* n. sp. (Acoela, Platyhelminthes) and the use of muscle patterns in taxonomy. Invertebrate Biology 118, 8–17.

Hooper, J. N. A.; Van Soest, R. W. M. (eds.). 2002a. Systema Porifera: a guide to the classification of sponges, vol. 1. Kluwer Academic/Plenum, New York/Boston.

Hooper, J. N. A.; Van Soest, R. W. M. (eds.). 2002b. Systema Porifera: a guide to the classification of sponges, vol. 2. Kluwer Academic/Plenum, New York/Boston.

Hopkins, M. J.; Smith, A. B. 2015. Dynamic evolutionary change in post-Paleozoic echinoids and the importance of scale when interpreting changes in rates of evolution. Proceedings of the National Academy of Sciences of the USA 112, 3758–3763.

Hori, H.; Osawa, S. 1987. Origin and evolution of organisms as deduced from 5S ribosomal RNA sequences. Molecular Biology and Evolution 4, 445–472.

Horikawa, D. D.; Cumbers, J.; Sakakibara, I.; Rogoff, D.; Leuko, S.; Harnoto, R.; Arakawa, K.; Katayama, T.; Kunieda,

Janies, D. A.; Mooi, R. D. 1999. *Xyloplax* is an asteroid. In: Candia Carnevali, M. D., Bonasoro, F. (eds.), Echinoderm Research 1998. A.A. Balkema, Rotterdam-Brookfield, pp. 311–316.

Janies, D. A.; Voight, J. R.; Daly, M. 2011. Echinoderm phylogeny including *Xyloplax*, a progenetic asteroid. Systematic Biology 60, 420–438.

Janies, D. A.; Witter, Z.; Linchangco, G. V.; Foltz, D. W.; Miller, A. K.; Kerr, A. M.; Jay, J.; Reid, R. W.; Wray, G. A. 2016. EchinoDB, an application for comparative transcriptomics of deeply-sampled clades of echinoderms. BMC Bioinformatics 17, 48.

Janssen, R.; Budd, G. E. 2010. Gene expression suggests conserved aspects of Hox gene regulation in arthropods and provides additional support for monophyletic Myriapoda. EvoDevo 1, 4.

Janssen, R.; Eriksson, B. J.; Tait, N. N.; Budd, G. E. 2014. Onychophoran Hox genes and the evolution of arthropod Hox gene expression. Frontiers in Zoology 11, 22.

Janssen, R.; Eriksson, J. B.; Budd, G. E.; Akam, M.; Prpic, N.-M. 2010. Gene expression patterns in onychophorans reveal that regionalization predates limb segmentation in panarthropods. Evolution and Development 12, 363–372.

Janssen, R.; Jörgensen, M.; Prpic, N.-M.; Budd, G. E. 2015. Aspects of dorso-ventral and proximo-distal limb patterning in onychophorans. Evolution and Development 17, 21–33.

Janssen, R.; Wennberg, S. A.; Budd, G. E. 2009. The hatching larva of the priapulid worm *Halicryptus spinulosus*. Frontiers in Zoology 6, 8.

Jaspers, C.; Haraldsson, M.; Bolte, S.; Reusch, T. B. H.; Thygesen, U. H.; Kiorboe, T. 2012. Ctenophore population recruits entirely through larval reproduction in the central Baltic Sea. Biology Letters 8, 809–812.

Jebram, D. H. A. 1992. The polyphyletic origin of the "Cheilostomata" (Bryozoa). Zeitschrift für zoologische Systematik und Evolutionsforschung 30, 46–52.

Jefferies, R. P. S. 1986. The Ancestry of the Vertebrates. British Museum (Natural History), London.

Jefferies, R. P. S. 1991. Two types of bilateral symmetry in the Metazoa: chordate and bilaterian. In: Bock, G. R.; Marsh, J. (eds.), Biological Asymmetry and Handedness. Wiley, Chichester, pp. 94–120.

Jeffery, W. R.; Swalla, B. J. 1997. Tunicates. In: Gilbert, S. F.; Raunio, A. M. (eds.), Embryology. Constructing the organism. Sinauer, Sunderland, MA, pp. 331–364.

Jékely, G. s. r.; Paps, J.; Nielsen, C. 2015. The phylogenetic position of ctenophores and the origin(s) of nervous systems. EvoDevo 6, 1.

Jell, J. S. 1984. Cambrian cnidarians with mineralized skeletons. Palaeontographica Americana 54, 105–109.

Jenner, R. A. 1999. Metazoan phylogeny as a tool in evolutionary biology: Current problems and discrepancies in application. Belgian Journal of Zoology 129, 245–262.

Jenner, R. A. 2000. Evolution of animal body plans: the role of metazoan phylogeny at the interface between pattern and process. Evolution and Development 2, 208–221.

Jenner, R. A. 2001. Bilaterian phylogeny and uncritical recycling of morphological data sets. Systematic Biology 50, 730–742.

Jenner, R. A. 2002. Boolean logic and character state identity: pitfalls of character coding in metazoan cladistics. Contributions to Zoology 71, 67–91.

Jenner, R. A. 2003. Unleashing the force of cladistics? Metazoan phylogenetics and hypothesis testing. Integrative and Comparative Biology 43, 207–218.

Jenner, R. A. 2004. Towards a phylogeny of the Metazoa: evaluating alternative phylogenetic positions of Platyhelminthes, Nemertea, and Gnathostomulida, with a critical reappraisal of cladistic characters. Contributions to Zoology 73, 3–163.

Jenner, R. A.; Scholtz, G. 2005. Playing another round of metazoan phylogenetics: Historical epistemology, sensitivity analysis, and the position of Arthropoda within Metazoa on the basis of morphology. In: Koenemann, S.; Jenner, R. A. (eds.), Crustacean Issues 16: Crustacea and Arthropod Relationships. Festschrift for Frederick R. Schram. Taylor and Francis, Boca Raton, pp. 355–385.

Jenner, R. A.; Schram, F. R. 1999. The grand game of metazoan phylogeny: rules and strategies. Biological Reviews 74, 121–142.

Jensen, S.; Gehling, J. G.; Droser, M. L. 1998. Ediacara-type fossils in Cambrian sediments. Nature 393, 567–569.

Jespersen, Å. 1987. Ultrastructure of the protonephridium in *Prostoma graecense* (Bohmig) (Rhynchocoela, Enopla, Hoplonemertini). Zoologica Scripta 16, 181–189.

Jespersen, Å.; Lützen, J. 1989. Ultrastructure of the protonephridial terminal organ in the terrestrial nemertean *Geonemertes pelaensis* (Rhynchocoela, Enopla, Hoplonemertini). Acta Zoologica 70, 157–162.

Jex, A. R.; Liu, S.; Li, B.; Young, N. D.; Hall, R. S.; Li, Y.; Yang, L.; Zeng, N.; Xu, X.; Xiong, Z.; et al. 2011. *Ascaris suum* draft genome. Nature 479, 529–533.

Jex, A. R.; Nejsum, P.; Schwarz, E. M.; Hu, L.; Young, N. D.; Hall, R. S.; Korhonen, P. K.; Liao, S.; Thamsborg, S.; Xia, J.; et al. 2014. Genome and transcriptome of the porcine whipworm *Trichuris suis*. Nature Genetics 46, 701–706.

Jiménez-Guri, E.; Okamura, B.; Holland, P. W. H. 2007a. Origin and evolution of a myxozoan worm. Integrative and Comparative Biology 47, 752–758.

Jiménez-Guri, E.; Paps, J.; García-Fernàndez, J.; Saló, E. 2006. *Hox* and *ParaHox* genes in Nemertodermatida, a basal bilaterian clade. International Journal of Developmental Biology 50, 675–679.

Jiménez-Guri, E.; Philippe, H.; Okamura, B.; Holland, P. W. 2007b. *Buddenbrockia* is a cnidarian worm. Science 317, 116–118.

Jo, J.; Oh, J.; Lee, H.-G.; Hong, H.-H.; Lee, S.-G.; Cheon, S.; Kern, E. M. A.; Jin, S.; Cho, S.-J.; Park, J.-K.; et al. 2017. Draft genome of the sea cucumber *Apostichopus japonicus* and genetic polymorphism among color variants. GigaScience 6, 1–6.

Jockusch, E. L.; Smith, F. W. 2015. Hexapoda: Comparative aspects of later embryogenesis and metamorphosis. In: Wanninger, A. (ed.), Evolutionary Developmental Biology of Invertebrates 5: Ecdysozoa III: Hexapoda. Springer, Vienna, pp. 111–208.

Joester, D.; Brooker, L. R. 2016. The chiton radula: A model system for versatile use of iron oxides. In: Faivre, D. (ed.),

Iron Oxides: From Nature to Applications. Wiley–VCH Verlag GmbH KGaA, Weinheim, pp. 177–205.

Joffe; B. I., Kotikova, E. A. 1988. Nervous system of *Priapulus caudatus* and *Halicryptus spinulosus* (Priapulida). Trudy Zoologicheskogo Instituta 183, 52–77.

Joffe, B. I.; Valiejo Roman; K. M., Birstein; V. Y., Troitsky; A. V. 1995. 5S rRNA sequences of 12 species of flatworms: Implications for the phylogeny of the Platyhelminthes. Hydrobiologia 305; 37–43.

Joffe; B. I.; Wikgren, M. 1995. Immunocytochemical distribution of 5-HT (serotonin) in the nervous-system of the gastrotrich *Turbanella cornuta*. Acta Zoologica 76, 7–9.

John, C. C. 1933. Habits, structure, and development of *Spadella cephaloptera*. Quarterly Journal of Microscopical Science 75, 625–696.

Johnsen, S. 1997. Identification and localization of a possible rhodopsin in the echinoderms *Asterias forbesi* (Asteroidea) and *Ophioderma brevispinum* (Ophiuroidea). Biological Bulletin 193, 97–105.

Johnson, C. H. 2010. Effects of selfing on offspring survival and reproduction in a colonial simultaneous hermaphrodite (*Bugula stolonifera*, Bryozoa). Biological Bulletin 219, 27–37.

Johnson, C. H.; Winston, J. E.; Woollacott, R. M. 2012. Western Atlantic introduction and persistence of the marine bryozoan *Tricellaria inopinata*. Aquatic Invasions 7, 295–303.

Johnson, C. H.; Woollacott, R. M. 2010. Larval settlement preference maximizes genetic mixing in an inbreeding population of a simultaneous hermaphrodite (*Bugula stolonifera*, Bryozoa). Molecular Ecology 19, 5511–5520.

Johnstone, I. L. 1994. The cuticle of the nematode *Caenorhabditis elegans*: A complex collagen structure. BioEssays 16, 171–178.

Jondelius, U. 1992a. Adhesive glands in the Pterastericolidae (Plathelminthes, Rhabdocoela). Zoomorphology 111, 229–238.

Jondelius, U. 1992b. Sperm morphology in the Pterastericolidae (Platyhelminthes, Rhabdocoela): Phylogenetic implications. Zoologica Scripta 21, 223–230.

Jondelius, U. 1998. Flatworm phylogeny from partial 18S rDNA sequences. Hydrobiologia 383, 147–154.

Jondelius, U.; Larsson, K.; Raikova, O. 2004. Cleavage in *Nemertoderma westbladi* (Nemertodermatida) and its phylogenetic significance. Zoomorphology 123, 221–225.

Jondelius, U.; Ruiz-Trillo, I.; Baguñà, J.; Riutort, M. 2002. The Nemertodermatida are basal bilaterians and not members of the Platyhelminthes. Zoologica Scripta 31, 201–215.

Jondelius, U.; Wallberg, A.; Hooge, M.; Raikova, O. I. 2011. How the worm got its pharynx: Phylogeny, classification and Bayesian assessment of character evolution in Acoela. Systematic Biology 60, 845–871.

Jones, B. W.; Nishiguchi, M. K. 2004. Counterillumination in the Hawaiian bobtail squid, *Euprymna scolopes* Berry (Mollusca: Cephalopoda). Marine Biology 144, 1151–1155.

Jones, M. L. 1985. On the Vestimentifera, new phylum: six new species, and other taxa from hydrothermal vents and elsewhere. Bulletin of the Biological Society of Washington 6, 117–158.

Jönsson, K. I.; Hygum, T. L.; Andersen, K. N.; Clausen, L. K.; Møbjerg, N. 2016. Tolerance to gamma radiation in the marine heterotardigrade, *Echiniscoides sigismundi*. PLoS One 11, e0168884.

Jönsson, K. I.; Levine, E. B.; Wojcik, A.; Haghdoost, S.; Harms-Ringdahl, M. 2018. Environmental adaptations: Radiation tolerance. In: Schill, R. O. (ed.), Water Bears: The Biology of Tardigrades. Zoological Monographs, vol. 2. Springer Nature Switzerland, Cham, pp. 311–330.

Jönsson, K. I.; Rabbow, E.; Schill, R. O.; Harms-Ringdahl, M.; Rettberg, P. 2008. Tardigrades survive exposure to space in low Earth orbit. Current Biology 18, R729–R731.

Jørgensen, A.; Faurby, S.; Hansen, J. G.; Møbjerg, N.; Kristensen, R. M. 2010. Molecular phylogeny of Arthrotardigrada (Tardigrada). Molecular Phylogenetics and Evolution 54, 1006–1015.

Jørgensen, A.; Kristensen, R. M. 2004. Molecular phylogeny of Tardigrada—investigation of the monophyly of Heterotardigrada. Molecular Phylogenetics and Evolution 32, 666–670.

Jørgensen, A.; Kristensen, R. M.; Møbjerg, N. 2018. Phylogeny and integrative taxonomy of Tardigrada. In: Schill, R. O. (ed.) Water Bears: The Biology of Tardigrades. Zoological Monographs, vol. 2. Springer Nature Switzerland, Cham, pp. 95–114.

Jørgensen, A.; Møbjerg, N. 2014. Notes on the cryptobiotic capability of the marine arthrotardigrades *Styraconyx haploceros* (Halechiniscidae) and *Batillipes pennaki* (Batillipedidae) from the tidal zone in Roscoff, France. Marine Biology Research 11, 214–217.

Jørgensen, A.; Møbjerg, N.; Kristensen, R. M. 2007. A molecular study of the tardigrade *Echiniscus testudo* (Echiniscidae) reveals low DNA sequence diversity over a large geographical area. Journal of Limnology 66, 77–83.

Jörger, K. M.; Neusser, T. P.; Brenzinger, B.; Schrödl, M. 2014. Exploring the diversity of mesopsammic gastropods: How to collect, identify, and delimitate small and elusive sea slugs? American Malacological Bulletin 32, 290–307.

Joysey, K. A. 1959. Probable cirripede, phoronid, and echionoid burrows within a Cretaceous echinoid test. Palaeontology 1, 397–400.

Juliano, C. E.; Yajima; M., Wessel, G. M. 2010. Nanos functions to maintain the fate of the small micromere lineage in the sea urchin embryo. Developmental Biology 337, 220–232.

Junoy, J.; Andrade, S. C. S.; Giribet, G. 2010. Phylogenetic placement of a new hoplonemertean species commensal of ascidians. Invertebrate Systematics 24, 616–629.

Junoy, J.; Montalvo, S.; Roldán, C.; García-Corrales, P. 2000. Ultrastructural study of the bacillary, granular and mucoid proboscidial gland cells of *Riseriellus occultus* (Nemertini, Heteronemertini). Acta Zoologica 81, 235–242.

Just, J.; Kristensen, R. M.; Olesen, J. 2014. *Dendrogramma*, new genus, with two new non-bilaterian species from the marine bathyal of southeastern Australia (Animalia, Metazoa incertae sedis)—with similarities to some medusoids from the Precambrian Ediacara. PLoS One 9, e102976.

Justine, J.-L. 1991. Phylogeny of parasitic Platyhelminthes: a critical-study of synapomorphies proposed on the basis of of specimens from Bermuda. Zoologica Scripta 5, 239–255.

Knauss, E. B. 1979. Indication of an anal pore in Gnathostomulida. Zoologica Scripta 8, 181–186.

Knaust, D. 2010. Remarkably preserved benthic organisms and their traces from a Middle Triassic (Muschelkalk) mud flat. Lethaia 43, 344–356.

Kniprath, E. 1981. Ontogeny of the molluscan shell field: A review. Zoologica Scripta 10, 61–79.

Kobayashi, M.; Furuya, H.; Holland, P. W. H. 1999. Dicyemids are higher animals. Nature 401, 762.

Kober, K. M.; Bernardi, G. 2013. Phylogenomics of strongylocentrotid sea urchins. BMC Evolutionary Biology 13, 88.

Koch, M.; Quast, B.; Bartolomaeus, T. 2014. Coeloms and nephridia in annelids and arthropods. In: Wägele, J. W.; Bartolamaeus, T. (eds.), Deep Metazoan Phylogeny: the Backbone of the Tree of Life. De Gruyter, Berlin, pp. 173–284.

Kocot, K. M.; Aguilera, F.; McDougall, C.; Jackson, D. J.; Degnan, B. M. 2016. Sea shell diversity and rapidly evolving secretomes: insights into the evolution of biomineralization. Frontiers in Zoology 13, 23.

Kocot, K. M.; Cannon, J. T.; Todt, C.; Citarella, M. R.; Kohn, A. B.; Meyer, A.; Santos, S. R.; Schander, C.; Moroz, L. L.; Lieb, B.; et al. 2011. Phylogenomics reveals deep molluscan relationships. Nature 447, 452–456.

Kocot, K. M.; Halanych, K. M.; Krug, P. J. 2013. Phylogenomics supports Panpulmonata: Opisthobranch paraphyly and key evolutionary steps in a major radiation of gastropod molluscs. Molecular Phylogenetics and Evolution 69, 764–771.

Kocot, K. M.; Struck, T. H.; Merkel, J.; Waits, D. S.; Todt, C.; Brannock, P. M.; Weese, D. A.; Cannon, J. T.; Moroz, L. L.; Lieb, B.; et al. 2017. Phylogenomics of Lophotrochozoa with consideration of systematic error. Systematic Biology 66, 256–282.

Kocot, K. M.; Tassia, M. G.; Halanych, K. M.; Swalla, B. J. 2018. Phylogenomics offers resolution of major tunicate relationships. Molecular Phylogenetics and Evolution 121, 166–173.

Kocot, K. M.; Todt, C.; Mikkelsen, N. T.; Halanych, K. M. 2019. Phylogenomics of Aplacophora (Mollusca, Aculifera) and a solenogaster without a foot. Proceedings of the Royal Society B: Biological Sciences 286, 20190115.

Koeppen, D.; Aurich, M.; Rampp, T. 2014. Medicinal leech therapy in pain syndromes: a narrative review. Wien Medizinische Wochenschrift 164, 95–102.

Kohn, A. B.; Citarella, M. R.; Kocot, K. M.; Bobkova, Y. V.; Halanych, K. M.; Moroz, L. L. 2012. Rapid evolution of the compact and unusual mitochondrial genome in the ctenophore, Pleurobrachia bachei. Molecular Phylogenetics and Evolution 63, 203–207.

Kokkinopoulou, M.; Spiecker, L.; Messerschmidt, C.; Barbeck, M.; Ghanaati, S.; Landfester, K.; Markl, J. 2015. On the ultrastructure and function of rhogocytes from the pond snail Lymnaea stagnalis. PLoS One 10, e0141195.

Komai, T. 1922. Studies on two aberrant ctenophores, Coeloplana and Gastrodes. Published by the Author, Kyoto.

Komai, T. 1941. A new remarkable sessile ctenophore. Proceedings of the Imperial Academy, Tokyo 17, 216–220.

Kon, T.; Nohara, M.; Yamanoue, Y.; Fujiwara, Y.; Nishida, M.; Nishikawa, T. 2007. Phylogenetic position of a whale-fall lancelet (Cephalochordata) inferred from whole mitochondrial genome sequences. BMC Evolutionary Biology 7, 127.

Kondo, K.; Kubo, T.; Kunieda, T. 2015. Suggested Involvement of PP1/PP2A activity and de novo gene expression in anhydrobiotic survival in a tardigrade, Hypsibius dujardini, by chemical genetic approach. PLoS One 10, e0144803.

Koroleva, A. G.; Chernyshev, A. V.; Kiril'chik, S. V.; Tasevska, O.; Kostoski, G.; Timoshkin, O. A. 2014. The first finding of freshwater nemerteans in Ohrid Lake (Macedonia) with some comments on the taxonomy of the genus Prostoma (Nemertea, Monostilifera). Biology Bulletin 41, 736–741.

Kostyuchenko, R. P.; Kozin, V. V.; Kupriashova, E. E. 2016. Regeneration and asexual reproduction in annelids: Cells, genes, and evolution. Biology Bulletin 43, 185–194.

Kotikova, E. A. 1998. Catecholaminergic neurons in the brain of rotifers. Hydrobiologia 387, 135–140.

Kotikova, E. A.; Raikova, O.; Reuter, M.; Gustafsson, M. K. S. 2005. Rotifer nervous system visualized by FMRFamide and 5-HT immunocytochemistry and confocal laser scanning microscopy. Hydrobiologia 546, 239–248.

Kotikova, E. A.; Raikova, O. I. 2008. Architectonics of the central nervous system of Acoela, Platyhelminthes, and Rotifera. Journal of Evolutionary Biochemistry and Physiology 44, 95–108.

Kotikova, E. A.; Raikova, O. I.; Flyatchinskaya, L. P.; Reuter, M.; Gustafsson, M. K. S. 2001. Rotifer muscles as revealed by phalloidin-TRITC staining and confocal scanning laser microscopy. Acta Zoologica 82, 1–9.

Kouchinsky, A.; Bengtson, S.; Runnegar, B.; Skovsted, C.; Steiner, M.; Vendrasco, M. 2012. Chronology of early Cambrian biomineralization. Geological Magazine 149, 221–251.

Koutsouveli, V.; Taboada, S.; Moles, J.; Cristobo, J.; Ríos, P.; Bertran, A.; Solà, J.; Avila, C.; Riesgo, A. 2018. Insights into the reproduction of some Antarctic dendroceratid, poecilosclerid, and haplosclerid demosponges. PLoS One 13, e0192267.

Koutsovoulos, G.; Kumar, S.; Laetsch, D. R.; Stevens, L.; Daub, J.; Conlon, C.; Maroon, H.; Thomas, F.; Aboobaker, A. A.; Blaxter, M. 2016. No evidence for extensive horizontal gene transfer in the genome of the tardigrade Hypsibius dujardini. Proceedings of the National Academy of Sciences of the USA 113, 5053–5058.

Kovalevskiĭ, A. O. 1867. Anatomy and history of the development of Phoronis [in Russian]. Supplement to the Bulletin of the Academie des Sciences, St. Petersbourg 11, 1–41.

Kowalevsky, A. O. 1867. Die Entwickelungsgeschichte des Amphioxus lanceolatus. Mémoires de l'Académie Impériale de Saint-Pétersbourg 11, 1–17.

Kozek, W. J.; Marroquin, H. F. 1977. Intracytoplasmic bacteria in Onchocerca volvulus. American Journal of Tropical Medicine and Hygiene 26, 663–678.

Koziol, U.; Jarero, F.; Olson, P. D.; Brehm, K. 2016. Comparative analysis of Wnt expression identifies a highly conserved developmental transition in flatworms. BMC Biology 14, 10.

Kozloff, E. N. 1969. Morphology of the orthonectid *Rhopalura ophiocomae*. Journal of Parasitology 55, 171–195.

Kozloff, E. N. 1972. Some aspects of development in *Echinoderes* (Kinorhyncha). Transctions of the American Microscopical Society 91, 119–130.

Kozloff, E. N. 1992. The genera of the phylum Orthonectida. Cahiers de Biologie Marine 33, 377–406.

Kozloff, E. N. 1994. The structure and origin of the plasmodium of *Rhopalura ophiocomae* (Phylum Orthonectida). Acta Zoologica 75, 191–199.

Kozloff, E. N. 1997. Studies on the so-called plasmodium of *Ciliocincta sabellariae* (Phylum Orthonectida), with notes on an associated microsporan parasite. Cahiers de Biologie Marine 38, 151–159.

Kozloff, E. N. 2007. Stages of development, from first cleavage to hatching, of an *Echinoderes* (phylum Kinorhyncha: class Cyclorhagida). Cahiers de Biologie Marine 48, 199–206.

Krapf, K.; Dunagan, T. T. 1987. Structural features of the protonephridia in female *Macracanthorhynchus hirudinaceus* (Acanthocephala). Journal of Parasitology 73, 1176–1181.

Kraus, Y.; Aman, A.; Technau, U.; Genikhovich, G. 2016. Prebilaterian origin of the blastoporal axial organizer. Nature Communications 7, 11694.

Krell, F.-T.; Cranston, P. S. 2004. Which side of the tree is more basal? Systematic Entomology 29, 279–281.

Krisko, A.; Leroy, M.; Radman, M.; Meselson, M. 2012. Extreme anti-oxidant protection against ionizing radiation in bdelloid rotifers. Proceedings of the National Academy of Sciences of the USA 109, 2354–2357.

Kristensen, R. M. 1981. Sense organs of two marine arthrotardigrades (Heterotardigrada, Tardigrada). Acta Zoologica 62, 27–41.

Kristensen, R. M. 1982. The first record of cyclomorphosis in Tardigrada based on a new genus and species from Arctic meiobenthos. Zeitschrift für zoologische Systematik und Evolutionsforschung 20, 249–270.

Kristensen, R. M. 1983. Loricifera, a new phylum with Aschelminthes characters from meiobenthos. Zeitschrift für zoologische Systematik und Evolutionsforschung 21, 163–180.

Kristensen, R. M. 1984. Nyt dyr—korsetdyret—opdaget. Naturens Verden 10, 357–367.

Kristensen, R. M. 1987. Generic revision of the Echiniscidae (Heterotardigrada), with a discussion of the origin of the family. In: Bertolani, R. (ed.), Biology of Tardigrades. Selected Symposia and Monographs U.Z.I. Mucchi, Modena, pp. 261–335.

Kristensen, R. M. 1991. Loricifera. In: Harrison, F. W.; Ruppert, E. E. (eds.), Microscopic anatomy of invertebrates, vol. 4: Aschelminthes. Wiley-Liss, New York, pp. 351–375.

Kristensen, R. M. 1992. Loricifera—a general biological and phylogenetic overview. Verhandlungen der Deutschen zoologischen Gesellschaft 84, 231–246.

Kristensen, R. M. 1995. Are Aschelminthes pseudocoelomate or acoelomate? In: Lanzavecchia, G.; Valvassori, R.; Candia Carnevali, M. D. (eds.), Body cavities: function and phylogeny. Selected Symposia and Monographs U.Z.I., Mucchi, Modena, pp. 41–43.

Kristensen, R. M. 2002. An introduction to Loricifera, Cycliophora, and Micrognathozoa. Integrative and Comparative Biology 42, 641–651.

Kristensen, R. M. 2003. Comparative morphology: Do the ultrastructural investigations of Loricifera and Tardigrada support the clade Ecdysozoa? In: Legakis, A.; Sfenthourakis, S.; Polymeni, R.; Thessalou-Legaki, M. (eds.), The new Panorama of Animal Evolution. Proceedings of the 18th Congress of Zoology. Pensoft, Sofia, pp. 467–477.

Kristensen, R. M. 2016. Loricifera. In: Brusca, R. C.; Moore, W.; Shuster, S. M. (eds.), Invertebrates, 3rd ed. Sinauer, Sunderland, MA, pp. 701–705.

Kristensen, R. M. 2017. Palaeontology: Darwin's dilemma dissolved. Nature Ecology and Evolution 1, 0076.

Kristensen, R. M.; Brooke, S. 2002. Phylum Loricifera. In: Young, C. M.; Sewell, M. A.; Rice, M. E. (eds.), Atlas of Marine Invertebrate Larvae. Academic Press, San Diego, pp. 179–187.

Kristensen, R. M.; Funch, P. 2000. Micrognathozoa: A new class with complicated jaws like those of Rotifera and Gnathostomulida. Journal of Morphology 246, 1–49.

Kristensen, R. M.; Heiner, T.; Higgins, R. P. 2007. Morphology and life cycle of a new loriciferan from the Atlantic coast of Florida with an emended diagnosis and life cycle of Nanaloricidae (Loricifera). Invertebrate Biology 126, 120–137.

Kristensen, R. M.; Neuhaus, B. 1999. The ultrastructure of the tardigrade cuticle with special attention to marine species. Zoologischer Anzeiger 238, 261–281.

Kristensen, R. M.; Neves, R. C.; Gad, G. 2013. First report of Loricifera from the Indian Ocean: a new *Rugiloricus*-species represented by a hermaphrodite. Cahiers de Biologie Marine 54, 161–171.

Kristensen, R. M.; Shirayama, Y. 1988. *Pliciloricus hadalis* (Pliciloricidae), a new loriciferan species collected from the Izu-Ogasawara Trench, western Pacific. Zoological Science 5, 875–881.

Kristof, A.; Maiorova, A. S. 2016. Annelida: Sipuncula. In: Schmidt-Rhaesa, A.; Harzsch, S.; Purschke, G. (eds.), Structure and Evolution of Invertebrate Nervous Systems. Oxford University Press, Oxford, pp. 248–253.

Kristof, A.; Wollesen, T.; Wanninger, A. 2008. Segmental mode of neural patterning in Sipuncula. Current Biology 18, 1129–1132.

Kröger, B.; Vinther, J.; Fuchs, D. 2011. Cephalopod origin and evolution: A congruent picture emerging from fossils, development and molecules. BioEssays 33, 602–613.

Kroh, A.; Mooi, R. 2018. World Echinoidea Database. Accessed at http://www.marinespecies.org/echinoidea on 2018-06-06.

Kudtarkar, P.; Cameron, R. A. 2017. Echinobase: an expanding resource for echinoderm genomic information. Database 2017, bax074.

Kumano, G.; Nishida, H. 2007. Ascidian embryonic development: An emerging model system for the study of cell fate specification in chordates. Developmental Dynamics 236, 1732–1747.

Kumar, S.; Koutsovoulos, G.; Kaur, G.; Blaxter, M. 2012. Toward 959 nematode genomes. Worm 1, 42–50.

Kuroda, R.; Endo, B.; Abe, M.; Shimizu, M. 2009. Chiral blastomere arrangement dictates zygotic left-right asymmetry pathway in snails. Nature 462, 790–794.

Kusche, K.; Ruhberg, H.; Burmester, T. 2002. A hemocyanin from the Onychophora and the emergence of respiratory proteins. Proceedings of the National Academy of Sciences of the USA 99, 10545–10548.

Kutikova, L. A. 1995. Larval metamorphosis in sessile rotifers. Hydrobiologia 313, 133–138.

Kuzmina, T. V.; Malakhov, V. V. 2011. The periesophageal celom of the articulate brachiopod Hemithyris psittacea (Rhynchonelliformea, Brachiopoda). Journal of Morphology 272, 180–190.

Kuzmina, T. V.; Malakhov, V. V. 2014. The accessory hearts of the articulate brachiopod Hemithyris psittacea. Zoomorphology 134, 25–32.

Kuzmina, T. V.; Malakhov, V. V.; Temereva, E. N. 2006. Anatomy of coelomic system in the articulate brachiopoda Hemithyris psittacea (Brachiopoda, Articulata). Zoologichesky Zhurnal 85, 1118–1127.

Kuzmina, T. V.; Temereva, E. N.; Malakhov, V. V. 2016. Structure of the main heart of the articulate brachiopod Hemithiris psittacea: Morphological evidence of dual function. Zoologischer Anzeiger 262, 1–9.

Kvist, S.; Chernyshev, A. V.; Giribet, G. 2015. Phylogeny of Nemertea with special interest in the placement of diversity from Far East Russia and northeast Asia. Hydrobiologia 760, 105–119.

Kvist, S.; Laumer, C. E.; Junoy, J.; Giribet, G. 2014. New insights into the phylogeny, systematics and DNA barcoding of Nemertea. Invertebrate Systematics 28, 287–308.

Kvist, S.; Oceguera-Figueroa, A.; Tessler, M.; Jiménez-Armenta, J.; Freeman, R. M.; Giribet, G.; Siddall, M. E. 2017. When predator becomes prey: investigating the salivary transcriptome of the shark-feeding leech Pontobdella macrothela (Hirudinea: Piscicolidae). Zoological Journal of the Linnean Society 179, 725–737.

Kvist, S.; Siddall, M. E. 2013. Phylogenomics of Annelida revisited: a cladistic approach using genome-wide expressed sequence tag data mining and examining the effects of missing data. Cladistics 29, 435–448.

Lacalli, T.; Stach, T. 2016. Acrania (Cephalochordata). In: Schmidt-Rhaesa, A.; Harzsch, S.; Purschke, G. (eds.), Structure and Evolution of Invertebrate Nervous Systems. Oxford University Press, Oxford, pp. 719–728.

Lacalli, T. C. 1997. The nature and origin of deuterostomes: some unresolved issues. Invertebrate Biology 116, 363–370.

Lacalli, T. C. 2002. Vetulicolians–are they deuterostomes? chordates? BioEssays 24, 208–211.

Lacalli, T. C. 2005. Protochordate body plan and the evolutionary role of larvae: old controversies resolved? Canadian Journal of Zoology 83, 216–224.

Lacalli, T. C.; Gilmour, T. H. J. 2001. Locomotory and feeding effectors of the tornaria larva of Balanoglossus biminiensis. Acta Zoologica 82, 117–126.

Lacaze-Duthiers, F.-J.-H. 1858. Histoire de l'organisation, du développement, des moeurs et des rapports zoologiques du Dentale. Victor Masson, Paris.

Lacorte, G. A.; Oliveira, I. d. S.; Da Fonseca, C. G. 2011. Phylogenetic relationships among the Epiperipatus lineages (Onychophora: Peripatidae) from the Minas Gerais State, Brazil. Zootaxa 2755, 57–65.

Ladurner, P.; Rieger, R. 2000. Embryonic muscle development of Convoluta pulchra (Turbellaria-Acoelomorpha, Platyhelminthes). Developmental Biology 222, 359–375.

Ladurner, P.; Rieger, R.; Baguñà, J. 2000. Spatial distribution and differentiation potential of stem cells in hatchlings and adults in the marine platyhelminth Macrostomum sp.: A bromodeoxyuridine analysis. Developmental Biology 226, 231–241.

Lafay, B.; Bouryesnault, N.; Vacelet, J.; Christen, R. 1992. An analysis of partial 28S ribosomal RNA sequences suggests early radiations of sponges. Biosystems 28, 139–151.

Lafferty, K. D. 1999. The evolution of trophic transmission. Parasitology Today 15, 111–115.

Laing, R.; Kikuchi, T.; Martinelli, A.; Tsai, I. J.; Beech, R. N.; Redman, E.; Holroyd, N.; Bartley, D. J.; Beasley, H.; Britton, C.; et al. 2013. The genome and transcriptome of Haemonchus contortus, a key model parasite for drug and vaccine discovery. Genome Biology 14, R88.

LaJeunesse, T. C.; Parkinson, J. E.; Gabrielson, P. W.; Jeong, H. J.; Reimer, J. D.; Voolstra, C. R.; Santos, S. R. 2018. Systematic revision of Symbiodiniaceae highlights the antiquity and diversity of coral endosymbionts. Current Biology 28, 2570–2580 e2576.

Lam, J.; Cheng, Y.-W.; Chen, W.-N. U.; Li; H.-H., Chen; C.-S., Peng; S.-E. 2017. A detailed observation of the ejection and retraction of defense tissue acontia in sea anemone (Exaiptasia pallida). PeerJ 5, e2996.

Lamarck, J. B. P. A. 1816. Histoire naturelle des animaux sans vertèbres, présentant les caractères généraux et particuliers de ces animaux, leur distribution, leurs classes, leurs familles, leurs genres, et la citation des principales espèces qui s'y rapportent, vol. 3. Baillière, Paris.

Lambert, J. D. 2008. Mesoderm in spiralians: the organizer and the 4d cell. Journal of Experimental Zoology (Mol. Dev. Evol.) 310, 15–23.

Lambert, J. D. 2010. Developmental patterns in spiralian embryos. Current Biology 20, R72–R77.

Lambert, K.; Bekal, S. 2002. Introduction to plant-parasitic nematodes. Plant Health Instructor.

Lambert, J. D.; Johnson, A. B.; Hudson, C. N.; Chan, A. 2016. Dpp/BMP2-4 mediates signaling from the D-quadrant organizer in a spiralian embryo. Current Biology 26, 2003–2010.

Lambert, J. D.; Nagy, L. M. 2001. MAPK signaling by the D quadrant embryonic organizer of the mollusc Ilyanassa obsoleta. Development 128, 45–56.

Lambshead, P. J.; Boucher, G. 2003. Marine nematode deep-sea biodiversity—hyperdiverse or hype? Journal of Biogeography 30, 475–485.

Lambshead, P. J. D. 1993. Recent developments in marine benthic biodiversity research. Oceanis 19, 5–24.

Lammert, V. 1985. The fine structure of protonephridia in Gnathostomulida and their comparison within Bilateria. Zoomorphology 105, 308–316.

Lamsdell, J. C. 2013. Revised systematics of Palaeozoic 'horseshoe crabs' and the myth of monophyletic Xiphosura. Zoological Journal of the Linnean Society 167, 1–27.

Land, M. F.; Nilsson, D.-E. 2012. Animal Eyes. 2nd ed. Oxford University Press, Oxford/New York.

Lander, E. S.; Linton, L. M.; Birren, B.; Nusbaum, C.; Zody, M. C.; Baldwin, J.; Devon, K.; Dewar, K.; Doyle, M.; FitzHugh, W.; et al. 2001. Initial sequencing and analysis of the human genome. Nature 409, 860–921.

Landing, E.; Antcliffe, J. B.; Brasier, M. D.; English, A. B. 2015. Distinguishing Earth's oldest known bryozoan (Pywackia, late Cambrian) from pennatulacean octocorals (Mesozoic–Recent). Journal of Paleontology 89, 292–317.

Landing, E.; Antcliffe, J. B.; Geyer, G.; Kouchinsky, A.; Bowser, S. S.; Andreas, A. 2018. Early evolution of colonial animals (Ediacaran Evolutionary Radiation–Cambrian Evolutionary Radiation–Great Ordovician Biodiversification Interval). Earth-Science Reviews 178, 105–135.

Landing, E.; English, A.; Keppie, J. D. 2010. Cambrian origin of all skeletalized metazoan phyla—Discovery of Earth's oldest bryozoans (Upper Cambrian, southern Mexico). Geology 38, 547–550.

Landman, N. H.; Mikkelsen, P. M.; Bieler, R.; Bronson, B. 2001. Pearls: A Natural History. H. N. Abrams in association with the American Museum of Natural History and the Field Museum, New York.

Landry, C.; Geyer, L. B.; Arakaki, Y.; Uehara, T.; Palumbi, S. R. 2003. Recent speciation in the Indo-West Pacific: rapid evolution of gamete recognition and sperm morphology in cryptic species of sea urchin. Proceedings of the Royal Society B: Biological Sciences 270, 1839–1847.

Lane, S. J.; Moran, A. L.; Shishido, C. M.; Tobalske, B. W.; Woods, H. A. 2018. Cuticular gas exchange by Antarctic sea spiders. Journal of Experimental Biology 221, jeb177568.

Lang, A. 1884. Die Polycladen (Seeplanarien) des Golfes von Neapel und der angrenzenden Meeres-Abschnitte. Wilhelm Engelmann, Leipzig.

Lang, K. 1953. Die Entwicklung des Eies von Priapulus caudatus Lam. und die systematische Stellung der Priapuliden. Arkiv för zoologi 5, 321–348.

Lanterbecq, D.; Rouse, G. W.; Milinkovitch, M. C.; Eeckhaut, I. 2006. Molecular phylogenetic analyses indicate multiple independent emergences of parasitism in Myzostomida (Protostomia). Systematic Biology 55, 208–227.

Lapan, E. A. 1975. Magnesium inositol hexaphosphate deposits in mesozoan dispersal larvae. Experimental Cell Research 94, 277–282.

Lapan, E. A.; Morowitz, H. J. 1974. Characterization of mesozoan DNA. Experimental Cell Research 83, 143–151.

Lapan, E. A.; Morowitz, H. J. 1975. The dicyemid Mesozoa as an integrated system for morphogenetic studies. I. Description, isolation and maintenance. Journal of Experimental Zoology 193, 147–159.

Larson, P. 2017. Brooding sea anemones (Cnidaria: Anthozoa: Actiniaria): paragons of diversity in mode, morphology, and maternity. Invertebrate Biology 136, 92–112.

Latire, T.; Legendre, F.; Bigot, N.; Carduner, L.; Kellouche, S.; Bouyoucef, M.; Carreiras, F.; Marin, F.; Lebel, J. M.; Galéra, P.; et al. 2014. Shell extracts from the marine bivalve Pecten maximus regulate the synthesis of extracellular matrix in primary cultured human skin fibroblasts. PLoS One 9, e99931.

Laumer, C. E. 2014. Isolated branches in the phylogeny of Platyhelminthes. PhD thesis, Department of Organismic and Evolutionary Biology, Harvard University, Cambridge, MA, pp. 216.

Laumer, C. E.; Bekkouche, N.; Kerbl, A.; Goetz, F.; Neves, R. C.; Sørensen, M. V.; Kristensen, R. M.; Hejnol, A.; Dunn; C. W., Giribet; G.; et al. 2015a. Spiralian phylogeny informs the evolution of microscopic lineages. Current Biology 25, 2000–2006.

Laumer, C. E.; Fernández, R.; Lemer, S.; Combosch, D. J.; Kocot, K.; Andrade, S. C. S.; Sterrer, W.; Sørensen, M. V.; Giribet, G. 2019. Revisiting metazoan phylogeny with genomic sampling of all phyla. Proceedings of the Royal Society B: Biological Sciences 286, 20190831.

Laumer, C. E.; Giribet, G. 2014. Inclusive taxon sampling suggests a single, stepwise origin of ectolecithality in Platyhelminthes. Biological Journal of the Linnean Society 111, 570–588.

Laumer, C. E.; Giribet, G. 2017. Phylogenetic relationships within Adiaphanida (phylum Platyhelminthes) and the status of the crustacean–parasitic genus Genostoma. Invertebrate Biology 136, 184–198.

Laumer, C. E.; Giribet, G.; Curini-Galletti, M. 2014. Prosogynopora riseri, gen. et spec. nov., a phylogenetically problematic lithophoran proseriate (Platyhelminthes : Rhabditophora) with inverted genital pores from the New England coast. Invertebrate Systematics 28, 309–325.

Laumer, C. E.; Gruber-Vodicka, H.; Hadfield, M. G.; Pearse, V. B.; Riesgo, A.; Marioni, J. C.; Giribet, G. 2018. Support for a clade of Placozoa and Cnidaria in genes with minimal compositional bias. eLife 7, e36278.

Laumer, C. E.; Hejnol, A.; Giribet, G. 2015b. Nuclear genomic signals of the "microturbellarian" roots of platyhelminth evolutionary innovation. eLife 4, e05503.

Lavrov, A. I.; Kosevich, I. A. 2014. Sponge cell reaggregation: Mechanisms and dynamics of the process. Russian Journal of Developmental Biology 45, 205–223.

Lavrov, D. V.; Adamski, M.; Chevaldonné, P.; Adamska, M. 2016. Extensive mitochondrial mRNA editing and unusual mitochondrial genome organization in calcaronean sponges. Current Biology 26, 86–92.

Lavrov, D. V.; Brown, W. M.; Boore, J. L. 2004. Phylogenetic position of the Pentastomida and (pan)crustacean relationships. Proceedings: Biological Sciences 271, 537–544.

Lavrov, D. V.; Forget, L.; Kelly, M.; Lang, B. F. 2005. Mitochondrial genomes of two demosponges provide insights into an early stage of animal evolution. Molecular Biology and Evolution 22, 1231–1239.

Lavrov, D. V.; Lang, B. F. 2005. Poriferan mtDNA and animal phylogeny based on mitochondrial gene arrangements. Systematic Biology 54, 651–659.

Lavrov, D. V.; Pett, W. 2016. Animal mitochondrial DNA as we do not know it: mt-genome organization and evolution in nonbilaterian lineages. Genome Biology and Evolution 8, 2896–2913.

Lavrov, D. V.; Pett, W.; Voigt, O.; Wörheide, G.; Forget, L.; Lang, B. F.; Kayal, E. 2013. Mitochondrial DNA of *Clathrina clathrus* (Calcarea, Calcinea): six linear chromosomes, fragmented rRNAs, tRNA editing, and a novel genetic code. Molecular Biology and Evolution 30, 865–880.

Layden, M. J.; Rentzsch, F.; Röttinger, E. 2016. The rise of the starlet sea anemone *Nematostella vectensis* as a model system to investigate development and regeneration. WIREs Developmental Biology 5, 408–428.

Lažetić, V.; Fay, D. S. 2017. Molting in *C. elegans*. Worm 6, e1330246.

Le, T. H.; Blair, D.; McManus, D. P. 2003. Complete DNA sequence and gene organization of the mitochondrial genome of the liverfluke, *Fasciola hepatica* L. (Platyhelminthes; Trematoda). Parasitology 123, 609–621.

Leasi, F.; Neves, R. C.; Worsaae, K.; Sorensen, M. V. 2012. Musculature of *Seison nebaliae* Grube, 1861 and *Paraseison annulatus* (Claus, 1876) revealed with CLSM: a comparative study of the gnathiferan key taxon Seisonacea (Rotifera). Zoomorphology 131, 185–195.

Leasi, F.; Norenburg, J. L. 2014. The necessity of DNA taxonomy to reveal cryptic diversity and spatial distribution of meiofauna, with a focus on Nemertea. PLoS One 9, e104385.

Leasi, F.; Pennati, R.; Ricci, C. 2009. First description of the serotonergic nervous system in a bdelloid rotifer: *Macrotrachela quadricornifera* Milne 1886 (Philodinidae). Zoologischer Anzeiger 248, 47–55.

Leasi, F.; Rothe, B. H.; Schmidt-Rhaesa, A.; Todaro, M. A. 2006. The musculature of three species of gastrotrichs surveyed with confocal laser scanning microscopy (CLSM). Acta Zoologica 87, 171–180.

Leasi, F.; Rouse, G. W.; Sørensen, M. V. 2011. A new species of *Paraseison* (Rotifera: Seisonacea) from the coast of California, USA. Journal of the Marine Biological Association of the U.K. 92, 959–965.

Lebrato, M.; Mendes, P. D.; Steinberg, D. K.; Cartes, J. E.; Jones, B. M.; Birsa, L. M.; Benavides, R.; Oschlies, A. 2013. Jelly biomass sinking speed reveals a fast carbon export mechanism. Limnology and Oceanography 58, 1113–1122.

Lechner, M. 1966. Untersuchungen zur Embryonalentwicklung des Rädertieres *Asplanchna girodi* de Guerne. Wilhelm Roux' Archiv für Entwicklungsmechanik der Organismen 157, 117–173.

Lee, B. P.; Messersmith, P. B.; Israelachvili, J. N.; Waite, J. H. 2011. Mussel-inspired adhesives and coatings. Annual Review of Materials Research 41, 99–132.

Lee, D. L. 2002. Life cycles. In: Lee, D. L. (ed.), The biology of nematodes. Taylor and Francis, London/New York, pp. 61–72.

Lee, P. N.; Callaerts, P.; De Couet, H. G.; Martindale, M. Q. 2003. Cephalopod *Hox* genes and the origin of morphological novelties. Nature 424, 1061–1065.

Lee, P. N.; Kumburegama, S.; Marlow, H. Q.; Martindale, M. Q.; Wikramanayake, A. H. 2007. Asymmetric developmental potential along the animal–vegetal axis in the anthozoan cnidarian, *Nematostella vectensis*, is mediated by Dishevelled. Developmental Biology 310, 169–186.

Lee, R. C.; Feinbaum, R. L.; Ambros, V. 1993. The C. elegans heterochronic gene *lin-4* encodes small RNAs with antisense complementarity to *lin-14*. Cell 75, 843–854.

Lee, W. L.; Reiswig, H. M.; Austin, W. C.; Lundsten, L. 2012. An extraordinary new carnivorous sponge, *Chondrocladia lyra*, in the new subgenus *Symmetrocladia* (Demospongiae, Cladorhizidae), from off of northern California, USA. Invertebrate Biology 131, 259–284.

Lefebvre, B.; Guensburg, T. E.; Martin, E. L. O.; Mooi, R.; Nardin, E.; Nohejlová, M.; Saleh, F.; Kouraïss, K.; El Hariri, K.; David, B. 2019. Exceptionally preserved soft parts in fossils from the Lower Ordovician of Morocco clarify stylophoran affinities within basal deuterostomes. Geobios 52, 27–36.

Legg, D. A.; Sutton, M. D.; Edgecombe, G. D. 2013. Arthropod fossil data increase congruence of morphological and molecular phylogenies. Nature Communications 4, 2485.

Lehnert, O.; Miller, J. F.; Cochrane, K. 1999. *Palaeobotryllus* and friends: Cambro-Ordovician record of probable ascidian tunicates. Acta Universitatis Carolinae Geologica 43, 447–450.

Leitz, T. 2016. Cnidaria. In: Schmidt-Rhaesa, A.; Harzsch, S.; Purschke, G. (eds.), Structure and Evolution of Invertebrate Nervous Systems. Oxford University Press, Oxford, pp. 26–47.

Lemaire, P. 2009. Unfolding a chordate developmental program, one cell at a time: Invariant cell lineages, short-range inductions and evolutionary plasticity in ascidians. Developmental Biology 332, 48–60.

Lemaire, P.; Piette, J. 2015. Tunicates: exploring the sea shores and roaming the open ocean. A tribute to Thomas Huxley. Open Biology 5, 150053.

Lemburg, C. 1995a. Ultrastructure of sense organs and receptor cells of the neck and lorica of the *Halicryptus spinulosus* larva (Priapulida). Microfauna Marina 10, 7–30.

Lemburg, C. 1995b. Ultrastructure of the introvert and associated structures of the larvae of *Halicryptus spinulosus* (Priapulida). Zoomorphology 115, 11–29.

Lemburg, C. 1998. Electron microscopical localization of chitin in the cuticle of *Halicryptus spinulosus* and *Priapulus caudatus* (Priapulida) using gold-labelled wheat germ agglutinin: phylogenetic implications for the evolution of the cuticle within the Nematelminthes. Zoomorphology 118, 137–158.

Lemburg, C. 1999. Ultrastrukturelle Untersuchungen an den Larven von *Halicryptus spinulosus* und *Priapulus caudatus*. Hypothesen zur Phylogenie der Priapulida und deren Bedeutung für die Evolution der Nemathelminthes. Cuvillier Verlag, Göttingen.

Lemche, H. 1957. A new living deep-sea mollusc of the Cambro-Devonian class Monoplacophora. Nature 179, 413–416.

Lemche, H.; Wingstrand, K. G. 1959. The anatomy of *Neopilina galatheae* Lemche, 1957 (Mollusca, Tryblidiacea). Galathea Report 3, 9–71.

Lemche, H.; Wingstrand, K. G. 1987. Notes on the anatomy of *Neopilina (Vema) ewingi* Clarke and Menzies, 1959 (Monoplacophora). Malacology Data Net (Ecosearch Series) 21, 15–17.

Lemer, S.; Bieler, R.; Giribet, G. 2019. Resolving the relationships of clams and cockles: dense transcriptome sampling drastically improves the bivalve tree of life. Proceedings of the Royal Society B: Biological Sciences, 20182684.

Lemer, S.; González, V. L.; Bieler, R.; Giribet, G. 2016. Cementing mussels to oysters in the pteriomorphian tree: A phylogenomic approach. Proceedings of the Royal Society B: Biological Sciences 283, 20160857.

Lemer, S.; Kawauchi, G. Y.; Andrade, S. C. S.; González, V. L.; Boyle, M. J.; Giribet, G. 2015. Re-evaluating the phylogeny of Sipuncula through transcriptomics. Molecular Phylogenetics and Evolution 83, 174–183.

Lengerer, B.; Pjeta, R.; Wunderer, J.; Rodrigues, M.; Arbore, R.; Scharer, L.; Berezikov, E.; Hess, M.; Pfaller, K.; Egger, B.; et al. 2014. Biological adhesion of the flatworm *Macrostomum lignano* relies on a duo-gland system and is mediated by a cell type-specific intermediate filament protein. Frontiers in Zoology 11, 12.

Lengfeld, T.; Watanabe, H.; Simakov, O.; Lindgens, D.; Gee, L.; Law, L.; Schmidt, H. A.; Özbek, S.; Bode, H.; Holstein, T. W. 2009. Multiple Wnts are involved in *Hydra* organizer formation and regeneration. Developmental Biology 330, 186–199.

Lentz, T. L.; Barrnett, R. J. 1965. Fine structure of the nervous system of *Hydra*. American Zoologist 5, 341–356.

Leonard, J. L.; Pearse, J. S.; Harper, A. B. 2002. Comparative reproductive biology of *Ariolimax californicus* and *A. dolichophallus* (Gastropoda; Stylommiatophora). Invertebrate Reproduction and Development 41, 83–93.

Leptin, M. 1991. *twist* and *snail* as positive and negative regulators during *Drosophila* mesoderm development. Genes and Development 5, 1568–1576.

Lesser, M. P.; Mazel, C. H.; Gorbunov, M. Y.; Falkowski, P. G. 2004. Discovery of symbiotic nitrogen-fixing cyanobacteria in corals. Science 305, 997–1000.

Levin, L. A. 1984. Multiple patterns of development in *Streblospio benedicti* Webster (Spionidae) from three coasts of North America. Biological Bulletin 166, 494–508.

Levin, L. A. 2010. Anaerobic metazoans: No longer an oxymoron. BMC Biology 8, 31.

Levin, M.; Anavy, L.; Cole, A. G.; Winter, E.; Mostov, N.; Khair, S.; Senderovich, N.; Kovalev, E.; Silver, D. H.; Feder, M.; et al. 2016. The mid-developmental transition and the evolution of animal body plans. Nature 531, 637–641.

Levine, M.; Tjian, R. 2003. Transcription regulation and animal diversity. Nature 424, 147–151.

Lewis, C.; Long, T. A. F. 2005. Courtship and reproduction in *Carybdea sivickisi* (Cnidaria: Cubozoa). Marine Biology 147, 477–483.

Lewis, E. B. 1978. A gene complex controlling segmentation in *Drosophila*. Nature 276, 565–570.

Lewis, Z. R.; Dunn, C. W. 2018. We are not so special. eLife 7, e38726.

Leys, S. P. 2015. Elements of a 'nervous system' in sponges. Journal of Experimental Biology 218, 581–591.

Leys, S. P.; Eerkes-Medrano, D. I. 2006. Feeding in a calcareous sponge: Particle uptake by pseudopodia. Biological Bulletin 211, 157–171.

Leys, S. P.; Hill, A. 2012. The physiology and molecular biology of sponge tissues. Advances in Marine Biology 62, 1–56.

Leys, S. P.; Riesgo, A. 2012. Epithelia, an evolutionary novelty of metazoans. Journal of Experimental Zoology, pt. B: Molecular and Developmental Evolution 314B, 438–447.

Li, J.; Celiz, A. D.; Yang, J.; Yang, Q.; Wamala, I.; Whyte, W.; Seo, B. R.; Vasilyev, N. V.; Vlassak, J. J.; Suo, Z.; et al. 2017a. Tough adhesives for diverse wet surfaces. Science 357, 378–381.

Li, J.; Volsteadt, M.; Kirkendale, L.; Cavanaugh, C. M. 2018a. Characterizing photosymbiosis between Fraginae bivalves and *Symbiodinium* using phylogenetics and stable isotopes. Frontiers in Ecology and Evolution 6, 45.

Li, L.; Connors, M. J.; Kolle, M.; England, G. T.; Speiser, D. I.; Xiao, X.; Aizenberg, J.; Ortiz, C. 2015a. Multifunctionality of chiton biomineralized armor with an integrated visual system. Science 350, 952–956.

Li, L.; Zhang, X.; Yun, H.; Li, G. 2017b. Complex hierarchical microstructures of Cambrian mollusk *Pelagiella*: insight into early biomineralization and evolution. Scientific Reports 7, 1935.

Li, T. M.; Chen, J.; Li, X.; Ding, X.-J.; Wu, Y.; Zhao, L. F.; Chen, S.; Lei, X.; Dong, M.-Q. 2013. Absolute quantification of a steroid hormone that regulates development in *Caenorhabditis elegans*. Analytical Chemistry 85, 9281–9287.

Li, W. X.; Zhang, D.; Boyce, K.; Xi, B. W.; Zou, H.; Wu, S. G.; Li, M.; Wang, G. T. 2017c. The complete mitochondrial DNA of three monozoic tapeworms in the Caryophyllidea: a mitogenomic perspective on the phylogeny of eucestodes. Parasites and Vectors 10, 314.

Li, Y.; Kocot, K. M.; Schander, C.; Santos, S. R.; Thornhill, D. J.; Halanych, K. M. 2015b. Mitogenomics reveals phylogeny and repeated motifs in control regions of the deep-sea family Siboglinidae (Annelida). Molecular Phylogenetics and Evolution 85, 221–229.

Li, Z.; Tiley, G. P.; Galuska, S. R.; Reardon, C. R.; Kidder, T. I.; Rundell, R. J.; Barker, M. S. 2018b. Multiple large-scale gene and genome duplications during the evolution of hexapods. Proceedings of the National Academy of Sciences of the USA 115, 4713–4718.

Liesenjohann, T.; Neuhaus, B.; Schmidt-Rhaesa, A. 2006. Head sensory organs of *Dactylopodola baltica* (Macrodasyida, Gastrotricha): A combination of transmission electron microscopical and immunocytochemical techniques. Journal of Morphology 267, 897–908.

Light, S. F. 1923. Amphioxus fisheries near the University of Amoy, China. Science 58, 57–60.

Lin, M. F.; Chou, W. H.; Kitahara, M. V.; Chen, C. L.; Miller, D. J.; Foret, S. 2016. Corallimorpharians are not "naked corals": insights into relationships between Scleractinia and Corallimorpharia from phylogenomic analyses. PeerJ 4, e2463.

Linchangco, G. V., Jr.; Foltz, D. W.; Reid, R.; Williams, J.; Nodzak, C.; Kerr, A. M.; Miller, A. K.; Hunter, R.; Wilson, N. G.; Nielsen, W. J.; et al. 2017. The phylogeny of extant starfish (Asteroidea: Echinodermata) including *Xyloplax*, based on comparative transcriptomics. Molecular Phylogenetics and Evolution 115, 161–170.

Lindberg, D. R. 2009. Monoplacophorans and the origin and relationships of mollusks. Evolution: Education and Outreach 2, 191–203.

Lindberg, D. R.; Ghiselin, M. T. 2003. Fact, theory and tradition in the study of molluscan origins. Proceedings of the California Academy of Sciences 54, 663–686.

Lindberg, D. R.; Sigwart, J. D. 2015. What is the molluscan osphradium? A reconsideration of homology. Zoologischer Anzeiger 256, 14–21.

Lindeque, P. K.; Parry, H. E.; Harmer, R. A.; Somerfield, P. J.; Atkinson, A. 2013. Next generation sequencing reveals the hidden diversity of zooplankton assemblages. PLoS One 8, e81327.

Lindgren, A. R.; Giribet, G.; Nishiguchi, M. K. 2004. A combined approach to the phylogeny of Cephalopoda (Mollusca). Cladistics 20, 454–486.

Lindsay, D. J.; Miyake, H. 2007. A novel benthopelagic ctenophore from 7,217 m depth in the Ryukyu Trench, Japan, with notes on the taxonomy of deep-sea cydippids. Plankton and Benthos Research 2, 98–102.

Linnaeus, C. 1758. Systema naturae per regna tria naturae, secundum classes, ordines, genera, species, cum characteribus, differentiis, synonymis, locis, vol. 1. Laurentii Salvii, Holmiae.

Liscovitch-Brauer, N.; Alon, S.; Porath, H. T.; Elstein, B.; Unger, R.; Ziv, T.; Admon, A.; Levanon, E. Y.; Rosenthal, J. J. C.; Eisenberg, E. 2017. Trade-off between transcriptome plasticity and genome evolution in cephalopods. Cell 169, 191–202.

Littlewood, D. T. J.; Bray, R. A. 2001. Interrelationships of the Platyhelminthes. Taylor and Francis, London.

Littlewood, D. T. J.; Olson, P. D. 2001. Small subunit rDNA and the Platyhelminthes: Signal, noise, conflict and compromise. In: Littlewood, D. T. J.; Bray, R. A. (eds.), Interrelationships of the Platyhelminthes. Taylor and Francis, London, pp. 262–278.

Littlewood, D. T. J.; Rohde, K.; Clough, K. A. 1999. The interrelationships of all major groups of Platyhelminthes: phylogenetic evidence from morphology and molecules. Biological Journal of the Linnean Society 66, 75–114.

Littlewood, D. T. J.; Smith, A. B. 1995. A combined morphological and molecular phylogeny for sea urchins (Echinoidea: Echinodermata). Philosophical Transactions of the Royal Society of London B: Biological Sciences 347, 213–234.

Littlewood, D. T. J.; Smith, A. B.; Clough, K. A.; Emson, R. H. 1997. The interrelationships of the echinoderm classes: morphological and molecular evidence. Biological Journal of the Linnean Society 61, 409–438.

Littlewood, D. T. J.; Telford, M. J.; Clough, K. A.; Rohde, K. 1998. Gnathostomulida—an enigmatic metazoan phylum from both morphological and molecular perspectives. Molecular Phylogenetics and Evolution 9, 72–79.

Liu, A. G.; Matthews, J. J.; Menon, L. R.; McIlroy, D.; Brasier, M. D. 2014a. *Haootia quadriformis* n. gen., n. sp., interpreted as a muscular cnidarian impression from the Late Ediacaran period (approx. 560 Ma). Proceedings of the Royal Society B: Biological Sciences 281, 20141202.

Liu, A. G.; McLlroy, D.; Brasier, M. D. 2010. First evidence for locomotion in the Ediacara biota from the 565 Ma Mistaken Point Formation, Newfoundland. Geology 38, 123–126.

Liu, C.; Li, S.; Huang, J.; Liu, Y.; Jia, G.; Xie, L.; Zhang, R. 2015a. Extensible byssus of *Pinctada fucata*: Ca²⁺-stabilized nanocavities and a thrombospondin-1 protein. Scientific Reports 5, 15018.

Liu, C.; Zhang, Y.; Ren, Y.; Wang, H.; Li, S.; Jiang, F.; Yin, L.; Qiao, X.; Zhang, G.; Qian, W.; et al. 2018. The genome of the golden apple snail *Pomacea canaliculata* provides insight into stress tolerance and invasive adaptation. GigaScience 7, 1–13.

Liu, J.; Ou, Q.; Han, J.; Li, J.; Wu, Y.; Jiao, G.; He, T. 2015b. Lower Cambrian polychaete from China sheds light on early annelid evolution. Science of Nature 102, 34.

Liu, P. J.; Xiao, S. H.; Yin, C. Y.; Chen, S. M.; Zhou, C. M.; Li, M. 2014b. Ediacaran acanthomorphic acritarchs and other microfossils from chert nodules of the Upper Doushantuo Formation in the Yangtze Gorges area, South China. Journal of Paleontology 88, 1–139.

Liu, Y.; Xiao, S.; Shao, T.; Broce, J.; Zhang, H. 2014c. The oldest known priapulid-like scalidophoran animal and its implications for the early evolution of cycloneuralians and ecdysozoans. Evolution and Development 16, 155–165.

Lockyer, A. E.; Olson, P. D.; Littlewood, D. T. J. 2003. Utility of complete large and small subunit rRNA genes in resolving the phylogeny of the Neodermata (Platyhelminthes): implications and a review of the cercomer theory. Biological Journal of the Linnean Society 78, 155–171.

Loesel, R.; Heuer, C. M. 2010. The mushroom bodies—prominent brain centres of arthropods and annelids with enigmatic evolutionary origin. Acta Zoologica 91, 29–34.

Loesel, R.; Nässel, D. R.; Strausfeld, N. J. 2002. Common design in a unique midline neuropil in the brains of arthropods. Arthropod Structure and Development 31, 77–91.

Loesel, R.; Wolf, H.; Kenning, M.; Harzsch, S.; Sombke, A. 2013. Architectural principles and evolution of the arthropod central nervous system. In: Minelli, A.; Boxshall, G.; Fusco, G. (eds.), Arthropod Biology and Evolution: Molecules, Development, Morphology. Springer, Heidelberg, pp. 299–342.

Long, J. A.; Stricker, S. A. 1991. Brachiopoda. Reproduction of marine invertebrates, vol. 6: Echinoderms and lophophorates. Boxwood Press, Pacific Grove, CA, pp. 47–84.

Long, K. A.; Nossa, C. W.; Sewell, M. A.; Putnam, N. H.; Ryan, J. F. 2016. Low coverage sequencing of three echinoderm genomes: the brittle star *Ophionereis fasciata*, the sea star *Patiriella regularis*, and the sea cucumber *Australostichopus mollis*. GigaScience 5, 20.

Lopanik, N.; Lindquist, N.; Targett, N. 2004. Potent cytotoxins produced by a microbial symbiont protect host larvae from predation. Oecologia 139, 131–139.

Lorenzen, S. 1981. Entwurf eines phylogenetischen Systems der freilebenden Nematoden. Veröffentlichungen des Institut für Meeresforschungen Bremerhaven, suppl. 7, 1–472.

Lorenzen, S. 1985. Phylogenetic aspects of pseudocoelomate evolution. In: Conway Morris, S.; George, J. D.; Gibson, R.; Platt, H. M. (eds.), The Origins and Relationships of Lower Invertebrates. Oxford University Press, Oxford, pp. 210–223.

Love, G. D.; Grosjean, E.; Stalvies, C.; Fike, D. A.; Grotzinger, J. P.; Bradley, A. S.; Kelly, A. E.; Bhatia, M.; Meredith, W.; Snape, C. E.; et al. 2009. Fossil steroids record the appearance of Demospongiae during the Cryogenian period. Nature 457, 718–721.

Lowe, C. J. 2008. Molecular genetic insights into deuterostome evolution from the direct-developing hemichordate Saccoglossus kowalevskii. Philosophical Transactions of the Royal Society B: Biological Sciences 363, 1569–1578.

Lowe, C. J.; Clarke, D. N.; Medeiros, D. M.; Rokhsar, D. S.; Gerhart, J. 2015. The deuterostome context of chordate origins. Nature 520, 456–465.

Lowe, C. J.; Terasaki, M.; Wu, M.; Freeman, R. M.; Runft, L.; Kwan, K.; Haigo, S.; Aronowicz, J.; Lander, E.; Gruber, C.; et al. 2006. Dorsoventral patterning in hemichordates: Insights into early chordate evolution. PLoS Biology 4, 1603–1619.

Lowe, C. J.; Wu, M.; Salic, A.; Evans, L.; Lander, E.; Stange-Thomann, N.; Gruber, C. E.; Gerhart, J.; Kirschner, M. 2003. Anteroposterior patterning in hemichordates and the origins of the chordate nervous system. Cell 113, 853–865.

Loxton, J.; Wood, C. A.; Bishop, J. D. D.; Porter, J. S.; Spencer Jones, M.; Nall, C. R. 2017. Distribution of the invasive bryozoan Schizoporella japonica in Great Britain and Ireland and a review of its European distribution. Biological Invasions 19, 2225–2235.

Lozano-Fernandez, J.; Carton, R.; Tanner, A. R.; Puttick, M. N.; Blaxter, M.; Vinther, J.; Olesen, J.; Giribet, G.; Edgecombe, G. D.; Pisani, D. 2016. A molecular palaeobiological exploration of arthropod terrestrialisation. Philosophical Transactions of the Royal Society B: Biological Sciences 371, 20150133.

Lu, T.-M.; Kanda, M.; Satoh, N.; Furuya, H. 2017. The phylogenetic position of dicyemid mesozoans offers insights into spiralian evolution. Zoological Letters 3, 6.

Ludeman, D. A.; Farrar, N.; Riesgo, A.; Paps, J.; Leys, S. P. 2014. Evolutionary origins of sensation in metazoans: functional evidence for a new sensory organ in sponges. BMC Evolutionary Biology 14, 3.

Lundin, K. 1997. Comparative ultrastructure of epidermal ciliary rootlets and associated structures in species of the Nemertodermatida and Acoela (Plathelminthes). Zoomorphology 117, 81–92.

Lundin, K. 1998. The epidermal ciliary rootlets of Xenoturbella bocki (Xenoturbellida) revisited: new support for a possible kinship with the Acoelomorpha (Platyhelminthes). Zoologica Scripta 27, 263–270.

Lundin, K. 1999. Symbiotic bacteria on the epidermis of species of the Nemertodermatida (Platyhelminthes, Acoelomorpha). Acta Zoologica 79, 187–191.

Lundin, K. 2000. Phylogeny of the Nemertodermatida (Acoelomorpha, Platyhelminthes). A cladistic analysis. Zoologica Scripta 29, 65–74.

Lundin, K. 2001. Degenerating epidermal cells in Xenoturbella bocki (phylum uncertain), Nemertodermatida and Acoela (Platyhelminthes). Belgian Journal of Zoology 131, 153–157.

Lundin, K.; Hendelberg, J. 1996. Degenerating epidermal bodies ("pulsatile bodies") in Meara stichopi (Plathelminthes, Nemertodermatida). Zoomorphology 116, 1–5.

Lundin, K.; Hendelberg, J. 1998. Is the sperm type of the Nemertodermatida close to that of the ancestral Platyhelminthes? Hydrobiologia 383, 197–205.

Lundin, K.; Schander, C.; Todt, C. 2009. Ultrastructure of epidermal cilia and ciliary rootlets in Scaphopoda. Journal of Molluscan Studies 75, 69–73.

Luo, H.; Hu, S.; Chen, L. 2001. New Early Cambrian chordates from Haikou, Kunming. Acta Geologica Sinica 75, 345–348.

Luo, Y.-J.; Kanda, M.; Koyanagi, R.; Hisata, K.; Akiyama, T.; Sakamoto, H.; Sakamoto, T.; Satoh, N. 2017. Nemertean and phoronid genomes reveal lophotrochozoan evolution and the origin of bilaterian heads. Nature Ecology and Evolution 2, 141–151.

Luo, Y.-J.; Takeuchi, T.; Koyanagi, R.; Yamada, L.; Kanda, M.; Khalturina, M.; Fujie, M.; Yamasaki, S.-i.; Endo, K.; Satoh, N. 2015a. The Lingula genome provides insights into brachiopod evolution and the origin of phosphate biomineralization. Nature Communications 6, 8301.

Luo, Y. J.; Satoh, N.; Endo, K. 2015b. Mitochondrial gene order variation in the brachiopod Lingula anatina and its implications for mitochondrial evolution in lophotrochozoans. Marine Genomics 24, 31–40.

Lüter, C. 1995. Ultrastructure of the metanephridia of Terebratulina retusa and Crania anomala (Brachiopoda). Zoomorphology 115, 99–107.

Lüter, C. 1996. The median tentacle of the larva of Lingula anatina (Brachiopoda) from Queensland, Australia. Australian Journal of Zoology 44, 355–366.

Lüter, C. 2000a. The origin of the coelom in Brachiopoda and its phylogenetic significance. Zoomorphology 120, 15–28.

Lüter, C. 2000b. Ultrastructure of larval and adult setae of Brachiopoda. Zoologischer Anzeiger 239, 75–90.

Lüter, C. 2007. Anatomy. In: Selden, P. A. (ed.), Treatise on Invertebrate Paleontology, part H: Brachiopoda, rev., vol. 6. The Geological Society of America and University of Kansas Press, Boulder and Lawrence, pp. 2321–2355.

Lüter, C. 2016a. Brachiopoda. In: Schmidt-Rhaesa, A.; Harzsch, S.; Purschke, G. (eds.), Structure and Evolution of Invertebrate Nervous Systems. Oxford University Press, Oxford, pp. 341–350.

Lüter, C. 2016b. Phylum Brachiopoda: the lamp shells. In: Brusca, R. C.; Moore, W.; Shuster, S. M. (eds.), Invertebrates, 3rd ed. Sinauer, Sunderland, MA, pp. 657–665.

Luttrell, S. M.; Glotting, K.; Ross, E.; Sánchez Alvarado, A.; Swalla, B. J. 2016. Head regeneration in hemichordates is not a strict recapitulation of development. Developmental Dynamics 245, 1159–1175.

Lutz, R. A.; Shank, T. M.; Fornari, D. J.; Haymon, R. M.; Lilley, M. D.; Vondamm, K. L.; Desbruyeres, D. 1994. Rapid growth at deep-sea vents. Nature 371, 663–664.

Lydeard, C.; Cummings, K. S. (eds.). 2019. Freshwater Mollusks of the World. Johns Hopkins University Press, Baltimore, MD.

Lyke, E. B.; Reed, C. G.; Woollacott, R. M. 1983. Origin of the cystid epidermis during the metamorphosis of three species of gymnolaemate bryozoans. Zoomorphology 102, 99–110.

Lyons, K. M. 1973. Collar cells in planula and adult tentacle ectoderm of the solitary coral Balanophyllia regia (111 Eupsammiidae). Zeitschrift für Zellforschung und mikroskopische Anatomie 145, 57–74.

Ma, J.-Y.; Buttler, C. J.; Taylor, P. D. 2014a. Cladistic analysis of the 'trepostome' Suborder Esthonioporina and the systematics of Palaeozoic bryozoans. Studi Trentini di Scienze Naturali 94, 153–161.

Ma, J. Y.; Taylor, P. D.; Xia, F. S.; Zhan, R. B. 2015a. The oldest known bryozoan: Prophyllodictya (Cryptostomata) from the lower Tremadocian (Lower Ordovician) of Liujiachang, south-western Hubei, central China. Palaeontology 58, 925–934.

Ma, X.; Aldridge, R. J.; Siveter, D. J.; Siveter, D. J.; Hou, X.; Edgecombe, G. D. 2014b. A new exceptionally preserved Cambrian priapulid from the Chengjiang Lagerstätte. Journal of Paleontology 88, 371–384.

Ma, X.; Edgecombe, G. D.; Hou, X.; Goral, T.; Strausfeld, N. J. 2015b. Preservational pathways of corresponding brains of a Cambrian euarthropod. Current Biology 25, 2969–2975.

Ma, X.; Edgecombe, G. D.; Legg, D. A.; Hou, X. 2014c. The morphology and phylogenetic position of the Cambrian lobopodian Diania cactiformis. Journal of Systematic Palaeontology 12, 445–457.

Ma, X.; Hou, X.; Baines, D. 2010. Phylogeny and evolutionary significance of vermiform animals from the Early Cambrian Chengjiang Lagerstätte. Science China Earth Sciences 53, 1774–1783.

Maas, A. 2013. Gastrotricha, Cycloneuralia and Gnathifera: the fossil record. In: Schmidt-Rhaesa, A. (ed.), Handbook of Zoology. Gastrotricha, Cycloneuralia and Gnathifera, vol. 1: Nematomorpha, Priapulida, Kinorhyncha, Loricifera. De Gruyter, Berlin/Boston, pp. 11–28.

Maas, A.; Huang, D.; Chen, J.; Waloszek, D.; Braun, A. 2007a. Maotianshan-Shale nemathelminths—morphology, biology, and the phylogeny of Nemathelminthes. Palaeogeography, Palaeoclimatology, Palaeoecology 254, 288–306.

Maas, A.; Waloszek, D. 2001. Cambrian derivatives of the early arthropod stem lineage, pentastomids, tardigrades and lobopodians—an 'Orsten' perspective. Zoologischer Anzeiger 240, 451–459.

Maas, A.; Waloszek, D.; Haug, J. T.; Mueller, K. J. 2009. Loricate larvae (Scalidophora) from the Middle Cambrian of Australia. Memoir of the Association of Australasian Palaeontologists 37, 281–302.

Maas, A.; Waloszek, D.; Haug, J. T.; Müller, K. J. 2007b. A possible larval roundworm from the Cambrian 'Orsten' and its bearing on the phylogeny of Cycloneuralia. Memoires of the Association of Australasian Palaeontologists 34, 499–519.

Machner, J.; Scholtz, G. 2010. A scanning electron microscopy study of the embryonic development of Pycnogonum litorale (Arthropoda, Pycnogonida). Journal of Morphology 271, 1306–1318.

Mactavish, T.; Stenton-Dozey, J.; Vopel, K.; Savage, C. 2012. Deposit-feeding sea cucumbers enhance mineralization and nutrient cycling in organically-enriched coastal sediments. PLoS One 7, e50031.

Magie, C. R.; Daly, M.; Martindale, M. Q. 2007. Gastrulation in the cnidarian Nematostella vectensis occurs via invagination not ingression. Developmental Biology 305, 483–497.

Mah, C. L. 2006. A new species of Xyloplax (Echinodermata: Asteroidea: Concentricycloidea) from the northeast Pacific: comparative morphology and a reassessment of phylogeny. Invertebrate Biology 125, 136–153.

Mah, C. L.; Blake, D. B. 2012. Global diversity and phylogeny of the Asteroidea (Echinodermata). PLoS One 7, e35644.

Mah, J. L.; Christensen-Dalsgaard, K. K.; Leys, S. P. 2014. Choanoflagellate and choanocyte collar-flagellar systems and the assumption of homology. Evolution and Development 16, 25–37.

Maida, M.; Coll, J. C.; Sammarco, P. W. 1994. Shedding new light on scleractinian coral recruitment. Journal of Experimental Marine Biology and Ecology 180, 189–202.

Malakhov, V. V. 1994. Nematodes. Smithsonian Institution Press, Washington, D.C.

Malakhov, V. V.; Berezinskaya, T. L.; Solovyev, K. A. 2005. Fine structure of sensory organs in chaetognaths. 1. Ciliary fence receptors, ñiliary tuft receptors and ciliary loop. Invertebrate Zoology 2, 67–77.

Malakhov, V. V.; Temereva, E. N. 1999. Embryonic development of phoronid Phoronis ijimai (Lophophorata, Phoronida): Two sources of coelomic mesoderm. Doklady Akademii Nauk 365, 574–576.

Maldonado, L. L.; Assis, J.; Araújo, F. M.; Salim, A. C.; Macchiaroli, N.; Cucher, M.; Camicia, F.; Fox, A.; Rosenzvit, M.; Oliveira, G.; et al. 2017. The Echinococcus canadensis (G7) genome: a key knowledge of parasitic platyhelminth human diseases. BMC Genomics 18, 204.

Maldonado, M. 2004. Choanoflagellates, choanocytes, and animal multicellularity. Invertebrate Biology 123, 1–22.

Maldonado, M.; Aguilar, R.; Bannister, R. J.; Bell, J. J.; Conway, K. W.; Dayton, P. K.; Díaz, C.; Gutt, J.; Kelly, M.; Kenchington, E. L. R.; et al. 2016. Sponge grounds as key marine habitats: A synthetic review of types, structure, functional roles, and conservation concerns. In: Rossi, S.; Bramanti, L.; Gori, A.; Orejas, C. (eds.), Marine Animal Forests. Springer, Cham, pp. 1–39.

Maldonado, M.; Aguilar, R.; Blanco, J.; García, S.; Serrano, A.; Punzón, A. 2015. Aggregated clumps of lithistid sponges: A singular, reef-like bathyal habitat with relevant paleontological connections. PLoS One 10, e0125378.

Maldonado, M.; Carmona, M. C.; Velásquez, Z.; Puig, A.; Cruzado, A.; López, A.; Young, C. M. 2005. Siliceous sponges as a silicon sink: An overlooked aspect of benthopelagic coupling in the marine silicon cycle. Limnology and Oceanography 50, 799–809.

Maldonado, M.; Riesgo, A.; Bucci, A.; Rützler, K. 2010. Revisiting silicon budgets at a tropical continental shelf: Silica standing stocks in sponges surpass those in diatoms. Limnology and Oceanography 55, 2001–2010.

Maletz, J. 2014a. Graptolite reconstructions and interpretations. Paläontologische Zeitschrift 89, 271–286.

Maletz, J. 2014b. Hemichordata (Pterobranchia, Enteropneusta) and the fossil record. Palaeogeography, Palaeoclimatology, Palaeoecology 398, 16–27.

Mallatt, J.; Craig, C. W.; Yoder, M. J. 2012. Nearly complete rRNA genes from 371 Animalia: Updated structure-based alignment and detailed phylogenetic analysis. Molecular Phylogenetics and Evolution 64, 603–617.

Mallatt, J.; Giribet, G. 2006. Further use of nearly complete 28S and 18S rRNA genes to classify Ecdysozoa: 37 more arthropods and a kinorhynch. Molecular Phylogenetics and Evolution 40, 772–794.

Mallatt, J.; Holland, N. 2013. *Pikaia gracilens* Walcott: Stem chordate, or already specialized in the Cambrian? Journal of Experimental Zoology, pt. B: Molecular and Developmental Evolution 320b, 247–271.

Mamkaev, Y. V.; Kostenko, A. G. 1991. On the phylogenetic significance of sagittocysts and copulatory organs in acoel turbellarians. Hydrobiologia 227, 307–314.

Mángano, M. G.; Buatois, L. A. 2014. Decoupling of body-plan diversification and ecological structuring during the Ediacaran-Cambrian transition: evolutionary and geobiological feedbacks. Proceedings of the Royal Society B: Biological Sciences 281, 20140038.

Manni, L.; Pennati, R. 2016. Tunicata. In: Schmidt-Rhaesa, A.; Harzsch, S.; Purschke, G. (eds.), Structure and Evolution of Invertebrate Nervous Systems. Oxford University Press, Oxford, pp. 699–718.

Manton, S. M. 1972. The evolution of arthropodan locomotory mechanisms, pt. 10. Locomotory habits, morphology and the evolution of the hexapod classes. Zoological Journal of the Linnean Society 51, 203–400.

Manuel, M.; Kruse, M.; Müller, W. E. G.; Le Parco, Y. 2000. The comparison of ß-thymosin homologues among Metazoa supports an arthropod-nematode clade. Journal of Molecular Evolution 51, 378–381.

Manylov, O. G. 1995. Regeneration in Gastrotricha–I. Light microscopical observations on the regeneration in *Turbanella* sp. Acta Zoologica 76, 1–6.

Manylov, O. G.; Vladychenskaya, N. S.; Milyutina, I. A.; Kedrova, O. S.; Korokhov, N. P.; Dvoryanchikov, G. A.; Aleshin, V. V.; Petrov, N. B. 2004. Analysis of 18S rRNA gene sequences suggests significant molecular differences between Macrodasyida and Chaetonotida (Gastrotricha). Molecular Phylogenetics and Evolution 30, 850–854.

Marchioro, T.; Rebecchi, L.; Cesari, M.; Hansen, J. G.; Viotti, G.; Guidetti, R. 2013. Somatic musculature of Tardigrada: phylogenetic signal and metameric patterns. Zoological Journal of the Linnean Society 169, 580–603.

Marcus, E. 1928a. Zur Embryologie der Tardigraden. Verhandlungen der Deutschen zoologischen Gesellschaft 32, 134–146.

Marcus, E. 1928b. Zur vergleichenden Anatomie und Histologie der Tardigraden. Zoologische Jahrbücher, Abteilung für allgemeine Zoologie und Physiologie der Tiere 45, 99–158.

Marcus, E. 1929. Zur Embryologie der Tardigraden. Zoologische Jahrbücher, Abteilung für Anatomie und Ontogenie der Tiere 50, 333–384.

Marcus, E. 1938. Bryozoarios marinhos brasileiros. II. Boletins da Faculdade de Philosophia, Sciencias e Letras, Universidade de São Paulo, Série Zoologia 2, 1–137.

Marcus, E. 1939. Bryozoarios marinhos brasileiros. III. Boletins da Faculdade de Philosophia, Sciencias e Letras, Universidade de São Paulo, Série Zoologia 3, 111–354.

Marcus, E. 1958. On the evolution of the animal phyla. Quarterly Review of Biology 33, 24–58.

Marin, F.; Le Roy, N.; Marie, B. 2012. The formation and mineralization of mollusk shell. Frontiers in Bioscience S4, 1099–1125.

Mark Welch, D. B. 2000. Evidence from a protein-coding gene that acanthocephalans are rotifers. Invertebrate Biology 119, 17–26.

Mark Welch, D. B.; Mark Welch, J. L.; Meselson, M. 2008. Evidence for degenerate tetraploidy in bdelloid rotifers. Proceedings of the National Academy of Sciences of the USA 105, 5145–5149.

Mark Welch, D. B.; Meselson, M. 2000. Evidence for the evolution of bdelloid rotifers without sexual reproduction or genetic exchange. Science 288, 1211–1215.

Märkel, K.; Mackenstedt, U.; Röser, U. 1992. The sphaeridia of sea urchins: ultrastructure and supposed function (Echinodermata, Echinoida). Zoomorphology 112, 1–10.

Marlétaz, F.; Gilles, A.; Caubit, X.; Perez, Y.; Dossat, C.; Samain, S.; Gyapay, G.; Wincker, P.; Le Parco, Y. 2008. Chaetognath transcriptome reveals ancestral and unique features among bilaterians. Genome Biology 9, R94.

Marlétaz, F.; Le Parco, Y.; Liu, S.; Peijnenburg, K. T. C. A. 2017. Extreme mitogenomic variation in natural populations of chaetognaths. Genome Biology and Evolution 9, 1374–1384.

Marlétaz, F.; Martin, E.; Perez, Y.; Papillon, D.; Caubit, X.; Lowe, C. J.; Freeman, B.; Fasano, L.; Dossat, C.; Wincker, P., et al. 2006. Chaetognath phylogenomics: a protostome with deuterostome-like development. Current Biology 16, R577–R578.

Marlétaz, F.; Peijnenburg, K. T. C. A.; Goto, T.; Satoh, N.; Rokhsar, D. S. 2019. A new spiralian phylogeny places the enigmatic arrow worms among gnathiferans. Current Biology 29, 312–318.

Marley, N. J.; McInnes, S. J.; Sands, C. J. 2011. Phylum Tardigrada: A re-evaluation of the Parachela. Zootaxa 2819, 51–64.

Marlow, H. Q.; Srivastava, M.; Matus, D. Q.; Rokhsar, D.; Martindale, M. Q. 2009. Anatomy and development of the nervous system of *Nematostella vectensis*, an anthozoan cnidarian. Developmental Neurobiology 69, 235–254.

Proceedings of the Royal Society B: Biological Sciences 282, 20142396.

Misof, B.; Liu, S.; Meusemann, K.; Peters, R. S.; Donath, A.; Mayer, C.; Frandsen, P. B.; Ware, J.; Flouri, T.; Beutel, R. G.; et al. 2014. Phylogenomics resolves the timing and pattern of insect evolution. Science 346, 763–767.

Mitchell, C. E.; Melchin, M. J.; Cameron, C. B.; Maletz, J. 2013. Phylogenetic analysis reveals that *Rhabdopleura* is an extant graptolite. Lethaia 46, 34–56.

Miyazaki, K. 2002. On the shape of foregut lumen in sea spiders (Arthropoda: Pycnogonida). Journal of the Marine Biological Association of the U.K. 82, 1037–1038.

Miyazawa, H.; Yoshida, M.; Tsuneki, K.; Furuya, H. 2012. Mitochondrial genome of a Japanese placozoan. Zoological Science 29, 223–228.

Møbjerg, N.; Halberg, K. A.; Jørgensen, A.; Persson, D.; Bjørn, M.; Ramløv, H.; Kristensen, R. M. 2011. Survival in extreme environments—on the current knowledge of adaptations in tardigrades. Acta Physiologica 202, 409–420.

Møbjerg, N.; Jørgensen, A.; Eibye-Jacobsen, J.; Agerlich Halberg, K.; Persson, D. K.; Møbjerg Kristensen, R. 2007. New records on cyclomorphosis in the marine eutardigrade *Halobiotus crispae* (Eutardigrada: Hypsibiidae). Journal of Limnology 66, suppl. 1, 132–140.

Møbjerg, N.; Jørgensen, A.; Kristensen, R. M.; Neves, R. C. 2018. Morphology and functional anatomy. In: Schill, R. O. (ed.) Water Bears: The Biology of Tardigrades. Zoological Monographs, vol. 2. Springer Nature Switzerland, Cham, pp. 57-94.

Moens, T.; Braeckman, U.; Derycke, S.; Fonseca, G.; Gallucci, F.; Gingold, R.; Guilini, K.; Ingels, J.; Leduc, D.; Vanaverbeke, J.; et al. 2014. Ecology of free-living marine nematodes. In: Schmidt-Rhaesa, A. (ed.), Handbook of Zoology. Gastrotricha, Cycloneuralia and Gnathifera, vol. 2: Nematoda. De Gruyter, Berlin/Boston, pp. 109–152.

Moller, P. C.; Ellis, R. A. 1974. Fine structure of the excretory system of *Amphioxus (Branchiostoma floridae)* and its response to osmotic stress. Cell and Tissue Research 148, 1–9.

Monge-Nájera, J.; Morera-Brenes, B. 2015. Velvet worms (Onychophora) in folklore and art: Geographic pattern, types of cultural reference and public perception. British Journal of Education, Society and Behavioural Science 10, 1–9.

Monks, S. 2001. Phylogeny of the Acanthocephala based on morphological characters. Systematic Parasitology 48, 81–116.

Monniot, C.; Monniot, F. 1990. Revision of the class Sorberacea (benthic tunicates) with descriptions of seven new species. Zoological Journal of the Linnean Society 99, 239–290.

Monteiro, A. S.; Okamura, B.; Holland, P. W. 2002. Orphan worm finds a home: *Buddenbrockia* is a myxozoan. Molecular Biology and Evolution 19, 968–971.

Monteiro, A. S.; Schierwater, B.; Dellaporta, S. L.; Holland, P. W. 2006. A low diversity of ANTP class homeobox genes in Placozoa. Evolution and Development 8, 174–182.

Montgomery. 1904. The development and structure of the larva of *Paragordius*. Proceedings of the Academy of Natural Sciences of Philadelphia 56, 738–755.

Montgomery, J. C.; Jeffs, A.; Simpson, S. D.; Meekan, M.; Tindle, C. 2006. Sound as an orientation cue for the pelagic larvae of reef fishes and decapod crustaceans. Advances in Marine Biology 51, 143–196.

Monticelli, F. S. 1893. *Treptoplax reptans* n.g., n.sp. Atti dell´ Academia dei Lincei, Rendiconti 5, 39–40.

Mooi, R. 2016. Phylum Echinodermata. In: Brusca, R. C.; Moore, W.; Shuster, S. M. (eds.), Invertebrates, 3rd ed. Sinauer, Sunderland, MA, pp. 968–1003.

Mooi, R.; David, B.; Marchand, D. 1994. Echinoderm skeletal homologies: classical morphology meets modern phylogenetics. In: David, B.; Guille, A.; Féral, J.-P.; Roux, M. (eds.), Echinoderms through Time. Balkema, Rotterdam, pp. 87–95.

Mooi, R.; David, B.; Wray, G. A. 2005. Arrays in rays: terminal addition in echinoderms and its correlation with gene expression. Evolution and Development 7, 542–555.

Moor, B. 1983. Organogenesis. The Mollusca, vol. 3: Development. Academic Press, New York, pp. 123–177.

Moore, J.; Gibson, R.; Jones, H. D. 2001. Terrestrial nemerteans thirty years on. Hydrobiologia 456, 1–6.

Moreno, E.; De Mulder, K.; Salvenmoser, W.; Ladurner, P.; Martinez, P. 2010. Inferring the ancestral function of the posterior Hox gene within the Bilateria: controlling the maintenance of reproductive structures, the musculature and the nervous system in the acoel flatworm *Isodiametra pulchra*. Evolution and Development 12, 258–266.

Moreno, E.; Nadal, M.; Baguñà, J.; Martinez, P. 2009. Tracking the origins of the bilaterian Hox patterning system: insights from the acoel flatworm *Symsagittifera roscoffensis*. Evolution and Development 11, 574–581.

Morera-Brenes, B.; Monge-Nájera, J. 2010. A new giant species of placented worm and the mechanism by which onychophorans weave their nets (Onychophora: Peripatidae). Revista de Biología Tropical 58, 1127–1142.

Morgan, T. H. 1898. Experimental studies of the regeneration of *Planaria maculata*. Archiv für Entwickelungsmechanik der Organismen 7, 364–397.

Moritz, L.; Wesener, T. 2018. *Symphylella patrickmuelleri* sp. nov. (Myriapoda: Symphyla): The oldest known Symphyla and first fossil record of Scolopendrellidae from Cretaceous Burmese amber. Cretaceous Research 84, 258–263.

Morokuma, J.; Durant, F.; Williams, K. B.; Finkelstein, J. M.; Blackiston, D. J.; Clements, T.; Reed, D. W.; Roberts, M.; Jain, M.; Kimel, K.; et al. 2017. Planarian regeneration in space: Persistent anatomical, behavioral, and bacteriological changes induced by space travel. Regeneration 4, 85–102.

Moroz, L. L.; Kocot, K. M.; Citarella, M. R.; Dosung, S.; Norekian, T. P.; Povolotskaya, I. S.; Grigorenko, A. P.; Dailey, C.; Berezikov, E.; Buckley, K. M.; et al. 2014. The ctenophore genome and the evolutionary origins of neural systems. Nature 510, 109–114.

Moroz, L. L.; Kohn, A. B. 2016. Independent origins of neurons and synapses: insights from ctenophores. Philosophical Transactions of the Royal Society B: Biological Sciences 371, 20150041.

Morris, J.; Nallur, R.; Ladurner, P.; Egger, B.; Rieger, R.; Hartenstein, V. 2004. The embryonic development of the flatworm *Macrostomum* sp. Development Genes and Evolution 214, 220–239.

Morris, J. L.; Puttick, M. N.; Clark, J. W.; Edwards, D.; Kenrick, P.; Pressel, S.; Wellman, C. H.; Yang, Z.; Schneider, H.; Donoghue, P. C. J. 2018. The timescale of early land plant evolution. Proceedings of the National Academy of Sciences of the USA 115, E2274–E2283.

Morris, V. B. 2012. Early development of coelomic structures in an echinoderm larva and a similarity with coelomic structures in a chordate embryo. Development Genes and Evolution 222, 313–323.

Morris, V. B.; Selvakumaraswamy, P.; Whan, R.; Byrne, M. 2009. Development of the five primary podia from the coeloms of a sea star larva: homology with the echinoid echinoderms and other deuterostomes. Proceedings of the Royal Society B: Biological Sciences 276, 1277–1284.

Morris, V. B.; Selvakumaraswamy, P.; Whan, R.; Byrne, M. 2011. The coeloms in a late brachiolaria larva of the asterinid sea star *Parvulastra exigua*: deriving an asteroid coelomic model. Acta Zoologica 92, 266–275.

Morse, M. P. 1981. *Meiopriapulus fijiensis* n. gen., n. sp.: An interstitial priapulid from coarse sand in Fiji. Transactions of the American Microscopical Society 100, 239–252.

Mortensen, T. 1910. *Tjalfiella tristoma* n. g., n. sp. A sessile ctenophore from Greenland. Videnskabelige Meddelelser fra den Naturhistoriske Forening i Kjøbenhavn 17, 249–253.

Mortensen, T. 1912. The Danish Ingolf-Expedition: Ctenophora. Videnskabelige Meddelelser fra den Naturhistoriske Forening i Kjøbenhavn 5, 1–95.

Morton, B. 2008. The evolution of eyes in the Bivalvia: new insights. American Malacological Bulletin 26, 35–45.

Moussian, B. 2013. The arthropod cuticle. In: Minelli, A.; Boxshall, G.; Fusco, G. (eds.), Arthropod Biology and Evolution: Molecules, Development, Morphology. Springer, Heidelberg, pp. 171–196.

Moysiuk, J.; Smith, M. R.; Caron, J.-B. 2017. Hyoliths are Palaeozoic lophophorates. Nature 541, 394–397.

Muir, L. A.; Botting, J. P.; Carrera, M. G.; Beresi, M. 2013. Cambrian, Ordovician and Silurian non-stromatoporoid Porifera. Geological Society, London, Memoirs: Early Palaeozoic Biogeography and Palaeogeography 38, 81–95.

Müller, C. H. G.; Harzsch, S.; Perez, Y. 2018. Chaetognatha. In: Schmidt-Rhaesa, A. (ed.), Handbook of Zoology: Miscellaneous Invertebrates. De Gruyter, Berlin/Boston, pp. 163–282.

Müller, C. H. G.; Rieger, V.; Perez, Y.; Harzsch, S. 2014. Immunohistochemical and ultrastructural studies on ciliary sense organs of arrow worms (Chaetognatha). Zoomorphology 133, 167–189.

Müller, C. H. G.; Sombke, A.; Hilken, G.; Rosenberg, J. 2011. Chilopoda—Nervous system. In: Minelli, A. (ed.), Treatise on Zoology—Anatomy, Taxonomy, Biology, vol. 1: The Myriapoda. Brill, Leiden/Boston, pp. 235–278.

Müller, C. H. G.; Sombke, A.; Rosenberg, J. 2007. The fine structure of the eyes of some bristly millipedes (Penicillata, Diplopoda): Additional support for the homology of mandibulate ommatidia. Arthropod Structure and Development 36, 463–476.

Müller, J. 1846. Bericht über einige neue Thierformen der Nordsee. Archiv für Anatomie und Physiologie 13, 101–104.

Müller, K. J. 1977. *Palaeobotryllus* from Upper Cambrian of Nevada—a probable ascidian. Lethaia 10, 106–118.

Müller, K. J.; Walossek, D.; Zakharov, A. 1995. 'Orsten' type phosphatized soft-integument preservation and a new record from the Middle Cambrian Kuonamka Formation in Siberia. Neues Jahrbuch für Geologie und Paläontologie. 197, 101–118.

Müller, M. C. M. 2006. Polychaete nervous systems: Ground pattern and variations—cLS microscopy and the importance of novel characteristics in phylogenetic analysis. Integrative and Comparative Biology 46, 125–133.

Müller, M. C. M.; Schmidt-Rhaesa, A. 2003. Reconstruction of the muscle system in *Antygomonas* sp. (Kinorhyncha, Cyclorhagida) by means of phalloidin labelling and cLSM. Journal of Morphology 256, 103–110.

Müller, M. C. M.; Sterrer, W. 2004. Musculature and nervous system of *Gnathostomula peregrina* (Gnathostomulida) shown by phalloidin labeling, immunohistochemistry, and cLSM, and their phylogenetic significance. Zoomorphology 123, 169–177.

Mulligan, K. L.; Hiebert, T. C.; Jeffery, N. W.; Gregory, T. R. 2014. First estimates of genome size in ribbon worms (phylum Nemertea) using flow cytometry and Feulgen image analysis densitometry. Canadian Journal of Zoology 92, 847–851.

Mullis, K. B.; Faloona, F. A. 1987. Specific synthesis of DNA in vitro via a polymerase-catalyzed chain reaction. Methods in Enzymology 155, 335–350.

Mun, S.; Kim, Y.-J.; Markkandan, K.; Shin, W.; Oh, S.; Woo, J.; Yoo, J.; An, H.; Han, K. 2017. The whole-genome and transcriptome of the Manila clam (*Ruditapes philippinarum*). Genome Biology and Evolution 9, 1487–1498.

Mundy, C. N.; Babcock, R. C. 1998. Role of light intensity and spectral quality in coral settlement: Implications for depth-dependent settlement? Journal of Experimental Marine Biology and Ecology 223, 235–255.

Murdock, D. J.; Bengtson, S.; Marone, F.; Greenwood, J. M.; Donoghue, P. C. 2014. Evaluating scenarios for the evolutionary assembly of the brachiopod body plan. Evolution and Development 16, 13–24.

Murgarella, M.; Puiu, D.; Novoa, B.; Figueras, A.; Posada, D.; Canchaya, C. 2016. A first insight into the genome of the filter-feeder mussel *Mytilus galloprovincialis*. PLoS One 11, e0151561.

Murienne, J.; Daniels, S. R.; Buckley, T. R.; Mayer, G.; Giribet, G. 2014. A living fossil tale of Pangaean biogeography. Proceedings of the Royal Society B: Biological Sciences 281, 20132648.

Muscente, A. D.; Hawkins, A. D.; Xiao, S. H. 2015a. Fossil preservation through phosphatization and silicification in the Ediacaran Doushantuo Formation (South China): a comparative synthesis. Palaeogeography, Palaeoclimatology, Palaeoecology 434, 46–62.

Muscente, A. D.; Michel, F. M.; Dale, J. G.; Xiao, S. H. 2015b. Assessing the veracity of Precambrian 'sponge' fossils using in situ nanoscale analytical techniques. Precambrian Research 263, 142–156.

Muschiol, D.; Traunspurger, W. 2009. Life at the extreme: meiofauna from three unexplored lakes in the caldera of the

the antiquity of the classical cadherin/β-catenin complex. Proceedings of the National Academy of Sciences of the USA 109, 13046–13051.

Nickel, M.; Scheer, C.; Hammel, J. U.; Herzen, J.; Beckmann, F. 2011. The contractile sponge epithelium *sensu lato*—body contraction of the demosponge *Tethya wilhelma* is mediated by the pinacoderm. Journal of Experimental Biology 214, 1692–1698.

Nielsen, C. 1966. On the life-cycle of some Loxosomatidae (Entoprocta). Ophelia 3, 221–247.

Nielsen, C. 1967a. The larvae of *Loxosoma pectinaricola* and *Loxosomella elegans* (Entoprocta). Ophelia 4, 203–206.

Nielsen, C. 1967b. Metamorphosis of the larva of *Loxosomella murmanica* (Nilus) (Entoprocta). Ophelia 4, 85–89.

Nielsen, C. 1971. Entoproct life-cycles and the entoproct/ectoproct relationship. Ophelia 9, 209–341.

Nielsen, C. 1985. Animal phylogeny in the light of the trochaea theory. Biological Journal of the Linnean Society 25, 243–299.

Nielsen, C. 1995. Animal evolution, interrelationships of the living phyla. Oxford University Press, Oxford.

Nielsen, C. 1999. Origin of the chordate central nervous system—and the origin of chordates. Development Genes and Evolution 209, 198–205.

Nielsen, C. 2001. Animal evolution: interrelationships of the living phyla. 2nd ed. Oxford University Press, Oxford.

Nielsen, C. 2002. The phylogenetic position of Entoprocta, Ectoprocta, Phoronida, and Brachiopoda. Integrative and Comparative Biology 42, 685–691.

Nielsen, C. 2003. Proposing a solution to the Articulata-Ecdysozoa controversy. Zoologica Scripta 32, 475–482.

Nielsen, C. 2004. Trochophora larvae: cell-lineages, ciliary bands, and body regions. 1. Annelida and Mollusca. Journal of Experimental Zoology (Mol. Dev. Evol.) 302B, 35–68.

Nielsen, C. 2005. Trochophora larvae: cell-lineages, ciliary bands and body regions. 2. Other groups and general discussion. Journal of Experimental Zoology (Mol. Dev. Evol.) 304B, 401–447.

Nielsen, C. 2008. Six major steps in animal evolution: are we derived sponge larvae? Evolution and Development 10, 241–257.

Nielsen, C. 2010. After all: *Xenoturbella* is an acoelomorph! Evolution and Development 12, 241–243.

Nielsen, C. 2012a. Animal evolution: interrelationships of the living phyla. 3rd ed. Oxford University Press, Oxford.

Nielsen, C. 2012b. How to make a protostome. Invertebrate Systematics 26, 25–40.

Nielsen, C. 2013. The triradiate sucking pharynx in animal phylogeny. Invertebrate Biology 132, 1–13.

Nielsen, C. 2015. Evolution of deuterostomy—and origin of the chordates. Biological Reviews 92, 316–325.

Nielsen, C. 2016a. Phylum Bryozoa: the moss animals. In: Brusca, R. C.; Moore, W.; Shuster, S. M. (eds.), Invertebrates, 3rd ed. Sinauer, Sunderland, MA, pp. 644–657.

Nielsen, C. 2016b. Phylum Entoprocta: The entoprocts. In: Brusca, R. C.; Moore, W.; Shuster, S. M. (eds.), Invertebrates, 3rd ed. Sinauer, Sunderland, MA, pp. 603–609.

Nielsen, C. 2018a. Origin of the trochophora larva. Royal Society Open Science 5, 180042.

Nielsen, C. 2018b. Why not 'Eurystomata'? Bulletin of the International Bryozoology Association 14, 12–13.

Nielsen, C. 2019. Was the ancestral panarthropod mouth ventral or terminal? Arthropod Structure and Development 49, 152–154.

Nielsen, C.; Brunet, T.; Arendt, D. 2018. Evolution of the bilaterian mouth and anus. Nature Ecology and Evolution 2, 1358–1376.

Nielsen, C.; Haszprunar, G.; Ruthensteiner, B.; Wanninger, A. 2007. Early development of the aplacophoran mollusc *Chaetoderma*. Acta Zoologica 88, 231–247.

Nielsen, C.; Hay-Schmidt, A. 2007. Development of the enteropneust *Ptychodera flava*: ciliary bands and nervous system. Journal of Morphology 268, 551–570.

Nielsen, C.; Jespersen, Å. 1997. Entoprocta. In: Harrison, F. W.; Woollacott, R. M. (eds.), Microscopic Anatomy of Invertebrates, vol. 13: Lophophorates, Entoprocta, and Cycliophora. Wiley-Liss, pp. 13–43.

Nielsen, C.; Nørrevang, A. 1985. The trochaea theory: an example of life cycle phylogeny. In: Conway Morris, S.; George, J. D.; Gibson, R., Platt, H. M. (eds.), The Origin and Relationships of Lower Invertebrates. Oxford University Press, Oxford, pp. 28–41.

Nielsen, C.; Rostgaard, J. 1976. Structure and function of an entoproct tentacle with a discussion of ciliary feeding types. Ophelia 15, 115–140.

Nielsen, C.; Scharff, N.; Eibye-Jacobsen, D. 1996. Cladistic analyses of the animal kingdom. Biological Journal of the Linnean Society 57, 385–410.

Nielsen, C., Worsaae, K. 2010. Structure and occurrence of cyphonautes larvae (Bryozoa, Ectoprocta). Journal of Morphology 271, 1094–1109.

Nijhout, H. F. 2013. Arthropod developmental endocrinology. In: Minelli, A.; Boxshall, G.; Fusco, G. (eds.), Arthropod Biology and Evolution: Molecules, Development, Morphology. Springer, Heidelberg, pp. 123–148.

Nikitin, M. 2015. Bioinformatic prediction of *Trichoplax adhaerens* regulatory peptides. General and Comparative Endocrinology 212, 145–155.

Nikulina, E. A.; Hanel, R.; Schäfer, P. 2007. Cryptic speciation and paraphyly in the cosmopolitan bryozoan *Electra pilosa*—impact of the Tethys closing on species evolution. Molecular Phylogenetics and Evolution 45, 765–776.

Nilsson, D.-E.; Gislén, L.; Coates, M. M.; Skogh, C.; Garm, A. 2005. Advanced optics in a jellyfish eye. Nature 435, 201–205.

Nilsson, D. E. 1994. Eyes as optical alarm systems in fan worms and ark clams. Philosophical Transactions of the Royal Society of London B: Biological Sciences 346, 195–212.

Nishiguchi, M. K.; Lopez, J. E.; Boletzky, S. V. 2004. Enlightenment of old ideas from new investigations: more questions regarding the evolution of bacteriogenic light organs in squids. Evolution and Development 6, 41–49.

Nishiguchi, M. K.; Mapes, R. H. 2008. Cephalopoda. In: Ponder, W. F.; Lindberg, D. R. (eds.), Phylogeny and Evolution of the Mollusca. University of California Press, Berkeley, pp. 163–199.

Noguchi, Y.; Endo, K.; Tajima, F.; Ueshima, R. 2000. The mitochondrial genome of the brachiopod *Laqueus rubellus*. Genetics 155, 245–259.

Nohara, M.; Nishida, M.; Miya, M.; Nishikawa, T. 2005. Evolution of the mitochondrial genome in Cephalochordata as inferred from complete nucleotide sequences from two *Epigonichthys* species. Journal of Molecular Evolution 60, 526–537.

Nomaksteinsky, M.; Röttinger, E.; Dufour, H. D.; Chettouh, Z.; Lowe, C. J.; Martindale, M. Q.; Brunet, J.-F. 2009. Centralization of the deuterostome nervous system predates chordates. Current Biology 19, 1264–1269.

Norén, M.; Jondelius, U. 1997. *Xenoturbella*'s molluscan relatives . . . Nature 390, 31–32.

Norenburg, J. 1985. Structure of the nemertine integument with consideration of its ecological and phylogenetic significance. American Zoologist 25, 37–51.

Normark, B. B. 2003. The evolution of alternative genetic systems in insects. Annual Review of Entomology 48, 397–423.

Nørrevang, A. 1964. Choanocytes in the skin of *Harrimania kupfferi* (Enteropneusta). Nature 204, 398–399.

Nørrevang, A.; Wingstrand, K. G. 1970. On the occurrence and structure of choanocyte-like cells in some echinoderms. Acta Zoologica 51, 249–270.

Nosenko, T.; Schreiber, F.; Adamska, M.; Adamski, M.; Eitel, M.; Hammel, J.; Maldonado, M.; Müller, W. E.; Nickel, M.; Schierwater, B.; et al. 2013. Deep metazoan phylogeny: When different genes tell different stories. Molecular Phylogenetics and Evolution 67, 223–233.

Nossa, C. W.; Havlak, P.; Yue, J. X.; Lv, J.; Vincent, K. Y.; Brockmann, H. J.; Putnam, N. H. 2014. Joint assembly and genetic mapping of the Atlantic horseshoe crab genome reveals ancient whole genome duplication. GigaScience 3, 9.

Noto, T.; Endoh, H. 2004. A "chimera" theory on the origin of dicyemid mesozoans: evolution driven by frequent lateral gene transfer from host to parasite. Biosystems 73, 73–83.

Novikova, E. L.; Bakalenko, N. I.; Nesterenko, A. Y.; Kulakova, M. A. 2013. Expression of Hox genes during regeneration of nereid polychaete *Alitta (Nereis) virens* (Annelida, Lophotrochozoa). EvoDevo 4, 14.

Nowak, M. A.; Tarnita, C. E.; Wilson, E. O. 2010. The evolution of eusociality. Nature 466, 1057–1062.

Nowell, R. W.; Almeida, P.; Wilson, C. G.; Smith, T. P.; Fontaneto, D.; Crisp, A.; Micklem, G.; Tunnacliffe, A.; Boschetti, C.; Barraclough, T. G. 2018. Comparative genomics of bdelloid rotifers: Insights from desiccating and nondesiccating species. PLoS Biology 16, e2004830.

Nübler-Jung, K.; Arendt, D. 1999. Dorsoventral axis inversion: Enteropneust anatomy links invertebrates to chordates turned upside down. Journal of Zoological Systematics and Evolutionary Research 37, 93–100.

Nyholm, K. G.; Nyholm, P. G. 1982. Spermatozoa and spermatogenesis in Homalorhagha Kinorhyncha. Journal of Ultrastructure Research 78, 1–12.

O'Donnell, M. J.; George, M. N.; Carrington, E. 2013. Mussel byssus attachment weakened by ocean acidification. Nature Climate Change 3, 587–590.

O'Hara, T. D.; Hugall, A. F.; MacIntosh, H.; Naughton, K. M.; Williams, A.; Moussalli, A. 2016. *Dendrogramma* is a siphonophore. Current Biology 26, R457–R458.

O'Hara, T. D.; Hugall, A. F.; Thuy, B.; Moussalli, A. 2014. Phylogenomic resolution of the class Ophiuroidea unlocks a global microfossil record. Current Biology 24, 1874–1879.

Oakley, T. H.; Wolfe, J. M.; Lindgren, A. R.; Zaharoff, A. K. 2013. Phylotranscriptomics to bring the understudied into the fold: monophyletic Ostracoda, fossil placement, and pancrustacean phylogeny. Molecular Biology and Evolution 30, 215–233.

Obst, M.; Funch, P. 2006. The microhabitat of *Symbion pandora* (Cycliophora) on the mouthparts of its host *Nephrops norvegicus* (Decapoda: Nephropidae). Marine Biology 148, 945–951.

Obst, M.; Funch, P.; Giribet, G. 2005. Hidden diversity and host specificity in cycliophorans: a phylogeographic analysis along the North Atlantic and Mediterranean Sea. Molecular Ecology 14, 4427–4440.

Obst, M.; Funch, P.; Kristensen, R. M. 2006. A new species of Cycliophora from the mouthparts of the American lobster, *Homarus americanus* (Nephropidae, Decapoda). Organisms, Diversity and Evolution 6, 83–97.

Obst, M.; Nakano, H.; Bourlat, S. J.; Thorndyke, M. C.; Telford, M. J.; Nyengaard, J. R.; Funch, P. 2011. Spermatozoon ultrastructure of *Xenoturbella bocki* (Westblad 1949). Acta Zoologica 92, 109–115.

Ochietti, S.; Cailleux, A. 1969. Comparaison des Conodontes et des mâchoires de Gnathostomulides. Comptes rendus de l'Académie des Sciences Paris 268, 2664–2666.

Oelofsen, B. W.; Loock, J. C. 1981. A fossil cephalochordate from the early Permian Whitehill formation of South-Africa. South African Journal of Science 77, 178–180.

Ogino, K.; Tsuneki, K.; Furuya, H. 2010. Unique genome of dicyemid mesozoan: Highly shortened spliceosomal introns in conservative exon/intron structure. Gene 449, 70–76.

Ogino, K.; Tsuneki, K.; Furuya, H. 2011. Distinction of cell types in *Dicyema japonicum* (phylum Dicyemida) by expression patterns of 16 genes. Journal of Parasitology 97, 596–601.

Ohama, T.; Kumazaki, T.; Hori, H.; Osawa, S. 1984. Evolution of multicellular animals as deduced from 5S rRNA sequences: a possible early emergence of the Mesozoa. Nucleic Acids Research 12, 5101–5108.

Oji, T.; Dornbos, S. Q.; Yada, K.; Hasegawa, H.; Gonchigdorj, S.; Mochizuki, T.; Takayanagi, H.; Iryu, Y. 2018. Penetrative trace fossils from the late Ediacaran of Mongolia: early onset of the agronomic revolution. Royal Society Open Science 5, 172250.

Okamura, B.; Curry, A.; Wood, T. S.; Canning, E. U. 2002. Ultrastructure of *Buddenbrockia* identifies it as a myxozoan and verifies the bilaterian origin of the Myxozoa. Parasitology 124, 215–223.

Okamura, B.; Gruhl, A. 2016. Myxozoa + *Polypodium*: A common route to endoparasitism. Trends in Parasitology 32, 268–271.

Okamura, B.; Gruhl, A.; Bartholomew, J. L., eds. 2015. Myxozoan Evolution, Ecology and Development. Springer, Cham.

Okusu, A. 2002. Embryogenesis and development of *Epimenia babai* (Mollusca Aplacophora). Biological Bulletin 203, 87–103.

Okusu, A.; Schwabe, E.; Eernisse, D. J.; Giribet, G. 2003. Towards a phylogeny of chitons (Mollusca, Polyplacophora) based on combined analysis of five molecular loci. Organisms, Diversity and Evolution 3, 281–302.

Oliveira, I. d. S.; Bai, M.; Jahn, H.; Gross, V.; Martin, C.; Hammel, J. U.; Zhang, W.; Mayer, G. 2016. Earliest onychophoran in amber reveals Gondwanan migration patterns. Current Biology 26, 2594–2601.

Oliveira, I. d. S.; Mayer, G. 2013. Apodemes associated with limbs support serial homology of claws and jaws in Onychophora (velvet worms). Journal of Morphology 274, 1180–1190.

Oliveira, I. d. S.; Read, V. M. S. J.; Mayer, G. 2012. A world checklist of Onychophora (velvet worms), with notes on nomenclature and status of names. ZooKeys 211, 1–70.

Oliveira, I. d. S.; Schaffer, S.; Kvartalnov, P. V.; Galoyan, E. A.; Plako, I. V.; Weck-Heimann, A.; Geissler, P.; Ruhberg, H.; Mayer, G. 2013. A new species of *Eoperipatus* (Onychophora) from Vietnam reveals novel morphological characters for the South-East Asian Peripatidae. Zoologischer Anzeiger 252, 495–510.

Oliveira, I. S.; Lacorte, G. A.; Fonseca, C. G.; Wieloch, A. H.; Mayer, G. 2011. Cryptic speciation in Brazilian *Epiperipatus* (Onychophora: Peripatidae) reveals an underestimated diversity among the peripatid velvet worms. PLoS One 6, e19973.

Oliver, P. G.; Taylor, J. D. 2012. Bacterial symbiosis in the Nucinellidae (Bivalvia: Solemyida) with descriptions of two new species. Journal of Molluscan Studies 78, 81–91.

Olson, P. D.; Caira, J. N.; Jensen, K.; Overstreet, R. M.; Palm, H. W.; Beveridge, I. 2010. Evolution of the trypanorhynch tapeworms: Parasite phylogeny supports independent lineages of sharks and rays. International Journal for Parasitology 40, 223–242.

Olson, P. D.; Cribb, T. H.; Tkach, V. V.; Bray, R. A.; Littlewood, D. T. J. 2003. Phylogeny and classification of the Digenea (Platyhelminthes: Trematoda). International Journal for Parasitology 33, 733–755.

Olson, P. D.; Littlewood, D. T. J. 2002. Phylogenetics of the Monogenea—evidence from a medley of molecules. International Journal for Parasitology 32, 233–244.

Olson, P. D.; Littlewood, D. T. J.; Bray, R. A.; Mariaux, J. 2001. Interrelationships and evolution of the tapeworms (Platyhelminthes: Cestoda). Molecular Phylogenetics and Evolution 19, 443–467.

Olson, P. D.; Zarowiecki, M.; Kiss, F.; Brehm, K. 2012. Cestode genomics—progress and prospects for advancing basic and applied aspects of flatworm biology. Parasite Immunology 34, 130–150.

Opperman, C. H.; Bird, D. M.; Williamson, V. M.; Rokhsar, D. S.; Burke, M.; Cohn, J.; Cromer, J.; Diener, S.; Gajan, J.; Graham, S.; et al. 2008. Sequence and genetic map of *Meloidogyne hapla*: A compact nematode genome for plant parasitism. Proceedings of the National Academy of Sciences of the USA 105, 14802–14807.

Orrhage, L.; Müller, M. C. M. 2005. Morphology of the nervous system of Polychaeta (Annelida). Hydrobiologia 535/536, 79–111.

Ortega-Hernández, J. 2015. Lobopodians. Current Biology 25, R873–R875.

Ortega-Hernández, J. 2016. Making sense of 'lower' and 'upper' stem-group Euarthropoda, with comments on the strict use of the name Arthropoda von Siebold, 1848. Biological Reviews 91, 255–273.

Ortega-Hernández, J.; Janssen, R.; Budd, G. E. 2017. Origin and evolution of the panarthropod head—A palaeobiological and developmental perspective. Arthropod Structure and Development 46, 354–379.

Ortega-Hernández, J.; Janssen, R.; Budd, G. E. 2019. The last common ancestor of Ecdysozoa had an adult terminal mouth. Arthropod Structure and Development 49, 155–158.

Osawa, S.; Jukes, T. H.; Watanabe, K.; Muto, A. 1992. Recent evidence for evolution of the genetic code. Microbiological Reviews 56, 229–264.

Osborn, K. J.; Gebruk, A. V.; Rogacheva, A.; Holland, N. D. 2013. An externally brooding acorn worm (Hemichordata, Enteropneusta, Torquaratoridae) from the Russian Arctic. Biological Bulletin 225, 113–123.

Osborn, K. J.; Kuhnz, L. A.; Priede, I. G.; Urata, M.; Gebruk, A. V.; Holland, N. D. 2012. Diversification of acorn worms (Hemichordata, Enteropneusta) revealed in the deep sea. Proceedings of the Royal Society B: Biological Sciences 279, 1646–1654.

Osborn, K. J.; Rouse, G. W. 2008. Multiple origins of pelagicism within Flabelligeridae (Annelida). Molecular Phylogenetics and Evolution 49, 386–392.

Osca, D.; Irisarri, I.; Todt, C.; Grande, C.; Zardoya, R. 2014. The complete mitochondrial genome of *Scutopus ventrolineatus* (Mollusca: Chaetodermomorpha) supports the Aculifera hypothesis. BMC Evolutionary Biology 14, 197.

Osca, D.; Templado, J.; Zardoya, R. 2015. Caenogastropod mitogenomics. Molecular Phylogenetics and Evolution 93, 118–128.

Osigus, H.-J.; Rolfes, S.; Herzog, R.; Kamm, K.; Schierwater, B. 2019. *Polyplacotoma mediterranea* is a new ramified placozoan species. Current Biology 29, R148–R149.

Ostrovsky, A. 2013. Evolution of sexual reproduction in marine invertebrates: example of gymnolaemate bryozoans. Springer, Dordrecht/Heidelberg.

Ostrovsky, A. N. 2008. *External* versus *internal* and *self- cross-*. Fertilization in Bryozoa: transformation of the view and evolutionary considerations. In: Wyse Jackson, P. N.; Spencer Jones, M. E. (eds.), Annals of Bryozoology 2: Aspects of the History of Research on Bryozoans. International Bryozoology Association, Dublin, pp. 103–115.

Ostrovsky, A. N.; Gordon, D. P.; Lidgard, S. 2009. Independent evolution of matrotrophy in the major classes of Bryozoa: transitions among reproductive patterns and their ecological background. Marine Ecology Progress Series 378, 113–124.

Ott, J.; Bright, M.; Bulgheresi, S. 2004. Symbioses between marine nematodes and sulfur-oxidizing chemoautotrophic bacteria. Symbiosis 36, 103–126.

Ou, Q.; Conway Morris, S.; Han, J.; Zhang, Z.; Liu, J.; Chen, A.; Zhang, X.; Shu, D. 2012a. Evidence for gill slits and a pharynx in Cambrian vetulicolians: implications for the early evolution of deuterostomes. BMC Biology 10, 81.

Ou, Q.; Han, J.; Zhang; Z., Shu, D.; Sun, G., Mayer, G. 2017. Three Cambrian fossils assembled into an extinct body plan of cnidarian affinity. Proceedings of the National Academy of Sciences of the USA 114, 8835–8840.

Ou, Q.; Liu, J. N.; Shu, D. G.; Han, J.; Zhang, Z. F.; Wan, X. Q.; Lei, Q. P. 2011. A rare onychophoran-like lobopodian from the Lower Cambrian Chengjiang Lagerstätte, Southwestern China, and its phylogenetic implications. Journal of Paleontology 85, 587–594.

Ou, Q.; Shu, D.; Mayer, G. 2012b. Cambrian lobopodians and extant onychophorans provide new insights into early cephalization in Panarthropoda. Nature Communications 3, 1261.

Ou, Q.; Xiao, S.; Han, J.; Sun, G.; Zhang, F.; Zhang, Z.; Shu, D. 2015. A vanished history of skeletonization in Cambrian comb jellies. Science Advances 1, e1500092.

Owre, H. B. 1973. A new chaetognath genus and species, with remarks on the taxonomy and distribution of others. Bulletin of Marine Science 23, 948–963.

Pabst, E. A.; Kocot, K. M. 2018. Phylogenomics confirms monophyly of Nudipleura (Gastropoda: Heterobranchia). Journal of Molluscan Studies 84, 259–265.

Pacheco, M. L. A. F.; Galante, D.; Rodrigues, F.; Leme Jde, M.; Bidola, P.; Hagadorn, W.; Stockmar, M.; Herzen, J.; Rudnitzki, I. D.; Pfeiffer, F.; et al. 2015. Insights into the skeletonization, lifestyle, and affinity of the unusual Ediacaran fossil Corumbella. PLoS One 10, e0114219.

Pagani, M.; Ricci, C.; Redi, C. A. 1993. Oogenesis in Macrotrachela quadricornifera (Rotifera, Bdelloidea). I. Germarium eutely, karyotype and DNA content. Hydrobiologia 255, 225–230.

Page, L. R. 2006. Early differentiating neuron in larval abalone (Haliotis kamtschatkana) reveals the relationship between ontogenetic torsion and crossing of the pleurovisceral nerve cords. Evolution and Development 8, 458–467.

Paknia, O.; Schierwater, B. 2015. Global habitat suitability and ecological niche separation in the phylum Placozoa. PLoS One 10, e0140162.

Palanisamy, S. K.; Rajendran, N. M.; Marino, A. 2017. Natural products diversity of marine ascidians (Tunicates; Ascidiacea) and successful drugs in clinical development. Natural Products and Bioprospecting 7, 1–111.

Pallas, P. S. 1774. Spicilegia Zoologica, vol 1. Fascicle 10, vol 1. G. A. Lange, Berlin.

Palmberg, I.; Reuter, M. 1992. Sensory receptors in the head of Stenostomum leucops. I. Presumptive photoreceptors. Acta Biologica Hungarica 43, 259–267.

Pang, K.; Ryan, J. F.; Baxevanis, A. D.; Martindale, M. Q. 2011. Evolution of the TGF-ß signaling pathway and its potential role in the ctenophore, Mnemiopsis leidyi. PLoS One 6, e24152.

Pani, A. M.; Mullarkey, E. E.; Aronowicz, J.; Assimacopoulos, S.; Grove, E. A.; Lowe, C. J. 2012. Ancient deuterostome origins of vertebrate brain signalling centres. Nature 483, 289–294.

Papillon, D.; Perez, Y.; Caubit, X.; Le Parco, Y. 2004. Identification of chaetognaths as protostomes is supported by the analysis of their mitochondrial genome. Molecular Biology and Evolution 21, 2122–2129.

Paps, J.; Baguñà, J.; Riutort, M. 2009a. Bilaterian phylogeny: A broad sampling of 13 nuclear genes provides a new Lophotrochozoa phylogeny and supports a paraphyletic basal Acoelomorpha. Molecular Biology and Evolution 26, 2397–2406.

Paps, J.; Baguñà, J.; Riutort, M. 2009b. Lophotrochozoa internal phylogeny: new insights from an up-to-date analysis of nuclear ribosomal genes. Proceedings of the Royal Society B: Biological Sciences 276, 1245–1254.

Paps, J.; Holland, P. W. H. 2018. Reconstruction of the ancestral metazoan genome reveals an increase in genomic novelty. Nature Communications 9, 1730.

Paps, J.; Riutort, M. 2012. Molecular phylogeny of the phylum Gastrotricha: new data brings together molecules and morphology. Mol Phylogenet Evol 63, 208–212.

Pardos, F.; Benito, J. 1988. Ultrastructure of the branchial sacs of Glossobalanus minutus (Enteropneusta) with special reference to podocytes. Archives of Biology 99, 351–363.

Pardos, F.; Kristensen, R. M. 2013. First record of Loricifera from the Iberian Peninsula, with the description of Rugiloricus manuelae sp. nov., (Loricifera, Pliciloricidae). Helgoland Marine Research 67, 623–638.

Park, J.-K.; Kim, K.-H.; Kang, S.; Kim, W.; Eom, K. S.; Littlewood, D. T. 2007. A common origin of complex life cycles in parasitic flatworms: evidence from the complete mitochondrial genome of Microcotyle sebastis (Monogenea: Platyhelminthes). BMC Evolutionary Biology 7, 11.

Park, J.-K.; Rho, H. S.; Kristensen, R. M.; Kim, W.; Giribet, G. 2006. First molecular data on the phylum Loricifera—an investigation into the phylogeny of Ecdysozoa with emphasis on the positions of Loricifera and Priapulida. Zoological Science 23, 943–954.

Park, T.-S.; Kihm, J.-H.; Woo, J.; Park, C.; Lee, W. Y.; Smith, M. P.; Harper, D. A. T.; Young, F.; Nielsen, A. T.; Vinther, J. 2018. Brain and eyes of Kerygmachela reveal protocerebral ancestry of the panarthropod head. Nature Communications 9, 1019.

Park, T.-y.; Woo, J.; Lee, D.-J.; Lee, D.-C.; Lee, S.-b.; Han, Z.; Chough, S. K.; Choi, D. K. 2011. A stem-group cnidarian described from the mid-Cambrian of China and its significance for cnidarian evolution. Nature Communications 2, 442.

Parkefelt, L.; Ekström, P. 2009. Prominent system of RFamide immunoreactive neurons in the rhopalia of box jellyfish (Cnidaria: Cubozoa). Journal of Comparative Neurology 516, 157–165.

Parkhaev, P. Y. 2008. The Early Cambrian radiation of Mollusca. In: Ponder, W. F.; Lindberg, D. R. (eds.), Phylogeny and Evolution of the Mollusca. University of California Press, Berkeley, pp. 33–69.

Parkhaev, P. Y. 2017. Origin and the early evolution of the phylum Mollusca. Paleontological Journal 51, 663–686.

Parry, L.; Tanner, A.; Vinther, J. 2014. The origin of annelids. Palaeontology 57, 1091–1103.

Parry, L.; Vinther, J.; Edgecombe, G. D. 2015. Cambrian stem-group annelids and a metameric origin of the annelid head. Biology Letters 11, 20150763.

Parry, L. A.; Boggiani, P. C.; Condon, D. J.; Garwood, R. J.; Leme, J. d. M.; McIlroy, D.; Brasier, M. D.; Trindade, R.; Campanha, G. A. C.; Pacheco, M. L. A. F.; et al. 2017. Ichnological evidence for meiofaunal bilaterians from the terminal Ediacaran and earliest Cambrian of Brazil. Nature Ecology and Evolution 1, 1455–1464.

Parry, L. A.; Edgecombe, G. D.; Eibye-Jacobsen, D.; Vinther, J. 2016. The impact of fossil data on annelid phylogeny inferred from discrete morphological characters. Proceedings of the Royal Society B: Biological Sciences 283, 20161378.

Parry, L. A.; Eriksson, M. E.; Vinther, J. 2019. The annelid fossil record. In: Purschke, G.; Böggemann, M.; Westheide, W. (eds.), Handbook of Zoology. Annelida, vol. 1. Basal Groups and Pleistoannelida, Sedentaria I. De Gruyter, Berlin/Boston, pp. 69–88.

Pascal, P.-Y.; Bellemare, C.; Sterrer, W.; Boschker, H. T. S.; Gonzalez-Rizzo, S.; Gros, O. 2015. Diet of *Haplognathia ruberrima* (Gnathostomulida) in a Caribbean marine mangrove. Marine Ecology 36, 246–257.

Pass, G. 2000. Accessory pulsatile organs: Evolutionary innovations in insects. Annual Review of Entomology 45, 495–518.

Pass, G.; Tögel, M.; Krenn, H.; Paululat, A. 2015. The circulatory organs of insect wings: Prime examples for the origin of evolutionary novelties. Zoologischer Anzeiger 256, 82–95.

Passamaneck, Y.; Halanych, K. M. 2006. Lophotrochozoan phylogeny assessed with LSU and SSU data: Evidence of lophophorate polyphyly. Molecular Phylogenetics and Evolution 40, 20–28.

Pastrana, C. C.; DeBiasse, M. B.; Ryan, J. F. 2019. Sponges lack ParaHox genes. Genome Biology and Evolution 11, 1250–1257.

Paterson, J. R.; García-Bellido, D. C.; Lee, M. S. Y.; Brock, G. A.; Jago, J. B.; Edgecombe, G. D. 2011. Acute vision in the giant Cambrian predator *Anomalocaris* and the origin of compound eyes. Nature 480, 237–240.

Pawlowski, J.; Montoya Burgos, J. I.; Fahrni, J. F.; Wuest, J.; Zaninetti, L. 1996. Origin of the Mesozoa inferred from 18S rRNA gene sequences. Molecular Biology and Evolution 13, 1128–1132.

Paxton, H. 2005. Molting polychaete jaws—ecdysozoans are not the only molting animals. Evolution and Development 7, 337–340.

Paxton, H. 2009. Phylogeny of Eunicida (Annelida) based on morphology of jaws. Zoosymposia 2, 241–264.

Paxton, H.; Eriksson, M. E. 2012. Ghosts from the past—ancestral features reflected in the jaw ontogeny of the polychaetous annelids *Marphysa fauchaldi* (Eunicidae) and *Diopatra aciculata* (Onuphidae). GFF 134, 309–316.

Payne, J. L.; Heim, N. A.; Knope, M. L.; McClain, C. R. 2014. Metabolic dominance of bivalves predates brachiopod diversity decline by more than 150 million years. Proceedings of the Royal Society B: Biological Sciences 281, 20133122.

Pearman, J. K.; Leray, M.; Villalobos, R.; Machida, R. J.; Berumen, M. L.; Knowlton, N.; Carvalho, S. 2018. Cross-shelf investigation of coral reef cryptic benthic organisms reveals diversity patterns of the hidden majority. Scientific Reports 8, 8090.

Pearse, V. B. 2002. Prodigies of propagation: the many modes of clonal replication in boloceroidid sea anemones (Cnidaria, Anthozoa, Actiniaria). Invertebrate Reproduction and Development 41, 201–213.

Pearse, V. B.; Voigt, O. 2007. Field biology of placozoans (*Trichoplax*): distribution, diversity, biotic interactions. Integrative and Comparative Biology 47, 677–692.

Pechenik, J. A. 2010. Biology of the Invertebrates. 6th ed. McGraw-Hill, Boston.

Pechmann, M.; Khadjeh, S.; Sprenger, F.; Prpic, N.-M. 2010. Patterning mechanisms and morphological diversity of spider appendages and their importance for spider evolution. Arthropod Structure and Development 39, 453–467.

Peck, L. S.; Hayward, P. J.; Spencer-Jones, M. E. 1995. A pelagic bryozoan from Antarctica. Marine Biology 123, 757–762.

Pecoits, E.; Konhauser, K. O.; Aubet, N. R.; Heaman, L. M.; Veroslavsky, G.; Stern, R. A.; Gingras, M. K. 2012. Bilaterian burrows and grazing behavior at >585 million years ago. Science 336, 1693–1696.

Pedersen, K. J.; Pedersen, L. R. 1986. Fine structural observations on the extracellular matrix (ECM) of *Xenoturbella bocki* Westblad, 1949. Acta Zoologica 67, 103–113.

Pedersen, K. J.; Pedersen, L. R. 1988. Ultrastructural observations on the epidermis of *Xenoturbella bocki* Westblad, 1949, with a discusion of epidermal cytoplasmic filament systems of Invertebrates. Acta Zoologica 69, 231–246.

Peel, J. S. 2006. Scaphopodization in Palaeozoic molluscs. Palaeontology 49, 1357–1364.

Peel, J. S. 2010a. Articulated hyoliths and other fossils from the Sirius Passet Lagerstätte (early Cambrian) of North Greenland. Bulletin of Geosciences 85, 385–394.

Peel, J. S. 2010b. A corset-like fossil from the Cambrian Sirius Passet Lagerstätte of North Greenland and its implications for cycloneuralian evolution. Journal of Paleontology 84, 332–340.

Peel, J. S. 2017. A problematic cnidarian (*Cambroctoconus*; Octocorallia?) from the Cambrian (Series 2–3) of Laurentia. Journal of Paleontology 91, 871–882.

Peel, J. S.; Stein, M.; Kristensen, R. M. 2013. Life cycle and morphology of a Cambrian stem-lineage loriciferan. PLoS One 8, e73583.

Pemberton, A. J.; Hansson, L. J.; Craig, S. F.; Hughes, R. N.; Bishop, J. D. D. 2007. Microscale genetic differentiation in a sessile invertebrate with cloned larvae: investigating the role of polyembryony. Marine Biology 153, 71–82.

Peña-Contreras, Z.; Mendoza-Briceno, R. V.; Miranda-Contreras, L.; Palacios-Pru, E. L. 2008. Synaptic dimorphism in onychophoran cephalic ganglia. Revista de Biología Tropical 55, 261–267.

Pennerstorfer, M.; Scholtz, G. 2012. Early cleavage in *Phoronis muelleri* (Phoronida) displays spiral features. Evolution and Development 14, 484–500.

Percival, E. 1944. A contribution to the life-history of the brachiopod *Terebratella inconspicua* Sowerby. Transactions and Proceedings of the Royal Society of New Zealand 74, 1–23.

Perea-Atienza, E.; Gavilan, B.; Chiodin, M.; Abril, J. F.; Hoff, K. J.; Poustka, A. J.; Martinez, P. 2015. The nervous system of Xenacoelomorpha: a genomic perspective. Journal of Experimental Biology 218, 618–628.

Pereira, R. B.; Andrade, P. B.; Valentão, P. 2016. Chemical diversity and biological properties of secondary metabolites from sea hares of *Aplysia* genus. Marine Drugs 14, 39.

Perez, Y.; Rieger, V.; Martin, E.; Müller, C. H. G.; Harzsch, S. 2013. Neurogenesis in an early protostome relative: Progenitor cells in the ventral nerve center of chaetognath hatchlings are arranged in a highly organized geometrical pattern. Journal of Experimental Zoology, pt. B: Molecular and Developmental Evolution 320B, 179–193.

Pérez-Porro, A. R.; Navarro-Gómez, D.; Uriz, M. J.; Giribet, G. 2013. A NGS approach to the encrusting Mediterranean sponge *Crella elegans* (Porifera, Demospongiae, Poecilosclerida): transcriptome sequencing, characterization and overview of the gene expression along three life cycle stages. Molecular Ecology Resources 13, 494–509.

Perseke, M.; Bernhard, D.; Fritzsch, G.; Brümmer, F.; Stadler, P. F.; Schlegel, M. 2010. Mitochondrial genome evolution in Ophiuroidea, Echinoidea, and Holothuroidea: insights in phylogenetic relationships of Echinodermata. Molecular Phylogenetics and Evolution 56, 201–211.

Perseke, M.; Hankeln, T.; Weich, B.; Fritzsch, G.; Stadler, P. F.; Israelsson, O.; Bernhard, D.; Schlegel, M. 2007. The mitochondrial DNA of *Xenoturbella bocki*: genomic architecture and phylogenetic analysis. Theory in Biosciences 126, 35–42.

Perseke, M.; Hetmank, J.; Bernt, M.; Stadler, P. F.; Schlegel, M.; Bernhard, D. 2011. The enigmatic mitochondrial genome of *Rhabdopleura compacta* (Pterobranchia) reveals insights into selection of an efficient tRNA system and supports monophyly of Ambulacraria. BMC Evolutionary Biology 11, 134.

Persson, D.; Halberg, K. A.; Jørgensen, A.; Ricci, C.; Møbjerg, N.; Kristensen, R. M. 2011. Extreme stress tolerance in tardigrades: surviving space conditions in low earth orbit. Journal of Zoological Systematics and Evolutionary Research 49, 90–97.

Persson, D. K.; Halberg, K. A.; Jørgensen, A.; Møbjerg, N.; Kristensen, R. M. 2012. Neuroanatomy of *Halobiotus crispae* (Eutardigrada: Hypsibiidae): Tardigrade brain structure supports the clade Panarthropoda. Journal of Morphology 273, 1227–1245.

Persson, D. K.; Halberg, K. A.; Jørgensen, A.; Møbjerg, N.; Kristensen, R. M. 2013. Brain anatomy of the marine tardigrade *Actinarctus doryphorus* (Arthrotardigrada). Journal of Morphology 275, 173–190.

Peters, S. E.; Gaines, R. R. 2012. Formation of the 'Great Unconformity' as a trigger for the Cambrian explosion. Nature 484, 363–366.

Peters, W. 1972. Occurrence of chitin in Mollusca. Comparative Biochemistry and Physiology 41B, 541–550.

Petersen, C. P.; Reddien, P. W. 2009a. Wnt signaling and the polarity of the primary body axis. Cell 139, 1056–1068.

Petersen, C. P.; Reddien, P. W. 2009b. A wound-induced Wnt expression program controls planarian regeneration polarity. Proceedings of the National Academy of Sciences of the USA 106, 17061–17066.

Petersen, J. K. 2007. Ascidian suspension feeding. Journal of Experimental Marine Biology and Ecology 342, 127–137.

Peterson, K. J.; Cameron, R. A.; Davidson, E. H. 1997. Set-aside cells in maximal indirect development: evolutionary and developmental significance. BioEssays 19, 623–631.

Peterson, K. J.; Cotton, J. A.; Gehling, J. G.; Pisani, D. 2008. The Ediacaran emergence of bilaterians: congruence between the genetic and the geological fossil records. Philosophical Transactions of the Royal Society B: Biological Sciences 363, 1435–1443.

Peterson, K. J.; Eernisse, D. J. 2001. Animal phylogeny and the ancestry of bilaterians: inferences from morphology and 18S rDNA gene sequences. Evolution and Development 3, 170–205.

Petrov, A. A.; Zaitseva, O. V. 2012. Muscle fibers in the central nervous system of nemerteans: Spatial organization and functional role. Journal of Morphology 273, 870–882.

Petrov, N. B.; Aleshin, V. V.; Pegova, A. N.; Ofitserov, M. V.; Slyusarev, G. S. 2010. New insight into the phylogeny of Mesozoa: Evidence from the 18S and 28S rRNA genes. Moscow University Biological Sciences Bulletin 65, 167–169.

Petrov, N. B.; Pegova, A. N.; Manylov, O. G.; Vladychenskaya, N. S.; Mugue, N. S.; Aleshin, V. V. 2007. Molecular phylogeny of Gastrotricha on the basis of a comparison of the 18S rRNA genes: Rejection of the hypothesis of a relationship between Gastrotricha and Nematoda. Molecular Biology 41, 445–452.

Petrov, N. B.; Vladychenskaya, N. S. 2005. Phylogeny of molting protostomes (Ecdysozoa) as inferred from 18S and 28S rRNA gene sequences. Molecular Biology 39, 503–513.

Pett, W.; Adamski, M.; Adamska, M.; Francis, W. R.; Eitel, M.; Pisani, D.; Wörheide, G. 2019. The role of homology and orthology in the phylogenomic analysis of metazoan gene content. Molecular Biology and Evolution 36, 643–649.

Pett, W.; Lavrov, D. V. 2015. Cytonuclear interactions in the evolution of animal mitochondrial tRNA metabolism. Genome Biology and Evolution 7, 2089–2101.

Pett, W.; Ryan, J. F.; Pang, K.; Mullikin, J. C.; Martindale, M. Q.; Baxevanis, A. D.; Lavrov, D. V. 2011. Extreme mitochondrial evolution in the ctenophore *Mnemiopsis leidyi*: Insight from mtDNA and the nuclear genome. Mitochondrial DNA 22, 130–142.

Philip, G. K.; Creevey, C. J.; McInerney, J. O. 2005. The Opisthokonta and the Ecdysozoa may not be clades: stronger support for the grouping of plant and animal than for animal and fungi and stronger support for the Coelomata than Ecdysozoa. Molecular Biology and Evolution 22, 1175–1184.

Philippe, H.; Brinkmann, H.; Copley, R. R.; Moroz, L. L.; Nakano, H.; Poustka, A. J.; Wallberg, A.; Peterson, K. J.; Telford,

M. J. 2011. Acoelomorph flatworms are deuterostomes related to *Xenoturbella*. Nature 470, 255–258.

Philippe, H.; Derelle, R.; Lopez, P.; Pick, K.; Borchiellini, C.; Boury-Esnault, N.; Vacelet, J.; Renard, E.; Houliston, E.; Quéinnec, E.; et al. 2009. Phylogenomics revives traditional views on deep animal relationships. Current Biology 19, 1–17.

Philippe, H.; Lartillot, N.; Brinkmann, H. 2005. Multigene analyses of bilaterian animals corroborate the monophyly of Ecdysozoa, Lophotrochozoa and Protostomia. Molecular Biology and Evolution 22, 1246–1253.

Picciani, N.; Kerlin, J. R.; Sierra, N.; Swafford, A. J. M.; Ramirez, M. D.; Roberts, N. G.; Cannon, J. T.; Daly, M.; Oakley, T. II. 2018. Prolific origination of eyes in Cnidaria with co-option of non-visual opsins. Current Biology 28, 2413–2419.

Pick, K. S.; Philippe, H.; Schreiber, F.; Erpenbeck, D.; Jackson, D. J.; Wrede, P.; Wiens, M.; Alié, A.; Morgenstern, B.; Manuel, M.; et al. 2010. Improved phylogenomic taxon sampling noticeably affects nonbilaterian relationships. Molecular Biology and Evolution 27, 1983–1987.

Pilato, G.; Binda, M. G.; Biondi, O.; D'Urso, V.; Lisi, O.; Marletta, A.; Maugeri, S.; Nobile, V.; Rapazzo, G.; Sabella, G.; et al. 2005. The clade Ecdysozoa, perplexities and questions. Zoologischer Anzeiger 244, 43–50.

Piraino, S.; Boero, F.; Aeschbach, B.; Schmid, V. 1996. Reversing the life cycle: medusae transforming into polyps and cell transdifferentiation in *Turritopsis nutricula* (Cnidaria, Hydrozoa). Biological Bulletin 190, 302–312.

Pisani, D.; Carton, R.; Campbell, L. I.; Akanni, W. A.; Mulville, E.; Rota-Stabelli, O. 2013. An overview of arthropod genomics, mitogenomics, and the evolutionary origins of the arthropod proteome. In: Minelli, A.; Boxshall, G.; Fusco, G. (eds.), Arthropod Biology and Evolution: Molecules, Development, Morphology. Springer, Heidelberg, pp. 41–61.

Pisani, D.; Feuda, R.; Peterson, K. J.; Smith, A. B. 2012. Resolving phylogenetic signal from noise when divergence is rapid: a new look at the old problem of echinoderm class relationships. Molecular Phylogenetics and Evolution 62, 27–34.

Pisani, D.; Pett, W.; Dohrmann, M.; Feuda, R.; Rota-Stabelli, O.; Philippe, H.; Lartillot, N.; Wörheide, G. 2015. Genomic data do not support comb jellies as the sister group to all other animals. Proceedings of the National Academy of Sciences of the USA 112, 15402–15407.

Plachetzki, D. C.; Fong, C. R.; Oakley, T. H. 2012. Cnidocyte discharge is regulated by light and opsin-mediated phototransduction. BMC Biology 10, 17.

Platnick, N. I. 2009. Platnick. In: Knapp, S.; Wheeler, Q. (eds.), Letters to Linnaeus. Linnean Society of London, London, pp. 199–203.

Plazzi, F.; Puccio, G.; Passamonti, M. 2016. Comparative large-scale mitogenomics evidences clade-specific evolutionary trends in mitochondrial DNAs of Bivalvia. Genome Biology and Evolution 8, 2544–2564.

Plazzi, F.; Ribani, A.; Passamonti, M. 2013. The complete mitochondrial genome of *Solemya velum* (Mollusca: Bivalvia) and its relationships with Conchifera. BMC Genomics 14, 409.

Plese, B.; Lukic-Bilela, L.; Bruvo-Madaric, B.; Harcet, M.; Imesek, M.; Bilandzija, H.; Cetkovic, H. 2012. The mitochondrial genome of stygobitic sponge *Eunapius subterraneus*: mtDNA is highly conserved in freshwater sponges. Hydrobiologia 687, 49–59.

Podar, M.; Haddock, S. H. D.; Sogin, M. L.; Harbison, G. R. 2001. A molecular phylogenetic framework for the phylum Ctenophora using 18S rRNA genes. Molecular Phylogenetics and Evolution 21, 218–230.

Podsiadlowski, L.; Braband, A.; Mayer, G. 2008. The complete mitochondrial genome of the onychophoran *Epiperipatus biolleyi* reveals a unique transfer RNA set and provides further support for the Ecdysozoa hypothesis. Molecular Biology and Evolution 25, 42–51.

Podsiadlowski, L.; Mwinyi, A.; Lesný, P.; Bartolomaeus, T. 2014. Mitochondrial gene order in Metazoa—theme and variations. In: Wägele, J. W.; Bartholomaeus, T. (eds.), Deep metazoan phylogeny: The backbone of the tree of life. New insights from analyses of molecules, morphology, and theory of data analysis. De Gruyter, Berlin/Boston, pp. 459–472.

Poinar, G., Jr. 1996. Fossil velvet worms in Baltic and Dominican amber: Onychophoran evolution and biogeography. Science 273, 1370–1371.

Poinar, G., Jr. 1999. *Paleochordodes protus* n.g., n.sp (Nematomorpha, Chordodidae), parasites of a fossil cockroach, with a critical examination of other fossil hairworms and helminths of extant cockroaches (Insecta: Blattaria). Invertebrate Biology 118, 109–115.

Poinar, G., Jr. 2000. Fossil onychophorans from Dominican and Baltic amber: *Tertiapatus dominicanus* n.g., n.sp. (Tertiapatidae n.fam.) and *Succinipatopsis balticus* n.g., n.sp. (Succinipatopsidae n.fam.) with a proposed classification of the subphylum Onychophora. Invertebrate Biology 119, 104–109.

Poinar, G., Jr. 2003a. A rhabdocoel turbellarian (Platyhelminthes, Typhloplanoida) in Baltic amber with a review of fossil and sub-fossil platyhelminths. Invertebrate Biology 122, 308–312.

Poinar, G., Jr. 2003b. First fossil record of nematode parasitism of ants; a 40 million year tale. Parasitology 125, 457–459.

Poinar, G., Jr. 2014a. Evolutionary history of terrestrial pathogens and endoparasites as revealed in fossils and subfossils. Advances in Biology 2014, 1–29.

Poinar, G., Jr. 2014b. Palaeontology of nematodes. In: Schmidt-Rhaesa, A. (ed.), Handbook of Zoology. Gastrotricha, Cycloneuralia and Gnathifera, vol. 2: Nematoda. De Gruyter, Berlin/Boston, pp. 173–178.

Poinar, G., Jr.; Acra, A.; Acra, F. 1994. Earliest fossil nematode (Mermithidae) in Cretaceous Lebanese amber. Fundamental and Applied Nematology 17, 475–477.

Poinar, G., Jr.; Brockerhoff, A. M. 2001. *Nectonema zealandica* n. sp. (Nematomorpha: Nectonematoidea) parasitising the purple rock crab *Hemigrapsus edwardsi* (Brachyura: Decapoda) in New Zealand, with notes on the prevalence of infection and host defence reactions. Systematic Parasitology 50, 149–157.

Poinar, G., Jr.; Buckley, R. 2006. Nematode (Nematoda: Mermithidae) and hairworm (Nematomorpha: Chordodidae) parasites in Early Cretaceous amber. Journal of Invertebrate Pathology 93, 36–41.

Poinar, G., Jr.; Kerp, H.; Hass, H. 2008. *Palaeonema phyticum* gen. n., sp n. (Nematoda : Palaeonematidae fam. n.), a Devonian nematode associated with early land plants. Nematology 10, 9–14.

Poinar, G., Jr.; Ricci, C. 1992. Bdelloid rotifers in Dominican amber: Evidence for parthenogenetic continuity. Experientia 48, 408–410.

Pojeta, J., Jr. 2000. Cambrian Pelecypoda (Mollusca). American Malacological Bulletin 15; 157–166.

Pojeta, J., Jr.; Runnegar, B.; Morris, N. J.; Newell, N. D. 1972. Rostroconchia: a new class of bivalved mollusks. Science 177, 264–267.

Pollock, L. W. 1975. Tardigrada. In: Giese, A. C.; Pearse, J. S. (eds.), Reproduction of Marine Invertebrates, vol. 2: Entoprocts and Lesser Coelomates. Academic Press, London, pp. 43–54.

Polz, M. F.; Felbeck, H.; Novak, R.; Nebelsick, M.; Ott, J. A. 1992. Chemoautotrophic, sulfur-oxidizing symbiotic bacteria on marine nematodes: morphological and biochemical characterization. Microbial Ecology 24, 313–329.

Ponder, W. F.; Lindberg, D. R. 1997. Towards a phylogeny of gastropod molluscs: an analysis using morphological characters. Zoological Journal of the Linnean Society 119, 83–265.

Ponder, W. F.; Lindberg, D. R. 2008a. Molluscan evolution and phylogeny: An introduction. In: Ponder, W. F.; Lindberg, D. R. (eds.), Phylogeny and Evolution of the Mollusca. University of California Press, Berkeley, pp. 1–17.

Ponder, W. F.; Lindberg, D. R. (eds.). 2008b. Phylogeny and Evolution of the Mollusca. University of California Press, Berkeley.

Pontin, R. M. 1964. A comparative account of the protonephridia of *Asplanchna* (Rotifera) with special reference to the flame bulbs. Journal of Zoology, 511–525.

Ponton, F.; Lebarbenchon, C.; Lefèvre, T.; Biron, D. G.; Duneau, D.; Hughes, D. P.; Thomas, F. 2006. Parasite survives predation on its host. Nature 440, 756.

Popiel, I. 1986. The reproductive biology of schistosomes. Parasitology Today 2, 10–15.

Popova, O. V.; Mikhailov, K. V.; Nikitin, M. A.; Logacheva, M. D.; Penin, A. A.; Muntyan, M. S.; Kedrova, O. S.; Petrov, N. B.; Panchin, Y. V.; Aleoshin, V. V. 2016. Mitochondrial genomes of Kinorhyncha: *trnM* duplication and new gene orders within animals. PLoS One 11, e0165072.

Por, F. D.; Bromley, H. J. 1974. Morphology and anatomy of *Maccabeus tentaculatus* (Priapulida : Seticoronaria). Journal of Zoology 173, 173–197.

Porter, S. M. 2004. Halkieriids in Middle Cambrian phosphatic limestones from Australia. Journal of Paleontology 78, 574–590.

Porter, S. M. 2008. Skeletal microstructure indicates chancelloriids and halkieriids are closely related. Palaeontology 51, 865–879.

Pouchkina-Stantcheva, N. N.; McGee, B. M.; Boschetti, C.; Tolleter, D.; Chakrabortee, S.; Popova, A. V.; Meersman, F.; Macherel, D.; Hincha, D. K.; Tunnacliffe, A. 2007. Functional divergence of former alleles in an ancient asexual invertebrate. Science 318, 268–271.

Powell, D.; Subramanian, S.; Suwansa-Ard, S.; Zhao, M.; O'Connor, W.; Raftos, D.; Elizur, A. 2018. The genome of the oyster *Saccostrea* offers insight into the environmental resilience of bivalves. DNA Research 25, 655–665.

Pray, F. A. 1965. Studies on the early development of the rotifer *Monostyla cornuta* Müller. Transactions of the American Microscopical Society 84, 210–216.

Prendini, L. 2001. Species or supraspecific taxa as terminals in cladistic analysis? Groundplans versus exemplars revisited. Systematic Biology 50, 290–300.

Presnell, J. S.; Vandepas, L. E.; Warren, K. J.; Swalla, B. J.; Ameniya, C. T.; Browne, W. E. 2016. The presence of a functionally tripartite through-gut in Ctenophora has implications for metazoan character trait evolution. Current Biology 26, 2814–2820.

Proestou, D. A.; Goldsmith, M. R.; Twombly, S. 2008. Patterns of male reproductive success in *Crepidula fornicata* provide new insight for sex allocation and optimal sex change. Biological Bulletin 214, 194–202.

Prokop, J.; Nel, A.; Hoch, I. 2005. Discovery of the oldest known Pterygota in the Lower Carboniferous of the Upper Silesian Basin in the Czech Republic (Insecta: Archaeorthoptera). Geobios 38, 383–387.

Protasio, A. V.; Tsai; I. J., Babbage, A.; Nichol, S.; Hunt, M.; Aslett, M. A.; De Silva, N.; Velarde, G. S.; Anderson, T. J.; Clark, R. C.; et al. 2012. A systematically improved high quality genome and transcriptome of the human blood fluke *Schistosoma mansoni*. PLoS Neglected Tropical Diseases 6, e1455.

Purcell, J. E. 2012. Jellyfish and ctenophore blooms coincide with human proliferations and environmental perturbations. Annual Review of Marine Science 4, 209–235.

Purcell, J. E.; Milisenda, G.; Rizzo, A.; Carrion, S. A.; Zampardi, S.; Airoldi, S.; Zagami, G.; Guglielmo, L.; Boero, F.; Doyle, T. K.; et al. 2015. Digestion and predation rates of zooplankton by the pleustonic hydrozoan *Velella velella* and widespread blooms in 2013 and 2014. Journal of Plankton Research 37, 1056–1067.

Purchon, R. D. 1960. Phylogeny in the Lamellibranchia. In: Purchon, R. D. (ed.), Proceedings of the Centenary and Bicentenary Congress of Biology, Singapore, Dec. 1958. University of Malaya Press, Singapore, pp. 69–82.

Purschke, G. 1997. Ultrastructure of nuchal organs in polychaetes (Annelida)—new results and review. Acta Zoologica 78, 123–143.

Purschke, G. 2003. Is *Hrabeiella periglandulata* (Annelida, "Polychaeta") the sister group of Clitellata? Evidence from an ultrastructural analysis of the dorsal pharynx in *H. periglandulata* and *Enchytraeus minutus* (Annelida, Clitellata). Zoomorphology 122, 55–66.

Purschke, G.; Arendt, D.; Hausen, H.; Müller, M. C. 2006. Photoreceptor cells and eyes in Annelida. Arthropod Structure and Development 35, 211–230.

Purschke, G.; Bleidorn, C.; Struck, T. 2014. Systematics, evolution and phylogeny of Annelida—a morphological perspective. Memoirs of Museum Victoria 71, 247–269.

Purschke, G.; Hessling, R.; Westheide, W. 2000. The phylogenetic position of the Clitellata and the Echiura—on the problematic assessment of absent characters. Journal of

Zoological Systematics and Evolutionary Research 38, 165–173.

Purschke, G.; Müller, M. C. 2006. Evolution of body wall musculature. Integrative and Comparative Biology 46, 497–507.

Purschke, G.; Westheide, W.; Rohde, D.; Brinkhurst, R. O. 1993. Morphological reinvestigation and phylogenetic relationship of *Acanthobdella peledina* (Annelida, Clitellata). Zoomorphology 113, 91–101.

Purschke, G.; Wolfrath, F.; Westheide, W. 1997. Ultrastructure of the nucal organ and cerebral organ in *Onchnesoma squamatum* (Sipuncula, Phascolionidae). Zoomorphology 117, 23–31.

Purschke, G. n. 2016. Annelida: Basal groups and Pleistoannelida. In: Schmidt-Rhaesa, A.; Harzsch, S.; Purschke, G. (eds.), Structure and Evolution of Invertebrate Nervous Systems. Oxford University Press, Oxford, pp. 254–312.

Putnam, N. H.; Butts, T.; Ferrier, D. E.; Furlong, R. F.; Hellsten, U.; Kawashima, T.; Robinson-Rechavi, M.; Shoguchi, E.; Terry, A.; Yu, J. K.; et al. 2008. The amphioxus genome and the evolution of the chordate karyotype. Nature 453, 1064–1071.

Putnam, N. H.; Srivastava, M.; Hellsten, U.; Dirks, B.; Chapman, J.; Salamov, A.; Terry, A.; Shapiro, H.; Lindquist, E.; Kapitonov, V. V.; et al. 2007. Sea anemone genome reveals ancestral eumetazoan gene repertoire and genomic organization. Science 317, 86–94.

Quattrini, A. M.; Faircloth, B. C.; Dueñas, L. F.; Bridge, T. C. L.; Brugler, M. R.; Calixto-Botía, I. F.; DeLeo, D. M.; Forêt, S.; Herrera, S.; Lee, S. M. Y.; et al. 2017. Universal target-enrichment baits for anthozoan (Cnidaria) phylogenomics: New approaches to long-standing problems. Molecular Ecology Resources 18, 281–295.

Raff, R. A.; Field, K. G.; Olsen, G. J.; Giovannoni, S. J.; Lane, D. J.; Ghiselin, M. T.; Pace, N. R.; Raff, E. C. 1989. Metazoan phylogeny based on analysis of 18S ribosomal RNA. In: Fernhölm, B.; Bremer, K.; Jörnvall, H. (eds.), The hierarchy of life. Molecules and morphology in phylogenetic analysis. Excerpta Medica, Amsterdam/New York/Oxford, pp. 247–261.

Rahm, G. 1937a. Eine neue Tardigraden-Ordnung aus den heissen Quellen von Unzen, Insel Kyushu, Japan. Zoologischer Anzeiger 120, 65–71.

Rahm, G. 1937b. A new ordo of tardigrades from the hot springs of Japan (Furu-Yu Section, Unzen). Annotationes Zoologicae Japonenses 16, 345–352.

Rahman, I. A.; Zamora, S.; Falkingham, P. L.; Phillips, J. C. 2015. Cambrian cinctan echinoderms shed light on feeding in the ancestral deuterostome. Proceedings of the Royal Society B: Biological Sciences 282, 20151964.

Rähr, H. 1979. The circulatory system of amphioxus (*Branchiostoma lanceolatum* [Pallas]): A light-microscopic investigation based on intravascular injection technique. Acta Zoologica 60, 1–18.

Raikova, E. V. 1994. Life cycle, cytology, and morphology of *Polypodium hydriforme*, a coelenterate parasite of the eggs of acipenseriform fishes. Journal of Parasitology 80, 1–22.

Raikova, E. V.; Ibragimov, A. Y.; Raikova, O. I. 2007. Muscular system of a peculiar parasitic cnidarian *Polypodium hydriforme*: A phalloidin fluorescence study. Tissue and Cell 39, 79–87.

Raikova, O. I.; Flyatchinskaya, L. P.; Justine, J.-L. 1998a. Acoel spermatozoa: ultrastructure and immunocytochemistry of tubulin. Hydrobiologia 383, 207–214.

Raikova, O. I.; Reuter, M.; Gustafsson, M. K. S.; Maule, A. G.; Halton, D. W.; Jondelius, U. 2004a. Basiepidermal nervous system in *Nemertoderma westbladi* (Nemertodermatida): GYIRFamide immunoreactivity. Zoology (Jena) 107, 75–86.

Raikova, O. I.; Reuter, M.; Gustafsson, M. K. S.; Maule, A. G.; Halton, D. W.; Jondelius, U. 2004b. Evolution of the nervous system in *Paraphanostoma* (Acoela). Zoologica Scripta 33, 71–88.

Raikova, O. I.; Reuter, M.; Jondelius, U.; Gustafsson, K. S. 2000a. The brain of the Nemertodermatida (Platyhelminthes) as revealed by anti-FMRFamide immunostainings. Tissue and Cell 32, 358–365.

Raikova, O. I.; Reuter, M.; Jondelius, U.; Gustafsson, K. S. 2000b. An immunocytochemical and ultrastructural study of the nervous and muscular systems of *Xenoturbella westbladi* (Bilateria inc. sed.). Zoomorphology 120, 107–118.

Raikova, O. I.; Reuter, M.; Kotikova, E. A.; Gustafsson, M. K. S. 1998b. A commissural brain! The pattern of 5-HT immunoreactivity in Acoela (Plathelminthes). Zoomorphology 118, 69–77.

Raikova, O. I.; Tekle, Y. I.; Reuter, M.; Gustafsson, M. K. S.; Jondelius, U. 2006. Copulatory organ musculature in *Childia* (Acoela) as revealed by phalloidin fluorescence and confocal microscopy. Tissue and Cell 38, 219–232.

Ramazzotti, G. 1962. Il phylum Tardigrada. Memorie dell'Istituto italiano di idrobiologia dott. Marco De Marchi 14, 1–595.

Ramm, S. A. 2017. Exploring the sexual diversity of flatworms: Ecology, evolution, and the molecular biology of reproduction. Molecular Reproduction and Development 84, 120–131.

Ramsköld, L.; Chen, Y.-J. 1998. Cambrian lobopodians: Morphology and phylogeny. In: Edgecombe, G. D. (ed.), Arthropod Fossils and Phylogeny. Columbia University Press, New York, pp. 107–150.

Rao, T. R.; Sarma, S. S. S. 1985. Mictic and amictic modes of reproduction in the rotifer *Brachionus patulus* Mueller. Current Science 54, 499–501.

Reardon, W.; Chakrabortee, S.; Pereira, T. C.; Tyson, T.; Banton, M. C.; Dolan, K. M.; Culleton, B. A.; Wise, M. J.; Burnell, A. M.; Tunnacliffe, A. 2010. Expression profiling and cross-species RNA interference (RNAi) of desiccation-induced transcripts in the anhydrobiotic nematode *Aphelenchus avenae*. BMC Molecular Biology 11, 6.

Rebecchi, L.; Altiero, T.; Guidetti, R.; Cesari, M.; Bertolani, R.; Negroni, M.; Rizzo, A. M. 2009. Tardigrade resistance to space effects: first results of experiments on the LIFE-TARSE mission on FOTON-M3 (September 2007). Astrobiology 9, 581–591.

Reddien, P. W. 2013. Specialized progenitors and regeneration. Development 140, 951–957.

Reddien, P. W. 2018. The cellular and molecular basis for planarian regeneration. Cell 175, 327–345.

Redl, E.; Scherholz, M.; Todt, C.; Wollesen, T.; Wanninger, A. 2014. Development of the nervous system in Solenogastres (Mollusca) reveals putative ancestral spiralian features. EvoDevo 5, 48.

Reed, C. G. 1991. Bryozoa. Reproduction of Marine Invertebrates, vol. 6: Echinoderms and Lophophorates. Boxwood Press, Pacific Grove, CA, pp. 85–245.

Reger, J. F. 1969. Studies on the fine structure of muscle fibres and contained crystalloids in basal socket muscle of the entoproct, *Barentsia gracilis*. Journal of Cell Science 4, 305–325.

Regier, J. C.; Shultz, J. W.; Zwick, A.; Hussey, A.; Ball, B.; Wetzer, R.; Martin, J. W.; Cunningham, C. W. 2010. Arthropod relationships revealed by phylogenomic analysis of nuclear protein-coding sequences. Nature 463, 1079–1083.

Rehkämper, G.; Storch, V.; Alberti, G.; Welsch, U. 1989. On the fine structure of the nervous system of *Tubiluchus philippinensis* (Tubiluchidae, Priapulida). Acta Zoologica 70, 111–120.

Rehkämper, G.; Welsch, U.; Dilly, P. N. 1987. Fine structure of the ganglion of *Cephalodiscus gracilis* (Pterobranchia, Hemichordata). Journal of Comparative Neurology 259, 308–315.

Rehm, P.; Meusemann, K.; Borner, J.; Misof, B.; Burmester, T. 2014. Phylogenetic position of Myriapoda revealed by 454 transcriptome sequencing. Molecular Phylogenetics and Evolution 77, 25–33.

Rehm, P.; Pick, C.; Borner, J.; Markl, J.; Burmester, T. 2012. The diversity and evolution of chelicerate hemocyanins. BMC Evolutionary Biology 12, 19.

Reich, A.; Dunn, C.; Akasaka, K.; Wessel, G. 2015. Phylogenomic analyses of Echinodermata support the sister groups of Asterozoa and Echinozoa. PLoS One 10, e0119627.

Reise, K. 1981. Gnathostomulida abundant alongside polychaete burrows. Marine Ecology Progress Series 6, 329–333.

Reisinger, E. 1924. Die Gattung Rhynchoscolex. Zeitschrift für Morphologie und Ökologie der Tiere 1, 1–37.

Reisinger, E. 1925. Untersuchungen am Nervensystem der Bothrioplana semperi Braun. (Zugleich ein Beitrag zur Technik der Vitalen Nervenfärbung und zur Vergleichenden Anatomie des Plathelminthennervensystems). Zeitschrift für Morphologie und Ökologie der Tiere 5, 119–149.

Reisinger, E. 1960. Was ist *Xenoturbella*? Zeitschrift für Wissenschaftliche Zoologie 164, 188–198.

Reisinger, E. 1961. Allgemeine Morphologie der Metazoa. Fortschritte der Zoologie 13, 1–82.

Reiswig, H. M. 1971. Particle feeding in natural populations of three marine demosponges. Biological Bulletin 141, 568–591.

Remane, A. 1933. Aschelminthen I. Buch: Rotatorien, Gastrotrichen und Kinorhynchen (continued). Bronn's Klassen und Ordnungen des Tier-Reichs, Leipzig, pp. 449–516.

Remane, A. 1936. Gastrotricha. In: Bronn, H. G. (ed.), Klassen und Ordnungen des Tierreiches. Gastrotricha und Kinorhyncha, vol. 4: Vermes II. Abteilung Askelminthes, Trochelminthes. Akademische Verlagsgesellschaft, Leipzig, pp. 1–242.

Remane, A. 1963. The evolution of the Metazoa from colonial flagellates *vs.* plasmodial ciliates. In: Dougherty, C. E. (ed.), The Lower Metazoa. Comparative Biology and Phylogeny. California University Press, Berkeley, pp. 23–32.

Retallack, G. J. 2013. Comment on "Trace fossil evidence for Ediacaran bilaterian animals with complex behaviors" by Chen et al. (Precambrian Research 224 [2013] 690–701). Precambrian Research 231, 383–385.

Reuter, M.; Raikova, O. I.; Gustafsson, M. K. 1998. An endocrine brain? The pattern of FMRF-amide immunoreactivity in Acoela (Plathelminthes). Tissue and Cell 30, 57–63.

Rex, M. A.; McClain, C. R.; Johnson, N. A.; Etter, R. J.; Allen, J. A.; Bouchet, P.; Warén, A. 2005. A source-sink hypothesis for abyssal biodiversity. American Naturalist 165, 163–178.

Reynolds, P. D.; Okusu, A. 1999. Phylogenetic relationships among families of the Scaphopoda (Mollusca). Zoological Journal of the Linnean Society 126, 131–153.

Ribeiro, A. R.; Barbaglio, A.; Benedetto, C. D.; Ribeiro, C. C.; Wilkie, I. C.; Carnevali, M. D.; Barbosa, M. A. 2011. New insights into mutable collagenous tissue: correlations between the microstructure and mechanical state of a sea-urchin ligament. PLoS One 6, e24822.

Ribes, M.; Coma, R., Gili, J.-M. 1999. Natural diet and grazing rate of the temperate sponge *Dysidea avara* (Demospongiae, Dendroceratida) throughout an annual cycle. Marine Ecology Progress Series 176, 179–190.

Ricci, C. 2016. Bdelloid rotifers: 'sleeping beauties' and 'evolutionary scandals', but not only. Hydrobiologia 796, 277–285.

Richards, G. S.; Simionato, E.; Perron, M.; Adamska, M.; Vervoort, M.; Degnan, B. M. 2008. Sponge genes provide new insight into the evolutionary origin of the neurogenic circuit. Current Biology 18, 1156–1161.

Richmond, R. H. 1985. Reversible metamorphosis in coral planula larvae. Marine Ecology Progress Series 22, 181–185.

Richter, D. J.; Fozouni, P.; Eisen, M. B.; King, N. 2018. Gene family innovation, conservation and loss on the animal stem lineage. eLife 7, e34226.

Richter, S.; Loesel, R.; Purschke, G.; Schmidt-Rhaesa, A.; Scholtz, G.; Stach, T.; Vogt, L.; Wanninger, A.; Brenneis, G.; Doring, C.; et al. 2010. Invertebrate neurophylogeny: suggested terms and definitions for a neuroanatomical glossary. Frontiers in Zoology 7, 29.

Richter, S.; Stein, M.; Frase, T.; Szucsich, N. U. 2013. The arthropod head. In: Minelli, A.; Boxshall, G.; Fusco, G. (eds.), Arthropod Biology and Evolution: Molecules, Development, Morphology. Springer, Heidelberg, pp. 223–240.

Ridgway, I. D.; Richardson, C. A. 2011. *Arctica islandica*: the longest lived non colonial animal known to science. Reviews in Fish Biology and Fisheries 21, 297–310.

Riedl, R. 1959. Beiträge zur Kenntnis der *Rhodope veranii*, Teil I. Geschichte und Biologie. Zoologischer Anzeiger 163, 107–122.

Riedl, R. J. 1969. Gnathostomulida from America. Science 163, 445–452.

Rieger, G. E.; Rieger, R. M. 1977. Comparative fine structure study of the gastrotrich cuticle and aspects of cuticle evolution within the Aschelminthes. Zeitschrift für zoologische Systematik und Evolutionsforschung 15, 81–124.

Rieger, R. M. 1976. Monociliated epidermal cells in Gastrotricha: significance for concepts of early metazoan evolution. Zeitschrift für zoologische Systematik und Evolutionsforschung 14, 198–226.

Rieger, R. M. 1980. A new group of interstitial annelids (Lobatocerebridae, nov. fam.) and its significance for metazoan phylogeny. Zoomorphology 95, 41–84.

Rieger, R. M. 1981. Morphology of the Turbellaria at the ultrastructural level. Hydrobiologia 84, 213–229.

Rieger, R. M. 1984. Evolution of the cuticle in the lower Eumetazoa. In: Bereiter-Hahn, J.; Matoltsy, A. G.; Richards, K. S. (eds.), Biology of the Integument, vol. 1: Invertebrates. Springer-Verlag, Berlin, pp. 389–399.

Rieger, R. M. 1986. Asexual reproduction and the turbellarian archetype. Hydrobiologia 132, 35–45.

Rieger, R. M. 1991a. *Jennaria pulchra*, nov. gen. nov. spec., eine den psammobionten Anneliden nahestehende Gattung aus dem Kustengrundwasser von North Carolina. Berichte des Naturwissenschaftlich-Medizinischen Vereins in Innsbruck 78, 203–215.

Rieger, R. M. 1991b. Neue Organisationstypen aus der Sandlückenraumfauna: die Lobatocerebriden und *Jennaria pulchra*. Verhandlungen der Deutschen zoologischen Gesellschaft 84, 247–259.

Rieger, R. M.; Legniti, A.; Ladurner, P.; Reiter, D.; Asch, E.; Salvenmoser, W.; Schürmann, W.; Peter, R. 1999. Ultrastructure of neoblasts in microturbellaria: Significance for understanding stem cells in free-living Platyhelminthes. Invertebrate Reproduction and Development 35, 127–140.

Rieger, R. M.; Mainitz, M. 1977. Comparative fine structural study of the body wall in Gnathostomulida and their phylogenetic position between Platyhelminthes and Aschelminthes. Zeitschrift für zoologische Systematik und Evolutionsforschung 15, 9–35.

Rieger, R. M.; Purschke, G. 2005. The coelom and the origin of the annelid body plan. Hydrobiologia 535/536, 127–137.

Rieger, R. M.; Ruppert, E.; Rieger, G. E.; Schoepfer-Sterrer, C. 1974. On the fine structure of gastrotrichs with description of *Chordodasys antennatus* sp.n. Zoologica Scripta 3, 219–237.

Rieger, R. M.; Sterrer, W. 1975a. New spicular skeletons in Turbellaria, and the occurrence of spicules in marine meiofauna. Zeitschrift für zoologische Systematik und Evolutionsforschung 13, 207–278.

Rieger, R. M.; Sterrer, W. 1975b. New spicular skeletons in Turbellaria, and the occurrence of spicules in marine meiofauna. II. Zeitschrift für zoologische Systematik und Evolutionsforschung 13, 249–278.

Rieger, R. M.; Tyler, S. 1995. Sister-group relationship of Gnathostomulida and Rotifera-Acanthocephala. Invertebrate Biology 114, 186–188.

Rieger, R. M.; Tyler, S.; Smith, J. P. S., III; Rieger, G. E. 1991. Platyhelminthes: Turbellaria. In: Harrison, F. W.; Bogitsh, B. J. (eds.), Microscopic Anatomy of Invertebrates, vol. 3: Platyhelminthes and Nemertinea. Wiley-Liss, New York, pp. 7–140.

Rieger, V.; Perez, Y.; Müller, C. H. G.; Lacalli, T.; Hansson, B. S.; Harzsch, S. 2011. Development of the nervous system in hatchlings of *Spadella cephaloptera* (Chaetognatha), and implications for nervous system evolution in Bilateria. Development, Growth and Differentiation 53, 740–759.

Rieger, V.; Perez, Y.; Müller, C. H. G.; Lipke, E.; Sombke, A.; Hansson, B. S.; Harzsch, S. 2010. Immunohistochemical analysis and 3D reconstruction of the cephalic nervous system in Chaetognatha: insights into the evolution of an early bilaterian brain? Invertebrate Biology 129, 77–104.

Riemann, O.; Ahlrichs, W. H. 2010. The evolution of the protonephridial terminal organ across Rotifera with particular emphasis on *Dicranophorus forcipatus*, *Encentrum mucronatum* and *Erignatha clastopis* (Rotifera: Dicranophoridae). Acta Zoologica 91, 199–211.

Riemann, O.; Kieneke, A. 2008. First record of males of *Encentrum mucronatum* Wulfert, 1936 and *Encentrum martes* Wulfert, 1939 (Rotifera: Dicranophoridae) including notes on males across Rotifera Monogononta. Zootaxa, 63–68.

Riesgo, A. 2010. Phagocytosis of sperm by follicle cells of the carnivorous sponge *Asbestopluma occidentalis* (Porifera, Demospongiae). Tissue and Cell 42, 198–201.

Riesgo, A.; Andrade, S. C. S.; Sharma, P. P.; Novo, M.; Pérez-Porro, A. R.; Vahtera, V.; González, V. L.; Kawauchi, G. Y.; Giribet, G. 2012. Comparative description of ten transcriptomes of newly sequenced invertebrates and efficiency estimation of genomic sampling in non-model taxa. Frontiers in Zoology 9, 33.

Riesgo, A.; Farrar, N.; Windsor, P. J.; Giribet, G.; Leys, S. P. 2014a. The analysis of eight transcriptomes from all Porifera classes reveals surprising genetic complexity in sponges. Molecular Biology and Evolution 31, 1102–1120.

Riesgo, A.; Maldonado, M. 2009. An unexpectedly sophisticated, V-shaped spermatozoon in Demospongiae (Porifera): reproductive and evolutionary implications. Biological Journal of the Linnean Society 97, 413–426.

Riesgo, A.; Maldonado, M.; López-Legentil, S.; Giribet, G. 2015. A proposal for the evolution of cathepsin and silicatein in sponges. Journal of Molecular Evolution 80, 278–291.

Riesgo, A.; Novo, M.; Sharma, P. P.; Peterson, M.; Maldonado, M.; Giribet, G. 2014b. Inferring the ancestral sexuality and reproductive condition in sponges (Porifera). Zoologica Scripta 43, 101–117.

Riesgo, A.; Taylor, C.; Leys, S. P. 2007. Reproduction in a carnivorous sponge: the significance of the absence of an aquiferous system to the sponge body plan. Evolution and Development 9, 618–631.

Riisgård, H. U. 2002. Methods of ciliary filter feeding in adult *Phoronis muelleri* (phylum Phoronida) and in its free-swimming actinotroch larva. Marine Biology 141, 75–87.

Rimskaya-Korsakova, N. N.; Kristof, A.; Malakhov, V. V.; Wanninger, A. 2016. Neural architecture of *Galathowenia oculata* Zach, 1923 (Oweniidae, Annelida). Frontiers in Zoology 13, 5.

Ringrose, J. H.; van den Toorn, H. W.; Eitel, M.; Post, H.; Neerincx, P.; Schierwater, B.; Altelaar, A. F.; Heck, A. J. 2013. Deep proteome profiling of *Trichoplax adhaerens* reveals remarkable features at the origin of metazoan multicellularity. Nature Communications 4, 1408.

Rink, J. C.; Vu, H. T.-K.; Sánchez Alvarado, A. 2011. The maintenance and regeneration of the planarian excretory system are regulated by EGFR signaling. Development 138, 3769–3780.

Riser, N. W. 1988. *Arhynchonemertes axi* gen. n., sp. n. (Nemertinea)—an insight into basic acoelomate bilaterian organology. Forschritte der Zoologie 36, 367–373.

Riutort, M.; Field, K. G.; Raff, R. A.; Baguñà, J. 1993. 18S rRNA sequences and phylogeny of Platyhelminthes. Biochemical Systematics and Ecology 21, 71–77.

Riutort, M.; Field, K. G.; Turbeville, J. M.; Raff, R. A.; Baguñà, J. 1992. Enzyme electrophoresis, 18S rRNA sequences, and levels of phylogenetic resolution among several species of freshwater planarians (Platyhelminthes, Tricladida, Paludicola). Canadian Journal of Zoology 70, 1425–1439.

Rizzo, A. M.; Altiero, T.; Corsetto, P. A.; Montorfano, G.; Guidetti, R.; Rebecchi, L. 2015. Space flight effects on antioxidant molecules in dry tardigrades: The TARDIKISS experiment. Biomed Research International 2015, 167642.

Robertson, H. E.; Lapraz, F.; Egger, B.; Telford, M. J.; Schiffer, P. H. 2017. The mitochondrial genomes of the acoelomorph worms *Paratomella rubra*, *Isodiametra pulchra* and *Archaphanostoma ylvae*. Scientific Reports 7, 1847.

Robilliard, G. A.; Dayton, P. K. 1972. A new species of platyctenean ctenophore, *Lyrocteis flavopallidus* sp. nov., from McMurdo Sound, Antarctica. Canadian Journal of Zoology 50, 47–52.

Rodríguez, E.; Barbeitos, M.; Daly, M.; Gusmão, L. C.; Häussermann, V. 2012. Toward a natural classification: phylogeny of acontiate sea anemones (Cnidaria, Anthozoa, Actiniaria). Cladistics 28, 375–392.

Rodriguez-Pascual, F.; Slatter, D. A. 2016. Collagen cross-linking: insights on the evolution of metazoan extracellular matrix. Scientific Reports 6, 37374.

Rogozin, I. B.; Wolf, Y. I.; Carmel, L.; Koonin, E. V. 2007. Ecdysozoan clade rejected by genome-wide analysis of rare amino acid replacements. Molecular Biology and Evolution 24, 1080–1090.

Rohde, K. 1991. The evolution of protonephridia of the Platyhelminthes. Hydrobiologia 227, 315–321.

Rohde, K. 2001. Protonephridia as phylogenetic characters. In: Littlewood, D. T. J.; Bray, R. A. (eds.), Interrelationships of the Platyhelminthes. Taylor and Francis, London, pp. 203–216.

Rohde, K.; Watson, N. A. 1998. The terminal protonephridial complex of *Haplopharynx rostratus* (Platyhelminthes, Haplopharyngida). Acta Zoologica 79, 329–333.

Roper, C. F. E.; Shea, E. K. 2013. Unanswered questions about the giant squid *Architeuthis* (Architeuthidae) illustrate our incomplete knowledge of coleoid cephalopods. American Malacological Bulletin 31, 109–122.

Rosa, M. T.; Oliveira, D. S.; Loreto, E. L. S. 2017. Characterization of the first mitochondrial genome of a catenulid flatworm: *Stenostomum leucops* (Platyhelminthes). Journal of Zoological Systematics and Evolutionary Research 55, 98–105.

Rosen, B. R. 2000. Algal symbiosis, and the collapse and recovery of reef communities: Lazarus corals across the K-T boundary. In: Culver, S. J.; Rawson, P. F. (eds.), Biotic Response to Global Change: The Last 145 Million Years. Cambridge University Press, Cambridge, pp. 164–180.

Rosenberg, J.; Müller, C. H. G.; Hilken, G. 2011. Chilopoda—Endocryne system. In: Minelli, A. (ed.), Treatise on Zoology—Anatomy, Taxonomy, Biology. The Myriapoda, vol. 1. Brill, Leiden/Boston, pp. 197–215.

Rosengarten, R. D.; Sperling, E. A.; Moreno, M. A.; Leys, S. P.; Dellaporta, S. L. 2008. The mitochondrial genome of the hexactinellid sponge *Aphrocallistes vastus*: Evidence for programmed translational frameshifting. BMC Genomics 9, 33.

Rota-Stabelli, O.; Campbell, L.; Brinkmann, H.; Edgecombe, G. D.; Longhorn, S. J.; Peterson, K. J.; Pisani, D.; Philippe, H.; Telford, M. J. 2011. A congruent solution to arthropod phylogeny: phylogenomics, microRNAs and morphology support monophyletic Mandibulata. Proceedings of the Royal Society B: Biological Sciences 278, 298–306.

Rota-Stabelli, O.; Daley, A. C.; Pisani, D. 2013. Molecular timetrees reveal a Cambrian colonization of land and a new scenario for ecdysozoan evolution. Current Biology 23, 392–398.

Rota-Stabelli, O.; Kayal, E.; Gleeson, D.; Daub, J.; Boore, J. L.; Telford, M. J.; Pisani, D.; Blaxter, M.; Lavrov, D. V. 2010. Ecdysozoan mitogenomics: Evidence for a common origin of the legged invertebrates, the Panarthropoda. Genome Biology and Evolution 2, 425–440.

Rothe, B. H.; Schmidt-Rhaesa, A. 2008. Variation in the nervous system in three species of the genus *Turbanella* (Gastrotricha, Macrodasyida). Meiofauna Marina 16, 175–184.

Rothe, B. H.; Schmidt-Rhaesa, A. 2009. Architecture of the nervous system in two *Dactylopodola* species (Gastrotricha, Macrodasyida). Zoomorphology 128, 227–246.

Rothe, B. H.; Schmidt-Rhaesa, A. 2010. Structure of the nervous system in *Tubiluchus troglodytes* (Priapulida). Invertebrate Biology 129, 39–58.

Rothe, B. H.; Schmidt-Rhaesa, A.; Kieneke, A. 2011. The nervous system of *Neodasys chaetonotoideus* (Gastrotricha: Neodasys) revealed by combining confocal laser-scanning and transmission electron microscopy: evolutionary comparison of neuroanatomy within the Gastrotricha and basal Protostomia. Zoomorphology 130, 51–84.

Rothe, B. H.; Schmidt-Rhaesa, A.; Todaro, M. A. 2006. The general muscular architecture in *Tubiluchus troglodytes* (Priapulida). Meiofauna Marina 15, 79–86.

Röttinger, E.; Lowe, C. J. 2012. Evolutionary crossroads in developmental biology: hemichordates. Development 139, 2463–2475.

Rouse, G. W. 1999a. Bias? What bias? The evolution of downstream larval-feeding in animals. Zoologica Scripta 29, 213–236.

Rouse, G. W. 1999b. Trochophore concepts: ciliary bands and the evolution of larvae in spiralian Metazoa. Biological Journal of the Linnean Society 66, 411–464.

Rouse, G. W. 2005. Annelid sperm and fertilization biology. Hydrobiologia 535/536, 167–178.

Rouse, G. W. 2016. Phylum Annelida. The segmented (and some unsegmented) worms. In: Brusca, R. C.; Moore, W.; Shuster, S. M. (eds.), Invertebrates, 3rd ed. Sinauer, Sunderland, MA, pp. 531–602.

Rouse, G. W.; Fauchald, K. 1997. Cladistics and polychaetes. Zoologica Scripta 26, 139–204.

Rouse, G. W.; Goffredi, S. K.; Vrijenhoek, R. C. 2004. Osedax: bone-eating marine worms with dwarf males. Science 305, 668–671.

Rouse, G. W.; Jermiin, L. S.; Wilson, N. G.; Eeckhaut, I.; Lanterbecq, D.; Oji, T.; Young, C. M.; Browning, T.; Cisternas, P.; Helgen, L. E.; et al. 2013. Fixed, free, and fixed: the fickle phylogeny of extant Crinoidea (Echinodermata) and their Permian-Triassic origin. Molecular Phylogenetics and Evolution 66, 161–181.

Rouse, G. W.; Wilson, N. G.; Carvajal, J. I.; Vrijenhoek, R. C. 2016. New deep-sea species of Xenoturbella and the position of Xenacoelomorpha. Nature 530, 94–97.

Rouse, G. W.; Wilson, N. G.; Worsaae, K.; Vrijenhoek, R. C. 2015. A dwarf male reversal in bone-eating worms. Current Biology 25, 236–241.

Rouse, G. W.; Worsaae, K.; Johnson, S. B.; Jones, W. J.; Vrijenhoek, R. C. 2008. Acquisition of dwarf male "Harems" by recently settled females of Osedax roseus n. sp. (Siboglinidae; Annelida). Biological Bulletin 214, 67–82.

Rowe, F. W. E.; Baker, A. N.; Clark, H. E. S. 1988. The morphology, development and taxonomic status of Xyloplax Baker, Rowe, and Clark (1986) (Echinodermata: Concentricycloidea), with the description of a new species. Proceedings of the Royal Society of London B: Biological Sciences 223, 431–439.

Roy, K.; Jablonski, D.; Valentine, J. W.; Rosenberg, G. 1998. Marine latitudinal diversity gradients: Tests of causal hypotheses. Proceedings of the National Academy of Sciences of the USA 95, 3699–3702.

Royuela, M.; Astier, C.; Grandier-Vazeille, X.; Benyamin, Y.; Fraile, B.; Paniagua, R.; Duvert, M. 2003. Immunohistochemistry of chaetognath body wall muscles. Invertebrate Biology 122, 74–82.

Rubinstein, N. D.; Feldstein, T.; Shenkar, N.; Botero-Castro, F.; Griggio, F.; Mastrototaro, F.; Delsuc, F.; Douzery, E. J. P.; Gissi, C.; Huchon, D. 2013. Deep sequencing of mixed total DNA without barcodes allows efficient assembly of highly plastic ascidian mitochondrial genomes. Genome Biology and Evolution 5, 1185–1199.

Ruhberg, H.; Daniels, S. R. 2013. Morphological assessment supports the recognition of four novel species in the widely distributed velvet worm Peripatopsis moseleyi sensu lato (Onychophora : Peripatopsidae). Invertebrate Systematics 27, 131–145.

Ruiz-Ramos, D. V.; Weil, E.; Schizas, N. V. 2014. Morphological and genetic evaluation of the hydrocoral Millepora species complex in the Caribbean. Zoological Studies 53, 4.

Ruiz-Trillo, I.; Burger, G.; Holland, P. W. H.; King, N.; Lang, B. F.; Roger, A. J.; Gray, M. W. 2007. The origins of multicellularity: a multi-taxon genome initiative. Trends in Genetics 23, 113–118.

Ruiz-Trillo, I.; Paps, J. 2015. Acoelomorpha: earliest branching bilaterians or deuterostomes? Organisms Diversity and Evolution 16, 391–399.

Ruiz-Trillo, I.; Paps, J.; Loukota, M.; Ribera, C.; Jondelius, U.; Baguñà, J.; Riutort, M. 2002. A phylogenetic analysis of myosin heavy chain type II sequences corroborates that Acoela and Nemertodermatida are basal bilaterians. Proceedings of the National Academy of Sciences of the USA 99, 11246–11251.

Ruiz-Trillo, I.; Riutort, M.; Fourcade, H. M.; Baguñà, J.; Boore, J. L. 2004. Mitochondrial genome data support the basal position of Acoelomorpha and the polyphyly of the Platyhelminthes. Molecular Phylogenetics and Evolution 33, 321–332.

Ruiz-Trillo, I.; Riutort, M.; Littlewood, D. T. J.; Herniou, E. A.; Baguñà, J. 1999. Acoel flatworms: earliest extant bilaterian metazoans, not members of Platyhelminthes. Science 283, 1919–1923.

Runnegar, B. 1981. Muscle scars, shell form and torsion in Cambrian and Ordovician univalved molluscs. Lethaia 14, 311–322.

Ruppert, E. E. 1978. The reproductive system of gastrotrichs. II. Insemination in Macrodasys: a unique mode of sperm transfer in Metazoa. Zoomorphologie 89, 207–228.

Ruppert, E. E. 1982. Comparative ultrastructure of the gastrotrich pharynx and the evolution of myoepithelial foreguts in Aschelminthes. Zoomorphology 99, 181–200.

Ruppert, E. E. 1991a. Gastrotricha. In: Harrison, F. W.; Ruppert, E. E. (eds.), Microscopic Anatomy of Invertebrates, vol. 4: Aschelminthes. Wiley-Liss, New York, pp. 41–109.

Ruppert, E. E. 1991b. Introduction to the aschelminth phyla: A consideration of mesoderm, body cavities, and cuticle. In: Harrison, F. W.; Ruppert, E. E. (eds.), Microscopic Anatomy of Invertebrates, vol. 4: Aschelminthes. Wiley-Liss, New York, pp. 1–17.

Ruppert, E. E. 1996. Morphology of Hatschek's nephridium in larval and juvenile stages of Branchiostoma virginiae (Cephalochordata). Israel Journal of Zoology 42, S161–S182.

Ruppert, E. E.; Balser, E. J. 1986. Nephridia in the larvae of hemichordates and echinoderms. Biological Bulletin 171, 188–196.

Ruppert, E. E.; Fox, R. S.; Barnes, R. D. 2004. Invertebrate zoology, a functional evolutionary approach. Thomson, Brooks/Cole, Belmont, CA.

Ruppert, E. E.; Travis, P. B. 1983. Hemoglobin-containing cells of Neodasys (Gastrotricha, Chaetonotida). I. Morphology and unltrstructure. Journal of Morphology 175, 57–64.

Rusin, L. Y.; Malakhov, V. V. 1998. Free-living marine nematodes possess no euteLy. Doklady Akademii Nauk 361, 132–134.

Ruthensteiner, B. 1997. Homology of the pallial and pulmonary cavity of gastropods. Journal of Molluscan Studies 63, 353–367.

Ryan, J. F.; Pang, K.; Program, N. C. S.; Mullikin, J. C.; Martindale, M. Q.; Baxevanis, A. D. 2010. The homeodomain complement of the ctenophore *Mnemiopsis leidyi* suggests that Ctenophora and Porifera diverged prior to the Para-Hoxozoa. EvoDevo 1, 9.

Ryan, J. F.; Pang, K.; Schnitzler, C. E.; Nguyen, A. D.; Moreland, R. T.; Simmons, D. K.; Koch, B. J.; Francis, W. R.; Havlak, P.; Smith, S. A.; et al. 2013. The genome of the ctenophore *Mnemiopsis leidyi* and its implications for cell type evolution. Science 342, 1242592.

Ryan, K.; Lu, Z.; Meinertzhagen, I. A. 2016. The CNS connectome of a tadpole larva of *Ciona intestinalis* (L.) highlights sidedness in the brain of a chordate sibling. eLife 5, e16962.

Ryan, T. J.; Grant, S. G. 2009. The origin and evolution of synapses. Nature Reviews Neuroscience 10, 701–712.

Sacks, M. 1955. Observations on the embryology of an aquatic gastrotrich, *Lepidodermella squamata* (Dujardin, 1841). Journal of Morphology 96, 473–495.

Saiki, R. K.; Gelfand, D. H.; Stoffel, S.; Scharf, S. J.; Higuchi, R.; Horn, G. T.; Mullis, K. B.; Erlich, H. A. 1988. Primer-directed enzymatic amplification of DNA with a thermostable DNA polymerase. Science 239, 487–491.

Saiz, J. I.; Bustamante, M.; Tajadura, J. 2016. A census of deep-water sipunculans (Sipuncula). Marine Biodiversity 48, 449–464.

Sakai, M.; Sakaizumi, M. 2012. The complete mitochondrial genome of *Dugesia japonica* (Platyhelminthes; order Tricladida). Zoological Science 29, 672–680.

Saldarriaga, J. F.; Voss-Foucart, M.-F.; Compère, P.; Goffinet, G.; Storch, V.; Jeuniaux, C. 1995. Quantitative estimation of chitin and proteins in the cuticle of five species of Priapulida. Sarsia 80, 67–71.

Salinas-Saavedra, M.; Rock, A. Q.; Martindale, M. Q. 2018. Germ layer-specific regulation of cell polarity and adhesion gives insight into the evolution of mesoderm. eLife 7, e36740.

Salvini-Plawen, L. v. 1972. Cnidaria as food-sources for marine invertebrates. Cahiers de Biologie Marine 13, 385–400.

Salvini-Plawen, L. v. 1980. Was ist eine Trochophora? Eine Analyse der Larventypen mariner Protostomier. Zoologisches Jahrbücher, Abteilung Anatomie 103, 389–423.

Salvini-Plawen, L. v. 1986. Systematic notes on *Spadella* and on the Chaetognatha in general. Zeitschrift für zoologische Systematik und Evolutionsforschung 24, 122–128.

Salvini-Plawen, L. v.; Bartolomaeus, T. 1995. Mollusca: Mesenchymata with a "coelom." In: Lanzavecchia, G.; Valvassori, R.; Candia Carnevali, M. D. (eds.), Body cavities: function and phylogeny. Selected Symposia and Monographs U.Z.I., Mucchi, Modena, pp. 75–92.

Salvini-Plawen, L. v.; Steiner, G. 1996. Synapomorphies and plesiomorphies in higher classification of Mollusca. In: Taylor, J. D. (ed.), Origin and evolutionary radiation of the Mollusca. Oxford University Press, Oxford, pp. 29–51.

Samadi, L.; Steiner, G. 2010. Expression of *Hox* genes during the larval development of the snail, *Gibbula varia* (L.)—further evidence of non-colinearity in molluscs. Development Genes and Evolution 220, 161–172.

Samanta, M. P.; Tongprasit, W.; Istrail, S.; Cameron, R. A.; Tu, Q.; Davidson, E. H.; Stolc, V. 2006. The transcriptome of the sea urchin embryo. Science 314, 960–962.

Sampaio-Costa, C.; Chagas-Junior, A.; Baptista, R. L. C. 2009. Brazilian species of Onychophora with notes on their taxonomy and distribution. Zoologia 26, 553–561.

Sánchez Alvarado, A. 2003. The freshwater planarian *Schmidtea mediterranea*: embryogenesis, stem cells and regeneration. Current Opinion in Genetics and Development 13, 438–444.

Sánchez, N.; Yamasaki, H.; Pardos, F.; Sørensen, M. V.; Martínez, A. 2016. Morphology disentangles the systematics of a ubiquitous but elusive meiofaunal group (Kinorhyncha: Pycnophyidae). Cladistics 32, 479–505.

Sanger, F.; Nicklen, S.; Coulsen, A. R. 1977. DNA sequencing with chain terminating inhibitors. Proceedings of the National Academy of Sciences of the USA 74, 5463–5468.

Sanmarco, P. W. 1982. Polyp bail-out: An escape response to environmental stress and a new means of reproduction in corals. Marine Ecology Progress Series 10, 57–65.

Sansom, R. S.; Gabbott, S. E.; Purnell, M. A. 2010. Non-random decay of chordate characters causes bias in fossil interpretation. Nature 463, 797–800.

Santagata, S. 2002. Structure and metamorphic remodeling of the larval nervous system and musculature of *Phoronis pallida* (Phoronida). Evolution and Development 4, 28–42.

Santagata, S. 2008a. Evolutionary and structural diversification of the larval nervous system among marine bryozoans. Biological Bulletin 215, 3–23.

Santagata, S. 2008b. The morphology and evolutionary significance of the ciliary fields and musculature among marine bryozoan larvae. Journal of Morphology 269, 349–364.

Santagata, S. 2011. Evaluating neurophylogenetic patterns in the larval nervous systems of brachiopods and their evolutionary significance to other bilaterian phyla. Journal of Morphology 272, 1153–1169.

Santagata, S. 2015a. Brachiopoda. In: Wanninger, A. (ed.), Evolutionary Developmental Biology of Invertebrates 2: Lophotrochozoa (Spiralia). Springer, Vienna, pp. 263–277.

Santagata, S. 2015b. Ectoprocta. In: Wanninger, A. (ed.), Evolutionary Developmental Biology of Invertebrates 2: Lophotrochozoa (Spiralia). Springer, Vienna, pp. 247–262.

Santagata, S. 2015c. Phoronida. In: Wanninger, A. (ed.), Evolutionary Developmental Biology of Invertebrates 2: Lophotrochozoa (Spiralia). Springer, Vienna, pp. 231–245.

Santagata, S.; Cohen, B. L. 2009. Phoronid phylogenetics (Brachiopoda; Phoronata): evidence from morphological cladistics, small and large subunit rDNA sequences, and mitochondrial *cox1*. Zoological Journal of the Linnean Society 157, 34–50.

Santagata, S.; Zimmer, R. L. 2002. Comparison of the neuromuscular systems among actinotroch larvae: systematic and evolutionary implications. Evolution and Development 4, 43–54.

Santerre, C.; Sourdaine, P.; Marc, N.; Mingant, C.; Robert, R.; Martinez, A.-S. 2013. Oyster sex determination is influenced by temperature—first clues in spat during first gonadic differentiation and gametogenesis. Comparative

Biochemistry and Physiology, pt. A: Molecular and Integrative Physiology 165, 61–69.

Sarà, M.; Liaci, L. 1964. Symbiotic association between Zooxanthellae and two marine sponges of the genus *Cliona*. Nature 203, 321.

Sasakura, Y.; Ogura, Y.; Treen, N.; Yokomori, R.; Park, S.-J.; Nakai, K.; Saiga, H.; Sakuma, T.; Yamamoto, T.; Fujiwara, S.; et al. 2016. Transcriptional regulation of a horizontally transferred gene from bacterium to chordate. Proceedings of the Royal Society B: Biological Sciences 283, 20161712.

Sasson, D. A.; Ryan, J. F. 2016. The sex lives of ctenophores: the influence of light, body size, and self-fertilization on the reproductive output of the sea walnut, *Mnemiopsis leidyi*. PeerJ 4, e1846.

Sato, A.; Bishop, J. D.; Holland, P. W. 2008. Developmental biology of pterobranch hemichordates: history and perspectives. Genesis 46, 587–591.

Satoh, N. 2003. The ascidian tadpole larva: comparative molecular development and genomics. Nature Reviews Genetics 4, 285–295.

Satoh, N.; Rokhsar, D.; Nishikawa, T. 2014. Chordate evolution and the three-phylum system. Proceedings of the Royal Society B: Biological Sciences 281, 20141729.

Savic, A. G.; Guidetti, R.; Turi, A.; Pavicevic, A.; Giovannini, I.; Rebecchi, L.; Mojovic, M. 2015. Superoxide anion radical production in the tardigrade *Paramacrobiotus richtersi*, the first electron paramagnetic resonance spin-trapping study. Physiological and Biochemical Zoology 88, 451–454.

Schachat, S. R.; Labandeira, C. C.; Saltzman, M. R.; Cramer, B. D.; Payne, J. L.; Boyce, C. K. 2018. Phanerozoic pO_2 and the early evolution of terrestrial animals. Proceedings of the Royal Society B: Biological Sciences 285, 20172631.

Schack, C. R.; Gordon, D. P.; Ryan, K. G. 2018. Classification of cheilostome polymorphs. In: Wyse Jackson, P. N.; Spencer Jones, M. E. (eds.), Annals of Bryozoology 6: aspects of the history of research on bryozoans. International Bryozoology Association, pp. 85–134.

Schaefer, K. 2000. The adoral sense organ in protobranch bivalves (Mollusca): comparative fine structure with special reference to *Nucula nucleus*. Invertebrate Biology 119, 188–214.

Schaeffer, B. 1987. Deuterostome monophyly and phylogeny. Evolutionary Biology 21, 179–235.

Schärer, L.; Littlewood, D. T. J.; Waeschenbach, A.; Yoshida, W.; Vizoso, D. B. 2011. Mating behavior and the evolution of sperm design. Proceedings of the National Academy of Sciences of the USA 108, 1490–1495.

Schell, T.; Feldmeyer, B.; Schmidt, H.; Greshake, B.; Tills, O.; Truebano, M.; Rundle, S. D.; Paule, J.; Ebersberger, I.; Pfenninger, M. 2017. An annotated draft genome for *Radix auricularia* (Gastropoda, Mollusca). Genome Biology and Evolution 9, 1–8.

Schellenberg, J.; Reichert, J.; Hardt, M.; Schmidtberg, H.; Kämpfer, P.; Glaeser, S. P.; Schubert, P.; Wilke, T. 2019. The precursor hypothesis of sponge kleptocnidism: Development of nematocysts in *Haliclona cnidata* sp. nov. (Porifera, Demospongiae, Haplosclerida). Frontiers in Marine Science 5, 509.

Scheltema, A. H. 1993. Aplacophora as progenetic aculiferans and the coelomate origin of mollusks as the sister taxon of Sipuncula. Biological Bulletin 184, 57–78.

Scheltema, A. H.; Ivanov, D. L. 2002. An aplacophoran postlarva with iterated dorsal groups of spicules and skeletal similarities to Paleozoic fossils. Invertebrate Biology 121, 1–10.

Scheltema, A. H.; Kerth, K.; Kuzirian, A. M. 2003. Original molluscan radula: comparisons among Aplacophora, Polyplacophora, Gastropoda, and the Cambrian fossil *Wiwaxia corrugata*. Journal of Morphology 257, 219–245.

Scherholz, M.; Redl, E.; Wollesen; T.; Todt, C.; Wanninger, A. 2013. Aplacophoran mollusks evolved from ancestors with polyplacophoran-like features. Current Biology 23, 2130–2134.

Schiemann, S. M.; Martín-Durán, J. M.; Børve, A.; Vellutini, B. C.; Passamaneck, Y. J.; Hejnol, A. 2017. Clustered brachiopod Hox genes are not expressed collinearly and are associated with lophotrochozoan novelties. Proceedings of the National Academy of Sciences of the USA 114, E1913–E1922.

Schierenberg, E.; Sommer, R. J. 2014. Reproduction and development in nematodes. In: Schmidt-Rhaesa, A. (ed.), Handbook of Zoology. Gastrotricha, Cycloneuralia and Gnathifera, vol. 2: Nematoda. De Gruyter, Berlin/Boston, pp. 61–108.

Schierwater, B.; Eitel, M. 2015. Placozoa. In: Wanninger, A. (ed.), Evolutionary Developmental Biology of Invertebrates 1: Introduction, Non-Bilateria, Acoelomorpha, Xenoturbellida, Chaetognatha. Springer, Vienna, pp. 107–114.

Schierwater, B.; Eitel, M.; Jakob, W.; Osigus, H. J.; Hadrys, H.; Dellaporta, S. L.; Kolokotronis, S. O.; DeSalle, R. 2009. Concatenated analysis sheds light on early metazoan evolution and fuels a modern "urmetazoon" hypothesis. PLoS Biology 7, e20.

Schiffer, P. H.; Kroiher, M.; Kraus, C.; Koutsovoulos, G. D.; Kumar, S.; Camps, J. I.; Nsah, N. A.; Stappert, D.; Morris, K.; Heger, P.; et al. 2013. The genome of *Romanomermis culicivorax*: revealing fundamental changes in the core developmental genetic toolkit in Nematoda. BMC Genomics 14, 923.

Schiffer, P. H.; Robertson, H. E.; Telford, M. J. 2018. Orthonectids are highly degenerate annelid worms. Current Biology 28, 1970–1974 e1973.

Schill, R. O. (ed.) 2018. Water Bears: The Biology of Tardigrades. Zoological Monographs, vol. 2. Springer Nature Switzerland, Cham.

Schilthuizen, M., Davison, A. 2005. The convoluted evolution of snail chirality. Naturwissenschaften 92, 504–515.

Schlegel, M.; Weidhase, M.; Stadler, P. F. 2014. Deuterostome phylogeny—a molecular perspective. In: Wägele, J. W.; Bartholomaeus, T. (eds.), Deep metazoan phylogeny: The backbone of the tree of life. New insights from analyses of molecules, morphology, and theory of data analysis. De Gruyter, Berlin/Boston, pp. 413–424.

Schlesinger, A.; Zlotkin, E.; Kramarsky-Winter, E.; Loya, Y. 2009. Cnidarian internal stinging mechanism. Proceedings of the Royal Society B: Biological Sciences 276, 1063–1067.

Schlichter, D. 1991. A perforated gastrovascular cavity in the symbiotic deep-water coral *Leptoseris fragilis*: A new strategy to optimize heterotrophic nutrition. Helgoländer Meeresuntersuchungen 45, 423–443.

Schmerler, S.; Wessel, G. M. 2011. Polar bodies—more a lack of understanding than a lack of respect. Molecular Reproduction and Development 78, 3–8.

Schmidt, C.; Martínez Arbizu, P. 2015. Unexpectedly higher metazoan meiofauna abundances in the Kuril–Kamchatka Trench compared to the adjacent abyssal plains. Deep-Sea Research, pt. 2. Topical Studies in Oceanography 111, 60–75.

Schmidt, G. 1985. Development and life cycles. In: Crompton, D.; Nickol, B. (eds.), Biology of the Acanthocephala. Cambridge University Press, Cambridge, pp. 273–305.

Schmidt-Rhaesa, A. 1996a. The nervous system of *Nectonema munidae* and *Gordius aquaticus* with implications for the ground pattern of the Nematomorpha. Zoomorphology 116, 133–142.

Schmidt-Rhaesa, A. 1996b. Ultrastructure of the anterior end in three ontogenetic stages of *Nectonema munidae* (Nematomorpha). Acta Zoologica 77, 267–278.

Schmidt-Rhaesa, A. 1996c. Zur Morphologie, Biologie und Phylogenie der Nematomorpha—Untersuchungen an *Nectonema munidae* und *Gordius aquaticus*. Cuvillier Verlag, Göttingen.

Schmidt-Rhaesa, A. 1997. Ultrastructural features of the female reproductive system and female gametes of *Nectonema munidae* Brinkmann 1930 (Nematomorpha). Parasitology Research 83, 77–81.

Schmidt-Rhaesa, A. 1998. Muscular ultrastructure in *Nectonema munidae* and *Gordius aquaticus* (Nematomorpha). Invertebrate Biology 117, 37–44.

Schmidt-Rhaesa, A. 2002. Two dimensions of biodiversity research exemplified by Nematomorpha and Gastrotricha. Integrative and Comparative Biology 42, 633–640.

Schmidt-Rhaesa, A. 2004. Ultrastructure of an integumental organ with probable sensory function in *Paragordius varius* (Nematomorpha). Acta Zoologica 85, 15–19.

Schmidt-Rhaesa, A. 2005. Morphogenesis of *Paragordius varius* (Nematomorpha) during the parasitic phase. Zoomorphology 124, 33–46.

Schmidt-Rhaesa, A. 2007. The Evolution of Organ Systems. Oxford University Press, Oxford.

Schmidt-Rhaesa, A. 2013a. Nematomorpha. In: Schmidt-Rhaesa, A. (ed.), Handbook of Zoology. Gastrotricha, Cycloneuralia and Gnathifera, vol. 1: Nematomorpha, Priapulida, Kinorhyncha, Loricifera. De Gruyter, Berlin/ Boston, pp. 29–145.

Schmidt-Rhaesa, A. 2013b. Priapulida. In: Schmidt-Rhaesa, A. (ed.), Handbook of Zoology. Gastrotricha, Cycloneuralia and Gnathifera, vol. 1: Nematomorpha, Priapulida, Kinorhyncha, Loricifera. De Gruyter, Berlin/Boston, pp. 147–180.

Schmidt-Rhaesa, A. (ed.). 2014. Handbook of Zoology: Gastrotricha, Cycloneuralia and Gnathifera, vol. 2: Nematoda. De Gruyter, Berlin.

Schmidt-Rhaesa, A. (ed.). 2015. Handbook of Zoology: Gastrotricha, Cycloneuralia and Gnathifera, vol. 3: Gastrotricha and Gnathifera. De Gruyter, Berlin.

Schmidt-Rhaesa, A. 2016. Gnathostomulida. In: Schmidt-Rhaesa, A.; Harzsch, S.; Purschke, G. (eds.), Structure and Evolution of Invertebrate Nervous Systems. Oxford University Press, Oxford, pp. 118–121.

Schmidt-Rhaesa, A.; Bartolomaeus, T.; Lemburg, C.; Ehlers, U.; Garey, J. R. 1998. The position of the Arthropoda in the phylogenetic system. Journal of Morphology 238, 263–285.

Schmidt-Rhaesa, A.; Biron, D. G.; Joly, C.; Thomas, F. 2005. Host–parasite relations and seasonal occurrence of *Paragordius tricuspidatus* and *Spinochordodes tellinii* (Nematomorpha) in Southern France. Zoologischer Anzeiger 244, 51–57.

Schmidt-Rhaesa, A.; Henne, S. 2016. Cycloneuralia (Nematoda, Nematomorpha, Priapulida, Kinorhyncha, Loricifera). In: Schmidt-Rhaesa, A.; Harzsch, S.; Purschke, G. n. (eds.), Structure and Evolution of Invertebrate Nervous Systems. Oxford University Press, Oxford, pp. 368–382.

Schmidt-Rhaesa, A.; Panpeng, S.; Yamasaki, H. 2017. Two new species of *Tubiluchus* (Priapulida) from Japan. Zoologischer Anzeiger 267, 155–167.

Schmidt-Rhaesa, A.; Pohle, G.; Gaudette, J.; Burdett-Coutts, V. 2012. Lobster (*Homarus americanus*), a new host for marine horsehair worms (*Nectonema agile*, Nematomorpha). Journal of the Marine Biological Association of the United Kingdom 93, 631–633.

Schmidt-Rhaesa, A.; Rothe, B. H. 2006. Postembryonic development of dorsoventral and longitudinal musculature in *Pycnophyes kielensis* (Kinorhyncha, Homalorhagida). Integrative and Comparative Biology 46, 144–150.

Schmidt-Rhaesa, A.; Rothe, B. H. 2014. Brains in Gastrotricha and Cycloneuralia—a comparison. In: Wägele, J. W.; Bartolomaeus, T. (eds.), Deep Metazoan Phylogeny: The Backbone of the Tree of Life. De Gruyter, Berlin, pp. 93–104.

Schnabel, R.; Bischoff, M.; Hintze, A.; Schulz, A. K.; Hejnol, A.; Meinhardt, H.; Hutter, H. 2006. Global cell sorting in the *C. elegans* embryo defines a new mechanism for pattern formation. Developmental Biology 294, 418–431.

Schnabel, R.; Hutter, H.; Moerman, D.; Schnabel, H. 1997. Assessing normal embryogenesis in *Caenorhabditis elegans* using a 4D microscope: variability of development and regional specification. Developmental Biology 184, 234–265.

Schneider, A. 1862. Ueber die Metamorphose der *Actinotrocha branchiata*. Archiv für Anatomie, Physiologie und Wissenschaftliche Medicin 1862, 47–65.

Schnitzler, C. E.; Pang, K.; Powers, M. L.; Reitzel, A. M.; Ryan, J. F.; Simmons, D.; Tada, T.; Park, M.; Gupta, J.; Brooks, S. Y.; et al. 2012. Genomic organization, evolution, and expression of photoprotein and opsin genes in *Mnemiopsis leidyi*: a new view of ctenophore photocytes. BMC Biology 10, 107.

Schokraie, E.; Hotz-Wagenblatt, A.; Warnken, U.; Mali, B.; Frohme, M.; Förster, F.; Dandekar, T.; Hengherr, S.; Schill, R. O.; Schnölzer, M. 2010. Proteomic analysis of tardigrades: towards a better understanding of molecular mechanisms by anhydrobiotic organisms. PLoS One 5, e9502.

World Database. Accessed at http://www.marinespecies.org/ascidiacea. doi:10.14284/353

Shenkar, N.; Gordon, T. 2015. Gut-spilling in chordates: evisceration in the tropical ascidian *Polycarpa mytiligera*. Scientific Reports 5, 9614.

Shibata, N.; Rouhana, L.; Agata, K. 2010. Cellular and molecular dissection of pluripotent adult somatic stem cells in planarians. Development, Growth and Differentiation 52, 27–41.

Shibazaki, Y.; Shimizu, M.; Kuroda, R. 2004. Body handedness is directed by genetically determined cytoskeletal dynamics in the early embryo. Current Biology 14, 1462–1467.

Shigeno, S.; Sasaki, T.; Haszprunar, G. 2007. Central nervous system of *Chaetoderma japonicum* (Caudofoveata, Aplacophora): Implications for diversified ganglionic plans in early molluscan evolution. Biological Bulletin 213, 122–134.

Shimizu, K.; Amano, T.; Bari, M. R.; Weaver, J. C.; Arima, J.; Mori, N. 2015. Glassin, a histidine-rich protein from the siliceous skeletal system of the marine sponge *Euplectella*, directs silica polycondensation. Proceedings of the National Academy of Sciences of the USA 112, 11449–11454.

Shimizu, K.; Luo, Y.-J.; Satoh, N.; Endo, K. 2017. Possible co-option of engrailed during brachiopod and mollusc shell development. Biology Letters 13, 20170254.

Shimotori, T.; Goto, T. 2001. Developmental fates of the first four blastomeres of the chaetognath *Paraspadella gotoi*: relationship to protostomes. Development, Growth and Differentiation 43, 371–382.

Shinn, G. L. 1997. Chaetognatha. In: Harrison, F. W.; Ruppert, E. E. (eds.), Microscopic Anatomy of Invertebrates, vol. 15: Hemichordata, Chaetognatha, and the Invertebrate chordates. Wiley-Liss, New York, pp. 103–220.

Shinzato, C.; Shoguchi, E.; Kawashima, T.; Hamada, M.; Hisata, K.; Tanaka, M., Fujie; M., Fujiwara; M., Koyanagi; R.; Ikuta, T.; et al. 2011. Using the *Acropora digitifera* genome to understand coral responses to environmental change. Nature 476, 320–323.

Shirley, T. C.; Storch, V. 1999. *Halicryptus higginsi* n.sp. (Priapulida)—a giant new species from Barrow, Alaska. Invertebrate Biology 118, 404–413.

Shoguchi, E.; Shinzato, C.; Kawashima, T.; Gyoja, F.; Mungpakdee, S.; Koyanagi, R.; Takeuchi, T.; Hisata, K.; Tanaka, M.; Fujiwara, M.; et al. 2013. Draft assembly of the *Symbiodinium minutum* nuclear genome reveals dinoflagellate gene structure. Current Biology 23, 1399–1408.

Shomrat, T.; Levin, M. 2013. An automated training paradigm reveals long-term memory in planarians and its persistence through head regeneration. Journal of Experimental Biology 216, 3799–3810.

Shu, D.; Conway Morris, S.; Han, J.; Hoyal Cuthill, J. F.; Zhang, Z.; Cheng, M.; Huang, H.; Zhang, X.-G. 2017. Multi-jawed chaetognaths from the Chengjiang Lagerstätte (Cambrian, Series 2, Stage 3) of Yunnan, China. Palaeontology 60, 763–772.

Shu, D.; Conway Morris, S.; Zhang, Z. F.; Liu, J. N.; Han, J.; Chen, L.; Zhang, X. L.; Yasui, K.; Li, Y. 2003a. A new species of yunnanozoan with implications for deuterostome evolution. Science 299, 1380–1384.

Shu, D.-G.; Conway Morris, S.; Han, J.; Li, Y.; Zhang, X.-L.; Hua, H.; Zhang, Z.-F.; Liu, J.-N.; Guo, J.-F.; Yao, Y.; et al.

2006. Lower Cambrian vendobionts from China and early diploblast evolution. Science 312, 731–734.

Shu, D.-G.; Conway Morris, S.; Han, J.; Zhang, Z.-F.; Liu, J.-N. 2004. Ancestral echinoderms from the Chengjiang deposits of China. Nature 430, 422–428.

Shu, D.-G.; Conway Morris, S.; Han, J.; Zhang, Z.-F.; Yasui, K.; Janvier, P.; Chen, L.; Zhang, X.-L.; Liu, J.-N.; Li, Y.; et al. 2003b. Head and backbone of the Early Cambrian vertebrate *Haikouichthys*. Nature 421, 526–529.

Shu, D.-G.; Conway Morris, S.; Zhang, X.-L. 1996. A *Pikaia*-like chordate from the Lower Cambrian of China. Nature 384, 157–158.

Shu, D.-G.; Conway Morris, S.; Zhang, Z.-F.; Han, J. 2010. The earliest history of the deuterostomes: the importance of the Chengjiang Fossil-Lagerstätte. Proceedings of the Royal Society B: Biological Sciences 277, 165–174.

Shu, D.-G.; Luo, H.-L.; Conway Morris, S.; Zhang, X.-L.; Hu, S.-X.; Chen, L.; Han, J.; Zhu, M.; Li, Y.; Chen, L.-Z. 1999. Lower Cambrian vertebrates from South China. Nature 402, 42–46.

Shu, D. G.; Chen, L.; Han, J.; Zhang, X. L. 2001a. An Early Cambrian tunicate from China. Nature 411, 472–473.

Shu, D. G.; Conway Morris, S.; Han, J.; Chen, L.; Zhang, X. L.; Zhang, Z. F.; Liu, H. Q.; Li, Y.; Liu, J. N. 2001b. Primitive deuterostomes from the Chengjiang Lagerstätte (Lower Cambrian, China). Nature 414, 419–424.

Shultz, J. W. 2001. Gross muscular anatomy of *Limulus polyphemus* (Xiphosura, Chelicerata) and its bearing on evolution in the Arachnida. Journal of Arachnology 29, 283–303.

Shultz, J. W. 2007. A phylogenetic analysis of the arachnid orders based on morphological characters. Zoological Journal of the Linnean Society 150, 221–265.

Shunatova, N. N.; Nielsen, C. 2002. Putative sensory structures in marine bryozoans. Invertebrate Biology 121, 262–270.

Shuster, C. N., Jr.; Barlow, R. B.; Brockmann, H. J. (eds.). 2003. The American horseshoe crab. Harvard University Press, Cambridge, MA.

Siddall, M. E.; Martin, D. S.; Bridge, D.; Desser, S. S.; Cone, D. K. 1995. The demise of a phylum of protists: phylogeny of Myxozoa and other parasitic cnidaria. Journal of Parasitology 81, 961–967.

Siddall, M. E.; Min, G.-S.; Fontanella, F. M.; Phillips, A. J.; Watson, S. C. 2011. Bacterial symbiont and salivary peptide evolution in the context of leech phylogeny. Parasitology 138, 1815–1827.

Sidri, M.; Milanese, M.; Bummer, F. 2005. First observations on egg release in the oviparous sponge Chondrilla nucula (Demospongiae, Chondrosida, Chondrillidae) in the Mediterranean Sea. Invertebrate Biology 124, 91–97.

Siebert, S.; Robinson, M. D.; Tintori, S. C.; Goetz, F.; Helm, R. R.; Smith, S. A.; Shaner, N.; Haddock, S. H. D.; Dunn, C. W. 2011. Differential gene expression in the siphonophore *Nanomia bijuga* (Cnidaria) assessed with multiple Next-Generation Sequencing workflows. PLoS One 6, e22953.

Sielaff, M.; Schmidt, H.; Struck, T. H.; Rosenkranz, D.; Welch, D. B. M.; Hankeln, T.; Herlyn, H. 2016. Phylogeny of Syndermata (syn. Rotifera): Mitochondrial gene order verifies epizoic Seisonidea as sister to endoparasitic Acanthocephala within monophyletic Hemirotifera. Molecular Phylogenetics and Evolution 96, 79–92.

Sigl, R.; Steibl, S.; Laforsch, C. 2016. The role of vision for navigation in the crown-of-thorns seastar, *Acanthaster planci*. Scientific Reports 6, 30834.

Signorovitch, A.; Hur, J.; Gladyshev, E.; Meselson, M. 2015. Allele sharing and evidence for sexuality in a mitochondrial clade of bdelloid rotifers. Genetics 200, 581–590.

Signorovitch, A. Y.; Buss, L. W.; Dellaporta, S. L. 2007. Comparative genomics of large mitochondria in placozoans. PLoS Genetics 3, e13.

Signorovitch, A. Y.; Dellaporta, S. L.; Buss, L. W. 2006. Caribbean placozoan phylogeography. Biological Bulletin 211, 149–156.

Sigwart, J. D.; Lindberg, D. R. 2015. Consensus and confusion in molluscan trees: Evaluating morphological and molecular phylogenies. Systematic Biology 64, 384–395.

Sigwart, J. D.; Sumner-Rooney, L. H. 2016. Mollusca: Caudofoveata, Monoplacophora, Polyplacophora, Scaphopoda, and Solenogastres. In: Schmidt-Rhaesa, A.; Harzsch, S.; Purschke, G. (eds.), Structure and Evolution of Invertebrate Nervous Systems. Oxford University Press, Oxford, pp. 172–189.

Sigwart, J. D.; Sumner-Rooney, L. H.; Schwabe, E.; Hess, M.; Brennan, G. P.; Schrödl, M. 2014. A new sensory organ in "primitive" molluscs (Polyplacophora: Lepidopleurida), and its context in the nervous system of chitons. Frontiers in Zoology 11, 7.

Sigwart, J. D.; Sutton, M. D. 2007. Deep molluscan phylogeny: synthesis of palaeontological and neontological data. Proceedings of the Royal Society B: Biological Sciences 274, 2413–2419.

Sikes, J. M.; Bely, A. E. 2008. Radical modification of the A–P axis and the evolution of asexual reproduction in *Convolutriloba* acoels. Evolution and Development 10, 619–631.

Sikes, J. M.; Bely, A. E. 2010. Making heads from tails: Development of a reversed anterior–posterior axis during budding in an acoel. Developmental Biology 338, 86–97.

Sikes, J. M.; Newmark, P. A. 2013. Restoration of anterior regeneration in a planarian with limited regenerative ability. Nature 500, 77–80.

Silén, L. 1954. Developmental biology of Phoronidea of the Gullmar Fiord area (West Coast of Sweden). Acta Zoologica Stockholm 35, 215–257.

Simakov, O.; Kawashima, T.; Marlétaz, F.; Jenkins, J.; Koyanagi, R.; Mitros, T.; Hisata, K.; Bredeson, J.; Shoguchi, E.; Gyoja, F.; et al. 2015. Hemichordate genomes and deuterostome origins. Nature 527, 459–465.

Simakov, O.; Marlétaz, F.; Cho, S. J.; Edsinger-Gonzales, E.; Havlak, P.; Hellsten, U.; Kuo, D.-H.; Larsson, T.; Lv, J.; Arendt, D.; et al. 2013. Insights into bilaterian evolution from three spiralian genomes. Nature 493, 526–531.

Simion, P.; Bekkouche, N.; Jager, M.; Quéinnec, E.; Manuel, M. 2014. Exploring the potential of small RNA subunit and ITS sequences for resolving phylogenetic relationships within the phylum Ctenophora. Zoology 118, 102–114.

Simion, P.; Philippe, H.; Baurain, D.; Jager, M.; Richter, D. J.; Di Franco, A.; Roure, B.; Satoh, N.; Queinnec, E.; Eres-

kovsky, A.; et al. 2017. A large and consistent phylogenomic dataset supports sponges as the sister group to all other animals. Current Biology 27, 958–967.

Simmons, D. K.; Martindale, M. Q. 2016. Ctenophora. In: Schmidt-Rhaesa, A.; Harzsch, S.; Purschke, G. (eds.), Structure and Evolution of Invertebrate Nervous Systems. Oxford University Press, Oxford, pp. 48–55.

Simone, L. R. L. 2011. Phylogeny of the Caenogastropoda (Mollusca), based on comparative morphology. Arquivos de Zoologia 42, 161–323.

Simpson, S. D.; Meekan, M. G.; Jeffs, A.; Montgomery, J. C.; McCauley, R. D. 2008. Settlement-stage coral reef fish prefer the higher-frequency invertebrate-generated audible component of reef noise. Animal Behaviour 75, 1861–1868.

Simpson, T. L. 1984. Gamete, embryo, larval development. The Cell Biology of Sponges. Springer, Berlin, pp. 341–413.

Singh, T. R.; Tsagkogeorga, G.; Delsuc, F.; Blanquart, S.; Shenkar, N.; Loya, Y.; Douzery, E. J. P.; Huchon, D. 2009. Tunicate mitogenomics and phylogenetics: peculiarities of the *Herdmania momus* mitochondrial genome and support for the new chordate phylogeny. BMC Genomics 10, 534.

Skovsted, C. B.; Balthasar, U.; Brock, G. A.; Paterson, J. R. 2009a. The tommotiid *Camenella reticulosa* from the early Cambrian of South Australia: Morphology, scleritome reconstruction, and phylogeny. Acta Palaeontologica Polonica 54, 525–540.

Skovsted, C. B.; Brock, G. A.; Paterson, J. R.; Holmer, L. E.; Budd, G. E. 2008. The scleritome of *Eccentrotheca* from the Lower Cambrian of South Australia: Lophophorate affinities and implications for tommotiid phylogeny. Geology 36, 171–174.

Skovsted, C. B.; Brock, G. A.; Topper, T. P.; Paterson, J. R.; Holmer, L. E. 2011. Scleritome construction, biofacies, biostratigraphy and systematics of the tommotiid *Eccentrotheca helenia* sp. nov. from the early Cambrian of South Australia. Palaeontology 54, 253–286.

Skovsted, C. B.; Clausen, S.; Álvaro, J. J.; Ponlevé, D. 2014. Tommotiids from the Early Cambrian (Series 2, Stage 3) of Morocco and the evolution of the tannuolinid scleritome and setigerous shell structures in stem group brachiopods. Palaeontology 57, 171–192.

Skovsted, C. B.; Holmer, L. E.; Larsson, C. M.; Högström, A. E. S.; Brock, G. A.; Topper, T. P.; Balthasar, U.; Stolk, S. P.; Paterson, J. R. 2009b. The scleritome of *Paterimitra*: an Early Cambrian stem group brachiopod from South Australia. Proceedings of the Royal Society B: Biological Sciences 276, 1651–1656.

Skuballa, J.; Taraschewski, H.; Petney, T. N.; Pfäffle, M.; Smales, L. R. 2010. The avian acanthocephalan *Plagiorhynchus cylindraceus* (Palaeacanthocephala) parasitizing the European hedgehog (*Erinaceus europaeus*) in Europe and New Zealand. Parasitology Research 106, 431–437.

Slater, B. J.; Harvey, T. H. P.; Guilbaud, R.; Butterfield, N. J. 2017. A cryptic record of Burgess Shale-type diversity from the early Cambrian of Baltica. Palaeontology 60, 117–140.

Slatko, B. E.; Taylor, M. J.; Foster, J. M. 2010. The *Wolbachia* endosymbiont as an anti-filarial nematode target. Symbiosis 51, 55–65.

Southcott, R. V.; Lange, R. T. 1971. Acarine and other micro-fossils from the Maslin Eocene, South Australia. Records of the South Australian Museum 16, 1–21.

Soviknes, A. M.; Glover, J. C. 2008. Continued growth and cell proliferation into adulthood in the notochord of the appendicularian *Oikopleura dioica*. Biological Bulletin 214, 17–28.

Speiser, D. I.; Eernisse, D. J.; Johnsen, S. 2011a. A chiton uses aragonite lenses to form images. Current Biology 21, 665–670.

Speiser, D. I.; Loew, E. R.; Johnsen, S. 2011b. Spectral sensitivity of the concave mirror eyes of scallops: potential influences of habitat, self-screening and longitudinal chromatic aberration. Journal of Experimental Biology 214, 422–431.

Spengel, J. W. 1932. Planctosphaera pelagica. Report on the Scientific Results of the "Michael Sars" North Atlantic Deep-Sea Expedition 1910 5, 1–27.

Sperling, E. A.; Frieder, C. A.; Raman, A. V.; Girguis, P. R.; Levin, L. A.; Knoll, A. H. 2013. Oxygen, ecology, and the Cambrian radiation of animals. Proceedings of the National Academy of Sciences of the USA 110, 13446–13451.

Sperling, E. A.; Peterson, K. J.; Pisani, D. 2009. Phylogenetic-signal dissection of nuclear housekeeping genes supports the paraphyly of sponges and the monophyly of Eumetazoa. Molecular Biology and Evolution 26, 2261–2274.

Sperling, E. A.; Pisani, D.; Peterson, K. J. 2007. Poriferan paraphyly and its implications for Precambrian palaeobiology. In: Vickers-Rich, P.; Komarower, P. (eds.), The Rise and Fall of the Ediacaran Biota. Geological Society of London Special Publications 286, London, pp. 355–368.

Sperling, E. A.; Pisani, D.; Peterson, K. J. 2011. Molecular paleobiological insights into the origin of the Brachiopoda. Evolution and Development 13, 290–303.

Sperling, E. A.; Robinson, J. M.; Pisani, D.; Peterson, K. J. 2010. Where's the glass? Biomarkers, molecular clocks, and microRNAs suggest a 200-Myr missing Precambrian fossil record of siliceous sponge spicules. Geobiology 8, 24–36.

Sperling, E. A.; Rosengarten, R. D.; Moreno, M. A.; Dellaporta, S. L. 2012. The complete mitochondrial genome of the verongid sponge *Aplysina cauliformis*: implications for DNA barcoding in demosponges. Hydrobiologia 687, 61–69.

Sperling, E. A.; Stockey, R. G. 2018. The temporal and environmental context of early animal evolution: Considering all the ingredients of an "explosion." Integrative and Comparative Biology 58, 605–622.

Sperling, E. A.; Vinther, J. 2010. A placozoan affinity for *Dickinsonia* and the evolution of late Proterozoic metazoan feeding modes. Evolution and Development 12, 201–209.

Spiridonov, S. E.; Schmatko, V. Y. 2013. Morphological and molecular characterization of *Gordionus alpestris* (Nematomorpha) from the North-West Caucasus, Russia. Helminthologia, Bratislava, 50, 67–72.

Sprinkle, J. 1973. Morphology and evolution of blastozoan echinoderms. Special Publication of the Museum of Comparative Zoology, 1–284.

Sprinkle, J.; Wilbur, B. C. 2005. Deconstructing helicoplacoids: reinterpreting the most enigmatic Cambrian echinoderms. Geological Journal 40, 281–293.

Spruyt, N.; Delarbre, C.; Gachelin, G.; Laudet, V. 1998. Complete sequence of the amphioxus (*Branchiostoma lanceolatum*) mitochondrial genome: relations to vertebrates. Nucleic Acids Research 26, 3279–3285.

Squires, L. N.; Rubakhin, S. S.; Wadhams, A. A.; Talbot, K. N.; Nakano, H.; Moroz, L. L.; Sweedler, J. V. 2010. Serotonin and its metabolism in basal deuterostomes: insights from *Strongylocentrotus purpuratus* and *Xenoturbella bocki*. Journal of Experimental Biology 213, 2647–2654.

Srivastava, M.; Begovic, E.; Chapman, J.; Putnam, N. H.; Hellsten, U.; Kawashima, T.; Kuo, A.; Mitros, T.; Salamov, A.; Carpenter, M. L.; et al. 2008. The *Trichoplax* genome and the nature of placozoans. Nature 454, 955–960.

Srivastava, M.; Mazza-Curll, K. L.; van Wolfswinkel, J. C.; Reddien, P. W. 2014. Whole-body acoel regeneration is controlled by Wnt and Bmp-Admp signaling. Current Biology 24, 1107–1113.

Srivastava, M.; Simakov, O.; Chapman, J.; Fahey, B.; Gauthier, M. E. A.; Mitros, T.; Richards, G. S.; Conaco, C.; Dacre, M.; Hellsten, U.; et al. 2010. The *Amphimedon queenslandica* genome and the evolution of animal complexity. Nature 466, 720–U723.

Stach, T. 2008. Chordate phylogeny and evolution: a not so simple three-taxon problem. Journal of Zoology 276, 117–141.

Stach, T. 2014. Deuterostome phylogeny—a morphological perspective. In: Wägele, J. W.; Bartholomaeus, T. (eds.), Deep metazoan phylogeny: The backbone of the tree of life. New insights from analyses of molecules, morphology, and theory of data analysis. De Gruyter, Berlin/Boston, pp. 425–457.

Stach, T. 2016a. Hemichordata. In: Schmidt-Rhaesa, A.; Harzsch, S.; Purschke, G. (eds.), Structure and Evolution of Invertebrate Nervous Systems. Oxford University Press, Oxford, pp. 689–698.

Stach, T. 2016b. *Xenoturbella*. In: Schmidt-Rhaesa, A.; Harzsch, S.; Purschke, G. (eds.), Structure and Evolution of Invertebrate Nervous Systems. Oxford University Press, Oxford, pp. 62–66.

Stach, T.; Braband, A.; Podsiadlowski, L. 2010. Erosion of phylogenetic signal in tunicate mitochondrial genomes on different levels of analysis. Molecular Phylogenetics and Evolution 55, 860–870.

Stach, T.; Dupont, S.; Israelson, O.; Fauville, G.; Nakano, H.; Kanneby, T.; Thorndyke, M. 2005. Nerve cells of *Xenoturbella bocki* (phylum uncertain) and *Harrimania kupfferi* (Enteropneusta) are positively immunoreactive to antibodies raised against echinoderm neuropeptides. Journal of the Marine Biological Association of the United Kingdom 85, 1519–1524.

Stach, T.; Gruhl, A.; Kaul-Strehlow, S. 2012. The central and peripheral nervous system of *Cephalodiscus gracilis* (Pterobranchia, Deuterostomia). Zoomorphology 131, 11–24.

Stach, T.; Turbeville, J. M. 2002. Phylogeny of Tunicata inferred from molecular and morphological characters. Molecular Phylogenetics and Evolution 25, 408–428.

Stach, T.; Winter, J.; Bouquet, J.-M.; Chourrout, D.; Schnabel, R. 2008. Embryology of a planktonic tunicate reveals traces of sessility. Proceedings of the National Academy of Sciences of the USA 105, 7229–7234.

Stanley, G. D.; Stürmer, W. 1983. The first fossil ctenophore from the Lower Devonian of West Germany. Nature 303, 518–520.

Stanley, G. D.; Stürmer, W. 1987. A new fossil ctenophore discovered by X-rays. Nature 328, 61–63.

Starunov, V. V.; Dray, N.; Belikova, E. V.; Kerner, P.; Vervoort, M.; Balavoine, G. 2015. A metameric origin for the annelid pygidium? BMC Evolutionary Biology 15, 25.

Starunov, V. V.; Voronezhskaya, E. E.; Nezlin, L. P. 2017. Development of the nervous system in *Platynereis dumerilii* (Nereididae, Annelida). Frontiers in Zoology 14, 27.

Stasek, C. R. 1963. Synopsis and discussion of the association of ctenidia and labial palps in the bivalved Mollusca. Veliger 6, 91–97.

Stechmann, A.; Schlegel, M. 1999. Analysis of the complete mitochondrial DNA sequence of the brachiopod *Terebratulina retusa* places Brachiopoda within the protostomes. Proceedings of the Royal Society B: Biological Sciences 266, 2043–2052.

Stegner, M. E. J.; Fritsch, M.; Richter, S. 2014. The central complex in Crustacea. In: Wägele, J. W.; Bartholomaeus, T. (eds.), Deep metazoan phylogeny: The backbone of the tree of life. New insights from analyses of molecules, morphology, and theory of data analysis. De Gruyter, Berlin/Boston, pp. 361–384.

Stegner, M. E. J.; Richter, S. 2011. Morphology of the brain in *Hutchinsoniella macracantha* (Cephalocarida, Crustacea). Arthropod Structure and Development 40, 221–243.

Stein, L. D.; Bao, Z.; Blasiar, D.; Blumenthal, T.; Brent, M. R.; Chen, N.; Chinwalla, A.; Clarke, L.; Clee, C.; Coghlan, A.; et al. 2003. The genome sequence of *Caenorhabditis briggsae*: a platform for comparative genomics. PLoS Biology 1, 166–192.

Steinböck, O. 1930–1931. Ergebnisse einer von E. Reisinger und O. Steinböck mit Hilfe des RaskOrsted Fonds durchgeführten Reise in Grönland 1926. 2. *Nemertoderma bathycola* nov. gen. nov. spec., eine eigenartige Turbellarie aus der Tiefe der Diskobay; nebst einem Beitrag zur Kenntnis des Nemertinenepithels. Videnskabelige meddelelser fra Dansk naturhistorisk forening i København 90, 47–84.

Steinmetz, P. R.; Kraus, J. E.; Larroux, C.; Hammel, J. U.; Amon-Hassenzahl, A.; Houliston, E.; Wörheide, G.; Nickel, M.; Degnan, B. M.; Technau, U. 2012. Independent evolution of striated muscles in cnidarians and bilaterians. Nature 487, 231–234.

Stemple, D. L. 2005. Structure and function of the notochord: an essential organ for chordate development. Development 132, 2503–2512.

Sterrer, W. 1968. Beiträge zur Kenntnis der Gnathostomulida. I. Anatomie und Morphologie des Genus *Pterognathia* Sterrer. Arkiv för zoologi 22, 1–125.

Sterrer, W. 1972. Systematics and evolution within the Gnathostomulida. Systematic Zoology 21, 151–173.

Sterrer, W. 1976. *Tenuignathia rikerae* nov. gen., nov. spec., a new gnathostomulid from the West Atlantic. Internationale Revue der Gesamten Hydrobiologie 61, 249–259.

Sterrer, W. 1992. Clausognathiidae, a new family of Gnathostomulida from Belize. Proceedings of the Biological Society of Washington 105, 136–142.

Sterrer, W. 1998a. Gnathostomulida from the (sub)Tropical Northwestern Atlantic. Studies on the Natural History of the Caribbean Region 74, 1–178.

Sterrer, W. 1998b. New and known Nemertodermatida (Platyhelminthes-Acoelomorpha)—a revision. Belgian Journal of Zoology 128, 55–92.

Sterrer, W.; Mainitz, M.; Rieger, R. M. 1985. Gnathostomulida: enigmatic as ever. In: Conway Morris, S.; George, J. D.; Gibson, R.; Platt, H. M. (eds.), The origins and relationships of lower invertebrates. Systematics Association/ Clarendon Press, Oxford, pp. 181–199.

Sterrer, W.; Sørensen, M. V. 2015. Phylum Gnathostomulida. In: Schmidt-Rhaesa, A. (ed.), Handbook of Zoology: Gastrotricha, Cycloneuralia and Gnathifera, vol. 3: Gastrotricha and Gnathifera. De Gruyter, Berlin, pp. 135–196.

Stewart, H.; Westlake, H. E.; Page, L. R. 2014. Rhogocytes in gastropod larvae: developmental transformation from protonephridial terminal cells. Invertebrate Biology 133, 47–63.

Stiasny-Wijnhoff, G. 1936. Die Polystilifera der Siboga-Expedition. Siboga Expeditie 22, 1–214.

Stock, C. W. 2001. Stromatoporoidea, 1926–2000. Journal of Paleontology 75, 1079–1089.

Stock, S. P. 2016. Phylum Nematoda: Roundworms and Threadworms. In: Brusca, R. C.; Moore, W.; Shuster, S. M. (eds.), Invertebrates, 3rd ed. Sinauer, Sunderland, MA, pp. 671–686.

Stöger, I.; Kocot, K. M.; Poustka, A. J.; Wilson, N. G.; Ivanov, D.; Halanych, K. M.; Schrödl, M. 2016. Monoplacophoran mitochondrial genomes: convergent gene arrangements and little phylogenetic signal. BMC Evolutionary Biology 16, 274.

Stöger, I.; Schrödl, M. 2013. Mitogenomics does not resolve deep molluscan relationships (yet?). Molecular Phylogenetics and Evolution 69, 376–392.

Stöger, I.; Sigwart, J. D.; Kano, Y.; Knebelsberger, T.; Marshall, B. A.; Schwabe, E.; Schrödl, M. 2013. The continuing debate on deep molluscan phylogeny: evidence for Serialia (Mollusca, Monoplacophora plus Polyplacophora). Biomed Research International 2013, 407072.

Stöhr, S.; O'Hara, T. D.; Thuy, B. 2012. Global diversity of brittle stars (Echinodermata: Ophiuroidea). PLoS One 7, e31940.

Stokes, A. N.; Ducey, P. K.; Neuman-Lee, L.; Hanifin, C. T.; French, S. S.; Pfrender, M. E.; Brodie, E. D., III; Brodie, E. D., Jr. 2014. Confirmation and distribution of tetrodotoxin for the first time in terrestrial invertebrates: Two terrestrial flatworm species (*Bipalium adventitium* and *Bipalium kewense*). PLoS One 9, e100718.

Stolarski, J.; Kitahara, M. V.; Miller, D. J.; Cairns, S. D.; Mazur, M.; Meibom, A. 2011. The ancient evolutionary origins of Scleractinia revealed by azooxanthellate corals. BMC Evolutionary Biology 11, 316.

Stolfi, A.; Brown, F. 2015. Tunicata. In: Wanninger, A. (ed.), Evolutionary Developmental Biology of Invertebrates 6: Deuterostomia. Springer, Vienna, pp. 135–204.

Stolfi, A.; Lowe, E. K.; Racioppi, C.; Ristoratore, F.; Brown, C. T.; Swalla, B. J.; Christiaen, L. 2014. Divergent mechanisms regulate conserved cardiopharyngeal development and gene expression in distantly related ascidians. eLife 3, e03728.

Stollewerk, A.; Chipman, A. D. 2006. Neurogenesis in myriapods and chelicerates and its importance for understanding arthropod relationships. Integrative and Comparative Biology 46, 195–206.

Storch, V. 1991. Priapulida. In: Harrison, F. W.; Ruppert, E. E. (eds.), Microscopic Anatomy of invertebrates, vol. 4: Aschelminthes. Wiley-Liss, New York, pp. 333–350.

Storch, V.; Higgins, R. P. 1989. Ultrastructure of developing and mature spermatozoa of *Tubiluchus corallicola* (Priapulida). Transactions of the American Microscopical Society 108, 45–50.

Storch, V.; Higgins, R. P.; Anderson, P.; Svavarsson, J. 1995. Scanning and transmission electron microscopic analysis of the introvert of *Priapulopsis australis* and *Priapulopsis bicaudatus* (Priapulida). Invertebrate Biology 114, 64–72.

Storch, V.; Higgins, R. P.; Morse, M. P. 1989a. Internal anatomy of *Meiopriapulus fijiensis* (Priapulida). Transactions of the American Microscopical Society 108, 245–261.

Storch, V.; Higgins, R. P.; Morse, M. P. 1989b. Ultrastructure of the body wall of *Meiopriapulus fijiensis* (Priapulida). Transactions of the American Microscopical Society 108, 319–331.

Storch, V.; Higgins, R. P.; Rumohr, H. 1990. Ultrastructure of introvert and pharynx of *Halicryptus spinulosus* (Priapulida). Journal of Morphology 206, 163–171.

Storch, V.; Welsch, U. 1972. Über Bau und Entstehung der Mantelrandstachen von *Lingula unguis* L. (Brachiopoda). Zeitschrift für wissenschaftiche Zoologie 183, 181–189.

Storch, V.; Welsch, U. 1974. Epitheliomuscular cells in *Lingula unguis* (Brachiopoda) and *Branchiostoma lanceolatum* (Acrania). Cell and Tissue Research 154, 543–545.

Storch, V.; Welsch, U. 1976. Elektronenmikroskopische und enzymhistochemische Untersuchungen über Lophophor und Tentakeln von *Lingula unguis* L. (Brachiopoda). Zoologische Jahrbücher für Anatomie und Ontogenie der Tiere 96, 225–237.

Stork, N. E. 1996. Tropical forest dynamics: the faunal components. Monographiae Biologicae 74, 1–20.

Stouthamer, R.; Luck, R. F.; Hamilton, W. D. 1990. Antibiotics cause parthenogenetic *Trichogramma* (Hymenoptera, Trichogrammatidae) to revert to sex. Proceedings of the National Academy of Sciences of the USA 87, 2424–2427.

Strader, M. E.; Aglyamova, G. V.; Matz, M. V. 2018. Molecular characterization of larval development from fertilization to metamorphosis in a reef-building coral. BMC Genomics 19, 17.

Strader, M. E.; Davies, S. W.; Matz, M. V. 2015. Differential responses of coral larvae to the colour of ambient light guide them to suitable settlement microhabitat. Royal Society Open Science 2, 150358.

Strand, M.; Hedstrom, M.; Seth, H.; McEvoy, E. G.; Jacobsson, E.; Göransson, U.; Andersson, H. S.; Sundberg, P. 2016. The bacterial (*Vibrio alginolyticus*) production of tetrodotoxin in the ribbon worm *Lineus longissimus*–just a false positive? Marine Drugs 14, 63.

Strand, M.; Norenburg, J.; Alfaya, J. E.; Fernández-Álvarez, F. Á.; Andersson, H. S.; Andrade, S. C. S.; Bartolomaeus, T.; Beckers, P.; Bigatti, G.; Cherneva, I.; et al. 2018. Nemertean taxonomy—Implementing changes in the higher ranks, dismissing Anopla and Enopla. Zoologica Scripta 48, 118–119.

Strausfeld, N. J. 1998. Crustacean—insect relationships: the use of brain characters to derive phylogeny amongst segmented invertebrates. Brain Behavior and Evolution 52, 186–206.

Strausfeld, N. J. 2005. The evolution of crustacean and insect optic lobes and the origins of chiasmata. Arthropod Structure and Development 34, 235–256.

Strausfeld, N. J. 2009. Brain organization and the origin of insects: an assessment. Proceedings of the Royal Society B: Biological Sciences 276, 1929–1937.

Strausfeld, N. J. 2012. Arthropod brains: evolution, functional elegance, and historical significance. Belknap Press of Harvard University Press, Cambridge, MA.

Strausfeld, N. J.; Andrew, D. R. 2011. A new view of insect–crustacean relationships inferred from neural cladistics. Arthropod Structure and Development 40, 276–280.

Strausfeld, N. J.; Buschbeck, E. K.; Gomez, R. S. 1995. The arthropod mushroom body: its functional roles, evolutionary enigmas and mistaken identities. In: Breidbach, O.; Kutsch, W. (eds.), The Nervous Systems of Invertebrates: an Evolutionary and Comparative Approach. Birkhäuser, Basel, pp. 349–381.

Strausfeld, N. J.; Ma, X.; Edgecombe, G. D. 2016a. Fossils and the evolution of the arthropod brain. Current Biology 26, R989–R1000.

Strausfeld, N. J.; Ma, X.; Edgecombe, G. D.; Fortey, R. A.; Land, M. F.; Liu, Y.; Cong, P.; Hou, X. 2016b. Arthropod eyes: The early Cambrian fossil record and divergent evolution of visual systems. Arthropod Structure and Development 45, 152–172.

Strausfeld, N. J.; Strausfeld, C. M.; Stowe, S.; Rowell, D.; Loesel, R. 2006a. The organization and evolutionary implications of neuropils and their neurons in the brain of the onychophoran *Euperipatoides rowelli*. Arthropod Structure and Development 35, 169–196.

Strausfeld, N. J.; Strausfeld, C. M.; Loesel, R.; Rowell, D.; Stowe, S. 2006b. Arthropod phylogeny: onychophoran brain organization suggests an archaic relationship with a chelicerate stem lineage. Proceedings of the Royal Society B: Biological Sciences 273, 1857–1866.

Stretton, A. O. W. 1976. Anatomy and development of the somatic musculature of the nematode *Ascaris*. Journal of Experimental Biology 64, 773–788.

Stretton, A. O. W.; Fishpool, R. M.; Southgate, E.; Donmoyer, J. E.; Walrond, J. P.; Moses, J. E. R.; Kass, I. S. 1978. Structure and physiological activity of the motoneurons of the nematode *Ascaris*. Proceedings of the National Academy of Sciences of the USA 75, 3493–3497.

Stricker, S. A. 1984. Styletogenesis in nemertean worms: The ultrastructure of organelles involved in intracellular calcification. Journal of Morphology 179, 119–134.

Stricker, S. A.; Cavey, M. J.; Cloney, R. A. 1985. Tetracycline labeling studies of calcification in nemertean worms. Transactions of the American Microscopical Society 104, 232–241.

Stricker, S. A.; Reed, C. G. 1985a. The ontogeny of shell secretion in *Terebratalia transversa* (Brachiopoda, Articulata), I: Development of the mantle. Journal of Morphology 183, 233–250.

Stricker, S. A.; Reed, C. G. 1985b. The ontogeny of shell secretion in *Terebratalia transversa* (Brachiopoda, Articulata). II. Formation of the protegulum and juvenile shell. Journal of Morphology 183, 251–271.

Stricker, S. A.; Reed, C. G.; Zimmer, R. L. 1988. The cyphonautes larva of the marine bryozoan *Membranipora membranacea*. II. Internal sac, musculature, and pyriform organ. Canadian Journal of Zoology 66, 384–398.

Struck, T. H. 2011. Direction of evolution within Annelida and the definition of Pleistoannelida. Journal of Zoological Systematics and Evolutionary Research 49, 340–345.

Struck, T. H.; Fisse, F. 2008. Phylogenetic position of Nemertea derived from phylogenomic data. Molecular Biology and Evolution 25, 728–736.

Struck, T. H.; Paul, C.; Hill, N.; Hartmann, S.; Hösel, C.; Kube, M.; Lieb, B.; Meyer, A.; Tiedemann, R.; Purschke, G.; et al. 2011. Phylogenomic analyses unravel annelid evolution. Nature 471, 95–98.

Struck, T. H.; Schult, N.; Kusen, T.; Hickman, E.; Bleidorn, C.; McHugh, D.; Halanych, K. M. 2007. Annelid phylogeny and the status of Sipuncula and Echiura. BMC Evolutionary Biology 7, 57.

Struck, T. H.; Westheide, W.; Purschke, G. 2002. Progenesis in Eunicida ("Polychaeta," Annelida)—separate evolutionary events? Evidence from molecular data. Molecular Phylogenetics and Evolution 25, 190–199.

Struck, T. H.; Wey-Fabrizius, A. R.; Golombek, A.; Hering, L.; Weigert, A.; Bleidorn, C.; Klebow, S.; Iakovenko, N.; Hausdorf, B.; Petersen, M.; et al. 2014. Platyzoan paraphyly based on phylogenomic data supports a non-coelomate ancestry of Spiralia. Molecular Biology and Evolution 31, 1833–1849.

Stunkard, H. W. 1982. Mesozoa. In: Parker, S. P. (ed.), Synopsis and classification of living organisms, vol. 1. McGraw-Hill, New York, pp. 853–855.

Suga, H.; Chen, Z.; de Mendoza, A.; Sebé-Pedrós, A.; Brown, M. W.; Kramer, E.; Carr, M.; Kerner, P.; Vervoort, M.; Sánchez-Pons, N.; et al. 2013. The *Capsaspora* genome reveals a complex unicellular prehistory of animals. Nature Communications 4, 2325.

Suga, K.; Mark Welch, D. B.; Tanaka, Y.; Sakakura, Y.; Hagiwara, A. 2008. Two circular chromosomes of unequal copy number make up the mitochondrial genome of the rotifer *Brachionus plicatilis*. Molecular Biology and Evolution 25, 1129–1137.

Sulston, J. E.; Schierenberg, E.; White, J. G.; Thomson, J. N. 1983. The embryonic cell lineage of the nematode *Caenorhabditis elegans*. Developmental Biology 100, 64–119.

Sumner-Rooney, L.; Rahman, I. A.; Sigwart, J. D.; Ullrich-Lüter, E. 2018. Whole-body photoreceptor networks are independent of 'lenses' in brittle stars. Proceedings of the Royal Society B: Biological Sciences 285, 20172590.

Sumner-Rooney, L. H.; Schrödl, M.; Lodde-Bensch, E.; Lindberg, D. R.; Hess, M.; Brennan, G. P.; Sigwart, J. D. 2015. A neurophylogenetic approach provides new insight to the evolution of Scaphopoda. Evolution and Development 17, 337–346.

Sumner-Rooney, L. H.; Sigwart, J. D. 2015. Is the Schwabe organ a retained larval eye? Anatomical and behavioural studies of a novel sense organ in adult *Leptochiton asellus* (Mollusca, Polyplacophora) indicate links to larval photoreceptors. PLoS One 10, e0137119.

Sumrall, C. D.; Wray, G. A. 2007. Ontogeny in the fossil record: diversification of body plans and the evolution of "aberrant" symmetry in Paleozoic echinoderms. Paleobiology 33, 149–163.

Sun, H.; Smith, M. R.; Zeng, H.; Zhao, F.; Li, G.; Zhu, M. 2018. Hyoliths with pedicles illuminate the origin of the brachiopod body plan. Proceedings of the Royal Society B: Biological Sciences 285, 20181780.

Sun, J.; Mu, H.; Ip, J. C. H.; Li, R.; Xu, T.; Accorsi, A.; Sánchez Alvarado, A.; Ross, E.; Lan, Y.; Sun, Y.; et al. 2019. Signatures of divergence, invasiveness and terrestralization revealed by four apple snail genomes. Molecular Biology and Evolution 36, 1507–1520.

Sun, J.; Zhang, Y.; Xu, T.; Zhang, Y.; Mu, H.; Zhang, Y.; Lan, Y.; Fields, C. J.; Hui, J. H. L.; Zhang, W.; et al. 2017. Adaptation to deep-sea chemosynthetic environments as revealed by mussel genomes. Nature Ecology and Evolution 1, 121.

Sun, S. 2006. On nemerteans with a branched proboscis from Zhanjiang, China. Journal of Natural History 40, 943–965.

Sun, W.-Y.; Shen, C.-Y.; Sun, S.-C. 2016. The complete mitochondrial genome of *Tetrastemma olgarum* (Nemertea: Hoplonemertea). Mitochondrial DNA. pt. A 27, 1086–1087.

Sun, W.-Y.; Xu, D.-L.; Chen, H.-X.; Shi, W.; Sundberg, P.; Strand, M.; Sun, S.-C. 2014. Complete mitochondrial genome sequences of two parasitic/commensal nemerteans, *Gononemertes parasita* and *Nemertopsis tetraclitophila* (Nemertea: Hoplonemertea). Parasites and Vectors 7, 273.

Sundaram, M. V.; Buechner, M. 2016. The *Caenorhabditis elegans* excretory system: A model for tubulogenesis, cell fate specification, and plasticity. Genetics 203, 35–63.

Sundberg, P. 1990. Gibson's reclassification on the enoplan nemerteans (Enopla, Nemertea): a critique and cladistic analysis. Zoologica Scripta 19, 133–140.

Sundberg, P.; Andrade, S. C. S.; Bartolomaeus, T.; Beckers, P.; Von Döhren, J.; Gibson, R.; Giribet, G.; Herrera-Bachiller, A.; Junoy, J.; Kajihara, H.; et al. 2016. The future of nemertean taxonomy (phylum Nemertea)—a proposal. Zoologica Scripta 46, 579–582.

Sundberg, P.; Hylbom, R. 1994. Phylogeny of the nemertean subclass Palaeonemertea (Anopla, Nemertea). Cladistics 10, 347–402.

Sundberg, P.; Strand, M. 2007. *Annulonemertes* (phylum Nemertea): when segments do not count. Biology Letters 3, 570–573.

Sundberg, P.; Turbeville, J. M.; Härlin, M. S. 1998. There is no support for Jensen's hypothesis of nemerteans as ancestors to the vertebrates. Hydrobiologia 365, 47–54.

Temereva, E. N.; Kuzmina, T. V. 2017. The first data on the innervation of the lophophore in the rhynchonelliform brachiopod *Hemithiris psittacea*: what is the ground pattern of the lophophore in lophophorates? BMC Evolutionary Biology 17, 172.

Temereva, E. N.; Malakhov, V. V. 2003. The organization and origin of the circulatory system of Phoronida (Lophophorata). Doklady Biological Sciences 389, 166–169.

Temereva, E. N.; Malakhov, V. V. 2006. Development of excretory organs in *Phoronopsis harmeri* (Phoronida): From protonephridium to nephromixium. Zoologichesky Zhurnal 85, 915–924.

Temereva, E. N.; Malakhov, V. V. 2007. Embryogenesis and larval development of *Phoronopsis harmeli* Pixell, 1912 (Phoronida): dual origin of the coelomic mesoderm. Invertebrate Reproduction and Development 50, 57–66.

Temereva, E. N.; Malakhov, V. V. 2009a. Microscopic anatomy and ultrastructure of the nervous system of *Phoronopsis harmeri* Pixell, 1912 (Lophophorata: Phoronida). Russian Journal of Marine Biology 35, 388–404.

Temereva, E. N.; Malakhov, V. V. 2009b. On the organization of the lophophore in phoronids (Lophophorata: Phoronida). Russian Journal of Marine Biology 35, 479–489.

Temereva, E. N.; Malakhov, V. V. 2010. Filter feeding mechanism in the phoronid *Phoronopsis harmeri* (Phoronida, Lophophorata). Russian Journal of Marine Biology 36, 109–116.

Temereva, E. N.; Malakhov, V. V. 2011. Organization of the epistome in *Phoronopsis harmeri* (Phoronida) and consideration of the coelomic organization in Phoronida. Zoomorphology 130, 121–134.

Temereva, E. N.; Neldyudov, B. V. 2017. A new phoronid species, *Phoronis savinkini* sp n., from the South China Sea and an analysis of the taxonomic diversity of Phoronida. Zoologichesky Zhurnal 96, 1285–1308.

Temereva, E. N.; Tsitrin, E. B. 2015. Modern data on the innervation of the lophophore in *Lingula anatina* (Brachiopoda) support the monophyly of the lophophorates. PLoS One 10, e0123040.

Temkin, M. H. 1994. Gamete spawning and fertilization in the gymnolaemate bryozoan *Membranipora membranacea*. Biological Bulletin 187, 143–155.

Teuchert, G. 1968. Zur Fortpflanzung und Entwicklung der Macrodasyoidea (Gastrotricha). Zeitschrift für Morphologie der Tiere 63, 343–418.

Teuchert, G. 1973. Die Feinstruktur des Protonephridialsystems von *Turbanella cornuta* Remane, einem marinen Gastrotrich der Ordnung Macrodasyoidea. Zeitschrift für Zellforschung und Mikroskopische Anatomie 136, 277–289.

Teuchert, G. 1976. Sinneseinrichtungen bei *Turbanella cornuta* Remane (Gastrotricha). Zoomorphologie 83, 193–207.

Teuchert, G. 1977. The ultrastructure of the marine gastrotrich *Turbanella cornuta* Remane (Macrodasoydea) and its functional and phylogenetic importance. Zoomorphologie 88, 189–246.

Thacker, R. W.; Freeman, C. J. 2012. Sponge-microbe symbioses: Recent advances and new directions. Advances in Marine Biology 62, 57–111.

Thacker, R. W.; Hill, A. L.; Hill, M. S.; Redmond, N. E.; Collins, A. G.; Morrow, C. C.; Spicer, L.; Carmack, C. A.; Zappe, M. E.; Pohlmann, D.; et al. 2013. Nearly complete 28S rRNA gene sequences confirm new hypotheses of sponge evolution. Integrative and Comparative Biology 53, 373–387.

Thayer, C. W. 1985. Brachiopods versus mussels: competition, predation, and palatability. Science 228, 1527–1528.

Theologidis, I.; Fodelianakis, S.; Gaspar, M. B.; Zouros, E. 2008. Doubly uniparental inheritance (DUI) of mitochondrial DNA in *Donax trunculus* (Bivalvia: Donacidae) and the problem of its sporadic detection in Bivalvia. Evolution 62, 959–970.

Thiel, D.; Bauknecht, P.; Jékely, G.; Hejnol, A. 2017. An ancient FMRFamide-related peptide-receptor pair induces defence behaviour in a brachiopod larva. Open Biology 7, 170136.

Thollesson, M.; Norenburg, J. L. 2003. Ribbon worm relationships: a phylogeny of the phylum Nemertea. Proceedings: Biological Sciences 270, 407–415.

Thomas, F.; Schmidt-Rhaesa, A.; Martin, G.; Manu, C.; Durand, P.; Renaud, F. 2002. Do hairworms (Nematomorpha) manipulate the water seeking behaviour of their terrestrial hosts? Journal of Evolutionary Biology 15, 356–361.

Thomas, R. 2018. Remarkably preserved chaetae of *Pelagiella* document the occurrence of torsion in development of an early Cambrian stem gastropod and support the lophotrochozoan afilation of the Mollusca. 5th International Paleontological Congress—Paris, 9th–13th July 2018, Paris, p. 311.

Thompson, G. A.; Dinofrio, E. O.; Alder, V. A. 2013. Structure, abundance and biomass size spectra of copepods and other zooplankton communities in upper waters of the Southwestern Atlantic Ocean during summer. Journal of Plankton Research 35, 610–629.

Thompson, I. 1979. Errant polychaetes (Annelida) from the Pennsylvanian Essex fauna of northern Illinois. Palaeontographica Abteilung A Palaeozoologie-Stratigraphie 163, 169–199.

Thompson, T. E. 1965. Epidermal acid-secretion in some marine polyclad Turbellaria. Nature 206, 954–955.

Thuesen, E. V.; Goetz, F. E.; Haddock, S. H. D. 2010. Bioluminescent organs of two deep-sea arrow worms, *Eukrohnia fowleri* and *Caecosagitta macrocephala*, with further observations on bioluminescence in chaetognaths. Biological Bulletin 219, 100–111.

Thuesen, E. V.; Kogure, K. 1989. Bacterial production of tetrodotoxin in four species of Chaetognatha. Biological Bulletin 176, 191–194.

Thuesen, E. V.; Kogure, K.; Hashimoto; K.; Nemoto, T. 1988. Poison arrowworms: A tetrodotoxin venom in the marine phylum Chaetognatha. Journal of Experimental Marine Biology and Ecology 116, 249–256.

Thuy, B.; Stöhr, S. 2016. A new morphological phylogeny of the Ophiuroidea (Echinodermata) accords with molecular evidence and renders microfossils accessible for cladistics. PLoS One 11, e0156140.

Tilic, E.; Bartolomaeus, T. 2016. Structure, function and cell dynamics during chaetogenesis of abdominal uncini in

Sabellaria alveolata (Sabellariidae, Annelida). Zoological Letters 2, 1.

Tilic, E.; Pauli, B.; Bartolomaeus, T. 2017. Getting to the root of fireworms' stinging chaetae–chaetal arrangement and ultrastructure of *Eurythoe complanata* (Pallas, 1766) (Amphinomida). Journal of Morphology 278, 865–876.

Todaro, M. A.; Dal Zotto, M.; Jondelius, U.; Hochberg, R.; Hummon, W. D.; Kånneby, T.; Rocha, C. E. F. 2012. Gastrotricha: A marine sister for a freshwater puzzle. PLoS One 7, e31740.

Todaro, M. A.; Kånneby, T.; Dal Zotto, M.; Jondelius, U. 2011. Phylogeny of Thaumastodermatidae (Gastrotricha: Macrodasyida) inferred from nuclear and mitochondrial sequence data. PLoS One 6, e17892.

Todaro, M. A.; Littlewood, D. T. J.; Balsamo, M.; Herniou, E. A.; Cassanelli, S.; Manicardi, G.; Wirz, A.; Tongiorgi, P. 2003. The interrelationships of the Gastrotricha using nuclear small rRNA subunit sequence data, with an interpretation based on morphology. Zoologischer Anzeiger 242, 145–156.

Todaro, M. A.; Telford, M. J.; Lockyer, A. E.; Littlewood, D. T. J. 2006. Interrelationships of the Gastrotricha and their place among the Metazoa inferred from 18S rRNA genes. Zoologica Scripta 35, 251–259.

Todd, J. A. 2000. The central role of ctenostomes in bryozoan phylogeny. In: Herrera Cubilla, A.; Jackson, J. B. (eds.), Proceedings of the 11th International Bryozoology Association Conference, Smithsonian Tropical Research Institute, Balboa, Republic of Panama, pp. 104–135.

Todd, J. A.; Taylor, P. D. 1992. The first fossil entoproct. Naturwissenschaften 79, 311–314.

Todt, C.; Buchinger, T.; Wanninger, A. 2008. The nervous system of the basal mollusk *Wirenia argentea* (Solenogastres): a study employing immunocytochemical and 3D reconstruction techniques. Marine Biology Research 4, 290–303.

Todt, C.; Wanninger, A. 2010. Of tests, trochs, shells, and spicules: Development of the basal mollusk *Wirenia argentea* (Solenogastres) and its bearing on the evolution of trochozoan larval key features. Frontiers in Zoology 7, 6.

Tokioka, T. 1965. The taxonomical outline of Chaetognatha. Publications of the Seto Marine Biological Laboratory 12, 335–357.

Topper, T. P.; Guo, J.; Clausen, S.; Skovsted, C. B.; Zhang, Z. 2019. A stem group echinoderm from the basal Cambrian of China and the origins of Ambulacraria. Nature Communications 10, 1366.

Topper, T. P.; Strotz, L. C.; Holmer, L. E.; Caron, J.-B. 2015. Survival on a soft seafloor: life strategies of brachiopods from the Cambrian Burgess Shale. Earth-Science Reviews 151, 266–287.

Torruella, G.; de Mendoza, A.; Grau-Bove, X.; Anto, M.; Chaplin, M. A.; Del Campo, J.; Eme, L.; Pérez-Cordón, G.; Whipps, C. M.; Nichols, K. M.; et al. 2015. Phylogenomics reveals convergent evolution of lifestyles in close relatives of animals and fungi. Current Biology 25, 2404–2410.

Toth, L. T.; Aronson, R. B.; Vollmer, S. V.; Hobbs, J. W.; Urrego, D. H.; Cheng, H.; Enochs, I. C.; Combosch, D. J.; van Woesik, R.; Macintyre, I. G. 2012. ENSO drove 2500-year collapse of Eastern Pacific coral reefs. Science 337, 81–84.

Travis, P. B. 1983. Ultrastructural study of body wall organization and Y-cell composition in the Gastrotricha. Zeitschrift für zoologische Systematik und Evolutionsforschung 21, 52–68.

Treibergs, K. A. 2019. How does a polymorphic colony divide labor among its modules? Colonial development in the marine invertebrate, *Bugulina stolonifera*. PhD thesis, Department of Organsmic and Evolutionary Biology, Harvard University, Cambridge, MA.

Treonis, A. M.; Wall, D. H. 2005. Soil nematodes and desiccation survival in the extreme arid environment of the Antarctic Dry Valleys. Integrative and Comparative Biology 45, 741–750.

Trewick, S. A. 2000. Mitochondrial DNA sequences support allozyme evidence for cryptic radiation of New Zealand *Peripatoides* (Onychophora). Molecular Ecology 9, 269–281.

Tsagkogeorga, G.; Turon, X.; Galtier, N.; Douzery, E. J. P.; Delsuc, F. 2010. Accelerated evolutionary rate of housekeeping genes in tunicates. Journal of Molecular Evolution 71, 153–167.

Tsagkogeorga, G.; Turon, X.; Hopcroft, R. R.; Tilak, M. K.; Feldstein, T.; Shenkar, N.; Loya, Y.; Huchon, D.; Douzery, E. J. P.; Delsuc, F. 2009. An updated 18S rRNA phylogeny of tunicates based on mixture and secondary structure models. BMC Evolutionary Biology 9, 187.

Tsai, I. J.; Zarowiecki, M.; Holroyd, N.; Garciarrubio, A.; Sanchez-Flores, A.; Brooks, K. L.; Tracey, A.; Bobes, R. J.; Fragoso, G.; Sciutto, E.; et al. 2013. The genomes of four tapeworm species reveal adaptations to parasitism. Nature 496, 57–63.

Tsaousis, A. D.; Martin, D. P.; Ladoukakis, E. D.; Posada, D.; Zouros, E. 2005. Widespread recombination in published animal mtDNA sequences. Molecular Biology and Evolution 22, 925–933.

Tsujimoto, M.; Imura, S.; Kanda, H. 2016. Recovery and reproduction of an Antarctic tardigrade retrieved from a moss sample frozen for over 30 years. Cryobiology 72, 78–81.

Turbeville, J. M. 1986. An ultrastructural analysis of coelomogenesis in the hoplonemertine *Prosorhochmus americanus* and the polychaete *Magelona* sp. Journal of Morphology 187, 51–56.

Turbeville, J. M. 1991. Nemertinea. In: Harrison, F. W.; Bogitsh, B. J. (eds.), Microscopic anatomy of invertebrates, vol. 3: Platyhelminthes and Nemertinea. Wiley-Liss, New York, pp. 285–328.

Turbeville, J. M.; Field, K. G.; Raff, R. A. 1992. Phylogenetic position of phylum Nemertini, inferred from 18S rRNA sequences: molecular data as a test of morphological character homology. Molecular Biology and Evolution 9, 235–249.

Turbeville, J. M.; Pfeifer, D. M.; Field, K. G.; Raff, R. A. 1991. The phylogenetic status of arthropods, as inferred from 18S rRNA sequences. Molecular Biology and Evolution 8, 669–686.

Turbeville, J. M.; Ruppert, E. E. 1983. Epidermal muscles and peristaltic burrowing in *Carinoma tremaphoros*

(Nemertini): correlates of effective burrowing without segmentation. Zoomorphology 103, 103–120.

Turbeville, J. M.; Ruppert, J. E. 1985. Comparative ultrastructure and the evolution of nemertines. American Zoologist 25, 53–71.

Turbeville, J. M.; Smith, D. M. 2007. The partial mitochondrial genome of the *Cephalothrix rufifrons* (Nemertea, Palaeonemertea): Characterization and implications for the phylogenetic position of Nemertea. Molecular Phylogenetics and Evolution 43, 1056–1065.

Turner, R. L.; Meyer, C. E. 1980. Salinity tolerance of the brackish-water echinoderm *Ophiophragmus filograneus* (Ophiuroidea). Marine Ecology Progress Series 2, 249–256.

Turon, X.; López-Legentil, S. 2004. Ascidian molecular phylogeny inferred from mtDNA data with emphasis on the Aplousobranchiata. Molecular Phylogenetics and Evolution 33, 309–320.

Turpeenniemi, T. A.; Hyvärinen, H. 1996. Structure and role of the renette cell and caudal glands in the nematode *Sphaerolaimus gracilis* (Monhysterida). Journal of Nematology 28, 318–327.

Tyler, S. 1988. The role of function in determination of homology and convergence—examples from invertebrate adhesive organs. Fortschritte der Zoologie 36, 331–347.

Tyler, S.; Hooge, M. 2004. Comparative morphology of the body wall in flatworms (Platyhelminthes). Canadian Journal of Zoology 82, 194–210.

Tyler, S.; Hooge, M. D. 2001. Musculature of *Gnathostomula armata* Riedl 1971 and its ecological significance. P.S.Z.N.: Marine Ecology 22, 71–83.

Tyler, S.; Melanson, L. A.; Rieger, R. M. 1980. Adhesive organs of the Gastrotricha. II. The organs of *Neodasys*. Zoomorphologie 95, 17–26.

Tyler, S.; Rieger, G. E. 1980. Adhesive organs of the Gastrotricha. I. Duo-gland organs. Zoomorphologie 95, 1–15.

Tyler, S.; Rieger, R. M. 1975. Uniflagellate spermatozoa in Nemertoderma (Turbellaria) and their phylogenetic significance. Science 188, 730–732.

Tyler, S.; Rieger, R. M. 1977. Ultrastructural evidence for the systematic position of the Nemertodermatida (Turbellaria). Acta Zoologica Fennica 154, 193–207.

Tzetlin, A. B.; Filippova, A. V. 2005. Muscular system in polychaetes (Annelida). Hydrobiologia 535/536, 113–126.

Ukmar-Godec, T.; Kapun, G.; Zaslansky, P.; Faivre, D. 2015. The giant keyhole limpet radular teeth: A naturally-grown harvest machine. Journal of Structural Biology 192, 392–402.

Uliano-Silva, M.; Dondero, F.; Dan Otto, T.; Costa, I.; Lima, N. C. B.; Americo, J. A.; Mazzoni, C. J.; Prosdocimi, F.; Rebelo, M. de F. 2018. A hybrid-hierarchical genome assembly strategy to sequence the invasive golden mussel, *Limnoperna fortunei*. GigaScience 7, 1–10.

Ullrich-Lüter, E. M.; Dupont, S.; Arboleda, E.; Hausen, H.; Arnone, M. I. 2011. Unique system of photoreceptors in sea urchin tube feet. Proceedings of the National Academy of Sciences of the USA 108, 8367–8372.

Ungerer, P.; Scholtz, G. 2008. Filling the gap between identified neuroblasts and neurons in crustaceans adds new support for Tetraconata. Proceedings of the Royal Society B: Biological Sciences 275, 369–376.

Ungerer, P.; Scholtz, G. 2009. Cleavage and gastrulation in *Pycnogonum litorale* (Arthropoda, Pycnogonida): morphological support for the Ecdysozoa? Zoomorphology 128, 263–274.

Uribe, J. E.; Kano, Y.; Templado, J.; Zardoya, R. 2016. Mitogenomics of Vetigastropoda: insights into the evolution of pallial symmetry. Zoologica Scripta 45, 145–159.

Ussow, M. 1887. Ein neue Form von Süfswasser-Cölenteraten. Morphologisches Jahrbuch 12, 137–153.

Uyeno, D.; Lasley, R. M., Jr.; Moore, J. M.; Berumen, M. L. 2015. New records of *Lobatolampea tetragona* (Ctenophora: Lobata: Lobatolampeidae) from the Red Sea. Marine Biodiversity Records 8, e33.

Vacelet, J. 2002. Recent 'Sphinctozoa,' Order Verticillitida, Family Verticillitidae Steinmann, 1882. In: Hooper, J. N. A.; Van Soest, R. W. M.; Willenz, P. (eds.), Systema Porifera. Springer, Boston, pp. 1097–1098.

Vacelet, J.; Boury-Esnault, N. 1995. Carnivorous sponges. Nature 373, 333–335.

Vacelet, J.; Boury-Esnault, N.; Fiala-Medioni, A.; Fisher, C. R. 1995. A methanotrophic carnivorous sponge. Nature 377, 296.

Valentine, J. W. 2004. On the Origin of Phyla. University of Chicago Press, Chicago.

Valentine, J. W.; Collins, A. G. 2000. The significance of moulting in ecdysozoan evolution. Evolution and Development 2, 152–156.

Van Beneden, É. 1876. Recherches sur les *Dicyémides*, survivants actuels d'un embranchement des *Mésozoires*. Bulletins de l'Académie royale des sciences, des lettres et des beaux-arts de Belgique 42, 35–97.

van Cleave, H. J. 1932. Eutely or cell constancy in its relation to body size. Quarterly Review of Biology 7, 59–67.

van den Biggelaar, J. A. M.; Dictus, W. J. A. G.; van Loon, A. E. 1997. Cleavage patterns, cell-lineages and cell specification are clues to phyletic lineages in Spiralia. Seminars in Cell and Developmental Biology 8, 367–378.

van der Kooi, C. J.; Matthey-Doret, C.; Schwander, T. 2017. Evolution and comparative ecology of parthenogenesis in haplodiploid arthropods. Evolution Letters 1, 304–316.

van Deurs, B. 1974. Pycnogonid sperm. An example of inter- and intraspecifical axonemal variation. Cell and Tissue Research 149, 105–111.

Van Dover, C. L.; Humphris, S. E.; Fornari, D.; Cavanaugh, C. M.; Collier, R.; Goffredi, S. K.; Hashimoto, J.; Lilley, M. D.; Reysenbach, A. L.; Shank, T. M.; et al. 2001. Biogeography and ecological setting of Indian Ocean hydrothermal vents. Science 294, 818–823.

Van Iten, H.; de Moraes Leme, J.; Coelho Rodrigues, S.; Guimarães Simoes, M. 2005. Reinterpretation of a conulariid-like fossil from the Vendian of Russia. Palaeontology 48, 619–622.

Van Iten, H.; de Moraes Leme, J.; Simões, M. G.; Marques, A. C.; Collins, A. G. 2006. Reassessment of the phylogenetic position of conulariids (?Ediacaran–Triassic) within the subphylum Medusozoa (phylum Cnidaria). Journal of Systematic Palaeontology 4, 109–118.

Van Iten, H.; Marques, A. C.; Leme, J. d. M.; Pacheco, M. L. A. F.; Simões, M. G. 2014. Origin and early diversification of the phylum Cnidaria Verrill: major developments in the analysis of the taxon's Proterozoic–Cambrian history. Palaeontology 57, 677–690.

van Megen, H.; van den Elsen, S.; Holterman, M.; Karssen, G.; Mooyman, P.; Bongers, T.; Holovachov, O.; Bakker, J.; Helder, J. 2009. A phylogenetic tree of nematodes based on about 1200 full-length small subunit ribosomal DNA sequences. Nematology 11, 927–S927.

van Oppen, M. J. H.; McDonald, B. J.; Willis, B.; Miller, D. J. 2001. The evolutionary history of the coral genus Acropora (Scleractinia, Cnidaria) based on a mitochondrial and a nuclear marker: Reticulation, incomplete lineage sorting, or morphological convergence? Molecular Biology and Evolution 18, 1315–1329.

Van Soest, R. W. M.; Boury-Esnault, N.; Vacelet, J.; Dohrmann, M.; Erpenbeck, D.; De Voogd, N. J.; Santodomingo, N.; Vanhoorne, B.; Kelly, M.; Hooper, J. N. 2012. Global diversity of sponges (Porifera). PLoS One 7, e35105.

Van Steenkiste, N.; Davison, P.; Artois, T. 2010. Bryoplana xerophila n. g. n. sp., a new limnoterrestrial microturbellarian (Platyhelminthes, Typhloplanidae, Protoplanellinae) from epilithic mosses, with notes on its ecology. Zoological Science 27, 285–291.

Vannier, J. 2012. Gut contents as direct indicators for trophic relationships in the Cambrian marine ecosystem. PLoS One 7, e52200.

Vannier, J.; Calandra, I.; Gaillard, C.; Żylińska, A. 2010. Priapulid worms: pioneer horizontal burrowers at the Precambrian-Cambrian boundary. Geology 38, 711–714.

Vannier, J.; Liu, J. N.; Lerosey-Aubril, R.; Vinther, J.; Daley, A. C. 2014. Sophisticated digestive systems in early arthropods. Nature Communications 5, 3641.

Vannier, J.; Steiner, M.; Renvoisé, E.; Hu, S.-X.; Casanova, J.-P. 2007. Early Cambrian origin of modern food webs: evidence from predator arrow worms. Proceedings of the Royal Society B: Biological Sciences 274, 627–633.

Velandia-Huerto, C. A.; Gittenberger, A. A.; Brown, F. D.; Stadler, P. F.; Bermúdez-Santana, C. I. 2016. Automated detection of ncRNAs in the draft genome sequence of a colonial tunicate: the carpet sea squirt Didemnum vexillum. BMC Genomics 17, 691.

Velasco-Castrillón, A.; McInnes, S. J.; Schultz, M. B.; Arróniz-Crespo, M.; D'Haese, C. A.; Gibson, J. A. E.; Adams, B. J.; Page, T. J.; Austin, A. D.; Cooper, S. J. B.; et al. 2015. Mitochondrial DNA analyses reveal widespread tardigrade diversity in Antarctica. Invertebrate Systematics 29, 578–590.

Vellutini, B. C.; Hejnol, A. 2016. Expression of segment polarity genes in brachiopods supports a non-segmental ancestral role of engrailed for bilaterians. Scientific Reports 6, 32387.

Vellutini, B. C.; Martín-Durán, J. M.; Hejnol, A. 2017. Cleavage modification did not alter blastomere fates during bryozoan evolution. BMC Biology 15, 33.

Vendrasco, M. J.; Checa, A. G.; Kouchinsky, A. V. 2011. Shell microstructure of the early bivalve Pojetaia and the independent origin of nacre within the Mollusca. Palaeontology 54, 825–850.

Vendrasco, M. J.; Wood, T. E.; Runnegar, B. N. 2004. Articulated Palaeozoic fossil with 17 plates greatly expands disparity of early chitons. Nature 429, 288–291.

Venter, J. C.; Adams, M. D.; Myers, E. W.; Li, P. W.; Mural, R. J.; Sutton, G. G.; Smith, H. O.; Yandell, M.; Evans, C. A.; Holt, R. A.; et al. 2001. The sequence of the human genome. Science 291, 1304–1351.

Verdes, A.; Gruber, D. F. 2017. Glowing worms: Biological, chemical, and functional diversity of bioluminescent annelids. Integrative and Comparative Biology 57, 18–32.

Verdes, A.; Simpson, D.; Holford, M. 2018. Are fireworms venomous? Evidence for the convergent evolution of toxin homologs in three species of fireworms (Annelida, Amphinomidae). Genome Biology and Evolution 10, 249–268.

Vermeij, G. J. 1987. Evolution and escalation: an ecological history of life. Princeton University Press, Princeton.

Vermeij, M. J. A. 2009. Floating corallites: a new ecophenotype in scleractinian corals. Coral Reefs 28, 987.

Vermeij, M. J. A.; Bakm, R. P. M. 2002. Corals on the move: rambling of Madracis pharensis polyps early after settlement. Coral Reefs 21, 262–263.

Vermeij, M. J. A.; Marhaver, K. L.; Huijbers, C. M.; Nagelkerken, I.; Simpson, S. D. 2010. Coral larvae move toward reef sounds. PLoS One 5, e10660.

Veronico, P.; Gray, L. J.; Jones, J. T.; Bazzicalupo, P.; Arbucci, S.; Cortese, M. R.; Di Vito, M.; De Giorgi, C. 2001. Nematode chitin synthases: gene structure, expression and function in Caenorhabditis elegans and the plant parasitic nematode Meloidogyne artiella. Molecular Genetics and Genomics 266, 28–34.

Verrill, A. E. 1879. Notice of recent additions to the marine invertebrates of the northeastern coast of America, pt. 1: Annelida, Gephyraea, Nemertina, Nematoda, Polyzoa, Tunicata, Mollusca, Anthozoa, Echinodermata, Porifera. Proceedings of the United States National Museum 2, 165–205.

Vidal-Dupiol, J.; Zoccola, D.; Tambutté, E.; Grunau, C.; Cosseau, C.; Smith, K. M.; Freitag, M.; Dheilly, N. M.; Allemand, D.; Tambutté, S. 2013. Genes related to ion-transport and energy production are upregulated in response to CO2-driven pH decrease in corals: new insights from transcriptome analysis. PLoS One 8, e58652.

Vila-Farré, M.; Álvarez-Presas, M.; Achatz, J. G. 2013. First record of Oligochoerus limnophilus (Acoela, Acoelomorpha) from British waters. Arxius de Miscel·lània Zoològica 11, 153–157.

Vinogradov, A. V. 1996. New fossil freshwater bryozoans from the Asiatic part of Russia and Kazakhstan. Paleontological Journal 30, 284–292.

Vinther, J. 2009. The canal system in sclerites of Lower Cambrian Sinosachites (Halkieriidae: Sachitida): Significance for the molluscan affinities of the sachitids. Palaeontology 52, 689–712.

Vinther, J. 2014. A molecular palaeobiological perspective on aculiferan evolution. Journal of Natural History 48, 2805–2823.

Vinther, J. 2015. The origins of molluscs. Palaeontology 58, 19–34.

Vinther, J.; Eibye-Jacobsen, D.; Harper, D. A. 2011. An Early Cambrian stem polychaete with pygidial cirri. Biology Letters 7, 929–932.

Vinther, J.; Parry, L.; Briggs, D. E. G.; Van Roy, P. 2017. Ancestral morphology of crown-group molluscs revealed by a new Ordovician stem aculiferan. Nature 542, 471–474.

Vinther, J.; Parry, L. A. 2019. Bilateral jaw elements in *Amiskwia sagittiformis* bridge the morphological gap between gnathiferans and chaetognaths. Current Biology 29, 881–888.

Vinther, J.; Porras, L.; Young, F. J.; Budd, G. E.; Edgecombe, G. D. 2016. The mouth apparatus of the Cambrian gilled lobopodian *Pambdelurion whittingtoni*. Palaeontology 59, 841–849.

Vinther, J.; Sperling, E. A.; Briggs, D. E. G.; Peterson, K. J. 2012. A molecular palaeobiological hypothesis for the origin of aplacophoran molluscs and their derivation from chiton-like ancestors. Proceedings of the Royal Society B: Biological Sciences 279, 1259–1268.

Vinther, J.; Van Roy, P.; Briggs, D. E. 2008. Machaeridians are Palaeozoic armoured annelids. Nature 451, 185–188.

Voight, J. R. 2005. First report of the enigmatic echinoderm *Xyloplax* from the North Pacific. Biological Bulletin 208, 77–80.

Voigt, O.; Adamska, M.; Adamski, M.; Kittelmann, A.; Wencker, L.; Wörheide, G. 2017. Spicule formation in calcareous sponges: Coordinated expression of biomineralization genes and spicule-type specific genes. Scientific Reports 7, 45658.

Voigt, O.; Collins, A. G.; Pearse, V. B.; Pearse, J. S.; Ender, A.; Hadrys, H.; Schierwater, B. 2004. Placozoa—no longer a phylum of one. Current Biology 14, R944–945.

Voigt, O.; Erpenbeck, D.; Wörheide, G. 2008. Molecular evolution of rDNA in early diverging Metazoa: first comparative analysis and phylogenetic application of complete SSU rRNA secondary structures in Porifera. BMC Evolutionary Biology 8, 69.

Vollmer, S. V.; Palumbi, S. R. 2002. Hybridization and the evolution of reef coral diversity. Science 296, 2023–2025.

von Dassow, G.; Emlet, R. B.; Maslakova, S. A. 2013. How the pilidium larva feeds. Frontiers in Zoology 10, 47.

von Döhren, J. 2011. The fate of the larval epidermis in the Desor-larva of *Lineus viridis* (Pilidiophora, Nemertea) displays a historically constrained functional shift from planktotrophy to lecithotrophy. Zoomorphology 130, 189–196.

von Döhren, J.; Bartolomaeus, T. 2007. Ultrastructure and development of the rhabdomeric eyes in *Lineus viridis* (Heteronemertea, Nemertea). Zoology (Jena) 110, 430–438.

von Erlanger, R. 1895. Beiträge zur Morphologie der Tardigraden. I. Zur Embryologie eines Tardigraden: *Macrobiotus macronyx* Dujardin. Morphologisches Jahrbuch 22, 491–513.

von Haffner, K. 1942. Untersuchungen über das Urogenitalsystem der Acanthocephalen. II Teil: Das Urogenitalsystem von Gigantorhynchus echinodiscus Diesing. Zeitschrift für Morphologie und Oekologie der Tiere Berlin 38, 295–316.

von Haffner, K. 1950. Organisation und systematische Stellung der Acanthocephalen. Zoologischer Anzeiger 145, 243–274.

von Nickisch-Rosenegk, M.; Brown, W. M.; Boore, J. L. 2001. Complete sequence of the mitochondrial genome of the tapeworm *Hymenolepis diminuta*: Gene arrangements indicate that platyhelminths are eutrochozoans. Molecular Biology and Evolution 18, 721–730.

von Reumont, B. M.; Campbell, L. I.; Jenner, R. A. 2014a. *Quo vadis* venomics? A roadmap to neglected venomous invertebrates. Toxins 6, 3488–3551.

von Reumont, B. M.; Campbell, L. I.; Richter, S.; Hering, L.; Sykes, D.; Hetmank, J.; Jenner, R. A.; Bleidorn, C. 2014b. A polychaete's powerful punch: Venom gland transcriptomics of *Glycera* reveals a complex cocktail of toxin homologs. Genome Biology and Evolution 6, 2406–2423.

von Reumont, B. M.; Jenner, R. A.; Wills, M. A.; Dell'Ampio, E.; Pass, G.; Ebersberger, I.; Meyer, B.; Koenemann, S.; Iliffe, T. M.; Stamatakis, A.; et al. 2012. Pancrustacean phylogeny in the light of new phylogenomic data: support for Remipedia as the possible sister group of Hexapoda. Molecular Biology and Evolution 29, 1031–1045.

von Siebold, C. T. W.; Stannius, H. 1848. Lehrbuch der vergliechenden Anatomie der Wirbellosen Tiere. Verlag von Veit and Comp., Berlin.

von Wenck, W. 1914. Entwicklungsgeschichtliche Untersuchungen an Tardigraden (Macrobiotus lacustris Duj.). Zoologische Jahrbücher, Abteilung für Anatomie und Ontogenie der Tiere 37, 465–514.

Voronezhskaya, E. E.; Croll, R. P. 2016. Mollusca: Gastropoda. In: Schmidt-Rhaesa, A.; Harzsch, S.; Purschke, G. (eds.), Structure and Evolution of Invertebrate Nervous Systems. Oxford University Press, Oxford, pp. 196–221.

Vortsepneva, E.; Tzetlin, A.; Purschke, G.; Mugue, N.; Hass-Cordes, E.; Zhadan, A. 2008. The parasitic polychaete known as *Asetocalamyzas laonicola* (Calamyzidae) is in fact the dwarf male of the spionid *Scolelepis laonicola* (comb. nov.) Invertebrate Biology 127, 403–416.

Voskoboynik, A.; Neff, N. F.; Sahoo, D.; Newman, A. M.; Pushkarev, D.; Koh, W.; Passarelli, B.; Fan, H. C.; Mantalas, G. L.; Palmeri, K. J.; et al. 2013. The genome sequence of the colonial chordate, *Botryllus schlosseri*. eLife 2, e00569.

Voskoboynik, A.; Weissman, I. L. 2015. *Botryllus schlosseri*, an emerging model for the study of aging, stem cells, and mechanisms of regeneration. Invertebrate Reproduction and Development 59, 33–38.

Wächtler, K.; Dreher-Mansur, M. C.; Richter, T. 2001. Larval types and early postlarval biology in naiads (Unionoida). Ecological Studies 145, 93–125.

Wada, H. 1998. Evolutionary history of free-swimming and sessile lifestyles in urochordates as deduced from 18S rDNA molecular phylogeny. Molecular Biology and Evolution 15, 1189–1194.

Wada, H.; Satoh, N. 1994. Details of the evolutionary history from invertebrates to vertebrates, as deduced from the sequences of 18S rDNA. Proceedings of the National Academy of Sciences of the USA 91, 1801–1804.

Waeschenbach, A.; Cox, C. J.; Littlewood, D. T. J.; Porter, J. S.; Taylor, P. D. 2009. First molecular estimate of cyclostome bryozoan phylogeny confirms extensive homoplasy among skeletal characters used in traditional taxonomy. Molecular Phylogenetics and Evolution 52, 241–251.

Waeschenbach, A.; Porter, J. S.; Hughes, R. N. 2012a. Molecular variability in the *Celleporella hyalina* (Bryozoa; Cheilostomata) species complex: evidence for cryptic speciation from complete mitochondrial genomes. Molecular Biology Reports 39, 8601–8614.

Waeschenbach, A.; Taylor, P. D.; Littlewood, D. T. J. 2012b. A molecular phylogeny of bryozoans. Molecular Phylogenetics and Evolution 62, 718–735.

Waeschenbach, A.; Telford, M. J.; Porter, J. S.; Littlewood, D. T. J. 2006. The complete mitochondrial genome of *Flustrellidra hispida* and the phylogenetic position of Bryozoa among the Metazoa. Molecular Phylogenetics and Evolution 40, 195–207.

Waeschenbach, A.; Vieira, L. M.; Reverter-Gil, O.; Souto Derungs, J.; Nascimento, K. B.; Fehlauer-Ale, K. H. 2015. A phylogeny of Vesiculariidae (Bryozoa, Ctenostomata) supports synonymization of three genera and reveals possible cryptic diversity. Zoologica Scripta 44, 667–683.

Wägele, H.; Ballesteros, M.; Avila, C. 2006. Defensive glandular structures in opisthobranch molluscs—From histology to ecology. Oceanography and Marine Biology: An Annual Review 44, 197–276.

Wägele, J. W.; Erikson, T.; Lockhart, P. J.; Misof, B. 1999. The Ecdysozoa: Artifact or monophylum? Journal of Zoological Systematics and Evolutionary Research 37, 211–223.

Wägele, J. W.; Kück, P. 2014. Arthropod phylogeny and the origin of Tracheata (= Atelocerata) from Remipedia–like ancestors. In: Wägele, J. W.; Bartholomaeus, T. (eds.), Deep Metazoan Phylogeny: The backbone of the tree of life. New insights from analyses of molecules, morphology, and theory of data analysis. De Gruyter, Berlin/Boston, pp. 285–341.

Wägele, J. W.; Misof, B. 2001. On quality of evidence in phylogeny reconstruction: a reply to Zrzavý's defence of the 'Ecdysozoa' hypothesis. Journal of Zoological Systematics and Evolutionary Research 39, 165–176.

Waggoner, B. M.; Poinar, G. O., Jr. 1993. Fossil habrotrochid rotifers in Dominican amber. Experientia 49, 354–357.

Wagner, D. E.; Wang, I. E.; Reddien, P. W. 2011. Clonogenic neoblasts are pluripotent adult stem cells that underlie planarian regeneration. Science 332, 811–816.

Waite, J. H.; Broomell, C. C. 2012. Changing environments and structure—property relationships in marine biomaterials. Journal of Experimental Biology 215, 873–883.

Walker, N. S.; Fernández, R.; Sneed, J. M.; Paul, V. J.; Giribet, G.; Combosch, D. J. 2019. Differential gene expression during substrate probing in larvae of the Caribbean coral *Porites astreoides*. Molecular Ecology. doi:10.1111/mec.15265.

Wallace, R. L. 1987. Coloniality in the phylum Rotifera. Hydrobiologia 147, 141–155.

Wallace, R. L. 2002. Rotifers: Exquisite metazoans. Integrative and Comparative Biology 42, 660–667.

Wallace, R. L.; Ricci, C., Melone, G. 1996. A cladistic analysis of pseudocoelomate (aschelminth) morphology. Invertebrate Biology 115, 104–112.

Wallberg, A.; Curini-Galletti, M.; Ahmadzadeh, A.; Jondelius, U. 2007. Dismissal of Acoelomorpha: Acoela and Nemertodermatida are separate early bilaterian clades. Zoologica Scripta 36, 509–523.

Wallberg, A.; Thollesson, M.; Farris, J. S.; Jondelius, U. 2004. The phylogenetic position of the comb jellies (Ctenophora) and the importance of taxonomic sampling. Cladistics 20, 558–578.

Waller, T. R. 1998. Origin of the molluscan class Bivalvia and a phylogeny of major groups. In: Johnston, P. A.; Haggart, J. W. (eds.), Bivalves: An eon of evolution—Palaeobiological studies honoring Norman D. Newell. University of Calgary Press, Calgary, pp. 1–45.

Wang, B.; Collins, J. J., III; Newmark, P. A. 2013. Functional genomic characterization of neoblast-like stem cells in larval *Schistosoma mansoni*. eLife 2, e00768.

Wang, B.; Lee, J.; Li, P.; Saberi, A.; Yang, H.; Liu, C.; Zhao, M.; Newmark, P. A. 2018. Stem cell heterogeneity drives the parasitic life cycle of *Schistosoma mansoni*. eLife 7, e35449.

Wang, C.; Grohme, M. A.; Mali, B.; Schill, R. O.; Frohme, M. 2014. Towards decrypting cryptobiosis—analyzing anhydrobiosis in the tardigrade *Milnesium tardigradum* using transcriptome sequencing. PLoS One 9, e92663.

Wang, S.; Zhang, J.; Jiao, W.; Li, J.; Xun, X.; Sun, Y.; Guo, X.; Huan, P.; Dong, B.; Zhang, L.; et al. 2017a. Scallop genome provides insights into evolution of bilaterian karyotype and development. Nature Ecology and Evolution 1, 0120.

Wang, X.; Han, J.; Vannier, J.; Ou, Q.; Yang, X.; Uesugi, K.; Sasaki, O.; Komiya, T.; Sevastopulo, G. 2017b. Anatomy and affinities of a new 535-million-year-old medusozoan from the Kuanchuanpu Formation, South China. Palaeontology 60, 853–867.

Wang, X.; Lavrov, D. V. 2007. Mitochondrial genome of the homoscleromorph *Oscarella carmela* (Porifera, Demospongiae) reveals unexpected complexity in the common ancestor of sponges and other animals. Molecular Biology and Evolution 24, 363–373.

Wang, X. Y.; Chen, W. J.; Huang, Y.; Sun, J. F.; Men, J. T.; Liu, H. L.; Luo, F.; Guo, L.; Lv, X. L.; Deng, C. H.; et al. 2011. The draft genome of the carcinogenic human liver fluke *Clonorchis sinensis*. Genome Biology 12, R107.

Wang, Y.; Wang, X.; Huang, Y. 2008. Megascopic symmetrical metazoans from the Ediacaran Doushantuo Formation in the Northeastern Guizhou, South China. Journal of China University of Geosciences 19, 200–206.

Wangensteen, O. S.; Palacín, C.; Guardiola, M.; Turon, X. 2018. DNA metabarcoding of littoral hard-bottom communities: high diversity and database gaps revealed by two molecular markers. PeerJ 6, e4705.

Wanninger, A. 2004. Myo-anatomy of juvenile and adult loxosomatid Entoprocta and the use of muscular body plans for phylogenetic inferences. Journal of Morphology 261, 249–257.

Wanninger, A. 2005. Immunocytochemistry of the nervous system and the musculature of the chordoid larva of

Symbion pandora (Cycliophora). Journal of Morphology 265, 237–243.

Wanninger, A. 2016a. Kamptozoa (Entoprocta). In: Schmidt-Rhaesa, A.; Harzsch, S.; Purschke, G. (eds.), Structure and Evolution of Invertebrate Nervous Systems. Oxford University Press, Oxford, pp. 166–171.

Wanninger, A. 2016b. Mollusca: Bivalvia. In: Schmidt-Rhaesa, A.; Harzsch, S.; Purschke, G. (eds.), Structure and Evolution of Invertebrate Nervous Systems. Oxford University Press, Oxford, pp. 190–195.

Wanninger, A.; Fuchs, J.; Haszprunar, G. 2007. Anatomy of the serotonergic nervous system of an entoproct creeping-type larva and its phylogenetic implications. Invertebrate Biology 126, 268–278.

Wanninger, A.; Koop, D.; Degnan, B. M. 2005. Immunocytochemistry and metamorphic fate of the larval nervous system of *Triphyllozoon mucronatum* (Ectoprocta: Gymnolaemata: Cheilostomata). Zoomorphology 124, 161–170.

Wanninger, A.; Ruthensteiner, B.; Haszprunar, G. 2000. Torsion in *Patella caerulea* (Mollusca, Patellogastropoda): Ontogenetic process, timing, and mechanisms. Invertebrate Biology 119, 177–187.

Wanninger, A.; Wollesen, T. 2015. Mollusca. In: Wanninger, A. (ed.), Evolutionary Developmental Biology of Invertebrates 2: Lophotrochozoa (Spiralia). Springer, Vienna, pp. 103–153.

Wanninger, A.; Wollesen, T. 2019. The evolution of molluscs. Biological Reviews 94, 102–115.

Ward, P.; Dooley, F.; Barord, G. J. 2016. *Nautilus*: biology, systematics, and paleobiology as viewed from 2015. Swiss Journal of Palaeontology 135, 169–185.

Warén, A.; Bengtson, S.; Goffredi, S. K.; Van Dover, C. L. 2003. A hot-vent gastropod with iron sulfide dermal sclerites. Science 302, 1007.

Warner, B. G.; Chengalath, R. 1988. Holocene fossil *Habrotrocha angusticollis* (Bdelloidea: Rotifera) in North America. Journal of Paleolimnology 1, 141–147.

Warner, F. D. 1969. The fine structure of the protonephridia in the rotifer *Asplanchna*. Journal of Ultrastructure Research 29, 499–524.

Warren, L. V.; Pacheco, M. L. A. F.; Fairchild, T. R.; Simões, M. G.; Riccomini, C.; Boggiani, P. C.; Cáceres, A. A. 2012. The dawn of animal skeletogenesis: Ultrastructural analysis of the Ediacaran metazoan *Corumbella werneri*. Geology 40, 691–694.

Warwick, R. M. 2000. Are loriciferans paedomorphic (progenetic) priapulids ? Vie et Milieu 50, 191–193.

Watabe, N.; Pan, C.-M. 1984. Phosphatic shell formation in atremate brachiopods. American Zoologist 24, 977–985.

Watson, J. D.; Crick, F. H. 1953. Molecular structure of nucleic acids; a structure for deoxyribose nucleic acid. Nature 171, 737–738.

Watson, N. A.; Rohde, K. 1993. Ultrastructural evidence for an adelphotaxon (sister group) to the Neodermata (Platyhelminthes). International Journal for Parasitology 23, 285–289.

Watson, N. A.; Rohde, K. 1995. Sperm and spermiogenesis of the "Turbellaria" and implications for the phylogeny of the phylum Platyhelminthes. In: Jamieson, B. G.; Ausió, J.;

Justine, J.-L. (eds.), Advances in Spermatozoal Phylogeny and Taxonomy. Mémoires du Muséum National d'Histoire Naturelle, Paris, pp. 37–54.

Watson, N. A.; Rohde, K.; Jondelius, U. 1993. Ultrastructure of sperm and spermiogenesis of *Pterastericola astropectinis* (Platyhelminthes, Rhabdocoela, Pterastericolidae). Parasitology Research 79, 322–328.

Weaver, J. C. 2010. Analysis of an ultra hard magnetic biomineral in chiton radular teeth. Materials Today 13, 42–52.

Webby, B. D.; Elias, R. J.; Young, G. A.; Neuman, B. E. E.; Kaljo, D. 2004. Corals. In: Webby, B. D.; Droser, M. L.; Percival, I. G. (eds.), The great Ordovician biodiversification event. Columbia University Press, New York, pp. 124–146.

Weber, R. E.; Fänge, R. 1980. Oxygen equilibrium of *Priapulus* hemerythrin. Experientia 36, 427–428.

Weber, R. E.; Fänge, R.; Rasmussen, K. K. 1979. Respiratory significance of priapulid hemerythrin. Marine Biology Letters 1, 87–97.

Webster, B. L.; Copley, R. R.; Jenner, R. A.; Mackenzie-Dodds, J. A.; Bourlat, S. J.; Rota-Stabelli, O.; Littlewood, D. T. J.; Telford, M. J. 2006. Mitogenomics and phylogenomics reveal priapulid worms as extant models of the ancestral Ecdysozoan. Evolution and Development 8, 502–510.

Webster, B. L.; Mackenzie-Dodds, J. A.; Telford, M. J.; Littlewood, D. T. J. 2007. The mitochondrial genome of *Priapulus caudatus* Lamarck (Priapulida : Priapulidae). Gene 389, 96–105.

Weigert, A.; Bleidorn, C. 2016. Current status of annelid phylogeny. Organisms Diversity and Evolution 16, 345–362.

Weigert, A.; Golombek, A.; Gerth, M.; Schwarz, F.; Struck, T. H.; Bleidorn, C. 2016. Evolution of mitochondrial gene order in Annelida. Molecular Phylogenetics and Evolution 94, 196–206.

Weigert, A.; Helm, C.; Meyer, M.; Nickel, B.; Arendt, D.; Hausdorf, B.; Santos, S. R.; Halanych, K. M.; Purschke, G.; Bleidorn, C.; et al. 2014. Illuminating the base of the annelid tree using transcriptomics. Molecular Biology and Evolution 31, 1391–1401.

Weisblat, D. A.; Shankland, M. 1985. Cell lineage and segmentation in the leech. Philosophical Transactions of the Royal Society of London B: Biological Sciences 312, 39–56.

Weiss, M. J. 2001. Widespread hermaphroditism in freshwater gastrotrichs. Invertebrate Biology 120, 308–341.

Welch, V. L.; Vigneron, J. P.; Parker, A. R. 2005. The cause of colouration in the ctenophore *Beroë cucumis*. Current Biology 15, R985–R986.

Wełnicz, W.; Grohme, M. A.; Kaczmarek, Ł.; Schill, R. O.; Frohme, M. 2011. Anhydrobiosis in tardigrades—The last decade. Journal of Insect Physiology 57, 577–583.

Welsch, U.; Rehkämper, G. 1987. Podocytes in the axial organ of echinoderms. Journal of Zoology 213, 45–50.

Weltner, W. 1907. Spongillidenstudien V. Zur Biologie von Ephydatia fluviatilis und die Bedeutung der Amöbocyten für die Spongilliden. Archiv für Naturgeschichte 73, 273–286.

Wendt, D. E.; Woollacott, R. M. 1999. Ontogenies of phototactic behavior and metamorphic competence in larvae of three species of *Bugula* (Bryozoa). Invertebrate Biology 118, 75–84.

Wennberg, S. A.; Janssen, R.; Budd, G. E. 2009. Hatching and earliest larval stages of the priapulid worm *Priapulus caudatus*. Invertebrate Biology 128, 157–171.

Wesenberg-Lund, C. 1930. Contributions to the Biology of the Rotifera. II. The Periodicity and Sexual Periods. A. F. Host and Son, Copenhagen.

Wesener, T.; Moritz, L. 2018. Checklist of the Myriapoda in Cretaceous Burmese amber and a correction of the Myriapoda identified by Zhang (2017). Check List 14, 1131–1140.

Wesołowska, W.; Wesołowski, T. 2014. Do *Leucochloridium* sporocysts manipulate the behaviour of their snail hosts? Journal of Zoology 292, 151–155.

West, R. R. 2011. Treatise Online, pt. E (rev.), vol. 4, chap. 2c: Classification of the fossil and living hypercalcified chaetetid-type Porifera (Demospongiae). Treatise Online 22, 1–24. https://doi.org/10.17161/to.v0i0.4139.

Westblad, E. 1937. Die Turbellarien-gattung Nemertoderma Steinböck. Acta Societatis pro Fauna et Flora Fennica 60, 45–89.

Westblad, E. 1949. *Xenoturbella bocki* n.g., n.sp., a peculiar, primitive Turbellarian type. Arkiv för zoologi 1, 3–29.

Westheide, W. 1987. Progenesis as a principle in meiofauna evolution. Journal of Natural History 21, 843–854.

Westheide, W.; Rieger, R. M. 1996. Spezielle Zoologie. Erster Teil: Einzeller und Wirbellose Tiere. Gustav Fischer Verlag, Stuttgart.

Westheide, W.; Rieger, R. M. (eds.). 2007. Spezielle Zoologie. Teil 1: Einzeller und Wirbellose Tiere. Spektrum Akademischer Verlag, Stuttgart.

Wey-Fabrizius, A. R.; Herlyn, H.; Rieger, B.; Rosenkranz, D.; Witek, A.; Mark Welch, D. B.; Ebersberger, I.; Hankeln, T. 2014. Transcriptome data reveal syndermatan relationships and suggest the evolution of endoparasitism in Acanthocephala via an epizoic stage. PLoS One 9, e88618.

Wey-Fabrizius, A. R.; Podsiadlowski, L.; Herlyn, H.; Hankeln, T. 2013. Platyzoan mitochondrial genomes. Molecular Phylogenetics and Evolution 69, 365–375.

Weygoldt, P.; Paulus, H. F. 1979. Untersuchungen zur Morphologie, Taxonomie und Phylogenie der Chelicerata. II. Cladogramme und die Entfaltung der Chelicerata. Zeitschrift für zoologische Systematik und Evolutionsforschung 17, 177–200.

Wheeler, W. C.; Giribet, G. 2016. Molecular data in systematics: a promise fulfilled, a future beckoning. In: Williams, D.; Schmitt, M.; Wheeler, Q. (eds.), The Future of Phylogenetic Systematics: The Legacy of Willi Hennig. Cambridge University Press/Systematics Association, Cambridge, pp. 329–343.

Whelan, N. V.; Kocot, K. M.; Moroz, L. L.; Halanych, K. M. 2015. Error, signal, and the placement of Ctenophora sister to all other animals. Proceedings of the National Academy of Sciences of the USA 112, 5773–5778.

Whelan, N. V.; Kocot, K. M.; Moroz, T. P.; Mukherjee, K.; Williams, P.; Paulay, G.; Moroz, L. L.; Halanych, K. M. 2017. Ctenophore relationships and their placement as the sister group to all other animals. Nature Ecology and Evolution 1, 1737–1746.

Whelan, N. V.; Kocot, K. M.; Santos, S. R.; Halanych, K. M. 2014. Nemertean toxin genes revealed through transcriptome sequencing. Genome Biology and Evolution 6, 3314–3325.

Whitington, P. M. 2007. The evolution of arthropod nervous systems: insights from neural development in the Onychophora and Myriapoda. In: Kaas, J. H. (ed.), Evolution of nervous systems: a comprehensive reference, vol. 1: Theories, development, invertebrates. Elsevier Academic Press, Amsterdam, pp. 317–336.

Whittaker, J. R. 1997. Cephalochordates, the lancelets. In: Gilbert, S. F.; Raunio, A. M. (eds.), Embryology. Constructing the organism. Sinauer, Sunderland, MA, pp. 365–381.

Whittle, R. J.; Hunter, A. W.; Cantrill, D. J.; McNamara, K. J. 2018. Globally discordant Isocrinida (Crinoidea) migration confirms asynchronous Marine Mesozoic Revolution. Communications Biology 1, 46.

Wicht, H.; Lacalli, T. C. 2005. The nervous system of amphioxus: structure, development, and evolutionary significance. Canadian Journal of Zoology 83, 122–150.

Wiedermann, A. 1995. Zur Ultrastruktur des Nervensystems bei *Cephalodasys maximus* (Macrodasyida, Gastrotricha). Microfauna Marina 10, 173–233.

Wild, E.; Wollesen, T.; Haszprunar, G.; Hess, M. 2015. Comparative 3D microanatomy and histology of the eyes and central nervous systems in coleoid cephalopod hatchlings. Organisms Diversity and Evolution 15, 37–64.

Wilkie, I. C. 2002. Is muscle involved in the mechanical adaptability of echinoderm mutable collagenous tissue? Journal of Experimental Biology 205, 159–165.

Willems, W. R.; Wallberg, A.; Jondelius, U.; Littlewood, D. T. J.; Backeljau, T.; Schockaert, E. R.; Artois, T. J. 2006. Filling a gap in the phylogeny of flatworms: relationships within the Rhabdocoela (Platyhelminthes), inferred from 18S ribosomal DNA sequences. Zoologica Scripta 35, 1–17.

Williams, A. 1997. Brachiopoda: Introduction and integumentary system. In: Harrison, F. W.; Woollacott, R. M. (eds.), Microscopic Anatomy of Invertebrates, vol. 13: Lophophorates, Entoprocta, and Cycliophora. Wiley-Liss, New York, pp. 237–296.

Williams, A.; Carlson, S. J.; Brunton, C. H. C.; Holmer, L. E.; Popov, L. 1996. A supra-ordinal classification of the Brachiopoda. Philosophical Transactions of the Royal Society of London B: Biological Sciences 351, 1171–1193.

Williams, S. T. 2017. Molluscan shell colour. Biological Reviews 92, 1039–1058.

Wills, M. A.; Gerber, S.; Ruta, M.; Hughes, M. 2012. The disparity of priapulid, archaeopriapulid and palaeoscolecid worms in the light of new data. Journal of Evolutionary Biology 25, 2056–2076.

Wilson, C. B. 1900. The habits and early development of *Cerebratulus lacteus* (Verrill). Quarterly Journal of Microscopical Sciences 43, 97–198.

Wilson, C. G.; Sherman, P. W. 2010. Anciently asexual bdelloid rotifers escape lethal fungal parasites by drying up and blowing away. Science 327, 574–576.

Wilson, E. B. 1892. The cell-lineage of *Nereis*. A contribution to the cytogeny of the annelid body. Journal of Morphology 6, 361–481.

Wilson, N. G.; Rouse, G. W.; Giribet, G. 2010. Assessing the molluscan hypothesis Serialia (Monoplacophora +

Polyplacophora) using novel molecular data. Molecular Phylogenetics and Evolution 54, 187–193.

Wilson, W. H. 1991. Sexual reproductive modes in polychaetes: Classification and diversity. Bulletin of Marine Science 48, 500–516.

Wilts, E. F.; Wulfken, D.; Ahlrichs, W. H. 2010. Combining confocal laser scanning and transmission electron microscopy for revealing the mastax musculature in *Bryceella stylata* (Milne, 1886) (Rotifera: Monogononta). Zoologischer Anzeiger 248, 285–298.

Wingstrand, K. G. 1972. Comparative spermatology of a pentastomid, *Raillietiella hemidactyli*, and a branchiuran crustacean, *Argulus foliaceus*, with a discussion of pentastomid relationships. Det Kongelige Danske Videnskabernes Selskabs Biologiske Skrifter 19, 1–72.

Winnepenninckx, B.; Backeljau, T.; De Wachter, R. 1995a. Phylogeny of protostome worms derived from 18S rRNA sequences. Molecular Biology and Evolution 12, 641–649.

Winnepenninckx, B.; Backeljau, T.; Kristensen, R. M. 1998. Relations of the new phylum Cycliophora. Nature 393, 636–638.

Winnepenninckx, B.; Backeljau, T.; Mackey, L. Y.; Brooks, J. M.; De Wachter, R.; Kumar, S.; Garey, J. R. 1995b. 18S rRNA data indicate that Aschelminthes are polyphyletic in origin and consist of at least three distinct clades. Molecular Biology and Evolution 12, 1132–1137.

Winston, J. E. 1984. Why bryozoans have avicularia—a review of the evidence. American Museum Novitates 2789, 1–26.

Winston, J. E. 1986. Victims of avicularia. Marine Ecology 7, 193–199.

Winston, J. E. 1995. Ectoproct diversity of the Indian River coastal lagoon. Bulletin of Marine Science 57, 84–93.

Wirkner, C. S.; Richter, S. 2010. Evolutionary morphology of the circulatory system in Peracarida (Malacostraca; Crustacea). Cladistics 26, 143–167.

Wirkner, C. S.; Tögel, M.; Pass, G. 2013. The arthropod circulatory system. In: Minelli, A.; Boxshall, G.; Fusco, G. (eds.), Arthropod Biology and Evolution: Molecules, Development, Morphology. Springer, Heidelberg, pp. 343–391.

Wirz, A.; Pucciarelli, S.; Miccli, C.; Tongiorgi, P.; Balsamo, M. 1999. Novelty in phylogeny of Gastrotricha: evidence from 18S rRNA gene. Molecular Phylogenetics and Evolution 13, 314–318.

Wissocq, J. 1970. Evolution de la musculature longitudinale dorsale et ventrale au cours de la stolonisation de *Syllis amica* Quatrefages (Annélide Polychète). 1. Muscles du vet asexue et muscles du stolon. Journal de Microscopie 9, 355–388.

Witek, A.; Herlyn, H.; Ebersberger, I.; Mark Welch, D. B.; Hankeln, T. 2009. Support for the monophyletic origin of Gnathifera from phylogenomics. Molecular Phylogenetics and Evolution 53, 1037–1041.

Wolenski, F. S.; Bradham, C. A.; Finnerty, J. R.; Gilmore, T. D. 2013. NF-*k*B is required for cnidocyte development in the sea anemone *Nematostella vectensis*. Developmental Biology 373, 205–215.

Wolf, K.; Markiw, M. E. 1984. Biology contravenes taxonomy in the Myxozoa: New discoveries show alternation of invertebrate and vertebrate hosts. Science 225, 1449–1452.

Wolf, Y. I.; Rogozin, I. B.; Koonin, E. V. 2004. Coelomata and not Ecdysozoa: evidence from genome-wide phylogenetic analysis. Genome Research 14, 29–36.

Wolff, C.; Scholtz, G. 2008. The clonal composition of biramous and uniramous arthropod limbs. Proceedings of the Royal Society B: Biological Sciences 275, 1023–1028.

Wolff, C.; Tinevez, J.-Y.; Pietzsch, T.; Stamataki, E.; Harich, B.; Guignard, L.; Preibisch, S.; Shorte, S.; Keller, P. J.; Tomancak, P.; et al. 2018. Multi-view light-sheet imaging and tracking with the MaMuT software reveals the cell lineage of a direct developing arthropod limb. eLife 7, e34410.

Wolff, G.; Strausfeld, Nicholas J. 2016. The insect brain: A commentated primer. In: Schmidt-Rhaesa, A.; Harzsch, S.; Purschke, G. (eds.), Structure and Evolution of Invertebrate Nervous Systems. Oxford University Press, Oxford, pp. 597–639.

Wolff, G. H.; Thoen, H. H.; Marshall, J.; Sayre, M. E.; Strausfeld, N. J. 2017. An insect-like mushroom body in a crustacean brain. eLife 6, e29889.

Wolff, T. 1960. The hadal community, an introduction. Deep-Sea Research 6, 95–124.

Wollesen, T. 2016. Mollusca: Cephalopoda. In: Schmidt-Rhaesa, A.; Harzsch, S.; Purschke, G. (eds.), Structure and Evolution of Invertebrate Nervous Systems. Oxford University Press, Oxford, pp. 222–240.

Wollesen, T.; Loesel, R.; Wanninger, A. 2009. Pygmy squids and giant brains: mapping the complex cephalopod CNS by phalloidin staining of vibratome sections and whole-mount preparations. Journal of Neuroscience Methods 179, 63–67.

Wollesen, T.; Rodríguez Monje, S. V.; Luiz de Oliveira, A.; Wanninger, A. 2018. Staggered Hox expression is more widespread among molluscs than previously appreciated. Proceedings of the Royal Society B: Biological Sciences 285, 20181513.

Wong, Y. H.; Ryu, T.; Seridi, L.; Ghosheh, Y.; Bougouffa, S.; Qian, P.-Y.; Ravasi, T. 2014. Transcriptome analysis elucidates key developmental components of bryozoan lophophore development. Scientific Reports 4, 6534.

Wong, Y. H.; Wang, H.; Ravasi, T.; Qian, P.-Y. 2012. Involvement of Wnt signaling pathways in the metamorphosis of the bryozoan *Bugula neritina*. PLoS One 7, e33323.

Wood, R. 1991. Non-spicular biomineralization in calcified demosponges. In: Reitner, J.; Keupp, H. (eds.), Fossil and Recent sponges. Springer-Verlag, Berlin/Heidelberg/New York, pp. 322–340.

Wood, R.; Liu, A. G.; Bowyer, F.; Wilby, P. R.; Dunn, F. S.; Kenchington, C. G.; Hoyal Cuthill, J. F.; Mitchell, E. G.; Penny, A. 2019. Integrated records of environmental change and evolution challenge the Cambrian explosion. Nature Ecology and Evolution 3, 528–538.

Wood, T. S. 2005. *Loxosomatoides sirindhornae*, new species, a freshwater kamptozoan from Thailand (Entoprocta). Hydrobiologia 544, 27–31.

Wood, T. S.; Lore, M. B. 2005. The higher phylogeny of Phylactolaemate bryozoans inferred from 18S ribosomal DNA sequences. In: Moyano, H. I.; Cancino, J. M.; Wyse Jackson, P. N. (eds.), Bryozoan studies 2004: proceedings of the Thirteenth International Bryozoology Association Conference, Concepción, Chile, 11–16 January 2004. A.A. Balkema, Leiden/London, pp. 361–367.

Woollacott, R. M.; Zimmer, R. L. 1971. Attachment and metamorphosis of the cheilo-ctenostome bryozoan *Bugula neritina* (Linné). Journal of Morphology 134, 351–382.

Woollacott, R. M.; Zimmer, R. L. 1972. Fine structure of a potential photoreceptor organ in the larva of *Bugula neritina* (Bryozoa). Zeitschrift für Zellforschung und Mikroskopische Anatomie 123, 458–469.

Woollacott, R. M.; Zimmer, R. L. 1978. Metamorphosis of cellularioid bryozoans. In: Chia, F. S.; Rice, M. E. (eds.), Proceedings of the symposium on settlement and metamorphosis of marine invertebrate larvae: American Zoological Society Meeting, Toronto, Ontario, Canada, December 27–28, 1977. Elsevier, New York/Oxford, pp. 49–63.

Wörheide, G. 2006. Low variation in partial cytochrome oxidase subunit I (COI) mitochondrial sequences in the coralline demosponge *Astrosclera willeyana* across the Indo-Pacific. Marine Biology 148, 907–912.

Wörheide, G.; Dohrmann, M.; Erpenbeck, D.; Larroux, C.; Maldonado, M.; Borchiellini, C.; Lavrov, D. 2012. Deep phylogeny and evolution of sponges (Phylum Porifera). Advances in Marine Biology 61, 1–78.

Worsaae, K.; Giribet, G.; Martínez, A. 2018. The role of progenesis in the diversification of the interstitial annelid lineage Psammodrilidae. Invertebrate Systematics 32, 774–793.

Worsaae, K.; Kristensen, R. M. 2016. Phylum Micrognathozoa: The micrognathozoans. In: Brusca, R. C.; Moore, W.; Shuster, S. M. (eds.), Invertebrates, 3rd ed. Sinauer, Sunderland, MA, pp. 626–632.

Worsaae, K.; Rouse, G. W. 2008. Is *Diurodrilus* an annelid? Journal of Morphology 269, 1426–1455.

Worsaae, K.; Rouse, G. W. 2010. The simplicity of males: Dwarf males of four species of *Osedax* (Siboglinidae; Annelida) investigated by confocal laser scanning microscopy. Journal of Morphology 271, 127–142.

Worsaae, K.; Sterrer, W.; Kaul-Strehlow, S.; Hay-Schmidt, A.; Giribet, G. 2012. An anatomical description of a miniaturized acorn worm (Hemichordata, Enteropneusta) with asexual reproduction by paratomy. PLoS One 7, e48529.

Wray, G. A. 1997. Echinoderms. In: Gilbert, S. F.; Raunio, A. M. (eds.), Embryology: Constructing the Organism. Sinauer, Sunderland, MA, pp. 309–329.

Wright, J. C. 1988. The tardigrade cuticle. I. Fine structure and the distribution of lipids. Tissue and Cell 20, 745–758.

Wright, J. C. 1989. The tardigrade cuticle II. Evidence for a dehydration-dependent permeability barrier in the intracuticle. Tissue and Cell 21, 263–279.

Wright, T. S. 1856. Description of two tubicolar animals. Proceedings of the Royal Physical Society of Edinburgh 1, 165–167.

Wu, T. H.; Ayres, E.; Bardgett, R. D.; Wall, D. H.; Garey, J. R. 2011. Molecular study of worldwide distribution and diversity of soil animals. Proceedings of the National Academy of Sciences of the USA 108, 17720–17725.

Wudarski, J.; Simanov, D.; Ustyantsev, K.; de Mulder, K.; Grelling, M.; Grudniewska, M.; Beltman, F.; Glazenburg, L.; Demircan, T.; Wunderer, J.; et al. 2017. Efficient transgenesis and annotated genome sequence of the regenerative flatworm model *Macrostomum lignano*. Nature Communications 8, 2120.

Wulfken, D.; Wilts, E. F.; Martínez-Arbizu, P.; Ahlrichs, W. H. 2010. Comparative analysis of the mastax musculature of the rotifer species *Pleurotrocha petromyzon* (Notommatidae) and *Proales tillyensis* (Proalidae) with notes on the virgate mastax type. Zoologischer Anzeiger 249, 181–194.

Xiao, S.; Knoll, A. H. 2000. Phosphatized animal embryos from the Neoproterozoic Doushantuo Formation at Weng'An, Guizhou, South China. Journal of Paleontology 74, 767–788.

Xiao, S.; Yuan, X.; Knoll, A. H. 2000. Eumetazoan fossils in terminal Proterozoic phosphorites? Proceedings of the National Academy of Sciences of the USA 97, 13684–13689.

Xiao, S. H.; Laflamme, M. 2009. On the eve of animal radiation: phylogeny, ecology and evolution of the Ediacara biota. Trends in Ecology and Evolution 24, 31–40.

Yager, J. 1989. The male reproductive system, sperm, and spermatophores of the primitive, hermaphroditic, remipede crustacean *Speleonectess benjamini*. Invertebrate Reproduction and Development 15, 75–81.

Yamada, K.; Ojika, M.; Kigoshi, H.; Suenaga, K. 2010. Cytotoxic substances from two species of Japanese sea hares: chemistry and bioactivity. Proceedings of the Japan Academy, Series B86, 176–189.

Yamaguchi, E.; Dannenberg, L. C.; Amiel, A. R.; Seaver, E. C. 2016. Regulative capacity for eye formation by first quartet micromeres of the polychaete *Capitella teleta*. Developmental Biology 410, 119–130.

Yamasaki, H.; Fujimoto, S.; Miyazaki, K. 2015. Phylogenetic position of Loricifera inferred from nearly complete 18S and 28S rRNA gene sequences. Zoological Letters 1, 18.

Yamasaki, H.; Hiruta, S. F.; Kajihara, H. 2013. Molecular phylogeny of kinorhynchs. Molecular Phylogenetics and Evolution 67, 303–310.

Yamasu, T. 1991. Fine-structure and function of ocelli and sagittocysts of acoel flatworms. Hydrobiologia 227, 273–282.

Yang, H.; Hochberg, R. 2018. Ultrastructure of the extracorporeal tube and "cement glands" in the sessile rotifer *Limnias melicerta* (Rotifera: Gnesiotrocha). Zoomorphology 137, 1–12.

Yang, J.; Ortega-Hernández, J.; Butterfield, N. J.; Liu, Y.; Boyan, G. S.; Hou, J. B.; Lan, T.; Zhang, X. G. 2016. Fuxianhuiid ventral nerve cord and early nervous system evolution in Panarthropoda. Proceedings of the National Academy of Sciences of the USA 113, 2988–2993.

Yang, J.; Ortega-Hernández, J.; Gerber, S.; Butterfield, N. J.; Hou, J. B.; Lan, T.; Zhang, X.-g. 2015. A superarmored lobopodian from the Cambrian of China and early disparity in the evolution of Onychophora. Proceedings of the National Academy of Sciences of the USA 112, 8678–8683.

Yang, J.; Ortega-Hernández, J.; Legg, D. A.; Lan, T.; Hou, J.-b.; Zhang, X.-g. 2018. Early Cambrian fuxianhuiids from China reveal origin of the gnathobasic protopodite in euarthropods. Nature Communications 9, 470.

Yang, J.; Smith, M. R.; Lan, T.; Hou, J-b.; Zhang, X.-g. 2014. Articulated *Wiwaxia* from the Cambrian Stage 3 Xiaoshiba Lagerstätte. Scientific Reports 4, 4643.

Yasui, K.; Reimer, J. D.; Liu, Y.; Yao, X.; Kubo, D.; Shu, D.; Li, Y. 2013. A diploblastic radiate animal at the dawn of Cambrian diversification with a simple body plan: distinct from Cnidaria? PLoS One 8, e65890.

Yin, L. M.; Zhu, M. Y.; Knoll, A. H.; Yuan, X. L.; Zhang, J. M.; Hu, J. 2007. Doushantuo embryos preserved inside diapause egg cysts. Nature 446, 661–663.

Yin, Z.; Zhu, M.; Davidson, E. H.; Bottjer, D. J.; Zhao, F.; Tafforeau, P. 2015. Sponge grade body fossil with cellular resolution dating 60 Myr before the Cambrian. Proceedings of the National Academy of Sciences of the USA 112, E1453–E1460.

Yochem, J.; Tuck, S.; Greenwald, I.; Han, M. 1999. A gp330/megalin-related protein is required in the major epidermis of *Caenorhabditis elegans* for completion of molting. Development 126, 597–606.

Yokobori, S.; Iseto, T.; Asakawa, S.; Sasaki, T.; Shimizu, N.; Yamagishi, A.; Oshima, T.; Hirose, E. 2008. Complete nucleotide sequences of mitochondrial genomes of two solitary entoprocts, *Loxocorone allax* and *Loxosomella aloxiata*: implications for lophotrochozoan phylogeny. Molecular Phylogenetics and Evolution 47, 612–628.

Yokobori, S.; Watanabe, Y.; Oshima, T. 2003. Mitochondrial genome of *Ciona savignyi* (Urochordata, Ascidiacea, Enterogona): Comparison of gene arrangement and tRNA genes with *Halocynthia roretzi* mitochondrial genome. Journal of Molecular Evolution 57, 574–587.

Yonge, C. M. 1979. Cementation in bivalves. In: van der Spoel, S.; van Bruggen, A. C.; Lever, J. (eds.), Pathways in Malacology. Bohn, Scheltema and Holkema, Utrecht, pp. 83–106.

Yoshida, Y.; Koutsovoulos, G.; Laetsch, D. R.; Stevens, L.; Kumar, S.; Horikawa, D. D.; Ishino, K.; Komine, S.; Kunieda, T.; Tomita, M.; et al. 2017. Comparative genomics of the tardigrades *Hypsibius dujardini* and *Ramazzottius varieornatus*. PLoS Biology 15, e2002266.

Young, G. A.; Hagadorn, J. W. 2010. The fossil record of cnidarian medusae. Palaeoworld 19, 212–221.

Young, N. D.; Jex, A. R.; Li, B.; Liu, S.; Yang, L.; Xiong, Z.; Li, Y.; Cantacessi, C.; Hall, R. S.; Xu, X.; et al. 2012. Whole-genome sequence of *Schistosoma haematobium*. Nature Genetics 44, 221–225.

Younossi-Hartenstein, A.; Hartenstein, V. 2000. The embryonic development of the polyclad flatworm *Imogine mcgrathi*. Development Genes and Evolution 210, 383–398.

Yuan, D.; Nakanishi, N.; Jacobs, D. K.; Hartenstein, V. 2008. Embryonic development and metamorphosis of the scyphozoan *Aurelia*. Development Genes and Evolution 218, 525–539.

Yuan, J.; Shaham, S.; Ledoux, S.; Ellis, H. M.; Horvitz, H. R. 1993. The *C. elegans* cell death gene *ced-3* encodes a protein similar to mammalian interleukin-1ß-converting enzyme. Cell 75, 641–652.

Zadesenets, K.; Ershov, N.; Berezikov, E.; Rubtsov, N. 2017a. Chromosome evolution in the free-living flatworms: First evidence of intrachromosomal rearrangements in karyotype evolution of *Macrostomum lignano* (Platyhelminthes, Macrostomida). Genes 8, 298.

Zadesenets, K. S.; Schärer, L.; Rubtsov, N. B. 2017b. New insights into the karyotype evolution of the free-living flatworm *Macrostomum lignano* (Platyhelminthes, Turbellaria). Scientific Reports 7, 6066.

Zadesenets, K. S.; Vizoso, D. B.; Schlatter, A.; Konopatskaia, I. D.; Berezikov, E.; Schärer, L.; Rubtsov, N. B. 2016. Evidence for karyotype polymorphism in the free-living flatworm, *Macrostomum lignano*, a model organism for evolutionary and developmental biology. PLoS One 11, e0164915.

Zamora, S.; Rahman, I. A. 2015. Deciphering the early evolution of echinoderms with Cambrian fossils. Palaeontology 57, 1105–1119.

Zamora, S.; Rahman, I. A.; Smith, A. B. 2012. Plated Cambrian bilaterians reveal the earliest stages of echinoderm evolution. PLoS One 7, e38296.

Zantke, J.; Wolff, C.; Scholtz, G. 2008. Three-dimensional reconstruction of the central nervous system of *Macrobiotus hufelandi* (Eutardigrada, Parachela): implications for the phylogenetic position of Tardigrada. Zoomorphology 127, 21–36.

Zapata, F.; Goetz, F. E.; Smith, S. A.; Howison, M.; Siebert, S.; Church, S. H.; Sanders, S. M.; Ames, C. L.; McFadden, C. S.; France, S. C.; et al. 2015. Phylogenomic analyses support traditional relationships within Cnidaria. PLoS One 10, e0139068.

Zapata, F.; Wilson, N. G.; Howison, M.; Andrade, S. C. S.; Jörger, K. M.; Schrödl, M., Goetz; F. E.; Giribet, G.; Dunn, C. W. 2014. Phylogenomic analyses of deep gastropod relationships reject Orthogastropoda. Proceedings of the Royal Society B: Biological Sciences 281, 20141739.

Zardus, J. D.; Morse, M. P. 1998. Embryogenesis, morphology and ultrastructure of the pericalymma larva of *Acila castrensis* (Bivalvia: Protobranchia: Nuculoida). Invertebrate Biology 117, 221–244.

Zarrella, I.; Herten, K.; Maes, G. E.; Tai, S.; Yang, M.; Seuntjens, E.; Ritschard, E. A.; Zach, M.; Styfhals, R.; Sanges, R.; et al. 2019. The survey and reference assisted assembly of the *Octopus vulgaris* genome. Scientific Data 6, 13.

Zattara, E. E.; Fernández-Álvarez, F. A.; Hiebert, T. C.; Bely, A. E.; Norenburg, J. L. 2019. A phylum-wide survey reveals multiple independent gains of head regeneration ability in Nemertea. Proceedings of the Royal Society B: Biological Sciences 286, 20182524.

Zelinka, C. 1890. Die Gasterotrichen. Eine monographische Darstellung ihrer Anatomie, Biologie und Systematik. Zeitschrift fur Wissenschaftliche Zoologie 49, 209–384.

Zelinka, K. 1928. Monographie der Echinodera. W. Engelmann, Leipzig.

Zhang, G.; Fang, X.; Guo, X.; Li, L.; Luo, R.; Xu, F.; Yang, P.; Zhang, L.; Wang, X.; Qi, H.; et al. 2012. The oyster genome reveals stress adaptation and complexity of shell formation. Nature 490, 49–54.

Zhang, H.; Maas, A.; Waloszek, D. 2017a. New material of scalidophoran worms in Orsten-type preservation from the Cambrian Fortunian Stage of South China. Journal of Paleontology 92, 14–25.

Zhang, H.; Xiao, S.; Liu, Y.; Yuan, X.; Wan, B.; Muscente, A. D.; Shao, T.; Gong, H.; Cao, G. 2015. Armored kinorhynch-like scalidophoran animals from the early Cambrian. Scientific Reports 5, 16521.

Zhang, X.; Liu, W.; Isozaki, Y.; Sato, T. 2017b. Centimeter-wide worm-like fossils from the lowest Cambrian of South China. Scientific Reports 7, 14504.

Zhang, Z.; Holmer, L. E. 2013. Exceptionaly preserved brachiopods from the Chengjiang Lagerstätte (Yunnan, China): Perspectives on the Cambrian explosion of metazoans. Science Foundation in China 21, 66–80.

Zhang, Z.; Holmer, L. E.; Skovsted, C. B.; Brock, G. A.; Budd, G. E.; Fu, D.; Zhang, X.; Shu, D.; Han, J.; Liu, J.; et al. 2013. A sclerite-bearing stem group entoproct from the early Cambrian and its implications. Scientific Reports 3, 1066.

Zhang, Z.-F.; Li, G.-X.; Holmer, L. E.; Brock, G. A.; Balthasar, U.; Skovsted, C. B.; Fu, D.-J.; Zhang, X.-L.; Wang, H.-Z.; Butler, A., et al. 2014. An early Cambrian agglutinated tubular lophophorate with brachiopod characters. Scientific Reports 4, 4682.

Zhang, Z.-F.; Shu, D.-G.; Han, J.; Liu, J.-N. 2004. Lophophore anatomy of Early Cambrian linguloids from the Chengjiang Lagerstätte, Southwest China. Carnets de Géologie/Notebooks on Geology Letter 2004/04, 1–7.

Zhang, Z.-Q. (Ed). 2011. Animal biodiversity: An outline of higher-level classification and survey of taxonomic richness. Magnolia Press, Auckland.

Zhao, F.; Smith, M. R.; Yin, Z.; Zeng, H.; Li, G.; Zhu, M. 2017. Orthrozanclus elongata n. sp. and the significance of sclerite-covered taxa for early trochozoan evolution. Scientific Reports 7, 16232.

Zhao, F. C.; Bottjer, D. J.; Hu, S. X.; Yin, Z. J.; Zhu, M. Y. 2013. Complexity and diversity of eyes in Early Cambrian ecosystems. Scientific Reports 3, 2751.

Zhao, Y.; Vinther, J.; Parry, L. A.; Wei, F.; Green, E.; Pisani, D., Hou, X.; Edgecombe, G. D.; Cong, P. 2019. Cambrian sessile, suspension feeding stem-group ctenophores and evolution of the comb jelly body plan. Current Biology 29, 1112–1125.

Zheng, H.; Zhang, W.; Zhang, L.; Zhang, Z.; Li, J.; Lu, G.; Zhu, Y.; Wang, Y.; Huang, Y.; Liu, J.; et al. 2013. The genome of the hydatid tapeworm Echinococcus granulosus. Nature Genetics 45, 1168–1175.

Zhou, Y.; Zheng, H. J.; Chen, Y. Y.; Zhang, L.; Wang, K.; Guo, J.; Huang, Z.; Zhang, B.; Huang, W.; Jin, K.; et al. 2009. The Schistosoma japonicum genome reveals features of host-parasite interplay. Nature 460, 345–351.

Zhu, M.-Y.; Zhao, Y.-L.; Chen, J.-Y. 2002. Revision of the Cambrian discoidal animals Stellostomites eumorphus and Pararotadiscus guizhouensis from South China. Geobios 35, 165–185.

Zhuravlev, A. Y.; Wood, R. A.; Penny, A. M. 2015. Ediacaran skeletal metazoan interpreted as a lophophorate. Proceedings of the Royal Society B: Biological Sciences 282, 20151860.

Ziegler, A.; Schröder, L.; Ogurreck, M.; Faber, C.; Stach, T. 2012. Evolution of a novel muscle design in sea urchins (Echinodermata: Echinoidea). PLoS One 7, e37520.

Zimmer, R. L. 1967. The morphology and function of accessory reproductive glands in the lophophores of Phoronis vancouverensis and Phoronopsis harmeri. Journal of Morphology 121, 159–178.

Zimmer, R. L. 1991. Phoronida. In: Pearse, J. S.; Pearse, V. B.; Giese, A. C. (eds.), Reproduction of marine invertebrates, vol. 6: echinoderms and lophophorates. Boxwood Press, Pacific Grove, CA, pp. 1–45.

Zimmer, R. L. 1997. Phoronids, brachiopods, and bryozoans, the lophophorates. In: Gilbert, S. F.; Raunio, A. M. (eds.), Embryology: constructing the organism. Sinauer, Sunderland, MA, pp. 279–305.

Zimmer, R. L.; Woollacott, R. M. 1977a. Metamorphosis, ancestrulae and coloniality in bryozoan life cycles. In: Woollacott, R. M.; Zimmer, R. L. (eds.), Biology of Bryozoans. Academic Press, New York/San Francisco/London, pp. 91–142.

Zimmer, R. L.; Woollacott, R. M. 1977b. Structure and classification of gymnolaemate larvae. In: Woollacott, R. M.; Zimmer, R. L. (eds.), Biology of Bryozoans. Academic Press, New York, San Francisco/London, pp. 57–89.

Zimmer, R. L.; Woollacott, R. M. 1989a. Intercoronal cell complex of larvae of the bryozoan Watersipora arcuata (Cheilostomata: Ascophora). Journal of Morphology 199, 151–164.

Zimmer, R. L.; Woollacott, R. M. 1989b. Larval morphology of the bryozoan Watersipora arcuata (Cheilostomata: Ascophora). Journal of Morphology 199, 125–150.

Zimmer, R. L.; Woollacott, R. M. 1993. Anatomy of the larva of Amathia vidovici (Bryozoa, Ctenostomata) and phylogenetic significance of the vesiculariform larva. Journal of Morphology 215, 1–29.

Zouros, E.; Ball, A. O.; Saavedra, C.; Freeman, K. R. 1994. Mitochondrial DNA inheritance. Nature 368, 818.

Zrzavý, J. 2003. Gastrotricha and metazoan phylogeny. Zoologica Scripta 32, 61–81.

Zrzavý, J.; Hypša, V. 2003. Myxozoa, Polypodium, and the origin of the Bilateria: The phylogenetic position of "Endocnidozoa" in light of the rediscovery of Buddenbrockia. Cladistics 19, 164–169.

Zrzavý, J.; Hypša, V.; Tietz, D. F. 2001. Myzostomida are not annelids: Molecular and morphological support for a clade of animals with anterior sperm flagella. Cladistics 17, 170–198.

Zrzavý, J.; Mihulka, S.; Kepka, P.; Bezděk, A.; Tietz, D. 1998. Phylogeny of the Metazoa based on morphological and 18S ribosomal DNA evidence. Cladistics 14, 249–285.

Zwarycz, A. S.; Nossa, C. W.; Putnam, N. H.; Ryan, J. F. 2016. Timing and scope of genomic expansion within Annelida: Evidence from homeoboxes in the genome of the earthworm Eisenia fetida. Genome Biology and Evolution 8, 271–281.

INDEX